PROCESS TECHNOLOGY AND FLOWSHEETS

PROCESS TECHNOLOGY AND FLOWSHEETS

Edited by

Vincent Cavaseno

and

The Staff of Chemical Engineering

McGraw-Hill Publishing Co., New York, N.Y.

Printed in the United States of America

Library of Congress Cataloging in Publication Data
Main entry under title:

Process technology and flowsheets.

Articles which appeared in Chemical engineering over the last five years.
1. Chemical processes. I. Chemical engineering.
TP155.7.P76 660.2′08 79-12117
ISBN 0-07-010741-6

Contents

PROCESS TECHNOLOGY AND FLOWSHEETS

Introduction

The seventies have been a time of challenge for the chemical process industries. Government regulations, higher energy costs, environmental considerations, and waning hydrocarbon feedstocks have changed the ground rules of chemical production and chemical process design.

One positive aspect of this generally unpleasant situation, however, is that it has provided the world's chemical engineers with a unique opportunity for innovation. Indeed, one of the primary goals of CHEMICAL ENGINEERING'S editorial policy has been to track and report on developments of innovative technology: processes using unconventional feedstocks, techniques of turning wastes into dollars, and refinements designed to squeeze the most out of each Btu.

This book presents a selection of the most significant processes that have appeared in the pages of *CE* over the last five years. Most have been taken from the Process Technology section, while the remainder come from news and engineering feature-articles. For each article, the date of original publication is shown.

At the end of the book, we include a special section devoted to recent winners of *CE's* biennial Kirkpatrick Chemical Engineering Award. This award is given to developers of processes judged by a panel of prominent engineering educators to be the most significant additions to the body of chemical engineering technology.

Vincent Cavaseno

Section I
COAL PROCESSING AND CONVERSION

Coal-based processes evaluated in Germany

A conference on coal conversion methods takes a look at the status of conventional technology and assesses the prospects of advanced gasification routes, including underground gasification, and two promising liquefaction processes.

☐ Coal specialists gathered in Düsseldorf, West Germany, this January under the aegis of the United Nations' Economic Commission for Europe to attend a four-day symposium aimed at reviewing the status of gasification and liquefaction processes, and evaluating coal's prospects as a rival of crude oil and natural gas.

Most experts agreed that gasification now has an edge because of the wide range of exising processes that are constantly being improved. Even discounting this technological lead, gasification is likely to dominate the picture on purely economic grounds in industrialized countries of high population density and declining natural-gas reserves. For example, in the U.S. and Western Europe, gasification will become firmly established within the next decade.

Speaking in general terms, prospects for liquefaction are not so immediately inviting, but many speakers at the symposium pinpointed this alternative as a logical complement or successor to oil refining over the medium term. In contrast, the probable role of gasification in the same period is to supplant dwindling reserves of natural gas.

Two options look especially apt for the developing countries with large coal reserves but small domestic demand: the coal can either be shipped elsewhere, or liquefied at home. The latter option, followed by hydrogenation to a relatively heavy fuel oil, seems the best way to obtain higher added value and return on resources.

Coalplexes—i.e., chemical complexes based on coal feedstock, and using both gasification and liquefaction techniques—received the endorsement of many experts in Düsseldorf. They stressed the fact that South Africa's Sasol project—a success despite much criticism from outsiders—points the way to a future coal-based chemical industry.

APPLYING PRESSURE—At Düsseldorf, conventional gasification technology was reviewed but the main emphasis was on development of second-generation processes in Europe and the U.S., using high-pressure operation to boost yields. In addition, a third generation of techniques, which use nuclear-process heat as energy input for plants, received plenty of attention.

The well-known Lurgi and Koppers-Totzek gasification methods are still undergoing further refinement. In the past decade, Lurgi has improved its pressurized units, increasing their capacities from about 15,000 to 80,000 m^3/h. The Koppers-Totzek route is being adapted from operation at "near atmospheric" pressure to functioning at 15–30 bars.

A Davy Powergas development, the high-temperature Winkler (HTW) process, is being modified by Rheinische Braunkohlenwerke (Cologne) and Aachen Technical University (Aachen). Tests carried out at 1.5, 3, 6 and 9 bars show that output goes up in proportion to the square root of absolute pressure, so investigators are trying a pressurized reactor and increased reaction temperatures.

The HTW route produces gas with low CO_2 and H_2O contents, suitable for use in direct reduction of iron ores. Preheated air at 850°C, or oxygen at 300°C, complement the reaction heat supplied by combustion of part of the coal. To prevent sintering of the ash at the temperature of the bed (950–1,100°C), lime or dolomite is added; this step also provides desulfurization.

MORE ON GASIFICATION—Rheinische Braunkohlenwerke is also working on what it calls the RO process, which employs a tube-furnace to handle lignite. Raw coal is fed to a fluidized bed, then gasified in the horizontal tubes of the reactor. These tubes are indirectly heated, and temperatures can be controlled to favor the gasification reactions, which take place in zones along the pipes. At pressures of up to 40 bars, the process yields gas with a methane content of 3.5% (wet basis). A 200-kg/h pilot plant that takes part of the product gas for process heating has been used for tests, but the German firm is now building another pilot unit of twice that capacity.

Attendants at the symposium also heard of a revival of interest in the Rummel-Otto slagging gasifier (ROG), developed and commercialized in the 1950s by Dr. C. Otto & Co. (Bochum, West Germany). The ROG is a three-stage vertically mounted reactor, with a slag bath in the bottom section. In this chamber, which is about 2 m high, all the fine particles of the feed are gasified.

Powdered coal and the gasification medium—a mixture of oxygen and steam—enter the first stage

Originally published March 29, 1976

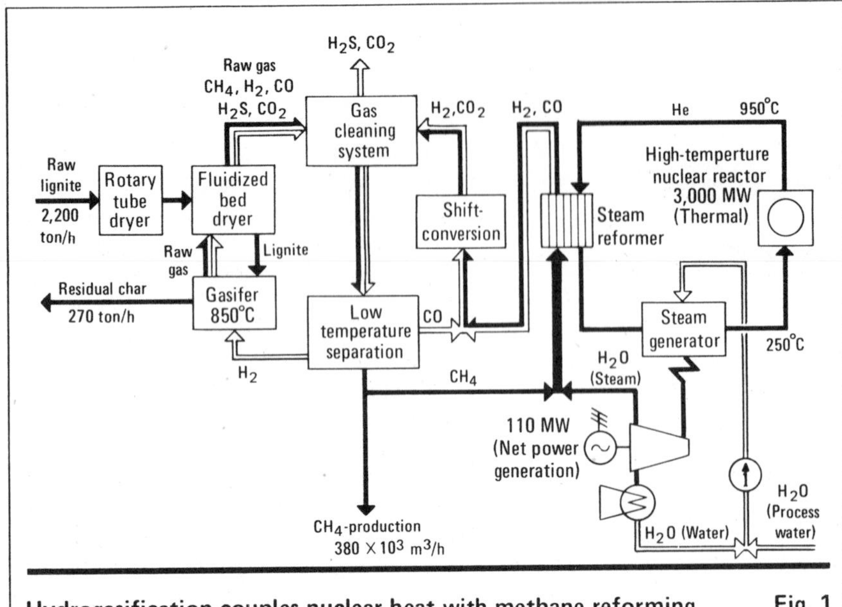

Hydrogasification couples nuclear heat with methane reforming Fig. 1

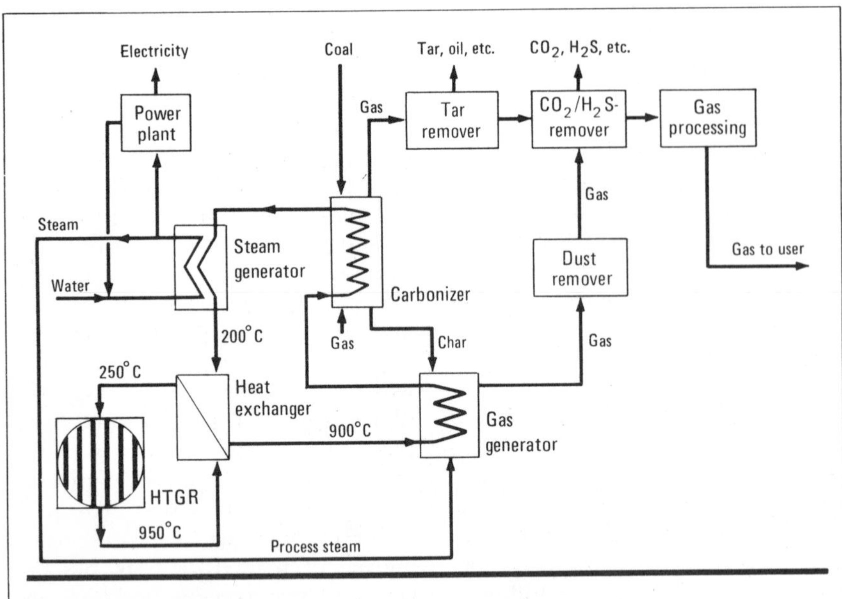

Plant combining HTGR and steam gasification of coal Fig. 2

through tangential tuyeres. A narrow throat connects this first stage with the second one (where entrained, fine coke particles are gasified), and as the rotating gas-stream (at 1,500–1,700°C) passes through it, large coke particles and slag droplets separate centrifugally and flow back to the slag bath.

In the third, uppermost, section, recycle gas or char is injected to quench the ascending gas stream and solidify ash droplets. Exit gas, at 800–900°C, carries 10–30% of the fresh coal-feed in the form of ash and coke, which are removed and recycled. The company plans to operate a new demonstration unit at 25 atm, producing 20,000 m^3/h of raw gas.

THIRD-GENERATION ROUTES—Hydrogasification and steam gasification, two third-generation techniques that use nuclear process heat, are also under study in West Germany. Their advantage lies in the promise of total conversion of the coal feed: Even in the advanced non-nuclear gasification plants, 30-40% of the feed is used to generate process heat, whereas the third-generation plants will pack virtually all of the coal's energy into the product gas. They depend, however, on the successful development of a high-temperature, gas-cooled nuclear reactor (HTGR).

Both German projects envision a 3,000-MW (thermal) HTGR, with a high-temperature stream of helium as energy carrier. Rheinische Braunkohlenwerke is working on a hydrogasification process (Fig. 1). Predried hard coal or lignite is gasified under pressure with hydrogen in a fluidized bed to obtain a methane-rich raw gas. Unreacted hydrogen is separated and recycled; to provide the rest of the hydrogen requirements, part of the product methane goes to a steam reformer to produce CO and H_2. The HTGR's helium coolant provides the heat needed for this endothermic reaction.

In a higher-efficiency variant of the process, hydrogen for the methane-forming reaction is produced in the gasification reactor by injecting a preheated, steam-containing gasification stream. The residual, low-sulfur char that is left over can go to power stations.

Bergbau-Forschung GmbH, at Essen, is studying steam-based gasification. Ground coal is precarbonized and devolatilized (Fig. 2), and the resulting low-temperature char is gasified, using superheated steam. Nuclear process heat supplies the needed energy. The HTGR's primary helium loop is heat-exchanged with a secondary helium cycle, heating the latter to 900°C. This stream then feeds through indirect heater tubes in the gasifier, and through heater tubes in the pretreater. Whatever heat is left over in the secondary helium stream is used to raise steam.

LIQUEFACTION TECHNOLOGY—Two interesting possibilities were described by the research arm of Britain's National Coal Board (Cheltenham, U.K.). Although both have similarities to known solvent-refined coal (SRC) processes, the British entries differ in that they maintain separate extraction and hydrogenation stages, thereby producing a range of "cuts."

One technique employs anthracene oil as solvent in a continuous recycle loop. Crushed dried coal is slurried with the oil, then pumped to

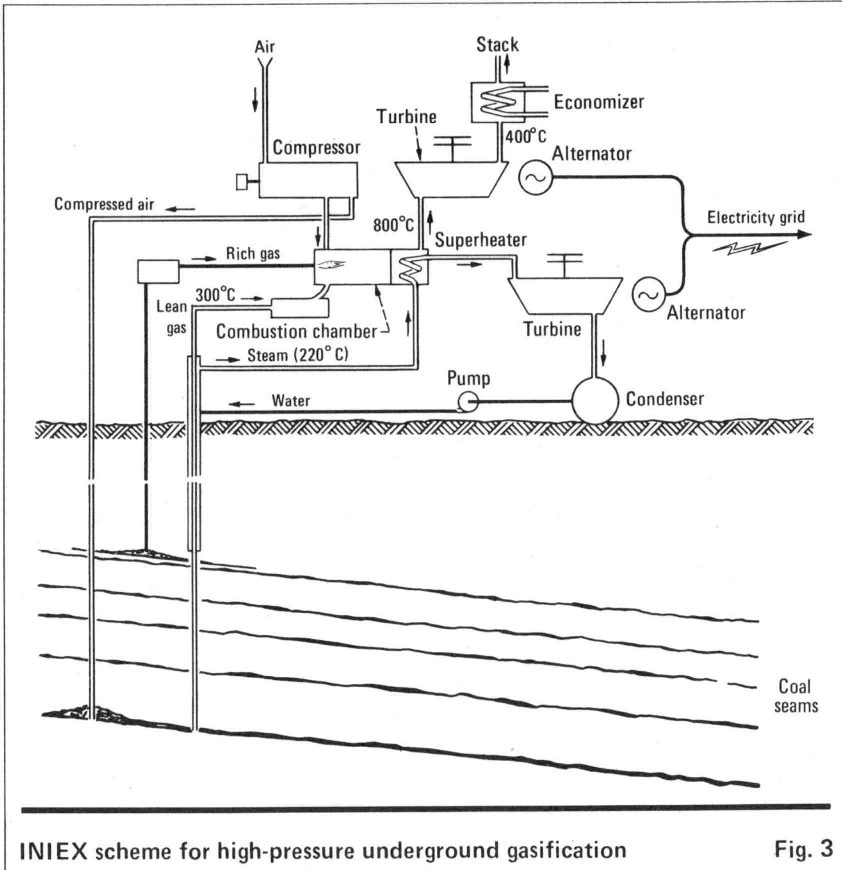

INIEX scheme for high-pressure underground gasification **Fig. 3**

a digester operating at 370–450°C. Up to 85% of the coal dissolves; once filtered and solvent-stripped, the resulting liquid is fed into a catalytic hydrogenation reactor.

This unit operates at hydrogen partial pressures of 200–340 bars, and temperatures of 400–480°C. Gaseous and liquid phases separate at high temperatures; the gaseous phase is cooled, scrubbed at high pressure, and recycled with addition of makeup hydrogen. Light oils recovered from the aqueous liquor go to the fractionation stage along with the liquid fraction.

Fractionation yields light- and middle-oil products. Some of the middle and heavy oils are recycled as solvent.

Experts say this process may prove to be competitive with crude-oil refining for production of aromatic feedstocks, but there is still a great deal of development work to be done, especially on hydrogenation catalysts. This technique is also under study in other countries, including the Soviet Union and Poland.

The British coal board is looking at a liquefaction route that employs compressed gases at near-critical temperatures to obtain a coal extract. Coal-tar- or petroleum-naphtha-derived gases at about 100 atm and 300–400°C could typically extract up to one-third by weight of the coal; the low-melting-point product is recovered by reducing the exit gas stream's pressure.

An economic study by U.S. company Air Products & Chemicals—based on NCB data—shows that manufacturing costs for a 3-million-metric-ton/yr plant are comparable with those published for existing liquefaction processes. Capital costs, at 1974 prices, are estimated at $140 million. Solid residue from the process is said to be ideally suited for fluidized-bed combustion in power generation facilities

IN-SITU GASIFICATION—Among the most interesting fringe ideas reviewed in Düsseldorf was *in-situ* gasification of coal. P. Ledent, a director of Belgium's Institut National des Industries Extractives (INIEX), stated that it holds great promise, at least for deep coal-deposits. Although a number of low-pressure processes have been exploited, the high-pressure routes needed for deep deposits have yet to be proven.

Soviet technology is currently operational at several sites—one of them Angren, in Uzbekistan, 120 km southeast of Tashkent. The U.S. firm Texas Utilities Services, Inc. (Dallas), has already licensed the Soviet knowhow and will use it to gasify a deep lignite deposit.

Still, and despite ambitious plans in the late 1950s, Soviet underground gasification facilities produce energy equivalent to only 200,000 metric tons/yr of coal. Costs are high, so the gas must be consumed locally; prices are not competitive with either natural gas or conventionally mined lignite.

Two new, more-promising schemes originate in the U.S. One of them is under development at Morgantown, W.Va., by the U.S. Bureau of Mines. It uses new directional drilling techniques to exploit thin seams that contain low-volatile matter. The drilling method, which involves boring parallel holes across the seams, will serve to produce a longwall generator.

Deep deposits are also the target of a second U.S. project, under study by the Lawrence Livermore Laboratory (Livermore, Calif.). Explosives introduced through drill holes will hopefully fragment about 600 tons of coal per ton of explosive. After access holes fitted with casings are drilled at the top and bottom of the fragmented seam, a fire will be started at the top of the formation.

Gasification will proceed in a downward direction, aided by injection of a mixture of water and oxygen at 35–70 bars. The high pressure favors formation of CO_2 and methane; after removal of water vapor and CO_2, a high-quality pipeline gas results. U.S. deposits eligible for this type of treatment could supply 30×10^{13} m^3 of gas—about 300 times annual U.S. consumption.

EUROPEAN CONTRIBUTION—German researchers at Rheinische Braunkohlenwerke and Aachen Technical University have also studied high-pressure underground gasification. Their process involves periodic variation of gas pressure in the underground gas-generation zone, aimed at broadening the reaction area and promoting gas-solid reactions. Conditions (40–60 bars,

800°C) favor direct hydrogenation of carbon to methane; some of the methane product will be hydrated to generate hydrogen needed for the underground reaction.

INIEX has its own *in-situ* technique, which can also collect mine gas (mostly methane) released from upper seams during gasification of lower ones (Fig. 3). A network of holes is bored down to the lowest seam of a deposit, and then successive seams are worked in an upward direction. Gas recovery is aided by injecting air at high pressure, with a backpressure of 15–25 bars maintained at the gas-outlet bores.

INIEX envisages either a constant-pressure operation, or cycling between 30–50 bars and 15–25 bars. A power plant sited at the surface would be able to use lean gas and steam produced by cooling the exit bore-holes, as well as high-Btu gas (methane) to offset changes in the calorific value of the lean gas.

Ledent stresses the advantages of such high-pressure, deep-deposit ventures—e.g., seams at depths of 600–700 m can be exploited, and the hydrostatic pressure of overlying strata contains the gas and eliminates pollution risks. One major hurdle is the high cost of drilling, but this can be outweighed by the favorable factors.

In contrast, low-pressure *in-situ* gasification is useful only in thick, shallow seams that can be easily exploited by conventional mining. There are also environmental problems—up to 30% of the gas may leak, and underground water-sources may be contaminated.—PETER SAVAGE, *European Editor, London.*

Coal research: momentum builds in Germany, U.K.

Development of coal gasification and liquefaction processes is moving ahead rapidly in those countries. Several technologies will be demonstrated within five years.

☐ Efforts to develop new coal-utilization technology are hitting high gear in West Germany and the U.K., with about a dozen projects entering the pilot-plant or demonstration-plant stage. Benefits for the chemical process industries are promised by coal gasification and liquefaction, as these technologies should ease shortfalls of petroleum and natural gas that are likely to occur in the late 1990s.

Because of the longterm nature of the development programs—most will not prove their worth before the mid-1980s—engineering contractors cannot expect an immediate bonanza. And experts throughout Europe are cautious about the prospects for individual processes, aware that in the U.S., at least, many processes have already fallen by the wayside.

Not surprisingly, most of the effort in development is centered in Western Europe's two largest coal-producing nations—the U.K. and West Germany. Research funds now come predominantly from two sources: national governments and the Paris-based International Energy Agency. But later this year, researchers may get additional substantial backing from the European Commission, which is planning a $100-million subsidy scheme for promising projects. (A final decision on the list of beneficiaries is not expected before December.)

GERMAN GASIFIERS—Nine weighty projects are underway in Germany, which originally developed a variety of coal gasification and synthesis techniques in the 1930s.

Government funding of the new German processes is relatively generous. The German Ministry for Research and Technology will put up 81% of the projected $62.5-million outlays for liquefaction and 71% of the planned $126 million to be spent on gasification between 1977 and 1980. The remainder will come from the International Energy Agency, the companies themselves, and possibly the European Commission.

The current German processes in the works range from straightforward improvements of established Lurgi-type gasifiers to sophisticated new

Originally published August 18, 1978

systems that use nuclear process heat for gasifying coal (for more details of some of these coal-conversion processes, see Coal-Based Processes Evaluated in Germany, pp. 3–6). They include:

■ *A pressurized, high-temperature Winkler gasifier.* This fluidized-bed system, which operates at 1,100°C and 10 atm, has been onstream since early this year at a 1-mt/h pilot plant. The developer, Rheinische Braunkohlenwerke AG, now plans to spend the equivalent of $73.5 million on a 6-yr project to build a demonstration plant that will gasify 15 to 20 mt/h of coal. Construction will begin in 1980.

The process involves adding lime or dolomite to the reactor bed to prevent sintering of the ash at the high temperatures used. Besides, the additives desulfurize the reaction gases.

■ *A 100-atm version of a Lurgi gasifier.* A joint effort of Ruhrgas AG, Ruhrkohle AG and STEAG, an electric utility, this fixed-bed gasifier will be demonstrated in a 70- to 150-mt/d plant due onstream at Dorsten (in the Ruhr) next spring. Investment costs will exceed $50 million.

■ *An entrained-phase gasifier.* Ruhrkohle and Ruhrchemie AG are involved in a project to demonstrate Texaco's entrained-phase gasification process. A 6-mt/h plant will be started up next April, operating at 1,300 to 1,500°C and 70 atm.

The process uses a wide range of coals, and produces a synthesis gas that contains 30 to 40% hydrogen, 50 to 60% carbon monoxide, and less than 1% methane. Of the $15 million investment, the German government is contributing $9 million, Ruhrkohle about $4 million, and Ruhrchemie approximately $2 million.

British development uses supercritical gas extraction of coal

■ *A liquid-slag gasifier.* This method, which uses a liquid-slag bath in the base of the gasification vessel, will be tried out in a demonstration plant feeding on 10 mt/h of coal to turn out 20,000 m^3/h of gas. Cost of building the plant and for one year's operation is put at $25 million, 75% of which will be government-financed. At presstime, it was due onstream shortly at Voelklingen, where it will be run by process developers Saarbergwerke AG and Dr. C. Otto GmbH.

■ *An improved I.G. Farben process.* Saarbergwerke is also planning a pilot plant for coal hydrogenation, taking off from old I.G. Farben liquefaction technology. A 6-mt/h unit, scheduled to be commissioned at Voelklingen in late 1979, will produce a coal oil for further processing into chemical feedstocks and fuels. The pilot unit will cost $8.5 million to build, and $5 million to run in its first two years of operation.

Saarbergwerke is aiming to streamline the I.G. Farben process, which saw its best days during World War II, when it produced all of Germany's aviation and diesel fuel, and 50% of its vehicle gasoline. Saarbergwerke hopes to upgrade the technology by cutting the operating pressure to 300 atm and replacing a centrifuge step with vacuum distillation.

■ *A catalytic hydrogenation process.* Bergbau-Forschung GmbH (Essen) has built a 20-kg/h pilot plant for catalytic hydrogenation of coal, also based on established I.G. Farben technology. The unit operates at 300 atm and 480°C. Together with chemical firm Veba-Chemie AG, Bergbau-Forschung plans to have a $70-million, 200-mt/d demonstration plant onstream at Bottrop in 1980.

Both Bergbau-Forschung and Saarbergwerke admit that there is some overlap in their work. Saarbergwerke, in fact, is part owner of Bergbau-Forschung. The reason that both groups are pressing on with their similar developments is that they have different patrons: the State of North Rhine Westphalia is funding Bergbau-Forschung's venture and the Bonn government, 75% of Saarbergwerke's.

■ *An advanced Koppers-Totzek gasifier.* Developers Deutsche Shell AG and Krupp-Koppers GmbH will bring onstream a 6-mt/h demonstration plant at Hamburg this autumn that will feature an entrained-phase gasifier based on the established atmospheric-pressure Koppers-Totzek process—but will operate at 30 atm and 1,500 to 1,950°C.

■ *A nuclear-heated gasifier.* A group of German researchers are working on linking coal gasification with high-temperature nuclear heat. The consortium, which includes Bergbau-Forschung, Rheinische Braunkohlenwerke, and various nuclear-energy firms, has an $87-million project in the works. The goal: to develop a pilot plant for steam gasification of lignite and hard coal, using heat from a planned, high-temperature nuclear reactor that would operate at 900°C.

According to Rheinische Braunkohlenwerke, a 0.2-mt/h facility is underway. What is lacking at present is the high-temperature reactor. Although a definite schedule is mapped out for the coal-process developments, the nuclear aspects of the project are clouded with uncertainty. Only small experimental reactors have been tested so far, while a 3,000 MW reactor is needed for commercialization of the concept. The consortium hopes to link the two parts of the development by the early 1990s.

Right now, it is difficult to determine which of these projects will be commercialized. According to a representative of the German Chemical Industry Assn., it will probably be ten years or so before the winners and losers are sorted out. Indeed, many observers note that a host of problems of materials selection, waste disposal and process optimization need to be solved before such coal-based processes can be fully scaled up.

LIQUIDS FAVORED IN U.K.—While German research is largely devoted to gasification projects, British efforts slightly favor liquefaction technology. (Gasification, however, figures heavily in some other projects aimed at power generation.)

In late May, the U.K.'s Dept. of Energy announced that it recommended government funding of three separate projects—extraction of coal, using liquid solvents; extraction using supercritical gas; and a "composite" gasifier based on an existing British Gas Corp. slagging gasifier. This recommendation was accepted, with the U.K. government now planning to cover (over 8 years) about two thirds of the three projects' expected $83-million combined costs.

The two liquefaction plans are being carried out by the National Coal Board at its Coal Research Establishment (Stoke Orchard, U.K.).

The first, liquid-solvent extraction (LSE), has much in common with U.S. and German solvent-refined coal technologies. The solvent used is a heavy oil produced from the coal itself; it is continuously recycled, making the process self-sufficient in solvent. In operation, a coal slurry is contacted with the solvent at about 400°C and a few atmospheres pressure. After filtering the ash, the coal extract is fed to a second, separate hydrocracking stage. This stage operates at 210 atm and 420°C, and uses virtually standard petroleum refining technology. After hydrocracking, the gases (mostly hydrogen) are scrubbed and recycled. The liquids are fractionated to produce a premium-quality light oil (boiling point 200°C), a middle-weight oil (boiling point 200° to 250°C), and a recycle stream of heavy oil.

The project currently consists of a pilot plant that produces 0.2 mt/d of liquids suitable for refining into a range of products, including kerosene and gasoline. The next stage of development is a $31-million project that

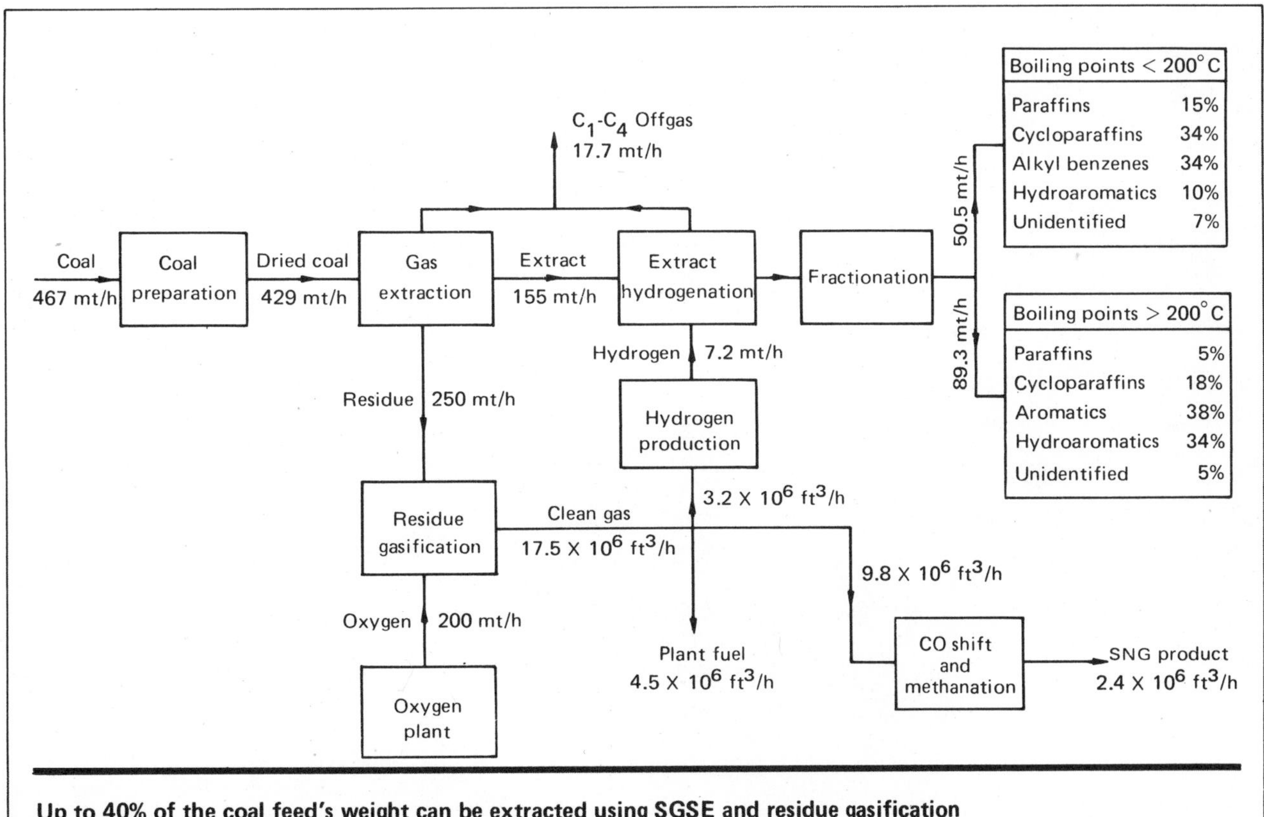

Up to 40% of the coal feed's weight can be extracted using SGSE and residue gasification

will include a pilot plant to produce 13 mt/d of liquids from 24 mt/d of coal. Startup is expected in 1982 or 1983. Design work is currently underway, and should be finished next year.

The Coal Research Establishment is now attempting to assess the kind of finishing needed to make the liquid extract—mostly aromatics—suitable as an auto fuel or a chemical-process-industries feedstock. But CRE admits that the economics of the process cannot become viable until the price of oil is twice that of coal, on the same basis.

GAS AS SOLVENT—The second coal-liquefaction technology is NCB's novel supercritical gas solvent-extraction (SGSE) process. SGSE uses a light organic solvent (currently toluene) at 350 to 450°C and 100 to 200 atm to remove the lighter components of the coal, leaving a reactive char as residue. A 5-kg/h pilot plant, built by Woodall-Duckham Ltd., and funded by the European Commission, was brought onstream last autumn. Next step in the development is to build a 1-mt/h, $28.9-million pilot plant in 1983, to assess the process in greater detail.

Up to 40% of the coal's weight can be extracted using SGSE, followed by residue gasification (see flowsheet). Like the LSE process, a downstream hydrogenation stage is employed; hydrogen from the gasifer might eventually be used as feed.

NCB has yet to decide on how to operate the process—economics look more favorable if a smaller percentage of the coal is extracted, at least from a gasoline-production standpoint. Beyond the 1-mt/h pilot plant, a 50 to 100-mt/h demonstration unit is envisioned for startup in 1990.

The third project that will get U.K. government funding is less futuristic: a $23.1-million venture that will use what developer British Gas Corp. calls a "composite" gasifier. This unit is designed to use a full range of coal sizes, unlike BGC's slagging gasifier, which cannot handle fines (see pp. 11 for details on the slagging system). The "composite" gasifier consists of a modified slagging gasifier with an entrained-phase gasification stage above the bed. BGC plans to build a 100-mt/d pilot unit in 1981 or 1982; early design work is now underway.

Peter R. Savage, European Editor
Jon Fedler, World News (Bonn)

Slagging gasifier aims for SNG market

Using a slagging concept, this gasifier boasts lower steam requirements, higher efficiency and greater output than conventional fixed-bed units.

Peter R. Savage, European Editor

☐ Partners from three different countries have contributed to developing coal-gasification technology that has netted a $24-million design contract from the U.S. Energy Research and Development Administration (ERDA). Awarded in May of this year to Conoco Coal Development Co., a subsidiary of Continental Oil Co. (Houston, Tex.), the contract covers the design of a demonstration-scale coal-gasification plant using a fixed-bed slagging gasifier originally developed by British Gas Corp. (London). The plant, to be sited in eastern Ohio, will use 3,800 tons/d of coal and will produce about 60 million std ft³/d of substitute natural gas (SNG). It will cost about $225 million to build, in 1975 dollars (the year of proposal to ERDA). With the joint approval of ERDA and Conoco, the project will move into a 2½-year construction phase following the 22-month design schedule.

The technology has largely been proven in a $10-million research project—funded by Conoco and a group of 15 U.S. oil and gas firms, with the active collaboration of Lurgi Mineraloeltechnik GmbH (Frankfurt/Main, West Germany)—that has been carried out over the past three years at the Westfield (U.K.) Development Center of British Gas Corp. The slagging gasifier—an extensively modified version of the standard Lurgi fixed-bed unit—is the most important component in the technology package, having a lower steam requirement, a higher efficiency and a greater output than conventional gasifiers, according to the companies.

EARLY WORK—Use of the slagging concept in such a unit had attracted the attention of researchers as early as the 1930s, and the Gas Council—British Gas Corp.'s predecessor—carried out development trials on an experimental basis at its Midlands Research Station (Solihull, U.K.) during the late 1950s and on a 100-ton/d pilot-scale unit in the early 1960s. The trials—which used a modified gasifier shell purchased from Lurgi, which had also performed tests—demonstrated the advantages and feasibility of the slagging operation. British Gas shelved proposals to build a prototype plant following the discovery of offshore natural-gas reserves.

The technology to be used in the U.S. project design will be essentially similar to that proven in two projects at Westfield: (1) the development of the slagging gasifier itself, and (2) the demonstration-scale conversion of lean gas from Lurgi gasifiers to SNG, using a methanation process devised by Conoco engineers, based on British Gas Corp. technology.

SLAGGING PRINCIPLES—A principle idea in the slagging type of gasifier is to provide a minimum of steam to the unit—ash from the process then melts and is run off as a liquid slag. In common with the conventional Lurgi gasifier, coal is fed into the unit through a lock hopper and a distributor (see Fig. 1). The coal is first preheated and distilled by heat exchange with the exit product-gases and vapor, which leave the system through a quench cooler. Next, the coal is gasified in steam and oxygen that are fed into the base of the vessel. The fuel bed rests on a hearth from which the slag is run off through a slag tap into a slag quench-chamber, and finally into a slag lock-hopper located below.

Performance data for pilot slagging-gasifier

Coal	
Type: Donisthorpe	
Size: (1/2-1 in.)	
Operating conditions	
Oxygen rate:	39,650 std ft³/h
Steam/oxygen ratio:	1.10 mol/mol
Products	
Gas composition (vol%)	
CO_2	2.55
O_2	0.10
C_nH_{2n}	0.45
H_2	28.05
CO	61.20
C_nH_{2n+2}	7.65
Calorific value: 374 Btu/ft³	
Gas output: 5.1 million ft³/d	
Tar: 15.4 gal/ton (dry, ash-free)	
Liquor: 2 gal/million Btu	
Consumption	
Coal: 981 lb/(ft²)(h)	
Steam: 25.6 lb/million Btu	
Oxygen: 552 ft³/million Btu	
Gas production:	11.3 million Btu/(ft²)(h)

Source: British Gas Corp.

In the coal-gasification process, oxygen in the oxygen/steam mixture is consumed rapidly in the lower, combustion zone of the unit, producing a rapid rise in temperature. This initiates decomposition of the steam in reaction with the carbon of the coal:

$$C + H_2O \rightarrow CO + H_2$$

The above reaction, being endothermic, absorbs heat liberated by combustion. Meanwhile, two exothermic reactions are also taking place:

$$CO + H_2O \rightarrow CO_2 + H_2$$
$$C + 2H_2 \rightarrow CH_4$$

In a conventional Lurgi gasifier, which uses excess steam to avoid the

Originally published September 12, 1977

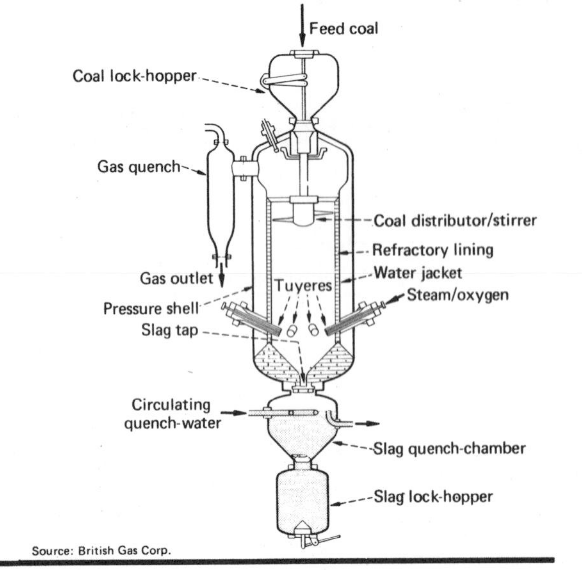

The slagging gasifier uses a minimum of live steam Fig. 1

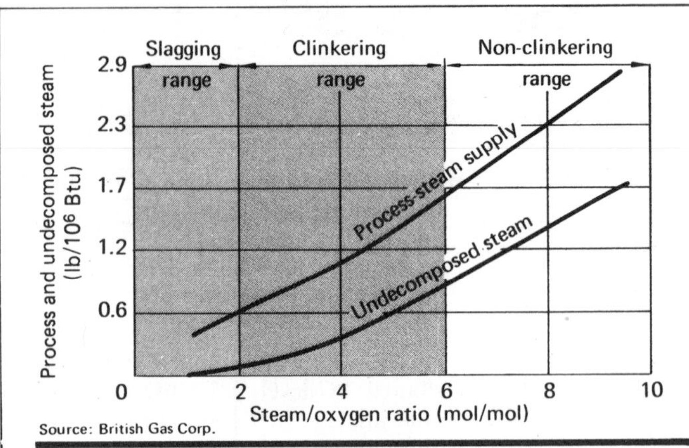

The slagging method uses process-steam efficiently Fig. 2

formation of clinker (the steam keeps the temperature in the combustion zone down, preventing ash from fusing into slag), steam rates are typically 6 to 10 mols of steam per mol of oxygen, and possibly higher with a low-reactivity fuel having a low-melting-point ash. For slagging gasification, however, the steam/oxygen ratio is between 1 and 1.5 mol/mol, and the influence of the coal's reactivity at the temperatures involved becomes insignificant, according to British Gas Corp. The main determining factor becomes slag quality, for handling purposes.

The graph (Fig. 2) shows the effect of varying the steam/oxygen ratio on the process steam supply and the steam remaining undecomposed. Under slagging conditions, practically all of the steam is used in gasifying the fuel.

HIGH OUTPUT—Aside from clinkering effects, gasifier output is determined by the rate at which product gases and undecomposed steam can be passed through without excessive carryover of fuel: A higher output is possible when undecomposed steam content is low. As a result of this, and the higher temperatures in the gasification zone of the bed, slagging units have a higher output than those operating under nonclinkering conditions.

British Gas feels that a slagging gasifier theoretically can have a gasification rate up to triple that of a standard Lurgi unit. Efficiency of a slagging gasifier (as measured by the ratio of the thermal value of gas produced to that of the feed coal and steam-raising coal) is figured at 76.2% compared with a Lurgi gasifier's 68.3%.

The performance data for the Westfield gasifier are confidential, but British Gas claims they "relate closely" to those of Westfield's predecessor—the Solihull pilot-plant unit—if account is taken of the differences in the quality of coals processed. In trials of the Solihull gasifier, a 3-ft-dia. unit, gas output ran at 142,000 std m^3/d (about 5.1 million std ft^3/d), using a 20-bar (about 300-psi) operating pressure. Other performance data are shown in the table.

FEED COMPOSITION—The Westfield trials, says British Gas, have demonstrated that, under slagging conditions, oxygen consumption shows very little variation with changing reactivity of the type of coal supplied. The absence of undecomposed steam from a slagging gasifier also results in the heat capacity of the product gas per unit mass of coal being "appreciably lower" than experienced under Lurgi gasification conditions. Thus, the temperature of the gas produced is lower, contributing to the higher efficiency of the process.

However, the exit gas temperature becomes sensitive to the ash- and moisture-content of the coal, since these inerts have to be heated up at the top of the fuel bed by the exit gases. British Gas notes that the effects of high ash or moisture content can be countered by drying of the coal, or by injection of tar or powdered coal into the fuel bed through tuyeres.

This, says British Gas, provides an outlet for surplus tar or fine coal that might otherwise not be salable, and is another advantage of the Westfield development.

COSTS AND PROJECTIONS—In 1975 dollars, the total capital cost stated for the Ohio demonstration plant is $225 million. For a commercial plant of 250 million std ft^3/d, a figure of $900 million is quoted. British Gas says that operating and maintenance costs for the gasification plants are only available in a combined form: $3.50/million Btu (again, on a 1975 dollar basis).

British Gas points out that the plants have only three effluents: the slag—which can be used as a road-surfacing material—and carbon dioxide and the aqueous liquor. The firm states that a U.S. facility would probably incorporate a biological treatment stage and recover water from the aqueous liquor.

Lurgi gasifiers will not be totally displaced by the slagging-gasifier development—British Gas sees it as complementing the standard Lurgi unit by extending the range of coals suitable for fixed-bed, steam-and-oxygen gasification into the less-reactive, lower-fusion-point types. Only those coals with high (15 to 20%) refractory-ash content are not well suited to use in the slagging gasifier. These coals are in the minority in the U.S., however.

Solvent-refined coal keeps flue gas clean

Removing sulfur from coal before combustion, by the solvent-refined-coal process, offers many boiler operators an attractive alternative to cleaning sulfur oxides from flue gas by scrubbing or other means.

Robert M. Jimeson, Federal Power Commission

☐ The best route for controlling sulfur oxides (SO_x) emissions is often not flue gas desulfurization, but rather combustion of low-sulfur, low-ash coal, produced by the solvent-refined-coal (SRC) process.

SRC is especially appropriate for an operator of a small-to-medium-size boiler who projects the load factor over the plant life at 65% or less, or desires to spend available capital on electric-generation equipment. No expensive flue-gas-desulfurization system is needed. With SRC, such a utility operator pays for pollution control only when fuel is being burned, as an add-on cost for coal, instead of incurring capital charges even during down periods.

Adoption of the intermittent control concept—requiring minimum SO_x emissions only during the 10% of the year when meteorological conditions are poor—would be tantamount to low-load-factor operation for any utility, regardless of the true load factor.

And a number of intangible advantages accrue with SRC: A utility has a positive guarantee that SO_x pollution will always be controlled, and its boiler runs independent of a flue-gas-desulfurization system and its reliability questions.

PROCESS BACKGROUND—The SRC process was first studied from 1962–1965, when Pittsburg and Midway Coal Mining Co. successfully performed small-scale pilot-plant work under the sponsorship of the U.S. Dept. of the Interior's Office of Coal Research.

In August 1972, the Southern Co. (an electric utility) and the Edison Electric Institute started construction of a 6-ton/d SRC pilot plant near Wilsonville, Ala. Catalytic, Inc., designed and built the $4-million plant, and in January 1974 began operations. The Electric Power Research Institute has since assumed the role of the Edison Electric Institute in the project.

Subsequent to the Wilsonville groundbreaking, the Office of Coal Research engineered the Pittsburg and Midway process, and began construction of a 50-ton/d pilot plant at Fort Lewis, Wash., in December 1972. Completion was in September 1974.

Cooperation in planning and information exchange takes place between the two separate units.

FLOW SEQUENCE—As shown in the flow diagram, the SRC process starts by slurrying crushed coal (90% through 200 mesh) with a recycling solvent, generated downstream from the coal itself. After pumping to 1,000–3,000 psi and addition of hydrogen, the three-phase mixture passes upward through a helical-tube preheater fired with liquefied petroleum gas. Contact time is 1.6 min and velocity 6–7 ft/s.

The heated material flows into a single-stage dissolver, normally held at 815–875°F and 1,700–2,400 psi. During a 40-min residence time, 92–94% of the coal reacts to form dissolved coal, a solvent fraction, and some gas. About 60% of the organic sulfur in the coal converts to hydrogen sulfide.

Products exiting the dissolver are partially cooled but kept at high pressure. A separator splits off unreacted hydrogen, hydrocarbon gases, water vapor and hydrogen sulfide from the liquid. In a commercial plant, hydrogen would recycle, hydrogen sulfide would serve as feedstock for elemental sulfur production, and the hydrocarbon gases would be another byproduct.

Liquid from the high-pressure separator, meanwhile, is reduced to 115 psi at 600°F, which flashes off more gas. The remaining liquid passes to a pressure-leaf filter for minerals separation. Filtration at a flowrate of 24 gph/ft^2 removes virtually all of the pyritic (inorganic) sulfur, as well as ash-forming substances and undissolved carbon, from the feed coal. Only 0.1 wt.% minerals is left behind in the filtrate. Trials with hydrocyclones for this key separation step have shown no advantages over the filter; disk and solid-bowl centrifuges are yet to be evaluated.

The filter residue contains a substantial quantity of absorbed solvent as well as minerals, and thus passes to a rotary-vacuum-drum dryer for solvent recovery before disposal.

The filtrate, consisting of solvent and dissolved coal, is reheated to 650°F and flashed in a vacuum distillation column operating at 2 psi abs. The middle distillate cut, with a boiling range of 350–780°F, serves as the process solvent.

Overhead product from the vacuum still is a light distillate (100–350°F boiling range), amounting to 10–15 wt.% of the coal charge. This light fraction contains 15–20 wt.% phenol and cresylic acids, which can be separated as valuable chemical byproducts. Hydrotreating and reforming, alternatively, could convert the light distillate into a re-

Originally published March 1, 1976

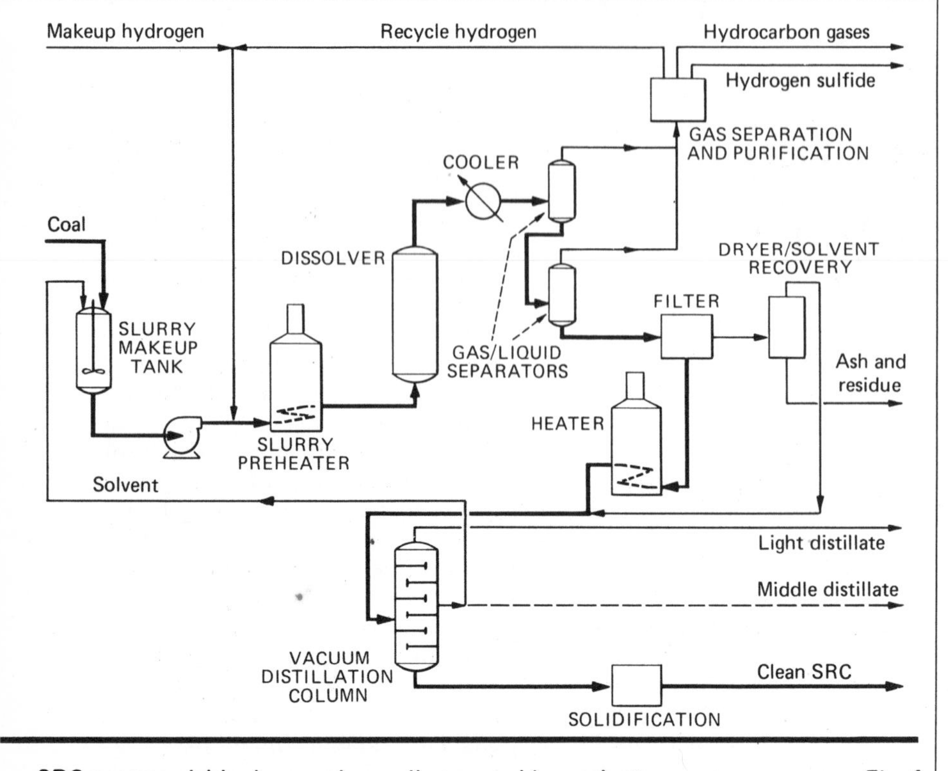

SRC process yields clean coal as well as several byproducts — Fig. 1

Add-on cost for SRC varies with operating conditions

Economic basis	
Feedrate, tons/d coal	20,000
Coal price, ¢/million Btu	80
Depreciable plant life, yr	20
Debt/equity ratio	90/10
Interest on debt, %	10
Return on investment, %	20
Federal income tax, %	52
Total capital investment, $ million	467.6
Annual revenue required, $ million	246.1
Annual revenue from SRC, $ million	172.3
SRC output, tons/yr	3,675,000
Operating variables	
Hydrogen consumption, wt.% of feed coal	2.0–3.0
Filtration rate, gph/ft^2	10–36
Solvent/coal ratio	1.75:1–2.5:1
Add-on cost, $/ton of SRC	**19.91–22.40**
Add-on cost, $/million Btu	**0.62–0.70**

formate rich in benzene, toluene and xylenes.

The bottoms stream from the vacuum still, in any case, is the solvent-refined coal product. It can remain as a hot liquid for burning in an adjacent power plant; or, cooling to below 300°F can solidify the stream for storage and subsequent shipping or handling. A vibrating water-cooled plate coil has recently been installed for solidification to replace the original water quenching, which was simpler but turned out a low-density, highly porous product that retained water too easily.

If desired, hydrocracking and reforming can convert SRC into light and heavy distillates, or a synthetic crude oil.

OPERATING RESULTS—During 1974 and the first half of 1975, the Wilsonville pilot plant confirmed its baseline operating conditions. The entire unit has run over a wide range of conditions, at coal rates of 50–130% of design, and for continuous periods as long as 75 days. Process variables tailored for about a half-dozen different coal types were studied during the rest of 1975.

Coal conversion rates generally range from 85–95%, and yields of solid SRC product from 50–70% of the moisture- and ash-free coal charge. Sulfur in the SRC product is 0.6–0.9%, depending on operating conditions and coal type, and ash content is 0.06–0.16%.

The product is brittle. Its gross heating value runs about 16,000 Btu/lb regardless of coal type, versus raw coal's 11,500–12,000 Btu/lb.

ECONOMICS—Continuous study of the economics of the SRC process indicates that capital requirements jumped tremendously from 1973 to 1975, from nearly $300 million to about $470 million for a 20,000-ton/d feedrate. This represents increased costs for materials and construction labor, not process changes.

An add-on cost of processing coal via the SRC process has been calculated. This add-on cost, plus the cost of raw coal, conversion efficiency losses, and credits for liquid and gaseous fuel byproducts determines the overall selling price of SRC.

The estimates of add-on cost in the tables are at the extreme limits of operating conditions. The most likely cost is 69¢/million Btu, or about $22/ton of SRC.

FUTURE PROSPECTS—An economically viable plant must process 10,000–20,000 tons/d of coal. Such a plant would have a number of parallel trains, because of equipment limitations and the undesirability of split-flow handling of slurries.

The next step in commercial development should thus be construction of a full-scale process train, probably 2,000–2,500 tons/d. This demonstration facility would itself not yield an economic return on investment but could later be integrated into a commercial clean-fuels plant.

Another alternative, which could lead to earlier commercialization, would use SRC as an intermediate for the manufacture of petrochemicals and carbon products. A single process train might indeed be economically justified, thus overcoming objections to building a demonstration train as only a research project.

The author

Robert M. Jimeson is assistant director for fuel resources in the Office of Energy Systems of the U.S. Federal Power Commission (Washington, DC 20426). Formerly, he was FPC's assistant advisor on environmental quality, and has worked for other federal agencies and private industry in various administrative, technical and engineering capacities. He holds degrees from Pennsylvania State, George Washington, and Stanford Universities (and has taught in the former two). He is a registered professional engineer and a member of several societies.

Coal cleaning readies for wider sulfur-removal role

New physical and chemical techniques to eliminate sulfur from coal prior to its combustion promise attractive alternatives to stackgas desulfurization. And the routes may win a role in coal-gasification, too.

☐ Away from the din of complaints about the expense and efficacy of flue-gas SO_x treatment processes, an old but modernized standby—coal cleaning—is gaining new adherents.

The term "coal cleaning" refers to any physical or chemical process that removes impurities (including sulfur) from coal prior to combustion, coking, conversion or other use. Some cleaning techniques handle dry coal; others run a wet feed. Almost all require comminution of the coal to as fine as 200 mesh before treatment. However, the basic structure of the ground coal feed remains intact throughout the processing. In this respect, coal cleaning differs from more-radical routes such as conversion or solvent refining (for an example, see pp. 13–14).

Coal-cleaning techniques already treat about half of the coal produced in the U.S., with most of this product going for steelmaking. Only about 40% of the coal destined for electricity generating stations receives such beneficiation—mostly to remove mineral matter, with sulfur elimination only a secondary aim.

However, a combination of events, both current and upcoming, could give a healthy boost to coal cleaning's prospects.

Development of new physical and chemical cleaning methods is progressing slowly but surely. In particular, some chemical leaching systems now being groomed promise much higher desulfurization rates than do current, conventional physical processes, which only remove gross pyrite particles. None of the chemical routes has yet reached commercialization but, perhaps signaling a trend, the U.S. Environmental Protection Agency recently awarded TRW Inc. (Redondo Beach, Calif.) a $2.4-million contract. The EPA money will fund the first process-development-scale unit to employ the firm's technique for chemical elimination of pyrite. And other chemical routes still in development attack organic sulfur as well as pyrite.

Also enhancing coal-cleaning prospects are the fast-approaching EPA 1977 deadlines (extended from the original 1975 time limit) for enforcement of state implementation plans for control of flue-gas emissions from existing sources. Coal cleaning may rate as the quickest and least capital-intensive route to compliance.

Further in the future, another lift may come from firms hoping to gasify coal. They could turn to coal cleaning as an adjunct to the basic conversion step. Though the economics of this combination is open to wide debate, many experts feel that smaller, cheaper gasification units, operating at less-severe conditions, could result if ash and sulfur loads were relieved by prior coal preparation (see box on p. 17).

POINTING THE WAY—Mine and utility operators need not look to tomorrow for the benefits of coal cleaning, according to a recent report by the Commerce Technical Advisory Board (CTAB), a blue-ribbon panel of experts commissioned by the U.S. Dept. of Commerce (Washington, D.C.). CTAB found that in the U.S. Northeast, at least, the incremental costs of coal cleaning in conjunction with flue-gas desulfurization should amount to 3–5 mills/kWh more than present generator operating tabs—a lower range than the 4–6 mills/kWh increment needed to meet regulations via flue-gas desulfurization alone. The savings stem mainly from reductions in lime and limestone required for scrubbing, as well as cuts in the volume of the resultant sludge.

CTAB's analysis covers conventional treatment techniques. An assessment of more-sophisticated cleaning methods, including chemical routes, should come out of another study soon to be funded by EPA. The agency recently received proposals from private groups for the 2–4-yr contract. In its request for these proposals, EPA focused on seven main topics: physical techniques for pyrite removal; chemical coal-cleaning routes; fine-coal dewatering and handling; coal-preparation requirements of synthetic coal-conversion processes; pollution hazards posed by coal-cleaning systems themselves; coal-cleaning-plant product and byproduct breakdowns; and an economic analysis of the various options.

In addition, the Electric Power Research Institute (EPRI) of Palo Alto, Calif., is interested enough in coal-cleaning to fund a demonstration plant based on a chemical route that gets at organic as well as pyritic sulfur. EPRI is now considering proposals, but hasn't yet decided on a dollar limit for the program.

PHYSICAL SEPARATION—The EPA study almost surely will find that plenty of room exists for more-severe desulfurization processes that can satisfy new clean-air standards. Present physical-separation techniques work adequately on only about 16–17% of the coal now mined in the U.S. The rest of the coal eludes successful treatment either because of too high a ratio of chemically bound organic sulfur to the inorganic pyritic portion, or too low a sulfur content to permit economic handling.

Nevertheless, work continues to broaden the applicability of physical separation techniques for coal desulfurization. These center on taking advantage of the differences in specific gravity between coal and pyrite, using jigs (devices in which pulsations of a liquid through a mixed bed aid separation), cyclones and, more recently, concentrating tables as well as single-stage froth-flotation systems.

The U.S. Bureau of Mines' Pittsburgh Energy Research Center (Bruceton, Pa.) and several equip-

Originally published March 1, 1976

ment makers are studying potentially better physical-separation techniques. One Bureau-developed system—two-stage froth flotation—is nearing its first full-scale test at a coal mine in western Pennsylvania.

Some of the methods under investigation border on the exotic. For example, the Bureau of Mines has looked at electrophoretic separation—wherein coal particles migrate toward a positive electrode at a faster rate than pyrite particles do—but has found disappointing economics for the route.

MAGNETIC IDEA—Instead of using an electric field, as in electrophoresis, separation in a magnetic field might prove more feasible. At least that's what Aquafine Corp. (Brunswick, Ga.) hopes.

The company is working to adapt its "high-extraction magentic filters," originally developed for kaolin refining, to coal cleaning. The units would only remove pyrite fractions, but at greater efficiency than conventional physical-treatment systems. The firm expects potential pyrite removal to run as much as 50% for dry, 200-mesh coal, and perhaps as high as 90% from (more magnetically permeable) wet streams. Of particular importance, says Aquafine, the filter boasts the ability to scavenge submicron-size particles, as demonstrated in five commercial kaolin applications.

Two projects involving magnetic separators are in progress. One, underway at Auburn U. (Auburn, Ala.) with financing by the National Science Foundation (NSF) of Washington, D.C., is testing a pilot-scale Aquafine unit for the desulfurization of solvent-refined coal from the Wilsonville, Ala., plant sponsored by EPRI and Southern Co. (Atlanta, Ga.). (This tack differs considerably from conventional process philosophy, which calls for desulfurization before conversion.) The other project is taking place at Indiana U. (Bloomington, Ind.), under the direction of H. H. Murray. Also funded by the Electric Power Research Institute, the work is being conducted concurrently with an NSF-sponsored study at the school to determine the responsiveness of some 40 minerals to magnetic beneficiation.

Proponents of the magnetic approach plan to gather this month (Mar. 23–26) at Auburn for a symposium on desulfurization.

DIFFERENT SITE—In another departure from conventional coal-cleaning practice, General Public Utilities Corp. (Parsippany, N.J.) and New York Electric & Gas Corp. (Binghamton, N.Y.) plan next year to build a desulfurization installation next to a generating station that the two firms jointly own at Homer City, Pa., rather than to construct it as is usual at a mine site. The \$35-million plant will feature a "multistream system," employing cyclones. The facility will turn out coal desulfurized to levels suitable for existing boilers, as well as a lower-sulfur portion to meet new-source standards (1.2 lb SO_2/million Btu evolved, which is generally equivalent to about 0.8 wt-% sulfur in coal), specified for a 650-MW generating unit scheduled for completion during 1977.

GPU's research and development manager, James F. McConnell, calls the chosen two-stage process the most extensive type of cleaning to be used by utilities; it even goes beyond traditionally most-severe, metallurgical-coal, standards. In deciding upon the route, the two companies analyzed the economic tradeoff between "deep cleaning," as the process is known, and stackgas scrubbing. The precombustion technique won, though partly because plans already had been drawn for a smaller, less-severe physical-pretreatment unit to service the existing generator.

CHEMICAL PYRITE-REMOVAL—If physical separation cannot get enough pyritic sulfur out of coal, chemical treatment might. TRW and Kennecott Copper Co.'s Ledgemont Laboratories (Lexington, Mass.) both see promise for leaching systems.

TRW hopes to prove its process in an 8-ton/d unit, which should be completed late this year at the company's Capistrano, Calif., test site. The route works on ground or pulverized coal by leaching out pyrite with aqueous ferric sulfate at about 250-265°F and 50–100 psig. In the leaching step, pyrite forms a solution of ferrous sulfate and sulfuric acid, plus a solid phase of sulfur, which clings to the coal matrix. This sulfur separates with the coal from the spent leachate; volatilization then recovers the sulfur in elemental form. Meanwhile, oxygenation regenerates the leach solution to the ferric state.

TRW has devised two basic ver-

Coal cleaning seeks a role in conversion

While most of the talk about coal cleaning now centers on desulfurization of feed for utilities, the horizon for the technique may broaden to encompass coal-conversion processes, particularly gasification routes—which could become common in the next decade.

Opinions on the merits of teaming the two technologies remain divided. However, many experts think that proper cleaning of coal to produce a uniform feedstock for conversion would save on capital and operating costs by reducing the required size of the conversion reactor, and by cutting down on the heat needed to process inert mineral matter. Lesser benefits would accrue from the removal of sulfur values; conversion would still mandate further sulfur-removal equipment before admitting gases to a methanation reactor.

In the combined flowscheme, the coal preparation unit and the gasifier can benefit from being designed in tandem. Battelle, for one, is following this philosophy. The research institute is working on a modified version of its Hydrothermal chemical-desulfurization process (see text) specifically for use with gasification reactors. A Battelle official says that, besides removing unwanted inert matter from the gasifier feed, the modified cleaning system would make coal more reactive by "opening up its structure" and by adding certain metallic components from the solvent (needed primarily to leach sulfur compounds out of the raw coal) that may provide a catalytic effect during gasification.

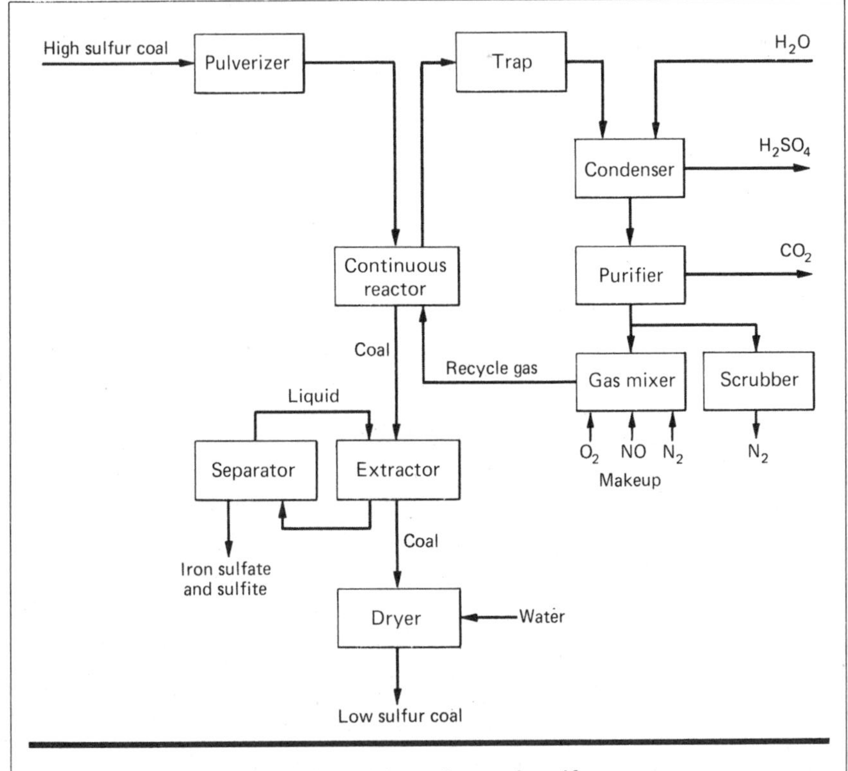

KVB process eliminates both pyritic and organic sulfur

sions of the process, one for fine coal of 8-mesh or less, and the other for coarser coal up to ⅜-in dia. The fines variant carries out the reaction and regeneration in the same vessel, while the coarse-coal route regenerates in a separate unit.

If all goes well with the Capistrano plant, the firm hopes to build a 50-ton/h fullscale unit, but design and construction wouldn't begin until 1980 at the earliest.

Ledgemont's efforts have not progressed as far. The research facility has investigated a process that employs oxygen directly. It oxidizes pyritic sulfur to sulfate, which goes into solution and can be removed by a solids-liquid separation step. Some coal is lost to oxidation but, claims Kennecott, the high pyrite-removal efficiency, which can total as much as 90%, offsets this penalty. Residence time for the oxidation reaction typically averages about 2 h.

(Such long residence times remain a major stumbling block in the development of chemical routes. But some experts hope to turn this disadvantage into an advantage: with the increasing interest in pipelining coal, these researchers foresee in-transit chemical processing. One University of Minnesota (Minneapolis) chemical engineering professor, Henry Tsuchiya, has even suggested bacteriological desulfurization in the pipeline. He has identified two strains of bacteria that might work, given a long enough residence time.)

Ledgemont was grooming the technique particularly for Kennecott's Peabody Coal subsidiary. However, the U.S. Dept. of Justice has ruled that the copper maker must divest itself of Peabody. So the laboratory has halted development of the desulfurization process.

GETTING OUT ORGANIC SULFUR— Pyrite adds up to a large part of the sulfur in coal, but organic sulfur compounds also are present to a significant extent. In fact, organic sulfur may account for up to 70% of the total sulfur in a coal feed, and seldom tallies less than 30%. The Bureau of Mines, Battelle Memorial Institute (Columbus, Ohio) and KVB Inc. (Tustin, Calif.) are working on ways to remove the organics. Atlantic Richfield Co.'s research group, located at Harvey, Ill., is also believed to be working on such a system.

The Bureau of Mines' effort has been limited until now. But there are signs that the Bruceton research center is placing greater emphasis on such systems, both in staff and spending.

So far, attention has focused on two processes: sodium hydroxide leaching at elevated temperatures, and a route involving steam and compressed air at about 300–400°F. Both systems can eliminate as much as 40% of the organic sulfur.

Battelle's Hydrothermal process, announced last June, uses much the same type of reaction chemistry as does the Bureau of Mines' sodium hydroxide route. The new technique takes out up to 99% of the pyritic sulfur and as much as 70% of the organics, claims the research institute.

The process entails five major steps: physical preparation to grind the feed coal to 200 mesh; contact with a leach solution; coal/leachate separation; coal drying; and leachate regeneration.

The leach solution contains sodium- and calcium hydroxides. The lean stream goes into the process in a ratio of about one part to ten parts of coal. After the spent solution's separation from the coal, sparging with carbon dioxide drives off sulfur as hydrogen sulfide, and regenerates the leachant to hydroxide form.

Battelle has lined up partial financial backing for a 50-ton/d, $33-million development unit that would handle high-sulfur coal. Mapco Inc. (Tulsa, Okla.) already has pledged about one-third of the building cost.

The Institute is working on a major simplification of the process design for use in combination with coal gasification (see box).

For its part, KVB Inc. espouses a sulfur-oxidation route featuring a gas mixture made up of oxygen, nitric oxide, nitrogen dioxide and nitrogen. At 1 atm and about 210°F, pyritic sulfur oxidizes to sulfate, while organic sulfur converts to sulfones or sulfoxides. Exposure to sodium hydroxide at 1–20 atm and slightly less than 210°F removes the organics.

KVB's chief chemist, Eugene D. Guth, says that in laboratory tests the firm has succeeded in taking out "substantially all of the inorganic sulfur and about half of the organic material." Development work is continuing, but a demonstration plant remains far off.

John C. Davis

Section II
FERTILIZER

Melt granulator featured in low-energy fertilizer route
Potash flotation method handles variable feed
Phosphoric acid purification: comparing the process choices
Enter: sulfur-coated urea, a slow-release fertilizer

Melt granulator featured in low-energy fertilizer route

Significant reductions in energy consumption, in investment and operating costs, and in dust- and fume-abatement problems, are achieved by a new technique for granulating fertilizers.

Robert G. Lee, Melvin M. Norton and Homer G. Graham, Tennessee Valley Authority

☐ A 400-ton/d demonstration plant for making fertilizers—a major innovation in fertilizer granulation—has now logged nearly two years of successful operating experience. The Tennessee Valley Authority (TVA) developed the process over a 10-yr span, starting with bench-scale work in 1964, and now runs the plant at its Muscle Shoals, Ala., National Fertilizer Development Center.

First outside user is Rasa Industries Ltd., which plans to start up a small 35,000-metric-ton/yr unit this November at Iwate-ken, Japan.

Because only anhydrous materials (rather than aqueous solutions or slurries) are fed to the melt granulator featured in the process, no drying is needed. Eliminating the dryer and its accessory dust collector and scrubbing equipment reduces investment and operating costs (exclusive of raw materials) by about 20% each, and also mitigates dust- and fume-abatement problems.

The process realizes a net energy savings as well. As much as 1.6 gal of fuel oil per ton of urea/ammonium-phosphate (UAP) product can be trimmed, a 50% savings over the conventional granulation approach. When the system alternatively produces ammonium polyphosphate (APP) fertilizer, energy usage falls about 3.3 gal of fuel oil per ton of product—a 95% reduction.

Dryerless operation is especially attractive for urea-based fertilizers because drying of these is difficult and requires a long retention time with low-temperature air.

MELT PREPARATION—The technique basically consists of cooling and crystallizing melts of urea and anhydrous APP (along with a recycle stream) in the granulator. Approximately equal parts of each are used for the UAP product (a 28-28-0 grade, in nitrogen-phosphorus-potassium content). No urea melt is needed for manufacturing the 11-57-0 APP fertilizer.

To prepare the urea melt for UAP, a 75% urea solution first flows through a steam-heated preevaporator (a vertical shell-and-tube heat exchanger). The stream then passes to the top of a rotating-disk evaporator containing eight sections. As the solution flows downwards, it is thrown out to the heat-exchange surfaces by the rotating disks. The resulting melt, concentrating to 99% urea and less than 1% water, exits at 285°F, and flows by gravity to the granulator.

A countercurrent stream of heated air meanwhile sweeps evaporated water overhead to a scrubber-condenser system.

Throughout the concentration system, all surfaces in contact with urea are 304L stainless steel.

For production of the APP melt, wet-process phosphoric acid (54% P_2O_5) is preheated to approximately 150°F, then introduced through a single spray-nozzle into the bottom of a spray reactor, measuring 3½ ft dia. by 20 ft high. An expanded section (5 ft dia. by 7 ft high) on top of the reactor provides space for entrainment separation; pumps recirculate most of the acid bottoms stream to three full-cone spray-nozzles spaced up through the reactor.

The circulating acid is partially neutralized to a pH of 1.5 (10% solution) by the ammonia/steam offgas from a pipe reactor. This equals an NH_3/H_3PO_4 mole ratio of about 0.4, which is near the maximum solubility of ammonium phosphate at the temperature (265°F) and concentration in the spray reactor.

Measurement of this pH, and adjustments in ammonia feedrate to hold pH very near to 1.5, are the main control criteria for the pipe reactor. This 6-in-dia, 10-ft-long unit takes in a metered stream of the hot, partially neutralized acid from the spray reactor, for reaction with gaseous 100°F ammonia. Polyphosphate content of the resulting APP melt is controlled to about 20-25% by regulating the feed-acid temperature. The polyphosphate content governs the readiness with which the melt crystallizes during granulation.

Originally published September 1, 1975

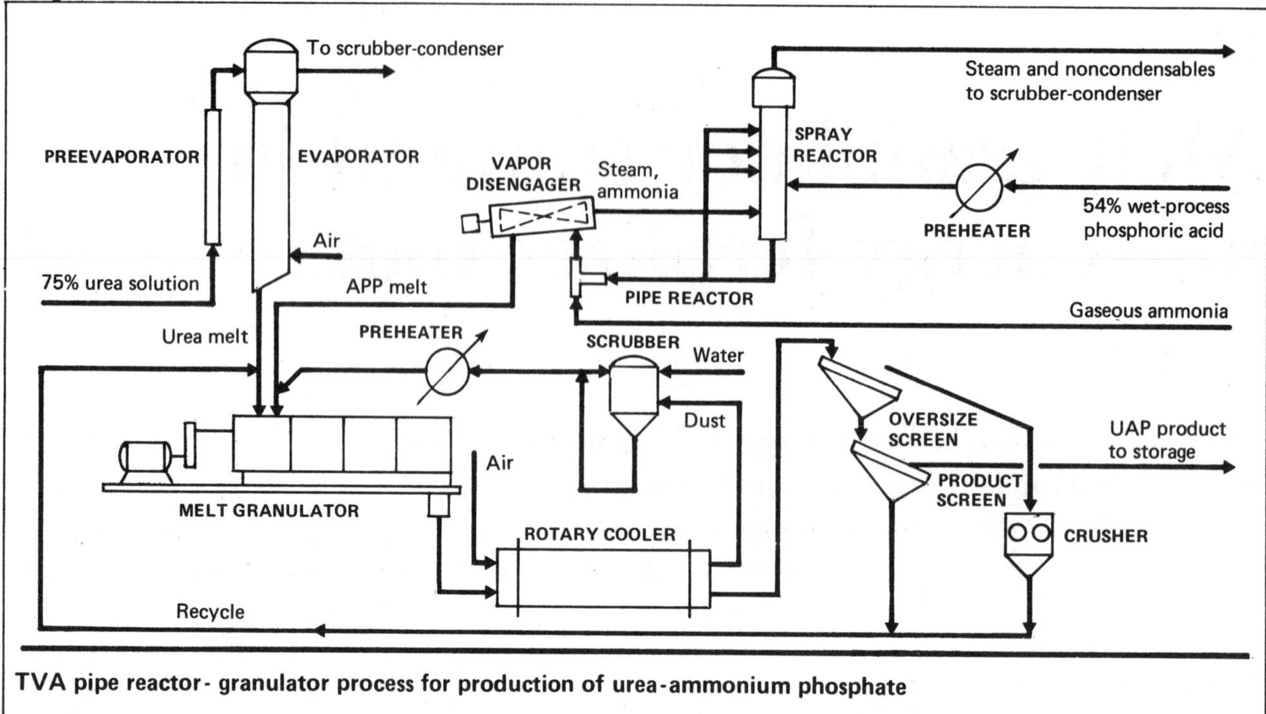

TVA pipe reactor- granulator process for production of urea-ammonium phosphate

The foamy APP melt discharges at 422°F into a vapor disengager (33 in dia. by 10 ft long), for separation of water vapor and unreacted ammonia. The disengager has a rotary helical blade, similar to a reel-type lawnmower, which is turned at 420 rpm by a 20-hp drive. The rotor keeps the melt on the wall of the disengager and conveys it to the discharge. However, it mainly shears the thixotropic melt to keep it fluid.

Gases evolved in the disengager, principally ammonia and steam, are drawn into the spray reactor for ammonia recovery. Overhead gases from the spray reactor pass to a scrubber-condenser that removes virtually all fluorine entering the system as phosphoric acid impurity.

The partially neutralized wet-process phosphoric acid is very corrosive at the high temperatures encountered in the reaction system. Tubes, tube sheets and head boxes for the acid preheater, for example, are all fabricated from 317L stainless steel.

The spray reactor shell can be either fiber-glass-reinforced polyester with fluorine-resistant liner, or Hastelloy alloy G. Other reactor materials include tantalum-clad 316 stainless for the recirculating pump, Teflon-lined carbon steel for recirculating lines, and Teflon for nozzles.

The disengager has a 317 stainless-steel shell, and a rotor of Hastelloy alloy C276, or 316L stainless. The pipe reactor is made from 316L stainless, schedule 40 pipe.

GRANULATION STEP—A recycle of undersize material and crushed oversize first enters the feed end of a pugmill granulator, at a rate of about 6 lb/lb of product. (Potash can be added with this recycle stream to manufacture NPK-grade fertilizers, which Rasa Industries will turn out.) The urea melt is next sprayed onto the recycle, followed by the APP melt, through an open trough. The pugmill in the demonstration plant measures 6 ft wide by 4 ft 10 in deep by 17 ft 6 in long (somewhat larger than needed). Material flowrates at design 400-ton/d UAP throughput are about 8.3 tons/h for urea, and 8.6 tons/h for the APP melt.

A process modification now being developed eliminates the vapor disengager and uses a drum melt-granulator, which is less expensive and more available and familiar to fertilizer operators than pugmills.

The APP melt is fed after the urea to improve binding of the crystallized urea into the granules and minimize dusting off of urea in the subsequent solids-handling equipment. The primary granulation-control parameters are recycle ratio, and APP polyphosphate content.

FINAL PROCESSING—The granulator discharges 160°F product into a countercurrent rotary cooler. Gases from the cooler are scrubbed by water and recirculating liquor. Part of the scrubbing solution, however, returns to the granulator at a point near the APP melt.

Electrically vibrated, single-deck units screen granules exiting the cooler. Oversize material passes to a chainmill crusher that operates either in a closed loop with the screens, or on a once-through basis as shown. Onsize granules (-6 to +12 mesh) are conveyed to storage without any further treatment.

Fertilizer consumers who have used the UAP and APP products in bulk blending and other applications have been well satisfied.

The authors

Robert G. Lee is a senior project leader in the Tennessee Valley Authority's Div. of Chemical Development (Muscle Shoals, AL 35660). A member of AIChE and ACS, he has been with TVA since earning his B.S. in chemical engineering from the University of Tennessee in 1962.

Melvin M. Norton is also a senior project leader in TVA's Div. of Chemical Development. He was graduated in 1950 from the University of North Alabama with a B.S. in chemistry, and has been with TVA since then.

Homer G. Graham is chief, nitrogen fertilizers branch, of TVA's Div. of Chemical Operations. A 27-yr veteran with TVA, he was graduated from the University of Florida in 1948 with a B.S in chemical engineering. He is a professional engineer in the state of Florida and a member of AIChE.

This article is based on a presentation before the 24th Annual Fertilizer Industry Round Table, Washington, D.C.

Potash flotation method handles variable feed

This anionic potash flotation process—the only one presently in commercial use—produces sulfate-of-potash fertilizer from Great Salt Lake brine. The technique can handle varying feed flowrates as well changing feed concentrations.

R. Bruce Tippin, *Great Salt Lake Minerals & Chemicals Corp.*

☐ Great Salt Lake Minerals & Chemicals Corp.'s (GSL) flotation facility beneficiates evaporation-pond deposits to a concentration (about 12−13% potassium) usable by an onsite chemical plant that produces fertilizer in the form of sulfate of potash, K_2SO_4. Flexible enough to operate with flowrates that vary by ±30% from its 120-ton/h nameplate capacity, the flotation unit adapts to inlet feeds that range from 2−8% K, recovering 75−90% of the available potassium, while rejecting 87% ± 7% of contaminating sodium chloride. The facility, which went onstream in January 1976, is the only anionic flotation process producing potash on a commercial scale.

GSL is located on the east shore of Utah's Great Salt Lake. The firm processes brine from the north arm of the lake to produce sulfate-of-potash fertilizer, salt cake, and magnesium-chloride and salt brine.

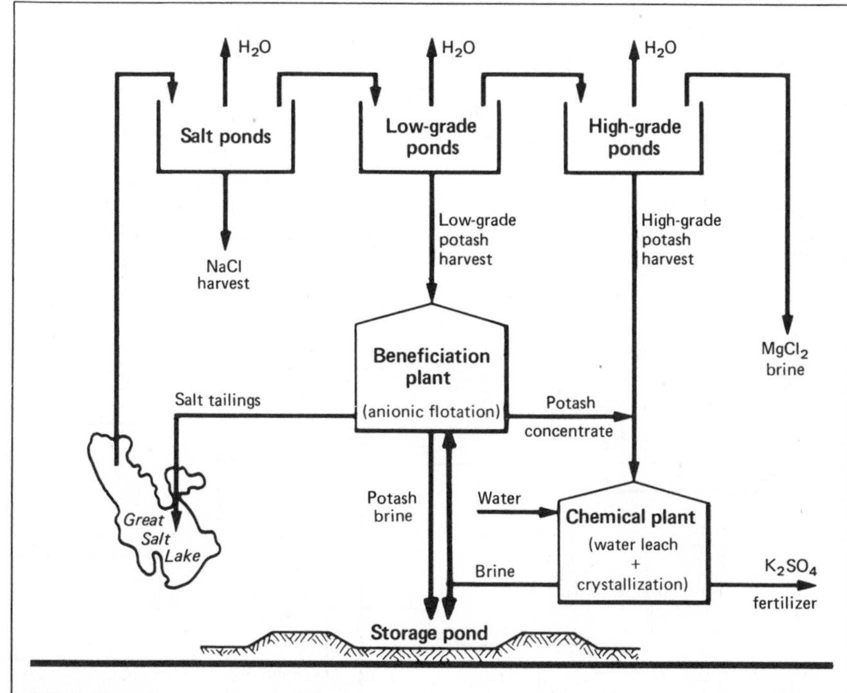

GSL's flotation process holds a central position in K_2SO_4 production Fig. 1

The company's complex (see schematic) consists of a solar evaporation system, and process plants that convert brine to the above products. The evaporation system comprises a 17,000-acre ponding operation where the various salts deposit during the summer evaporation season, and a harvesting system that recovers the deposits during the remainder of the year. High-grade (high-potassium-salt) harvests can be fed directly to the chemical plant. Low- or marginal-grade harvests, however, result in low recovery of K_2SO_4 there, because sodium chloride contamination limits recovery. And marginal-grade harvests occur with (unpredictable) regularity because vagaries of weather alter both the quantities of solar salt deposits and their compositions.

Thus, flotation beneficiates these marginal grade harvests, the resulting concentrate being blended with the high-grade harvest, which blend is fed directly to the chemical plant.

ANIONIC VS. CATIONIC FLOTATION— The Great Salt Lake is a sulfate, rather than a chloride lake, and the complex chemistry of lake brine results in a solar deposit that is a blend of minerals, containing primarily kainite ($KCl{\cdot}MgSO_4{\cdot}3H_2O$), schoenite ($K_2SO_4{\cdot}MgSO_4{\cdot}6H_2O$), and halite ($NaCl$), with some carnallite ($KCl{\cdot}MgCl_2{\cdot}6H_2O$) and epsomite ($MgSO_4{\cdot}7H_2O$). Other minor minerals may be deposited, depending upon the weather during the evaporation season.

Kainite, however, is a less-stable potash salt than is schoenite, and under the right time/temperature conditions, it converts to schoenite, with magnesium chloride brine as a reaction byproduct. In fact, potash deposits that may be 20% schoenite/80% kainite at the beginning of the harvest period often end up assaying at 80−100% schoenite at the end of the season.

Testing revealed that commonly used cationic flotation-reagents, al-

Originally published July 18, 1977

though selective for KCl-based potash salts, were not particularly effective in floating schoenite. Anionic reagents had the opposite effect.

GSL uses a short-chain, saturated-fatty-acid collector derived from coconut oil to float the schoenite away from the sodium chloride that remains in the brine. Although the reagent is quite specific to schoenite, the actual mechanism whereby flotation occurs is not fully understood.

The key then, to the success of the GSL beneficiation process, is the conversion of the harvest to schoenite. Although this will in time happen naturally, GSL "forces" the conversion by reacting the harvest with chemical-plant brine during grinding, conditioning and even flotation. And since neither kainite nor carnallite is found in flotation concentrates, it seems that conversion to schoenite takes place even in the flotation circuit itself.

THE PROCESS—Feed harvest comes from the low-grade ponds (see flowsheet), and consists of about 75% solids. Potash salts make up approximately 20–60% of the feed stream.

The harvest goes through a crusher, where particles are reduced to less than 1 in. in size, and then through a sampler into a ball-mill collecting-sump. Samplers located throughout the process provide specimens for analysis of the five ions of interest (K^+, Na^+, Mg^{+2}, Cl^- and SO_4^{-2}) and insolubles (such as clay, sand, etc.).

Brine from the chemical plant is added to the harvest in the collecting sump. The purpose of the brine is twofold: (1) it dilutes the harvest to 50% solids, the concentration at which the ball mill operates most efficiently, and (2) the brine "reacts" with kainite in the harvest to yield schoenite, the mineral form required for most-effective flotation. The heat developed in the ball mill during comminution aids in the conversion of the kainite to the required mineral form.

In the ball mill, the harvest is ground to less than 35 mesh. The ground material then flows to another sump where additional brine from the chemical plant is added. Once again, the purpose of the brine is both to abet the conversion to schoenite and to further dilute the stream to a thin slurry of about 20% solids. Recycled materials from the flotation cells also combine with the harvest at this point.

The thin slurry is pumped to a classifying cyclone that culls oversized material from the flowstream. The cyclone's underflow—a stream of about 55% solids, containing particles larger than 35 mesh—goes back into the collecting sump for recycle through the ball mill. The cyclone's overflow—properly-sized particles in a stream that is about 15% solids—flows through a sampler into the first of two agitated conditioning tanks.

The anionic reagent is added in the first tank; the collector adheres to the schoenite particles but not to the halite. The second conditioning tank simply provides holdup time, to ensure that the reagent has adsorbed onto the potash and to provide extra reaction time between the brine and any kainite still in the flowstream. The total holdup is about 5–10 min.

From the conditioning tanks, the harvest proceeds onward to the flotation cells.

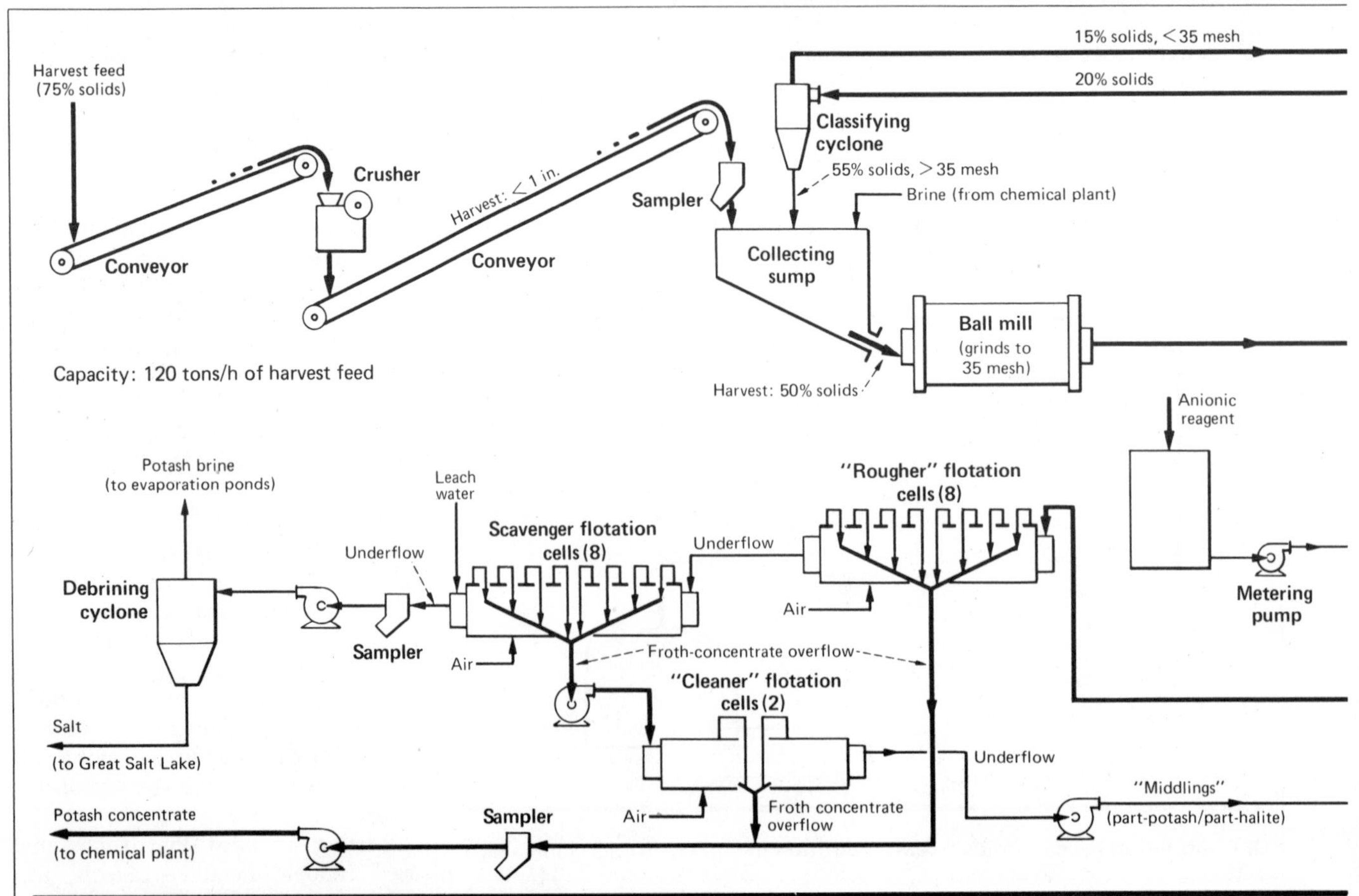

The flotation process is designed for flexible operation

HEART OF THE PROCESS—Actual beneficiation of the harvest takes place in a flotation system consisting of a "rougher" flotation circuit followed by a scavenger set of flotation tanks.

The flotation cells are standard units made from long, trough-like vessels divided into vertical compartments by baffles that do not extend to the base of the trough. Each compartment contains an agitator, from the bottom of which air is sparged. The air, preferentially adhering to the reagent-coated schoenite particles, increases the buoyancy of the particles so that they float to the surface of the compartment, forming a froth. The froth overflows the top of the compartment into a collecting trough, whence it progresses through the system.

The heavier, nonbuoyant halite particles remain in the solution left behind and underflow the baffles, being subjected to the aeration process in each compartment until the halite-laden stream exits the flotation cells.

Because of the specificity of the flotation system for schoenite, the froth concentrate that overflows from the "rougher" flotation circuit is essentially saltfree and can go directly to the chemical plant. The halite-laden underflow proceeds to the scavenger circuit for recovery of the last remaining potassium values.

The underflow from the scavenger unit—which may still contain small amounts of schoenite—is subjected to a water leach upon exiting from the flotation cell, in order to dissolve entrained potash. The leach stream then goes to a debrining cyclone, where the halite separates out as a solid, and the potash—in solution as a brine—goes to the evaporation ponds for use in the next evaporation season. The halite returns to the Great Salt Lake.

The froth-concentrate overflow from the scavenger unit proceeds to a subsequent cleaning unit because, with much of the schoenite having been removed in the rougher circuit, a considerable amount of halite gets entrapped in the froth. By slowing the air-sparging rate, and by controlling the speed of the agitators in the "cleaner" flotation cells, the halite drops out, resulting in a very-high-grade concentrate. The cleaner and rougher concentrates are then combined to make a final flotation-concentrate product that is pumped over to the chemical plant.

Underflow from the cleaner unit contains primarily "middlings"—part-schoenite/part-halite particles that cannot be separated by leaching or flotation. These middlings—most greater than 35 mesh—are recycled back to the ball mill for further grinding and another chance for potash recovery.

Fig. 2

FLEXIBLE PROCESS A MUST–GSL must closely monitor and review the economic balance of operating the flotation plant in conjunction with the chemical plant on the basis of the season's harvest tonnage and potassium distribution. This is because the flotation concentrate is blended with direct harvest to provide feed for the chemical plant. And the higher the grade of the flotation concentrate, the lower the grade of direct harvest that can be blended to result in an optimum chemical-plant feedstock.

Experience has shown that sometimes the scavenger froth-concentrates are sufficiently high-grade as to not require cleaning. The plant is so designed that the scavenger concentrate can be directly combined with the rougher concentrate and sent to the chemical plant. Conversely, the cleaner-unit tailings are often as low in potassium values as are the scavenger-unit tailings, and can be rejected directly to waste, rather than being recycled into the flotation circuit.

It is apparent from production runs that a 12%-K flotation concentrate is easily achievable and is not dependent on the grade of harvest feed. Flotation concentrates as high as 16% K have been produced periodically in the process, and it is not inconceivable that as operators become more experienced, the concentrates should average 14% potassium. (Schoenite theoretically contains 19% potassium.)

COST INFORMATION—The 120-ton/h facility cost approximately $2 million to build, and includes all equipment shown in the flowsheet, from the initial conveyor to the final dewatering cyclone. Operating cost factors break down into labor, chemicals and utilities.

The flotation unit requires only one operator per shift, and the process operates for three shifts a day. Chemical costs run about 50–75¢/lb of reagent, and reagent use is at the rate of ¼–½ lb/ton of feed. Utilities usage is primarily electricity, since no water—other than leach water—is used, and the process needs no heat input. Electrical consumption is approximately 12.5 kWh/ton of feed.

OPERATING NOTE—GSL's flotation process produces a very tough and voluminous froth, which poses a problem in pumping the concentrate from the flotation plant to the chemical plant, in that the pumps have a hard time handling the froth volume at the design rate. The use of brine sprays, air impingement and baffles can partially control the froth, but it is still necessary to add a defoamer to the concentrate to deaerate it for pumping.

The author

R. Bruce Tippin is Manager of Research and Development for Great Salt Lake Minerals & Chemicals Corp. (P.O. Box 1190, Ogden, UT 84402). He received his Ph.D. in Minerals Engineering from the U. of Minnesota. He holds a B.S. From New Mexico School of Mines and an M.S. from the U. of Alabama.

Phosphoric acid purification: comparing the process choices

Solvent extraction techniques are already commercial for purifying wet-process phosphoric acid. Other types of processes, however, are under laboratory or pilot-plant development at TVA; indirect purification by a urea-phosphate pyrolysis method is especially promising.

John F. McCullough, Tennessee Valley Authority

☐ In the fertilizer industry, the major reason for purifying wet-process phosphoric acid is to produce clear solution fertilizers that can be stored for extended periods without precipitation of solids.

Today's solution fertilizers are prepared either from unpurified wet-process acid, or from wet-process acid purified in respect to magnesium only. These are being successfully used, but storage life is shortened markedly by the impurities.

The need for purified acids for preparing truly reliable solution fertilizers, furthermore, is expected to become even more pressing in the future, because of a general decline in phosphate rock quality.

The degree of purification required for solution fertilizers has not been defined precisely—it varies from acid to acid. Generally, however, the removal of about 60% of the metallic and fluorine impurities from typical Florida acid should be sufficient. Higher degrees of purification probably are needed for acids prepared from North Carolina and western rocks because of their high magnesium content.

FOUR PROCESS TYPES—Purification processes designed specifically for the fertilizer industry (and not to turn out higher purity, higher priced product for other industrial uses) fall into the following broad categories:

Solvent extraction—A partially miscible solvent extracts the major portion of acid, but little of the impurities, from the impure aqueous acid. Back-extraction with water recovers phosphoric acid from the recycled solvent. Small amounts of solvent entering the aqueous phase are usually recovered by distillation.

Solvent precipitation—Acid is treated with a completely miscible solvent, usually in combination with alkalis or ammonia, to cause impurities to precipitate, mostly as phosphate salts. The solds are separated, and the solvent distilled from the liquid phase and recycled, leaving purified acid as residue.

Indirect purification—Acid is not produced *per se*, but is separated from the impurities as a salt or an organic adduct, which is converted into the end-product.

Clarification—Suspended solid impurities are separated from the phosphate by settling or flotation—the former has long been a routine practice in the fertilizer industry.

SOLVENT EXTRACTION—One such process has been developed by Israel's IMI-Institute for Research & Development, and industrially implemented by Fertilizantes Fosfatados Mexicanos.

The feedstock is 54% wet-process phosphoric acid. Extraction and phase separation are made at low temperature in mixer-settlers, and the clean acid is released from the bulk of the solvent into water at about ambient temperature. Dissolved solvent is recovered from the pure and impure aqueous streams by steam-stripping under vacuum. From 60–70% of the P_2O_5 is recovered as purified acid having a concentration of 48–51% P_2O_5.

About 97% or more of the metals, and about 88% of the fluorine, is removed from the usual Florida acid. Although organic matter is preferentially concentrated in the raffinate, the acid still blackens on concentration. To produce clear acid, a proprietary oxidation method is used.

A solvent extraction technique using *n*-heptanol, developed by USS Agri-Chemicals, has also been implemented on a plant scale. The low solubility of heptanol in aqueous phosphoric acid makes it unnecessary to recover solvent from the acid, and the absence of heating and cooling steps further simplifies the process and decreases capital costs. Solvent losses due to solubility and entrainment per ton of P_2O_5 are only 7–12 lb with green acid, and 12–17 lb with black acid.

The process is fed with 54% acid, and uses mixer-settlers for extraction and phase separation. The concentration of the product acid is about 44% P_2O_5, and it typically contains 80% of the input acid. Typical impurity rejections are 80–90% Mg, 70–80% Al, 50–60% Fe, and 75–85% F. The process accepts black acid, but a solvent cleanup step must be

* This article is based on a presentation at the 26th Annual Meeting of the Fertilizer Industry Round Table, Oct. 26-28, 1976, Atlanta, Ga.

Originally published December 6, 1976

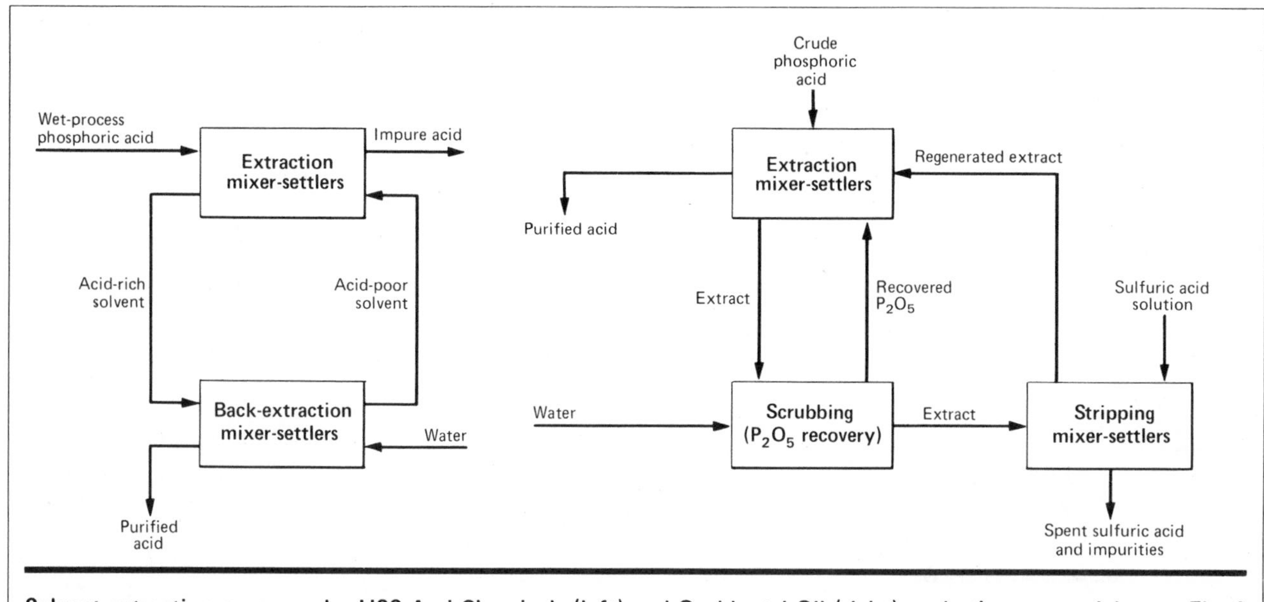

Solvent extraction processes by USS Agri-Chemicals (left) and Occidental Oil (right) are both commercial **Fig. 1**

added to prevent buildup of tarlike organics. Green acid provides smoother and more-economical plant operation. (See Fig. 1.)

The S-X process developed by Garrett Research, and successfully tested on a plant scale by Occidental Oil, is unique in that it selectively extracts Mg and Ca from the acid by liquid ion exchange. The extractant is sulfonic acid dissolved in kerosene. The objective of the process: upgrading wet-process phosphoric acid made from low-grade rock, without the coproduction of impure byproduct acid. About 94–97% of the input P_2O_5 is indeed recovered as purified acid.

Crude acid first contacts extractant in mixer-settlers to remove Mg and incidental amounts of other metals. Entrained P_2O_5 is recovered by water-scrubbing from the extract, which is regenerated by contact with sulfuric acid to remove metals. The spent sulfuric acid solution carrying the metal impurities is treated with lime and discarded. (Also Fig. 1.)

SOLVENT PRECIPITATION—A process developed through the laboratory stage and patented by TVA treats merchant-grade acid with methanol and a small amount of ammonia, to precipitate most of the impurities as metal ammonium phosphates and fluorine compounds. The slurry is centrifuged and the filtrate distilled to remove methanol and water, leaving purified superphosphoric acid. Methanol is separated from water in a fractionating tower, and reused. The solids are washed with recirculating methanol and dried. Solvent is lost only by mechanical means; chemical losses are negligible.

Purification increases with use of more methanol and/or ammonia, but filtration rates drop off. About 3.5 lb of methanol and 0.04 lb of ammonia per lb of P_2O_5 are optimum amounts for good filtration and the removal of about 90% or more of Fe, Al and F, and up to 70% of the Mg, from the usual Florida acid. Unfortunately, none of the carbon is removed during the process and the product acid is black.

The solids typically contain 10% of the input P_2O_5 and have a 5-48-0 grade (nitrogen-P_2O_5-K_2O content). All of the P_2O_5 is citrate-soluble and 50% is water soluble. Therefore, the byproduct is a potential fertilizer.

The most costly part of the process is distillation; however, this is offset by the high concentration of the product acid (% P_2O_5) and the 90/10 split between product and raffinate streams.

Another such process developed through the laboratory stage by TVA treats the feed acid with acetone and a small amount of ammonia. This forms two liquid phases: The light, acetone-rich phase contains the purified acid, while the heavy, acetone-poor phase contains most of the impurities and about 25% of the acid.

The phases are quickly and cleanly separated in a settler. All of the acetone and part of the water is distilled from the acetone-rich phase to yield concentrated purified acid. The acetone is separated from the water in a fractionation column and reused. The acetone-poor phase is ammoniated and dried to form impure diammonium phosphate, while recovering acetone. Acetone is lost only mechanically.

The optimum amounts of acetone and ammonia, for both smooth operation and the removal of 90% or more of the impurities from the usual acid, are 3.4 lb of acetone and 0.037 lb of ammonia per lb of P_2O_5. Under these conditions about 75% of the acid comes out in the purified product. Only part of the carbonaceous matter is removed, so the concentrated acid is black. The byproduct diammonium phosphate has a typical grade of 16-44-0; essentially all of the P_2O_5 is citrate-soluble and about 80% is water-soluble.

As with the methanol process, distillation is the major expense. Compared with the methanol process, the acetone/ammonia method has the advantages of liquid-liquid separation and lower energy costs for distillation, and the disadvantages of a lower product-to-raffinate split and the use of a more expensive solvent.

INDIRECT PURIFICATION—The

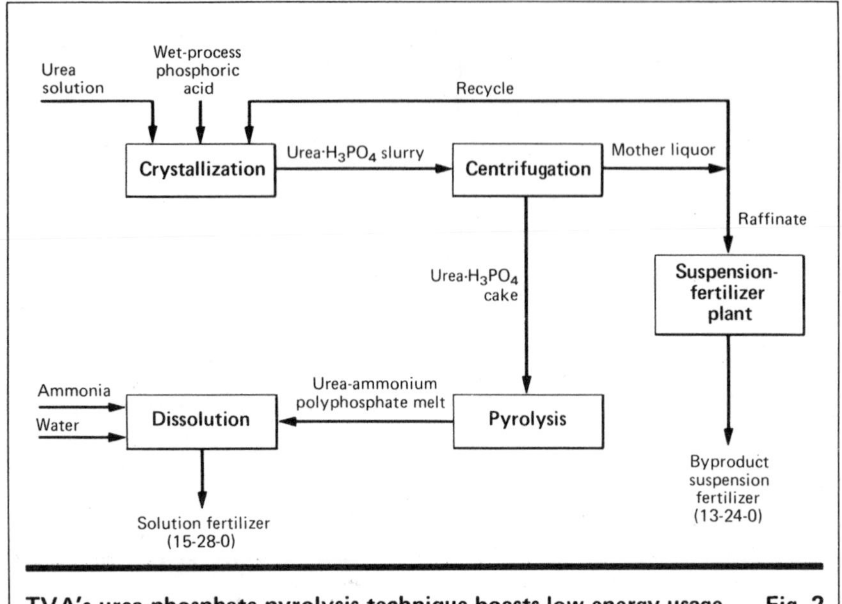

TVA's urea-phosphate pyrolysis technique boasts low energy usage Fig. 2

process detailed here has been under pilot-plant development by TVA; similar ones have been groomed by Pechiney Ugine Kuhlmann in France and BASF in Germany.

Urea solution (99+%), 54% wet-process phosphoric acid, and recycle filtrate combine in a reactor to precipitate pure urea phosphate (Fig. 2). Mother liquor is centrifuged from the crystals; part of the liquor recycles to the reactor to fluidize the slurry, while the rest is processed into 13-24-0 suspension fertilizer. The urea phosphate cake is pyrolyzed at 250+°F to form a melt of ammonium polyphosphate and undecomposed urea. The melt is then dissolved in water, and ammonia is added as needed for pH adjustment to form 15-28-0 solution.

The typical solution is of satisfactory clarity. It contains 85% of the phosphate but only 15% of the original impurities. However, clarity can be a problem when the solution contains a greater proportion of the impurities.

The major advantage of the urea-phosphate pyrolysis process is its low energy requirement, due to the complete dewatering of the urea phosphate and the limited heat needed for pyrolysis. Other pluses are the removal of carbonaceous matter, and process simplicity. A disadvantage is the use of relatively expensive urea nitrogen for condensation of the phosphate.

A process for preparing phosphoric acid from purified urea phosphate by treatment with nitric acid has also been developed through the laboratory stage by TVA. A similar procedure has been patented by the French company Azote et Produits Chimiques.

Purified urea phosphate (prepared as described above) is treated with concentrated nitric acid to form crystalline urea nitrate and phosphoric acid. The urea nitrate slurry is filtered, and the cake washed with nitric acid (subsequently used for the reaction with urea phosphate). The filtrate is purified phosphoric acid; part is recycled to the reactor to fluidize the slurry, and the rest is concentrated to superphosphoric acid.

The urea nitrate cake is neutralized with aqueous ammonia to form urea ammonium nitrate solution. Trace amounts of phosphate and impurities in this stream do not affect its stability.

The process can use nitric acid concentrations from 50–72%, but as high a concentration as is economically feasible should be chosen, to maximize the phosphoric acid concentration. Substantially all of the P_2O_5 and impurities in the starting urea phosphate comes out in the purified acid product. It is clear, amber in color, and typically contains 85% of the P_2O_5 and 15% of the impurities in the wet-process phosphoric acid.

The urea nitrate process has higher capital and operating costs than the urea-phosphate pyrolysis process. However, this is at least partially offset by manufacture of the more versatile acid and by lower raw material costs (since the cost of urea and nitric acid can be charged to urea ammonium nitrate production).

CLARIFICATION—A new process for the clarification of black solution fertilizers has been developed through the pilot-plant stage by TVA. It starts by mixing warm, freshly prepared, black solution fertilizer with long-chain aliphatic amines and quaternary ammonium chlorides. A pound of each flotation agent is used per ton of solution. In a separator, finely dispersed black matter floats to the surface for withdrawal along with 15% of the solution. This stream is cooled and processed into suspension fertilizer. The clarified product, containing 85% of the input solution, is withdrawn from the bottom of the separator and cooled. The clarification process removes only solid carbonaceous material; other impurities remain in the clarified product in the same proportion as in the original black liquid fertilizer.

ECONOMIC COMPARISON—TVA has made preliminary estimates for the cost of solution fertilizers prepared from phosphoric acid purified by the methanol, acetone and urea-phosphate processes. Although product from the urea phosphate alternative came out slightly cheaper than the others, all product costs were essentially the same within the probable error of the estimates.

Technical factors such as process complexity, reliability and energy-cost trends were thus considered in selecting the most promising process. The urea-phosphate pyrolysis method is the least complex, and has by far the lowest energy requirement. Therefore, TVA has chosen this technique for further development in a large-scale pilot plant.

The author

John F. McCullough is a research chemist in the Fundamental Research Branch, Division of Chemical Development, Tennessee Valley Authority, Muscle Shoals, AL 35660. He holds B.S. and M.S. degrees in chemistry from Auburn University, and is a member of the American Chemical Soc.

Enter: Sulfur-Coated Urea A Slow-Release Fertilizer

Moving from development to commercialization is a method that uses sulfur to encapsulate nitrate fertilizers and thus slow down the release of nutrients.

Changing from air-atomized sulfur sprays to hydraulic atomization at high sulfur-pump pressures sounds like a simple matter, and it is. But that switch has doubled the capacity and also reduced sulfur dust formation by 90% in a sulfur-coated urea pilot plant at Tennessee Valley Authority's Fertilizer Development Center at Muscle Shoals, Ala.

The new TVA technology (see flowsheet) is going into a 5-ton/hr, $750,000 commercial plant that Canadian Industries Ltd. plans to bring onstream in mid-1975 at Courtright, Ont. TVA itself will build a 10-ton/hr unit by 1976 in order to meet the increasing demand by developing countries and others for sulfur-coated urea (SCU).

Pilot-plant or semicommercial units are also in operation in Japan and England—notably, Imperial Chemical Industries' 1.4-ton/hr plant at Billingham, Teasside, U.K.

Outlook for SCU—ICI has been employing a modified version of the TVA process to make SCU at Billingham since March 1972. Feed is urea prills, and air atomization is used for sulfur spraying. Pan-granulated urea (which TVA uses) or spherodized urea (C & I Girdler's product) would be preferred as the substrate, as both are larger and more-uniform than prills, but neither of these varieties is readily available in Britain.

ICI markets its SCU in the U.S. and elsewhere under the name GOLD-N and has had good acceptance for fertilizing lawns, parks, golf courses, etc. Only one application a year is required, as compared with perhaps three for uncoated urea or ammonia nitrate. ICI expects a bigger market for SCU also will develop for fertilizing specialty crops such as lettuce, brussels sprouts and cabbage. By applying a slow-release fertilizer at the time of planting, the

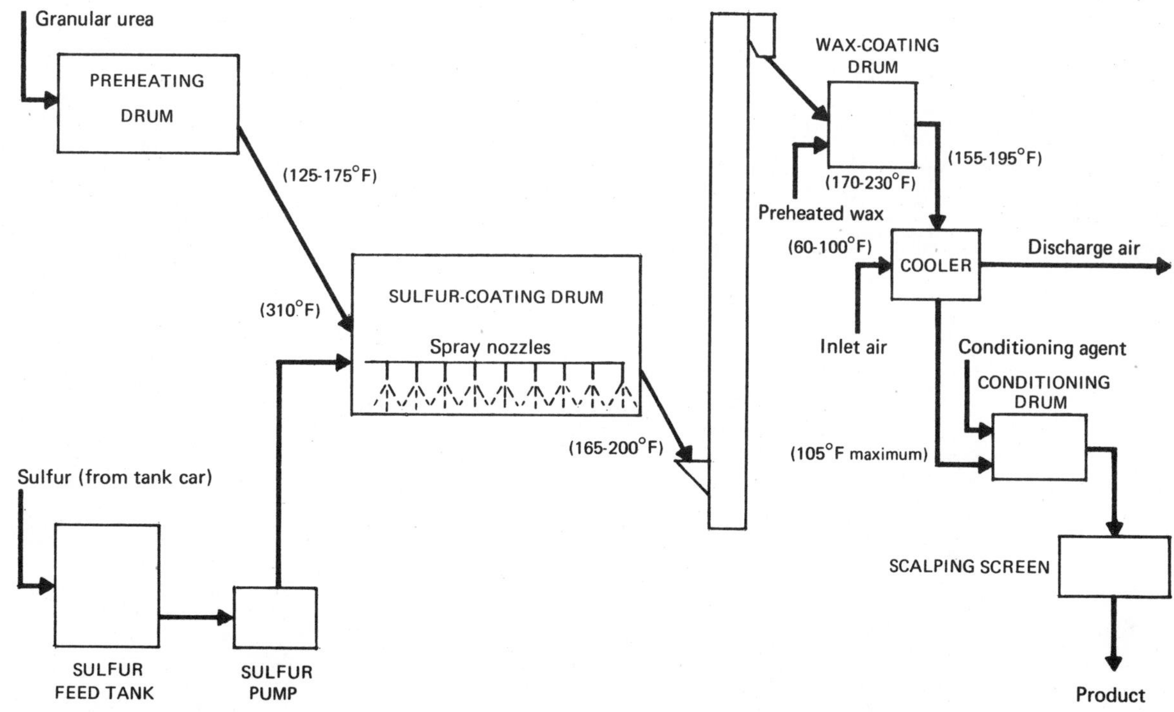

SULFUR-COATING system for urea, as developed by the Tennessee Valley Authority, starts by preheating the urea (radiant heaters are used in the present 1-ton/hr unit, but fluid-bed, hot-air heat will be used in a 10-ton/hr plant scheduled for 1976). After sulfur is applied (13-26%, by weight) in the coating drum, wax (2-3%, by weight) is sprayed on as a sealant. Next, a fluid-bed unit cools the material (to prevent sticking) and a diatomaceous earth is added (2-3%, by weight) to condition the product for storage. Finally, any agglomerates that form are removed.

Originally published December 23, 1974

farmer not only saves labor but also avoids leaf burning.*

(A competing slow-release nitrogen fertilizer is ureaform. A copolymer of urea and formaldehyde, the resin is a complicated mixture of cyclic and polymeric compounds, many of which are insoluble in water. In fact, only about 70% of ureaform ever reaches the plant. Moreover, TVA estimates SCUs at about half that of ureaform.

Cost Comparison—SCU costs 25-40% more than uncoated urea. But for grass and slow-maturing crops, says TVA and ICI, this is more than offset by savings in urea losses and less-frequent application.

The Hawaii Sugar Cane Planters Assn. has conducted a series of six tests which show that one application of SCU to sugar cane will yield the same as four to six of uncoated urea. Also, the last two applications of conventional fertilizer must be made by air, because the cane is then too thick for tractors.

TVA believes that there will be a big demand for SCU in the developing countries, especially for intermittently flooded rice, sugar cane, pineapple, watermelon, tomatoes, etc. Last October, TVA released estimated 1977 cost figures for SCU plants built in developing countries such as Taiwan, South Korea and the Philippines. For example, one 200-metric-ton/d SCU plant that uses prills from an existing ammonia-urea complex requires a $3,375,000 capital investment. Feeding $150/metric ton urea, total production cost of SCU would be $141/metric ton. With a 20% return on investment before taxes, the inplant price of SCU would be $151/metric ton for a 32-0-0-26S fertilizer. #

*Nitrogen compounds—ammonia, ammonium nitrate, urea, etc.—are very soluble and therefore subject to considerable loss by runoff, leaching or evaporation (urea hydrolyzes to ammonia and CO_2). Some plants such as lettuce take up nitrogen faster than needed and give off the excess through their leaves—a process called leaf burning.

Coating-Drum Design Was an Evolutionary Process

After studying the sulfur-encapsulation of soluble fertilizer granules for about 12 yr, the Tennessee Valley Authority built a 1-ton/hr pilot plant in 1972. Since that time, the heart of that system—the coating drum—has been extensively modified: A sulfur-dust-explosion hazard was solved by switching from air atomization to hydraulic sprays (sulfur-in-air levels dropped from 10 g/m to 1 g/m). This change also permitted the use of lifting flights and a collecting pan to create a falling curtain of urea and double the capacity of the drum to 2 tons/hr. The stop-action photo shows the falling curtain concept in action—a system that could not be used with air atomization because the excessive sulfur dust and mist produced severe sulfur buildup on the inside of the drum. TVA has also varied the sulfur-pump pressure and has found the optimum to be around 1,000 psi.

Section III
FOODS, FLAVORS AND FRAGRANCES

Fermentation Process Turns Whey Into Valuable Protein

A large-scale operation is now growing high-protein yeast on whey, converting a potential environmental contaminant into a needed food and feed product.

An "extremely promising" scheme for fermenting yeast on whey started up in January in a prototype but full-size commercial facility. That news was reported last month by Sheldon Bernstein, president of both Milbrew, Inc. and its Amber Laboratories Div. (330 South Mill St., Juneau, WI 53039), speaking before the National Conference on Management & Disposal of Residues from the Treatment of Industrial Wastewaters.*

Using a process groomed by Amber Laboratories with partial sponsorship from the U.S. Environmental Protection Agency, the brand-new plant is rated to turn out 5,000 tons/yr of yeast product from three times as much whey solids. Actual output can be several tons higher or lower, depending on product type and operating conditions.

Bernstein points out that disposal or utilization of whey has long been a problem (see box). Nearly 1 billion lb of whey solids go unutilized in the U.S. each year, representing about half the solids carried in the cheese industry's 30 billion lb/yr of liquid whey byproduct.

Medium Preparation—The fermentation process described by Bernstein starts off with acid or sweet whey in concentrated form (45-50% solids). This is diluted with water, raw whey (6-6.5% solids), or recycle condensate water, to an appropriate concentration of lactose—the carbohydrate substrate for yeast growth that comprises 64-72% of whey solids.

Other additions to the medium include anhydrous ammonia as the primary exogenous source of nitrogen, dilute phosphoric acid and yeast extract, as growth nutrients, and hydrochloric acid to adjust pH to 4.5.

After pasteurization by heating to 80°C for 45 min, the medium is cooled to fermentation temperature of 30°C and transferred to the fermenter. The commercial plant has a 15,000-gal deep-tank unit, fabricated from stainless steel, fully aerated, and jacketed for heat removal.

The yeast microorganism selected by Amber Laboratories is a strain of *Saccharomyces fragilis*. To provide adequate fermentation conditions, seed material amounting to 10-20% of the fermenter volume must also be added to the fermenter.

Fermentation Step—While both batch and semicontinuous fermentations can be run, the system operates more efficiently with continuous addition of fresh medium and withdrawal of fermented mass.

A continuous fermentation begins when the cell count in the fermenter broth reaches 1×10^9 cells/ml, and lactose concentration is 0.50-0.75%. (In one of the first runs with the new 15,000-gal unit, cell count came to that level after 12 h; addition and withdrawal rates were then kept at 1,250 gal/h.) Conversion of lactose to yeast is virtually complete, and results in 0.45-0.55 lb of yeast per lb of lactose.

By operating at low pH, large seed size and high cell count, the fermentation sidesteps contamination problems, and thus the need for sterile or aseptic equipment or techniques. Aeration requirements are reasonable. Neither foam control nor temperature control is troublesome.

Final Processing—Upon exiting the fermenter, the product mass passes through a holding tank to a three-stage evaporation. Concentration from the 8%-solids level of the fermentation broth to 27% solids occurs. Spray drying of the stream finally yields a yeast-fermentation-solubles product, approved by the U.S. Food and Drug Administration as an

Why Whey Is A Problem

Whey is a greenish-yellow liquid containing 6-6.5% solids and most of the water-soluble vitamins and minerals of the whole milk it is derived from during cheesemaking. In addition to 64-72% lactose, whey solids consist of 11-13% protein, 8-9% minerals, and small amounts of fat and lactic acid.

Since whey liquid is mostly water, hauling it any great distance for use or disposal is extremely expensive. The half that is processed for recovery of dry solids (a human and animal food) requires a large capital investment for evaporation equipment, says Amber Laboratories' Bernstein, and often processing costs are just barely recovered.

Disposing of raw whey causes trouble, too. Biochemical oxygen demand of the material runs 30,000-50,000 ppm, and with large cheese plants churning out up to 1 million lb/d, large and costly waste-treatment plants are needed.

*The conference proceedings, available for $25 from Information Transfer, Inc., 6110 Executive Blvd., Suite 750, Rockville, MD 20852, includes the entire text of Bernstein's presentation, on which this article is based.

Originally published March 17, 1975

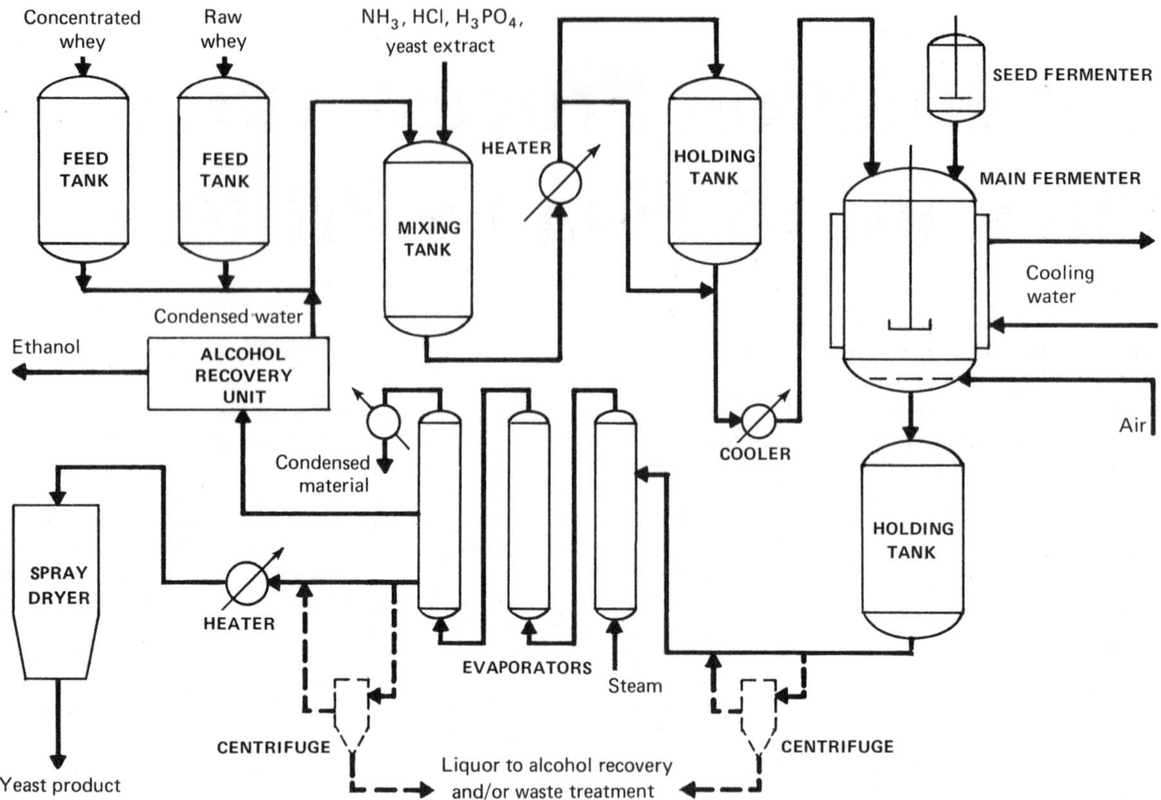

animal feed. Proximate analysis of the material shows a crude protein content of 35-50%, ash content of 15-20%, fat 2-3%, and moisture 3-4%.

The overhead from the evaporation system contains not only water but also ethyl alcohol, one of the metabolic products of fermentation. This is recovered as byproduct. The amount can in fact be increased (at the expense of cell yield) by modifying fermentation conditions, Bernstein notes, to an optimum level determined by such economic factors as recovery cost.

Water condensed during alcohol recovery, meanwhile, can be recycled to the medium-preparation section. This results in a closed-loop system and zero effluents for the process as a whole. Fermentation does not differ significantly whether tap water, raw whey or condensed water serves for diluting the medium, Bernstein says, and there is no apparent buildup of toxic materials.

Though with a penalty in fermentation yield, an FDA-approved food-grade yeast can also be produced. Centrifugation harvests this from the yeast-fermentation-solubles mass (shown in dotted lines in flow diagram). Supernatant streams from the centrifuges could be processed for alcohol recovery, then disposed of after waste treatment. The product yeast, designated Amber Nutrex, contains 45-55% crude protein, 6-10% ash, 2% fat, and 3-4% moisture.

Amino acid content of Amber Nutrex is shown in the table, versus the United Nations' Food and Agricultural Organization standard profile, and versus brewers and torula yeasts. Quality is good overall, but just as with other single-cell proteins, the product is somewhat deficient in sulfur-containing compounds such as methionine.

Bernstein figures a commercial plant for 4,000-10,000 tons/yr of product would cost in the range of $5-15 million. Production expenses for feed-grade material are estimated at 10-15¢/lb, a level that should be competitive with alternative sources.—NRI #

Amino Acid Content of Various Proteins

	% of Total Protein			
Amino Acid	FAO Profile	Amber Nutrex	Brewers Yeast	Torula Yeast
Lysine	4.2	6.9	6.8	8.5
Threonine	2.8	5.8	5.9	5.1
Methionine	2.2	1.6	1.5	1.5
Valine	4.2	5.4	4.7	5.6
Leucine	4.8	7.0	5.8	8.0
Isoleucine	4.2	4.0	3.6	6.4
Tyrosine	2.8	2.5	2.7	4.3
Phenylalanine	2.8	3.4	3.4	5.1
Tryptophan	1.4	1.4	1.1	—
Histidine	—	2.1	2.1	2.2

Current Developments in Fermentation

Fermentation is becoming ever more important to chemical engineers. Here is a summary of those applications that hold the greatest promise, plus highlights of the related chemical-engineering principles.

ARTHUR E. HUMPHREY, University of Pennsylvania

I believe society is heading toward a world community of energy imperialists and protein imperialists, where both protein and energy will be derived from many sources including oil, wastes, and renewable resources.

For some time, society has primarily considered only fixed resources for energy, and fossil fuels have been squandered in man's desire for private transportation, to the point where modern man is pictured as driving to his energy-poor house in a Rolls-Royce.

Similarly, the demand for status foods, i.e. red meat, has resulted in a squandering of precious reserves of cereal grains in many areas. A look at yield data in Table I indicates the wastefulness, i.e. low efficiency of crop carbohydrates, as these products are used in the production of status foods.

In effect, the world is rapidly approaching the point where man can no longer depend solely on agriculture, animal breeding and fishing for his food. He must turn to other forms of food synthesis. The energy and environmental crises will affect decisions regarding the development of these forms. Will society expend its precious fixed resources primarily for energy, or will those resources be used for chemicals and food production? What type of considerations—free-enterprise or societal—will dominate?

I personally feel that the time scale for the production of alternate protein sources depends solely upon (1) the extent of the energy crisis (because agriculture is now an energy-intensive process) and (2) the nature and extent of the next agricultural shortfall.

Also, we must consider the size and extent of the solid-waste problem in the U.S. (Table II). The biggest

Originally published December 9, 1974

Conversion of Sugar to Various Foods*— Table I

Item	g Food/100 g Sugar
Beef	8-12
Poultry	10-15
Milk	15-20
Yeast	45-55
Sweet syrup	90-100

*Hirst, E., Food - Related Energy Requirements, Science, 184, 134 (1974).

U.S. Solid-Waste Production* — Table II

Waste Type	10^6 Ton/Yr
Agricultural and food wastes	400
Manure	200
Urban refuse	150
Logging and other wood wastes	60
Industrial wastes	45
Municipal sewage solids	15
Miscellaneous organic wastes	70
Total	940

Note: Approximately half of this is cellulosic material, i.e., 0.5 billion tons.

*Steffgen, F. W., Project Rescue — Energy from Solid Wastes, Pittsburgh Energy Research Center, U.S. Bureau of Mines, Pittsburgh, Oct. 1972.

portion of wastes are agricultural in origin. The largest single, readily collectable wastes are those from large animal feedlots, i.e. manure.

In order to apply his skills to these alternatives, the new breed of chemical engineer must not only be an engineer, he must also be capable of factoring the human element into his process considerations. He will become much involved with the many life-science processes that are being reduced to practice through application of chemical engineering principles.

The popular press has reported many of these chemical-engineering activities, such as creating heart-lung machines to assist doctors in open-heart surgery. However, we shall here consider some of the life-science activities of chemical engineers that have received less public attention but are every bit as important in our daily lives. We shall consider fermentation—both in terms of the promise that this ancient art now holds for us in the production of protein and how protein production embodies some modern applications of chemical-engineering principles.

Mass Doubling Times — Table III

Organism	Time for One Mass Doubling
Bacteria and yeast	10-120 min
Mold and algae	2-6 h
Grass and some plants	1-2 wk
Chickens	2-4 wk
Pigs	4-6 wk
Cattle	1-2 mo
People	0.2-0.5 yr

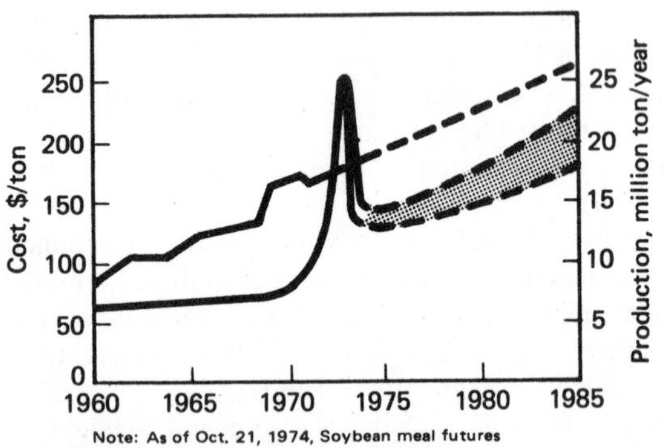

SOYBEAN MEAL PRICES AND PRODUCTION (FOB Decatur, Ill.) projected through 1985—Fig. 1

Food and Feed by Fermentation

One of the important new sources of food and animal-feed protein is that of single-cell protein (SCP).* SCP substrates include those derived from oil, wastes, and renewable resources. The idea of single-cell protein is not new. Torula yeast derived by aerobically fermenting wood wastes has been produced and used as food in Germany as far back as World War I. Only recently, however, has it been recognized that protein from single cells offers the best hope for major new protein supplies independent of agricultural land used.†

What are the advantages of SCP?

- The microorganisms do not depend on agricultural or climatic conditions, but are cultured in large fermentation vessels.
- They have rapid mass-doubling times.
- Genetic experimentation for protein improvement can be readily undertaken.
- Production of SCP is not limited by land surface or sunlight.

*The term SCP was coined at M.I.T. by Professor Carrol Wilson in May, 1966. He was seeking a general name plus an acronym to identify feed and food protein derived from single-cell microorganisms grown on various resources and wastes. Professor Wilson felt, and rightly so, that the use of terms such as "bacterial" or "microbial" had unpleasant connotations relative to food usage.

†The Advisory Committee of the United Nations Economic and Social Council on the Application of Science and Technology to Development (1967).

Amino-Acid Distribution of Petroleum-Derived SCP – Table IV

Microorganism	Substrate	Lysine	Methionine	Threonine	Tryptophan
Bacteria	Natural gas	5.3	3.4	4.5	–
Bacteria	*n*-Paraffin	6.5	2.0	4.0	0.9
Bacteria	*n*-Paraffin	7.0	1.8	4.9	1.4
Bacteria	Gas oil	4.3	1.2	4.0	1.2
Yeast	*n*-paraffin	7.0	1.2	3.9	0.5
Yeast	*n*-paraffin	7.0	1.8	4.9	1.4
Yeast	Gas oil	7.8	1.6	5.4	1.3
Yeast	Gas oil	6.8	1.3	5.6	2.3
Algae	Carbon dioxide+	4.6	1.4	4.6	1.4
Soy meal*	–	6.6	1.1	3.9	1.2

+Obtained by burning fossel fuels.
*For comparison purpose.

The advantage of rapid mass-doubling time is illustrated in Table III. SCP doubling times are on the order of minutes, compared with days for agricultural plants and weeks for animals. Genetic experimentation requires only a matter of days for development of new strains of organisms, so it is possible to select and develop strains that have varied amino-acid profiles, protein content and cell-wall digestibility (Table IV).

Bacterial SCP is generally comparable to fish meal, running around 60–75% crude protein; yeast SCP is more like soy meal, running 45–55% crude protein; and mycelial fungi SCP is usually somewhat lower in protein content.*

Many different carbon-containing substrates can be used for SCP production, including hydrocarbons, sugar and starches from agricultural products, and cellulosic material from agricultural and domestic wastes. Moreover, SCP can be utilized in numerous forms, including food, feeds, and protein isolates. As a consequence, there is almost an infinite variety of schemes for producing and utilizing this protein source.

Opinion is extremely diverse on the direction to be taken in developing SCP processes. Western European nations, Russia, Japan and the U.S. have plans for 100,000–1,000,000 tons/yr of feed protein from very complicated plants using the latest computer-control techniques (Table V). Countries in Asia, Africa and South America talk about village-level technology and systems that are capable of producing only 1,000–10,000 tons/yr of feed SCP.

*However, the threadlike mycelial fungi SCP has certain advantages for chemical-engineering reasons: It can be collected by simple filtration rather than costly centrifugation, as in the case of bacteria and yeast.

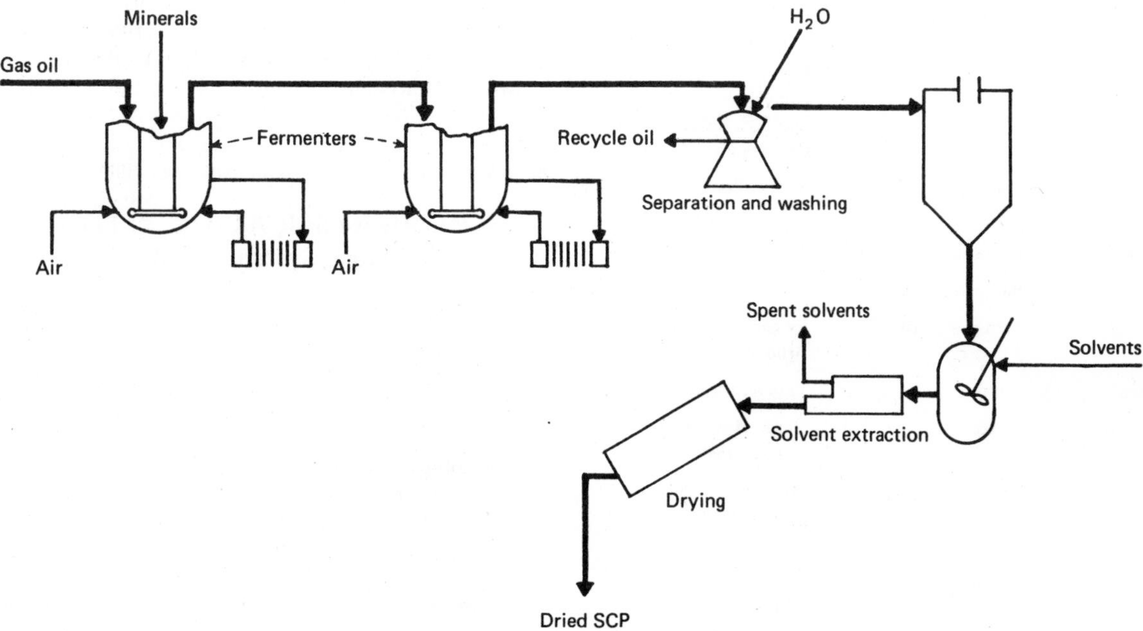

GAS OIL PROCESS uses fermenters and solvent extraction to make SCP—Fig. 2

Energy-rich countries with access to natural gas and oil are building plants to produce SCP from gas, oil and methanol. Countries with a strong agricultural base, such as the U.S., are utilizing solid wastes, including feedlot manure, seed-grass straw, and city refuse to produce SCP. Developing countries such as those in Central America are looking to local agricultural byproducts and wastes, such as molasses, bagasse, coffee pulp, etc.

An idea of the economic potential of SCP production can be obtained from soybean meal, an alternative source of protein (Fig. 1). Some economists are projecting that by 1985 soybean-meal prices, fob. Decatur, Ill., will be \$185–240/ton. This could mean European prices of \$300–350/ton for soybean meal, and \$410–465/ton for fish meal.

However, short-term effects, such as the 1972 worldwide shortfall and the 1973 Arab oil embargo, cloud the issues and make it difficult to produce sound economic advice with respect to SCP processes. If future agricultural shortfalls should occur, and if embargoes are repeated, Free World trade could be grossly restricted, and this could give rise to very local decisions with respect to SCP process feasibility.

Energy considerations could also affect the kind of SCP process. Aeration costs could become important if energy costs continue to rise. Hence, an anaerobic fermentation process, in which both alcohol and SCP feed are produced, could be the process of choice.

Let us consider these alternatives further according to whether the process is aerobic fermentation of oil and related products, solid wastes, refuse, agricultural wastes, or whether the fermentation is anaerobic.

SCP Processes Currently Being Developed Table V

Substrate	Organism	Companies Involved
n-Paraffin	Yeast or bacteria	British Petroleum Liquichimica Gulf Kanegafuchi U.S.S.R. Government Chinese Petroleum
Gas oil	Yeast or bacteria	British Petroleum U.S.S.R. Government
Methane	Bacteria	Shell
Methanol	Bacteria or yeast	ICI
Ethanol	Bacteria yeast	Exxon Amoco Food Co.
Sugars	Fungi	Tate and Lyle
Starch	Fungi	Rank, Hovis and McDougal
Cellulose	Bacteria	General Electric Co. Finnish P. & P. LSU – Bechtel
CO_2	Algae	Inst. Francais du Petrole

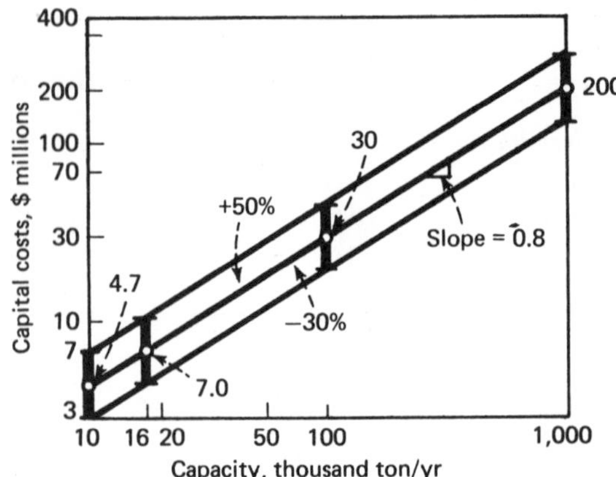

PLANT INVESTMENTS (1973) for making SCP from hydrocarbon prorate by the 0.8 power of capacity—Fig. 3

SCP From Oil

Large-scale production of SCP from oil and related products is a reality. Several 10,000–30,000-ton/yr plants are onstream, and at least three 100,000-ton/yr plants are being constructed. Five basic approaches are used: (1) growth of bacteria on natural gas, (2) growth of yeast or bacteria on methanol derived from natural gas, (3) growth of yeast on ethyl alcohol derived from chemically oxidized ethylene, (4) growth of bacteria or yeast on the gas-oil fraction of crude oil, and (5) growth of bacteria or yeast on purified n-alkanes separated from crude oil.

A typical process using gas oil is being operated at Lavera, France, by the Société Française des Petroles B.P. (Fig. 2). This plant's operating capacity is 17,000–20,000 tons/yr; its cost was approximately \$7.5 million. A 100,000-ton/yr plant using n-alkanes is expected to be onstream in Calabria, Italy, by the end of this year. This plant, owned by Liquichimica Biosintesi S.P.A., is projected to cost \$49 million, and is to yield a 15.8% return on investment from an SCP with 63% crude protein and competing with fish meal at \$424/ton (Table VI).

British Petroleum is reportedly considering the con-

SCP Costs for 100,000-Ton/Yr Plant for n-Alkane System – Table VI

Item	\$/Ton
Direct manufacturing costs	227.13
Sales expenses	24.00
Depreciation	47.18
Overhead	34.07
Total production costs	332.38
Before-tax profit	91.38
Price/ton	423.76
Total capital required	\$49,182,900
Return on investment =	15.8%

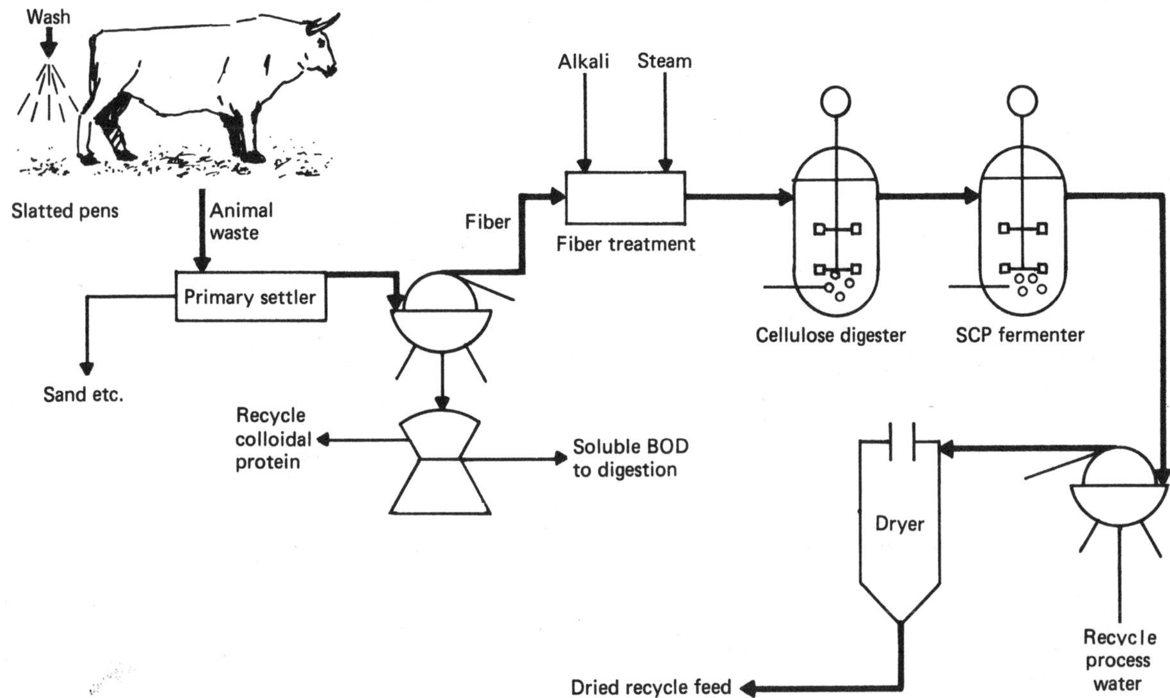

MANURE-CONVERTING PROCESS uses predigestion before fermentation—Fig. 4

struction of a 1-million-ton/yr plant. On the basis of a scaleup factor of 0.8, such a plant would cost around $200 million (Fig. 3). Obviously, such a scheme is only considered possible in one of the highly developed, energy-rich countries.

SCP From Solid Wastes

In the U.S. alone, it has been estimated that there are nearly 1 billion tons/yr of solid wastes. The bulk of this, as indicated in Table II, is agricultural, food, refuse, and feedlot wastes. The predominant component of this waste is cellulose, a material that is relatively difficult to ferment without some predigestion, by either enzymatic, chemical, or heat treatment.

Several solutions for disposing of manure have been suggested. Among those is a General Electric process for converting manure into recycle SCP feed (Fig. 4). Manure is first collected by either mechanical or hydraulic means and the fiber recovered by filtration. The washed fiber is predigested by a combination of alkali and heat treatment. The treated fibers are then microbially digested and converted to SCP in a one- or two-stage continuous process. The resulting microbial cells are filtered, washed and dried prior to being used as recycle feed. Among the difficulties with this process is that of producing a contaminant-free feed.

City Refuse

City refuse can be handled much the same way as manure (Fig. 5). The fibrous material is separated by hydroclassification, then filtered and washed, treated with hot alkali, and fermented to make SCP. At the present time, many large cities are simply dewatering the refuse, separating out the glass and metals, and then burning it along with coal in steam boilers. Because of the simplicity of this operation, plus the energy recovery, it is doubtful that city refuse has an immediate potential in SCP production.

Agricultural Wastes

Numerous crops yield low-value wastes (Table VII). Some of these are widely distributed and hence difficult to collect in economical volumes. For the most part, they occur in the developing countries having an agricultural

Examples of Low-Value Agricultural Wastes
Table VII

Crop	Type of Waste
Sugar cane	Bagasse
Sugar beet	Beet pulp
Coffee	Pulp, wash water
Carob	Husks
Dates	Date wastes
Potato	Peelings, starch water
Tomato	Pulp
Grass seed	Straw

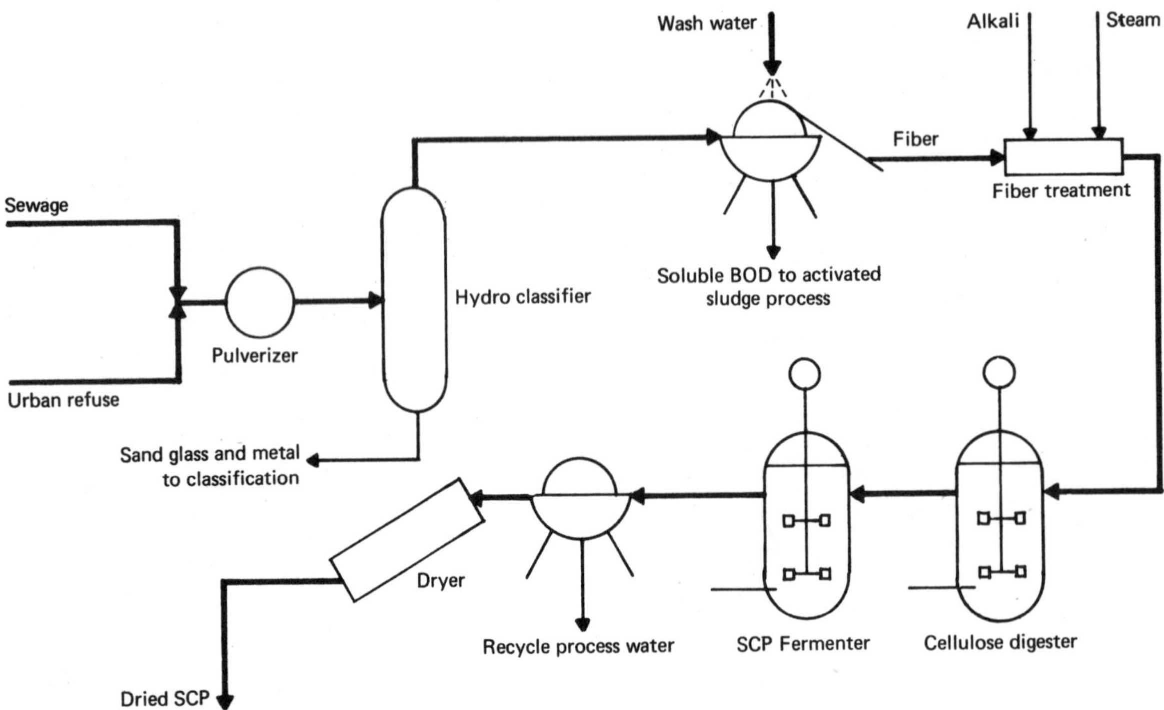

DOMESTIC-WASTE PROCESS separates cellulose for SCP production—Fig. 5

base. Examples are wastes from coffee and banana processes of Central America.

Contrasted with these limited-availability wastes are large-volume wastes, such as those from paper, sugar, and grass-seed production—for example, the waste liquor from sulfite wood pulping. Once sulfite waste liquors are steam stripped to remove SO_2, additions of various minerals (essentially fertilizer) make a fermentable liquor on which single-celled microorganisms can grow (Fig. 6).

A general process for the conversion of cellulosic wastes to SCP has been developed by Louisiana State University, initially for converting cane bagasse. This process is applicable to most cellulosic wastes. Bechtel Corp. is further developing it for worldwide application.

Anaerobic Processes

Anaerobic, rather than aerobic, fermentation presents an interesting alternative. One such process is the "Bactolac" feed developed by Hugh Henderson at Michigan State University. In this process, lactose from whey wastes is anaerobically fermented to lactate. This lactate is neutralized with ammonia to yield ammonium lactate, which can be used to provide up to 85% of the nitrogen for rumen diet. Rather than separate out the lactate, the whole broth is dried. Bacterial cells that grew on the whey wastes, converting the lactose to lactate, provide the needed source of organic nitrogen for the rumen. The cells plus ammonium lactate make a well-balanced nitro-

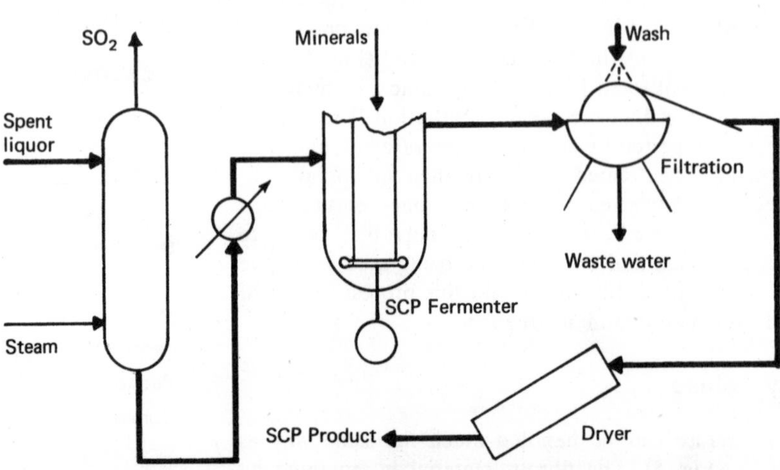

SULFITE LIQUOR PROCESS adds minerals to make pulp waste fermentable—Fig. 6

SCP Selling Price and ROCE* Starch and Agricultural Waste Systems — Table VIII

	Size	Price	ROCE,* %
High technology	9,000	325	4
	1,000	325	Loss
Low technology	9,000	325	14
	1,000	325	Loss

* Return on capital investment.

Village Level Technology SCP Processes Table IX

Plant	Case I	Case II	Case III
Working days/yr	100	200	300
Capacity, tons/day	84	170	250
Selling price, $/ton	325	325	325
Return on Investment, %	9	18	25

gen source that can virtually substitute for soybean in the rumen diet.

Yet-another viable alternative converts agricultural wastes to alcohol by anaerobic fermentation (Fig. 7). In this process, the yeast cells that grow on the waste, converting sugar and starches to alcohol, are utilized for feed and food purposes. The resulting alcohol can be used for chemical synthesis, for energy in terms of gasoline extender, or for producing food-grade SCP. A process for making food-grade SCP from ethanol was recently announced by American Oil Co.

Village-Level Technology

Meanwhile, Tate and Lyle, Ltd., a major world sugar company, is pushing the concept of village-level SCP systems to handle distributed agricultural wastes. These systems are constructed of plastic materials, operate non-aseptically, and utilize cheap labor rather than costly instrumentation for control. In contrast to high-level technology, which only pays within huge economies of scale (Table VIII), village-level technology can be economical on a small scale (Table IX).

Tate and Lyle believes that the idea of a cooperative SCP fermenter having roughly a 3-ton/day output is feasible as a small-village plant. Its people have designed

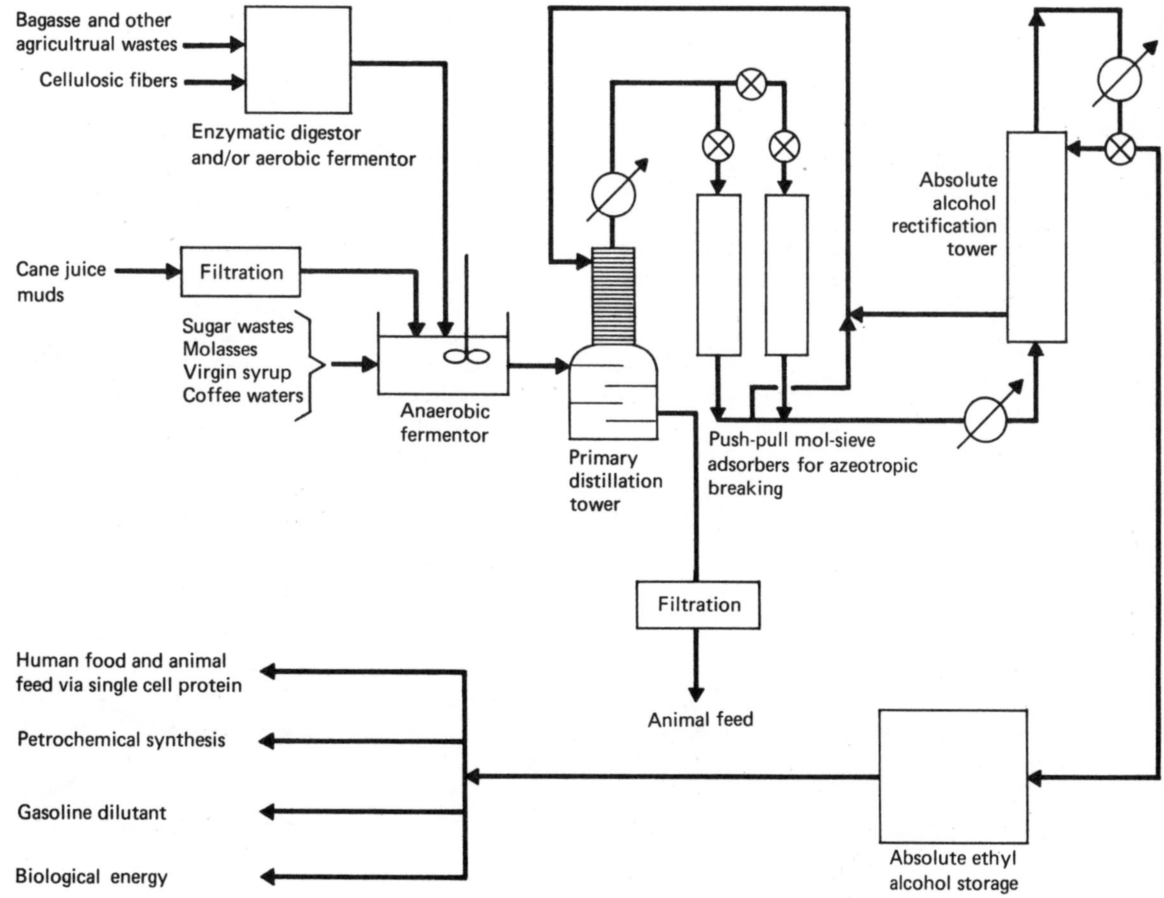

AGRICULTURAL WASTE PROCESS uses anaerobic fermentation to make C_2H_5-OH and SCP—Fig. 7

a plastic processing unit that can be erected in the field and that costs only around $50,000. In the hands of highly skilled technicians, this system will operate satisfactorily. Tests under field conditions need to be performed in order to verify the system's true value.

DESIGN CONSIDERATIONS

The Growth Process

Once the economic viability for an SCP system has been determined, many factors must be considered in the selection and design of the final process. Beyond considerations of product safety, product end-use and product handling characteristics, the process design is concerned with productivity (which is largely dependent on the organism's growth rate) and the yield (which is affected by the organism's maintenance requirements).

An organism may contain the genetic information to produce a wide variety of enzymes; however, only some enzymes are produced at all times, while others are greatly influenced by the substrate. Certain compounds interact with the substrate to repress the translation of genetic information for synthesis. (This is called repression.) Also, the substrate sometimes reacts with a compound that is a genetic-mechanism repressor to remove its action. (This is referred to as de-repression.)

These processes allow cells to regulate their enzyme content in direct response to the environment. They prevent the formation of excess end-product and superfluous enzymes.

Mutants can be selected that lack these genetic controls. Such mutants have been changed so that the genetic mechanism no longer is sensitive to a particular controlling metabolite. For industrial processes, such strains with faulty regulation, altered permeability, or metabilic deficiencies may be used to accumulate products.

Glutamic acid fermentation offers an example of such control. Monosodium glutamate is an important flavor-accentuating compound; in 1972, some 180,000 tons were produced, 90% of which was made by fermentation. The *Corynebacterium* used for this process is biotin-requiring and lacks the enzyme α-ketoglutarate dehydrogenase, which catalyzes the conversion of α-ketoglutarate to succinyl ~ CoA. And in the presence of ammonia, α-ketoglutarate is converted to glutamate (Fig. 8). Glutamate accumulates and is excreted by the cell as an end-product.

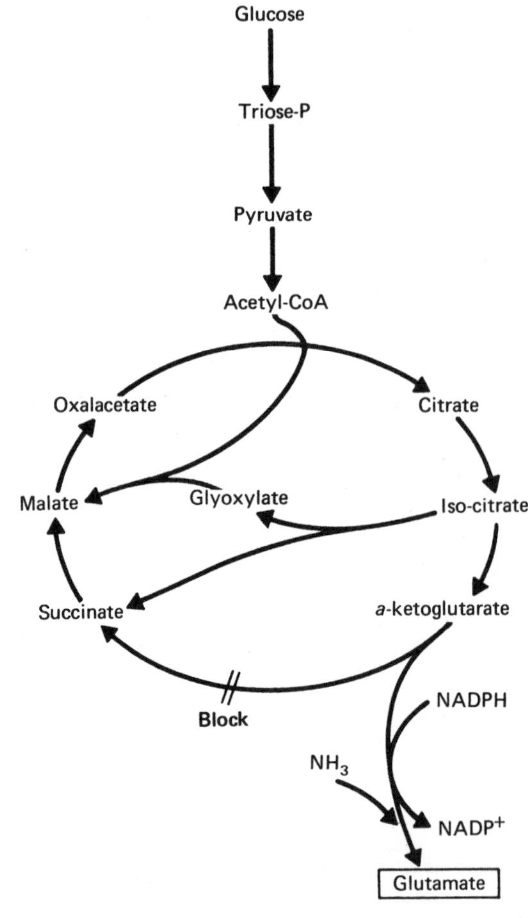

ENZYMATIC CONTROL of biosynthesis in the production of glutamate—Fig. 8

Enzyme action can also be inhibited by the system. This is best illustrated by a series of biosynthetic reactions, A to D, each controlled by a different enzyme, E_a, E_b, etc., and leading to an end product, P (Fig. 9). If the product is formed faster than needed for subsequent reactions, its concentration rises; and this inhibits the first enzyme E_a involved in the formation of the product. The effect, which is immediate and tends to decrease the formation of end product, is freely reversible should the concentration of the end-product fall.

A single product often inhibits the activity of the first

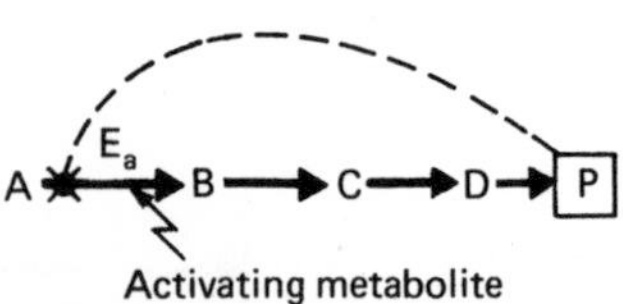

Product inhibits first enzyme, E_a, in series.

PRODUCT INHIBITION in unbranched series—Fig. 9

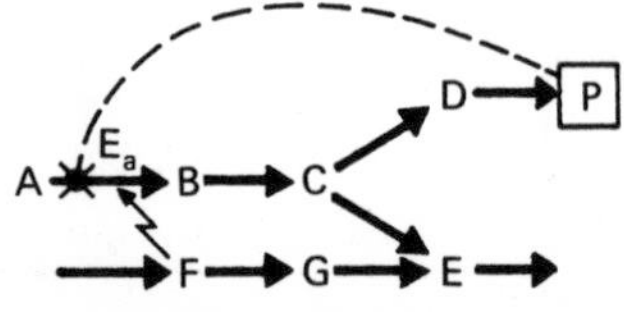

Product inhibits first enzyme, E_a, but this is partly offset by activating metabolite, F.

REGULATION by intermediates from related pathway—Fig. 10

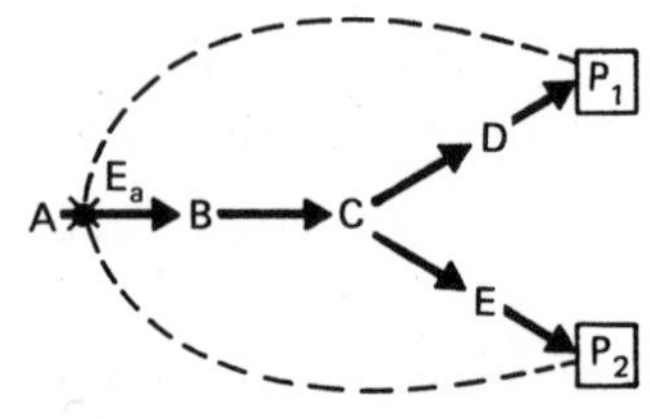

SYNERGISTIC INHIBITION by multiple products—Fig. 11

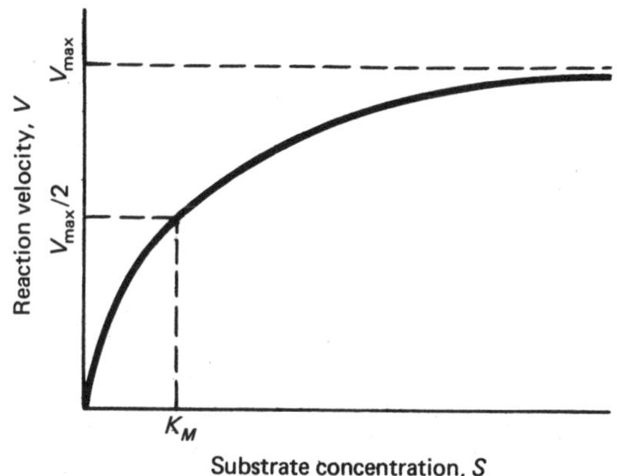

SUBSTRATE/REACTION-VELOCITY relation of a simple Michaelis-Menton kinetic model—Fig. 12

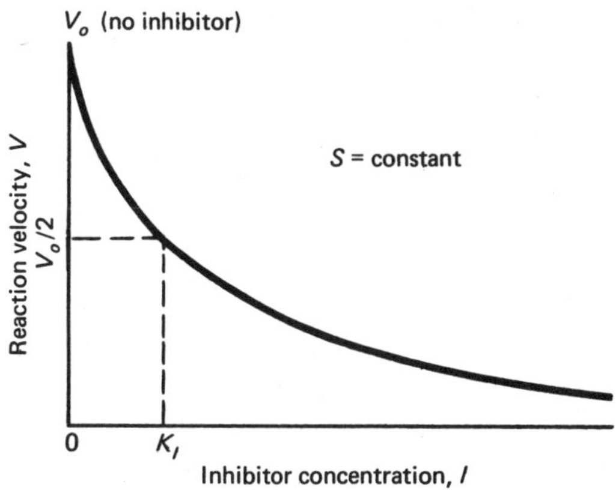

INHIBITOR EFFECT on enzyme kinetics as typified by organic acids—Fig. 13

enzyme (E_a) of a series, while intermediates of a related pathway may activate this enzyme (Fig. 10). With branched pathways leading to several products, total inhibition of E_a by a high concentration of one of the products would also act to stop growth of the other end-products.

In multivalent feedback inhibition (Fig. 11), the presence of all end-products is required to inhibit E_a. A variant of this type of control is found where each product inhibits the activity of E_a by a small amount; the inhibition is additive and synergistic when all the products of the pathway are present.

Kinetics

The literature dealing with the kinetics of biosystems is extremely extensive. No single review could possibly do justice to the whole field. Only those systems felt to be of greatest immediate concern to chemists and engineers, i.e. enzymes and single-cell activities, will be considered here.

For enzyme kinetics, a simple model proposed by Michaelis-Menton has been particularly useful. In this model, the enzyme (E) reversibly combines with the substrate (S) forming an enzyme-substrate complex (ES), which irreversibly decomposes to form product (P) and free enzyme:

$$E + S \underset{k_{-S}}{\overset{k_{+S}}{\rightleftharpoons}} ES \xrightarrow{k_{+P}} E + P \qquad (1)$$

If the substrate concentration is much greater than the enzyme concentration (i.e., $S_o >> E_o$), and if the decomposition of the enzyme-substrate complex, ES, is the rate-limiting step (i.e., $k_{+S} >> k_{+P}$), the relation between substrate and reaction velocity (V), shown in Fig. 12, is mathematically expressed as:

$$V = \frac{V_{max}(S)}{K_M + (S)} \qquad (2)$$

where $K_M = (k_{-S} + k_{+P})/k_{+S}$.

In general, K_M for the respiratory enzymes (those associated with sugar metabolism) is lower than K_M for the hydrolytic enzymes (those associated with primary substrate attack), as for example:

Enzyme	**Substrate**	**K_M Molarity**
Maltase	Maltose	2.1×10^{-1}
Sucrase	Sucrose	2.8×10^{-2}
Phosphatase	Glycerophosphate	3.0×10^{-3}
Lactic dehydrogenase	Pyruvate	3.5×10^{-5}

Enzymes are commonly inhibited by competitive actions or reversible noncompetitive actions. The competitive actions can be depicted as:

$$\begin{array}{l} \pm I \quad EI(\text{inactive}) \\ E \\ \pm S \quad ES \longrightarrow E + P \end{array} \qquad (3)$$

This leads to a reaction-rate expression of the form:

$$V = \frac{V_{max}(S)}{K_M + (S) + \frac{K_M}{K_I}(I)} \qquad (4)$$

where $K_I = k_{+I}/k_{-I}$.

Competitive inhibition is frequently found in reactions where the product inhibits the enzyme, i.e. P = I. An example is glucose, which is a competitive inhibitor of the action of invertase on sucrose.

The reversible noncompetitive actions can be depicted as:

$$\begin{array}{ccc} \pm I & EI & \pm S \\ E & & EIS \longrightarrow EI + S \\ \pm S & ES & \pm I \\ & \downarrow & \\ & E+P & \end{array} \qquad (5)$$

This model leads to a reaction-rate expression of the form:

$$V = \frac{V_{max}(S)}{K_M + (S)}\left[\frac{K_I}{K_I + (I)}\right] \qquad (6)$$

The effect of inhibitor on enzyme kinetics is shown in Fig. 13. This kind of inhibition is typical of the effect that we expect to be exhibited by organic acids such as acetate, propionate and lactate on the hydrolytic enzymes.

Immobilized enzymes are currently receiving quite a bit of attention as catalyst systems. Immobilized enzymes can be insoluble systems. They are formed by binding the enzyme to the surface of a nonporous solid or within a porous solid (Fig. 14).

With the enzymes bound to a nonporous solid, the reaction rate can be controlled by diffusion of substrate from the bulk solution to the enzymatically active surface, by the enzymatic reaction at the surface, or by diffusion of the reactant products back into the bulk of solution. The approximate behavior of such systems, when expressed in mathematical terms, can be represented by:

$$N_S = \frac{V_{max}E_S S_B}{S_B + K_M + \dfrac{V_{max}E_S}{D_S}\Delta X} \tag{7}$$

where N_S is the molar flux of substrate at the boundary layer, V_{max} is the maximum specific activity of the enzyme, E_S is the surface unit concentration of enzyme, S_B is the bulk concentration of substract, ΔX is the boundary layer thickness, and D_S is the substrate diffusivity. This equation leads to the kind of results shown in Fig. 15.

Enzymes bound to a porous solid bring about even more complicated situations. In addition to being controlled by the reaction and diffusion in the boundary layer, the reaction rate can also be controlled by the diffusion rate within the pore. Explicit solutions for these kinetics cannot be obtained. Rather, as in heterogeneous catalysis, an effectiveness factor can be defined for the reaction in terms of the ratio of the actual reaction rate to the rate at which the reaction would proceed if there were no diffusional limitations.

For single-cell kinetics, the growth rate (μ) of single cells can be expressed in terms of the cell concentration

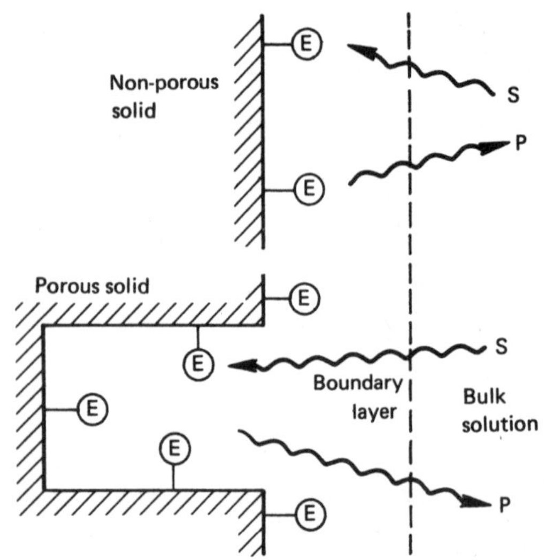

IMMOBILIZATION OF ENZYMES as accomplished by porous and nonporous solids—Fig. 14

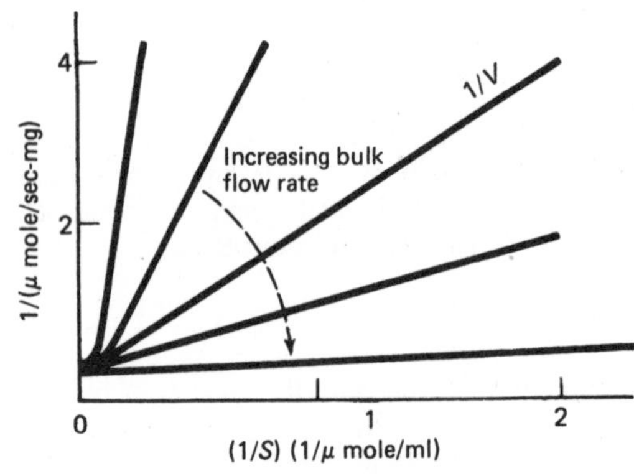

NONPOROUS-SOLID-BOUND ENZYMES typically exhibit this type of kinetic relationship—Fig. 15

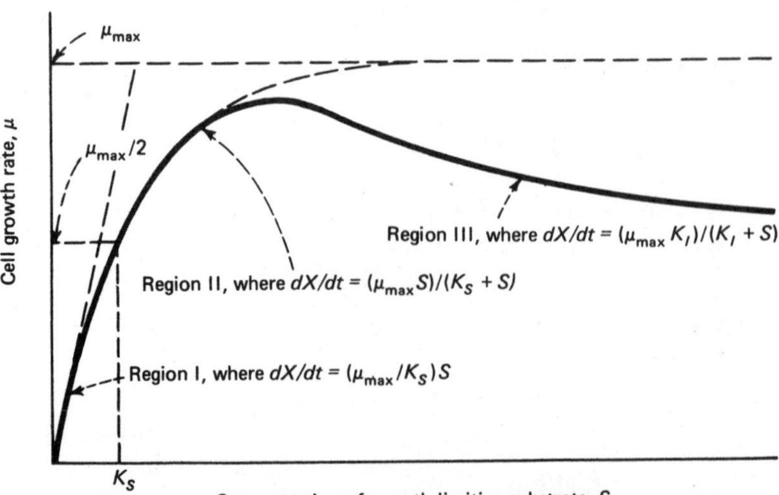

GROWTH/SUBSTRATE RELATIONSHIPS typically fall into one of three regions—Fig. 16

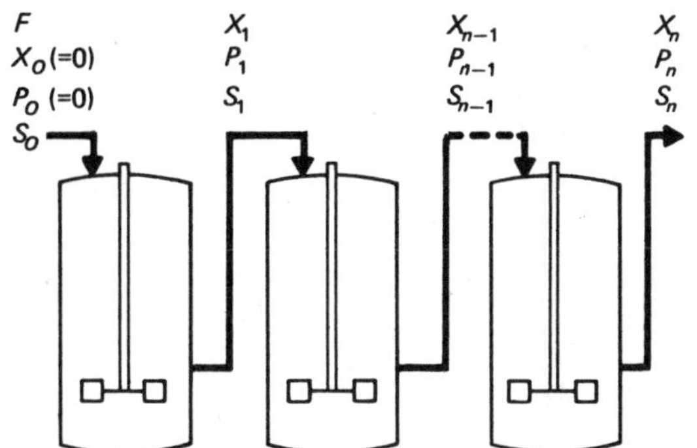

STEADY-STATE CONTINUOUS cultivation in a series of stirred reactors—Fig. 17

(X), the concentration of a growth-limiting substrate (S), and an inhibitor or predator (I):

$$\mu \equiv dX/Xdt = f(X, S, I) \tag{8}$$

There are three regions (Fig. 16) in which different models have been used to fit data for the growth-substrate relationships. Region I occurs where $S \lesseqgtr K_S$, and an essentially linear relationship exists between the growth rate and the substrate concentration. This behavior approximates many of the biological-waste-treatment processes, such as activated sludge. Region II occurs where the single limitation of simple-enzyme kinetics applies best. And Region III occurs at high substrate concentration, where the maximum growth rate is normally achieved, but where inhibition due to either metabolic growth products or to the substrate also occurs.

When no single substrate is limiting, growth the kinetics are much more complicated. This is also true when predation occurs or in unsteady-state growth, where structured cell function must be included in the kinetic equation to account for dynamic behavior.

Continuous Cultivation

With the exception of SCP processes such as the production of fodder yeast from sulfite-pulping waste, bakers yeast from molasses and activated sludge domestic waste, there has been little use of continuous fermentation in full-scale production processes. This for three reasons: the chances for deleterious mutations, technical difficulties of continuous aseptic operation, and lack of knowledge of microbial behavior. It is in the resolution of such problems that chemical engineering skills can perhaps exert a most beneficial influence.

Steady-state continuous cultivation theory has much in common with theories of continuous stirred reactors. Consider a series of equivolume vessels, n in total number (Fig. 17). The rate of flow of the medium through the vessels is F, and the volume of each vessel is V. The mass-balance equations regarding concentrations of cell mass (X), product (P) and limiting substrate (S) can be expressed, respectively, as follows:

$$V\frac{dX_n}{dt} = FX_{n-1} - FX_n + V\left(\frac{dX_n}{dt}\right)_{\text{Growth in } n\text{-th vessel}} \tag{9}$$

$$= FX_{n-1} - FX_n + V\mu_n X_n \tag{10}$$

$$\frac{dX_n}{dt} = D(X_{n-1} - X_n) + \mu_n X_n$$

Where:

$D = F/V =$ dilution rate, hr^{-1}

$= 1/(V/F) = 1/\bar{t}$

$=$ reciprocal of mean holding time (or retention time) of the flowing medium in each vessel

$\mu_n = \frac{1}{X_n}\frac{dX_n}{dt} =$ specific growth rate of cells in n-th vessel

subscripts n, $(n-1)$ = n-th, $(n-1)$-th vessel, respectively

Similarly . . .

$$\frac{dP_n}{dt} = \frac{F}{V}(P_{n-1} - P_n) + \left(\frac{dP_n}{dt}\right)_{\text{Production in } n\text{-th vessel}} \tag{11}$$

$$\frac{dP_n}{dt} = D(P_{n-1} - P_n) + Y_{P/X}\mu_n X_n \tag{12}$$

Where:

$Y_{P/X} = \Delta P/\Delta X =$ yield of product based on cell mass

Then . . .

$$\frac{dS_n}{dt} = \frac{F}{V}(S_{n-1} - S_n) + \left(\frac{dS_n}{dt}\right)_{\text{Consumption in } n\text{-th vessel}} \tag{13}$$

$$\frac{dS_n}{dt} = D(S_{n-1} - S_n) - \frac{1}{Y_{X/S}}\mu_n X_n$$

$$= D(S_{n-1} - S_n) - \frac{Y_{P/X}}{Y_{P/S}}\mu_n X_n \tag{14}$$

Where:

$Y_{X/S} = -\Delta X/\Delta S =$ yield of cell growth based on limiting substrate

$Y_{P/S} = -\Delta P/\Delta S =$ yield of product based on limiting substrate

$$= \left(\frac{\Delta P}{\Delta X}\right)\left(\frac{\Delta X}{-\Delta S}\right) = Y_{P/X} \cdot Y_{X/S}$$

The derivation of Eq. (9) to (14) assumes that the medium flowing from the $(n-1)$-th vessel into the n-th vessel is mixed instantaneously and completely with the

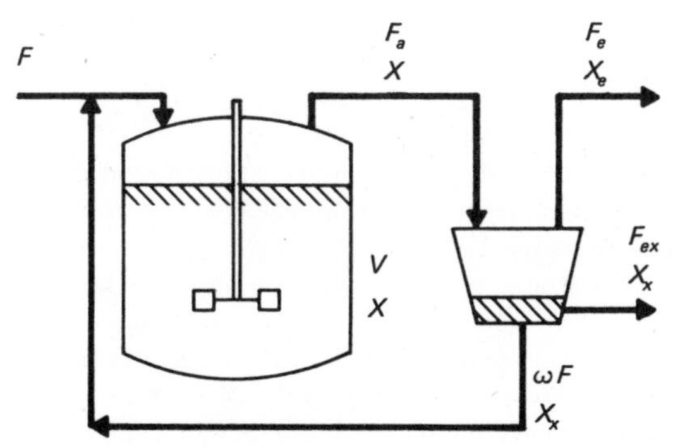

CONTINUOUS CULTIVATION as done in a single vessel with recycle—Fig. 18

contents of the n-th vessel. Values of $Y_{X/S}$, $Y_{P/X}$ and Y_{P-S} in these equations are all assumed to be constant, regardless of the number of vessels under consideration.

At steady-state conditions, the left-hand side of all Eq. (9) to (14) is zero. Then, from Eq. (10):

$$X_n = \frac{DX_{n-1}}{D - \mu_n}, \ (n \neq 1) \quad (15)$$

This equation is useful for estimating the value of X_n. Conversely, the value of μ_n in the steady state can be estimated from the cell concentrations, X_n, X_{n-1}, etc.

An interesting case of continuous cultivation is that of single vessels with a recycle of some of the microbial mass, as is done in the purification of sewage with activated sludge (Fig. 18). It is evident from this figure that:

$$F_a = (1 + \omega)F \quad (16)$$

$$F = F_e + F_{ex} \quad (17)$$

Assuming complete mixing, the mass balance for X in the reactor of Fig. 18 is:

$$V\frac{dX}{dt} = X_X \omega F - F_a X + V\left(\frac{dX}{dt}\right)_{Growth}$$

And at steady state:

$$\frac{1}{X}\left(\frac{dX}{dt}\right)_{Growth} = \frac{F_a}{V} - \frac{\omega F}{V}\cdot\frac{X_X}{X} \quad (18)$$

From Eq. (16) and (18):

$$\mu = (1 + \omega)D - \omega D\frac{X_X}{X}$$

$$= D\left\{1 + \omega\left(1 - \frac{X_X}{X}\right)\right\} \quad (19)$$

where, $D = F/V$.

Assuming a steady state in the separator vessel of Fig. 18, a mass balance around that vessel gives:

$$F_a = F_e\frac{X_e}{X} + \frac{X_X}{X}(F_{ex} + \omega F)$$

$$\frac{X_X}{X} = \frac{F_a - F_e\left(\frac{X_e}{X}\right)}{F_{ex} + \omega F} \quad (20)$$

Using Eq. (16) and (17), Eq. (20) can be rearranged as:

$$\frac{X_X}{X} = \frac{1 + \omega - \frac{F_e}{F}\cdot\frac{X_e}{X}}{1 + \omega - \frac{F_e}{F}} \quad (21)$$

Then from Eq. (19) and (21), we get:

$$\mu = D\left\{1 + \omega\left(1 - \frac{1 + \omega - \frac{F_e}{F}\cdot\frac{X_e}{X}}{1 + \omega - \frac{F_e}{F}}\right)\right\} \quad (22)$$

And if $X_o/X \doteqdot 0$, Eq. 25 reduces to:

$$\mu = D\left\{1 + \omega\left(1 - \frac{1 + \omega}{1 + \omega - \frac{F_e}{F}}\right)\right\} \quad (23)$$

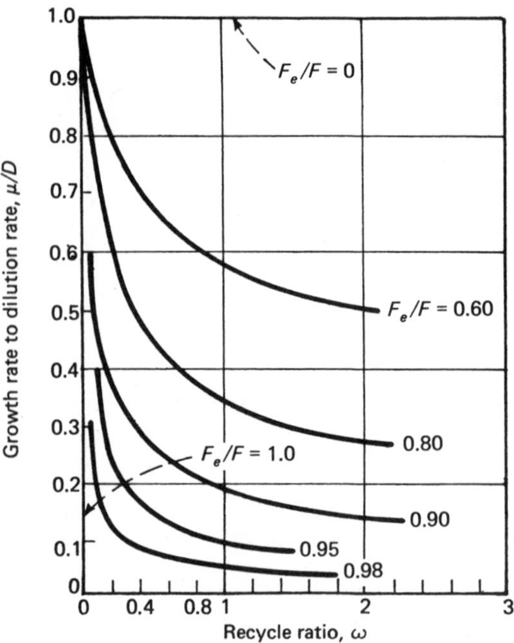

GROWTH AND DILUTION at steady state versus recycle in a single vessel—Fig. 19

This Eq. (23) gives relationships between μ and D in recycled vessels, as shown in Fig. 19, where μ/D is plotted versus ω, with parameters of F_e/F. This plot shows that steady-state operation can be realized, even if the dilution rate, D, in single vessels is larger than the specific growth rate, μ, of the cells, as long as the recycle ratio, ω, is selected according to the appropriate value of F_e/F.

An aeration basin for an activated-sludge process for sewage and industrial-waste treatment illustrates a typical case. Its dilution rate, D, is large compared to the specific growth rate, μ, of the sludge. However, recycle is the key variable for enhanced capacity of the system for BOD removal.

When F_e/F is reduced to zero, Eq. (26) reduces to nonrecycling. Here the value of μ must be equal to D for steady state to exist.

In addition to microbial problems, continuous cultivation is subject to certain inherent practical problems, including: (1) attaining homogeniety in the reactor, (2) maintaining sterility, and (3) maintaining the stability of the system.

The homogeniety of nutrients throughout a reactor becomes more important at low dilution rates, were the limiting substrate concentration is low. Good mixing of the broth, either with mechanical agitation or high aeration, becomes essential. Research projects on mixing highly viscuous or non-Newtonian fluids are lacking, but are needed if this problem is to be understood and solved in large-scale continuous cultivations. Good techniques are also needed for measuring the dissolved oxygen concentration in fermentation broth. Such techniques would allow aeration to be carried out with a minimal expenditure of energy.

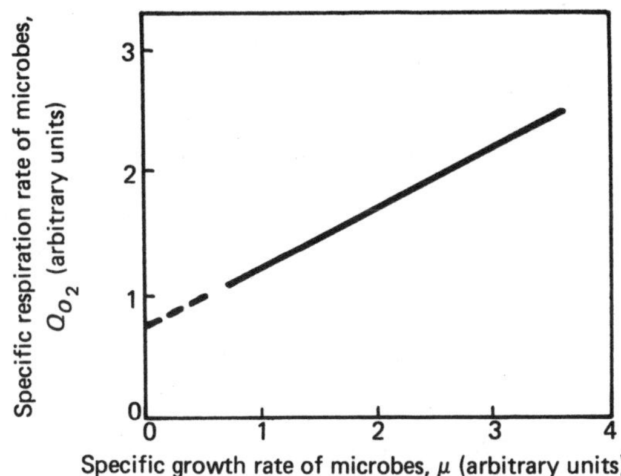

CELL-MAINTENANCE RESPIRATION as estimated from zero specific growth—Fig. 20

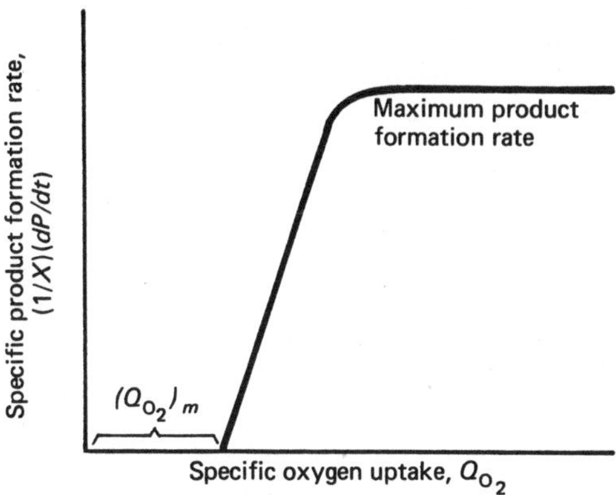

OXYGEN-UPTAKE/PRODUCTION RELATIONSHIP as typified by penicillin fermentation—Fig. 21

Oxygen Transfer of Fermentation Systems

In aerobic fermentations oxygen is a basic substrate that must be supplied for growth. As in enzymatic reactions, the relationship between oxygen concentration and growth is of a Michaelis-Menton type (see Fig. 12). The specific rate at which cells respire (Q_{O_2}) increases rapidly with an increase in the dissolved oxygen concentrations ($\bar{C}$) up to C_{crit}, which typically ranges 0.1-1.0 ppm for well-dispersed bacteria, yeast and fungi growing at 20–30°C. Beyond C_{crit}, the specific oxygen uptake rate increases only slightly with increasing oxygen concentrations.

If Q_{O_2} is plotted against the specific growth rate (μ) of the microbe, a linear correlation is obtained (Fig. 20). An intersection of this straight line with the ordinate is designated as $(Q_{O_2})_m$, which represents the oxygen required for cellular maintenance.

In mathematical terms, this relation is given by:

$$Q_{O_2}X = (Q_{O_2})_m X + (1/Y_{X/O_2})(dX/dt) \qquad (24)$$

where Y is the yield based on cell mass and t the time.

This type of correlation can be applied to almost any substrate involved in cell energy metabolism. It is justified from an energetic point of view and is supported by experimental steady-state data. Care must be used in applying this equation to transient conditions. It can only be used for situations at or near the steady-state equilibrium conditions.

When a product other than cell biomass is involved, the situation can be considerably more complex. Oxygen can be utilized for maintenance and growth as well as product formation. The simplest possible expression to represent the rate of oxygen utilization in this case is:

$$Q_{O_2}X = (Q_{O_2})_m + (1/Y_{X-O_2})^{(dX/dt)} + (1/Y_{P/O_2})^{(dP/dt)} \qquad (25)$$

This is equivalent to saying that oxygen utilization can be either growth related or cell-concentration related, or both. A typical relationship between oxygen uptake and product formation rates, as experienced in penicillin fermentation, is shown in Fig. 21.

Of course the oxygen transfer rate under steady-state conditions must be equal to the oxygen uptake rate, i.e.:

$$Q_{O_2}X = k_{la}(C^* - \bar{C})_{mean} \qquad (26)$$

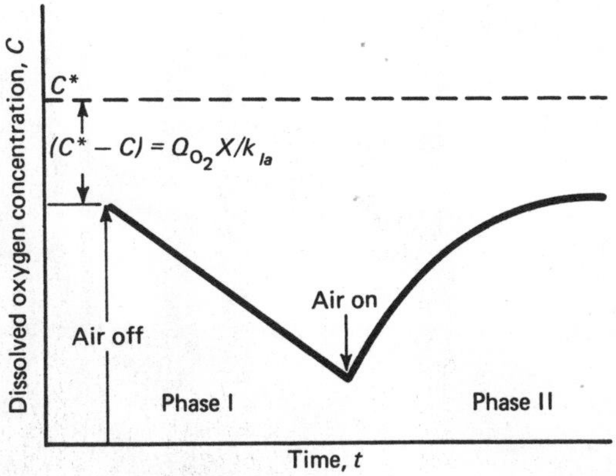

OXYGEN CONSUMPTION measured by alternating fermenter operation—Fig 22

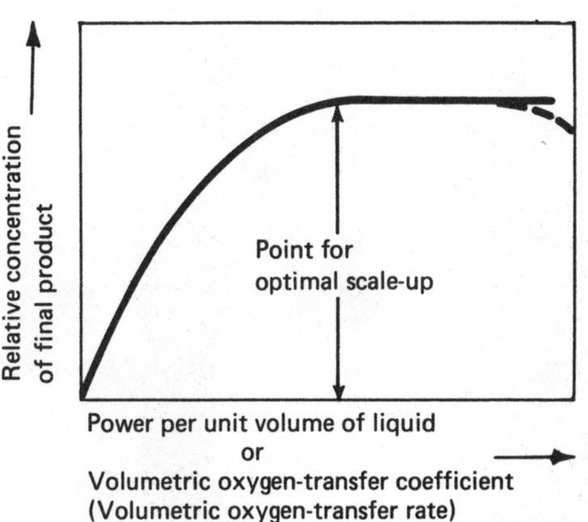

OPERATING REGION to be used for optimal scaleup of equipment—Fig. 23

where C^* = concentration of oxygen in the liquid that would be in equilibrium with the gas-bubble concentration and k_{la} = the volumetric oxygen transfer rate.

Utilizing this relationship, the volumetric oxygen transfer coefficient can be estimated, i.e.:

$$k_{la} = (Q_{O_2}X)/(C^* - \bar{C})_{mean} \quad (27)$$

In practice, the dissolved oxygen concentration ($\bar{C}$) is monitored by a membrane-covered galvanic probe. When the fermentation is equipped with a fast-responding probe (i.e. the response time is less than 6 seconds), then the fermenter can be used as a respirameter by making dynamic measurements of the oxygen concentration under aeration and non-aeration conditions. This is illustrated in Fig. 22.

The following equations represent these conditions:

air off, Phase I:

$$d\bar{C}/dt = -Q_{O_2}X \quad (28)$$

air on, Phase II:

$$d\bar{C}/dt = k_{la}(C^* - \bar{C})_{mean} - Q_{O_2}X \quad (29)$$

Scaleup

Many processes experience not only the problem of scaling up a new fermentation but also the translation of process-improvement data for well established fermentations from laboratory operations to plant scale. Most fermentations are scaled up on the basis of achieving similar oxygen transfer capabilities in the plant equipment to that proved optimal in bench scale.* The design equations are generally of a form:

$$P_g = K(P_o ND^3/F^{0.56})^{0.45} \quad (30)$$

and

$$k_{la} = K'(P_g/V)^{\alpha}(F)^{\beta} \quad (31)$$

where P_g = the gassed power input

P_o = the power input

N = impeller speed

D = impeller diameter

F = aeration rate

V = fermenter volume

and K, K', α and β are constants.

These relationships can be correlated with product concentration to set optimal design parameters (Fig. 23).

Once a plant is built, the conditions of agitation, aeration, mass (oxygen) transfer, and heat transfer are more or less set. Therefore, it has been suggested that the problem of translating process improvements is not one of scaleup but rather that of scale-down. Those environmental conditions achievable in plant-scale equipment should be scaled down to the pilot plant and screensize equipment (shaken flask) to insure that studies are carried out under conditions that can be duplicated.

A probable prerequisite for scaleup and/or scale-down is to make the environmental conditions identical be-

* A good review of this subject can be found in Wang, D. I-C., and Humphrey. A. E., "Developments in Agitation and Aeration of Fermentation Systems," Prog. in Ind. Microbiology, **8**, *1* (1968).

COMPUTER-CONTROLLED system for analyzing and controlling fermentation—Fig. 24

Gateway Measurements – Table X

Measurement	Result
pH	Acid product formation rate
Air flow rate In & out O_2 concentration	O_2 uptake rate
Air flow rate Out CO_2 concentration	CO_2 evolution rate
CO_2 evolution rate O_2 uptake rate	Respiratory quotient
Power input Air flow rate	O_2 transfer rate

tween plant and bench-scale equipment, but the identity does not always guarantee similar behavior of microorganisms. This situation remains very difficult to overcome. Obviously there is a great need for the biochemical engineer and the microbial physiologist to come together to solve the problem of designing microbial processes. This is beginning to happen.

Fermenter Instrumentation

Until recently, fermentation was more art than science; but as the mechanisms of gene replication and enzyme regulation have become known, the various environmental effects on a fermentation process have become somewhat clearer. We no longer need to limit fermentation control to that of temperature, pH, and aeration, but can now think in terms of sophisticated control systems for fermentation processes.

Indeed, we can even think in terms of utilizing computers to analyze and control our fermentations (Fig. 24), and dynamic optimization is just around the corner.

We still need new instruments. The most important sensor would be a reliable biomass monitoring device that could be sterilized. Also, there is need for a carbon substrate sensor and a nitrogen substrate sensor that can withstand repeated system sterilizations.

Indirect measurement via computers appears to be a viable alternative to measurement involving sampling. Certain sensor information can be combined to give additional information such as oxygen-uptake rate, carbon-dioxide-evolution rate, respiratory quotient, etc. These measurements can be thought of as "gateway" measurements because they make possible the calculation of additional information (Table X).

Further, the indirect measurement of a given component can be achieved by material-balancing that component around the fermenter. If a model for utilization of that component for biomass or product production is known, then either the biomass or product level can be estimated by computer summation or integration of the model.

Besides these uses of the computer, it has application in fermentation processes for continuous non-prejudical monitoring and (most important) continuous feedback control and dynamic optimization of the process.

Summary

Contemporary fermentation processes have many new opportunities largely centered around feed and feed production and the utilization of wastes. Economic viability of these processes depends on availability of raw material, investment requirements, and product demand. Knowledge of fermentation is approaching the point where meaningful kinetic models can be derived to represent microbial behavior and response to environmental changes. The design of large-scale continuous fermentation plants is feasible. And computer-coupled fermentation systems now provide a means for continuous and instantaneous system analysis, as well as a means for using this information to feedback-control the fermentation at its optimal operating point. With the advent of computer control, fermentation technology can begin moving into the modern age. It can no longer be considered an art. #

References

1. Aiba, S., Humphrey, A. E., Millis, N., "Biochemical Engineering," Academic Press, N.Y. (1973)
2. Matiles, R. I., and Tannenbaum, S., "Single Cell Protein," M.I.T. Press, Boston (1968)
3. Wang, D. I-C., and Humphrey, A. E., Developments in Agitation and Aeration of Fermentation Systems, Prog. in Ind. Microbiology, **8,** 1 (1968)
4. Gounelle de Pontanel, H., "Levures Cultures sur Alcanes," Centre de Recherches Foch, Paris (1972)
5. Wingard, L. B., "Enzyme Engineering," Interscience Publishers, John Wiley, N.Y. (1972)
6. Ministere de l'Agriculture, INRA, Proceedings at the 1st European Conference on Computer Control in Fermentation, Dijon, France, Sept. 3–5, 1973
7. Humphrey, A. E., Kinetics of Biological Systems—a Review, ACS Symposium Series, "Chemical Reaction Engineering," 1972 (no. 109, pp. 603)
8. Ryu, D. Y., and Humphrey, A. E., A Reassessment of Oxygen-Transfer Rate in Antibiotic Fermentation, J. Ferment. Technol., **50** (6) 424 (1972)

Meet the Author

Arthur E. Humphrey is professor of chemical and biochemical engineering, and dean of the College of Engineering and Applied Sciences, at University of Pennsylvania. A member of the National Academy of Engineering, he is a well-known author and holds several patents. He received the AIChE Professional Progress Award in 1972, and the Food and Bioengineering Award in 1973. He received his BSChE (1948) and MSChE (1950) from the University of Idaho, a MS (Food Tech.) (1960) from MIT and PhDChE (1953) from Columbia. He has been a Director of New Brunswick Scientific Co.

l-Menthol synthesis employs cheap, available feedstocks

The first to make *l*-menthol from a nonstereospecific feedstock, this process cuts raw materials costs and offers operating flexibility.

John C. Davis, *Senior Associate Editor*

☐ This technique to synthesize *l*-menthol is the first to make the material from a nonstereospecific feedstock, *m*-cresol. Previously, most *l*-menthol had been isolated from its natural source, Japanese mint (*Mentha arvensis*), with the remainder synthesized from stereospecific materials also isolated from nature, such as *alpha*- or *beta*-pinene, *d*-citronellal, or phellandrene, which comes from lemon oil.

Developed by Haarmann & Reimer GmbH (Holzminden, West Germany), the process was pioneered commercially at the firm's headquarters in Holzminden about four years ago, and improved with the startup of H&R's second plant, at Bushy Park, S.C., earlier this year. The route starts with a cheaper raw material than its predecessors, yields a consistently pure product, and has the flexibility to accept as alternative feedstocks any of several optically active or inactive isomers of menthol, turning everything into the levo-rotatory form, if desired.

Its $15-million U.S. plant may well come at an opportune time for H&R, one of Germany's oldest chemical firms, and one that virtually originated the synthetic flavors and fragrances business just over a century ago, with the invention and commercialization of a way to synthesize vanillin. First and foremost, the Bushy Park facility provides the company with a U.S. manufacturing base just as currency revaluations make German exports less competitive in the U.S.

In the longer run, H&R is looking forward to the growing importance of synthetic menthol, as opposed to the natural variety derived from *M. arvensis*. The shrub is grown mostly in Brazil, using a variety transplanted from Japan. In the short time since World War II, suitable acreage for cultivating this plant has been greatly diminished, primarily because *M. arvensis* is a demanding plant that exhausts virgin ground in only about four years. By the early 1970s, Brazilian cultivation had supplied about 80% of the world demand for menthol, but production there today can meet no more than 40% of needs, according to H&R's estimates.

The firm's synthetic-menthol capacity in Germany and in the U.S. can meet about 35% of world demand. The Holzminden plant churns out somewhat more than its 770,000-lb/yr nameplate capacity, and the Bushy Park unit is rated at 1.5 million lb/yr. This may not seem like a lot of product, but considering the fact that flavoring ingredients are present in foods, toothpastes, tobacco and drugs in amounts of one thousandth the total product weight, or less, their effect reaches far.

A RACEMIC MIXTURE—The H&R process breaks down basically into two steps: (1) the conversion of *m*-cresol feed material into geometrical and optical menthol isomers, and (2) the subsequent separation of *l*-menthol from the isomer mixture.

Cresol is first reacted with a propylating agent over a heterogeneous catalyst (see flowsheet), such that an

H&R's menthol plant at Bushy Park, S.C., uses *m*-cresol as its feed.

Originally published May 22, 1978

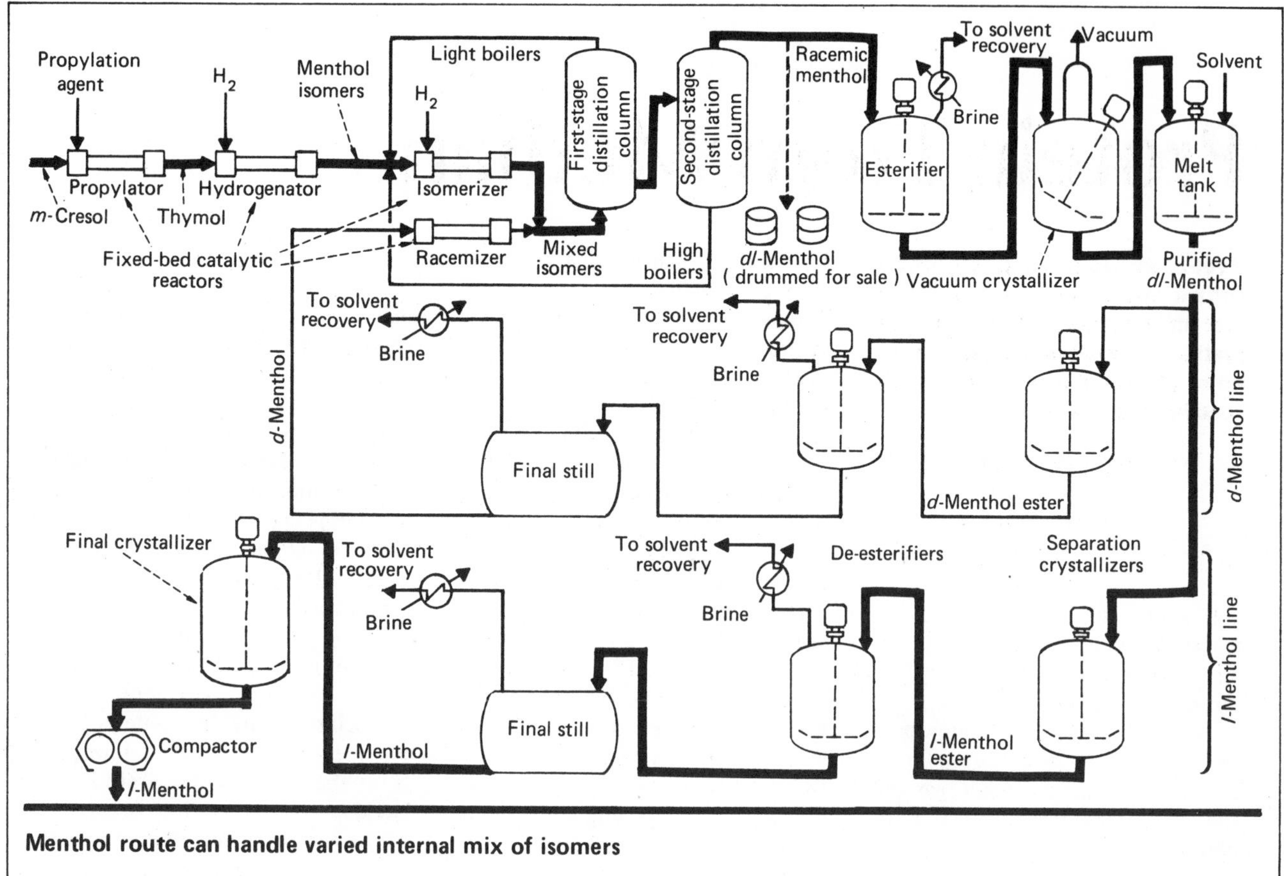

Menthol route can handle varied internal mix of isomers

isopropyl group attaches in the number 4 position on the cresol ring to form thymol. Subsequent hydrogenation results in a mixture of four geometric isomers, plus the optical variations of each (termed enantiomers), totaling eight different forms: *d*- and *l*-menthol, *d*- and *l*-isomenthol, *d*- and *l*-neomenthol, and *d*- and *l*-neoisomenthol.

The desired racemic mixture, *dl*-menthol, is actually produced in fairly low yield in H&R's synthesis—about 14%—but this is of little consequence, since the mixture is passed over another heterogeneous catalyst that further rearranges the other isomers, providing a final menthol concentration of about 50%. The sought-after *dl*-menthol is separated by distillation from the remaining undesired isomers, which are recycled over the catalyst until they finally rearrange into the proper isomer.

For some customers, the process stops here, and the racemic menthol (less costly than the purer *l*-menthol) is drummed and sold. For those wanting the pure levo-isomer, the mixture moves on to separation of *d*- and *l*- forms by "separation crystallization"—H&R's term for this type of fractionation.

SEPARATION—Until this step, all reactants pass through the system in a pure form in the liquid state. At this point, however, a proprietary solvent is introduced to form a menthol ester and to carry it through crystallization. Both the ester type and the solvent's identity are proprietary.

Following esterification of *dl*-menthol, the reaction stream is purified via vacuum crystallization and solvent drying. The purified stream is then melted and dissolved in a new solvent system, and split into two streams that are processed in parallel. The first step is separation crystallization, where temperatures are controlled to within 0.01°C.

With *d*- and *l*- forms separated in their respective ester fractions, both are put through parallel de-esterification and distillation, while product *l*-menthol is subjected to an additional final crystallization.

The purpose of this last step is not so much to provide greater purity, but to produce the crystals most desired by customers. Product purity is at least 99.9%, and this level is just about fully achieved in the prior distillation step.

The *d*-menthol, meanwhile, is routed back to the beginning of the process for racemization and eventual conversion into the *l*- and other isomers.

ANCILLARY SYSTEMS—A process-control computer is very important to the Bushy Park operation, especially in sequencing several of the batch operations employed. These include esterification, de-esterification (one for each menthol enantiomer), crude distillation of the de-esterified isomers, and a finishing distillation.

For environmental control, the top of each distillation column is outfitted with secondary brine heat-exchangers, in order to catch all the solvent. The exchangers condense vapors at −15°C. There is no aqueous process effluent; in fact, the whole process is nonaqueous.

The Bushy Park plant takes utilities from its sister operation next door, the Dyestuffs Div. of Mobay Chemical Co., to which H&R also sends waste effluent for treatment. H&R's relationship with Mobay is through both firms' German parent, Bayer AG (Leverkusen, West Germany).

Protein From Methanol

Pilot-scale tests are going on in Britain and Japan for processes that make single-cell protein from methanol, ammonia, air and nutrients. Methanol replaces the already-proven gas-oil feedstock.

MARK ROSENZWEIG
European Editor
SHOTA USHIO
McGraw-Hill World News, Tokyo

If pilot-scale tests with methanol feedstock continue to go well, the biosynthesis of proteins from hydrocarbon substrates could be well on its way to becoming a common process. The technique has already been commercially proven with gas oil and *n*-paraffin feedstocks (see p. 299).

The methanol alternative has a potentially big advantage going for it in light of current fuel shortages: The alcohol can be made cheaply from Middle Eastern or African natural gas that would otherwise be flared. It is not expected to be in the same tight demand as gas oil is for fuel.

Other advantages of methanol as a protein feedstock: It is water-soluble, contains no polycyclic aromatic compounds, and requires less oxygen than does methane. It has a lower boiling point to ease the separation of feedstock from the product stream. Investment is lower, too. ICI Ltd., one of the process developers, says the investment is about 10-15% lower than *n*-paraffin routes in the 100,000-metric-ton/yr. range. The ICI route also consumes slightly less power, but feedstock costs are about the same.

Not surprisingly, the two chief proponents of the protein-from-methanol scheme are big methanol producers themselves: ICI Ltd. and Mitsubishi Gas Chemical Co. ICI already has a 1,000-ton/yr. plant in operation at Teesside, U.K., and Mitsubishi is constructing a "large-scale pilot plant" due to come onstream in 1974. ICI expects to decide by midyear on whether to build a 100,000-metric-ton/yr. commercial scale unit at Teesside.

Enough is known of the two processes to discern some basic differences. These chiefly concern the design consequences of employing markedly different single-cell proteins that feed on methanol at different temperatures. ICI employs a bacterium of the genus *Pseudomonas* at a temperature of 37 C. in the fermenter. Mitsubishi, which still has not decided between yeast or bacteria, is looking at strains capable of withstanding temperatures up to 46 C. The higher operating temperature is important in Japan, where cooling water is more valuable, in order to gain a greater temperature gradient in the fermenter's cooling circuit.

The choice between yeast or bacteria has a lot to do with the design of dehydration equipment downstream of the fermenter, because tiny bacterial single-cell protein is too small to be immediately centrifuged. ICI has chosen a method of preconcentration of the fermenter effluent prior to centrifugation. And Mitsubishi is considering similar techniques in its work on bacterial strains.

Both developers are giving special attention to the fermenter design. ICI is unquestionably furthest along, with a patented Pressure Cycle Fermentor, consisting of upper and lower crossovers, a riser, and a downcomer (see figure), with no internal moving parts. Mitsubishi is believed to be studying a more common fermenter—an air-lift type with draft tubes—and has built a tower 5.6 m. high and 45 cm. dia. Both companies are designing their fermenters to ensure maximum oxygen transfer, a common fermentation-rate limitation.

ICI's Technology—In the ICI route, methanol plus essential nutrients such as phosphoric acid, potassium sulfate, sodium chloride and traces of iron, copper, zinc and molybdenum compounds are piped into a tank and diluted with recycled water. The mixture is then passed to a sterilization tank where all microorganisms are killed by heating with steam to about 100 C. The sterile stream then goes to the downcomer portion of the fermenter.

Meanwhile, ammonia is mixed with air, filtered to remove all microorganisms, and also sent to the downcomer. The high-integrity filters are arranged in parallel for easy maintenance while the unit is onstream.

The fermenter must be at least 80 ft. high, regardless of capacity. The unit can be made in any size, with a definite cost savings upon scaleup. The absence of internal moving parts saves on power and allows the unit to be easily sterilized.

Both liquid and gaseous streams are fed into the unit at various points. This, coupled with the fermenter geometry and the use of a number of stationary baffle-like devices, maintains balanced methanol, oxygen and carbon dioxide concentrations throughout. Temperature varies less than 1 C. Oxygen-transfer rates of 0.93 lb. O_2 (in air)/cu. ft. of vessel volume—the same as are claimed for BP Protein's gas-oil route—are claimed.

The medium fills the entire unit except for the top portion of the upper crossover, where carbon dioxide and excess air collect. A hydrostatic head of about 3 atm. in the riser (providing a high gas-absorption rate) gives way to lower atmospheric pressure in the upper crossover for CO_2 evolution.

Fermentation takes place at a pH of about 7, and a methanol concentration of a few ppm. Heat is pro-

Originally published January 7, 1974

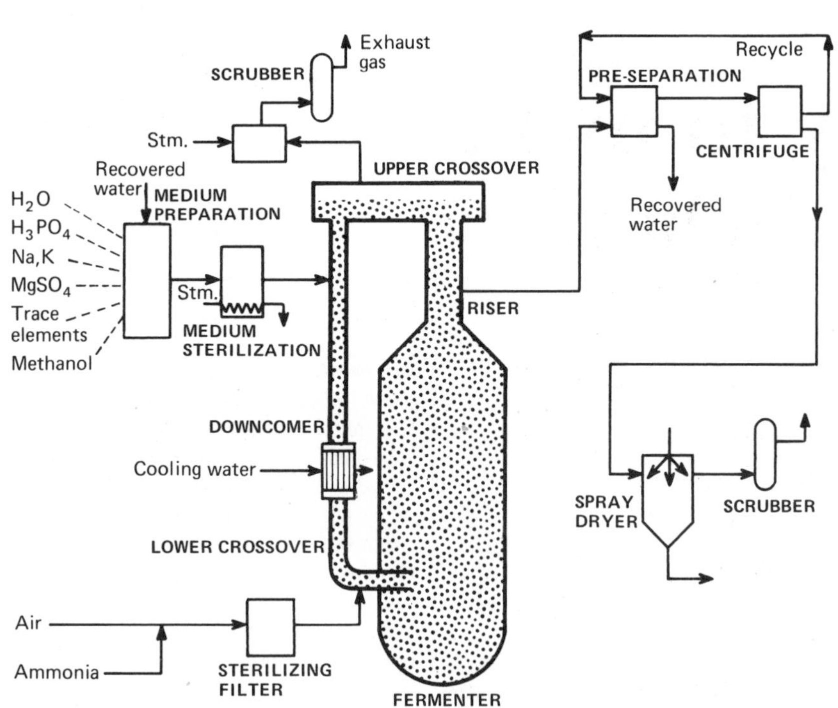

PRESSURE CYCLE FERMENTOR at ICI's Teesside plant is made of stainless steel. Photo of actual unit is shown at right.

duced, but the temperature is maintained by exchange against cooling water in the downcomer.

A product stream containing about 3% suspension of cellular dry matter is taken from near the top of the riser. ICI has devised a patented preconcentration system of agglomeration in a tank by combined chemical and thermal means. The chemicals used are as yet undisclosed, but are said to be a normal part of an animal's diet. The preconcentrated stream then goes to a flotation vessel, from which 10% dry cellular material is recovered from the top. This is sent to a centrifuge and brought up to about 20%. Water is recycled from both steps. The rest of the water is removed in a spray dryer (dryer offgas is scrubbed to prevent the escape of dust).

Whatever methanol is left in the fermenter product-stream is either devoured by the microorganisms during subsequent processing, or vaporized in the dryer. No methanol is left with the powdery product.

The pseudomonad product has 20-30% more protein on a weight basis than yeast, and also does not have yeast's deficiency in methionine, an essential amino acid, says ICI. It is, however, a bit below fishmeal in lysine, another essential amino acid.

Mitsubishi's Progress—Mitsubishi's investigation grew out of a joint project begun in 1969 with Japan's government-run Fermentation Research Institute. Eighteen continuous fermenters ranging in capacity from 1-50 l. have been built at Mitsubishi's Niigata Laboratory.

The company has found at least two yeasts that grow at 40-44 C. instead of the normal 30 C. without any significant increase in cell doubling-time. Mitsubishi's bacteria grow in the 38-42 C. range in "good" yields. The savings in cooling costs when using these strains, as opposed to those requiring 30 C., is about one-third, estimates the company.

Not much is known of Mitsubishi's method of bacteria preconcentration, although it is probably related to the company's work on preconcentration of yeasts, some details of which Mitsubishi has revealed at Japanese technical-society meetings. CHEMICAL ENGINEERING learned at one of these meetings that calcium chloride was used in tests as a flocculant at variable pH levels. "Considerable" cohesion and precipitation were caused by raising culture pH to 4 or more when a certain yeast was adopted. Researchers also report that cell volume is packed to 30% within 10 min. after the pH of the fermentation broth is set at 4, while cell concentration in the now-compressed cell layer exceeds 10%.

Acceptance Tests—Even though the product is meant as animal food, public acceptance and toxicity questions remain obstacles to commercialization. Safety trials are being run for both the ICI and Mitsubishi products. Mitsubishi says it is too soon to make definitive statements, since its animal tests are only a year in progress, but that so far no evidence of toxic substances in the protein product has been produced. The company gives the impression that it feels a couple of years are needed before it can establish the single-cell protein technology, including a confirmation of safety. Right now, however, the social climate for such a product in Japan is not too good: Dainippon Ink & Chemicals Inc. and Kanegafuchi Chemical Industry Co., which have both developed *n*-paraffin routes, still cannot build the 60,000-metric-ton/yr. plant they want in Japan because of the cool attitude of the Japanese public. #

A better way to make protein from whey?

Interest in using byproduct whey from cheesemaking to produce protein seems to be increasing. This may benefit a new route that is said to offer several advantages by relying for fermentation on stringy fungi instead of normally used yeast.

☐ Western European nations, eyeing their balance of payments, are concerned about their continued dependence on imports of U.S. soya for animal feed. One alternative, producing protein from hydrocarbons, remains surrounded by controversy. However, another option—protein-from whey—seems set to make inroads. And a French firm hopes to accelerate the trend with a new process it is now launching.

Indeed, the company, Heurtey SA (Paris), says that its technique is getting a warm welcome. And interest in whey-based protein processes is perking up generally, adds Fromageries Bel (Paris), which offers an established route.

The reason is straightforward: cheesemaking produces an enormous amount of whey. The French dairy industry alone turns out about 7 million m^3 of the byproduct annually, while Western Europe generates 23 million m^3/yr and world output totals around 60 million m^3/yr. Some of the whey—about half of that in France—is dried for animal-feed-supplement use, but the remainder often goes to waste and can pose a disposal problem. Upgrading the byproduct offers a way out of this dilemma. The Heurtey process, for instance, yields protein-rich fractions for human and animal consumption, as well as amino acids for pharmaceutical manufacture.

A number of companies have developed yeast-based fermentation routes to upgrade the whey. The roster includes Fromageries Bel, which now operates two plants, the first built in 1962 and the other commissioned last year, and Milbrew, Inc. (Juneau, Wis.), which started up a unit in January 1975. (For details on this latter facility, see p. 33.)

Heurtey takes a different approach. Its process uses a stringy fungus of the *Penicillium cyclopium* family for fermentation.

Through a subsidiary, Heurtey Equipments Entreprises (Paris), the firm recently won its first order—from a major French milk cooperative for an about-10-metric-tons/d protein plant, which will treat approximately 300 m^3/d of whey. Heurtey figures on a French market alone equivalent to 20 such plants (at around $6 million each) within the next ten years.

PLUSES—As explained by Pierre Henriet, a director of the Heurtey subsidiary, the use of stringy fungi rather than yeast brings a couple of specific benefits—faster protein production; and better distribution of amino acids in the product, requiring no additions to it of lysine, methionine or cysteine. Also, the product can be harvested by filtration, while yeast-based processes call for more-costly centrifuging and spray drying, contends the firm.

All of these advantages have long been recognized by researchers, but lack of a suitable fungal strain hardy enough for industrial applications has hindered commercialization.

Heurtey spent about $3.4 million on its research program, which was carried out at Ste. Industrielle de Recherches Biologiques in Toulouse

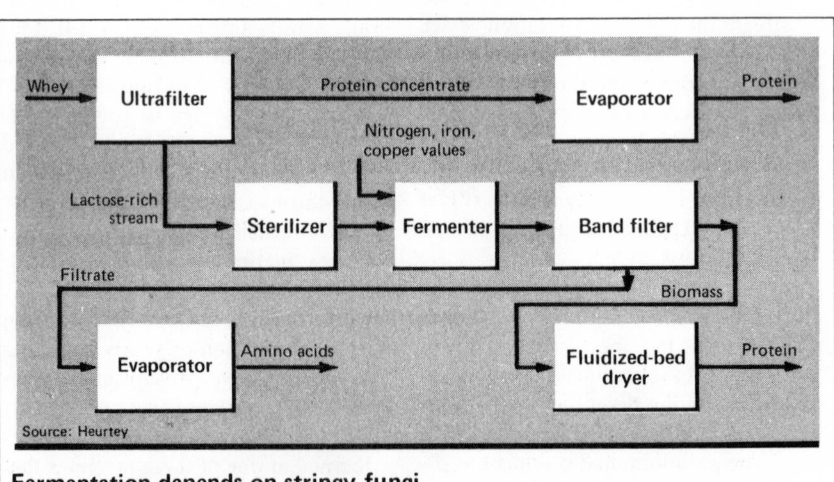

Fermentation depends on stringy fungi

Originally published March 13, 1978

over the past four years. The company boasts that the process could have been ready sooner had the firm not chosen to develop a pollution-free technique. Process discharge water is claimed pure enough for reuse in the plant, because the stream's biochemical oxygen demand is less than 160 mg/L.

HOW IT WORKS—The process has three main steps, as shown in the flow diagram below.

In the first stage, the whey undergoes ultrafiltration, to yield a protein fraction and a lactose-rich liquid stream. The protein is spray dried, and then sold for infant and animal-feed supplements.

Meanwhile, the liquid, which generally contains 38 to 53 g/L lactose, passes to a fermentation stage. Nitrogen, iron and copper values are added to an oxygenated fermenter, followed by the lactose stream. Fermentation temperature and pH vary case-by-case to suit the grade of whey. Typically, however, temperature runs about 25° C, and pH 3.0 to 4.5.

The biomass from the fermenter is continuously harvested, and is separated by a band filter into a protein-rich solid fraction and a filtrate.

Then, in the final stage, the protein fraction goes to a fluidized-bed dryer. Product is suitable for use in animal feeds. Meanwhile, filtrate passes to an evaporator, to recover amino acids. These can be used for making pharmaceuticals, says the firm.

Heurtey says that fermented protein should cost about 1.4¢/L to produce, and should fetch 1.6¢/L, depending on the market and other factors.

STATUS—Heurtey already has received inquiries from companies in North America, Africa and Eastern Europe, and is currently negotiating with five French firms. It expects two orders this year, in addition to the one in hand. That's from cooperative Centre-Lait (Aurillac), which bands together about 6,000 milk producers ranging from small farms upward.

Centre-Lait's plant, slated for Aurillac, will treat 300 m^3/d of whey to produce 3 mt/d of ultrafiltered protein, 7.5 mt/d of fungal protein, and 300 kg/d of amino acids. Final go-ahead on construction awaits a decision by the French government on its possible capital support for the project.

OTHER ACTION—Fromageries Bel, though not an engineering company like Heurtey, is also hopeful about selling its technology. Besides its internal use, at a pioneer, 2,500-mt/yr unit at Vendôme and an 1,800-mt/yr facility at Sablé, the process is already employed by Laiterie du Mont St. Michel at an 1,800-mt/yr plant at St.-Brice-en-Coglé. And Fromageries Bel expects to attract other clients—for instance, contracts for two more French plants within the next eighteen months—according to the firm's scientific director, Yves Vrignaud. In addition, talks are underway with companies in Scandinavia, Ireland, Canada and the U.S.

After all, explains Vrignaud, Bel's process is proven: "We can say what our production costs and sales prices are. There are no surprises." The company claims production tabs for its protein of 1.3¢/L and a selling price of 1.5¢/L for animal feed.

William Kosman, Paris
McGraw-Hill World News

Section IV
INORGANIC CHEMICALS

Ammonia from coal
Ammonia from coal: a technical/economic review
Hydrogen recovery unit ups NH_3-plant efficiency
New feeds and processes perk ammonia production
New solvent cuts costs of carbon monoxide recovery
Chloralkali plants win by teaming technology
New ion-exchange membrane stars in chloralkali plant
Recovering chlorine from HCl
Nitric acid route recovers more heat from tailgas
Strong-nitric-acid process features low utility costs
Recovery of active silica cuts costs of cryolite route
Soda ash plant exploits mineral-laden brine
Coal converts SO_2 to S
Selective oxidation in sulfuric and nitric acid plants: current practices

Ammonia from coal

As the price of natural gas goes up, and with increased restrictions on its use, it becomes preferable to make ammonia from coal. Here is a description of a process for making ammonia by using this widely available feedstock, together with a detailed analysis of economics.

David Netzer and James Moe, Fluor Engineers and Constructors, Inc.

☐ Making ammonia from coal is an old concept, but up to now the process has not been economical in the United States. The high cost of coal-fed plants coupled with the artificially low price of natural gas has made steam/methane-reforming plants the logical choice.

There is, of course, some point at which the relative costs of coal and natural gas will make coal-to-ammonia a practical choice, and this point is at hand, or nearly so. Although it is difficult to predict what price natural gas will command in the next few years, it is certain that it will increase above present levels. Some predictions indicate that the price of natural gas will rise to over $2.50 per million Btu. It also seems possible that new ammonia plants based on natural-gas feeds cannot be constructed in the United States; the political and social implications militating against this are severe, and probably cannot be overcome. Hess [*1*] made an interesting analysis of the economics of ammonia production and pointed out the illogic of constructing coal-to-SNG (substitute natural gas) plants and then using the SNG for making ammonia.

Several authors have presented costs for coal-to-ammonia plants, based upon commercially proven coal gasification technology [*2–5*]. Because of the strong efforts being made to develop new coal-gasification processes, it is reasonable to hope that one or more of them may offer some advantages for ammonia production. Fluor has been actively engaged in making engineering and feasibility studies of coal-conversion processes for the past six years, and decided to evaluate the new coal-gasification processes as they relate to making ammonia. Some of the desirable features of coal-to-ammonia can be summarized as follows:

- The process operates at high pressure, thus reducing downstream compression requirements.
- There are no troublesome byproducts.
- Coal delivered as slurry can be gasified directly.
- Any type of coal can be gasified.

The Texaco coal gasification process has these desirable features and therefore was selected as a promising contender. This process embodies commercially proven aspects of the Texaco oil-gasification process. A demon-

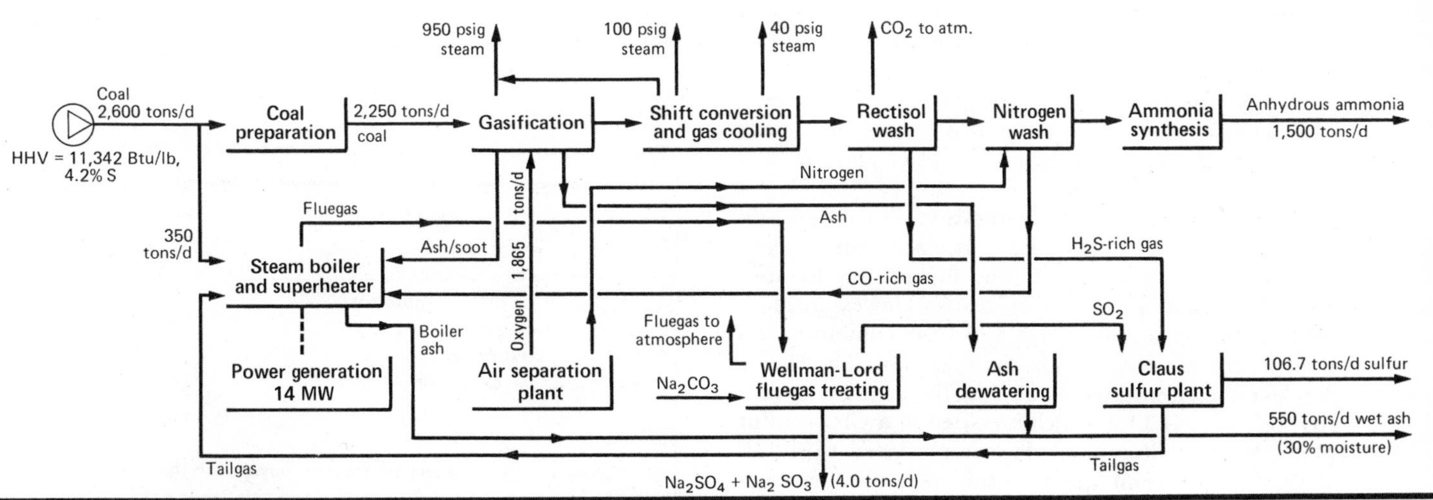

Flow diagram of the 1,500-ton/d coal-to-ammonia plant described in the text — Fig. 1

Originally published October 24, 1977

Analysis of coal used for ammonia plant feed — Table I

Material	Wt%
Carbon	67.0
Hydrogen	4.5
Oxygen	8.4
Sulfur	4.2
Nitrogen	1.3
Chlorine	0.01
Ash	14.6
	100.0
Free moisture,	6.0%

Higher heating value, *HHV*, (as received), 11,342 Btu/lb; *HHV* (dry), 12,022 Btu/lb.
Flow point of the ash in reducing atmosphere is 2,350°F.
Softening point of the ash in reducing atmosphere is 1,850°F.

Yield of various compounds from coal gasifier — Table II

Compound	Yield, mol%
Carbon monoxide	30.60
Hydrogen	23.80
Carbon dioxide	12.90
Methane	0.26
Nitrogen	0.36
Argon	0.11
Hydrogen sulfide	1.01
Carbon oxysulfide	0.067
Ammonia	0.09
Water	30.80
	100.00
Total mols/h	21,430

stration plant using coal as a feedstock for synthesis gas is currently under construction in Germany.

Process description

Fig. 1 is a block flow-diagram of a "grass roots" 1,500-ton/d plant. This plant generates its own electric power and requires only coal and water for operation. The net raw material and utilities requirements are:

Coal—2,600 tons/d (based on coal analysis in Table I)
Water—1,800 gal/min

In the coal preparation section, coal is first crushed to pass a ¾-in. screen. It then enters a wet grinding circuit and is ground to a fine mesh, forming a coal slurry of 55% (or higher) by weight. The slurry is pumped into the Texaco reactor, which operates at 850 psig and 2,400°F. The coal chemicals react to form light gases (shown in Table II) and the ash is slagged. Most of the slag is quenched in water at the bottom of the reactor, and removed via ash locks as a water slurry. The gas stream passes from the reactor to a quench chamber where the gas is cooled to 1,800°F by a combination of water quench and cool-gas recycle. At this temperature, the ash in the gas stream is below its fusion temperature. The hot gases pass through a waste-heat boiler, generating saturated steam at 950 psig, and on to a saturator, which also removes soot.

The CO in the soot-free synthesis gas is catalytically converted to CO_2 in water-gas shift reactors, producing H_2. The shifted gas is cooled and flows to the Rectisol acid-gas-removal process. The Rectisol process employs cold methanol to reduce H_2S and COS content to less than 1 ppm and CO_2 to less than 30 ppm. The H_2S is selectively concentrated to 35% in the acid-gas stream (the balance is CO_2), which is routed to a Claus sulfur plant. In the sulfur plant, H_2S is processed to sulfur.

Treated gas from the Rectisol wash unit enters a molecular sieve unit for removal of traces of CO_2 and methanol. The treated gas from the molecular sieve flows to a nitrogen wash unit where all the CO, CH_4 and argon and some hydrogen are scrubbed by liquid nitrogen. The CO, CH_4, argon and H_2, along with the nitrogen, are recovered as a low-pressure residue gas—HHV (higher heating value) about 160 Btu/std ft^3—and used as fuel gas in the plant.

Gaseous nitrogen with less than 5 ppm O_2 is provided by the air separation plant. Liquid nitrogen is formed in the nitrogen wash unit by expansion of nitrogen and by use of low-level refrigeration from the wash unit.

The overhead gas from the nitrogen scrubbing is a systhesis gas with less than 50 ppm (CH_4 + Ar) and less than 5 ppm (CO + O_2). Nitrogen is added to the gas in the desired ratio, and the gas is compressed to the pressure of the ammonia-synthesis loop, which operates at 3,150 psia. Water cooling and refrigeration are used to condense ammonia in the synthesis loop. The refrigeration system employed in the wash plant provides the necessary refrigeration for the loop.

The main power requirements of the plant are listed in Table III. Most of the drivers are powered by 850-psig, 900°F steam. Since heat recovery from the process streams is not sufficient for all the energy needs, additional high-pressure steam is produced in a boiler. This boiler is fired with a mixture of coal, unconverted car-

Power requirements for coal-to-ammonia plant — Table III

Unit	Power, hp
Synthesis gas makeup, recycle compressor	17,600
Refrigeration compressor	14,400
Air-separation-plant compressor	31,500
Oxygen compressor	15,500
Nitrogen compressor	12,000
Electric-power generator (14 MW)	18,700*
	109,700

*Provides all the necessary electric power for the plant.

Most of the power is provided by 850-psig, 900°F steam. The rest of it is provided by an excess of 100 psig saturated steam. All the motive steam is exhausted to 3½ in. Hg, abs. (120°F).

bon from the gasification unit, and residue gas from the nitrogen wash unit. The boiler can be operated independently of the other process units and is used to start up the plant. The overall energy requirement for the plant is 39.3 million Btu per ton of ammonia. For a lignite-based plant, it is about 4 to 6% higher.

A Wellman-Lord unit treats the flue gas from the boiler to remove SO_2, which is sent to the Claus plant for sulfur recovery. The Wellman-Lord unit incorporates a venturi scrubber for particulate removal, and SO_2 is absorbed by Na_2SO_3 to form $NaHSO_3$. $NaHSO_3$ is thermally regenerated to SO_2 and Na_2SO_3. Some oxidation to Na_2SO_4 occurs during the absorption.

Total sulfur emissions from the plant amount to 3.5 tons/d of SO_2. This is equivalent to 0.12 lb SO_2 per million Btu of coal fed to the plant or, based on the boiler firing rate, the emissions are 0.6 lb SO_2 per million Btu fired. The boiler stackgas includes the tailgas from the Claus plant, and it is not clear how the environmental regulations would apply. Nevertheless, the emission rate is comfortably below the EPA limitation for emission from coal-fired boilers. A small amount of Na_2SO_4 and Na_2SO_3 is purged and recovered as a dry solid. Na_2CO_3 is supplied to make up the sodium losses.

The plant is designed to have no discharge of liquid wastes. Blowdown water from the cooling tower is used to slurry the coal feed to the gasifiers because there are no particular purity specifications for the coal slurry. Solids contained in the incoming raw water are discharged with the ash.

Economics

The capital cost estimate was prepared, based on equipment factoring, for the coal-based plant. The ammonia selling price was determined by two methods; discounted cash flow (DCF) [6], and utility financing.

The main difference between the two financing methods is that in the DCF method 100% equity capital is assumed, while in the utility-financing, only a fraction of the capital is equity, with the balance being bonded indebtedness. As a result, the overall capital charges in utility financing are lower, and so is the selling price of the ammonia.

For DCF, the following basis was assumed:

- 20-yr project life.
- 16-yr sum-of-the-years'-digits depreciation of total plant investment.
- 100% equity capital.
- 12% discount cash-flow return rate–Case A.
- 15% discount cash-flow return rate–Case B.
- 48% federal income tax rate.
- 1st-quarter 1977 pricing was used, at an assumed central U.S. location.

For utility financing, the basis was:

- Straight-line depreciation over 25 years.
- Equity 33%, debt 67%.
- Return on equity, 12% compounded semi-annually.
- Interest rate, 9% compounded quarterly.
- Federal income tax, 48%.

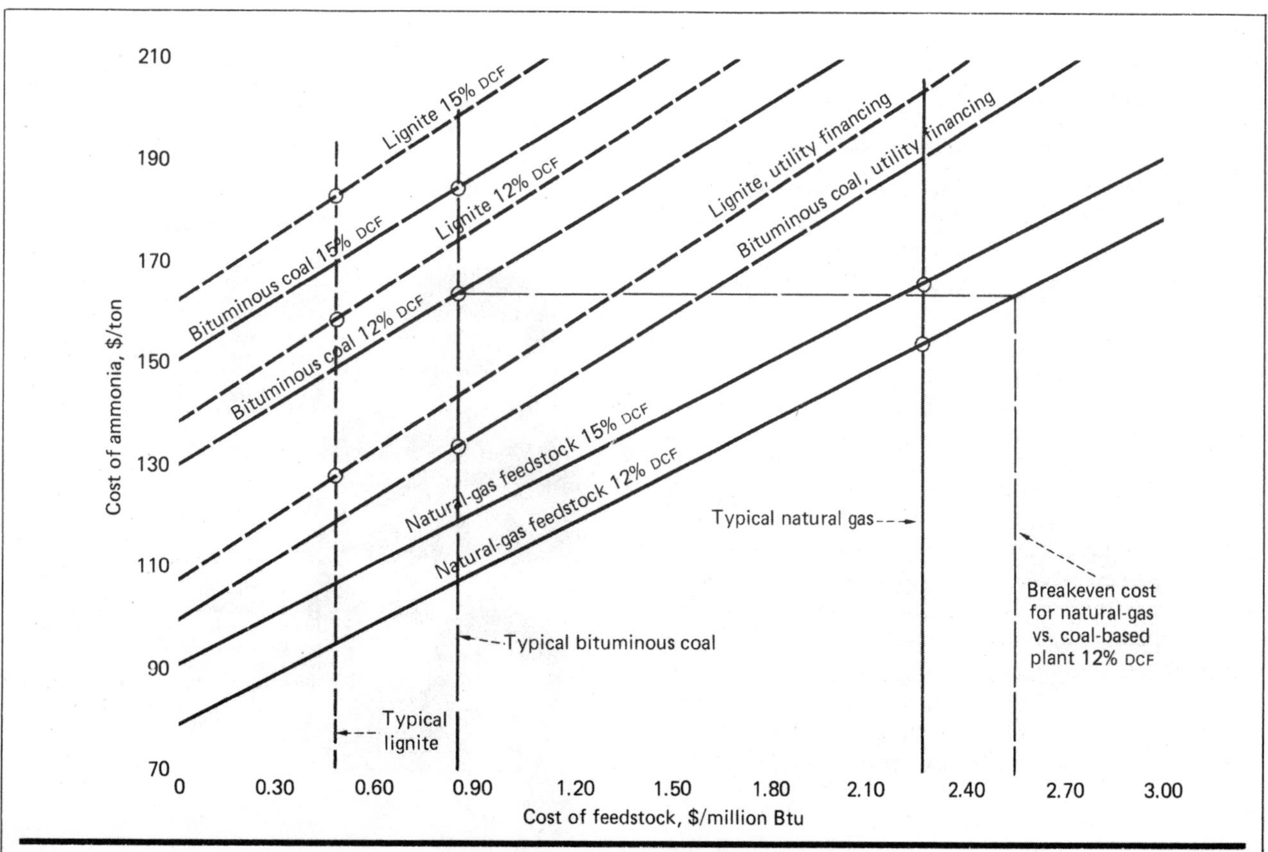

Relationship between the cost of various feedstocks and the cost of ammonia produced Fig. 2

Ammonia process economics, DCF method (All figures based on 1st quarter 1977) Table IV

	Coal, $ million	Natural gas, $ million
Plant investment	157.0*	90.0‡
Royalties	3.80	1.5
Construction interest	24.4	11.6
Capital charges, *I*	185.2	103.1
Working capital, *W*	2.0	3.4
Labor costs and ash disposal	2.55	1.84
Maintenance material and raw water	4.1	2.8
General administration	3.1	2.0
Property tax and insurance	3.8	2.3
Feedstock (coal @$0.85/million Btu, natural gas at $2.25/million Btu†	16.5	36.7
Electric power	—	1.1
Annual operating costs, *N*	30.05	46.74
Startup costs, *S*	6.93	3.93
General cost equation (at 0.9 ¶ service factor):		
Case A cost/ton NH_3 = 12% DCF $\dfrac{N + 0.247I + 0.133S + 0.23W}{1500\ (365)\ 0.9}$		
Ammonia price, $/ton	156.6	149.2
Case B cost/ton NH_3 = 15% DCF $\dfrac{N + 0.301I + 0.159S + 0.275W}{1500\ (365)\ 0.9}$ ¶		
Ammonia price, $/ton	177.4	160.1

*If the coal is fed to the plant as a slurry, this cost figure may be reduced by about 6 to 8%.

†Typical average for bituminous coal, and newly contracted natural gas in intrastate market. The energy requirement for natural-gas-based plant is estimated at 32.5 million Btu per ton of ammonia.

‡For naphtha feedstock the capital investment is approximately 7 to 10% higher and feedstock requirement is about 34 million Btu per ton of ammonia. For heavy oil feedstock, the capital investment is about 40-50% higher and feedstock requirement is about 37 million Btu per ton of ammonia.

¶ Factors in cost equations are derived in Ref. [*6*].

Table IV lists the main capital and operating costs.

In Fig. 2 the ammonia selling price is shown as a function of the feedstock cost. For example, bituminous coal at $0.85/million Btu and DCF of 12% will be competitive with a natural-gas-based plant with feedstock at $2.55/million Btu.

The ammonia price does not include a credit for 107 tons/d sulfur byproduct. A reasonable credit could be in the range of $1 to $3 per ton of ammonia.

It is apparent that coal-to-ammonia plants are more capital-intensive than their natural-gas counterparts. The use of Texaco gasification technology can reduce the capital cost of a coal-to-ammonia plant by over 15% and improve the thermal efficiency by about 15%, compared with older coal-to-ammonia technologies. Nevertheless, even using emerging gasification technologies, a greater amount of capital is needed to make ammonia from coal as compared to natural gas.

Making the ammonia plant as an increment to a SNG complex will further reduce the cost of the plant. If utility financing is assumed, as would apply to the SNG plant, the cost of ammonia can be significantly reduced (see Fig. 2).

If the gas-generation section is utility-financed while the ammonia synthesis is DCF-financed, the selling price of ammonia would fall between the two cases discussed previously.

The price of coal varies from $0.40 to over $1.20/ million Btu, depending upon how it is mined and upon the sulfur and ash content. For the purpose of gasifying coal, it is desirable to increase the heat content of the coal-slurry feed. Therefore, increased ash or chemically bonded moisture in the coal will have an adverse affect on the process. The sulfur content has virtually no impact on the capital and operating cost of the plant.

When lignites and high-ash coals are used, the cost of the ammonia plant will escalate on the order of 6 to 10% in most cases (mainly because of increased oxygen requirement and coal-handling facilities). The lower cost of lignite, however, which normally ranges from $0.40 to 0.60/million Btu, will more than offset the higher capital cost. This low cost, in fact, will make many types of lignite attractive as feedstocks.

References

1. Hess, Martin, Ammonia: Coal versus Gas, *Hydrocarbon Process.*, Nov. 1976, p. 97.
2. Ibid.
3. Buividas, L. G., others, Alternative Ammonia Feedstocks, *Chem. Eng. Progr.*, Oct. 1974, p. 21.
4. Franzen, J. E., and Goeke, E. K., Gasify Coal for Petrochemicals, *Hydrocarbon Process.*, Nov. 1976, p. 134.
5. Mitsak, D. M., others, Economics of the K-T Process, *Energy Comm.*, Vol. 1, No. 2 (1975).
6. Factored Estimate for Western Coals: Commercial Concepts, C. F. Braun report for American Gas Assn./Energy Research and Development Administration, E(4918)-2240, Oct. 1976.
7. Strelzoff, Samuel, Make Ammonia from Coal, *Hydrocarbon Process.*, Oct. 1974, p. 133.

The authors

David Netzer is a principal process engineer at the Houston Div. of Fluor Engineers and Constructors, Inc., 4620 North Braeswood, Houston, TX 77035. He received his B.S. in chemical engineering from the Technion (Israel Institute of Technology), Haifa, Israel, and is a registered professional engineer in the states of New York and Illinois.

James Moe works in the Process Engineering Dept. of Fluor's Southern California Div, Irvine, CA 92714. He was formerly associated with Girdler Catalysts, and Minerals and Chemicals Co. He holds a degree in chemical engineering from the University of Colorado.

Ammonia from coal: A technical/economic review

Concern over meeting U.S. fertilizer needs—because of the natural gas shortage—has spurred projects to produce ammonia from alternative feedstocks, mainly coal. This article looks at efforts being made in the U.S. and overseas.

Donald A. Waitzman, *Tennessee Valley Authority**

☐ At least one third of the food and fiber produced in the U.S. can be credited to the use of fertilizers, with nitrogen being the nutrient of major importance. Practically all nitrogen fertilizer is made from ammonia, which, in turn, is made from natural gas. But most experts predict that U.S. natural gas reserves will be depleted within 30 years or so. Between now and then, gas is expected to become increasingly costly, or unavailable, to those on interstate (regulated) supplies. And, although naphtha or fuel oil could replace natural gas as a feedstock, they too are scarce and expensive. Thus, coal seems to represent the only viable alternative for the foreseeable future.

Of current world ammonia capacity—about 77 million short tons of nitrogen equivalent—64% is based on natural gas, 13% on naphtha, and 12% on coal or coke, with the remaining 11% equally divided among other feedstock sources. In an effort to shift away from reliance on natural gas, three projects are underway in the U.S., all for production of ammonia from coal: one by the Tennessee Valley Authority (TVA) and two under the auspices of a branch of the U.S. Dept. of Energy (DOE), formerly known as the Energy Research and Development Administration. The three efforts complement one another, in that the DOE projects involve grass-roots facilities, while the TVA project is a retrofit to an existing plant.

In addition, there is a great deal of work being done around the world based on petroleum technology and German coal technology. At present, there are at least 14 coal-based ammonia plants in operation (none in the U.S.) and a larger number (some of which are in the U.S.) of petroleum-based, partial-oxidation units making NH_3.

In order to gain knowledge of the advantages and disadvantages of the various processes, members of TVA's Ammonia from Coal Projects staff traveled to West Germany to visit the process-development firms and to South Africa and India to meet with operators of some of the processes. The conclusion of the survey is that both the technical and economic aspects of coal-based ammonia production are unclear. No one gasification process will be applicable for all situations in the U.S., and the ammonia producer will have to consider several alternatives, depending on his own particular circumstances.

DOMESTIC EFFORTS—Once TVA dedicated itself to developing and demonstrating a method to make ammonia from coal, analysis showed that the cheapest and quickest way to accomplish that goal was to retrofit a coal gasification plant onto TVA's small, but modern, ammonia plant at Muscle Shoals, Ala. (see Fig. 1).

The plant is a 225-ton/d, natural-gas-feed, steam-reforming plant, completed in 1972. After the retrofit it will operate either at full capacity, with 60% of the feed being synthesis gas

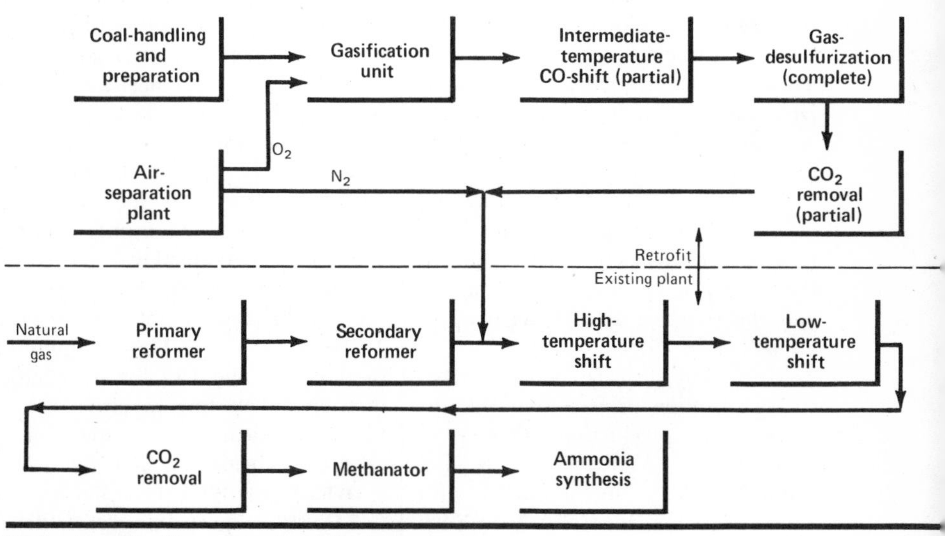

TVA will add a Texaco coal-gasification process to an existing ammonia plant — Fig. 1

*This article is based on a presentation at the 27th Annual Meeting of the Fertilizer Industry Round Table, held Oct. 25–27, 1977, at Washington, D.C.

Originally published January 30, 1978

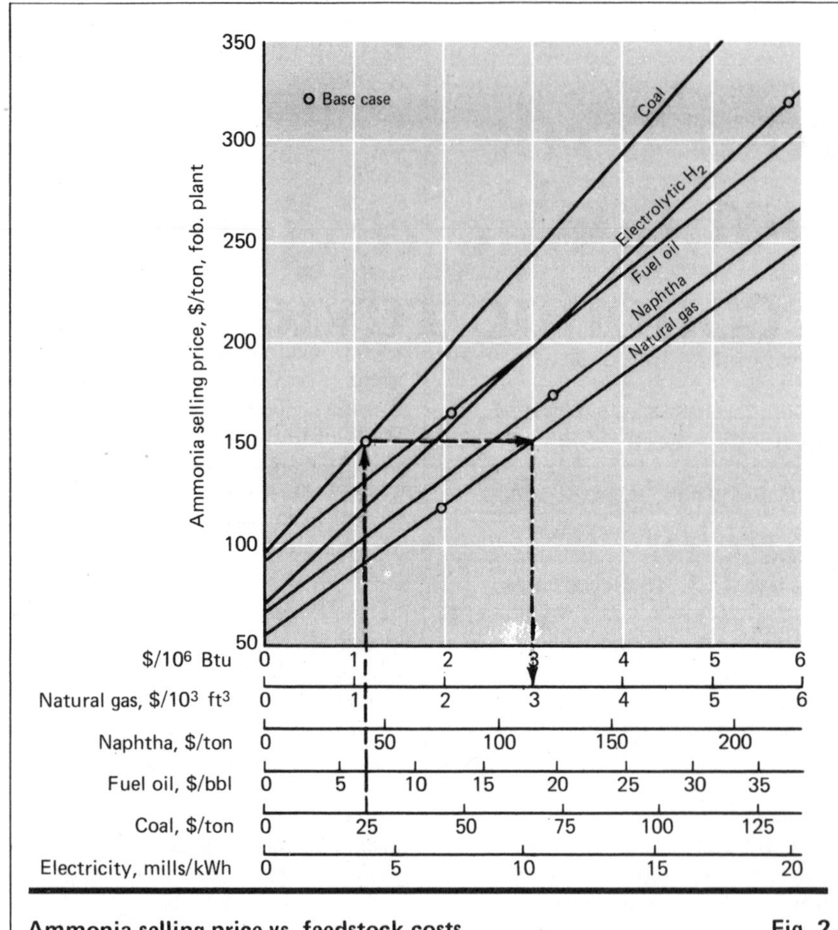

Ammonia selling price vs. feedstock costs **Fig. 2**

from coal, the remainder being natural gas, or at 60% capacity, using only coal-generated synthesis gas.

After a thorough review of the available processes, TVA concluded that the partial-oxidation process from Texaco Development Corp. (New York City)—originally commercialized using an oil feedstock—best met its criteria. Design conditions are: to gasify 168 tons/d of coal at a pressure of 490 lb/in^2 to produce 135 tons/d of ammonia. A final contract for engineering, procurement and erection of the facility is to be awarded shortly, and the complex should begin operations late in 1979.

Meanwhile, DOE is conducting two coal-gasification projects that will involve NH_3 production. One is a contract with W.R. Grace & Co. (New York City), in conjunction with Ebasco Services, Inc. (also New York City), while the other is an agreement to negotiate a contract with Air Products & Chemicals Inc. (Allentown, Pa.).

The Grace/Ebasco plant, according to DOE, would use the Texaco process, and would gasify 1,700 tons/d of coal at a pressure of 2,500 lb/in^2. The gas would be scrubbed, purified and sent to a new ammonia plant not funded by DOE. A feasibility study will be made first, and operation is scheduled for mid-1982 if a "go" decision is made by DOE—the Grace/Ebasco contract is one of two competing for funds available for only one plant. The other contract—which does not involve ammonia—is with Memphis (Tenn.) Light, Gas & Water Div.

Using the Koppers-Totzek process, the Air Products plant would gasify 1,210 tons/d of Texas lignite to produce H_2 and CO for distribution through an existing pipeline to Gulf Coast chemical firms. Operation is expected to begin late in 1981.

OVERSEAS WORK—TVA staffers visited West Germany and process-development firms Lurgi Kohle und Mineraloeltechnik GmbH (Frankfurt/Main), Krupp-Koppers GmbH (Essen) and Davy Powergas GmbH (Cologne), which offers the Winkler process. In addition, operators of the processes in Africa and Asia were also visited: South African Coal, Oil and Gas Corp. Ltd.'s (Sasol) plant at Sasolburg and African Explosives & Chemical Industries Ltd.'s (AE&CI) facility at Modderfontein, both in South Africa, and a unit of Fertilizer Corp. of India at Talcher, India.

■ *Lurgi*: Sasol uses the Lurgi process, and has 13 coal gasifiers.

Gasifier operating pressure is about 400 lb/in^2, and gasification temperature is about 2,200°F. The raw gas from the gasifier contains significant quantities of condensibles, such as ammonia, phenols and tars that Sasol recovers for fuel or sale.

Developers at Lurgi suggest that part of the condensibles, the "oil," can be hydrotreated over a catalyst to convert organic sulfur to H_2S, to reduce the CO content to 5%, and to reduce HCN to methane and ammonia. The resulting purified naphtha could be used as fuel.

The Lurgi process has the lowest oxygen requirement of all commercial gasifiers. Other conditions being equal, the need ranges from 0.11 std ft^3 of oxygen per std ft^3 of synthesis gas for active lignites, to 0.21 std ft^3/std ft^3 of synthesis gas for anthracites. In addition, Lurgi routes need an oxygen purity of only 90%, which should reduce the cost of air separation facilities by about 10%.

■ *Davy Powergas*: Davy has built 16 plants having a total of 36 Winkler gasifiers using a variety of coals. Twelve of the 16 were dedicated to ammonia production; three plants are still operational. All the gasifiers operated at near atmospheric pressure. However, Davy is now ready to offer a process operating at 45 lb/in^2. Among the reasons for choosing such a pressure:

- The pressure is sufficient to allow final particulate removal in venturi scrubbers, rather than in electrostatic precipitators.
- Major savings in capital cost and compression costs can be obtained at this pressure level without increasing the methane content of the synthesis gas. Also, at that pressure, the process uses 97 to 98% of the carbon values in the feed.

A variety of coals have been gasified in the Winkler units, but lignites and subbituminous coals are preferred. The Winkler is a nonslagging gasifier operating at temperatures lower than the ash-fusion temperature, but still high enough to prevent formation of methane and condensibles.

Davy feels that the major advantages for the Winkler gasifier are its simplicity and high capacity and that the nonslagging operation gives a high onstream efficiency. The large inventory of the fluidized bed provides a safety against oxygen breakthrough and tends to even out variations in feedstock quality. Additionally, the process is not sensitive to feed size-distribution—an advantage when gasifying low-grade, high-ash coals, since no pulverization is required.

■ *Krupp-Koppers*: Koppers stated that the Shell-Koppers development program at the Harburg, W. Germany, refinery of Deutsche Shell AG (Hamburg) involves the installation of a demonstration gasifier. Dry pulverized coal will be used as the feedstock to produce about 10 million std ft^3/d of synthesis gas.

All six Koppers-Totzek gasifiers at AE&CI's plant are normally used to produce 1,100 short tons/d of combined ammonia and methanol. However, full production can be maintained with only five units.

The coal is dried and pulverized, then carried by entrainment in a steam/oxygen stream into the gasifier, where reaction takes place at 2,900 to 3,600°F and slightly above atmospheric pressure.

Fertilizer Corp. of India's two plants, at Talcher and Ramagundam, are essentially identical and will produce 1,000 short tons/d of ammonia and 1,650 short tons/d of urea. Both facilities—presently under construction—will have three Koppers four-headed gasifiers. Design gasification temperature is 2,900°F at a pressure of about 16 in. H_2O. The Talcher plant is in the final phases of construction; the Ramagundam unit is about two months behind the former.

PROBLEMS—At each location visited, specific questions were asked about experiences pertaining to air and stream pollution and occupational safety and health.

Air pollution problems are being handled in varying degrees, depending on emission regulations in effect at the particular location. Generally speaking, air pollution controls in the U.S. will have to be much more extensive than those observed in the survey. Sasol has, until recently, released H_2S through a boiler stack, but it has now installed a sulfur-recovery unit. Coals used in most places visited were low-sulfur, but release of carbon monoxide to the atmosphere was allowed in some of the plants.

Water pollution in South Africa is rigidly controlled because water is scarce, and plant water effluents go into the community water-supply system. Close monitoring for heavy metals contamination has been practiced, but no corrective action has been needed.

Incidents with toxic fumes have been reported, ranging from inhalation of methanol fumes to fatalities from carbon monoxide and nitrogen asphyxiations. Effective corrective action seems to have been taken by installation of suitable fume-detection devices and unit modifications.

ASSESSMENT—Each coal gasification process studied has certain apparent advantages in given situations.

The Lurgi process would have prime application where there would be a use or market for a multiplicity of products consisting of synthesis gas, methane, oils, tars, phenols, etc. Its best utilization is with those coals that have noncaking characteristics.

The Koppers-Totzek process seems most applicable for coals having varying reactivities and ash-melting behavior and temperatures, and where coal-tar condensibles are not desired. A benefit is that the process yields high-pressure steam in a commercially proven waste-heat boiler.

The Winkler process performs well with certain coals, but not so well with others. Winkler does not produce condensibles, but it does yield a combustible char that must be disposed of. This would not be a problem in ammonia production, however, because the char could be used as steam-generation fuel.

The Texaco process has not yet advanced to commercial use with coal as a feedstock, but it does offer the potential of operating at elevated pressure, and it is expected to accept a variety of coals without producing condensibles.

Davy and Koppers are engaged in development work on their processes for gasification at elevated pressures; Lurgi has been involved in development of a slagging gasifier (see pp. 11–12.)

ECONOMIC OUTLOOK—Very little capital or operating cost information was available during the TVA team's visits. Information obtained was either inapplicable to the U.S. or was out of date. TVA has, however, prepared a series of conceptual designs and cost estimates on 1,000-short-ton/d grassroots ammonia plants.

The estimated cost of a 1,000-ton/d natural-gas, steam-reforming plant is about $75 million, and of a coal-based partial-oxidation ammonia plant, approximately $140 million. Estimated ammonia selling price, fob. plant, for 1,000-ton/d plants is shown on Fig. 2 for various feedstocks. Price includes cost of raw materials and chemicals, operating labor and supervision, utilities, maintenance, simple depreciation at 15 years, insurance, plant and administrative overheads, a 50-50 debt-equity capital structure, interest at 10% on borrowed capital, marketing, and a 14% after-tax return on the owner's equity.

Ammonia could be produced in a natural-gas steam-reforming plant, built in 1977, at a selling price of about $120/ton, using $2/1,000-$ft^3$ natural gas. The price for NH_3 from a coal-based plant would be about $150/ton, using $25/ton coal. It can be seen that if natural gas rises to $3/1,000 ft^3, coal would be competitive at $25/ton.

Naphtha, heavy oil and electrolytic hydrogen are also shown on the curve and are not competitive with coal at current costs of $13/bbl for fuel oil, $120/ton for naphtha, and 20 mills/kWh for electricity.

To bring all of this into perspective, prices for ammonia delivered to retail dealers in the Midwest are currently about $125 to $130 per ton, about $100/ton on the Gulf Coast, and less on some small shipments. Coal costs at Muscle Shoals are currently between $25 and $30 per ton. The cost would be about $17 to $23 per ton for a coal-based plant located at the coal mine (high-sulfur, bituminous).

For more information on ammonia-from-coal technology, see pp. 59–62 and 69–72.

The author

Donald A. Waitzman is projects manager for Ammonia from Coal Projects, Tennessee Valley Authority (National Fertilizer Development Center, Muscle Shoals, AL 35660). He holds a bachelor's degree in chemical engineering from Auburn University, and is a member of the American Institute of Chemical Engineers and the National Management Association. He is a registered professional engineer in Alabama.

Hydrogen recovery unit ups NH_3-plant efficiency

This recovery installation recycles most of the hydrogen contained in ammonia-plant purge gas back to the synthesis loop, for conversion to more ammonia. At the same time, the remaining waste fuel-gas stream supplements the natural-gas feed for the reformer burners.

Roy Banks, *Petrocarbon Developments, Inc.*

☐ The first cryogenic hydrogen-recovery unit specifically designed to recirculate hydrogen back to an ammonia plant's synthesis loop is presently operational at Vistron Corp.'s (Cleveland, Ohio) 1,500-ton/d plant at Lima, Ohio. And a second facility is being engineered to serve a total capacity of 1,750 tons/d from two ammonia plants at American Cyanamid Co.'s (Wayne, N.J.) Fortier, La., site.

In operation since April of this year, the Vistron recovery system has increased ammonia production by approximately 75 std tons/d—an attractive investment, since the installed cost of the recovery plant and ancillaries was less than $2 million. It is estimated that the energy cost of increasing ammonia-plant capacity by using larger production equipment would be about 35 million Btu/ton NH_3, compared with the cryogenic unit's cost of approximately 25 million Btu/ton NH_3 produced.

The plant processes approximately 11 million std ft^3/d of hydrogen-rich ammonia purge gas, and, after rejecting the bulk of inerts (argon and methane) contained therein, returns a 91%-hydrogen (by volume) stream back to the ammonia synthesis loop. The remaining waste fuel-gas stream—which still contains 35 to 40% of the heat content of the original purge gas—then goes to the primary reformer burners.

In the Vistron facility, recycling of the hydrogen overcame a bottleneck existing in the primary reformer (syngas production) unit, which limited production of hydrogen, and hence, ammonia. In a new plant, however, the recycled hydrogen boosts ammonia production capacity even further.

A typical Petrocarbon Developments recovery plant can accommodate feedstocks of varying compositions and

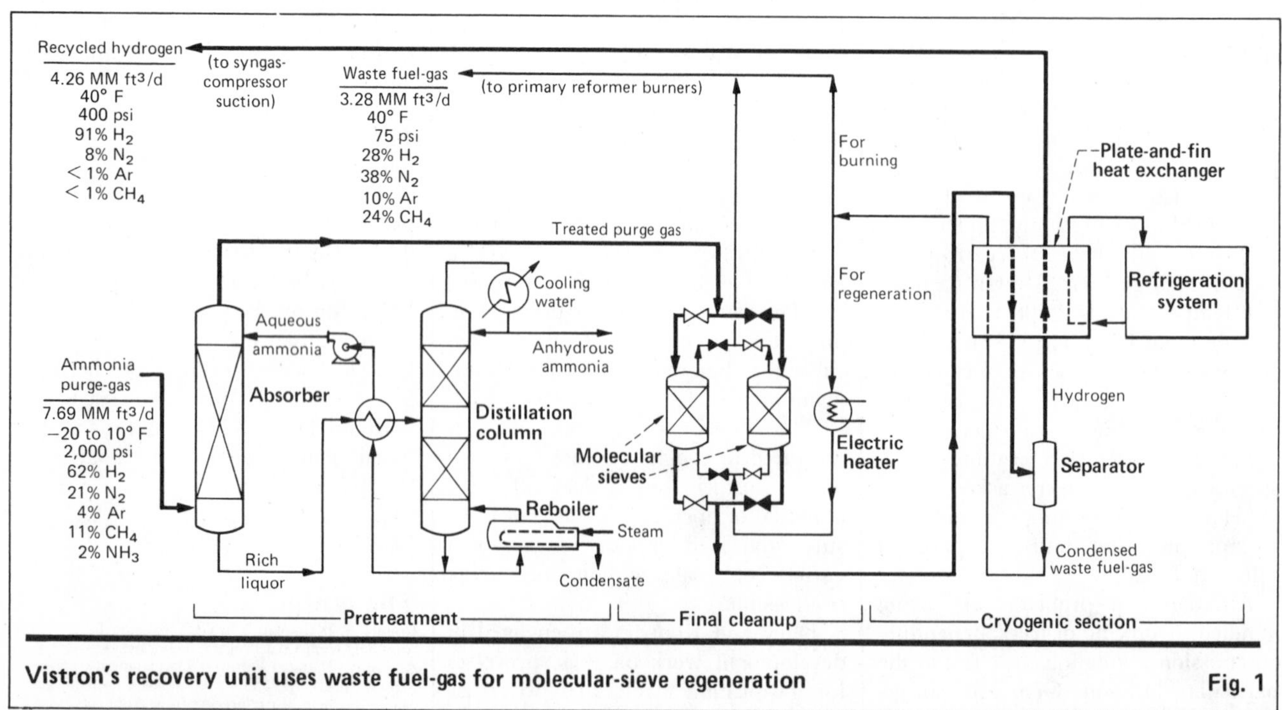

Vistron's recovery unit uses waste fuel-gas for molecular-sieve regeneration — Fig. 1

Originally published October 10, 1977

flowrates, and can turn down to below 40% of rated capacity.

Fully automatic in operation, the plant can be monitored continuously from a panelboard. Space requirements are modest; the facility can usually fit into an area of 2,000 to 3,000 $ft.^2$

The hydrogen recovery unit can be considered either as (1) a retrofit to upgrade capacity (as in the Vistron application), or to improve the overall thermal efficiency of the operating plant, or (2) as part of any new expansion in ammonia-plant capacity. When applied as a retrofit, the recovery unit can be installed without upsetting the operation of the existing ammonia facility. In addition, emergency shutdown of the recovery unit will cause negligible upset to ammonia production in the main plant.

HOW IT WORKS—The hydrogen recovery facility consists of three sections: (1) pretreatment, (2) final cleanup, and (3) cryogenic separation.

■ *Pretreatment*: The purpose of pretreatment is to remove ammonia from the purge-gas stream. Concentration of ammonia present in the purge gas can vary between 1 and 4% by volume, and all traces of this ammonia must be removed to avoid freezing up the cryogenic section. For the Vistron facility, the most reliable and economic method for removal of the ammonia was via a two-step process:

In the first step, a lean solution of aqueous ammonia washes the ammonia-rich purge gas in a packed-column absorber (see Fig. 1). This washing removes the bulk of the ammonia present. The resulting rich liquor is removed from the base of the column and sent, in the second step, to a packed distillation column where the ammonia is distilled—using a steam-heated reboiler—as anhydrous product. The recovered lean solution gets recycled back to the absorber column, and any water loss is made up by a small input of demineralized water.

■ *Final Cleanup*: The processed purge gas leaves the top of the absorber and passes through a two-bed molecular-sieve unit, where the final traces of NH_3 and H_2O are removed. The molecular sieve operates automatically on a fixed cycle; changeover is accomplished by pneumatically operated valves, actuated by a cam timer. No operator intervention is needed.

Regeneration gas for the off-stream molecular-sieve bed is provided by diverting a portion of either the waste fuel-gas stream or the hydrogen product stream exiting the cryogenic section of the process. The regeneration stream is heated electrically to approximately 550°F before regenerating the bed; on completion of the heating period, the same gas cools the bed back down before the bed goes back onstream.

Whichever gas is used (waste fuel-gas or hydrogen), it will contain traces of NH_3 and H_2O that have desorbed from the molecular sieve at a concentration of 0.5 to 1.0%. When hydrogen product is used as the regeneration gas (and it can be returned to the synloop containing some H_2O), this increases ammonia recovery by returning adsorbed ammonia to the loop and avoids the possibility of NO_x emissions that would result from the combustion of ammonia in the fuel-gas stream.

■ *Cryogenic Separation*: The final cryogenic section is enclosed within an insulant-filled, self-supporting plate-and-frame structure containing aluminum-alloy plate-and-fin heat exchangers, separators, and a refrigeration cycle to ensure operation of the process.

In operation, the purge gas exiting the molecular sieve passes through the plate-and-fin heat exchanger, where it is cooled, and enters a separator. The hydrogen product stream leaves the top of the separator, while the condensed waste fuel-gas exits at the bottom of the unit. A portion of the waste fuel-gas may be tapped off for use in regenerating the molecular-sieve beds, or all of the fuel gas may go to the reformer burners. That portion used for regeneration later rejoins the main waste fuel-gas stream.

The cryogenic section is completely shop-assembled, and is delivered to the site as a single unit. Equipment and pipework are fabricated using all-welded construction to eliminate mechanical joints as much as possible.

The refrigeration for the Vistron system is supplied by a closed-loop nitrogen cycle consisting of a two-stage, oilfree, reciprocating nitrogen compressor that feeds a radial-inflow gas-bearing expansion turbine. Thus, the Petrocarbon unit can be rapidly cooled down and brought onstream in usually less than eight hours.

EFFECT ON AMMONIA PLANT—The ammonia plant's operating mode does not have any significant effect on the operating conditions in the recovery unit. However, the recovery unit allows the ammonia producer to run his plant in response to prevailing economic conditions. Thus, he can:

■ Achieve maximum possible ammonia production during periods of high ammonia demand.

■ Maximize ammonia production while consuming his normal natural-gas allocation.

■ Extend run-times at design production levels, where they may have been previously limited by primary reformer-tube failures.

At the Vistron facility, as stated earlier, the front-end reformer was the limiting factor, whereas the process-air system and synthesis loop had capacity for more production. Hence, the primary justification for installation of the recovery unit was to make optimum use of recycled hydrogen, so as to debottleneck the plant and maximize ammonia production.

When the recycled hydrogen joins the syngas mixture, the combined stream is compressed through the syngas compressor, which then needs to achieve an additional throughput of approximately 5%. Some extra air must also be supplied to the secondary reformer to provide additional nitrogen (about 3%). A similar 5% capacity increase is also needed from the refrigeration unit to exploit the full potential increase in ammonia production of about 67 to 80 std tons/d.

If these additional capacity requirements can be accommodated, then the increase in ammonia output on a 1,500-std ton/d unit (at $150/ton) represents additional income of $3.0 to $3.6 million per year. The actual increase in ammonia production achievable depends on the operating constraints and catalyst activity within the ammonia plant and the resulting purge-gas flowrate. Against this revenue must be offset the extra power needed for the air and syngas compressors, and the additional fuel-gas requirements of the primary reformer. With these paid, a net revenue of over $1.0 million per year is expected, with natural gas at $2/1,000 ft^3.

ECONOMIC EVALUATION—Any method used for economic evaluation must inevitably be adjusted to particular company circumstances. The following analysis is based on a typical hydrogen recovery unit associated with

Rate of return after taxes for hydrogen recovery facility

Basis: Total plant erected cost: $2 million
Onstream time: 8,000 h/yr

	Annual cost ($)
Direct operating cost	
Utilities	
Electricity (120 kWh/h @ $0.02/kWh)	19,200
Steam (1.47 MM Btu/h @ $3.50/MMBtu)	41,200
Nitrogen purge (73.8 tons/yr @ $77/ton)	5,664
Cooling water (9,000 gal/h @ $0.05/gal)	5,150
Total recovery plant utilities	71,214
Additional natural gas for ammonia plant (53,750 std ft^3/h @ $2.00/1,000 std ft^3)	860,000
Total utilities cost	931,214
Operating labor (1/3 shift operator @ $12/h, including overheads, for 24 h/d, 365 d/yr)	35,040
Maintenance (3% of erected equipment cost)	60,000
Taxes and insurance (2% of erected equipment cost)	40,000
Total direct cost	1,066,254
Depreciable capital requirements	
Fixed capital requirement (non-annual cost)	2,000,000
Non-depreciable capital (molecular sieve, etc.)	negligible
Depreciation (@ 13 yrs, straight-line)	153,847
Average interest on depreciable (@ 12%/yr, avg.)	120,000
Total capital cost	273,847
Total direct cost	1,066,254
Total direct and capital costs	1,340,101
Overall economics	
Additional income (from NH_3 sales from increased NH_3 production: 17,233 tons/yr @ $150/ton)	2,585,000
Extra cost incurred to achieve additional income	1,340,101
Gross profit (before taxes)	1,244,899
Net profit (after taxes @ 50%)	622,450
Depreciation	153,847
Total cash flow	776,297
Rate of return (cash flow/fixed capital requirement)	38.8%/yr
Payback after taxes	2.6 yrs

Source: Petrocarbon Developments, Inc.

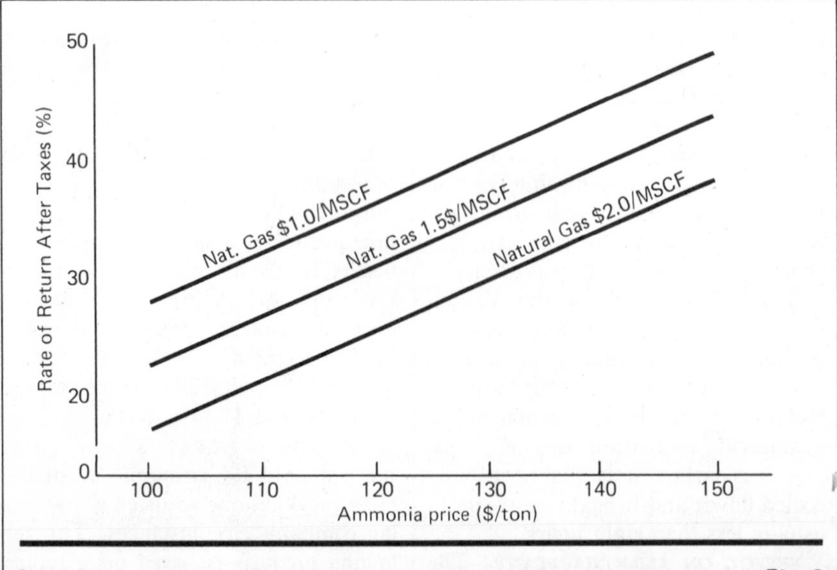

Sensitivity analysis: effect of ammonia price and natural-gas cost Fig. 2

a current-generation 1,150-std-ton/d ammonia plant when ammonia production is increased by 51.7 tons/d at the same primary-reformer input feed-gas consumption. Payback time is approximately 2½ years; rate of return is about 40%. When purge gas is available from an ammonia facility having a higher capacity, the scale factor applied to capital investment for the recovery unit is such that the payback time reduces even further.

The control panel for the hydrogen-recovery unit will be located in the main ammonia-plant control room, and one third of the shift operator's time will be charged against the recovery facility.

The material balance and utilities consumption shown in the analysis are based on a conservative 81% hydrogen recovery for a purge-gas stream where the recycled hydrogen is returned to the synthesis-gas compressor suction at a flowrate of 4.26 million std ft^3/d, with a purity of about 91% and a pressure of 400 lb/in^2, and where the fuel-gas main operates at 75 lb/in^2. The plant recovers ammonia, present in the purge gas, as anhydrous product at a pressure of about 250 lb/in^2. The actual hydrogen recovery achieved can usually be expected to approach 90 to 95%, and the increased ammonia production would rise in proportion.

The economic analysis is based on a capital cost of $2 million, and a hydrogen-recovery plant onstream-time of 8,000 h/yr.

SENSITIVITY ANALYSIS—To investigate the effect of variations in the selling price of ammonia on the above conclusions, the calculations were repeated for a range of ammonia prices between $120 and $150 per ton, and natural-gas prices between $1 and $2 per million std ft^3. These results are shown in Fig. 2.

The author

Roy Banks is technical manager, Petrocarbon Developments, Inc. (P.O. Box 58249, Houston, TX 77058), and is currently responsible for technical sales for the Cryogenic Business Group of Petrocarbon Developments in the U.S. He is a member of AIChE, and holds a Ph.D. and a B.Sc. in chemical engineering from Leeds University, England.

New feeds and processes perk ammonia production

As uncertainty grows over the availability and price of natural-gas feed, some producers are looking at routes that use coal and fuel oil, while others are developing energy-saving, high-efficiency plant designs.

☐ There is good reason to believe that the manufacture of ammonia, often called the highest volume petrochemical, will change considerably in coming years, at least in the U.S. and Europe.

U.S. ammonia producers, conscious of the fact that domestic supplies of natural-gas feed will be neither plentiful nor cheap in the 1980s, are making advances in ammonia-from-coal technology* (at first glance, this is a return to the past, since coal was one of the earliest NH_3 feedstocks).

Europe isn't facing such a pressing feedstock-supply problem (indeed, a swing toward natural gas—instead of naphtha—for older plants is underway). But European engineering and contracting firms, anxious to sell ammonia plants and knowhow abroad, are hard at work on routes that feature alternative feeds. And local ammonia producers, responding to a worldwide energy-crisis atmosphere, are seeking ways to cut energy consumption and increase the efficiency of traditional processes.

THE COAL OPTION—An unnamed site in western Kentucky will be the home of the first U.S. ammonia-from-coal facility. In late August of this year, W. R. Grace & Co. (New York City) announced that it would join the Energy Research and Development

*For more on ammonia from coal, see p. 59.

Originally published October 24, 1977

Administration (ERDA) in a feasibility study for a \$350-million project to produce ammonia from coal. The first phase will consist of a \$10-million design and engineering study scheduled to take two years. If successful, this will lead to construction of a demonstration plant that will use 1,700 tons/d of coal to make synthesis gas for a 1,200-ton/d ammonia plant.

ERDA is thinking of providing 47% of the necessary funds for the project, which will use Texaco Inc.'s (New York City) coal-gasification process to produce synthesis gas. An ERDA spokesman notes that although this coal-conversion technology has been around for some time, it has not been demonstrated at the scale of the plant under consideration. He adds that the W. R. Grace/ERDA plant is an example of second-generation technology that aims at higher-efficiency, lower-cost production in comparison with existing processes that are based on natural gas.

Whether ERDA shares the cost of the larger-scale demonstration facility ultimately depends on how well the initial Kentucky plant does in comparison with another joint project ERDA is involved in: a fuel-gas-from-coal plant to be built by Memphis Light, Gas and Water Div. (Memphis, Tenn.) along the Mississippi River.

Besides ERDA, there is another backer of the Texaco route. This summer, the Tennessee Valley Authority (TVA) revealed that it would use the firm's process in a much smaller project. TVA will build a \$46-million plant that will feed on 7 short tons/d of coal to produce synthesis gas as part of the raw material for a 225-short-ton/d ammonia plant currently fed solely by natural gas. The demonstration gasifier is scheduled to start operation in 1980.

WHAT IT CAN DO—A switch to coal from natural-gas feed could give some producers in the U.S. and Canada a strong competitive edge. "It will be very attractive to firms with relatively new plants in midwestern states or Canadian-prairie provinces to consider a conversion to coal in the 1980s. They may be able to get a premium on ammonia prices over prices charged by firms with interrupted gas supplies," says Frank Brown, ammonia-group manager for Humphreys & Glasgow Ltd. (London, U.K.). Brown has been battling for conversion to coal for the last couple of years, and Humphreys & Glasgow has already carried out several studies for interested firms in North America.

In a paper given at Eurochem, a symposium recently held (June 20-24) in Birmingham, U.K., Brown assessed a range of existing coal-gasification processes for suitability in conversion of existing ammonia plants. He concluded that the two most likely processes were the Texaco one—already adopted by Grace/ERDA and TVA—and the slagging gasifier developed by British Gas Corp. (London) from the older Lurgi process. One reason for his choice is that such medium- to high-pressure processes require lower coal consumption, and plant size and the number of necessary gasifiers are much reduced. For a 1,000-ton/d ammonia plant, a single Texaco gasifier operating at 55 bar would have adequate capacity.

The high pressure at which the Texaco process operates provides additional advantages, says Brown. For example, the synthesis-gas-generation stage can be operated at a constant pressure, obviating the need for additional compressors. Maintaining a high pressure at that stage also helps to cut down compression requirements in the ammonia-synthesis step.

According to Brown, in process selection, capital charges outweigh feedstock and utility costs. "The capital component is still the most important difference between gasification processes," he notes, "that is, until the price of coal exceeds \$30/metric ton."

Weighing the economics of natural-gas vs. coal use, he adds: "A coal-based plant using high-quality feed [with a caloric value of 5,000 kcal/kg] available at less than \$10/metric ton, and a natural-gas-fed facility using raw material priced at \$3/metric ton will produce ammonia at comparable costs. However, if the natural-gas/coal price-ratio widens faster than the capital-cost index, coal-based production has an edge."

Warning against natural-gas shortages, Brown cautions: "If natural-gas supplies are interrupted, each day's production loss is likely to cost the operator [of a 1,000-metric-ton/d ammonia plant] about \$100,000, including loss of profit and built-up capital charges."

CALL FOR FUEL OIL—Coal is not the only alternative to natural-gas feed. Naphtha, for instance, has been widely employed in Europe, though its use is declining there. Heavy fuel-oil, a low-price material, offers still another choice.

Two French firms—Société de la Grande Paroisse (Paris) and Heurtey S.A. (Paris)—have teamed to develop a reforming process that can use asphalt or heavy fuel-oil as feedstock for synthesis gas. Says Michel Bonnet, director of the engineering division of Grande Paroisse: "If it is successful, it could revolutionize the ammonia industry."

According to Roger Dumon, research and development director of Heurtey, the new route requires only half the investment needed for such partial-oxidation processes as Shell's or Texaco's. It uses a fluidized bed of chromium/cobalt/molybdenum catalyst on an alumina support, and has no coking problems. Heavy-fuel-oil fractions containing up to 2% sulfur can be employed as feed.

So far, the French developers have operated a pilot plant capable of producing 10,000 m^3/d of hydrogen; next year, they plan to build a 100,000-m^3/d unit. While these sizes are adequate for demonstrating the process to prospective customers in refining (who seek hydrogen-generation capacity for hydrocracking or desulfurization units), they are a far cry from the capacities needed in the ammonia business. (A commercial ammonia plant would call for hydrogen capacities of about 1 million m^3/d.)

While doubts remain that the process can be scaled up to such proportions, Bonnet is optimistic: "There is no reason why we can't reach those magnitudes," he says.

Allied Chemical Corp. (Morristown, N.J.) also has a fuel-oil development that could be used in ammonia production. Known as the VFO (for vaporized fuel oil) process, it is mainly used to fire natural-gas reformers, but Allied is trying to convert it into an ammonia-feedstock technique, according to William Hoehing, technical director of the agricultural division.

Hoehing warns that commercial application is still about five years away, and depends on finding a catalyst able to handle large amounts of fuel-oil feed.

FUEL-OIL ECONOMICS—Some experts, including Humphreys & Glasgow's Brown, believe that heavy fuel-oil feed does not promise a cost-saving bonanza. He notes: "We still get inquiries for naphtha reformers, and naphtha costs 2.5 to 3 times more than natural gas or fuel oil. But there aren't many serious requests for fuel-oil-based plants. They are capital intensive and less reliable, have all sorts of complications, and need more workers and operating skill."

Among the drawbacks mentioned by Brown: Such facilities need an air-separation plant, and various sulfur compounds must be removed from the synthesis gas because fuel oil, as a feedstock, is not particularly clean.

These disadvantages notwithstanding, one West German company—Veba Chemie AG—has been making ammonia from a fuel oil feedstock for some time.

The firm started up a 1,250-ton/d plant at Gelsenkirchen in 1972, and is currently building a 1,660-ton/d successor at Bruensbuttel as a joint venture with Denmark's Superfos A/S (which will have 40% equity). The new facility will go onstream next year. Lurgi AG (Frankfurt) did the engineering for the gasification unit, which features Shell's partial oxidation process.

Speaking from experience, a Veba executive estimates that investment costs for a heavy-fuel-oil-based ammonia plant are about 40% higher than for one feeding on natural gas.

In explaining its decision to go with fuel oil, Veba concedes that additional quantities of natural gas can still be contracted for in Germany, and that fuel-oil gasification is not competitive at the current cost of natural gas in Europe. However, the company expects this cost to climb considerably during the early 1980s, as existing natural-gas contracts expire.

Several European sources cite a different motive for Veba's actions. They point out that the firm, which is Germany's largest oil refiner, has plenty of high-sulfur, heavy fuel oil for which there is virtually no market. Absorbing it into ammonia production is one way to avoid a costly refinery overhaul.

EFFICIENCY BOOSTERS—Energy-saving measures are resulting in real improvements in ammonia technology, both in the U.S. and abroad. One example is Imperial Chemical Industries Ltd.'s (London, U.K.) new 300,000-metric-ton/yr ammonia plant that went onstream early this year in Billingham. The company worked closely with the contractor (Humphreys & Glasgow) to optimize plant design, using mathematical modeling of the steam system, via analog and digital techniques, to minimize excess steam produced.

According to project manager Robin Gadsby, the result is a stable plant with virtually no excess steam. ICI has chosen "the most efficient and reliable turbines available, and picked the best possible steam conditions." This called for, among other things, reheating the steam supplied to the condensing turbine to minimize pressure in the vacuum condenser, and hence, obtain maximum kilowatts per ton of steam per hour.

ICI also decided to get the most out of the reformer flue-gas, using it extensively to preheat combustion-air. Although this is a well-established procedure, Gadsby claims that ICI "pushed it farther," combining it with a low excess-combustion-air-feed to the reformer.

Relatively new in the U.S., air preheating is favored by Allied's Hoehing, who estimates that an investment in preheaters (cost: about $1.3 million) pays for itself in as little as two years. Hoehing believes that air preheating reduces energy consumption by about 0.75 million Btu/ton of ammonia produced.

OTHER DESIGNS—Bechtel Corp. (Houston, Tex.) claims to have reduced fuel consumption of ammonia facilities. In conventional processing, says Bechtel, about 60% of the natural-gas feed goes into making hydrogen for synthesis gas, and 40% is for fuel

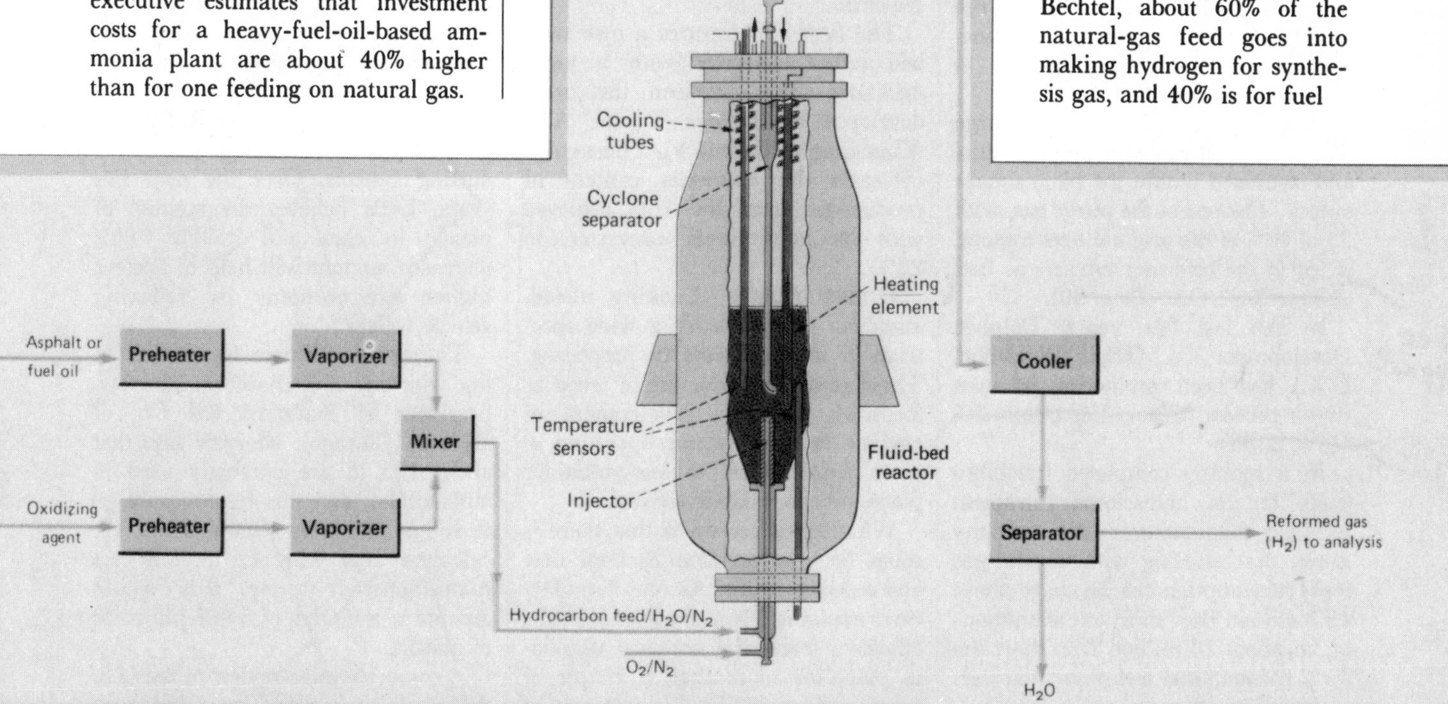

Flowsheet for French process uses specially designed reactor to produce hydrogen from heavy hydrocarbon feed **Fig. 1**

and utilities. The company's new design has a requirement of 28.2 million Btu/ton of ammonia; of this total, 10.2 million Btu is for fuel. Bechtel achieves this by means of a "combined cycle," in which a gas-turbine drive is used for one of the three large compressors, while steam turbines operate the other two.

Other companies have also upgraded their ammonia processes. European sources say that Pullman Kellogg (Houston, Tex.) has developed a new synthesis loop that runs at lower pressures, and that the firm has already offered this version to some prospective clients. There is also talk that Kellogg has tested the new design on a Verdigris, Okla., plant owned by Agrico Chemical Co. However, neither Kellogg nor Agrico will comment on this.

Also ready to unveil a significant ammonia-plant development is Humphreys & Glasgow. The new design is said to "substantially reduce" natural-gas consumption and capital investment. "It's quite unbeatable," says Brown, who nevertheless prefers to reserve further details for prospective clients.

A hydrogen purge-gas recovery system is attracting interest in the U.S. Developed by Petrocarbon Developments Inc. (Houston, Tex.), it can be designed into new plants or adapted to older ones. The system takes purge gas, separates the argon and methane, and recycles remaining hydrogen to the synthesis loop.

Ammonia separated from the purge gas is recovered as anhydrous product and combined within the plant's main output. The rest of the purge gas, with 35 to 40% of the original heat content, is fed to the reformer burners as fuel (see p. 66).

In the last few years, Dolphin Development Co. (High Wycombe, U.K.) has been promoting its own Breda process.

In a recently completed feasibility study for an undisclosed European fertilizer manufacturer, the company shows that, starting with natural-gas feed, consumption can be cut to about 15.2 million Btu/short ton of ammonia, vs. about 20 million Btu/short ton for a conventional technique that uses a secondary reformer.

According to technical director A. M. Dark, the Breda process can almost double the capacity of small existing plants. The shift conversion, carbon dioxide removal, and various other sections of a conventional plant can be easily adapted. And the only additional equipment required is extra steam-reforming capacity, a small cryogenic unit to supply pure nitrogen, and a centrifugal makeup-gas compressor and synthesis-loop section.

INTEGRATION SAVES—Italy's Snamprogetti S.p.A. (Milan) sees another way to save energy and investment in the manufacture of nitrogen fertilizer. Many ammonia producers use the petrochemical to make urea, and Snamprogetti believes that combining these operations into an ammonia/urea complex (as in its own process) can save 15% on investment and 16.8% on production costs. "This is an example," says Vincenzo Lagana, manager of the firm's special processes department, "of ways to eliminate parasite satellite plants and save money."

In the Snamprogetti flowscheme, ammonia reacts directly with carbon dioxide to make carbamate, and urea is the only end-product. This eliminates four expensive stages: carbon dioxide removal from converter gas; carbon dioxide compression; a cryogenic system to remove ammonia from synthesis-loop product gas; and liquid-ammonia storage and the equipment it requires.

The firm also features a new two-bed radial converter with a heat-exchange stage between the beds, developed by Haldor Topsoe A/S (Copenhagen, Denmark). This system increases the ammonia content of product-gas from the 16% achieved with its conventional converter, to 22%.

FUTURE TRENDS—Looking ahead, ammonia producers see a wide spectrum of improvements in processing. These range from the use of wood as feedstock, and the development of enzyme catalyst systems operating at room temperature, to less-ambitious plans to boost catalyst activity.

What seems certain is that there is room for improvement in both new and existing plants. As one manufacturer explains: "Nobody lives with the efficiency levels of a standard ammonia plant for more than a couple of years without spotting his own way of bettering the system."

Peter R. Savage

New solvent cuts costs of carbon monoxide recovery

Dramatic reductions in capital and operating costs, and a considerable number of technical advantages, are achieved by the new Cosorb process for carbon monoxide recovery.

Donald J. Haase, Tenneco Chemicals, Inc., Organics & Polymers Div.

☐ A new solvent and process—both referred to as Cosorb by developer Tenneco Chemicals—are now being introduced on a commercial scale.

The initial applications of the absorption/stripping technique will be for recovery of high-purity carbon monoxide (CO) from a refinery gas stream and from an acetylene-plant tailgas. Most of Cosorb's potential uses, however, lie in merchant production of CO for chemical synthesis, via conventional steam reforming or partial oxidation. Other applications are seen for preparation of ammonia synthesis gas and, longer term, for separation of CO from nitrogen in the product gases from air-fed coal gasifiers.

DEMONSTRATION Cosorb plant has prompted Dow's full-size unit.

Originally published August 4, 1975

A European firm will start up the refinery-gas Cosorb unit by year-end. During the first quarter of 1976, Dow Chemical Co. will bring the second unit onstream, at Freeport, Tex. This will feed on 14,000 normal m^3/h of tailgas and turn out a 99%-pure CO product. Some CO will be consumed internally to make phosgene as feedstock for toluene diisocyanate and eventually polyurethanes; the rest will be sold.

Dow has operated a small demonstration Cosorb unit since January (see photo). A demonstration plant with capacity of 100 tons/yr of CO has been operated by Tenneco Chemicals for two years to corroborate the simplicity of the Cosorb process and the noncorrosive behavior of the system under actual field conditions.

Tenneco Chemicals has qualified and approved several domestic and foreign firms to provide engineering and construction of units using the Cosorb process, but will directly license the process with each user.

SOLVENT IS THE STAR—The special Cosorb solvent consists of an active component, cuprous aluminum chloride ($CuAlCl_4$), in an aromatic base (toluene). The copper compound is in single phase with, and an integral part of, the stable solvent.

The solvent works by forming a chemical complex with CO (as opposed to physical absorption). Substances such as hydrogen, carbon dioxide, methane and nitrogen, generally found in synthesis gas, are chemically inert to the Cosorb solvent, but do exhibit some physical solubility in the aromatic base of the solvent.

The feedgas to a Cosorb unit usually requires a pretreatment. The most common impurity that must be removed beforehand is water, since the Cosorb solvent is nonaqueous and its active components will react quantitatively with any water present. Pretreatment with either molecular sieves or activated alumina can reduce water in the feedgas to less than 1 ppm.

The trace amount of water remaining reacts with the Cosorb solvent to produce quantitative amounts of hydrogen chloride (HCl)

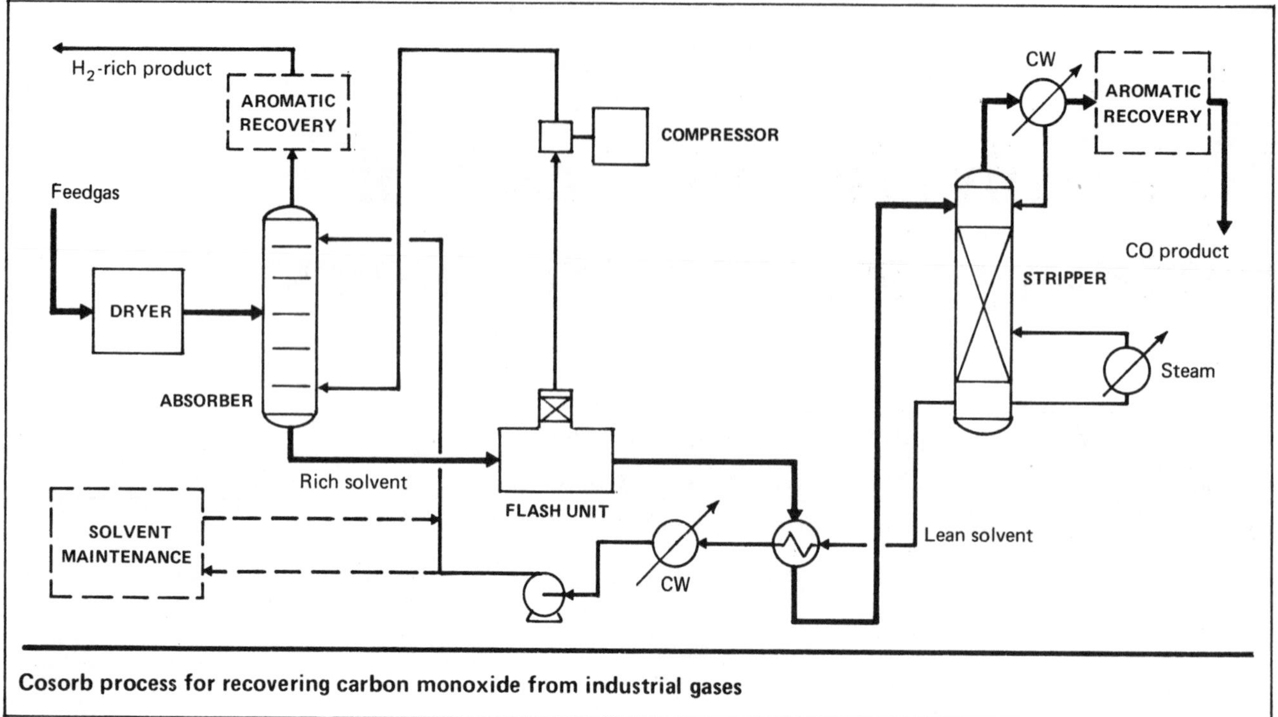

Cosorb process for recovering carbon monoxide from industrial gases

and a waste product soluble in the Cosorb solvent. The water-Cosorb reaction is advantageous in that the tailgas and CO product from the Cosorb unit come out absolutely dry; the ppm quantities of HCl produced in the dry atmosphere are thus noncorrosive.

Other components sometimes found in synthesis-gas streams, such as hydrogen sulfide, sulfur dioxide and ammonia, will have some reaction with the Cosorb solvent, and should be reduced prior to the Cosorb unit. But none of these impurities will cause a process shutdown, as water entering a cryogenic CO-recovery system might. They will only result in a calculable increase in solvent consumption.

ABSORPTION AND STRIPPING—The homogeneous Cosorb solvent contacts CO-laden feedgas in a countercurrent absorber, at ambient temperature and any practical pressure. It can chemically complex and recover more than 99 vol.% of the CO from the feedgas, usually leaving behind a hydrogen-rich coproduct stream.

Other common compounds in synthesis gas that have a physical solubility in the aromatic base of the Cosorb solvent are removed from the solvent in a following flash unit. This small stream can be recycled to the absorber, as shown in the flowsheet, or burned as fuel.

The CO-rich solvent from the flash unit is heated against recycling lean solvent, and passed to a stripper. Heat provided by low-pressure steam, and stripping action from the aromatic solvent-base, recover CO as an overhead product gas typically greater than 99 vol.% purity.

As shown in the flowsheet, aromatic recovery systems may be required on both the CO product stream and the absorber tailgas. These systems use either compression, refrigeration, activated carbon, or nonvolatile-liquid scrubbing.

Lean solvent exits the stripper at about 135°C, gives up heat to the stripper feed, and recycles to the absorber. A solvent maintenance section may be needed to regenerate solvent collected from accidental spills, and to remove impurities from the solvent that are produced by reactions with impurities in the feedgas to the Cosorb unit.

PROCESS BENEFITS—The Cosorb system has a number of advantages and attractions compared with competing technologies, i.e. cryogenic CO recovery, and absorption by ammoniacal copper liquor:

1. Cosorb recovers CO in almost quantitative yields and at high 99+% purity, while at the same time remaining able to produce a high-purity hydrogen stream. Neither competing route does as well on yield or purity.

2. Feeds containing such synthesis-gas components as hydrogen, carbon dioxide, methane and nitrogen can be handled by Cosorb without pretreatments. Cryogenic systems, by contrast, cannot separate CO from nitrogen nearly as well, because of the close boiling points of the two materials. Cosorb also tolerates water and other reactive impurities. Both water and carbon dioxide must be removed before cryogenic treatment to avoid solidification in the cold box.

3. Cosorb operation does not depend on high pressure or subambient temperature for effective removal of CO from gaseous mixtures. Operation is simple, especially compared with the copper liquor approach, which entails close control over cupric/cuprous ratio, recovery of ammonia from product gases, and high-pressure problems.

4. The Cosorb solvent has a high absorption capacity, low viscosity and high stability. It does not react with oxygen (though explosive mixtures of oxygen, solvent and feedgas

Comparison of carbon-monoxide recovery alternatives

Process	Cosorb	Copper liquor	Cryogenic
CO recovery, %	99+	96	87
CO purity, %	99.8	98.5	99.4
Capital requirement, $	Base	160%	170%
Utilities and chemicals (per million lb CO)			
Cooling water, million lb	45.38	153.6	17.08
Wash water, million lb	–	2.83	–
Steam, 50 psi, lb	1,282,000	3,354,000	–
Steam, 600 psi, lb	231,000	–	26,000
Power, kWh	33,960	66,850	187,148
Toluene, lb	84	–	–
Ammonia, lb	–	3,588	–
Lean MEA solution, lb	–	1,968,750	–
Nitrogen, std. ft^3	–	–	426,062
Total cost for utilities and chemicals, $	Base	220%	190%

should be avoided). And it does not share the problem in copper liquor systems of loss of active solution components by precipitation.

5. Because the Cosorb solvent is noncorrosive, carbon-steel construction can be used throughout a Cosorb plant while maintaining a corrosion rate within design limits (as experienced in the Tenneco Chemicals demonstration plant and in coupon tests). Both the cryogenic and copper liquor alternatives require special, more expensive metals. Such other materials of construction, however, are equally applicable for Cosorb if required by outside influences such as use of a corrosive stream as a heating or cooling medium.

6. CO production costs are less than with the competitive technologies, and total capital investment for a Cosorb plant is significantly less.

A quantitative comparison of the Cosorb process with cryogenic recovery and an ammoniacal copper liquor system, for the production of high-purity CO from a synthesis gas, is shown in the table.

The feedstream is based on the nitrogen-free conditions preferred for cryogenic separation (the same specification is used for all three cases), and includes the prior removal of carbon dioxide. Hydrogen accounts for about 71.5% of the feedgas, CO for about 25%, and methane for most of the rest.

It should be reemphasized that the Cosorb process does not require the removal of carbon dioxide. This is not considered within the comparison. In an overall optimization of the synthesis-gas plant and the Cosorb process, a simplified carbon-dioxide removal system would result in a further reduction in capital investment as well as operating expenses.

APPLICATIONS—Most potential uses of the Cosorb process are for producing high-purity CO for chemical synthesis. Other applications are seen in the preparation of ammonia-plant synthesis gas. There, using the Cosorb process in place of CO shift conversion and methanation, would yield from the ammonia synthesis gas a significant CO stream for chemicals production, increase the capacity of the ammonia synthesis loop by reduction of inerts, and reduce the loss of reactants through decreased purge rate.

The Cosorb process can remove and recover CO from the offgas of blast furnaces and phosphorus furnaces. It can be used to produce a feedgas for a hydroformylation unit, from a synthesis gas or an offgas stream, by properly adjusting the hydrogen-to-CO ratio. The Cosorb process is also an attractive candidate for removing environmentally unacceptable concentrations of CO from vent gases.

The simplicity of the nitrogen/CO separation with the Cosorb process introduces yet another potential area of significant use—coal gasification. The general objective of coal gasification is to minimize inerts in the product gas, the major inert under consideration being nitrogen. Unless a low-Btu gas (around 150 Btu/ft^3) is desired, the nitrogen must be removed.

The general solution has been to use air separation plants to separate nitrogen from the start, and then gasify with oxygen. This technique results in optimizing the coal gasification step for minimum oxygen consumption.

The Cosorb solution instead allows use of air for gasification, then separates product CO from nitrogen. The gasification step can thus be optimized for maximum conversion of carbon to gaseous fuel.

In addition, the integrated coal-gasification/Cosorb process permits operating the gasifier with very small quantities of steam. This results in a more efficient reaction temperature in the gasifier and a higher thermal efficiency. Several evaluations of this coal gasification scheme are being conducted by industry, and one such study is being sponsored by the U.S. government's Energy Research and Development Administration.

Considering an underground bed of coal as a gasifer, finally, the Cosorb process can produce an acceptable fuel and chemical feedstock from an *in situ* gasification of coal with air.

The author

Donald J. Haase is manager of Cosorb technology, Business Development and Planning Group, Organics & Polymers Div., Tenneco Chemicals, Inc. (1433 West Loop South, Suite 410, Houston, TX 77027). His professional work experience includes gas-solids adsorption technology, process computer control, project management, pilot plants, high-temperature reaction kinetics, process optimization, gas-liquid absorption technology, and process and economic evaluation. He earned his B.S., M.S. and Ph.D. in chemical engineering from the University of Texas. He is a member of AIChE and ACS, and a Registered Professional Engineer in Texas.

Chloralkali plants win by teaming technology

Energy requirements in chloralkali plant operation can be shaved by opting for a combination of mercury- and diaphragm-cell units, two major West German firms have rediscovered.

☐ Energy consumption in chloralkali production can be lowered by combining a diaphragm electrolysis plant with a mercury-cell plant, rather than installing units of only one type. So says Chemische Werke Huels AG (Marl), which last year supplemented its two existing mercury-cell plants with a new 125,000-metric-ton/year diaphragm plant. By using the salt crystallizing in the diaphragm plant's caustic lye concentrator as feedstock for the mercury-cell electrolyzers, CWH reportedly achieves energy savings that equal approximately 5% of the energy consumed by the diaphragm units.

The reason is fairly simple. A mercury electrolysis plant uses a closed-loop brine cycle, which needs solid salt to make up for decomposition losses in the electrolysis process. Consequently, mercury cells need a source of solid salt, usually provided by evaporating brine pumped to the plant from a solution-mined salt deposit.

In contrast, diaphragm electrolyzers are once-through units, with liquid brine feedstock. Diaphragm plants still feature an evaporator, though, because the caustic soda produced is too dilute and too salt-contaminated to be marketable. (The caustic stream contains 10 to 12% caustic soda and 16 to 18% salt. Evaporating to concentrate the caustic to 50% results in precipitation of salt, leaving only 1 to 1.2% salt in the concentrated stream.) Salt from the diaphragm plant's evaporator can be recycled, and often is, but dissolving it in water for reuse in diaphragm cells wastes the energy consumed to produce the solid salt.

CWH instead employs an elegant trick—it uses the solid salt as feedstock in its two mercury-cell plants, thus cutting energy consumption by evaporating less brine for the mercury cells. No brine evaporation at all would be needed if a producer had mercury and diaphragm cells in a capacity ratio of 1 to 0.9. (Its present set up is a non-ideal 13 to 5.)

Another major West German producer is also hopping on the combined-plant bandwagon. BASF AG (Ludwigshafen) is building a new diaphragm electrolysis plant that will

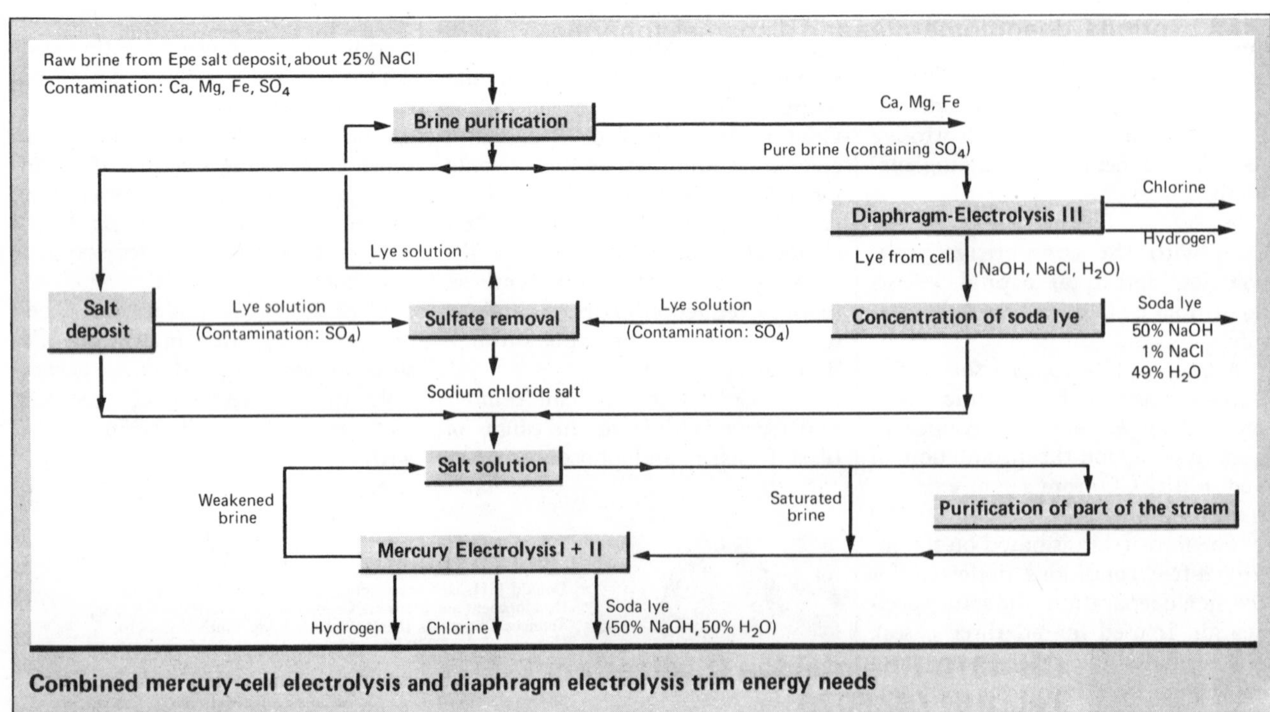

Combined mercury-cell electrolysis and diaphragm electrolysis trim energy needs

Originally published April 24, 1978

be integrated with existing mercury-cell capacity. This will boost chlorine capacity at Ludwigshafen from 200,000 mt/yr to 300,000 mt/yr, with diaphragm plants accounting for over half the total.

REDISCOVERY—Both CWH and BASF took their cue from a small diaphragm/mercury combination built by Dynamit Nobel AG at its Rheinfelden plant over 50 years ago.

Despite this early lead, there were no followers for some time. European chlorine producers were deterred by the higher energy consumption of diaphragm electrolysis, and instead concentrated on mercury-cell units, which still account for 80% of European capacity, according to CWH's estimates.

But in recent years, there has been a gradual switch. Energy consumption by diaphragm electrolysis units has been cut by the replacement of graphite anodes with metal anodes, and the development of new diaphragm types. At the same time, mercury-cell plants have to cope with steadily tightening environmental requirements concerning mercury losses.

ECONOMIC FACTORS—During the planning stage on the new project, CWH found little to choose between the diaphragm and the mercury-cell processes:

Mercury cell	**KWh/ton Cl_2**
Electrolysis	3,300
Steam for brine evaporation (1,300 kg low-pressure steam/ton Cl_2) equivalent to:	180
Electricity for brine evaporation	35
	3,515

Diaphragm cell	**KWh/ton Cl_2**
Electrolysis	2,650
Steam for lye evaporation (1,700 kg low-pressure steam/ton Cl_2) equivalent to:	500
Electricity for lye evaporation	30
	3,180

The lower energy consumption of diaphragm electrolyzers is offset by a difference in investment costs, CWH found. Overall costs are roughly equal, so environmental considerations tipped the scales in favor of a diaphragm plant.

The discovery of potential energy savings came later, when CWH's team, headed by Wolfgang Schade and Gerhard Schrewe, set to work on detailed plans and calculations for the 125,000-mt/yr diaphragm plant. Reviewing the advantages and disadvantages of chlorine plants elsewhere, they came upon Dynamit Nobel's venerable integrated diaphragm and mercury-cell plant.

CWH's design is full of energy-saving details. "Ten years ago, chlorine producers everywhere, including Europe, were more interested in low investment costs than in low energy consumption," Wolfgang Schade explains. "The oil crisis has triggered a re-thinking process."

In addition to higher energy prices, environmental problems, too, weigh more heavily now, and both aspects are reflected in the design.

HOW IT WORKS—Brine arriving at the plant by pipeline must first be purified. CWH uses a discontinuous process devised by Vereinigte Schweizerische Rheinsalinen, and offered by engineering firm Escher-Wyss. The process precipitates only the cations Ca, Mg and Fe. (Other processes are available which remove sulfate, too, but CWH's design leaves this to a later stage, resulting in a marketable sodium sulfate byproduct, and also cutting byproduct sludge volume in half by eliminating use of barium or calcium to precipitate sulfate ions).

Purified brine, at 310 g/L concentration, is fed to the diaphragm electrolysis plant, where approximately half of its salt content is decomposed by electrolysis.

The dilute, saline caustic soda stream is fed into a three-stage evaporator, which raises NaOH content from 11 to 50%, and reduces NaCl content from 16 to 1%. Crystallized salt from the evaporator is fed to the two mercury-cell plants. Since the new 125,000-mt/yr plant's salt output can meet only about one third of the salt requirements of the 325,000-mt/yr mercury-cell capacity at the Huels plant, additional salt has to be supplied by a brine evaporator, for which energy-conscious CWH picked a five-effect unit.

The brine still contains sulfate ion, so it is necessary to guard against sodium sulfate crystallization as a result of rising concentration in the evaporator. CWH's design manages this by continuously tapping off part of the liquid from the last evaporator stage into a sulfate-removal unit. (This two-stage evaporator takes advantage of sodium sulfate's decreasing solubility with increasing temperature. The liquid is fed first to a high-temperature stage to crystallize the sodium sulfate, the liquor then cycling back to a lower-temperature vacuum stage where sodium chloride is crystallized.)

A similar procedure applies to the lye evaporator of the diaphragm plant. To avoid sulfate contamination, the salt crystallizing in the higher-concentration stages is washed with condensate water and the remaining liquid, containing some sodium chloride—and plenty of sodium sulfate—is fed to the sulfate removal plant, along with the side stream from the brine evaporator.

TREND-SETTER?—Will this energy-saving combination of units become the preferred chloralkali production system of the future? Both CWH and BASF cautiously refrain from identifying any future trend.

It would certainly call for a host of new diaphragm plants in Europe, but, less likely, new mercury-cell plants in the U.S. (where 85% of capacity consists of diaphragm plants).

Neither German firm has a tight grip on the general process design used. A couple of U.S. firms are believed to have essentially similar combination plants, and chloralkali experts readily identify the merits of the idea. But it is in Europe that the greatest potential is available, because of the large number of mercury cells there.

A lot will depend on future environmental regulations, industry sources point out. The scheme is evidently a non-starter if mercury cells are eventually banned. But this is unlikely, because diaphragm-plant products—NaOH, chlorine and hydrogen—are less pure than those from mercury-cell plants.

While upcoming membrane-cell technology offers advantages in terms of purity, it is still being scaled up toward the size needed for commercial users.

New ion-exchange membrane stars in chlor-alkali plant

A special ion-exchange membrane, together with innovations in electrolyzer design and caustic-soda evaporation, results in a chlor-alkali manufacturing process that reduces costs for energy, other utilities, and capital investment.

Nicholas R. Iammartino, Associate Editor

☐ The world's first commercial membrane-cell chlor-alkali plant, rated to produce 40,000-metric tons/yr of caustic soda (expressed as 100% concentration), has been running successfully since April 1975, for Asahi Chemical Industry Co. at Nobeoka, Japan, on Kyusa island.

Commercial operation of the plant was detailed this April at the Centennial Meeting of the American Chemical Soc. in New York, by Asahi Chemical's Maomi Seko, member of the board, senior managing director, and manager of the firm's development department for the membrane chlor-alkali process.

The electrolyzers featured in the plant (see photo) initially used a perfluorosulfonic acid ion-exchange membrane (Nafion 315). Seko listed these typical operating results at design throughput: actual caustic-soda concentration, 17.6–21%; current density, 5 kA/m^2; current efficiency, 80.5%; electrolysis voltage, 4.20 V; and power consumption, 3,496 d.c.-kWh/(metric ton of 100% strong caustic and 0.89 metric tons of chlorine). Product quality has been equivalent to that from the mercury process, and easily superior to that of the diaphragm process, Seko says.

Recently, however, a perfluorocarboxylic acid membrane developed by Asahi Chemical replaced part of the Nafion, and plans are being laid to complete the switchover. The major advantage over Nafion of the new membrane, according to Seko, is its high 90+% current efficiency at high concentration of caustic soda from the electrolyzer.

Asahi Chemical has also specially designed the metal-anode electrolyzer, a brine-pretreatment system, and a heat-recovery evaporator. Japan's Denki Kagaku Kogyo licensed the technology package from Asahi Chemical last year, and Seko says more licensing inquiries are on hand.

The membrane basically acts as a separator between the anode and cathode compartments of an electrolyzer cell. Salt feedstock is supplied as brine to the anode side. Sodium ions migrate through the cation-exchange membrane to the cathode compartment. These sodium ions, together with hydroxide ions generated (along with hydrogen) from water at the cathode, form the caustic soda product. Meanwhile, the membrane prevents transfer of chloride ions from the anode compartment. Chlorine gas generates at the anode as coproduct.

MEMBRANE REQUIREMENTS—The criteria for membrane performance in this service include:

■ Chemical stability—The mem-

Large-capacity electrolyzer has run commercially for over a year.

* Asahi Chemical Industry Co., 1-2, 1-chome, Yurakucho, Chiyoda-ku, Tokyo, Japan.

Originally published June 21, 1976

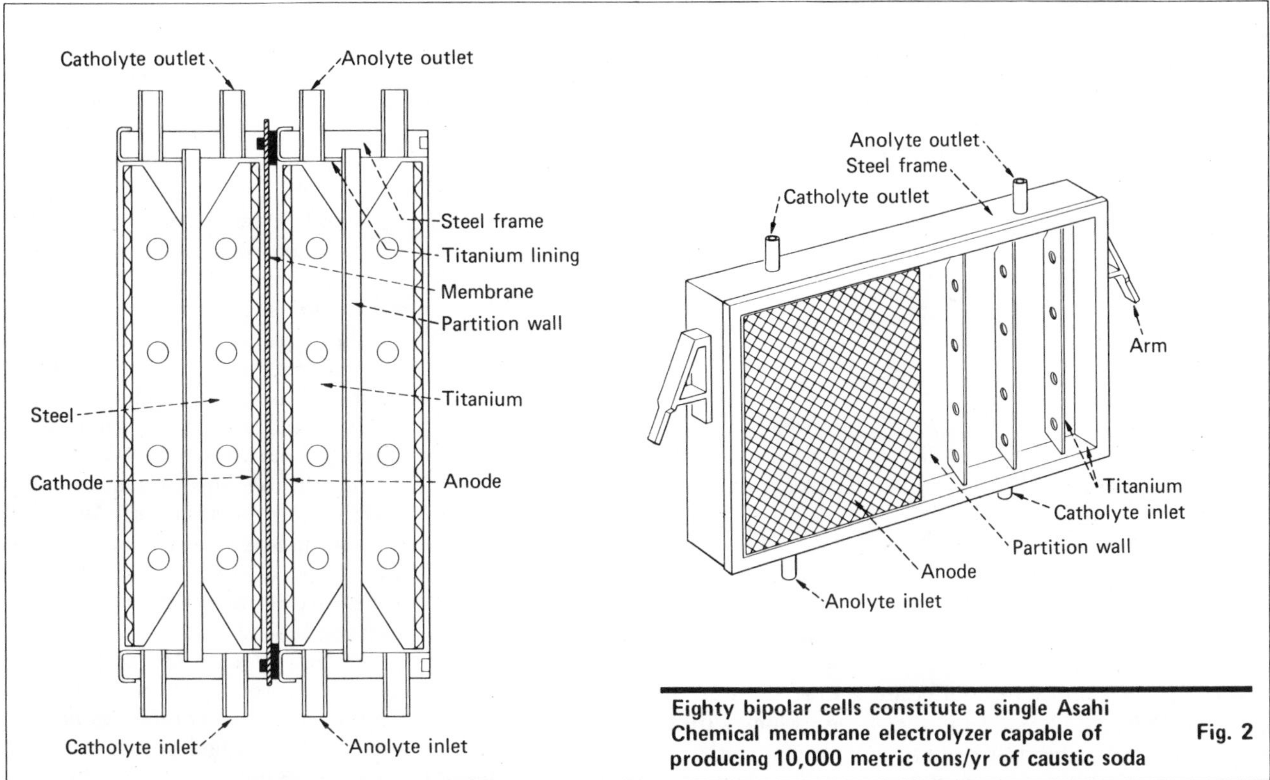

Eighty bipolar cells constitute a single Asahi Chemical membrane electrolyzer capable of producing 10,000 metric tons/yr of caustic soda **Fig. 2**

brane should not deteriorate in the presence of chlorine gas or hypochlorite ion at the high temperatures of electrolysis. Hydrocarbon-type cation-exchange membranes, for example, cannot be used for chlor-alkali cells because carbon-hydrogen bonds would break.

■ Dimensional stability—Size, flatness, etc., should be maintained despite the strong oxidation environment of electrolysis at high 80–120°C temperature.

■ Selectivity—Current efficiency should be at least 90%, even when making concentrated caustic.

■ Low resistance—Voltage drop across the membrane should be under 1 V, at high current density.

As typified by Nafion, Seko told the ACS audience, chemical stability can be achieved by selecting a perfluoro compound as the membrane, and dimensional stability comes from reinforcing the perfluoro compound with Teflon mesh.

However, he continued, perfluorosulfonic acid membranes cannot have high current efficiency when producing high-concentration caustic soda, because the hydrophilic sulfonic-acid groups lead to countermigration of hydroxide ions from the cathode to anode compartment. The maximum practical strength of caustic soda attainable with sulfonic acid membranes is thus limited to about 17.6% at a current efficiency of 80% (see Fig. 1).

Asahi Chemical's membrane, having carboxylic acid groups rather than sulfonic acid groups, does achieve high current efficiency—over 90% for caustic up to 40% strong.

Furthermore, the perfluorocarboxylic acid membrane has sufficient mechanical strength and toughness for preparation in large 4 x 8-ft sheets, which lend themselves to single large electrolyzer units capable of producing 10,000 metric tons/yr of caustic soda. Service life of the membrane is quite satisfactory. And voltage drop across it matches the 1-V maximum.

METAL ANODES—Cells using ion-exchange membranes require metal

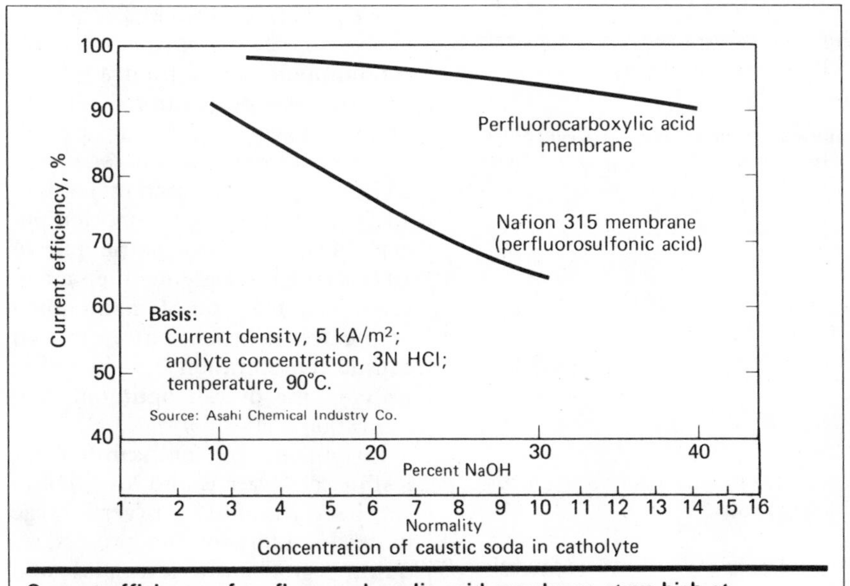

Current efficiency of perfluorocarboxylic acid membrane stays high at high caustic soda concentration, while efficiency of Nafion 315 drops **Fig. 1**

How performance compares for different chlor-alkali cells — Table I

Type of electrolyzer	Asahi Chemical membrane cell	Asbestos-diaphragm cell		Mercury cell, metal anode
		H4 type*	V1144 type*	
Current per cell, kA	10.8	150	72	277.5
Cells per electrolyzer	80	1	11	1
Total current, kA	864	150	792	277.5
Cell voltage, V	3.75	3.85	3.6	4.49
Electrolyzer voltage, V	300	3.85	39.6	4.49
Current efficiency, %	93	96.6	96	96.5
Power consumption, d.c.- kWh/metric-tonNaOH	2,703	2,671	2,513	3,119
Production rate per electrolyzer, metric tons/yr NaOH	10,070	1,816	9,529	3,356

*Data for Hooker Chemicals & Plastics Corp. H4 cell, and for PPG Industries, Inc., V1144 Glanor cell, obtained from Japan Soda Assn.'s "Soda Handbook—1975."

anodes having properties different from those needed by anodes in mercury cells or asbestos-diaphragm cells. Asahi Chemical has developed its own anode design and has used it successfully in its Nobeoka plant, Seko reported.

Since the membrane always contains strong caustic that contacts the anode locally during electrolysis, the anode must resist alkali corrosion. A three-component solid solution developed by Asahi Chemical to coat the anode provides this corrosion resistance. The solution contains more than 50 mol % ruthenium oxide, along with titanium dioxide and a metal oxide.

The coating also helps to limit oxygen formation and thus voltage drop. Membrane processes have an inherently high tendency to make unwanted oxygen, since the anodes often exist in an alkali condition locally because of diffusion of caustic through the membrane.

The coating has a long service life, too. This is necessary to minimize anode recoating, which is troublesome because anode and cathode are both welded into a unit assembly with other cell components.

ELECTROLYZER DESIGN—Asahi Chemical's membrane and anode are shown in a cell in Fig. 2.

A partition wall of explosion-bonded titanium and steel separates an anode-compartment/cathode-compartment pair from other pairs. Titanium covers the anode compartment, and steel forms the cathode compartment. All other components of both (except for Teflon gaskets) are metal, so the cell is mechanically accurate and withstands high operating temperatures of 90–120°C.

The effective area of a single electrode is 2.7 m^2 and a large 32 ft^2 of membrane is used in each cell. Voltage drop between adjacent electrodes is only about 3 mV even at high current density of 5 kA/m^2. Generation of chlorine and hydrogen on the electrodes does not influence electrode voltage. Adjacent anodes and cathodes are installed strictly in parallel, and the distance between electrodes maintained at 2–3 mm. Accordingly, total ohmic drop in the catholyte and anolyte is limited to 0.2 V even at 5 kA/m^2.

One electrolyzer has about 80 bipolar electrodes assembled in a single frame. It can produce 10,000 metric tons/yr of caustic. Table I, compiled by Seko, compares its specifications with those for asbestos-diaphragm and mercury cells.

COST SAVINGS—The lower cost of raw materials and utilities claimed by Seko for Asahi Chemical's membrane process is shown in Table II.

The substantial saving in steam consumption results from a special heat-recovery evaporator. This employs heat generated during electrolysis to concentrate caustic soda from 25–30% up to the market level of 48–50%, at a steam consumption under 0.4 metric tons/metric ton of 100%-NaOH. (Despite the membrane characteristic of maintaining high current efficiency at up to 40% caustic concentration from the electrolyzer, the overall optimum concentration is 20–30%.)

Investment cost for Asahi Chemical's electrolyzer is also low, due to compact assembly of very large units. In addition, the process requires significantly less land and building area than diaphragm-cell and mercury-cell processes.

Raw materials and utilities cost less for membrane process — Table II

Basis: Cost per metric ton of caustic soda (100% concentration) and 0.89 metric tons chlorine; Japanese location; April 1976.

Technology	Asahi Chemical membrane process	Asbestos-diaphragm process, expandable anode	Mercury process, metal anode
Salt ($21.34/metric ton)	32.46	33.82	32.12
Electrolysis power (2.88¢/a.c.- kWh)	80.24	68.28	92.48
Misc. chemicals	2.48	2.72	7.46
Motor power (2.88¢/a.c.- kWh)	3.17	6.05	2.48
Steam ($7.72/metric ton)	3.09	27.79	0.77
Misc. utilities	0.10	0.16	0.16
Misc. costs (lease, membrane, etc.)	15.00	10.63	5.32
Total cost per ton of NaOH,$	**136.54**	**149.45**	**140.79**

Recovering chlorine from HCl

The commercially proved Kel-Chlor process overcomes century-old problems that have held back recovery of chlorine from hydrogen chloride streams, by virtue of its extremely high conversion rate, reduced corrosion tendencies, and high energy efficiency.

Louis E. Bostwick, Pullman Kellogg

☐ The world's first plant using the Pullman Kellogg Kel-Chlor process to make chlorine from hydrogen chloride has now been operating successfully, with minimal process problems, since May 1974. The plant is an integral part of Du Pont Co.'s complex at Corpus Christi, Tex., feeding on HCl offgas from production units and recycling 600 tons/d of liquid chlorine. Easy and quick to start up and shut down, the Kel-Chlor unit has demonstrated its inherent operational stability by producing specification chlorine at a variety of output rates.

Developed over a ten-year period in collaboration with Du Pont, the route reacts oxygen with HCl over nitrogen oxide catalysts in a circulating stream of sulfuric acid, according to the reaction:

$$4\,HCl + O_2 \rightleftharpoons 2\,Cl_2 + 2\,H_2O \qquad (1)$$

The highly exothermic reaction finishes at 250–350°F. The water byproduct is removed by the circulating acid so as to favor equilibrium at conversion levels as high as 99.5%.

Thus, Kel-Chlor is an updated Deacon process, the 1870s-era route that used air and a copper chloride catalyst to convert HCl to chlorine. The Deacon route was abandoned because equilibrium considerations limited conversion to just 75%, while corrosion was severe, and energy consumption was high. Kel-Chlor, on the other hand, operates at high conversion as well as solves corrosion problems with available materials and cuts energy requirements to a small fraction.

PROCESS SEQUENCE—HCl feed-gas first enters into a stripper, where it countercurrently contacts hot aqueous sulfuric acid (about 80%). The acid contains water of reaction and the catalyst—the latter present mainly as nitrosylsulfuric acid ($HNSO_5$)—and strips the catalyst into the gas phase by the reaction:

$$HNSO_5 + HCl \rightleftharpoons NOCl + H_2SO_4 \qquad (2)$$

The oxygen feed-gas then removes any dissolved HCl from the acid, and the acid stream leaves the stripper carrying the water of reaction and only traces of $HNSO_5$.

The stripper offgas, composed of nitrogen oxychloride (NOCl), HCl and oxygen, enters a two-stage oxidizer and is heated by mixing with partially oxidized gas within the vessels. Temperature is chosen to sustain the following reactions:

$$2\,NOCl \rightleftharpoons 2\,NO + Cl_2 \qquad (3)$$

$$2\,NO + O_2 \rightleftharpoons 2\,NO_2 \qquad (4)$$

$$NO_2 + 2HCl \rightleftharpoons NO + Cl_2 + H_2O \qquad (5)$$

Excess heat from these exothermic reactions is removed in a steam-generation loop. The steam powers ejectors in the sulfuric acid vacuum-flash system (detailed later), and provides all needed process heat.

The oxidizer offgas is countercurrently contacted in an absorber by the circulating sulfuric-acid stream. Conditions favor the reverse of reaction (2), so NOCl that was not decomposed in the oxidizer is absorbed into the acid. The catalyst is thus regenerated. Most of the remaining HCl is oxidized to chlorine, by reaction (5). Additional catalyst is recovered by the reaction:

$$NO + NO_2 + 2\,H_2SO_4 \rightleftharpoons 2\,HNSO_5 + H_2O \qquad (6)$$

All the water of reaction exits from the bottom of the stripper in the circulating sulfuric acid, and is removed as overhead in an adiabatic vacuum flash. Most of the flashed acid recycles hot to the absorber; a small amount is cooled before entering the top section of the absorber, to reduce the exit-gas temperature.

The cooled absorber gases, mean-

Operating requirements for Kel-Chlor

Basis: Feed, anhydrous HCl gas; product, liquid chlorine; operating pressure, 4 atm abs. All figures per ton of chlorine product.

Chemicals	
HCl (100%), ton	1.030
Oxygen (99.5%), ton	0.227
NaOH (50%), lb	1
H_2SO_4 (98%), lb consumed	36
H_2SO_4 (80%), lb produced	42
Utilities	
Power, kWh	52
Cooling water (20°F rise), gal	12,900
Boiler feedwater, gal	24
Process water, gal	1
Miscellaneous	
Operating labor, men/shift (not including supervision)	2
Maintenance labor, % of capital investment/yr	2
Maintenance materials, % of capital investment/yr	3

Originally published October 11, 1976

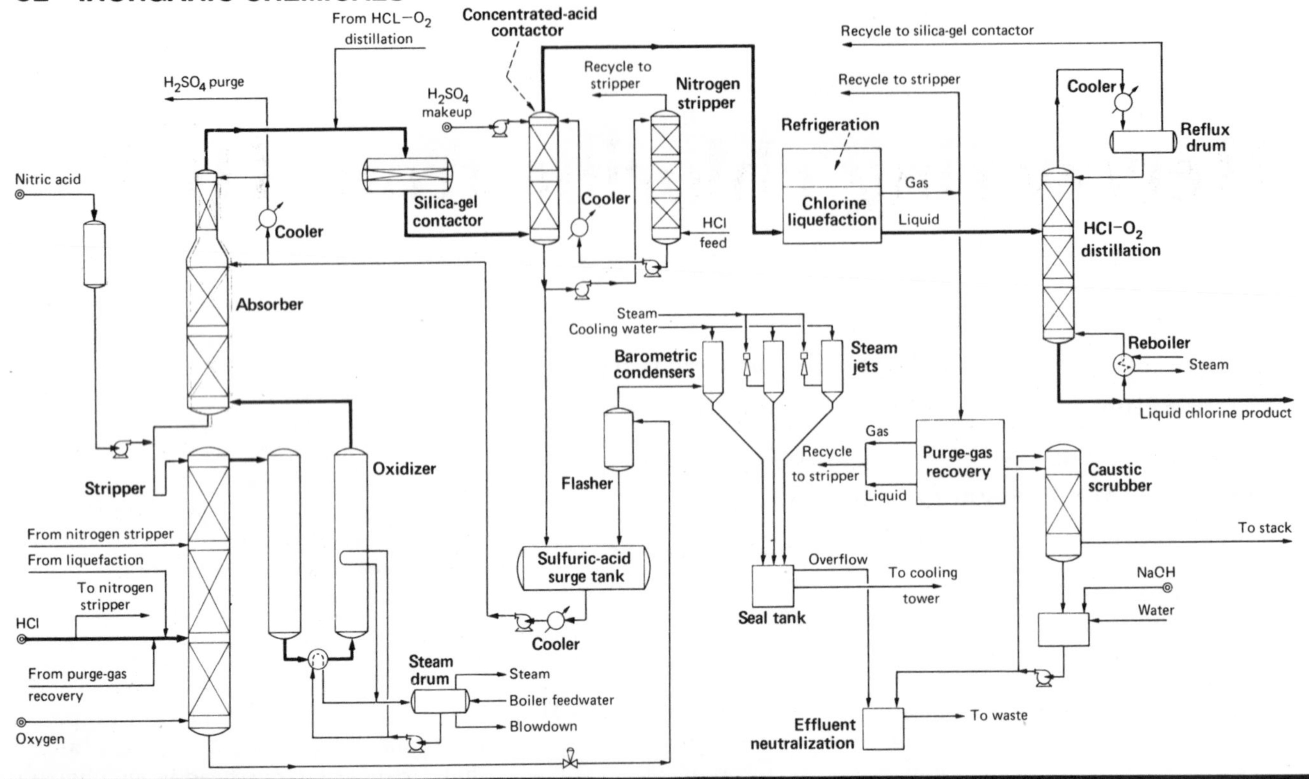

Kel-Chlor process features a circulating sulfuric-acid stream carrying a nitrogen oxides catalyst

while, pass through a silica gel contactor to convert any residual nitrosyl chloride (NO_2Cl), via:

$$No_2C1 + 2HC1 \rightleftharpoons NOCl + Cl_2 + H_2O \quad (7)$$

The NOCl and H_2O are absorbed by makeup H_2SO_4 in a contactor, desorbed from the acid in a nitrogen stripper, by a portion of the HCl feed via reaction (2); and recycled.

The dry gas is then refrigerated, condensing chlorine, which is stripped of dissolved oxygen and HCl by distillation. Noncondensables in the dry gas (HCl, oxygen and inerts) are recycled to the leadoff stripper, or purged to control the inerts level in the process. HCl and chlorine in the purge stream are recovered and recycled; the rest of the gases are vented through a caustic scrubber.

Process requirements for liquid chlorine output appear in the table. HCl recovery as chlorine, and utilization of oxygen, surpass 99.5%.

VARIATIONS—The Du Pont Kel-Chlor unit was designed for very high (99+%) overall conversion of HCl, with no recycle from chlorine liquefaction. Such high conversion in a single pass requires either a large oxidizer or high oxidizer pressure. Du Pont opted for high pressure—15 atm abs.—to keep the size of the oxidizer and other process vessels small. But this choice required tantalum linings for many vessels.

Capital investment and operating costs can be cut substantially at lower single-pass HCl conversion, if unconverted HC1 and oxygen are recycled. Here, pressure can be reduced to 4–8 atm abs., allowing use of steel vessels lined with a plastic membrane that is protected by acid-resistant brick. The lower conversion per pass ranges between 95–98%, to avoid an excessively large oxidizer. Overall HCl conversion and oxygen utilization remain above 99.5%.

When the presence of about 1.8% HCl by volume in the liquid chlorine product is tolerable, the HCl - O_2 product distillation can be replaced with simple oxygen stripping, with savings in investment and operating costs.

Kel-Chlor can also be designed to deliver chlorine gas. This eliminates liquefaction and purge-gas recovery, with substantial cost savings. The gas would contain less than 11% oxygen and is suitable for many chlorination reactions.

On the feedstock side, Kel-Chlor can be designed for aqueous HCl feed without prior separation of the HCl and H_2O. It can also process HCl containing hydrocarbon or chlorocarbon impurities, without drastic feed pretreatment.

Finally, Kel-Chlor can not only serve for waste HCl disposal, but also as an integral part of a manufacturing facility. It can replace oxychlorination in vinyl chloride plants to increase vinyl chloride yield, decrease or eliminate waste products, and reduce costs. In tolylene diisocyanate manufacture, Kel-Chlor can process byproduct HCl and thus virtually eliminate new chlorine consumption. In addition, processes that are so far judged uneconomical because of HCl output and its disposal problems, may now become feasible.

The author

Louis E. Bostwick is product manager for Kel-Chlor at Pullman Kellogg Div. of Pullman Inc. (1300 Three Greenway Plaza East, Houston, TX 77046). He has a B.Sc. in Chem. Eng. from Rensselaer Polytechnic Inst., holds several patents, and is a registered engineer in Florida.

Nitric Acid Route Recovers More Heat From Tailgas

Exchange of heat between tailgas headed for the stack and absorber overhead going to the combustion gas-cooling train is accomplished while avoiding the tailgas dewpoint.

LUIS M. MARZO
Española de Investigación
y Desarrollo, S.A.

The first nitric acid plant designed and erected with a process developed by Española de Investigación y Desarrollo, S.A. (Espindesa) has registered its first year of operation. Operated by Fertiberia S.A. at Castellón, Spain, the plant has a design capacity of 250 metric tons (based on 100% HNO_3) and produces 60% acid. It met all guarantees during the first test run, and since then runs of up to 50 days of continuous operation have been reached before catalyst replacement. A second plant of 270 metric tons/day capacity is now in design and scheduled for commissioning in 1974.

ABSORPTION TOWER has bleaching section in the bottom.

Originally published November 26, 1973

Basically the process is an optimization of conventional high-pressure processes, so that operating and investment costs are substantially reduced. In the more conventional routes, very little is done to recover the heat carried by the tailgas exhausted from the expander. The Espindesa process not only utilizes this heat in a steam economizer, but it recovers additional heat downstream of the economizer in an absorber offgas heater. The exhausted, cold tailgas finally goes to the stack at a temperature low enough to be not worth additional heat recovery but high enough as to be far from its dewpoint, which could cause condensation of nitric acid in the stack and piping. By a proper selection of the temperature profile, most of the materials of construction from the expander to the stack can be carbon steel.

Another savings of steam and investment is obtained with the ammonia superheater that cools the secondary air flowing from the compressor to the absorber in a carbon steel exchanger. This avoids duplication of an ammonia superheater with a separate air cooler. An ammonia detector is installed in the secondary air line to check possible leaks of NH_3 that could reach the absorption and bleaching section.

These and other aspects of the process are covered in Spanish patents and British Patent No. 1,278,828.

Investment and Operating Data

Investment*, $ million	3.4
Operating Requirements, per metric ton HNO_3	
Ammonia, lb.	630
Cooling water, gal.	35,000
Electricity**, kwh.	2.5
Steam export, lb. (14 psig.)	(2,550)
Naphtha, lb.	40.7

*For a turnkey plant in Spain, including engineering and royalties but excluding cooling-water tower and pumps.
**The plant only requires pumps for feeding condensate to the absorber and for the compressor set.

Acid up to 65% has been produced in the absorption system, and with a few more adjustments in the temperatures and compositions of the streams feeding the absorber, it seems possible to get even higher strengths.

Oxidation—As shown in the flow diagram, filtered air is compressed to 8.5 atm. and split into two streams: The small one (10% of the total) goes to the bleaching section of the absorption tower. The main stream is heated with process gases and mixed with the superheated ammonia. It then flows to the reactor where the mixed gases pass downward through the platinum/rhodium catalyst gauze, which is contained in a basket. Thanks to the intimate mixing of the ammonia and air streams and to special devices located at the reactor cover and basket that uniformly distribute the air-ammonia mixture, peak values of nitrogen oxides conversion of up to 96.2% are obtained. Weekly average values are in the order of 95.0%, which is closer to values obtained with medium pressure oxidation than with high pressure.

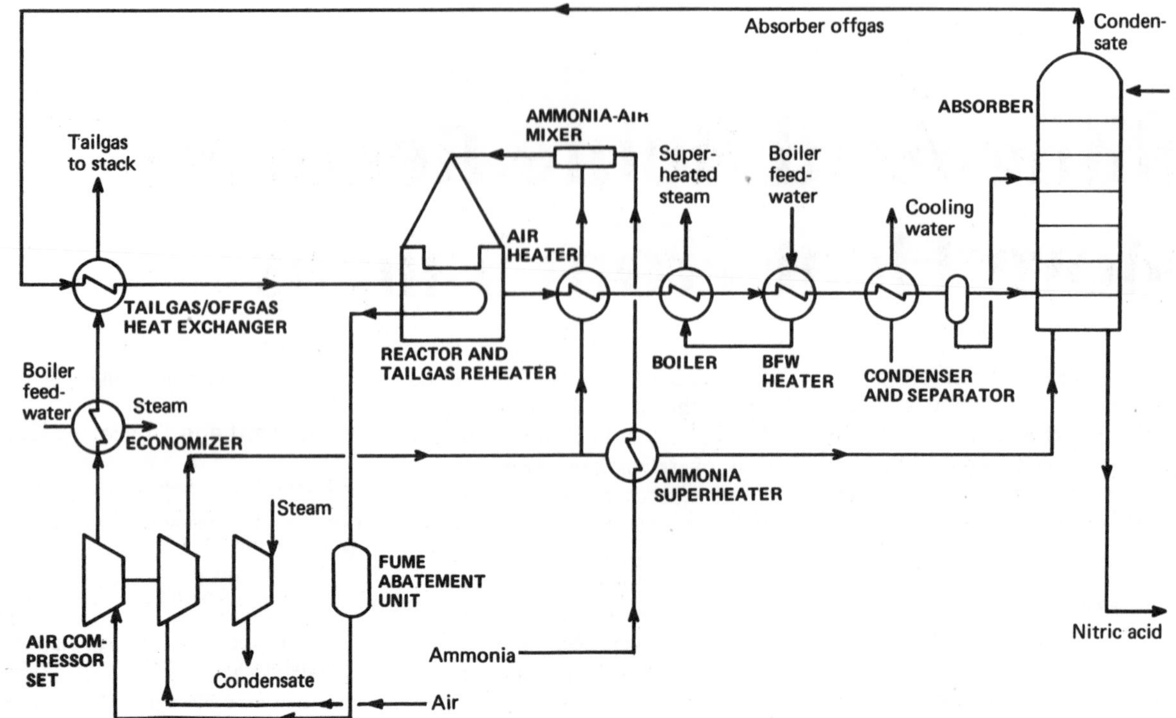

Catalyst changes are accomplished in just 55 min. from the loosening of the first bolt to the tightening of the last one. Moreover, by keeping the catalyst meshes below the level of the reactor flange, deformation problems that would otherwise appear in the flange after some time of operation are avoided.

Gases leave the catalyst at 920 C. and are immediately cooled in a series of heat exchangers, namely: a tailgas heater, the above-mentioned air heater, a boiler provided with a steam superheater, and a boiler-feedwater heater. Part of the superheated steam is used to move the steam turbine that drives the air compressor set, and the rest is exported outside the plant. Finally, the process gases pass through a platinum recovery filter and to the condenser where they reach the dewpoint and are split into a liquid stream (at a concentration of about 40% nitric acid) and a gaseous one (containing uncondensed nitric oxides). The liquid stream is fed to an intermediate tray of the absorption tower, and the gaseous stream enters the bottom of the absorption section of the tower.

Absorption—The tower is composed of sieve trays and makes extensive use of nitrogened stainless steels that permit a substantial decrease in shell thickness. Cost savings of up to 35% are obtained, compared to the conventional 304L stainless steel bubble-cap towers.

The use of sieve trays in nitric acid processes is favored by the flat nature of the curve for mass flow vs. gas velocity. This means that even at 40-50% loads, the velocity of the gas through the tray is almost the same as that of the 100% design load, and tray stability is reasonably sure.

From the process point of view, the tower is divided into two sections. The bottom section acts as a bleacher, and it produces clear acid with less than 100 ppm. of nitrous acid—without the use of a nitric acid heater.

The rest of the tower (37 trays) makes up the absorption section. When making 60% acid, the nitric oxides content of the tailgas is kept low, between 600 and 800 ppm.

After absorption, the offgas is reheated in two heat exchangers against hot tailgas entering the stack and hot process gases leaving the catalyst gauzes. At this point, the offgas temperature is about 450 C., and it enters the fume abatement unit for decolorization (i.e., reduction of red NO_2 to colorless NO). Liquid naphtha is the reducing agent. The temperature of the gas reaches 680 C. and its pressure is 6.5 atm. It then enters the gas expander, where more than 80% of its energy is recovered for air compression. The depressurized tailgas then passes through the economizer and, finally, the above-mentioned tailgas/offgas heater.

The thermal efficiency of the process is such that the actual amount of steam export, with fume abatement and a steam turbine-driven compressor, has reached 2,520 lb./metric ton of 100% acid. Additional operating data are given in the table.

During the design of the first plant, several problems had to be solved because the makeup water has a high chloride content. The circulating cooling water reached values of up to 500 ppm. chlorides. Austenitic stainless steels had to be avoided for services in contact with cooling water. The ferritic stainless steels selected made some of the construction steps difficult, but now after one year of operation no corrosion problem has appeared, and a careful inspection of critical items such as the cooler-condenser or absorption system has revealed no symptom of corrosion. ■

Meet the Author

Luis M. Marzo holds a Ph. D. in industrial chemistry from the University of Madrid. He is manager of development with Espindesa and he managed the team that developed the nitric acid process.

Strong-nitric-acid process features low utility costs

Commercially proven technique for producing strong nitric acid uses recycle loops and a special rectification step to eliminate the usual refrigeration equipment, oxygen plant and chemical dehydrating agents.

Lars Hellmer, Davy Powergas GmbH

☐ The world's premier commercial plant employing the SABAR (Strong Acid By Azeotropic Rectification) process has now turned in its first, and highly successful, year of operation. The facility started up last December near Tarragona, Spain, and produces 120 metric tons/d of 99% nitric acid for owner Unión Explosivos Río Tinto S.A.

A pair of recycle loops and a special rectification step are innovations of the process. These eliminate the refrigeration, oxygen and chemical dehydrating agents (e.g., sulfuric acid or magnesium nitrate) required by classical manufacturing techniques, and thus decrease utilities and operating expenses considerably. Atmospheric emissions have been less than 500 ppm in nitrogen oxides (NO_x) without any special abatement units.

Developer of the SABAR process (U.S. 3,676,065) is the former company Bamag Verfahrenstechnik GmbH, which under the new name Davy Powergas GmbH is now a member of the Davy International Ltd. family. Davy Powergas did a major part of the engineering for the new plant.

REACTION STEP—Starting off the process, liquid ammonia feedstock is vaporized and mixed with a filtered stream of air co-reactant. The blend reacts over a platinum-rhodium catalyst, at about atmospheric pressure and 850°C, to form nitric oxide (NO) and water:

$$4NH_3 + 5O_2 \rightarrow 4NO + 6H_2O$$

Reaction products pass through a waste-heat boiler, a tailgas preheater and a cooler/condenser to dissipate the heat generated. Virtually all of the water present condenses and is drained off as a 2–3 wt.% nitric acid solution (about 0.3 ton/ton of concentrated acid product). Operation at atmospheric pressure helps keep acid content down.

The NO-rich, anhydrous gas exiting the condenser is blended with a recycle of nitrogen dioxide (NO_2) and air, and then compressed. Compression energy comes partly from a tailgas expander; the balance is provided by electric motor or by a turbine driven by steam from the waste-heat boiler. At the higher pressure nearly all NO oxidizes to NO_2.

PRODUCT ACID SEQUENCE—After cooling of the gas, part of the NO_2 is chemically absorbed by an azeotropic acid of approximately 68 wt.% nitric acid (HNO_3). This brings acid concentration to superazeotropic, according to the formula:

$$4NO_2 + O_2 + 2H_2O \rightarrow 4HNO_3$$

The NO_2/air recycle portion of the total absorber feed gas is critical. It boosts partial pressure in the system to enhance absorption and thus

Nitric-acid plant in Spain premiered the SABAR process a year ago.

Originally published December 8, 1975

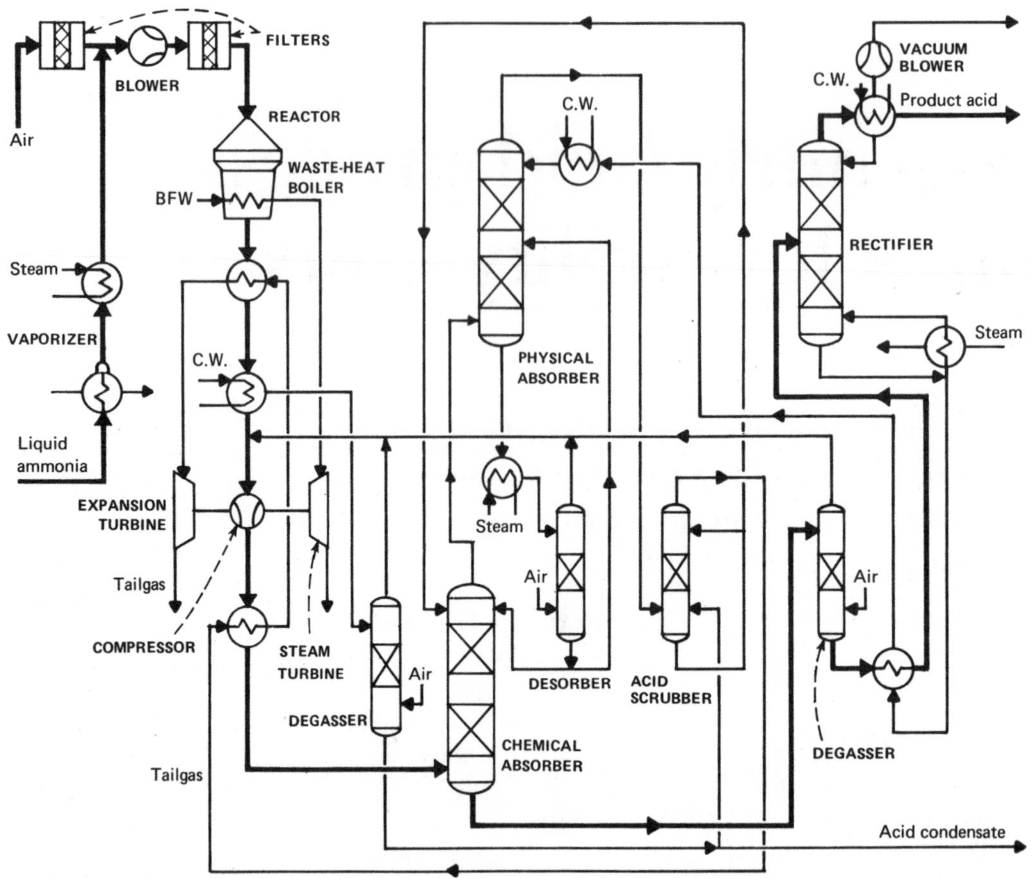

SABAR strong-nitric-acid process uses special recycle loops and rectification technique

keeps equipment size within reason.

The superazeotropic acid leaving the absorber (a countercurrent, packed-bed unit) is degassed by air, preheated, and fed to a rectification column. The acid separates under vacuum into a high-concentration overhead product about 99 wt.% strong, and an azeotropic bottoms.

TWO LOOPS—This bottoms stream from the rectifier serves as a physicial washing agent for NO_2 carried in the overhead gas from the chemical absorption column. The bottoms stream is then degassed by air and sent back to the chemical absorber to again become superazeotropic. The washed overhead gas meanwhile goes through a scrubber for removal of acid vapor, and exits the system through the expander as tailgas bearing less than 500 ppm NO_x.

The overhead vapors from each degasser—two in the acid loop and one treating the acid condensate—comprise the NO_2/air stream that is blended with the fresh reaction products just before compression.

OTHER FEATURES—Two equipment problems in particular were overcome in bringing the SABAR process to commercial status.

One arose in designing the multistage gas compressor. Equilibrium shifts that occur between NO_2 and N_2O_4 from stage to stage affect compression heat balances significantly. Extensive design calculations were needed to compensate.

Materials of construction for the rectification column were even more troublesome. The reboiler and bottom section of the column must handle boiling azeotropic acid; the upper section is in contact with even stronger acid vapors. For both, a proprietary stainless steel was developed to withstand corrosion.

Utilities consumption is about 10–40% lower for the SABAR process than for competing methods using refrigeration, oxygen or chemical dehydrating agents. Including engineering, license fee, equipment, civil facilities, erection and commissioning (but not offsites), capital costs for turnkey plants in a late-1975 average European location are \$4.4 million for a 50-metric-ton/d plant, \$6.4 million for 150 tons/d, and \$8.8 million for 270 tons/d.

Turndown ratio for a typical installation is 60–70%. Special design for surge characteristics of the gas compressor can achieve greater turndown if necessary. An acid product of medium strength, furthermore, can be manufactured with some process changes, i.e., adding the acid condensate and process water to the bottom stream of the rectification column, or to another point appropriate for the particular case.

SABAR raw material and utility requirements

Ammonia, kg	285
Catalyst, mg	65
Electricity, kWh	310
Steam, tons	0.25
Cooling water, m^3	220
Labor, men/shift	2

Basis: Per metric ton of 100% nitric acid; electricity and steam figures assume motor-driven compressors.

The author

Lars Hellmer is process design director for Davy Powergas GmbH (Aachener Strasse 958, D 5000 Cologne 41, Federal Republic of Germany). He holds a doctorate degree in chemical engineering from the Technical University in Berlin. Formerly, he was head of process design, research and development for Bamag Verfahrenstechnik.

Recovery of Active Silica Cuts Costs of Cryolite Route

Even though it uses waste fluosilicic acid, this process produces a salable byproduct form of silica that could result in a 14% reduction in overall costs.

ANTON REINHART, Wotag AG

Growing needs for cryolite by the aluminum and other industries are increasingly being met via synthetic feed-material. Waste fluosilicic acid is a suitable and cheap raw material for making cryolite, but conventional routes based upon this feedstock are burdened with production of a worthless byproduct, microcrystalline silica.

Now, however, there is a process* that coproduces a marketable form of silica. The byproduct credit, which is not claimed for the conventional routes, makes this process potentially much more economical for producing cryolite. A 22-ton/day plant is now operating in Yugoslavia.

The heart of the process is a novel, two-step method for the neutralization of fluosilicic acid by an ammoniacal solution. This enables the making of quality cryolite and amorphous, active silica, which is highly suited for use as a filler material. Credit for this byproduct means a production-cost reduction of about 14% (Table I). Other features of the process are the use of sodium carbonate, the limited requirements for both caustic soda and ammonia, and the ability of the process to be modified to accept other, locally available, raw materials.

*West German patent application 2121 152, Apr. 29, 1971, by IHP Prahovo, Yugoslavia. Swiss Patent Application 16505/72, Nov. 13, 1972, by Wotag AG, Zurich, Switzerland.

Relative Costs — Table I

Item	Units of Cost: Conventional Process	New Process
Raw materials	61	59
Active silica	–	(23)
Utilities	13	16
Labor and maintenance	10	13
Depreciation and interest	16	21
Total	**100**	**86**

Originally published December 10, 1973

Cryolite Production—In the process, caustic soda is charged to a batch reactor and heated to a temperature of about 205 F. Soda ash and aluminum hydroxide are added in stoichiometric ratio to form a sodium aluminate mixture:

$$2Al(OH)_3 + 4NaOH + Na_2CO_3 \rightarrow NaAlO_2 + Na_3AlO_3 + Na_2CO_3 + 5H_2O$$

The use of soda ash improves process economy by cutting the requirements for the more-expensive caustic soda. Moreover, the carbonate ions enhance the downstream filterability and purity of both cryolite and active silica. The aluminum hydroxide feed must be free from aluminum oxide contamination to assure a high yield of cryolite, Na_3AlF_6.

To form that product, the sodium aluminate reaction mixture is piped to another batch-reaction vessel, where an ammonium fluoride (NH_4F) solution obtained from a downstream filter is also added. At approximately 120 F., a suspension of 10 to 18% cryolite is produced according to the following reaction:

$$NaAlO_2 + Na_3AlO_3 + Na_2CO_3 + 12NH_4F \rightarrow 2Na_3AlF_6 + 12NH_3 + CO_2 + 6H_2O$$

By careful heating to around 210 F., a mixture of water, ammonia and carbon dioxide is evaporated. After partial condensation, the residual gas is fed to an absorber. There, ammonia is absorbed into a circulating aqueous ammoniacal stream.

Meanwhile, cryolite suspension from the reactor is continuously pumped to a rotary vacuum filter where most of the water is separated. The wet cryolite is then sent to a dryer/calciner, operated at 1,110 F., and the dry product is bagged and stored.

Making Silica—To produce a marketable byproduct, a solution of fluosilicic acid is cooled to a low temperature and continuously charged to two reactors in series. Here, it is neutralized by an approximately 24% ammoniacal solution containing some ammonium carbonate; this comprises the stream from the absorber brought up to the desired strength by the addition of fresh makeup ammonia. The exact strength of the solution is adjusted according to the concentration of the feed fluosilicic acid, which can vary from 10 to 25% in different plants.

Ammonium fluoride and active silica are formed according to this reaction:

$$H_2SiF_6 + 6NH_4OH \rightarrow 6NH_4F + SiO_2 + n \cdot H_2O$$

The sequence of neutralization exerts a prime influence on the quality and nature of the silica obtained, the active area of which can be adjusted to requirements within certain limits by the conditions selected.

The neutralization is carried out in two steps. In the first, the fluosilicic acid is neutralized to pH 4 to form ammonium fluosilicate. Overall temperature must not exceed 95 F., and local hot spots must be avoided. Heat of reaction is removed by an external graphite heat exchanger,

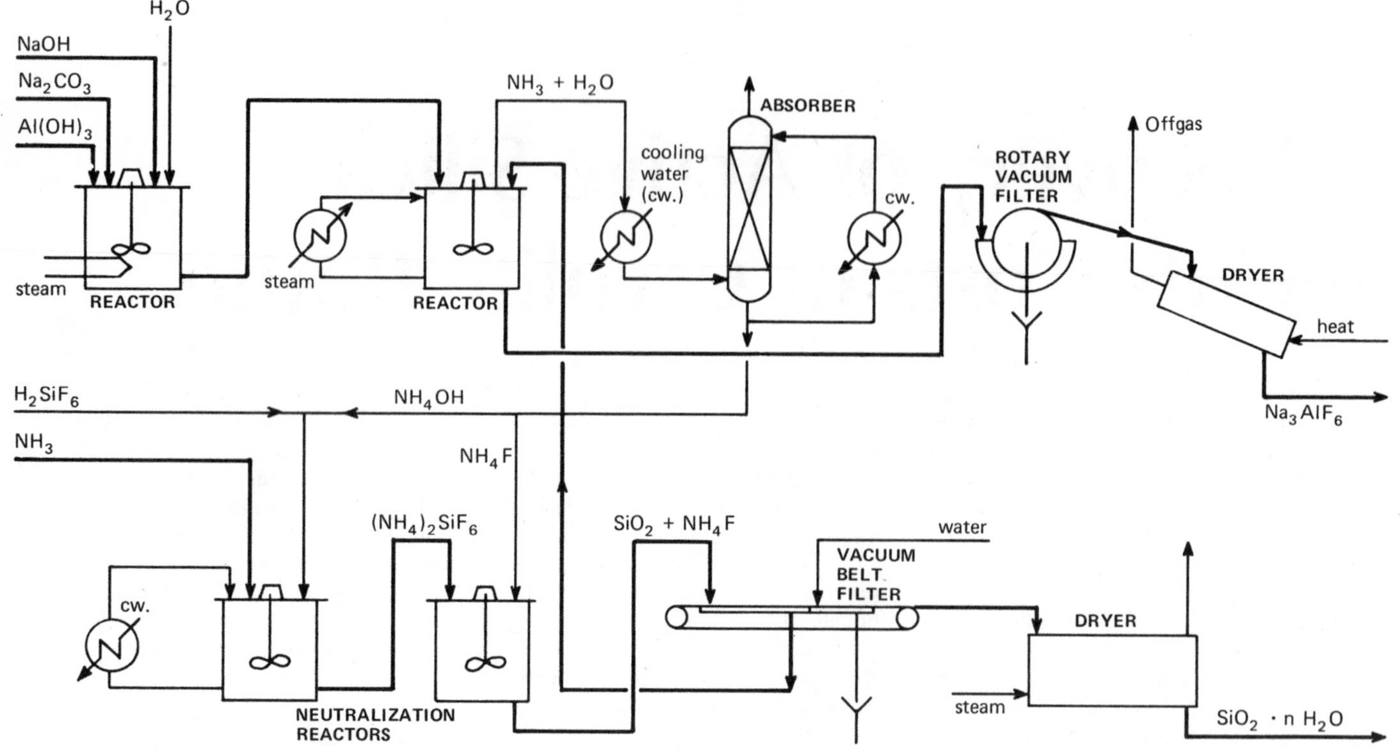

Process Requirements* — Table II

Item	Quantity
Investment (excluding buildings, storage, licensing and engineering)	$1,733,560
Raw materials per short ton of cryolite	
Fluosilicic acid (100% basis)	0.750 ton
Aluminum hydroxide (min. 95 wt. %)	0.415 ton
Caustic soda (min. 98 wt. %)	0.404 ton
Soda ash (min. 98 wt. %)	0.268 ton
Ammonia liquid (min. 99.5 wt. %)	0.118 ton
Utilities per short ton of cryolite	
Steam (min. 140 psi.)	8.0 tons
Fuel oil (10,000 kcal./kg.)	0.25 ton
Demineralized water (min. 60 psig.)	420 cu.ft.
Refrigerant (41 F.)	2.8×10^6 Btu.
Cooling water (min. 60 psig.)	7060 cu.ft.
Electricity (380/220 v.)	260 kwh.
Manpower	
Operators	7
Foreman	1

*For a plant producing 22 short tons/day of cryolite and 7 short tons/day of active silica.

and hot spots are thwarted by agitation and the proper design of the reactor. In the second step, the pH of the reaction mixture is raised to 9. Active silica is precipitated, with proper quality depending upon residence time, temperature, and the concentration of ammonium and carbonate ions.

The solid silica forms a suspension that is highly amorphous yet filterable. It is continuously pumped to a vacuum-belt filter. The ammonium fluoride solution is separated and recycled to the second reactor in the cryolite-production circuit. The ammonium-fluoride and ammoniacal solutions thus form a closed loop, limiting ammonia needs to loss makeup only.

The silica from the filter, after careful washing, has a water content of about 80%. It is conveyed to a belt dryer operating below 240 F. The dried product is then packaged and stored.

The only process effluents are filter wash-water, which contains a maximum of 1% ammonium hydroxide and ammonium flouride, and off-gases from the dryers and absorber, which do not normally require scrubbing.

Typical process requirements are summarized in Table II.

Materials of construction are carbon steel for the cryolite circuit, rubber-lined steel and graphite for fluosilicic-acid neutralization, and stainless steel for filtration and drying of silica. ■

Meet the Author

Anton Reinhart is manager of the Technical Dept. of Wotag AG (Zurich, Switzerland). He holds a Doctorate in Chemical Engineering from the Federal Institute of Technology in Zurich.

Soda ash plant exploits mineral-laden brine

Using a unique process—direct carbonation of brine—this new plant boasts greater efficiency than earlier facilities of the same type. It will employ what is believed to be California's first coal-fired power plant to supply process steam and electricity.

Gerald Parkinson, McGraw-Hill World News, Los Angeles

☐ In February, Kerr-McGee Chemical Corp.* (Oklahoma City, Okla.) will put onstream a new 1.3-million-ton/yr soda ash plant at Trona, Calif. The plant—located along the shore of Searles Lake, a "dry" lake consisting of massive salt strata sandwiched between layers of mud—is the largest yet built to yield soda ash by direct carbonation of natural brine. In fact, the direct carbonation process, itself, is peculiar to the Searles Lake operations of Kerr-McGee, where an existing plant is capable of producing 150,000 tons/yr of soda ash by direct carbonation. Two other plants in the area, now making 145,000 tons/yr of soda ash, will be phased out of that service.

Kerr-McGee decided to expand its capacity to help meet increased demand for soda ash caused by the closing of synthetic soda-ash plants. Strict pollution-control requirements have made the plants generally uneconomical to run; only three such plants are still operating in the U.S.

The new plant operates on a much bigger scale than earlier ones and incorporates a number of processing innovations, so that efficiency is improved. According to a Kerr-McGee spokesman, the new plant will yield 1.3 million tons/yr of soda ash, while employing only 25% of the number of pieces of equipment now used in the present facility, which makes just 1 million tons/yr of borax and potash products.

*Kerr-McGee Chemical Corp., Kerr-McGee Center, P.O. Box 25861, Oklahoma City, OK 73125

Processing innovations include:

- Carbonation under 13.5-lb/in^2 pressure—as opposed to atmospheric carbonation in earlier plants—which improves carbonation performance.
- Use of a coal-fired power plant to manufacture both process steam and electricity.
- Recovery and use in the carbonation process of the CO_2 from the power-plant flue gases. Kerr-McGee states that this will be more economical than current methods of obtaining carbon dioxide: purchase, direct injection of flue gas into the carbonators, and calcination of limestone.

THE THEORY—In the direct carbonation process, carbon dioxide reacts with sodium carbonate in brine to form sodium bicarbonate ($NaHCO_3$) in a mother liquor that contains the remaining brine salts. When the solution is cooled, the sodium bicarbonate crystallizes out. The crystals are then filtered out of the mother liquor and dried to drive off carbon dioxide and water, leaving sodium carbonate, or soda ash, (Na_2CO_3).

BRINE MINING—Brine for the plant will be pumped from about 300 ft below the surface of Searles Lake, a depth at which the sodium carbonate content of the brine is about 6.5% by weight, the design basis for the plant.

The brine's largest component is sodium chloride—about 16%, or half the mineral content (roughly 35%, by weight) of the brine. Other common minerals present include:

borax ($Na_2B_4O_7 \cdot 10\ H_2O$),
burkeite ($Na_6(CO_3)(SO_4)_2$),
hanksite ($Na_{22}K(SO_4)_9(CO_3)_2Cl$),
glaserite ($K_6Na_2(SO_4)_4$)
and trona ($Na_3H(CO_3)_2 \cdot 2\ H_2O$).

Mining is done by drilling a well and cementing a casing into the drill hole. Brine rises in the casing and is removed by well pumps. The Trona complex has about 150 wells in operation, and the brine from the different wells is blended to provide the desired feed to each plant.

The brine is pumped about five miles to the plant's 1-million-gal storage tank, through a 24-in.-dia. pipeline. Pumped at a pressure of 150 lb/in^2, the brine's temperature is about 80°F.

THE PROCESS—The sodium carbonate in the brine is converted to sodium bicarbonate by reaction with CO_2 gas, at about 13.5 lb/in^2, as mentioned earlier. Most of the CO_2 comes from recovery and recycle of carbon dioxide from the decomposition of sodium bicarbonate in steam-heated dryers further along in the process. Additional makeup CO_2 is recovered from boiler flue gas by monoethanolamine (MEA) extraction and is provided to the carbonation system as essentially pure carbon dioxide.

Since CO_2 forms a corrosive mixture when combined with brine, interiors of all vessels are lined with glass-reinforced polyester, and pipes are made of reinforced plastic, or stainless or rubber-lined steel.

The carbonated brine proceeds to the bicarbonate crystallizers where the sodium bicarbonate crystallizes out.

Originally published November 7, 1977

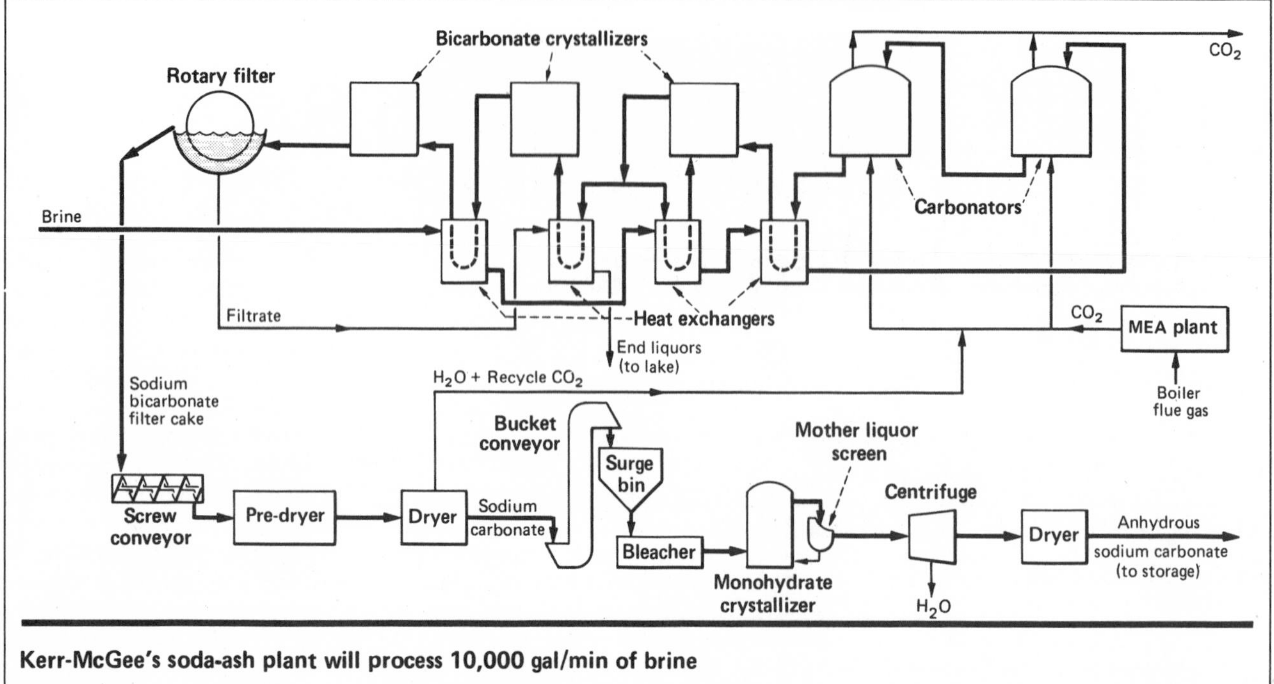

Kerr-McGee's soda-ash plant will process 10,000 gal/min of brine

The bicarbonate crystals are separated from the mother liquor by six 6-ft x 6-ft rotary vacuum filters, and are blown off the filter screens by an air stream onto a screw conveyor.

The conveyor carries the filter cake to steam-heated dryers where the water vapor and carbon dioxide are removed: the water vapor is condensed, the CO_2 recycled to the brine carbonators.

Impure sodium carbonate (light soda ash) leaving the dryers goes by bucket elevator to one of three 150-ton-capacity surge bins that provide in-process storage. Each of the surge bins feeds an associated gas- or oil-fired rotary bleacher where discoloring materials are burned off at a temperature of 850°F. The bleaching agent is sodium nitrate.

The bleached material is reslurried and recrystallized in a monohydrate crystallizer, where the light soda ash forms into large, dense crystals. Impurities remaining after the bleaching operation exit in the mother liquor following screening of the crystals.

Next, a two-stage centrifuge takes out all but 5% of the free moisture in the crystals. The remaining moisture and the water of hydration are removed in a rotary, 450-lb/in² steam-tube dryer. Following screening by a 16-mesh screen, the crystals are sent to storage. Lumps go to a roller mill for crushing.

The final product is anhydrous sodium carbonate having a density of about 62 lb/ft³, compared with a density of approximately 40 lb/ft³ for the light ash prior to crystallization. About 95% of the crystals will be greater than 100 mesh.

THE COST—Kerr-McGee spent about $1.5 million testing new ideas in a pilot plant for several years before and during construction of the new facility. And, out of the $200-million project cost, about $22 million is for pollution control equipment, and $16 million of that amount is for the power plant with its related pollution-control apparatus.

DETAILS—The power plant will generate both electricity and process steam. To accomplish this, it will have two 32-MW, single, noncondensing steam turbines. Steam will come from two 600,000-lb/h, 1,500-psi, coal-fired boilers. The tangential-fired boilers have overfire air to maintain a lower flame temperature and thereby meet NO_x emission standards. (Searles Lake is near the northern boundary of San Bernardino County, where air pollution regulations dictate that each boiler may release no more than 10 lb/h of particulates, 200 lb/h of sulfur as SO_2, or 140 lb/h of NO_x.)

Each boiler has an electrostatic precipitator rated at 98.5% efficiency, and a sodium carbonate flue-gas scrubber with an efficiency rating of 95%. The precipitators operate on the cold side of the boiler air-preheaters because hot-side precipitators could not guarantee compliance with pollution regulations, says a spokesman for Kerr-McGee.

However, the 300°F temperature on the cold side makes it difficult to collect fly ash. The Western coal (from New Mexico) that will be used has a sulfur content of only 0.7%, and particle ionization at 300°F will be insufficient for efficient fly-ash collection by the precipitators. To overcome this problem, Kerr-McGee will mix high-sulfur petroleum coke with the coal to raise the sulfur content to 1.5%, thus improving particle ionization and, in turn, fly-ash collection.

The two flue-gas scrubbers will use end-liquor from the soda ash plant to extract SO_2. Designed to operate in the sulfite/sulfate range, the scrubbers will discharge their spent liquors back to Searles Lake along with the rest of the plant end-liquor.

Scrubbed flue gases will be processed through two MEA plants for extraction of CO_2. Each plant will produce 300 tons/d of CO_2, and this will be the main source of makeup carbon dioxide in the closed-cycle plant. Liquid carbon dioxide will be stored for plant startup and for use in emergencies or for speeding up the carbonation of a completely fresh brine charge.

Coal Converts SO_2 to S

Coal acts as reductant in a new system that produces elemental sulfur from the sulfur dioxide in powerplant, metallurgical and chemical-processing offgases.

W. F. BISCHOFF, JR.
PETER STEINER
Foster Wheeler Energy Corp.

This month, an attractive new method for curbing sulfur dioxide (SO_2) emissions is scheduled for commercial startup.

Developed by the Research Division of Foster Wheeler Energy Corp., the Resox process has been chosen for demonstration at Gulf Power Co.'s Scholz steam plant at Chattahoochee, Fla. Input will be flue gas preconcentrated to about 20 vol.% SO_2; output will be marketable, high-purity liquid sulfur (S).

The only material needed for the SO_2 to S conversion is coal, and not scarce and costly natural gas or oil. Though previously considered as an SO_2 reductant, coal does the job in Resox in the presence of steam, a fresh approach that achieves a fast reaction rate at relatively low operating temperature.

The Resox process handles SO_2 streams produced by many of the techniques now offered for flue-gas cleanup. It works directly (and with especially good economics) for such metallurgical applications as offgas from copper-smelting reverberatory furnaces, where SO_2 concentration runs as low as 2 vol.%. And Resox serves also in chemical-processing, where SO_2 can range anywhere upward from Gulf Power's 20 vol.%.

Feed Preparation—The SO_2-bearing feed needs no special purification. Resox not only tolerates residual flyash in boiler applications but actually filters some particulate from the gas. Reduction in a steam atmosphere eliminates gas drying. The absence of catalysts sidesteps concern over trace contaminants.

Coal is prepared simply by crushing, in a unit tailored to produce "rice size" particles, about ¼-½ in. If fines content rises too high, the fines are screened out to avoid excessive pressure drop through the reactor bed. Foster Wheeler plans to soon test Resox with larger coals.

So far, anthracite coal is the preferred type. Its low volatiles content maintains a high-purity product sulfur. Bituminous coals high in volatiles, on the other hand, result in volatiles condensing along with the sulfur in the Resox condenser. If these must be used—e.g., so a utility can stockpile a single coal type for both Resox and burning—a devolatizer can first prepare the coal. A reactor now in development is aimed at accommodating even caking coals, by means of moving internals.

Reduction Reaction—The current Resox reactor has no moving parts, and is simply a shell completely filled with coal. The coal enters the reactor by gravity through a feeding tube and rotary airlock valve. It moves slowly downward in the vessel, exiting through a magnetic-drive vibratory feeder that controls bed level. Most exits as spent ash.

The SO_2 stream flows countercurrently, after injection with air at several ports distributed around the lower vessel circumference. At Gulf Power's unit, the influent contains about 20 vol.% SO_2,* 50-60 vol.% steam, with the remainder nitrogen and carbon dioxide. The overall reduction reaction is:

$$C + SO_2 \rightarrow CO_2 + S$$

Gas residence time in the atmospheric-pressure reactor runs 3-4 s, converting 90-95% of SO_2 to S.

Reaction temperature ranges between 1,100-1,500°F, near the higher figure for anthracite coal and near the lower end for bituminous coal. While a fast reaction between SO_2 *or* steam with coal calls for a much higher temperature, Foster Wheeler found a synergistic effect between the two reactions. Both are promoted, and SO_2 is reduced at a practical low temperature.

This 1,100-1,500°F temperature occurs only in the small zone of the coal bed near the gas inlet. Swings of ±50°F are easily tolerated. The air injected with the SO_2 gas burns some of the coal to provide this temperature control and, in some cases, burns larger amounts of coal in or-

*Gulf Power's influent is brought to this level by a Bergbau Forschung - Foster Wheeler (BF-FW) System for removing SO_2 from flue gas. This new technique first adsorbs SO_2 on a moving bed of activated char, then releases it as a concentrated gaseous stream during char regeneration.

PILOT PLANT for Resox has a 2 ft^3 reactor, capable of treating 1,200-1,500 actual ft^3/d of 20-30%-SO_2 gas.

Originally published January 6, 1975

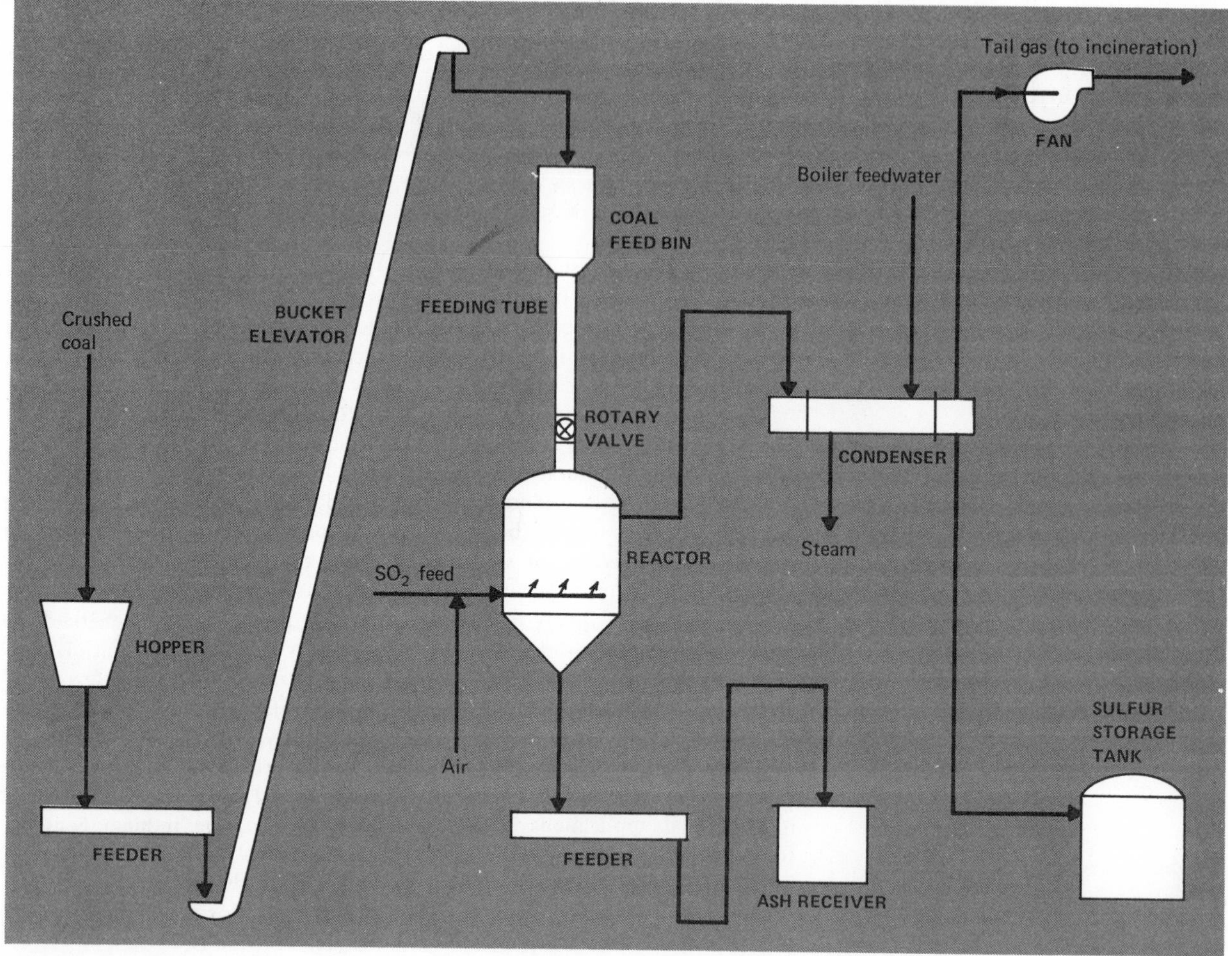

der to provide heat to the bed. (SO_2-bearing feeds from the BF-FW system need no extra heat generation, and other inlet gases can be heated at least partly by exchange with the reactor's hot product-gas.)

Three Outputs—After exiting from the top of the reactor, the gas stream passes to an inclined (about 5 deg.) shell-and-tube condenser similar to those in Claus sulfur-recovery units. Sulfur condenses on the tube side at 320°F, while 50-psi steam is generated from boiler feedwater on the shell side. The steam serves for line tracing and to maintain a 270°F temperature in the jacketed, insulated storage tank for liquid sulfur product. A net steam production typically results.

Noncondensing gases contain some unreacted SO_2, hydrogen sulfide and carbonyl sulfide, so are not acceptable for atmospheric venting. These can be incinerated completely to SO_2 in an external incinerator, in a boiler furnace, or within a BF-FW system. All the alternatives close the loop by directing the incinerated stream back to the front end of the pollution-control setup.

The ash stream most often is collected, cooled and discarded, e.g., in the ash pond traditionally used by utilities for flyash and bottoms ash.

Other Features—Turndown to 40% of design rate for Resox matches the load-swing capability of typical powerplant boilers. Relatively little coal and electricity are consumed—about 0.2 lb coal/lb inlet SO_2, and 0.02 kW/lb SO_2.

The Gulf Power unit can handle a total gas flow from the BF-FW system of 5,000-5,500 lb/h, SO_2 accounting for about 2,200 lb/h. The reactor measures 8 ft O.D. and stands 13 ft high from top to discharge-cone outlet. The whole Resox system takes up a 14 x 25-ft plot. The Gulf Power boiler is specified at 47.5 MW, and burns bituminous coals that average 3% S, 14% ash and 11,000 Btu/lb.

For a 500-MW powerplant burning 4.3%-S coal, estimated capital cost for Resox is $6.50/kW (including engineering, equipment and erection), about 10% of total pollution-control expense. Standard materials of construction help shave the cost. The reactor and most of the other equipment are of carbon steel, though conventional ceramic inserts are used at the feed end of the sulfur condenser, and stainless steel is used for the hot-flyash feeder and some hot-offgas piping.

Further, Resox offers an interesting flexibility to turn out a hydrogen sulfide product gas rather than a sulfur-rich stream, if an outlet for the former exists. Reaction temperature and water concentration are simply raised; no equipment modifications are required. #

Meet the Authors

W. F. Bischoff, Jr., is manager, Environmental Systems Dept., Equipment Div., for Foster Wheeler Energy Corp. (110 South Orange Ave., Livingston, NJ 07039). He holds a B.S. in mechanical engineering from the Newark College of Engineering, and is a member of the Air Pollution Control Assn. He has been working in pollution control for four years.

Peter Steiner is a research associate involved with development and evaluation of new processes, located at Foster Wheeler Energy Corp.'s John Blizard Research Center in Livingston. He holds a master's degree in physical chemistry from Technische Hochschule Stuttgart, West Germany, and is a member of the American Chemical Soc.

Selective oxidation in sulfuric and nitric acid plants: current practices

More-stringent pollution rules have forced operators to boost product conversion and recovery at the expense of energy efficiency.

*B. G. Mandelik and W. Turner, Pullman Kellogg Div., Pullman Inc.**

☐ Sulfuric and nitric acid find large-scale use in the preparation of fertilizers. Since it appears that the need for fertilizers will increase at a rate of about 6%/yr, acid demand will likely continue to be strong.

Heterogeneous catalytic oxidation using atmospheric air as the oxidant is the basic step in the manufacture of both acids.

These oxidation reactions are highly exothermic, and recovery of the reaction heat as high-pressure steam strongly affects the economy of plant operation. In both processes, there is so much high-pressure byproduct steam that all of the large process blowers and pumps can be driven by it. However, recently required pollution-control devices have diminished the amount of steam that can be exported to other users. And while modern sulfuric acid plants still show an excess of steam, an auxiliary energy supply has proved necessary in some nitric acid plants.

Incorporation of atmospheric pollution controls accounts for practically all of the major process and equipment-design modifications of recent years. Such changes in other areas have been limited, and consequently the design of major equipment items for both processes is fully developed and equipment is readily available.

Naturally, there are fundamental differences between the two processes. The nitric-acid-plant catalyst consists of a relatively small amount of alloys of precious metals; the sulfuric acid process uses large volumes of rather inexpensive metal oxides. The nitric acid reactor is compact, operating at high pressure and temperature, whereas the sulfuric acid plant converter is very large and operates at moderate temperatures, and at nearly atmospheric pressure.

Reviewed here are operating problems encountered in sulfuric and nitric acid plants, with special emphasis on those problems connected with either the catalysts or the reactors.

*Portions of this article were delivered at the Jan. 5, 1977 meeting of the Institution of Chemical Engineers in Runcorn, England.

Originally published April 25, 1977

SULFURIC ACID PLANTS

Simplified flowsheets for sulfuric acid plants are shown in Fig. 1. Atmospheric air is dried by circulating 93% (weight) acid; it is then used to burn sulfur in a furnace. Combustion gases cool in a boiler and pass into a four-stage converter where SO_2 is converted to SO_3. Temperature control is achieved by recovering the heat of reaction. SO_3-rich gases contact countercurrently with 98% sulfuric acid, which removes SO_3 from the gas. Unconverted SO_2 is discharged to atmosphere along with any remaining oxygen and nitrogen. In a double absorption-process, gases leaving the second or third catalyst bed pass through a first-stage absorption tower, return to the converter, and are finally discharged to the atmosphere after passing through a second absorption tower.

In a sulfuric acid plant, conversion of SO_2 to SO_3 is catalytic, whereas conversion of the raw materials to SO_2 is noncatalytic, when both elemental sulfur and metal sulfide ores are burned in air. It is essential to produce an SO_2-containing gas that has an excess of oxygen suitable for the catalytic conversion. When elemental sulfur is burned in air, O_2 to SO_2 volume ratios of 1.2 to 1.4 can be obtained, even with high SO_2 content in the converter feed gas. An 8.94% (volume) SO_2 gas would contain 12.06% free O_2, and would have a 1.35 O_2/SO_2 volume ratio.

Before the double-absorption process became standard in the industry, this ratio was chosen for high conversions. However, when sulfide ores were burned as the SO_2 source—with some of the oxygen being used to oxidize the metal sulfide to oxides—this resulted in lower feed-gas concentrations of SO_2.

In general, the strength of the converter feed gas is 6 to 10 vol.% SO_2, with O_2/SO_2 volume ratios of 1.2 to 1.3. In the double-absorption process, low O_2/SO_2 volume ratios in the first stages are acceptable, since after SO_3 is removed from the gas, a high O_2/SO_2 ratio is available in a subsequent conversion stage.

Oxidation step

The reaction of SO_2 with oxygen to form SO_3 is a highly exothermic, reversible reaction. The equilibrium composition depends on the initial gas composition and the final temperature. The rate at which equilibrium is attained is a function of catalyst activity, which is in turn related to gas composition and temperature:

$$\log K_p = 4{,}956/T - 4.678 \qquad (1)$$

$$K_p = \frac{P_{SO_3}}{P_{SO_2}\,(P_{O_2})^{1/2}} \qquad (2)$$

where T is given in °K and P is the component partial pressures in atm.

The mechanism by which a catalyst reacts SO_2 with O_2 to form SO_3 is still a matter of controversy. A widely accepted mathematical model for this reaction was suggested by Calderbank [1]. This model assumes that chemisorbed SO_2 retained on the catalyst reacts with oxygen molecules of the gas phase passing through the catalyst. This assumption can be represented by the following kinetic expression:

$$r = k\left(\frac{P_{SO_2}}{P_{SO_3}}\right)^{1/2}\left(P_{O_2} - \frac{P_{SO_3}}{P_{SO_2}K_p}\right)^2 \qquad (3)$$

where r is the reaction rate per unit volume of catalyst, K the reaction rate constant, and P the partial pressures of the corresponding gases. The rate-determining step is the rate of reaction between gaseous O_2 and chemisorbed SO_2 on the V_2O_5 catalyst. Both O_2 and SO_2 are chemisorbed, O_2 being weakly adsorbed with a heat of adsorption of 6.4 kcal/mole, and SO_2 strongly adsorbed with a heat of 28.8 kcal/mole. The number of active centers on the catalyst increases with temperature.

When the equilibrium and kinetic expressions are combined with the appropriate heat balances, the temperature and composition changes across a converter bed can be modeled. Constraints must be placed on the maximum allowable catalyst temperature, which also influences the optimum temperature for each bed of a multibed reactor. Major catalyst makers have made available computer programs for the solution of SO_2 conversion problems to catalyst users.

At present, many sulfuric acid plants must conform to a pollution standard limiting the concentration of atmospherically discharged SO_2 to about 250 ppm. However, this level corresponds to an overall conversion of 99.7%, and cannot be attained by simple equilibrium conversion. Even a 6% SO_2 feed-gas cannot yield a conversion higher than approximately 99.5%.

For this reason, a new process called "double absorption" was developed. SO_3 leaving the third converter stage is absorbed in sulfuric acid, and lean gas containing only SO_2, O_2 and N_2 is introduced into the last catalyst bed. High conversions can be obtained in such a gas, and the required reduction in SO_2 discharge is readily obtained. Absorption of SO_3 from converted gas may also take place between the second and third converter stages.

Oxidation catalysts

Catalytic conversion, as mentioned above, is usually conducted in four stages. Cooling between stages is required because the conversion of SO_2 to SO_3 is favored by lower temperatures.

For example, in a four-stage converter using an 8% SO_2 feed gas, the gas temperature in the first bed may rise from 419°C to 593°C, with a conversion of 75.7%. After cooling to 480°C, the gas passes to the second bed, where the temperature rises to 517°C and conversion is 91.5%. Cooled to 457°C, the gas flows to the third bed, where 96.3% is converted and the outlet temperature is 469°C. At the final bed, the temperature of the gas is 430°C, and only rises to 434°C. The ultimate conversion is nearly 98%.

Removal of heat between stages is usually achieved by a combination of steam generation, steam superheating, air heating, and SO_2-gas heating. This last method is always used for at least part of the heat removal in double-absorption acid plants; in plants that use cold SO_2 feed, SO_2 combustion gases are cooled to remove the bulk of the process water before they enter the converter. Cooling between stages can also be achieved by introduction of cold quench-gas.

The overall amount of catalyst needed in a sulfuric acid plant usually varies between 160 and 220 L per daily ton of H_2SO_4 production. The 160 to 220-L load is equivalent to 5.65 to 7.77 ft^3/(ton)(d) of H_2SO_4, and corresponds to a space velocity of 500 to 700 for an 8% SO_2 gas.

Conversions for a single-absorption process corresponding to these amounts of catalyst are shown in Fig. 2.

The usual distribution of catalyst among individual reactors for single-absorption systems is about 15%, 20%, 30% and 35% of the total volume; that for the double absorption systems is about 19%, 25%, 27% and 29%.

Catalysts used in this process are based either on platinum black or vanadium pentoxide. Original sulfuric-acid, contact-plant catalysts were based on platinum black. Good conversions were obtained with such catalysts, but poisoning by various compounds made the operation of the plants difficult. Some platinum black catalysts are still used, either (1) to raise the temperature of the feed gas from 350–370°C to 425°C with the heat of reaction so that it can be introduced into a vanadium catalyst bed, or (2) to allow operation of the last bed at a lower temperature for a better equilibrium. Pellets of platinum catalyst are usually larger than V_2O_5 pellets but because of their high activity require only one fourth the bed-depth needed with V_2O_5 pellets. Pressure drops through the converter are thus substantially lower.

Principal differences between brands of vanadium pentoxide catalysts stem from their shapes, and from the activators used by various manufacturers. V_2O_5 and potassium silicate or potassium oxide are always the basic components, but alkali-metal activators may vary considerably in composition. Some catalysts are extruded; others are pelletized as either tablets or spheres.

The active components of these catalysts are vanadium pentoxide and potassium oxide. The content of V_2O_5 in various commercial catalysts varies between 6 and 9.5%, and that of K_2O between 9 and 13.5%. V_2O_5 contents above 6% do not seem to have any significant

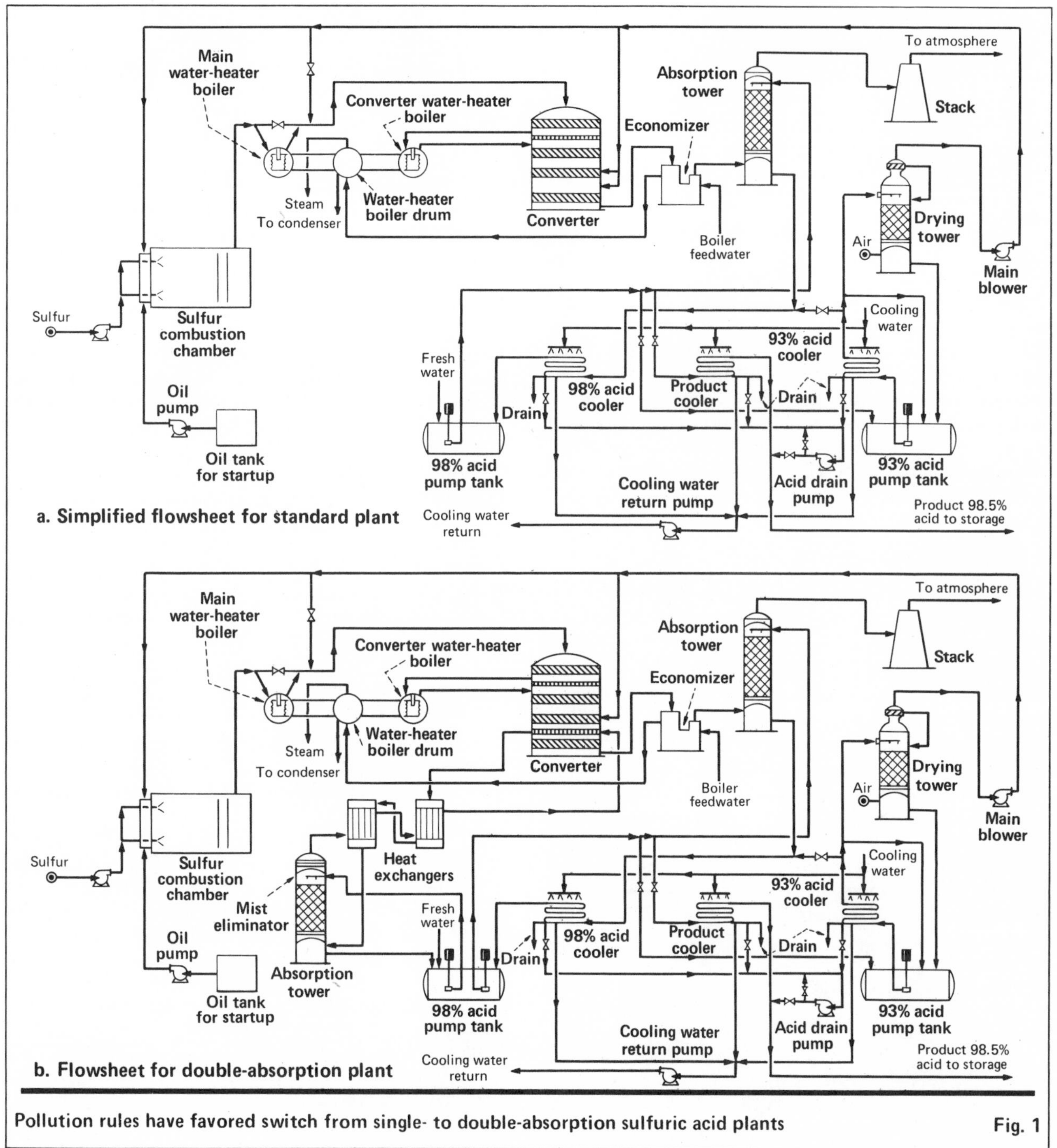

Pollution rules have favored switch from single- to double-absorption sulfuric acid plants — Fig. 1

effect on the rate of SO_2 conversion, but there is a definite connection between the life of the catalyst and the V_2O_5 content. Vanadium pentoxide is an oxidation catalyst. It has been surmised that the mechanism of its action upon SO_2 involves oxygen absorption by $V_2O_{4.34}$, which is oxidized to V_2O_5. (At 400°C, this reaction rate is proportional to the square root of time.) V_2O_5 next reacts with SO_2, oxidizing it to SO_3, and then reverts to $V_2O_{4.34}$.

K_2O content is also very important for the activity of the catalyst and affects its ignition (or activation) temperature. Substitution of rubidium or cesium for potassium has the advantage of lowering this temperature.

The type of support for the V_2O_5 and K_2O is critical, since porosity and strength decide the life of a catalyst. High silica content in the catalyst support provides these qualities.

Use of a more-rugged, lower-activity catalyst in the first and second beds is recommended by some manufacturers. Some recently improved, higher-activity catalysts (for example, ICI-33-2) exhibit ignition temperatures of 370° to 390°C. Maximum operating temperature of such a catalyst is 500°C, compared with 600°C for more-rugged standard catalysts.

Several attempts have been made to develop a pressure process for sulfuric acid manufacture. The main

Typical operating conditions for ammonia combustion

Pressure, atm	Gauze temp., °C	NH_3 content in feed gas, vol %	Yield, %	Catalyst Pt loss, g/ton HNO_3	Operating time, mo
1 (Low)	810-850	12.0-12.5	97.0-98.0	0.04-0.05	8-12
3 to 5 (Medium)	870-890	10.5-11.0	96.0-96.5	0.10-0.11	4-6
7 to 9 (High)	920-940	10.3-10.5	94.5-95.0	0.25-0.30	1.5-3

obstacle has been the design of a power-recovery expander. Sulfuric acid mist present in the stack gases is very corrosive and extremely difficult to remove from a gas stream. At present, a double-absorption plant operating at 30 psig is being offered in the U.S. As far as we know, an expander for H_2SO_4 stack gases has never been developed, so extreme are the corrosion problems associated with such a process.

Reaction trouble-spots

Operating problems in sulfuric acid plants are frequently caused by overheating of the catalyst in the first bed, due to reaction heat. The higher the SO_2 gas strength, the higher the danger of exceeding maximum allowable temperatures. To prevent this, adjustments are made to: the inlet temperature at the first stage; the depth of the catalyst bed installed in the converter; and

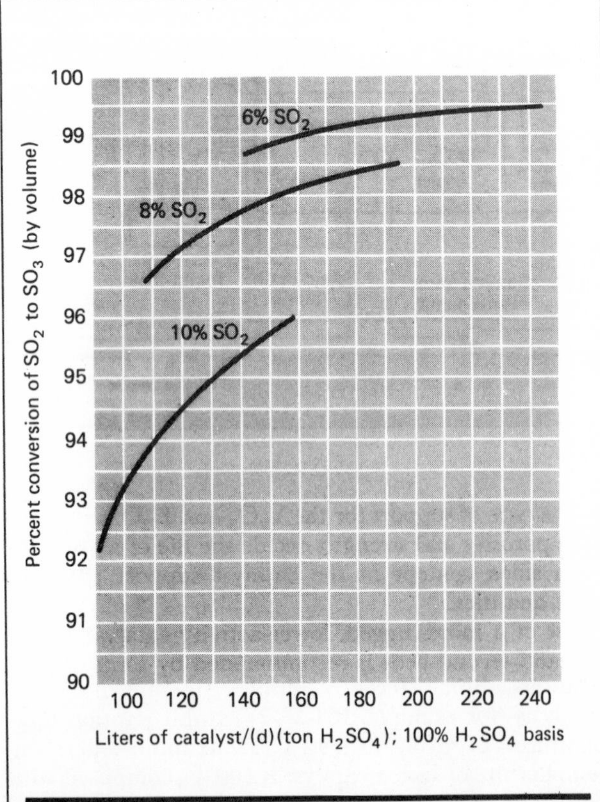

Catalyst demand for conversion of SO_2 to SO_3 in a four-bed reactor **Fig. 2**

the oxygen to sulfur-dioxide ratio of the gas (by volume). Overheating converts potassium sulfate to pyrosulfate, reducing the availability of potassium for catalytic conversion

Another source of catalyst damage is migration of V_2O_5 and K_2O into crusts formed in the beds by solids carried into the first converter with the SO_2 feed gas. The most troublesome contaminant is perhaps iron oxide, which can be present in many feed gases. Solids deposited between the catalyst particles form crusts that may become quite strong mechanically; these deposits increase the pressure drop, forcing a shutdown of the plant so that workers can break up the crust. Usually, the remaining catalyst particles will be damaged mechanically and show reduced contents of V_2O_5 and K_2O.

We know of a case where the pressure drop across the first catalyst bed increased from 5 to 90 in. H_2O within 9 mo of operation.

Sulfur ash may contain as much as 20% Fe (70% Fe_2O_3). Treatment of such sulfur for elimination of suspended solids is necessary prior to combustion. Spherical catalysts are said to show a higher resistance to formation of crusts and to allow a longer run between shutdowns for catalyst cleaning.

Catalyst beds can also be damaged by contact with water. Water can penetrate into the converter during a shutdown period, and long periods of contact with moist air affect the structure of the silica support. Once the mechanical strength of the catalyst is reduced, efficient operation becomes difficult because of the increased pressure drop. Weakened catalysts cannot be improved by spraying vanadium and potassium salt solutions on their surface.

In the double-absorption process, elimination of droplets from the stream leaving the first absorption tower is essential. Droplets of 98% acid caught on the catalyst can cause serious damage. Therefore, high-quality mist-eliminators are installed downstream of the first absorption tower. Similar damage may be caused by moisture contained in converter feed-gases.

Successful operation of a sulfuric acid plant requires good temperature control and maintenance of sufficient acid strengths in the tower circulation systems. The most important variables controlling sulfur dioxide conversion are the temperatures of the gas streams entering and leaving the individual catalyst beds. A suitable excess of oxygen for the targeted conversion must also be ensured.

Operational difficulties in a converter may sometimes result from failures in equipment that recovers the heat of SO_2 conversion from reaction gases; these snags may result not only in an improper temperature profile, but also in harmful mixing of reaction gases that are at various stages of conversion.

If a boiler is placed between the first and second converter beds, leakage from the high-pressure water side is a real danger. Because of this possibility, cooling between the second and third beds is preferably done by preheating combustion air or another gas stream, rather than by preheating boiler feedwater.

Of course, any leakage between the hot and cold sides of heat exchangers is always detrimental. Such leakages

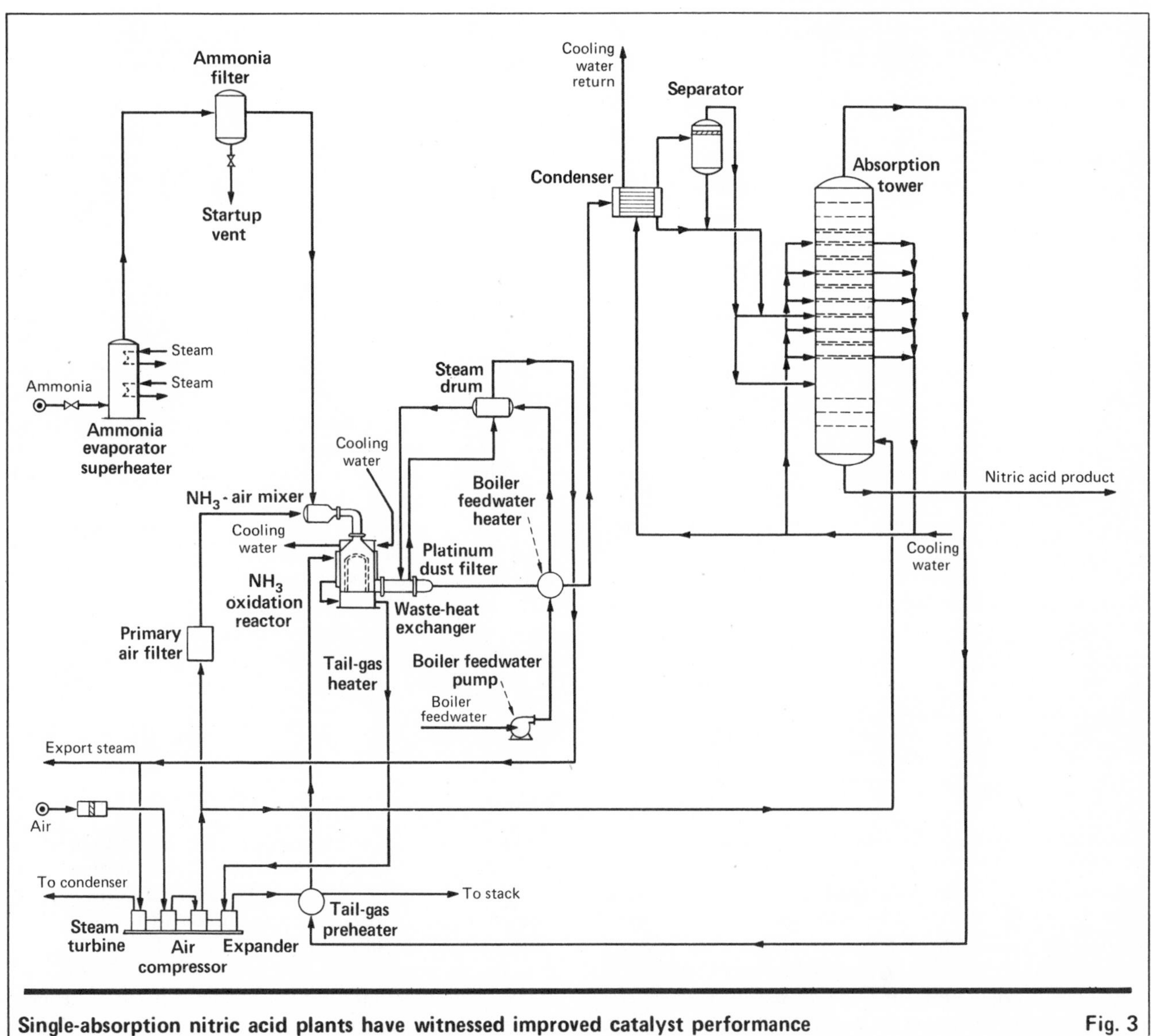

Single-absorption nitric acid plants have witnessed improved catalyst performance **Fig. 3**

usually stem from acid formation and subsequent corrosion, or from droplet carryover into the heat-exchanger equipment. In most cases, iron sulfate is formed in areas where gas velocities are low, or where equipment walls are cold. Avoiding deposition of this salt will help forestall corrosion.

Adequate heat-resisting alloys must be used in the converter catalyst supports and in the heat-exchange equipment located within the converter. The metallurgy of corrosion-resistant alloys for contact with sulfuric acid, SO_2 and SO_3 is highly developed, and manufacturers' guidance for specific applications is readily available.

NITRIC ACID PLANTS

A typical nitric-acid-plant flowsheet is shown in Fig. 3. Ammonia is vaporized by steam, mixed with compressed air and then reacted over a catalyst at around 900°C. Combustion gases containing NO are cooled and mixed with additional air to convert NO to NO_2, which is absorbed by a nitric acid solution in a tray tower. Water enters at the top of the tower; strong acid leaves at the bottom. There are basically two oxidation reactions that take place during ammonia combustion:

$$4\ NH_3 + 5\ O_2 = 4\ NO + 6\ H_2O \tag{4}$$

$$4\ NH_3 + 3\ O_2 = 2\ N_2 + 6\ H_2O \tag{5}$$

The feed gas usually contains between 8 and 12% of NH_3 by volume, at pressures between 1 and 8 atm. The lower flammability limit runs about 13% at 1 atm, and 11% at 8 atm. Because of these restrictions on ammonia concentration, enough oxygen must always be available for NH_3 oxidation to NO, via Eq. (4); under favorable operating conditions, up to 98% (volume) of NH_3 may be converted.

Converter operating variables are: catalyst condition; temperature; pressure; and gas composition. The rate of reaction is a function of the rate of diffusion of NH_3 molecules to the surface of Pt catalyst. High gas-velocity favors high yields, but the gas velocity must allow the necessary contact time with the catalyst and not exceed the value compatible with the mechanical strength of the activated platinum gauze.

Combustion catalysts

Normal conditions for ammonia combustion are shown in the table. The catalyst for the reaction is usually a platinum-alloy wire gauze containing 10% rhodium. Presence of Rh improves the mechanical strength and corrosion resistance of the wires. The catalyst is usually woven as an 80-mesh gauze using 0.003-in.-dia. wire, most often made of 90% Pt–10% Rh, or 90% Pt–5% Rh–5% Pd. Catalyst loading in 8-atm plants usually runs 2 troy oz/(ton)(d) of HNO_3; for atmospheric combustion, it is 4; for intermediate pressure, 1½.

The reaction temperature of the combustion is so high that it causes a loss of mechanical strength in the platinum. This strength is already lower under oxidation conditions prevailing in a nitric acid reactor. Furthermore, activated catalyst with an increased surface area is subject to higher erosion rates.

Because of the high cost of the platinum (at present, about $170 per troy oz), efforts are always made to recover as much of the metal as possible. The classical method is to provide a "platinum filter," whereby lowered flow velocity at around 260°C enables recovery of Pt particles collected on a ceramic fiber pad. Some Pt is also recovered by regular scraping of heat-exchanger areas on which it has deposited.

During a study of this problem in 1964–1966, the Tennessee Valley Authority (TVA) was able to recover up to 70% of Pt. About 50% of the Pt lost was recovered by the filter, and 20% from heat-exchange areas.

Improvement in Pt recovery was likewise achieved when Degussa introduced its gold "getter" gauzes into the reactor. These gauzes are removed from the pad at the end of each cycle and treated to recover Pt. "Getter" gauzes are fabricated in the same form as Pt gauzes; an assembly of six of the getter gauzes will recover about 40% of the total amount of the volatilized Pt and Rh. About one third of the Pt is lost from the getter gauzes through volatilization.

Whereas a 60% Pt recovery (a loss of 0.005 tr oz/ton HNO_3) was once considered good, a recovery of 85% (a loss of 0.002 tr oz/ton HNO_3) has been claimed with the gold getter assembly. Part of this improved performance is no doubt due to a higher pressure drop across the catalyst assembly, which produces a better flow distribution in the reactor and eliminates hot spots on the gauze.

The catalyzed oxidation reaction is extremely fast, requiring contact times of around 10^{-3} s. (Longer contact times may result in NO decomposition to N_2.) Such short times mean that the catalyst bed is only about ¼ in. thick, and therefore that the pressure drop across the bed is very low. It is impossible to locate a distribution plate above the catalyst, since it would be heated by radiation from the catalyst bed and would crack ammonia to N_2 and H_2. A distribution device that has proved successful is a cone, through which gases coming from the ammonia/air mixer expand. Gas distribution is not perfect—as can be seen from color changes in different areas of the gauze.

About 10 years ago, Engelhard Industries developed its Random Pack catalyst system, where about one-half of the normal Pt-Rh load is replaced with a porous pad made of a nickel-chrome alloy. The pressure drop across this pad should equal that of replaced catalyst gauzes, but because of the structure of the pad, a more uniform flow pattern across the catalyst assembly is achieved, reducing Pt losses by eliminating "hot spots." Matthey Bishop offers a similar device under the designation PGS System.

When gold getter gauzes are used with the Random Pack system, they are located between the catalyst and the porous pad. Such a combined assembly reduces Pt volatilization by 25% and yields a 40% recovery of volatilized Pt on the getter gauze. About 15% of the volatilized Pt is recovered on heat-exchange surfaces. Because of reduced losses, the amount of Pt found on the filter is much lower than in a plant with a catalyst consisting only of Pt-Rh gauzes. Overall losses of 0.002 to 0.004 tr oz per ton HNO_3 are claimed for the Random Pack Getter Gauze assembly.

It is difficult to compare the quality of the catalyst gauzes supplied by individual manufacturers. We must assume that all are of a uniform quality and perform equally well, and it is quite common that nitric acid plant operators will purchase their catalyst from several suppliers. Although the new catalyst prices are nearly uniform, costs of recovery of Pt and Rh from used gauzes may vary.

Other catalyst options

When the price of ammonia was low, at a time when ammonia plant feeds were priced at 15¢/million Btu, the price of platinum was only some 10% lower than now. Therefore, attempts were made to replace platinum with oxides of common metals such as Fe, Co, Bi, Cu, Zn. In the U.S., C&I Girdler, Inc. developed an Fe-Co catalyst that was used successfully in several plants. Similar catalysts were developed by ICI in Great Britain.

At present, the attraction of these catalysts is limited. While the cost of precious metals lost in the production of one ton of nitric acid is between $1 and $1.50, the cost of ammonia at $190/ton corresponds to $54 per ton of acid. Because these catalysts operate with 3 to 4% lower conversion efficiencies than platinum, their advantages are difficult to evaluate. The main improvement claimed for base-metal oxide catalysts is that the conversion efficiency remains unchanged, making longer cycles possible; whereas the conversion efficiency of Pt catalysts decreases over the length of the run.

Undesirable reactions

Losses in yield in a nitric acid plant may result from the following undesirable reactions:

- Burning of ammonia to nitrogen.
- Reaction of NO with ammonia to yield nitrogen.
- Cracking of ammonia to nitrogen and hydrogen.
- Decomposition of nitric oxide to nitrogen and oxygen.

The first reaction always takes place, but can be limited by proper plant design and catalyst care.

The second reaction can be suppressed by proper

mixing of the burner gas, and by keeping the entire area of the catalyst active.

The third is prevented by avoiding contact of ammonia with hot metal surfaces.

The last can be forestalled by cooling burner reaction-gases quickly and avoiding their contact with hot platinized metal surfaces.

Difficulties with conversion

Low conversion efficiencies may have many causes. One of the most common is accumulation of impurities, mainly iron oxide and oil, on the catalyst gauzes.

Ammonia is always in contact with steel equipment during its production and storage. Therefore, it is important to provide a filter in the vaporized-ammonia line. An atmospheric air filter must also be provided, and air lines, ahead of the mixer, are preferably made of aluminum or stainless steel.

The gauzes removed from the reactor for treatment are first brushed to remove nonadhering particles, and then pickled in 18% HCl. Some plants prefer pickling by condensing HCl on the gauzes for as long as 28 days.

Silica deposits can be removed from the gauze by treatment with hydrofluoric acid. This treatment must always be followed by HCl treatment to remove all traces of HF. Pt and Rh particles collected in pickling acid are recovered from the wash acid. Dissolved precious metals are usually recovered from pickling solutions by dissolving zinc in these acids.

Rhodium is less volatile than platinum. In some cases, low conversion may result from the higher rate of Pt loss from the gauze and the concomitant formation of rhodium shells. Rh forms oxides at temperatures below 900°C, leading to lower yields.

Ammonia conversions obtained in high-pressure nitric acid plants vary between 94 and 96%. Results are usually quite consistent. For instance, TVA made 38 conversion determinations during 1966 on two original Du Pont plants. Of those, 31 had values between 93 and 95%; 3 were below this range, and 4 above.

It is important to closely control gauze temperatures during operation. For high-pressure converters, 925°C is usually the best temperature. For medium- and low-pressure converters, operation temperatures run lower, as already mentioned. Temperatures are maintained by preheating combustion air going into the process stream, and by preheating ammonia with steam heat exchange. Gases leave the mixer at around 260°C.

Poor mixing of air with ammonia frequently causes low conversions, inducing hot spots on the gauze, from which high amounts of Pt are volatilized; it can also be responsible for cold spots through which unconverted ammonia passes to react with NO, yielding elemental nitrogen. Ammonia droplet carryover has a similar effect.

It is essential to avoid overheating the ammonia-air mixture to prevent cracking of NH_3 to N_2. The distribution cones above the catalyst are usually made of polished stainless steel, or are provided with external water cooling to prevent their being heated by radiation from the gauze, which may initiate undesirable NH_3 cracking on their surfaces.

The platinum catalyst holders must not be too massive, and should be as flat as possible. In a plant where the holders were 20 in. high, about 2% of ammonia was cracked by contact with the holders before it could reach the gauze.

Nitric oxide formed on the gauze decomposes if it contacts platinum for excessive periods of time. It is important to design the reactor so that the reaction gases cool below 800°C before they contact any areas on which Pt is deposited. Therefore, all reactors are arranged so that a heat-exchange area is located within inches of the bottom of the catalyst layer.

The new anti-pollution laws applied to nitric acid plants have not affected catalytic oxidation practices, but have forced improved efficiencies in absorption of nitric oxides. Better catalyst performance has been achieved recently through more-uniform distribution of gases passing over the catalyst and through enhanced recovery of volatilized catalyst particles. The nature of the platinum alloys used and the design of the mechanical supports have not changed drastically in recent years. In fact, the composition and the shape of the catalyst is standard for all manufacturers (quite distinct from the diversity among sulfuric acid catalysts). Thus the improved absorption system used in nitric acid plants is really a refinement of the single-absorption process, not affecting the basic catalytic oxidation techniques.

References

1. Calderbank, P. H., *Chem. Eng. Prog.*, Vol. 49. 1953, p. 585.
2. Duecker, W. W., and West, J. R., "The Manufacture of Sulfuric Acid", Reinhold Publishing Corp., New York, 1959.
3. Newman, D. J., and Klein, L. A., *Chem. Eng. Prog.,* Vol. 68, No. 4, 1972, p. 62.
4. Gillespie, G. R., and Goodfellow, D., *Chem. Eng. Prog.,* Vol. 70, No. 3, 1974, p. 81.
5. Edwards, W. M., Zuniga-Chaves, J. E., Worley, F. L., Jr., and Luss, Dan, *AIChE Journal,* Vol. 20, No. 3, 1974, p. 571.

The authors

Bernard G. Mandelik is Process Development Manager with the Pullman Kellogg Div. of Pullman Inc., 1300 Greenway Plaza East, Houston, TX 77046. In recent years, he has been concerned with the reduction of energy requirements for the production of ammonia and methanol, the removal of sulfur from industrial gases, and the design of nitric acid and ammonium nitrate plants. A registered professional engineer in New York, he belongs to the AIChE and the ACS, and holds a D. Sc. from the Technical University in Prague, Czechoslovakia.

William Turner is Director of Process Engineering, Kellogg International Corp., Wembley, England. A chartered Engineer and a Fellow of the Institution of Chemical Engineers, he received his B.Sc. Tech. in chemical engineering from the University of Manchester.

Section V
METALLURGICAL PROCESSES

Alumina Producers Look To Alternate Raw Materials

Lest they be caught in the same sort of raw-material squeeze as the oil companies recently experienced, alumina producers are looking at processes that can accept non-bauxite ores.

Aiming to squelch a concerted price-raising effort by bauxite-producing nations, U.S. and French aluminum manufacturers last month took the wraps off new alumina processes based on ores other than bauxite.* Many of the alternate ores are abundantly available in the manufacturing countries. Among the recent happenings:

- France's Pechiney-Ugine Kuhlmann disclosed in February details of a 20-ton/d alumina plant in southern France. The pilot plant will test the company's new H-Plus process for handling undisclosed non-bauxite ores.
- Earth Sciences, Inc., together with its joint-venture partners National Steel Corp. and Southwire Co., started up its 10-ton/d pilot plant in Golden, Colo., last November. This plant takes alunite* ore from Utah.
- **In March, Reynolds Metals Co.** held a week-long full-scale test on laterite* ore at Hurricane Creek, Ark. The demonstration took place just as the producing countries were meeting in Guinea to discuss prices and the possibility of an embargo.

In addition, the State of Georgia is pressing on with its two-year effort to interest manufacturers in the state's kaolin deposits by providing financial incentives to any company wishing to go ahead with the first major commercial development.

The Reynolds Metals test involved little more than running lower-grade ore from the Pacific Northwest through the company's existing Bayer process unit. The Bayer route is by far the industry workhorse, and uses caustic to extract alumina from silica in the ore. The aluminum hydroxide produced in this way is calcined to an alumina that is pure enough for feeding to potlines.

Reynolds is apparently betting that if bauxite prices go through the roof, new technology will not be necessary, and existing equipment can be converted to process U.S. laterite ores. Laterites contain 30 to 40% aluminum oxide. The chief penalties would be capacity reduction and the need for more settling area to accommodate the increased tailings.

Pechiney Project—The really big departure in established technology may be coming from the French. Though scanty in detail, the facts divulged by Pechiney tell of a concentrated sulfuric acid leach that produces aluminum sulfate from the ore. Under normal conditions, the ensuing crystallization is unsatisfactory, resulting in an impure alumina end-product. But Pechiney's general manager, Maurice Serpette, explains that the addition of hydrochloric acid during the crystallization stage produces an aluminum chloride that crystallizes well, and ultimately yields an alumina powder having ten times fewer impurities than Bayer-process alumina.

Pechiney is not talking much about just what ore it intends to use the most, though it has tested clays from the U.S. and coal shales from the north of France. One indication could be the company's recent deal with the Soviet Union to build a 1-million ton/yr plant there; the H-Plus technology has been discussed, says Pechiney, though so far there are no firm plans to exploit Russia's abundant nepheline ores.

The shales and clays used in the H-Plus process mean that Pechiney will have to treat at least twice as much raw mineral ore as it would with bauxite. Many bauxite ores have alumina contents of up to 50%, while clays such as those found in Georgia have about 38%. Pechiney says the minimum economic level at current prices is 18-20%.

The purity of the H-Plus alumina saves on downstream purification equipment, and of course there could be lower transportation costs if local ore were used. However, debits outnumber credits to the extent that the H-Plus route is about 10-15% more costly than the Bayer process. Energy costs are higher, as are material-handling expenses. The investment is likely to be 20-30% more,

*Geological terms used in this article: Bauxite is a hydrated alumina, $Al_2O_3.2H_2O$, with silica, clay and other impurities. Alunite is $KAl_3(SO_4)_2(OH)_6$. Kaolin rock contains the clay mineral kaolinite, $Al_2(Si_2O_5)(OH)_4$. Laterite is a far more general term for red rocks that contain hydroxides of aluminum.

PECHINEY's new pilot plant for the H-Plus scheme was preceded by this 1-ton/d unit in Gardanne, France.

Originally published April 29, 1974

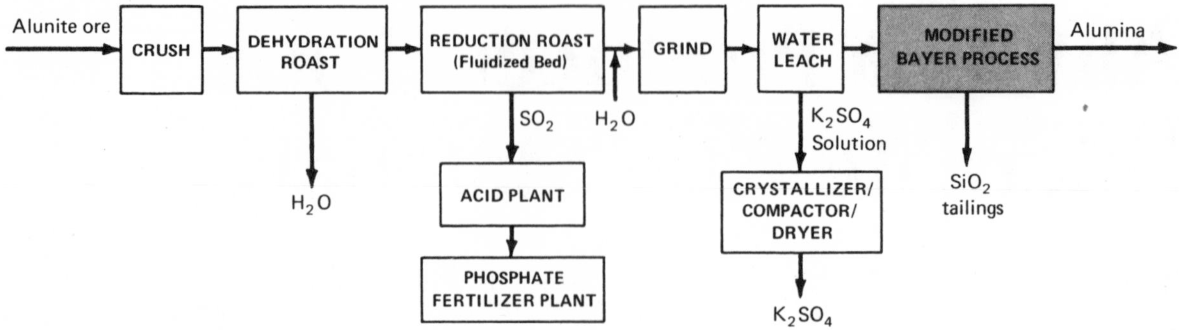

EARTH SCIENCES combines specially designed front-end with Bayer route.

says Pechiney, or between $400 and $500 per ton of installed capacity.

By next year, the company will start up a large "pre-industrial" prototype near Marseilles, which will produce 20 tons/d of alumina by the end of 1976. Pechiney engineers figure that if bauxite prices zoom or if supplies get tight, the firm could move on to a full-scale industrial plant of 200,000 or 300,000 tons annual capacity by 1978 or 1979.

Alunite Processing—Until recently, alunite was not considered a viable alternative to bauxite. However, the Soviet Union operates an alunite processing plant at Kirovabad, near the Caspian Sea, and Guanos y Fertilizantes de Mexico, S.A., has talked seriously about building a plant at Salamanca, Mexico, using a process developed at the University of Guanajuato. The Russians use a caustic-leach process, while the Mexican plant would employ an acid-reaction route.

The Earth Sciences approach also uses caustic, and the latter part of the route is billed as a modified Bayer process. Like the Mexican and Russian versions, it produces potassium sulfate and sulfuric acid as by-products.

Alunite is a hydrated sulfate of alumina and potassium uniformly distributed in a silica matrix. It constitutes 40% of the Utah ore being mined by Earth Sciences, while 50% is quartz and 10% is iron and other constituents. Overall, the alumina equivalent is 14.8% of the ore.

The ore's low grade is a relatively minor factor, the company says. More important is the uniform quality and abundance in the western U.S. In Beaver County, Utah, the company has drill-proved a deposit of more than 100 million tons. Exploratory drilling has also been done at four other peripheral locations in the county to define an ore zone covering more than 25 square miles. Total deposits in the five explored zones are estimated at 680 million tons, with the chance of much more.

Ore used by the pilot plant is hauled 23 miles by road to a railroad for shipping to the Golden pilot plant. Before shipping, it is crushed to 1 in. dia. by a jaw crusher. It is crushed further to about ⅛ in at Golden before being fed into a rotary kiln for dehydration at temperatures in the range of 400 to 700°C (but below the temperature at which sulfur would evolve).

The ore then goes to a fluidized-bed reactor, where reducing gases attack the sulfur fraction contained in the aluminum sulfate. The SO_2 so evolved would be collected in a wet scrubber and used to make acid.

Next, the material is ground to a 14 to 60-mesh powder, and water is injected for subsequent wet grinding. Still hot, the slurry flows to a series of three agitator tanks, where water leaches out the potassium sulfate. The K_2SO_4 solution is drawn off to a crystallizer/compactor/dryer unit. The remaining ore, mostly quartz and alumina, settles out.

The rest of the process is the so-called modified Bayer route, in which a sodium hydroxide solution leaches the alumina from the quartz. Since the bulk of the alunite structure has already been removed, all that is left is a loose network of alumina, which readily dissolves, leaving behind the solid quartz. In this route, the caustic is kept at about 95°C and atmospheric pressure. Normal Bayer routes require considerable pressure.

After removal of the quartz tailings from the pregnant caustic solution, the process stream is sent to four precipitation tanks, which are seeded with alumina hydrate to precipitate alumina. The remaining caustic solution is reused. The final step, calcining, is done batchwise.

From the data gained at the $3-million pilot plant, Earth Sciences and its partners hope to be able to define engineering details for a 500,000-ton/yr prototype. Such a plant would also produce 250,000 tons/yr of potassium sulfate and 450,000 tons/yr of sulfuric acid. The acid would most likely be used to acidulate phosphate rock from a mine that Earth Sciences plans to develop in southeastern Idaho.

Georgia Kaolin—Heavy interest in developing Georgia's kaolin deposits dates back to an April 1972 report by John F. Husted of the Georgia Institute of Technology, which spelled out the vulnerable U.S. position vis-à-vis aluminum raw materials. At that time, Husted estimated that domestic supplies were only 11% of annual need, and this figure was projected to go down to 5% by 1980. But a variety of economic factors hastened the decline to 5% by as early as late-1973.

At least two acidulation processes have been under investigation. Anaconda Co. has reportedly piloted a hydrochloric acid method of obtaining alumina from kaolin at a rate of 7 tons/d over an 18-mo test period. Others have looked at nitric acid treatment. Both routes are estimated to cost about $80/ton.—JCD#

Continuous copper-smelting process uses single vessel

The streamlined Noranda reactor for processing copper concentrates offers major fuel savings, simplified operation, and reduced sulfur-dioxide emissions. It is also highly flexible, accommodating various feeds, products and fuels.

Geoffrey D. Hallett, *Noranda Mines Ltd.*

☐ The first commercial plant using the Noranda process for continuous smelting and converting of copper concentrates has now operated for just over three years at Noranda Mines Ltd.'s Horne smelter at Noranda, Quebec, Canada, culminating a development effort that started in 1964.

The simple process combines into one reaction vessel all three stages of conventional smelting (roasting, reverberatory smelting, and converting). It produces either metallic copper ready for refining to anode copper, or a high-grade copper matte (70–80% copper) that is converted further before refining.

During its operating period, the $12-million Horne unit has exceeded its design capacity significantly. It has also fed on many different copper concentrates; has burned natural gas, oil and coal as fuels; and has used both air and oxygen-enriched air for oxidation.

The single-vessel concept reduces fuel usage, moreover, by recovering the heat liberated during oxidation of feed sulfides, for drying and melting incoming concentrates. Sulfur dioxide is produced continuously in a constant-volume and high-strength stream, which is more suitable for sulfuric acid manufacture than the weak gas from a reverberatory furnace, or the sporadic and changing-strength gas leaving a converter during iron blowing. When copper matte is the Noranda product, in fact, the matte has so little iron left that the iron blow is almost absent and converting proceeds without interruptions for skimming of slag.

A second installation featuring the Noranda process is now under construction for Kennecott Copper Corp. at Garfield, Utah. Scheduled for mid-1977 completion, the $280-million facility will have three Noranda reactors in parallel.

PROCESS DESCRIPTION—The Horne plant uses a single reactor (Fig. 1) of identical size to those being built for Kennecott. Pelletized copper concentrates and silica flux are fed onto the surface of the reactor bath, which is maintained in a highly turbulent state by air, or oxygen-enriched air, introduced via tuyeres. The exothermic oxidation reactions provide part of the heat for smelting. A burner at the feed end of

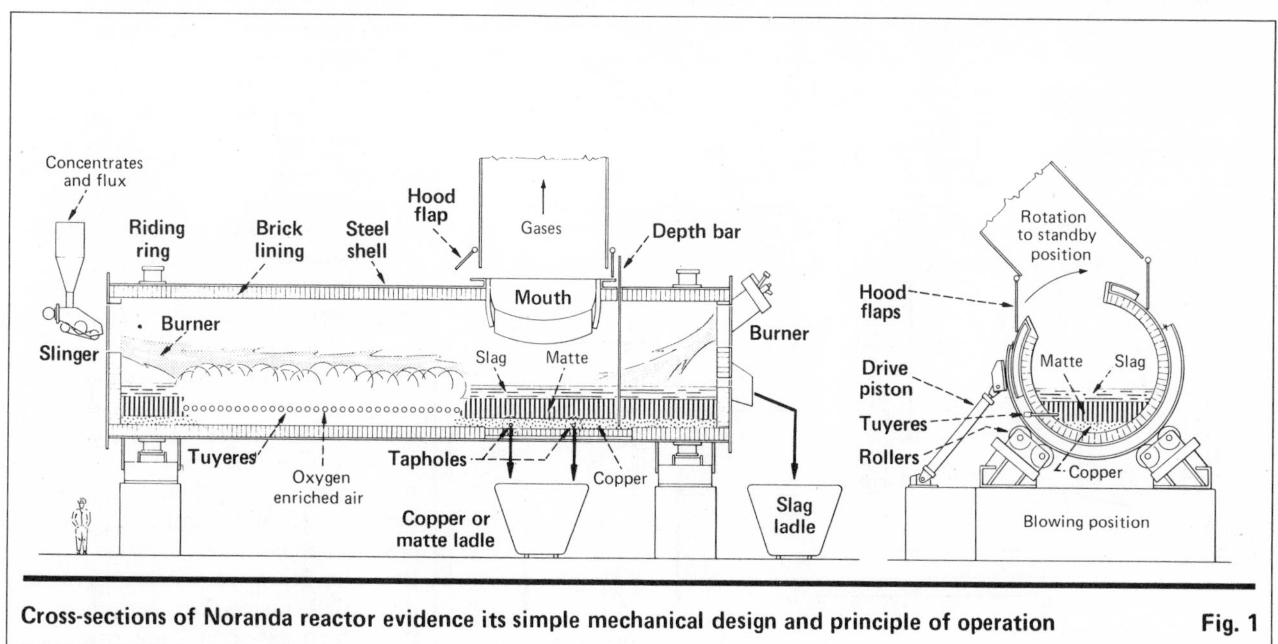

Cross-sections of Noranda reactor evidence its simple mechanical design and principle of operation **Fig. 1**

Originally published April 26, 1976

the reactor, or solid-fuel additions to the feed, supply the rest.

The usual copper reactions occur: iron sulfides and copper sulfides in the feed break down to form metallic copper or copper matte as the primary product, an iron-bearing slag, and an offgas containing SO_2.

When producing metallic copper, the slag is highly oxidized, and contains 20–30% magnetite (Fe_3O_4) in solid and entrained form and 8–12% copper. A high-grade matte coexists inside the reactor with the metallic copper and slag.

When producing matte, the slag contains less magnetite and copper; the levels range from 15–25% Fe_3O_4 and 3–7% Cu, depending on matte grade. The lower-magnetite slag is more fluid and allows operation of the reactor at a slag temperature of 2,200°F, compared to 2,250°F while making metallic copper.

In either case, the copper-bearing product settles to the bottom of the reactor for periodic withdrawal, while the slag is skimmed from the end of the reactor into ladles. As shown in the overall plant flowsheet (Fig. 2), the slag is slowly cooled, crushed, and sent to a concentrator, where flotation produces a 45%-Cu stream for recycle and an iron-rich tailings for discard.

Offgases leave the reactor through a water-cooled hood to an evaporative water-spray cooler, and then pass on to an electrostatic precipitator for dust removal. A fan exhausts the cleaned gases to the converter-aisle stack. Dust returns to the reactor through a pelletizer, and represents 2–5% of the dry weight of solid charge.

The basic control parameters of the Noranda process are (1) the oxygen/concentrate ratio, which is controlled at the value required to oxidize the feedstock sulfur and iron, and to give copper or matte of a desired grade, and (2) the flux/concentrate ratio, which is controlled to maintain the desired iron/silica ratio in the slag. The slag and metallic copper or matte are tapped so as to keep relatively constant levels of matte, slag and copper in the reactor. Dipping a steel bar into the bath serves as a simple hourly level check. The fuel input regulates temperature, and thus provides the required fluidity of the molten products.

EQUIPMENT DETAILS—The Noranda reactor at the Horne smelter is a 70-ft-long, 17-ft-dia. steel cylinder, supported on rollers so that a hydraulic drive-piston can rotate it through 48 deg to bring the tuyeres above the bath and stop the process. Direct-bonded chrome-magnesite brick serves as the shell lining, except in wear areas, where fused-cast magnesite-chrome brick is used.

The 40-ton ladles used to tap slag are transported by a rubber-tired carrier to the cooling area. Here, two ladles of slag are poured into a bed 30 × 15 × 1.5 ft. After about 24 h of cooling, a front-end loader breaks up the slag, which is then milled and concentrated for recycle.

Copper or matte is tapped intermittently through a 1½-in-dia. taphole, into a ladle that goes to the converter-anode aisle of the conventional smelter at Horne.

Offgases leave the reactor through a 12 × 8-ft mouth in the barrel of the reactor. The water-cooled hood covers the mouth when the tuyeres are in or out of the bath. This avoids the spill gas common to converter operations. To minimize air infiltration, movable flaps are installed on all four sides of the hood.

The evaporative cooler is bolted to the reactor hood. Sixteen air-atomized water sprays cool the gas to 700°F. (In other installations, a waste-heat boiler may be used.) The electrostatic-precipitation section consists of three parallel units and has an efficiency exceeding 98%.

OPERATING RESULTS—The Noranda reactor at the Horne site was built to expand existing facilities (three reverberatory furnaces and five converters). For almost two years following its March 1973 startup, it smelted near its design feedrate of 800 tons/d of concentrate, turning out copper metal. But early in 1975, copper-concentrate deliveries to the Horne facility began falling off, and a reverberatory furnace was closed down. This resulted in spare converting capacity, and allowed the Noranda reactor to switch to a high-grade matte product that could be processed further in the converters. Daily reactor throughput was thus raised to about 1,000 tons/d of concentrate, without modifying any equipment.

In July 1975, a second-hand oxygen plant rated for 90 tons/d came online for enriching oxidation air, and further increased the smelting rate of the Noranda reactor to 1,400 tons/d of concentrate.

Even with one reverberatory furnace closed down, there has at times been insufficient concentrate to operate at maximum smelting capacity. The Noranda reactor is closed down first during these periods, since it is much easier and less expensive to lay over than another reverberatory furnace.

The reactor has treated a wide va-

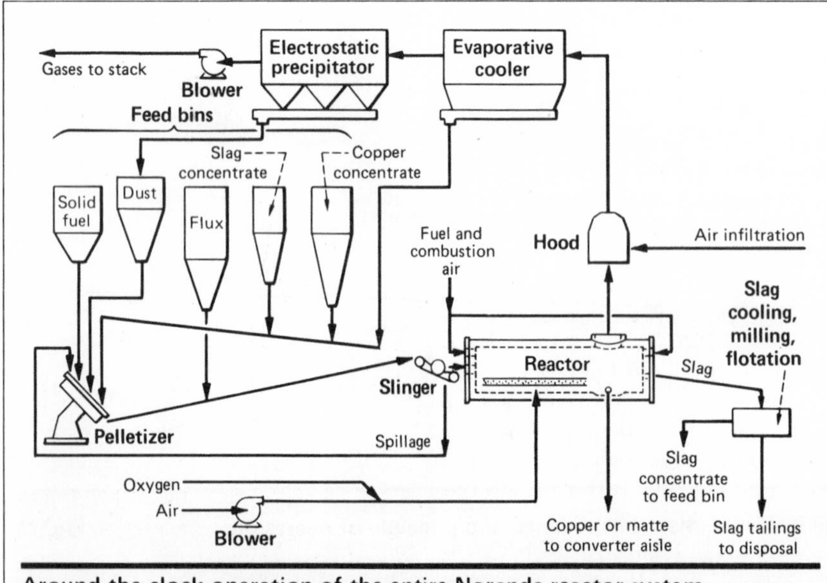

Around-the-clock operation of the entire Noranda-reactor system requires a total work force of about 50 people

Fig. 2

Reactor operating data for 10-d period at Horne smelter — Table I

Input	Dry tons	H_2O,%	Cu	Fe	SiO_2	S	Pb	Zn
			Composition (dry basis), %					
Copper concentrate	13,270	8.0	24.0	28.0	2.4	34.6	1.9	4.6
Slag concentrate	1,184	11.0	38.3	18.2	12.1	12.9	1.4	2.8
Crushed converter slag	453	–	19.0	27.7	21.8	0.6	6.0	1.5
Flux	2,820	3.5	–	4.0	70.5	–	–	–
Evaporative cooler dust and crushed reverts	173	–	41.3	17.4	5.5	15.1	2.5	3.3
Precipitator dust (tonnage estimated)	660	–	9.4	6.4	2.1	9.7	18.1	16.6
Output								
Matte	4,539	–	70.4	3.1	–	20.6	1.7	0.8
Slag	10,926	–	4.8	37.6	23.3	1.6	2.0	5.6
Evaporative cooler dust and reverts	173	–	41.3	17.4	5.5	15.1	2.5	3.3
Precipitator dust	660	–	9.4	6.4	2.1	9.7	18.1	16.6

Production data	
Instantaneous smelting rate, tons/d	1,425
Average blowing rate, scfm	46,840
Tonnage oxygen to tuyeres (at 96% oxygen), tons/d	84
Oxygen content of tuyere blast, %	23.4
Blowing time, %	93.2
Average fuel ratio, million Btu/ton of copper concentrate	2.67
Feed-end-burner firing rate, million Btu/h	88
Solid-fuel rate (21 million Btu/ton), tons/h	2.7

Projected operating data for Noranda reactors producing matte or copper, while burning different fuels — Table II

Product type	75% matte	75% matte	Copper	Copper
Copper-concentrate smelting rate, tons/d	2,100	2,400	1,870	2,050
Tonnage oxygen to tuyeres (at 96%), tons/d	415	415	415	415
Tuyere blowing rate, scfm	45,000	45,000	45,000	45,000
Oxygen content of tuyere blast, %	32.6	32.6	32.6	32.6
Feed-end-burner firing rate, million Btu/h	0	90	0	60
Solid-fuel rate, million Btu/h	45	0	29	0
Total fuel ratio, million Btu/ton Cu concentrate	0.63	1.0	0.50	0.82
Mouth offgas flowrate (dry basis), scfm	45,000	58,000	45,000	54,000
SO_2 at the mouth (dry basis), %	21.1	18.7	22.8	20.8
Expected air infiltration before scrubber, scfm	19,000	45,000	19,000	36,000
Annual throughput of copper concentrate, tons	604,000	690,000	514,000	563,000

riety of copper concentrates from some 25 sources, most of which were essentially chalcopyrite. These have wide ranges of analyses, namely: Cu, 20–28%; SiO_2, 0.8–9%; Fe, 25–36%; S, 25–35%; Pb, 0.1–8%; and Zn, 0.1–10%. Chalcocite concentrate, cement copper, refinery residues, and smelter reverts have also been processed at Horne.

The simple feed system to the reactor operates with few problems. When moisture in the concentrate exceeds 9%, no water addition to the pelletizer is required. Operating without the pelletizer increases dust load to the precipitator only slightly but results in dirtier dust recycle.

At the Horne smelter, the Noranda-reactor flux normally contains 67% silica, 4% iron and 12% alumina, and is crushed to less than ⅜ in. Flux as large as 1 in has been used. Mine flotation tailings, assaying 80% silica, have also served.

In a hot spot adjacent to the burner, just above the slag line opposite the tuyeres, the refractory has worn out rapidly. By using oxygen for converting, and thus a lower burner rate and smaller burner, it is anticipated that this hot spot will not reappear.

Refractory wear at the tuyere line is more severe when producing metallic copper than when making matte. After 110 d of making matte, the tuyere-line brick had lost only 1½ in of its original 15 in, compared with a 7-in loss after 70 d when producing copper.

Except for these areas, refractory wear is slight, especially below the slag line. Brick consumptions are less than for conventional smelting, which averages 4.84 lb of basic refractory per ton of concentrate at the Horne smelter. This compares to 3.3 lb when making copper and 2.57 lb when making matte via the Noranda reactor (which includes 0.2 lb for anode treatment, and 0.7 lb for converter treatment, of matte).

Predicted campaigns are 72 d for metallic copper production with a 16-d repair, and 120 d for matte production with a 20-d repair.

The burners operate on natural gas or oil. Coke or coal that may be added to the feed burns in the bath, using converter air for combustion. Bituminous and anthracite coals and coke fines have been used in the as-received state (generally smaller than 1½ in). Poor-quality coals with high ash, sulfur and moisture contents may also be burned.

With oxygen enrichment, the smelting rate rises, and, since there is less nitrogen carrying away heat in and diluting the offgas, fuel use falls and SO_2 strength rises.

Table I shows the most recent operating data for ten days during January 1976, when the Noranda reactor was producing a 70%-copper matte and using a small amount of oxygen enrichment. Table II shows predicted results for a reactor turning out copper or matte using different fuels.

The author

Geoffrey D. Hallett is assistant manager of the Continuous Smelting Div. of Noranda Mines Ltd. (P.O. Box 45, Commerce Court West, Toronto, Ont. M5L 1B6, Can.). He previously spent five years (of his seven with the firm) working in the Horne smelter at Noranda, Que. He attended the Royal School of Mines of London University, receiving B.Sc.(Eng.) and M.Sc.(Eng.) degrees in metallurgy, and earning an A.R.S.M. (Associate of the Royal School of Mines) and a D.I.C. (Diploma of Imperial College).

Cu/U ore-leaching route cuts pollution, trims costs

Less costly than agitation-leaching processes, and said to be as effective, this thin-layer leaching route reduces construction and operating costs, land and water requirements, and pollution problems. The methods may even help solve uranium-tailings woes.

Gerald Parkinson, Western Regional Editor

☐ Greatly reducing the chance of groundwater and airborne pollution, the TL (for "thin layer") process also eliminates a considerable amount of the large equipment associated with conventional leaching routes. Developed by Holmes & Narver, Inc.,* the process is said to be as effective, where applicable, as agitation leaching, but only about 50 to 70% as expensive to build, and 5 to 15% less costly to operate. Water requirements may be as little as one fourth the amount needed for agitation leaching, and tailings are produced in moist, solid, granular form instead of as a conventional slurry that contains 50% water, by weight. And since the TL route does not involve many fines, there is less likelihood of windblown dust.

The first commercial installation is under construction in Santiago, Chile, for Sociedad Minera Pudahuel, Ltda. Scheduled to go onstream late in 1979, the plant will handle 2,600 metric tons/d of copper ore, producing 45 metric tons/d of cathode copper by leaching, extraction and electrowinning operations.

In addition, Holmes & Narver has two TL pilot-plant projects underway with mining companies. A Latin American plant that started up last month will process 7 tons/d of copper ore, while a U.S. facility scheduled to go into operation by the end of the year will handle 25 tons/d of uranium ore.

THIN LAYERS—The TL process is so called because the ore is spread in layers 3 ft thick, or less, for leaching. After treatment with concentrated sulfuric acid—which liberates the copper or uranium—the crushed ore is allowed to cure for a day, then is spread over shallow beds for leaching, the shallowness of the beds permitting uniform contact between the leach liquor and the ore by minimizing compacting and channeling.

"We have run tests on 10 to 15 copper ores and 15 to 20 uranium ores," says Jay N. Palley, Holmes & Narver's vice-president for metallurgical plants, "and with few exceptions we have achieved recoveries as good as, or better than, agitation leaching."

FEWER WASTE WOES—In conventional copper and uranium processing, tailings are typically mixed with about 50 wt. % of water (about three times their volume) for disposal. And since less than 5% of the solid mined material is extracted in the process, the slurried tailings require extensive diked ponds—where the solids settle out of suspension and the liquids are reclaimed or evaporate.

But ponds—especially those for radioactive uranium tailings—pose environmental problems, such as seepage of liquids from the ponds into aquifers and into the soil, where they may be taken up into the food cycle. Dispersion of dried tailings solids (both by wind and man) is also of concern.

These problems are not insignificant. Los Alamos Scientific Laboratory (Los Alamos, N.M.) reports that, as of 1970, over 88 million tons of uranium mill tailings existed in the U.S., representing a volume of 1.7 billion ft^3 that would cover one square mile to a depth of sixty feet. Los Alamos estimates that by the year 2000 over 900 million tons of uranium tailings will have accumulated in New Mexico alone.

In the TL process, tailings have a moisture content of only about 14%, compared with a natural ore moisture of about 5 to 6%, a reflection of low water use in the process—as little as one fourth that used in agitation leaching. Because the tailings do not contain so much water, they have only about one fourth the volume of tailings generated in conventional routes. In addition, the granular TL tailings may be piled to a height of about 35 to 40 ft to save even more space, or returned to a worked-out portion of the mine.

And because the TL process requires that ores be crushed only to an approximately 3/8-in. size—as opposed to less than 65 mesh for copper and less than 28 mesh for uranium in conventional routes—there are fewer fines, minimizing the likelihood of windblown dust.

PROCESS DETAILS—Pudahuel's plant will exploit a 10-million-metric-ton ore body having an average grade of about 2.2% copper. Initially, about 80% of the copper in the ore will be acid soluble, while after three or four years, the mix will be about 50/50 soluble/nonsoluble. After about eight years, only 20% will be soluble. Plans provide for installation of a concentrator at some future date.

Ore will be crushed in a three-stage unit (see flowsheet), consisting of a jaw crusher, a secondary cone crusher and two tertiary cone crushers.

The crushed ore will proceed by conveyor to the leach plant, where it will enter a horizontal rotating drum. Here, concentrated sulfuric acid will spray onto the ore through jets at the

*Holmes & Narver, Inc., 999 Town and Country Rd., Orange, CA 92668.

Originally published April 24, 1978

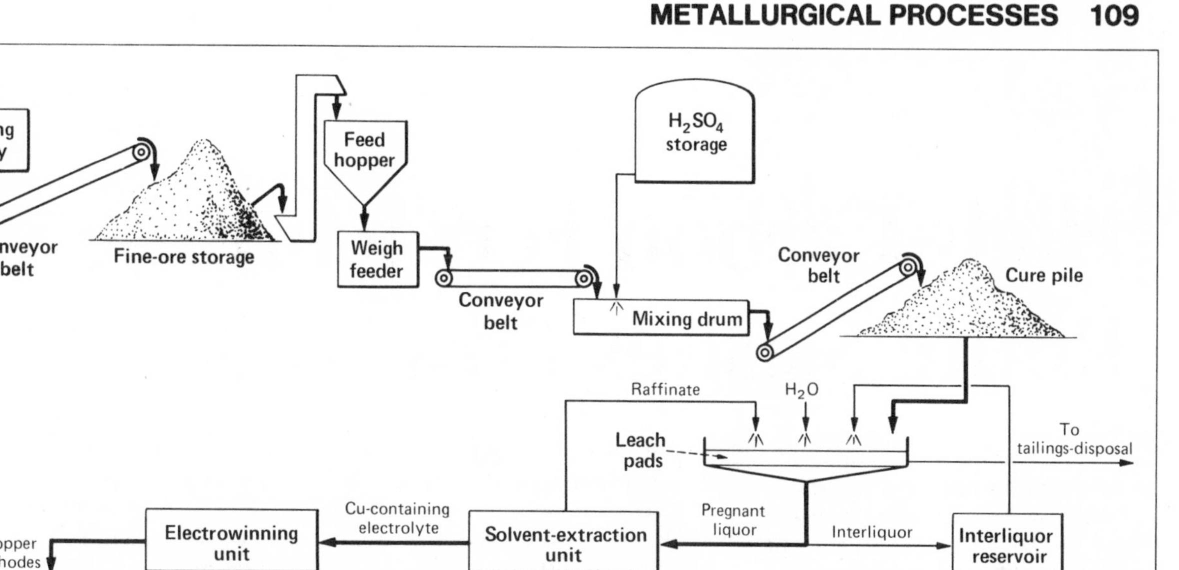

TL process finds its first commercial use in Chile

drum's inlet end. The acid will be controlled to maintain a moisture content of about 10% in the material, so there will be no free liquid. Ore passes through the drum in 60 s, adequate time for a thorough mix and some agglomeration of any fines.

Sulfuric-impregnated ore falls from the drum onto a conveyor that feeds one of four cure piles. One cure pile will be built up every 8 hours, so that the residence time for each pile will be 24 hours.

Front-end loaders will take material from a cure pile and deposit it on leach pads made of acid-resistant concrete. The Pudahuel unit will have 20 adjacent pads, each approximately 150 × 50 ft, with trench systems for collection of leach liquor.

Sprays, located on low walls on two sides of each pad, will apply leach liquor uniformly over the beds. The initial application will be of interliquor already used to leach another pad. This will be followed by raffinate recycled from the solvent-extraction plant. A liter (L) of raffinate will contain 7.5 to 9.0 g of sulfuric acid, regenerated from the extraction of copper in the solvent-extraction unit. In addition, a very small amount of ferric iron will be maintained in the liquor to enhance leaching of non-acid-soluble copper. The final wash of the ore will be accomplished with fresh water.

Cycle time on the beds will be six days at Pudahuel, but could vary elsewhere from two to six days, depending on the nature of the ore processed. The first liquor drawn from a bed may have as much as 25 g of copper per liter, but this concentration will fall rapidly. Liquor will be collected in reservoirs until the copper content is 5 to 6 g/L, approximately the optimum for the solvent-extraction unit.

After the final wash, the leached material will be loaded onto a conveyor and discharged at a tailings-disposal site.

EFFICIENT PROCESS—Holmes & Narver's Palley says: "Based on test data, we expect Pudahuel will initially recover about 90% of the total contained copper by leaching. When the firm gets to the mixed ore, tests indicate it can expect up to 75% recovery of the total copper, and for the final ore, we predict up to 50% recovery, just by leaching."

The route for uranium processing is basically the same, except that ferric iron is not used. Instead, sodium chlorate may be employed to enhance leaching.

Holmes & Narver has processed 0.06%- to 0.33%-uranium ores in an 8-ton/d demonstration plant at Atlas Minerals Corp.'s uranium mill at Moab, Utah. Says Palley: "We processed three different ores with parallel laboratory tests, and plant results verified lab results." A liter of pregnant solution containing 2 to 5 g of uranium (depending on the ore) was obtained. This compares with a typical solution of 1 g/L in conventional leach operations, according to the Holmes & Narver spokesman.

ECONOMICS—Costwise, the capital investment for a copper-producing TL leach plant is only about 50% that of an agitation leach facility, the comparison covering only the mill section and not ancillaries or the subsequent solvent-extraction and electrowinning sections. Savings in the mill section result from the elimination of grinding equipment. And because washing is done in the leach beds, countercurrent decantation and fines-separation systems (often needed for vat leaching routes) are obviated.

Savings can also be realized in both the solvent-extraction and tailings operations: In the former, because metal concentration in TL leach liquor can be controlled to a higher concentration than in agitation leaching, according to Holmes & Narver, reducing the size of the solvent-extraction plant needed by 40 to 60%, and in the latter, because of the diminished tailings-volume.

The leach, solvent-extraction and electrowinning sections of the Pudahuel project will cost about $30 million, for a capital investment of $1,875 per annual ton of copper produced (based on a production year of 355 days.)

While operating costs for both copper and uranium plants are expected to be 5 to 15% less than those for agitation leaching plants, according to Holmes & Narver, a grassroots uranium plant—excluding .only the tailings-disposal system—would cost about 70% as much as a plant that used agitation leaching.

High-Carbon Ferrochrome Route Slashes Power Use

Electric power consumption is cut in half compared with conventional methods because pellets of ore mixed with carbon are hot-charged into a closed-type electric furnace. Delicate temperature control is maintained.

KAZUO ICHIKAWA
Showa Denko K.K.

Like so many other manufacturers today, stainless steel producers are looking for processing techniques that cut power consumption as much as possible. For Showa Denko K.K., one search dates back to at least 1968 when it successfully developed a way of making high-carbon ferrochrome suitable as an addition agent in the production of stainless steel. A 60,000-ton/yr plant using this process and operated by Shunan Denko K.K., a Showa Denko subsidiary, cuts electric power consumption in half while doubling the productivity of the electric smelting furnace.

One of the main features of the process is a prereduction step ahead of the furnace, in which the chrome ore is pelletized along with coke, binder and flux, and roasted in a rotary kiln. Reaction between the coke and the ore within the pellet proceeds because of a hard film formed on the outside of the pellet that protects the contents from oxidation by the roasting atmosphere. Furthermore, the pellets are strong enough to be tumbled in the kiln to ensure the proper temperature control within the pellet bed that is crucial to the process.

By pelletizing the fine, low-grade chrome ore more commonly available around the world, it is possible to use a closed-type electric furnace. This enables a substantial reduction in unit power consumption, and in pollution-abatement costs that would otherwise be necessitated by handling the dusty ore in the open.

Because chromium has a stronger affinity for oxygen than does iron, it is not possible to use a carbon monoxide gas-reduction method to make reduced pellets of chrome ore. It is, therefore, necessary to have the carbonaceous reducing material come into microscopic contact with the ore and react locally at temperatures higher than 2,350°F.

To make the pellets, ore and carbonaceous material are pulverized to a fine homogeneous mixture. Most of the particles are less than 100 microns in size. This size is best suited for forming pellets of good solidity and reactivity. The pellets are then charged to the rotary kiln. Since the difference between the temperature required for reaction and the softening temperature of the pellets is small, special attention is given to controlling the flame and the furnace temperature when heating the kiln.

Sixty percent of the chromium and iron oxides in the ore are reduced in the kiln prior to charging the pellets to the electric furnace. This enables the economic use of the low-grade ores that are more plentiful than those of a high chromium/iron ratio.

In Flux—A key ingredient in the pellets is a flux consisting of silica and one or more of various metallic oxides. It is added in two stages, once before roasting, and a second time before smelting in the electric furnace. Addition prior to roasting aids in producing a strong pellet capable of holding its shape in the kiln. It also helps to accelerate the reduction reaction between chromium and iron oxides and carbon, partly because the flux helps destroy chromite crystal structures in the ore and thus improve the contact between ore and carbon particles. An added advantage to flux addition is that it allows a lower starting temperature in the roaster.

Silica concentration in the flux is between 20 and 45% on a weight basis. The other constituents are one or more of the following: lime, magnesia and alumina. The following formula specifies the correct mixture of the four possible components:

$$R = (1.0\,MgO + 0.72\,CaO)/1.0\,Al_2O_3$$

where R = 1.3 to 2.0, and the quantities of the flux materials are given as weight percents.

If the amount of silica is less than 20%, the silica will not melt enough to take part in the reaction, and the reaction rate will be slow. But silica in an amount greater than about 45% causes the pellet itself to melt, resulting in melt-adhesion of the pellet onto the walls of the rotary kiln.

During roasting, the flux forms a mixed-solid solution of forsterite ($2MgO \cdot SiO_2$) and spinel ($MgO \cdot Al_2O_3$). These do not inhibit the extraction, from the pellet, of carbon monoxide formed in the reduction reaction.

As applied to the pellet, the flux forms a film of a thickness ideally between 1.0 and 3.0 mm. A thinner film runs the risk of surface imperfections, while a thicker film does not yield any greater processing advantage.

Roasting—Roasting proceeds at a temperature of between 2,450 and 2,640°F in the presence of an oxidizing atmosphere in which air is sup-

Originally published April 1, 1974

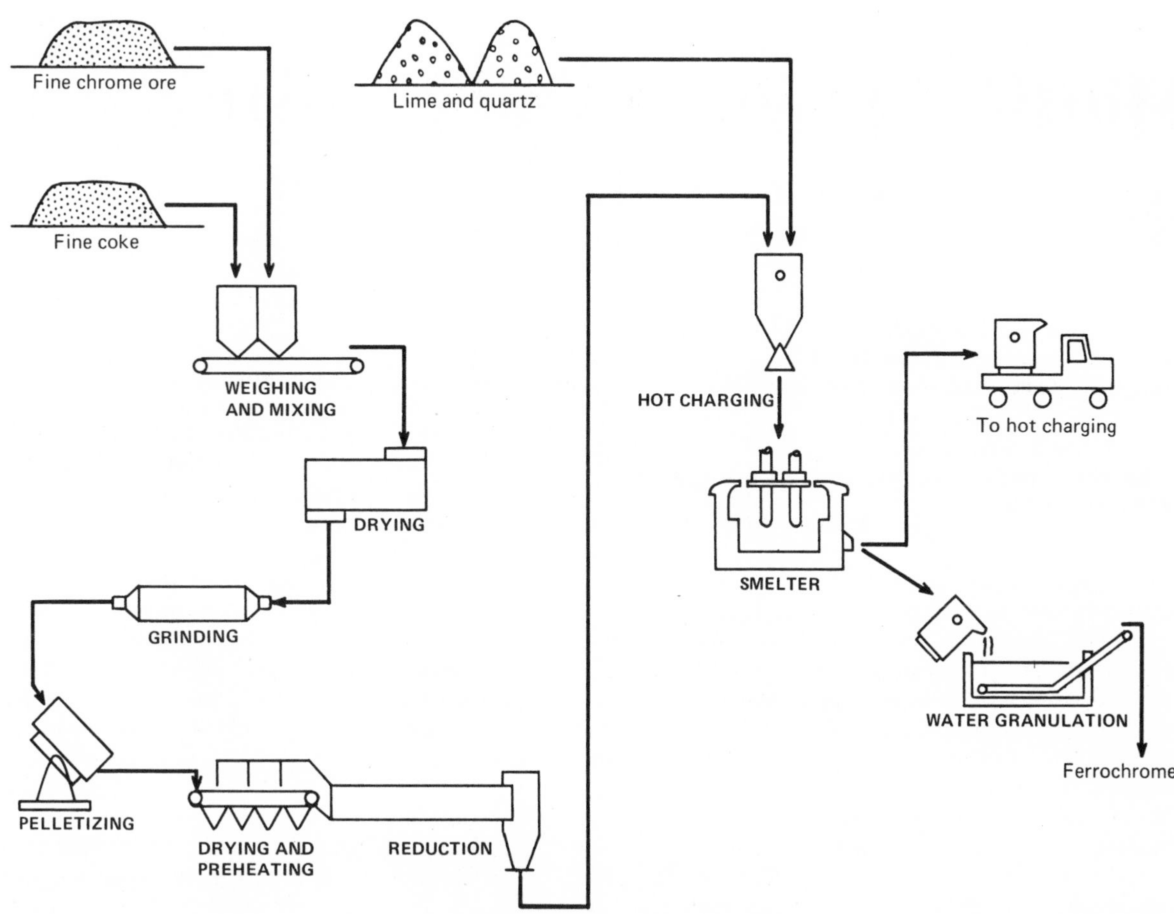

plied at 1.5 to 2.5 times the stoichiometric oxygen requirement for the fuel. Under these conditions, carbon monoxide evolved during the reaction is quickly oxidized. Meanwhile, the coated flux-components penetrate into the inner portion of the pellets to harden them.

By keeping the pellets hot after roasting, they can be charged directly into the electric furnace, thereby saving additional energy. Along the way, additional flux is added to the furnace charge, this time to ensure smooth completion of the smelting step.

Smelting—In this step, the reduction in the molten state of residual $Cr_2O_3 \cdot FeO$ and removal of slag is performed. Because the temperature must be higher than 2,900°F in this operation, a submerged-type electric smelting furnace is used. The hot pellets are charged continuously. The cell is completely closed.

The reduced hot pellets generate heat from oxidation when in contact with air, and they are easily sintered in a hopper or chute. Thus, special precautions are taken to prevent sintering during transportation, storing, and charging into the furnace.

The pellets are also easily subject to fusion inside the furnace, and the balance between fusion and reaction is very fine and difficult to maintain. In order to stabilize the reaction, a unique blend of raw materials and an electrical control method are employed.

The product obtained has the following analysis:

Chromium	52-60%
Carbon	about 8.2%
Silicon	about 2%
Iron	29-37%
Other	0.8%

Unit power consumption totals 2,000 to 2,100 kWh/ton of ferrochrome.

South African Plant—Under terms of an agreement signed in January, South Africa's Johannesburg Consolidated Investment (JCI) will be operating a 120,000-metric ton/yr plant at Lyndenburg, Transvaal, by 1976. The deal was made on a cash, rather than royalty, basis. The contract stipulates export of at least 30,000 metric tons/yr of this output to Showa Denko through Japanese trading companies. In this way, the Japanese firm reaps the double benefit of an energy savings and a chance to utilize South Africa's 2-billion-ton low-grade chromium ore deposits, the largest in the world. (Rhodesia has about 600 million tons of high-grade ore, but importation of this material into Japan is not possible under United Nations sanctions against Rhodesia.) #

Meet the Author

Kazuo Ichikawa is deputy manager of the Administration and Planning Dept., Metals & Alloys Div., Showa Denko K.K. He is a graduate of Tokyo Institute of Technology.

Iron-ore concentration process unlocks low-grade reserves

A just-commercialized process for concentrating iron ore—featuring selective-flocculation, desliming and flotation steps—sets the stage for use of vast low-grade reserves.

Arthur Zimmerman, McGraw-Hill World News

☐ The first commercial plant for concentrating fine-grain, nonmagnetic hematite ore into iron pellets was started up ten months ago by Cleveland-Cliffs Iron Co.*, at its Tilden mine near Ishpeming on Michigan's Upper Peninsula.

Though the firm has owned the 900-million-ton ore reserve for more than a century, mining has heretofore been on a small scale. Previous flotation schemes proposed for beneficiating Tilden's nonmagnetic ore were not commercially feasible because of high losses of the fine-size iron particles during silica-rejection steps.

The breakthrough that now avoids these losses is a selective-flocculation–desliming treatment, followed by froth flotation. Cleveland-Cliffs jointly developed the technique with the U.S. Bureau of Mines' Twin Cities Metallurgy Research Center (Minneapolis, Minn.).

About 70% of the iron content of the crude ore (a low-grade, 35%-iron, 45%-silica ore) is thus recovered in the product concentrate, and only 30% sent to tailings. The final pellets contain a high 65% iron, and 5% silica. Overall, the $200-million-plus facility will turn out 4 million long-tons/yr of pellets from 10 million long-tons/yr of crude ore.

Cleveland-Cliffs is well along with plans for an equal-size expansion at Tilden, with yet another 4 million long-tons/yr contemplated.

The process could also be applied at the firm's other reserves, like Tilden in the Michigan Marquette Range, which have similar ore characteristics. With some modifications, moreover, it could conceivably unlock even the vast nonmagnetic taconite reserves of the Western Mesabi Range in Minnesota.

SIZE REDUCTION—A crusher that reduces mined ore to chunks sized less than 12 in. starts off the process. Six parallel concentrator lines follow (only one is in the flowsheet).

Autogenous grinding mills use a tumbling action to reduce ore size.

* Cleveland-Cliffs Iron Co. (1460 Union Commerce Bldg., Cleveland, OH 44155) is manager for the Tilden Mining Co., a joint venture of Cleveland-Cliffs with Algoma Steel Corp., Jones & Laughlin Steel Corp., Sharon Steel Corp., Steel Co. of Canada, and Wheeling-Pittsburgh Steel Corp.

Originally published October 27, 1975

Deslime thickeners (above) separate extremely fine silica particles from starch-flocculated iron particles.

Flotation cells (left) remove most of the remaining silica impurity, again with the help of starch as flocculant.

Vacuum disk-filters (below) concentrate the iron-rich stream in preparation for final pelletizing operations.

A primary autogenous mill in each line grinds the crushed ore to the consistency of a coarse beach sand, in a 65%-solids aqueous solution. Sodium hydroxide added to the mill closely controls the pH of the water (10.5-11.0 pH), and sodium silicate is added to disperse fine siliceous materials in the ore, in preparation for desliming.

Cleveland-Cliffs has stayed on the conservative side in selecting the autogenous mills. The 27-ft-dia., 14.5-ft-long (inside chamber) units draw 5,720 hp with both motor drives operating. The firm could have chosen mills as large as 32–36 ft dia. (used elsewhere in the industry), which would have reduced the number needed but increased the impact of possible mechanical problems.

After separation of oversize particles for recycle, the onsize output from each primary mill feeds two pebble mills. These reduce 85% of the ore particles to less than 25μ, or 500 mesh (larger ones are again recycled). This grind is finer than now used in other iron-ore processes, emphasizing the need for an efficient, low-loss system for flocculation and desliming.

The two pebble mills might have been combined into a single larger unit. But Cleveland-Cliffs was again conservative, choosing the smaller models. Both measure 15 ft I.D. by 30 ft long, and draw 2,520 hp.

KEY SEPARATIONS—A commercial grade of corn starch mixes into the 8%-solids stream coming off the pebble mills to flocculate the iron oxide particles. The stream then passes into a pair of 55-ft-dia. deslime thickeners. Here, a slowly rising column of water carries the dispersed, extremely fine silica particles upwards and out as overflow slime, while the flocculated iron sinks for takeoff in the underflow.

The deslimers reject 20% of the total feed weight as overflow, which contains about one third of the input silica, and only 7% of the total iron from the crude ore. The 50%-solids underflow retains the rest of the iron, as a 41.8-wt.%-iron stream.

All six underflow streams are next merged to provide a single uniform stream; but, this is redivided for feeding into six flotation-cell lines. The 500-ft^3 cells, numbering 25 per line, are the largest in the industry.

CRUSHER PRODUCT BIN
CRUSHER FEED BIN
CONE CRUSHER
PEBBLE COLLECTING HOPPER
Over-flow
PEBBLE STORAGE BIN
CYCLONE
Starch flocculant
CONDITIONER
Lime flocculant
DESLIME THICKENER
PULP DISTRIBUTOR
Reuse water from pond
Caustic soda
Silicate dispersant
Crude ore
FEEDER
PRIMARY AUTOGENOUS MILL
SCREENS
Reuse water from pond
Reuse water from pond
PEBBLE MILL
MIXING TANK
Overflow (slime tailings)
FLOTATION FEED DISTRIBUTOR
Starch depressant
CONDITIONER
Amine collector material
ROUGHER FLOTATION FLIGHT
Polymer flocculant
Starch depressant
CONCENTRATE THICKENER
Polymer flocculant
Overflow
REUSE WATER POND
SCAVENGER FLOTATION FLIGHTS
Polymer flocculant
Reuse water
Middlings
SLURRY TANK
FILTER
Lime flocculant
DISTRIBUTOR
Steam
MIXING TANK
TAILINGS POND
Filter cake to pelletizing

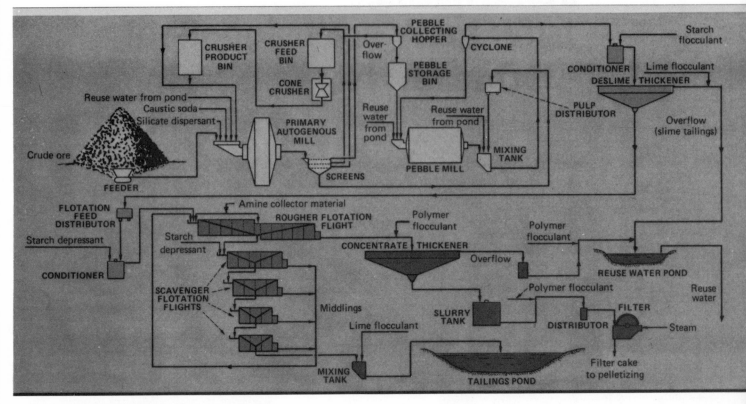

Tilden-mine process for iron-ore concentration

A cationic-amine collector material added in the first, rougher, flotation flight helps silica particles cling to the flotation air bubbles. Even though early work by the Bureau of Mines had emphasized an anionic approach, Cleveland-Cliffs chose the cationic method as a simpler, more-reliably-controlled system for its maiden plant. There are fewer reagents required and less opportunity for imbalances to develop.

Another starch stream, this acting as a depressant, is added prior to the rougher flight and to the first of four scavenger flights. Siliceous froth from the rougher flight passes to the scavengers and the final scavenger froth goes to a tailings pond for disposal, carrying with it about 23% of the iron from the crude ore. Middlings from the scavenger cells are recycled to the rougher stage.

FINAL PROCESSING—The 16.5%-solids, iron-rich stream from the cells is thickened to 70% solids in a 150-ft-dia. concentrator/thickener. Vacuum disk-filters heated with steam next turn the stream into a 10%-moisture filter cake.

A conventional pelletizing operation finally drum-rolls the cake (in a mix with bentonite) into green balls; drys and preheats these over a traveling-grate; hardens the pellets in a 2,400°F rotary kiln; and cools the pellets before transfer to storage.

MORE FEATURES—Recycling of process water aids materially in recovering reagents and controlling costs. Fully 95% of the 86,000 gpm of process water is recycled, the remaining 5% coming from a reservoir.

Though the design schedule had called for applying 2.7 lb of sodium hydroxide per long-ton of dry crude ore, for example, additions currently stand at only 1.0 lb. Sodium silicate is being added only intermittently to the recycle water, versus a design rate of 0.57 lb/long-ton of ore. And amine, calculated at 0.3 lb/long-ton, ranges between 0.15–0.2 lb.

Cleveland-Cliffs has experimented with varying additions of 2.5-wt.% starch solution into the deslime thickeners, originally set at 0.5 lb/long-ton, but is currently running lower. A planned 1.3 lb/long-ton addition of starch to the flotation stage is holding at or below design, as is the 3.4 lb/long-ton of lime into the deslimer overflow headed for the reuse-water pond.

Total power consumption at Tilden runs about 150 kWh/long-ton of pellets. Heat-energy consumption is 218,000 Btu/long-ton of filter cake for the steam filtering step, and 985,000 Btu/long-ton of pellets for drying and firing.

The firm sees possible refinements and improved efficiencies in the process, particularly in the choice of dispersants and flocculants. It continues to evaluate sodium tripolyphosphate and tannin products as dispersants; and tapioca starch, potato starch, various dextrins, and both cationic and anionic polymers as flocculants. All function in the process, but because of price, availability or efficiency were not chosen initially.

The author

Arthur Zimmerman has been chief of the McGraw-Hill Publications Co. World News bureau in Cleveland, Ohio, for the past 16 years. He specializes in editorial coverage of the chemical, plastics and metalworking industries. He graduated from Hiram College in Hiram, Ohio, with a B.A. in English, and has done graduate work in journalism at Kent State University.

Direct iron reduction: the role widens for natural-gas alternatives

Interest is growing in generating of reductants from fuel oil, coke-oven gas and other materials in routes that now rely on natural gas for the job. And present coal-based processes should gain ground. Nuclear-reactor-linked schemes, however, look far off.

☐ The harsh winter weather hobbling many areas of the U.S. points up at least one cold, hard fact: the vulnerability of natural-gas supplies in some sectors of the chemical process industries. Gas cutbacks and forced shutdowns of plants are reinforcing increasingly expressed qualms about the future availability and price of the gas. U.S. industry hardly worries alone—CPI companies in many parts of the world echo the apprehensions.

Firms are rethinking their reliance on natural gas, particularly for plants at the planning stage. This holds especially true for steelmakers eyeing direct-reduction routes to yield sponge iron from iron oxide.

Currently, most direct-reduction facilities worldwide rely on natural gas to provide the carbon-monoxide/hydrogen reducing gas used in the process. According to a study issued last September by the U.S. Dept. of Commerce, a total capacity of 22.7 million metric tons/yr is now operating, under construction or in contract globally. Of this amount, over 18 million metric tons/yr bank on natural gas, while the rest largely depend on coal. So, around 80% of capacity hinges on natural gas.

NEW OPTIONS—However, licensors of natural-gas-based direct-reduction technology are offering or grooming process variants in which other materials provide the necessary reducing gas. Alternatives include naphtha, fuel oils, coke-oven gas, electro-reduction iron-smelting-furnace offgas, and coal gasification.

For instance, Arthur G. McKee & Co. (Cleveland, Ohio) has looked at naphtha, fuel oils, coke-oven gas, and product from both developmental and proven coal-gasification techniques as sources of reducing gas for its Fior process. And, in fact, the firm is now studying—for a group including itself, Babcock & Wilcox, Northstar Steel, Timken, Keystone Steel and Wire, and Connors Steel—the feasibility of a 1.5-million-metric-tons-of-sponge-iron/yr plant, based on naphtha, #2 fuel oil or low-sulfur #6 fuel oil, for the U.S. Gulf Coast. A preliminary evaluation, completed last March, was favorable, and McKee expects a decision on the project during the first half of this year. Given a go-ahead then, the plant could be completed by around 1980.

Mexico's Hojalata y Lamina S.A. (Monterrey, Nuevo Leon) has also investigated naphtha, coke-oven gas and coal gasification for its HyL flowscheme. And the route's exclusive licensing agent, Pullman Inc.'s Pullman Swindell Div. (Pittsburgh, Pa.), signed a contract in 1975 with the government of Zambia for a 250,000-metric-tons-of-sponge-iron/yr plant, based on naphtha, at a site in the northern part of that nation. There reportedly have been some holdups on the government's side, and construction has not yet begun. However, the project is still alive, stresses Swindell. Some equipment has already been ordered, and the unit could go onstream by as early as 1979.

Armco Steel Corp.'s Armco International Div. (Middletown, Ohio) has appraised using naphtha, light and heavy fuel oils, and coke-oven gas in its now-natural-gas based system, in conjunction with its licensee Foster Wheeler Corp. (New York City). Indeed, Armco has prepared a proposal for a European plant running on coke-oven gas. This project is now in limbo, though. Armco has run into some problems with the furnace at its Houston, Tex., direct-reduction plant, and has suspended all licensing activities until the difficulties are sorted out.

MOVING AHEAD—Meanwhile, Midrex Corp. (Charlotte, N.C.) has a number of projects featuring alternatives to natural gas in various stages of development. The company, a U.S. subsidiary of West Germany's Korf Group, has studied fuel-oil-based systems for Brazil and South Africa, and has looked at naphtha, propane and refinery offgases, as well.

The Midrex process should get a commercial tryout with coke-oven gas in two 400,000-metric-tons-of-sponge/yr modules planned for Brazil. The firm has signed a letter of intent with Usinas Siderurgicas de Minas Gerais, S.A. (Belo Horizonte, Brazil) to construct the units at Ipatinga. No site work has yet been done and project timing remains uncertain, because the Brazilian government is reviewing the state-owned steel company's expansion plans. This review might be completed this year. In the meantime, Midrex is discussing another Brazilian plant, with private interests, based on offgas from electric-arc, low-shaft, iron-smelting furnaces.

Coal gasification may also play a role in the Midrex process. For instance, the company is now doing an evaluation for Turkey's Ege Demir Sanayii A.S. (Izmir) of a 400,000-metric-ton/yr installation based on lignite gasification. The study should likely be completed during the first quarter of this year.

Midrex notes that it will likely be in the 1980s before any of its proposed natural-gas-alternative projects could reach commercialization.

Back in the U.S., the State of Kentucky, which has sizable reserves of high-sulfur coal, is hoping to pave the way for use of the material in di-

Originally published February 28, 1977

rect reduction. The state is sponsoring a joint effort by Westinghouse Research Laboratories (Pittsburgh, Pa.) and the Institute for Mining and Minerals Research at the University of Kentucky (Lexington). The work now centers on a feasibility study on a moderate-pressure (about 10 atm) reduction using low-to-intermediate-heating-value gas derived from the coal. Results should be ready by around September. If all goes well, a second phase of the program, likely involving some hardware, would then take place.

IMPLICATIONS—Opting for an alternative to natural gas may likely change both process configuration and costs.

For example, coke-oven gas usually is considerably dirtier than natural gas, notes Midrex. Not only does the gas require cleaning, it also must undergo desulfurization to about 7 ppm (by volume) to be suitable for the reformer. However, the thermal load on the reformer is lower than for natural gas, because coke-oven gas contains only around 26% methane instead of 80%. This translates to a savings, says the firm, of approximately 1 Gigacalorie per metric ton of sponge iron.

Using coal- or oil-gasification, or offgas from low-shaft electric iron-smelting furnaces will change the Midrex flowscheme somewhat more substantially, as shown in the diagram below. The gas does not need reforming. Rather, after cleaning, it passes to a humidifier; this helps provide the desired hydrogen-to-carbon-monoxide ratio. A heater then brings the gas up to reduction temperature. Top gas from the reduction furnace is recycled via an amine scrubber, which removes carbon dioxide.

Russell Ellis, a business development specialist at McKee, puts the capital-cost impact of the alternatives into some perspective. Considering only the reducing-gas generation and fuel requirements of a gas-based direct-reduction process—which together typically account for about 30% of plant investment—he estimates that for a commercial-size installation at a U.S. or Venezuelan site: natural gas and coke-oven gas would rate about even; naphtha would incur a capital-cost multiplier of 1.1 (i.e., a 10% penalty); partial oxidation of high-sulfur fuel oil would come in at 2.1; and coal gasification would have a considerably higher tab.

Of course, the relative costs of the alternative feeds, or factors such as security of supply or politics, may be paramount in choosing among the options.

SOLID SUCCESS—Another alternative involves using coal directly without gasification. Several companies offer processes in which coal is fed into a kiln or reactor with the iron ore. A number of such solid-reductant plants are commercial. And the firms say that interest in the approach is increasing.

The SL/RN process rates as the most widely used of the solid-reductant routes. Developed jointly by Steel Co. of Canada (Toronto), Lurgi Chemie und Hütten-Technik GmbH (Frankfurt/Main), Republic Steel Corp. (Cleveland, Ohio) and NL Industries Inc. (New York City), the technique can employ lignite, high-volatile coal, sub-bituminous coal, coke breeze and anthracite for reducing fine ores, such as beach sand, as well as screened lump ore and pellets.

The process is already being utilized by seven firms around the world. Capacity of the plants ranges from 18,000 to 360,000 metric tons of sponge iron/yr. The first unit came onstream in 1969, and a new, 100,000-metric-ton/yr, plant is under construction for Siderperu at Chimbote, Peru. Overall capacity of all the Midrex kilns totals 3 million metric tons/yr.

In addition, Lurgi says that various "serious" lignite-based SL/RN projects, with a combined annual capacity of 1.8 million metric tons, are in preparation.

Another route that can use coal directly comes from the Reduction

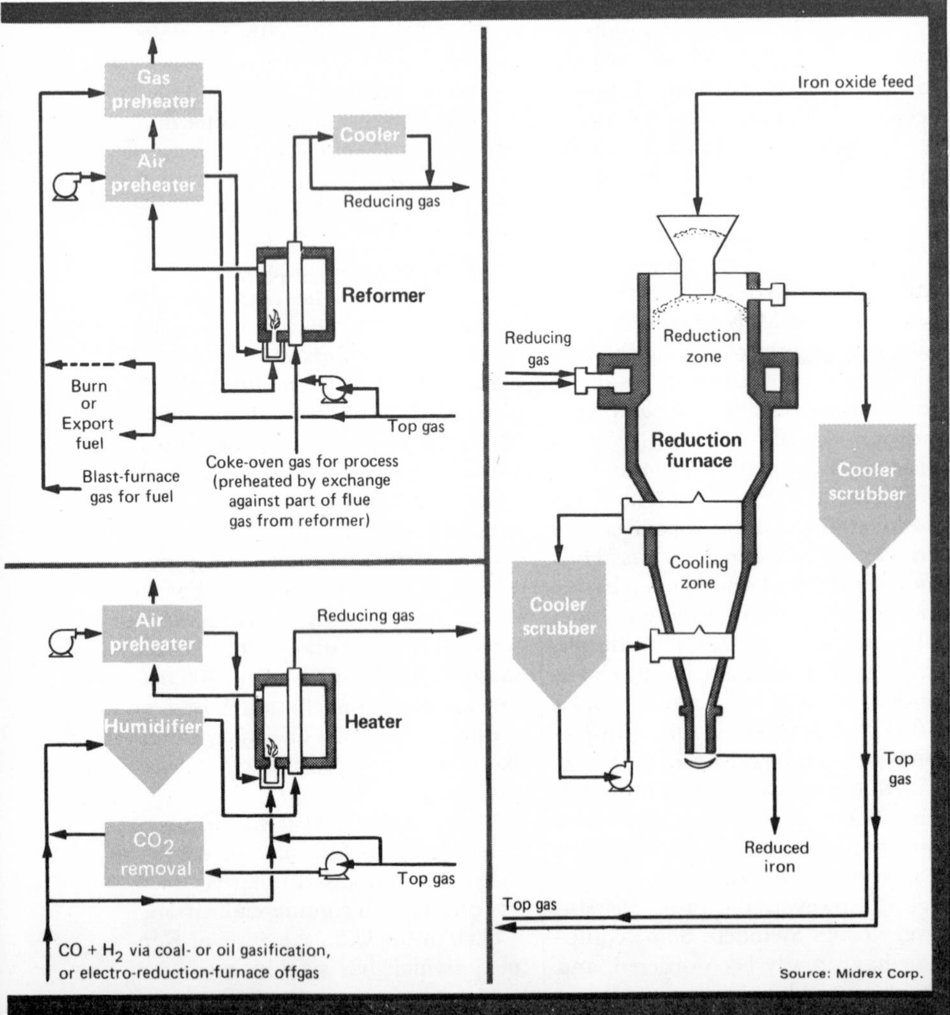

Switching to an alternative gas alters a direct-reduction flowsheet

Systems Div. of Allis-Chalmers (Milwaukee, Wis.). Its Accar process can operate entirely with natural gas, fuel oil or coal, or a combination of these. The company started up an approximately-52,000-metric-tons-of-iron/yr plant at Niagara Falls, Ont., in January 1973. Over the last few years, coal has been receiving more emphasis in that unit. And, in fact, the installation now uses a combination of natural gas and coal. (A larger, about-225,000-metric-ton/yr, facility was put onstream last May at Sudbury, Ont., for Sudbury Metals Co., a fifty-fifty joint venture of Allis-Chalmers and National Steel Co. of Canada. That unit is temporarily shut down. Timing of restart-up will largely depend on the price of scrap iron, which is a competitor to the plant's sponge-iron product. The facility has not used coal.)

Allis-Chalmers notes that it is negotiating three projects outside of North America. Each would employ coal for 80% or more of reductant/fuel requirements. The company hopes that the first order, for a 100,000-metric-ton/yr unit, will be signed this month. Then, the plant likely could be ready for late-1978 or early-1979 startup. The second deal, for a 150,000–200,000-metric-ton/yr installation, might come by midyear, while a pact for a 250,000-metric-ton/yr facility might follow by the end of 1977, says the firm. It adds that it is in the early stages of negotiation for a second North American joint venture.

Meanwhile, another solid-reductant process has reached commercialization in Italy. Kinglor-Metor SpA (Buttrio, Italy) placed its pioneer full-size plant in service last March for Ferriere Arvedi e Cia. SpA (Cremona, Italy). The unit boasts two furnaces of 20,000-metric-tons/yr capacity apiece. It was shut down after running for five months; its feed-hopper system is being redesigned to improve efficiency. The plant probably will go back into operation in early April, states Kinglor-Metor. Cremona uses bituminous coal, but reactive coke, sub-bituminous coal, anthracite and lignite also could be fed in.

Ferriere Arvedi is planning to double capacity of its installation, but no date has yet been set for the project. Kinglor-Metor says that ten other companies are "very" interested in the process. It particularly suits smaller capacity units, from 20,000–100,000 metric tons/yr, notes the firm, which expects to sell about a plant per year from now to 1985.

A coal-based process from West Germany's Fried. Krupp GmbH (Essen) already is running at a couple of plants. And the company is aiming for other sales.

NUCLEAR FALLING-OUT—A few years ago, the idea of using a high-temperature gas-cooled nuclear reactor to provide heat for reforming and reduction was voiced with high hopes. Now, however, the enthusiasm seems to have waned.

The American Iron and Steel Institute (Washington, D.C.) had set up a nuclear steelmaking taskforce, which studied the feasibility of nuclear heat for natural-gas-based and coal-gasification-based direct reduction. The taskforce now is dormant. And West Germany's Arbeitsgemeinschaft Nukleare Prozesswärme has moved away from the concept of an integrated reactor/steelmaking complex, because even a large steel plant would be much too small to utilize all of the heat produced by a nuclear reactor.

Only Japan is still actively pursuing the idea, but on a limited budget. Its Agency for Industrial Science and Technology hopes to have a test high-temperature heat exchanger and a reformer finished by April 1978.

Industry experts generally don't see the possibility of nuclear-linked direct reduction at least through the year 2000.

Mark D. Rosenzweig

Sponge-iron process combines flexibility, low costs

Newly commercial, this solid-reductant direct-reduction process can accommodate a wide variety of iron ores or pellets, reductants and fuels while turning out product of high quality. Intended for small-producer markets, the system features low capital and operating costs.

***Franco Colautti** and **Alessio Barbi**, Kinglor-Metor & C., S.p.A.*

☐ Sponge iron is a highly concentrated form of metal that can be fed directly to electric steelmaking furnaces. The sponge is produced by a process known as direct reduction, in which iron oxide ore is metallized by removal of oxygen by a reductant at temperatures below the melting point of the ore. Some observers feel that direct reduction will challenge—and perhaps replace—conventional coke-oven and blast-furnace systems, particularly in small facilities producing less than 1,000 tons/d of iron. Although the present low price of scrap metal—the traditional feedstock for electric steelmaking furnaces—doesn't encourage installation of direct-reduction facilities right now, the volatile nature of scrap prices could turn the situation around at any time.

Kinglor-Metor's solid-reductant process, born out of the need for mini steel-plants, is geared for production of 20,000 – 200,000 metric tons/yr of sponge iron. The first commercial installation is located in Cremona, Italy, at the works of Ferriere Arvedi & C. S.p.A. Consisting of two direct-reduction modules that produce 20,000 metric tons/yr each, the unit started up in September 1976, shut down for examination and raw-material-feed changes in January 1977, and restarted this month.

The process produces sponge iron with an adjustable metallization of 88 – 95%, and the Cremona unit's modularity makes production schedules very flexible. Advantages of the unit are low installation and operating costs. Installation costs run about $120/ton of output; production costs are dependent on costs of raw materials, but a range of about $85 – 93 per metric ton of sponge is projected. Manpower requirements are minimal, and energy needs are approximately 16 million Btu/metric ton.

FLEXIBLE—The process is versatile in that it can use a wide range of raw materials and fuels.

Iron-supplying raw materials may include natural or artificial iron-ore pellets or mill-scale briquettes. The recommended feedstock is hematite or limonite pellets ranging from 1/4 to 1 in. in size.

Although the Cremona facility uses bituminous coal as a reductant, suitable alternatives include reactive coke, and subbituminous and brown coal. The optimum grain size is again 1/4 to 1 in., and the reductant used should not have an ash-softening point below 1,200°C (2,192°F).

Fuels suitable for use with the process include—but are not limited to—natural gas, liquefied petroleum gas, producer gas, light fuel oil and substitute natural gas.

Fluxing materials, such as limestone, may be used, but are not needed if the reductant is low in sulfur.

The relative flexibility gained by use of a wide range of raw materials and fuels frees the process from many limitations of supply as well from instability of price.

STRAIGHTFORWARD FLOWSCHEME—The Kinglor-Metor reduction unit consists of a vertical-shaft furnace heated by one of the fuel alternatives listed above. The furnace surrounds six vertical silicon-carbide reactors that reduce the iron ore or pellets in the presence of a solid reductant at about 1,050°C (1,922°F).

In the process (see flowsheet), screened iron ore or pellets are mixed with a solid reductant composed of virgin and recirculated material and, if necessary, a fluxing agent. This mixture feeds into the top of the furnace, where it is preheated.

The preheated charge drops, at a controlled rate, into the reduction portion of the furnace, made up of the six reactors. Here the reductant combines with the oxygen of the iron oxide to produce a granular, spongelike metallic mass having a ferric content of 88 – 95%.

Following reduction, the sponge is cooled in a water-jacketed section of the unit and discharged into a screening operation, where product material and fines are separated. Both streams undergo a subsequent magnetic segregation: The fines are separated into nonmagnetics and sponge-iron fines, while the larger particles are segregated into final-product sponge iron and unreacted reductant, with the latter recirculated back to the mixture-preparation step.

RUGGED CONSTRUCTION—Each module, having a capacity of 20,000 metric tons/yr of sponge iron, is approximately 82 ft high, 40 ft long, and 20 ft wide. A metal structure supports the entire furnace and the extraction system; the reactors are self-

Originally published April 25, 1977

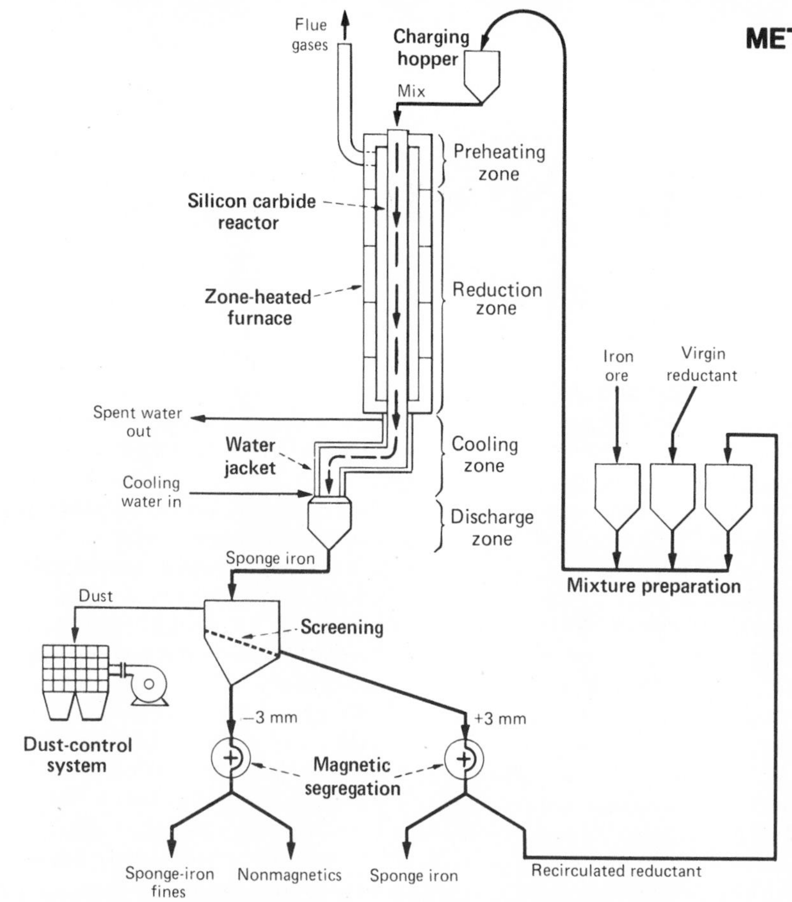

Kinglor-Metor units' capacities can be expanded modularly

supporting, transmitting their weight to the metal structure.

There are no particularly sensitive portions of the system, even though the silicon carbide reactors are exposed to temperatures over 1,100°C (2,012°F) on the heat-source side. In fact, examinations during the recent shutdown at Cremona showed that the reactors had not experienced any wear during the initial startup run.

The process has few moving parts, and most of its components are industrially proven and readily available in the marketplace.

Thus, the process is composed of simple components well within the reach of small and medium-sized steel manufacturers, assuring considerable reliability and, above all, safety of operation.

From an environmental standpoint, there is almost no consumption of water, hence no water pollution. Water consumption for cooling is only about 6.6 gal/min. A small amount of dust is generated during magnetic separation, but this is easily controlled.

PRODUCTIVITY—Up to now, the Cremona facility has been using ore pellets with an iron content of about 65%, a medium value. If an ore of higher Fe value were used, with reducibility being equal, the output would go up.

ECONOMICS—The direct-reduction process is an extremely simple modular installation costing approximately $120/metric ton of sponge-iron output. Subsequent modules would each run about $114/metric ton of output. Battery limits for the modules include: facilities for raw-material and product dressing and screening; the reduction units; sponge-cooling and dust-collection systems; control room and boards; and compressors and blowers. The limits exclude raw-material discharge and storage facilities, auxiliary buildings, and items, such as fencing, parking areas, or railway sidings.

The cost of sponge iron produced depends on the price of suitable ores, reductants and fuels, which may vary considerably for different locations, even within one country. Also relevant is the size of the operation—a larger plant purchasing larger quantities of materials and fuels will probably be able to obtain them at lower prices than a smaller unit; a larger output-volume will also result in a lower pro-rata charge for fixed overhead, although the effect on labor cost would be small.

The accompanying table gives a projection of the approximate production cost per ton of sponge iron for a unit turning out 18,000–20,000 metric tons/yr, with a minimum metallization of 88–90%. The plant runs 330 d/yr, 3 shifts/d, 7 d/wk. All prices are as of February 1977.

Sponge-iron production cost

Item	Quantity per metric ton of sponge iron	Unit price, in Italy ($ U.S.)	Cost per metric ton of sponge iron ($ U.S.)
Iron ore (~65% Fe)	1.350 tons	33.84/ton	45.68
Coal (bituminous)	0.364 ton	53.32/ton	19.39
Fuel (natural gas)	7,065 std ft^3	1.58/1,000 std ft^3	11.16
Electric power	68 kWh	0.035/kWh	2.38
Labor	1.2 man-h*	5.65/man-h	6.78*
Spare parts	$1.20	—	1.20
			86.59

Note: This estimate does not take into account depreciation, overhead and other general expenses.

*If the plant consists of more than one module, the labor required per ton of output decreases.

The authors

Franco Colautti is general manager of Kinglor-Metor & C., S.p.A. (Cas. Postale 113, 33042 Buttrio, Udine, Italy). After graduation from the Technical and Industrial Institute of Udine, he worked for several steel firms, in the areas of sponge-iron making and sponge-iron use in electric steelmaking furnaces.

Alessio Barbi is assistant general manager of Kinglor-Metor. He attended Parma University, graduating in industrial chemistry, and was formerly an assistant professor at Padova University.

Metals From Mn Nodules

A number of chemical processes can extract such metals as manganese, copper, nickel and cobalt from ocean manganese nodules.

NICHOLAS R. IAMMARTINO
Associate Editor

In their thrust to exploit manganese nodules commercially by the turn of the decade, potential ocean miners have now hurdled at least one key obstacle—finding a way to win metal values from the nodules.

Deepsea Ventures, Inc., Kennecott Copper Corp., and the Duisburger Kupferhütte AG - Bayer AG team have come out publicly with their process favorites. Other firms seem as far along, but are less willing to identify their choices, including Germany's AMR combine,* a four-member French consortium† and International Nickel Co. Most prefer hydrometallurgical techniques, with at least one continuing to study pyrometallurgical methods as well. Physical separations like flotation fail for nodules—the raw material is too fine-grained and metal values too intimately bound.

Having the process technology in-hand is, of course, a long stride in itself. In addition, more attention can now be given to such other aspects of the budding nodule business as international regulations and environmental studies, and prototype mining methods.

Hydrochlorination Approach—The most talked about process, a hydrochlorination method (see flowsheet), belongs to Deepsea Ventures.

After crushing and drying, the highly oxidized nodules (manganese dioxide, MnO_2, and oxides of the associated metals) react with hydrogen chloride (HCl) in a multihearth furnace. The reaction disrupts the MnO_2 crystal lattice to release metals as water-soluble chlorides, points out director of research Paul H. Cardwell. Goethite (iron hydrogen oxide) converts to ferric oxide, freeing cobalt values and forestalling HCl consumption by the iron.

As a secondary result, Cardwell cites production of chlorine from HCl, induced by the oxidizing capability of MnO_2. Byproduct HCl, e.g., from the chlorine industry, can thus serve as a source of chlorine, parallel with winning metals from nodules.

Water next leaches out 96+% of the soluble metal chlorides, yielding a liquor of manganese, nickel, copper and cobalt chlorides, with trace amounts of cadmium, molybdenum, vanadium and zinc. A solid residue of silicates, sulfates and metal oxides (mainly iron) remains behind for disposal as clean landfill.

A proprietary liquid-ion-exchange step separates the leach liquor into four streams. Aqueous solutions of nickel, copper and cobalt chlorides pass to electrolytic cells for metal-winning. After purification and crystallization, manganese chloride is converted to the metallic state by an undetailed, patent-pending process.

The overall system provides closed-loop recycles of leach solutions and HCl. Cardwell points out that recent modifications not only improve economics but also allow recovery of "three more metals." All the process operations have been proved in Deepsea Ventures' 1½-ton/d pilot plant. The firm is "approaching the point of engineering" a demonstration plant, but offers no predictions on size or timing.

Differential Extraction—Kennecott has similarly piloted its metals-extraction route, in a ½-1 ton/d unit

*Arbeitsgemeinschaft Meerestechnisch-Gewinnbare Rohstoffe, consisting of Metallgesellschaft AG, Preussag AG, Rheinische Braunkohlenwerke AG and Salzgitter AG.

†Consisting of Centre National pour l'Exploitation des Océans, Société le Nickel, Commissariat à l'Energie Atomique, and Bureau de Recherches Géologiques et Minières.

PILOT PLANT has proved Kennecott's nodule process.

Metals Separation With Lewatit TP 207 Resin

	Stream Concentration		
	Feed	Filtrate	Elute
Manganese, g/l	73	66-70	5-10
Nickel, mg/l	3,200	<3	24-27,000
Copper, mg/l	2,100	∿2	15-18,000
Cobalt, mg/l	1,000	<5	7-9,000
Iron, mg/l	800	Up to 300	Up to 4,000
Zinc, mg/l	500	<50	Up to 3,000
Aluminum, mg/l	50	<5	300

Source: Bayer AG

Originally published November 25, 1974

(see photo). It plans to decide by next year whether scaleup to a commercial-size plant (on the order of 3 million tons/d) is already possible, or should be ensured by first building and operating a 10-100-ton/d semicommercial plant. Additional data would smooth startup of the maiden facility, says Marne A. Dubs, director of Kennecott's ocean-mining efforts, and would also put product on the market for evaluation.

The firm's hydrometallurgical process takes a slurry of nodules direct to reduction, without any drying. Kennecott won't tell whether HCl, sulfur dioxide or another reducing material does the job, but claims the patent-pending operation has "never been used before."

An ammoniacal solution differentially extracts only copper and nickel from the MnO_2 molecule. Not extracting manganese and other metals is the hard part, explains Dubs. About 97% of the feed nodules end up as a stable MnO_2/clay tailings, which Dubs says can be dumped at sea without harming the environment, or stored on land to build up an artificial mineral deposit (of manganese, zinc, rare earths, etc.) for exploiting later on.

Liquid ion exchange separates copper and nickel streams from the extraction liquor, for final processing by electrolysis or other methods to high-purity metal or oxide products.

Virtually all chemicals and water recycle through the system, bringing both economic and ecologic pluses. The absence of nodule pretreatment keeps energy costs to a minimum.

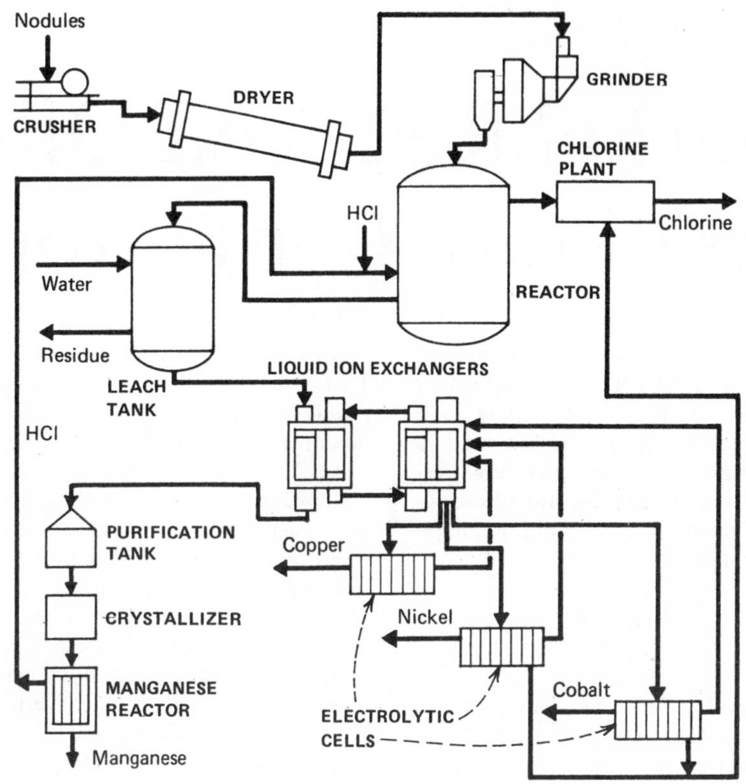

Solid Ion Exchange—The Duisburger/Bayer process contender turns out concentrated streams of a half dozen or so metals (see table).

After leaching ground nodules with sulfuric acid and gaseous sulfur dioxide to form metal sulfates, and filtering off insoluble residuals, the technique passes the stream over a bed of solid ion-exchange resin—Bayer's specially developed Lewatit TP 207. Metal ions other than manganese form complexes with, and adsorb on, the macroporous resin. In the original process, manganese values passed through the bed and out as a 98% stream. But recovery of pure manganese even from this proved too expensive, so Duisburger together with Germany's Krupp has now found a way to purge manganese prior to ion exchange, and thus slash that operation's cost.

A selective elution of the exhausted resin yields two offstreams—one a concentrated 60-g/copper solution for metal recovery by conventional methods (such as electrolytic deposition), and the other consisting of zinc, cobalt, nickel, iron, aluminum and residual manganese. The latter three metals in the combined flow are precipitated by the addition of calcium hydroxide (to 3.3 pH) and dosing with chlorine at 60-80°C. Filtrate treatment includes, sequentially, extraction of zinc with tributyl phosphate, extraction of cobalt with triisononylamine, and finally precipitation of nickel.

Other Efforts—Germany's AMR group hasn't yet selected a process, but has identified several leach, roast-leach and pure smelting (pyrometallurgical) candidates that it hopes to pick from within 12-18 months. International Nickel and the French consortium have groomed techniques of their own, but decline for now to reveal details.

However, the French group still has in the running two "classical" leach processes, one with ammonia, the other with sulfuric acid. The ammonia route calls for a larger capital investment, more operating personnel, and higher energy input. But it boasts a high nickel and copper extraction yield, is said to work with all types of nodules, and recycles almost all ammonia used. The process sequence breaks down nicely into a concentration stage (yielding an easily shipped material about 5% of original nodule volume) and an extraction stage— opening the door to build the first stage near the nodule field and the second elsewhere.

The sulfuric acid alternative doesn't recover as much copper and nickel, but does offer the chance to produce manganese if the market appears favorable. Acid consumption could be high, depending on nodule alkalinity, and overall process sensitivity to nodule composition is greater. The H_2SO_4 process generates large amounts of waste sulfates for disposal, furthermore, and seems less economical than the ammonia route. #

Nickel Reduced Directly In Rotary Converter

Canada's newest nickel refinery employs the first top-blown rotary converter to reduce sulfide feed to nickel. The converted feed is further refined via pressure carbonylation. Special nickel carbonyl decomposers yield a nickel pellet product.

JOHN C. DAVIS
Associate Editor

Peculiar feed and product characteristics prompted International Nickel Co. to employ the little-used pressure carbonyl process for its new 125-million-lb/yr nickel refinery, which started up last summer.

The Copper Cliff, Ont., plant handles an assortment of mineral concentrates and smelter or refinery intermediates, and directly produces a pellet product. Pellets are desirable in making alloys for the aircraft and electronics industries.

The new technology covers two major aspects: First, the process has a top-blown rotary converter for the pyrometallurgical desulfurization of the feed, with simultaneous direct reduction of nickel sulfide to crude nickel. Second, the converter product is reacted with carbon monoxide at about 70 atm in a carbonylation reactor to selectively evolve nickel- and iron-carbonyl vapors, which are drawn off and condensed. The reactor residue contains copper, cobalt and precious metals, which are sent to other Inco plants for refinement.

In addition to these two innovations, Inco has developed careful control procedures, and electrically heated carbonyl decomposers downstream of the carbonyl reactor and distillation column. The controlled decomposition of nickel carbonyl vapor on heated pellets permits the growth of the pelletized product. Two other products are made—a powdered nickel, and a powdered mix of iron and nickel, called ferronickel.

Converter—For years, nickel producers have been trying to develop direct nickel reduction in a large-scale converter, but to little avail. Nickel oxide usually forms a suffocating slag that kills the reaction. Inco has solved this problem by employing a top-blown rotary converter of the type already proven in the steel industry, which provides several advantages over the more conventional horizontal side-blown unit.

Reactants can be mixed batchwise in a turbulent bath to prevent localized hot spots and ensure even distribution of oxygen. This in turn enables operation at higher temperatures. High temperatures are needed as the sulfur content decreases, or else nickel oxide will accumulate, cutting oxygen efficiency such that slag formation is only hastened. Furthermore, NiO can react with refractory materials in the converter, and it increases the viscosity of the bath.

Temperature increases steadily while the converter charge consumes more and more sulfur. Starting with a sulfur content of roughly 20% and a temperature of about 1,380°C, the sulfur decreases to about 4% while temperature increases to at least 1,600°C. Enough sulfur is left with the charge to properly catalyze the subsequent carbonylation step.

Two 50-ton converters are in use at Copper Cliff. Each rotates at 40 rpm when charged. An entire day's output from the refinery (275 tons) can be handled by a single converter.

The molten product is tapped into a ladle, transported to an induction holding furnace, and charged to a water-quenched granulator. There,

OXYGEN is fed through lances to top of bath. Converter rotates steadily.

Originally published January 21, 1974

feed particles of proper size and activation are prepared for the carbonylation. The granules are dried prior to this step.

Carbonylation—The reaction of carbon monoxide and nickel to selectively remove nickel from the crude feed via vaporization was discovered before the turn of the century and first practiced by Inco at its Clydach, Wales, refinery soon thereafter. The Wales operation, however, runs at lower pressure. The higher pressure at Copper Cliff improves reaction kinetics and stabilizes the carbonyl to permit the higher temperature and fast reaction rate.

Because of the higher pressure, the iron in the feed also undergoes carbonylation, but it is separated from the nickel carbonyl in subsequent distillation. Cobalt, another major constituent of the feed, drops out as residue from the reactor, along with copper and precious metals.

The carbonyl process is very selective to nickel and iron, and is ideally suited for the Inco operation because of the wide variety of feed materials. These include native mineral concentrates, mattes, metallic intermediates, semirefined products, refinery residues, dusts and scrap.

The three carbonylation reactors, 12 ft dia. and 67 ft long, operate about 42 h on each charge and rotate continuously. Because of their gigantic mass—550 tons when fully charged—rotational force is applied by rollers in contact with massive wheels at both ends of the horizontal reactors.

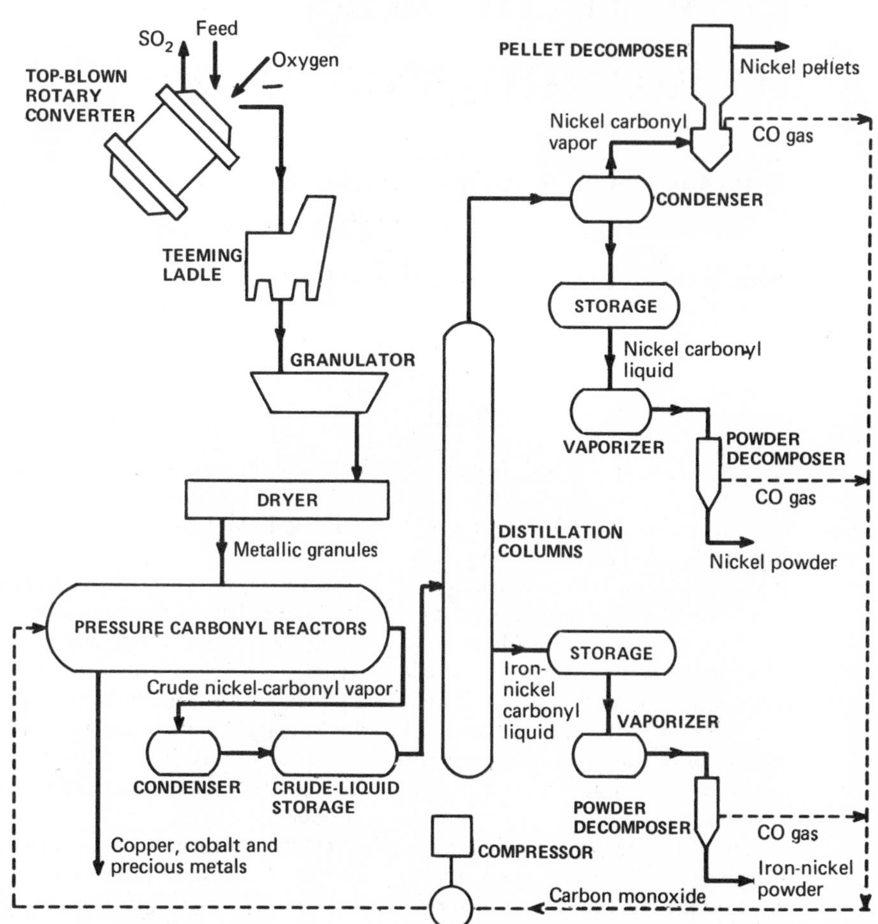

WHEN FINISHED, bath is 1,600 + °C.

Distillation, Decomposition—As the reaction proceeds, carbon monoxide recycled from downstream decomposition steps is fed in, and crude nickel carbonyl vapor is drawn off and sent to a condenser. The condensed crude is added to the middle of a distillation column, from which iron-nickel carbonyl liquid is obtained as bottoms, and nearly pure nickel carbonyl vapor is recovered as the overhead.

The overhead is then partially condensed. Liquid is revaporized and decomposed by conventional means to produce nickel powder, one of the three main products. The uncondensed portion goes to the special pellet decomposers. Meanwhile, the iron-nickel carbonyl liquid is also revaporized and decomposed to form a powder product.

The most up-to-date control equipment is provided, including a computer for data logging. The distillation column is controlled to produce a ratio of 30/70 nickel/iron carbonyl product in the distillation bottoms, and has the capability of adjusting this ratio to 50/50.

Extensive refrigeration equipment has been installed to cool the brine used in the carbonyl condensers. The process area also includes CO generation equipment capable of producing 7,000 ft³/h at a purity of 99.9%.

The decomposer area houses eighteen units for the decomposition of carbonyl into nickel pellets, eight units for the decomposition of carbonyl into nickel powder, and two to produce the iron-nickel powder.

The powder decomposers thermally shock the vaporized condensation products by rapidly heating them to about 200°C. The powder drops out, and CO gas comes off to be filtered, compressed and recycled.

The gaseous feed to the pellet decomposers is first diluted with CO and then put through a more gentle heating. Electric heat is used for careful temperature control. The gas contacts a slowly descending stream of preheated nickel pellets, which are continuously recycled within the decomposer until they reach the desired size (about ¼ in.).

All these products then go to an automated packaging plant, where they are put in 500-lb drums.

New technology tries to tap tungsten trove

A developmental technique may succeed in adding significantly to current U.S. tungsten reserves, which are not now meeting domestic demand.

☐ A new process for tungsten recovery, now under development by the U.S. Bureau of Mines' Metallurgy Research Center (Salt Lake City), may have a lasting effect on the U.S. supply/demand picture for that metal. If the technique becomes commercial, it will allow the extraction of tungsten from a previously untapped but bountiful source—the brines of Searles Lake, Calif. According to the Bureau, the lake contains an estimated 135 million lb of tungsten, compared with a total of 275 million lb for all known U.S. reserves.

Key to the new process is a metal-selective ion-exchange resin formulated by the Bureau. Dubbed HERF, it is polymerized hydroxyquinoline-ethylenediamine-resorcinol-formaldehyde. The resin's main asset is that it can work with low-grade feed—a brine containing as little as 70 ppm, or 0.07 g/L, of WO_3—and recover 95 to 100% of the tungsten. This is obtained as an iron-tungsten concentrate containing 45 to 55% WO_3, which can be sold to refiners.

Until now, no available technique has succeeded in economically recovering the metal from Searles Lake brine, even though those waters are already being commercially exploited for valuable compounds. Kerr-McGee Corp. owns lakeside plants for soda ash and various borax and potash products.

CUTTING IMPORTS—Recovery from Searles Lake would certainly reduce current U.S. reliance on tungsten imports. The nation's consumption of ore and concentrates, says the Bureau, totaled about 17.25 million lb last year, up from 16.1 million lb in 1976. Shipments from domestic mines reached only about 7 million lb; an equal amount had to be imported, while the rest (3.25 million lb) came mostly from stockpiles.

The main U.S. producers are Union Carbide Corp., which obtains ammonium paratungstate from scheelite ore mined in California and Nevada, and Amax Inc., whose tungsten output is a byproduct of its molybdenum mines located near Empire, Colo. About 68% of U.S. consumption is used for tungsten-carbide tools; 15% goes for alloy steels.

NO OVERNIGHT STAR—It has taken the Bureau a long time to develop the new method. Says project leader Paulette Altringer, "We started working on it seven years ago, and tested many commercial ion-exchange resins with no success. Eventually, we came up with a promising solvent-extraction system that we patented, but it did not turn out to be economically feasible. The new ion-exchange process is much more promising."

Last month, the Bureau started testing the method in a pilot plant, which consists of three 6-ft, 3-in.-dia. columns, each packed with HERF resin. In normal operation, two of the columns are hooked up in series for tungsten adsorption, while the other undergoes elution.

At startup, only two columns are used. When the first in the series is

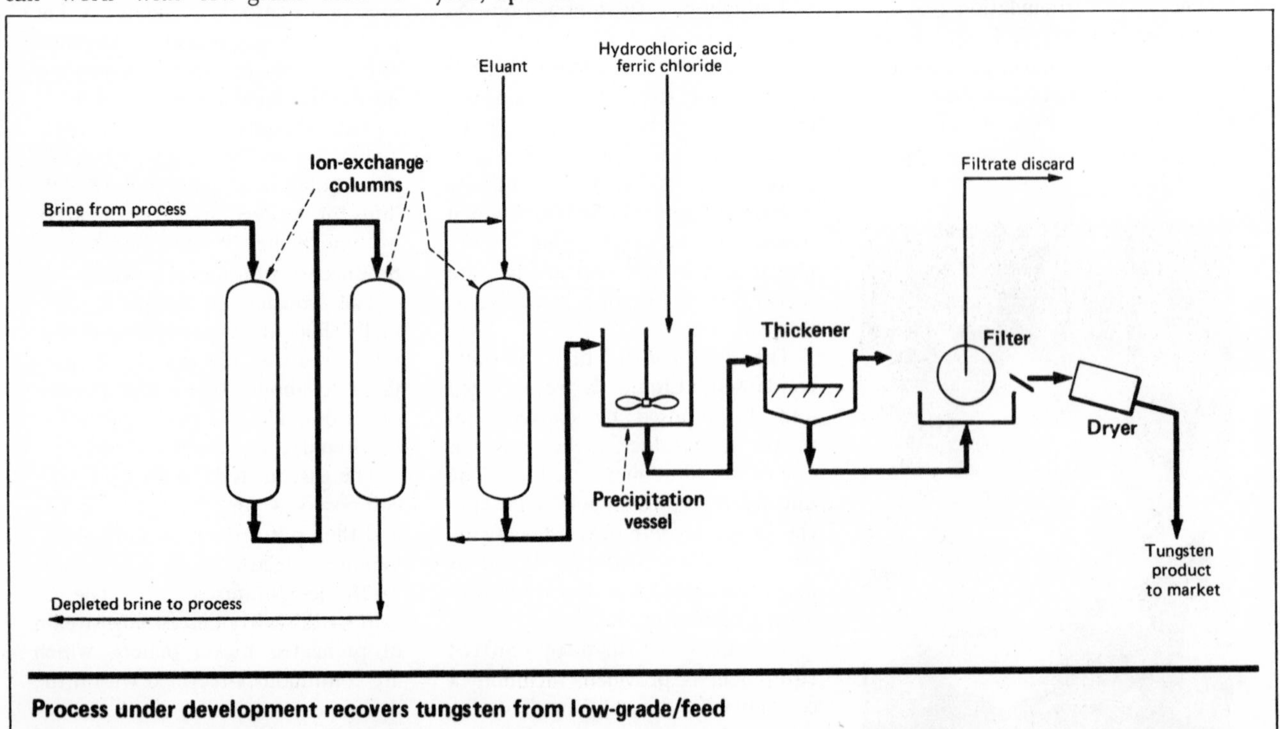

Process under development recovers tungsten from low-grade/feed

Originally published February 27, 1978

half loaded with tungsten, it is disconnected and eluted. The second column then becomes the first in the series, and the third column is hooked up behind it. Once the second column is loaded, it is disconnected for elution, and the first column is linked up behind the third.

ELUTION AND RECOVERY—Elution is carried out by a weak solution of soda ash, which does not seem to harm the HERF resin. In fact, resin degradation is claimed to be less than 10% after 1,200 cycles. "We are still studying the mechanics of both tungsten adsorption and elution," says Altringer, "they aren't completely understood at this time."

On a bench scale, the Bureau has operated the process at a feedrate of 5 gal/min per square foot of ion-exchange-column area; this is a rate typical of many industrial ion-exchange operations. The Bureau has been using granular resin in its laboratory work, but it is now working with beads, which it claims are considerably more efficient.

Tungsten is recovered from the soda ash eluate by coprecipitation with ferric iron. A ferric chloride solution is added in a ratio of approximately 1:2 parts WO_3. Subsequent filtration and drying yield an iron-tungsten concentrate suitable for sale to refiners.

At present, the filtration step is causing some trouble. "The iron/tungsten particles are so fine that they cling and clog the filter," says Altringer. "A commercial process would probably use a rotary filter." As an alternative, the Bureau is working on a solvent-extraction technique that would recover tungsten from the eluate in a purer form. But this work is still in the research stage.

Despite the obstacle, the Bureau remains optimistic. "At 5 gal/min and 60 bed-volumes, we recover 95 to 100% of the tungsten," says Altringer. "We don't yet know whether it would be more economical to extract that much metal from 60 bed-volumes or recover instead 70% from 100 bed-volumes. This is because we haven't completed our cost evaluation."

If and when the process reaches the commercial stage, a plant could be installed as part of Kerr-McGee's complex at Searles Lake. A company spokesman says Kerr-McGee is following the Bureau's work with interest, but there are no plans for commercialization at present, because the technique is still under development.

Kerr-McGee has developed and patented two methods for tungsten recovery from Searles Lake brine, but neither is economically feasible. One is a solvent-extraction route, in which brine is sulfidized to convert the tungsten into a thiotungstate compound. The metal is then extracted by quaternary amines. The other process uses oxine to precipitate tungsten trioxide from a concentrate.

Gerald Parkinson,
Los Angeles

Section VI
ORGANIC CHEMICALS

Combined ammonia urea process trims operating, capital costs

A single-train route that produces urea directly from natural gas eliminates some of the energy-consuming stages of conventional facilities, while replacing entire processing units with less-expensive equipment.

Vincenzo Laganà, Snamprogetti SpA

☐ By integrating ammonia and urea processes into a single train, Snamprogetti (Milan)—the chemical engineering subsidiary of the Italian state oil group, ENI—has trimmed 16.8% from production costs and 15% from plant investment. In the integrated unit, all the ammonia produced in the synthesis loop is turned into urea, resulting in a net saving of $8.93 per metric ton of prilled product, and a $16-million reduction in investment cost when urea is the end-product and natural gas is the feedstock.

Whereas conventional urea production requires two separate plants (one to produce liquid ammonia, yielding carbon dioxide as a byproduct, and one to produce urea), Snamprogetti's process eliminates expensive, energy-consuming stages. And it uses specialized equipment to replace entire processing units—for example, the process employs a falling-film absorber in place of a conventional carbon-dioxide-removal unit.

Snamprogetti has had an 80-metric-ton/d pilot plant in operation at the ANIC (a chemical producer and sister subsidiary in the ENI group) facility at Gela, in southern Italy, since 1973 and is currently negotiating the sales of a 250-metric-ton/d unit.

STREAMLINED ROUTE—In the ideal integration scheme, ammonia reacts directly with carbon dioxide to form carbamate, and then, urea. Thus, four stages common to conventional ammonia trains may be omitted: (1) decarbonation of converted gas; (2) CO_2 compression; (3) refrigeration for separating the liquid ammonia from unreacted gases leaving the synthesis reactor; and (4) liquid-ammonia storage facilities and auxiliary equipment.

The initial four process steps (see flowsheet)—feedstock desulfurization, primary reforming, secondary reforming, and high- and low-temperature CO-shift conversion—in the integrated route are common to those for conventional ammonia-production plants. The primary change comes with the elimination of CO_2-removal facilities. Because the requirement for heating a CO_2-removal reboiler is thus obviated, a high-heat-content converted gas is not needed, allowing the user to lower the steam/carbon ratio at the reforming-furnace inlet to 3:3.3 from 3.5:4.

Converted gas gets compressed to about 200 kg/cm^2 (2,845 lb/in^2) and conveyed into the bottom of a urea stripping tower. Here the NH_3 content and residual CO_2 from the urea solution are eliminated, the stripped gas proceeding to a CO_2 absorber.

Flowing from bottom to top through tubes, the gas reacts countercurrently with a lean water/ammonia/carbonate solution to form carbamate. Concentrated carbamate solution exiting the bottom of the CO_2 absorber is injected, via a venturi nozzle, into the urea reactor. Once the ammonia-rich decarbonated gas leaves the CO_2 absorber, it is cooled in a falling-film ammonia absorber, where part of the ammonia is condensed and recycled to the CO_2 absorber. Ammonia content is reduced to about 3%, by volume.

Decarbonated gas then goes to a methanator—where CO and traces of CO_2 are reduced to approximately 10 ppm—and then to the ammonia synthesis loop.

AMMONIA LOOP—In the synthesis loop, ammonia is absorbed down to 0.5%, volumetrically, by means of a countercurrent water stream in a falling-film absorber. This yields an 80-wt% aqueous-ammonia solution that goes to the urea reactor in the urea synthesis loop. (This NH_3 solution acts as the driving fluid through the venturi that feeds carbamate to the urea reactor.)

The gas leaving the ammonia absorber, along with the makeup syngas (from the syngas circulator following the methanator), goes to the dehydration unit, where it is quenched and scrubbed with liquid ammonia. Next, the gas is conveyed to the syngas circulator, which delivers it to the ammonia converter, a radial type with intermediate quenching. About 70% of the reaction heat is recovered.

The ammonia absorber is installed downstream from the ammonia converter, following a conventional heat-exchange train. The syngas—containing about 20% ammonia—from the converter enters the bottom of the absorber, whence exiting gas recycles to the dehydrator; meanwhile, as said previously, the aqueous-ammonia solution enters the urea reactor.

UREA SYNTHESIS—Both the aqueous ammonia and carbamate solutions are injected into the bottom of the urea reactor via the venturi nozzle. The NH_3/CO_2 and H_2O/CO_2 mole ratios at the reactor inlet are 5 and 1.1, respectively, while CO_2 conversion in the unit is approximately 70%.

When the product urea solution leaves the reactor, it enters the high-

Originally published January 2, 1978

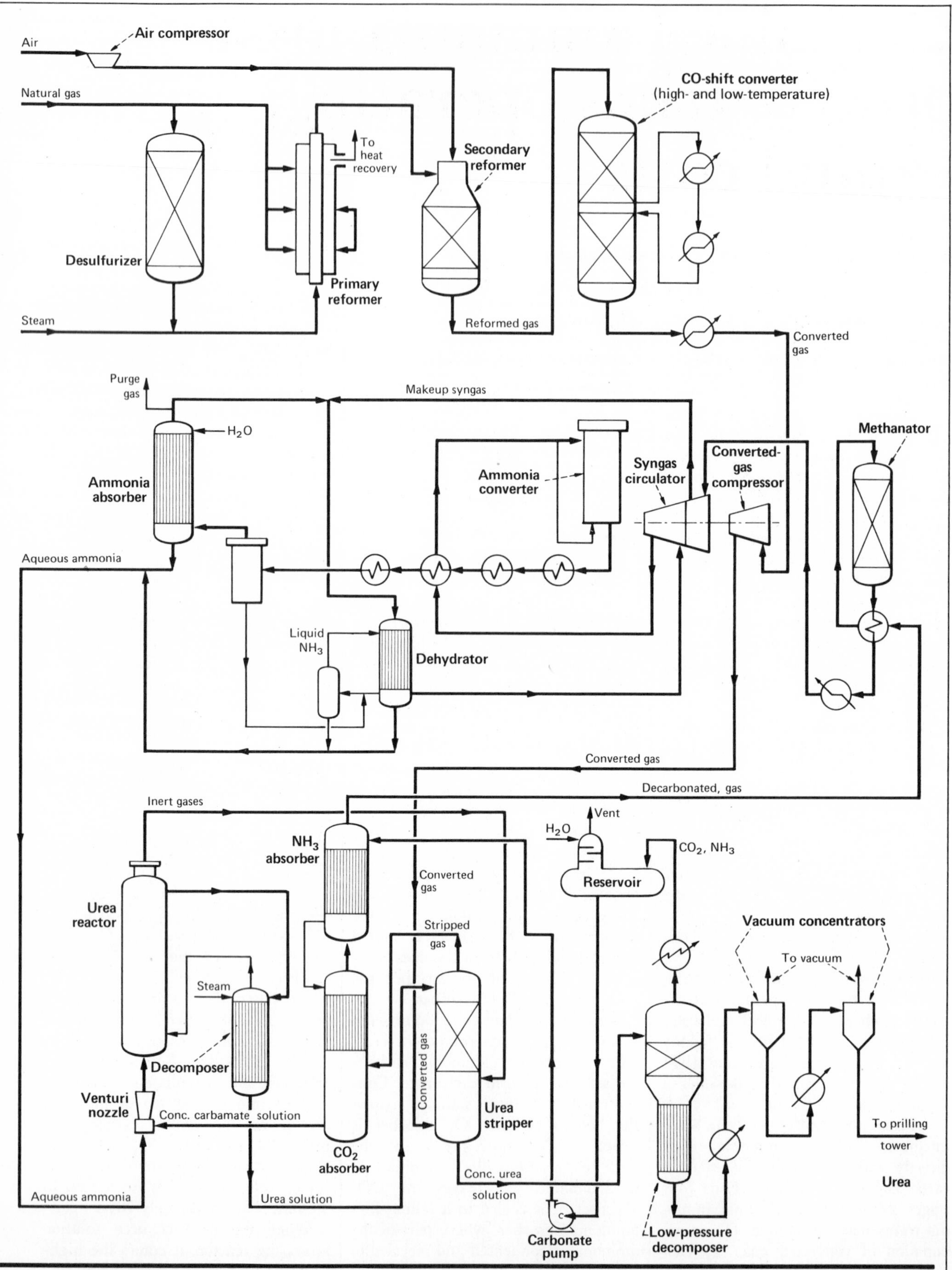

Snamprogetti's integrated ammonia/urea process reduces consumption and costs.

pressure carbamate decomposer, where condensing steam heats the solution to 210°C. In this manner, part of the carbamate and most of the contained ammonia are stripped off and recycled to the urea reactor.

From the carbamate decomposer bottom, the urea solution enters the urea stripping tower, where the stripping agent is converted gas. This purifies the urea solution of ammonia and residual CO_2, producing a concentrated urea stream, about 65% by weight. This stream goes directly to a low-pressure (approximately 4.5 kg/cm² or 64 lb/in²) recovery stage where the final traces of residual carbamate present in the solution are decomposed, with the resulting gases condensed and recycled as in conventional systems.

Once out of the low-pressure section, the urea solution enters two vacuum-concentration stages, of 0.3 kg/cm² (4.3 lb/in²) and 0.03 kg/cm² (0.43 lb/in²), respectively. At this point, the urea is ready for prilling.

WATER POLLUTION ELIMINATED—Snamprogetti has eliminated the wastewater pollution problem that often results from ammonia/urea plants by recycling, as steam, treated process condensates—such as those from the urea vacuum-concentration operations—to the primary reformer.

MACHINERY NEEDS—In the integrated process, high-pressure feedpumps conventionally found in the urea section are eliminated because the aqueous ammonia solution obtained in the ammonia synthesis loop is available at a high-enough pressure to enter the urea reactor directly.

The syngas compressor for small plants (having a capacity under 350 metric tons/d of urea) is of the reciprocating type. In such plants, a multiservice machine would be the best choice. A single frame would provide for: (1) natural-gas compression from network pressure up to 40 kg/cm² (570 lb/in²); (2) process-air compression from atmospheric pressure up to 36 kg/cm² (512 lb/in²); (3) syngas compression from 30 kg/cm² up to 260 kg/cm² (3,700 lb/in²); and (4) syngas recirculation from 255 kg/cm² (3,630 lb/in²) to 270 kg/cm² (3,843 lb/in²).

Integrated plants having a capacity over 350 metric tons/d of urea can use a centrifugal compressor that has syngas compression and circulation only, while air and methane employ individual machines.

SELF-SUFFICIENT PLANT—The integrated plant will be equipped with a steam-powered electric generator having enough capacity to make the facility entirely power-independent.

Four steam levels are envisaged for the plant. In one possible arrangement, boilers will produce steam at 100 kg/cm² (1,422 lb/in²) at 510°C, which will be conveyed to the turbines of the syngas and process air compressors. These turbines will be of the extraction and condensation type. Steam extracted at 27 kg/cm² (384 lb/in²) is sent in part to the urea plant and in part to drive the air compressor and the turbogenerator. Part of the steam is extracted from the turbogenerator saturated at 5.5 kg/cm² (78 lb/in²) and sent to the urea plant also. Finally, part of the process condensates recovered will be evaporated to produce steam also at 100 kg/cm², expanded in a second turbine driving the air compressor, and extracted from the turbine at a pressure of 40 kg/cm² (570 lb/in²) and 385°C to meet process needs, such as reforming.

However, individual steam-distribution systems will be designed to meet the client's needs.

OTHER HYDROCARBON FEEDS—In cases where the hydrocarbon feed to the reforming section has a carbon/hydrogen ratio higher than in natural gas—virgin naphtha, for example—the excess CO_2 present in the converted gas has to be removed to obtain the correct C/H proportion to balance the NH_3/CO_2 in the urea production.

Based on this concept, it is possible to design integrated ammonia/urea plants that use different feedstocks and that combine ammonia production with urea production. However, the scope of the integrated process decreases because as the pure ammonia requirement increases, the route tends to approach conventionality. The economic limit for the integrated route is to produce no more than 50% pure ammonia, with the remainder of the NH_3 transformed into urea.

COMBINED PRODUCTION—In cases where some ammonia end-product is needed, the integrated process must be modified. A conventional decarbonation unit must be installed after the CO-shift conversion section to remove the excess CO_2 not required for urea production. Following CO_2-removal, the gas stream is compressed, and it follows the steps of the integrated process, as described previously, until it reaches the ammonia absorber.

The ammonia solution obtained in the absorber is split into two streams: (1) one to the urea reactor, where the proper quantity of carbamate is formed to produce the desired amount of urea; and (2) the other to a distillation tower, from which is obtained the quantity of pure ammonia desired.

ECONOMICS—Table shows consumption and costs of raw materials and utilities for two cases: (A) a 1,000-ton/d ammonia/urea integrated process, and (B) ammonia (600 tons/d) and urea (1,000 tons/d) conventional processes, where the total amount of ammonia is converted into urea. Both cases refer to a plant having an auxiliary steam-generating facility and a turbine-driven generator to make the plants fully self-supporting so far as electrical power is concerned. Capital costs for both cases are also shown.

Consumption and production and capital costs

Consumption	Case A	Case B
Process natural gas (kcal)	3.44×10^6	3.40×10^6
Fuel natural gas (kcal)	1.56×10^6	2.55×10^6
Total natural gas (kcal)	5.00×10^6	5.95×10^6
Catalyst (5-yr life) ($U.S.)	0.45	0.45
Chemicals ($U.S.)	–	0.07
Electric power ($U.S.)	–	–
Steam ($U.S.)	–	–
Demineralized water (m³)	0.44	1.40
Cooling water (ΔT = 10°C), (m³)	210	265
Production costs ($U.S.)		
Natural gas	40.00	47.60
Catalyst	0.45	0.45
Chemicals	–	0.07
Demineralized water	0.17	0.55
Cooling water	3.36	4.24
Total raw materials and utilities costs	43.98	52.91
Capital costs ($U.S.)	90×10^6	106×10^6

Basis: 1 metric ton of prilled urea
Source: Snamprogetti SpA

Raw-materials and utilities prices were taken as follows:

- Process fuel and natural gas: $0.008/1,000 kcal ($2/1,000 ft³).
- Cooling water: $0.016/m³ ($0.06/1,000 gal).
- Demineralized water: $0.39/m³ ($1.48/1,000 gal).

The author

Vincenzo Laganà is engineer-in-charge of the special process department of Snamprogetti SpA (Casella Postale 4169, 20097 San Donato Milanese, Milan, Italy). He holds a degree in chemical engineering from Politecnico di Milano.

New processes will feed on mixed-butenes glut

Novel technology stands poised to handle byproduct supplies of mixed butenes, increasing the possibility of exploiting individual components of such mixtures. One question: Will the resulting products find a place in the market?

☐ Ethylene production from such heavy feeds as naphtha and gas oil will increase as U.S. natural-gas supplies dwindle. This in turn will yield plentiful supplies of byproduct mixed butenes, whose main traditional outlet has been in alkylation for high-octane gasoline. Now, chemical processors are searching for ways to make the most of the expected production, especially because the new supplies will significantly increase the already appreciable amounts of butenes coming from catalytic crackers.

New technology may be the answer. Processors now have available several separation methods and some isomerization techniques. These can open the door to further downstream development, yielding products suitable for chemical-feedstock use or for deriving a higher-octane alkylate.

ISOMERIZATION ROUTES—The main rationale for isomerization is to make it easier to fractionate isobutene from the mixed butenes. Of secondary importance is a possible need to increase the butene-2 yield.

Phillips Petroleum Co. (Bartlesville, Okla.) and Petro-Tex Corp. (Houston)—a subsidiary of Tenneco Inc.—recently opted for butene isomerization. A few months ago, Phillips started up a unit of its own design at its Borger, Tex., refinery. And Petro-Tex has announced plans to install a unit based on a process by Institut Français du Pétrole (IFP), Rueil Malmaison, France. (The IFP route works in the liquid phase, while the Phillips process operates in the vapor phase.) Another firm, UOP Inc. (Des Plaines, Ill.), offers a version of the vapor-phase route; the firm does not yet have a commercial installation to its credit.

When used for isobutene recovery, such processes replace cold sulfuric acid extraction—a traditional method that is falling into disfavor because of high energy consumption and adverse environmental effects.

This process substitution has, in fact, already started. Petro-Tex, which is one of four cold-acid plant operators in the U.S., plans to cut back on its cold-acid unit when the isomerization plant goes onstream, even though it will maintain some capacity to meet demands for isobutene dimer. And Neches Butane, a joint venture of Texas U.S.A. and B. F. Goodrich at Port Neches, Tex., says it will soon decide whether to replace its acid unit with a new isomerization plant.

PHYSICAL SEPARATION—At least two physical removal processes for butene-1 are being offered: a molecular-sieve technique developed by the Linde Div. of Union Carbide Corp. (New York City), and an adsorptive method by UOP. The second one is specific to butene-1, adsorbing it while rejecting isobutene and butene-2; the Linde process captures both normal-butene isomers.

Processors would have to follow up the Linde technique with a fractionator to recover pure butene-1. In any case, the method is also suitable for separating pure isobutene.

These processes are more energy-intensive than isomerization, but the end-product (butene-1) is a very valuable normal butene. The isomer shows potential as the monomer of polybutene-1, an emerging polyolefin. Shell Chemical Co., which recently bought from Witco Chemical the only polybutene-1 plant in the U.S., has announced a doubling of capacity in that Taft, La., facility.

Low-density polyethylene might also create more need for butene-1 as a comonomer in Ziegler-type process technology, which is becoming more popular because of its energy-saving advantages over converted free-radical chemistry.

OTHER OUTLETS—The available processes won't be of much use unless new butenes markets develop sufficiently to absorb the expected supplies. Besides the above-mentioned possibilities for butene-1, other probable applications include methyl *tert*-butyl ether (MTBE)—an octane boosting additive made in Europe by Italy's Anic and West Germany's Chemische Werke Hüls. One of MTBE's feedstocks is isobutene that may contain some amounts of normal butene and butane impurities.

Another possible consumer of isobutene is a new process to make methyl methacrylate. The original technique was developed by Japan's Asahi Glass Co.; as yet, there is no commercial plant in operation. However, Rohm and Haas Inc. (Philadelphia) has plans to use its own version of the process in a 300-million-lb/yr unit plant to be built in Texas by the early 1980s.

Of course, there are several traditional markets for high-purity butenes—e.g., butyl elastomers, polyisobutene, specialty chemicals—plus others for isobutene mixed with normal butenes (oil additives, sealants).

Promising as these applications sound, they still account for only 13% of U.S. mixed-butenes consumption, according to SRI International (formerly Stanford Research Institute), Palo Alto, Calif. Gasoline absorbs the rest.

However, it is now thought that gasoline might also profit from the new butenes-separation technology. HF alkylation, which accounts for one third of all U.S. alkylation capacity and is growing in popularity, has been shown to yield a higher-octane alkylate if butene components are first isomerized to butene-2.

At this point, there seems to be no clear answer as to who will be purifying the butenes streams, and what products they will be recovering. Says Vincent J. Guercio, of Chemical Technomics (Montclair, N.J.), a petrochemicals consulting firm: "Nobody goes after each isomer in particular. In

Originally published February 13, 1978

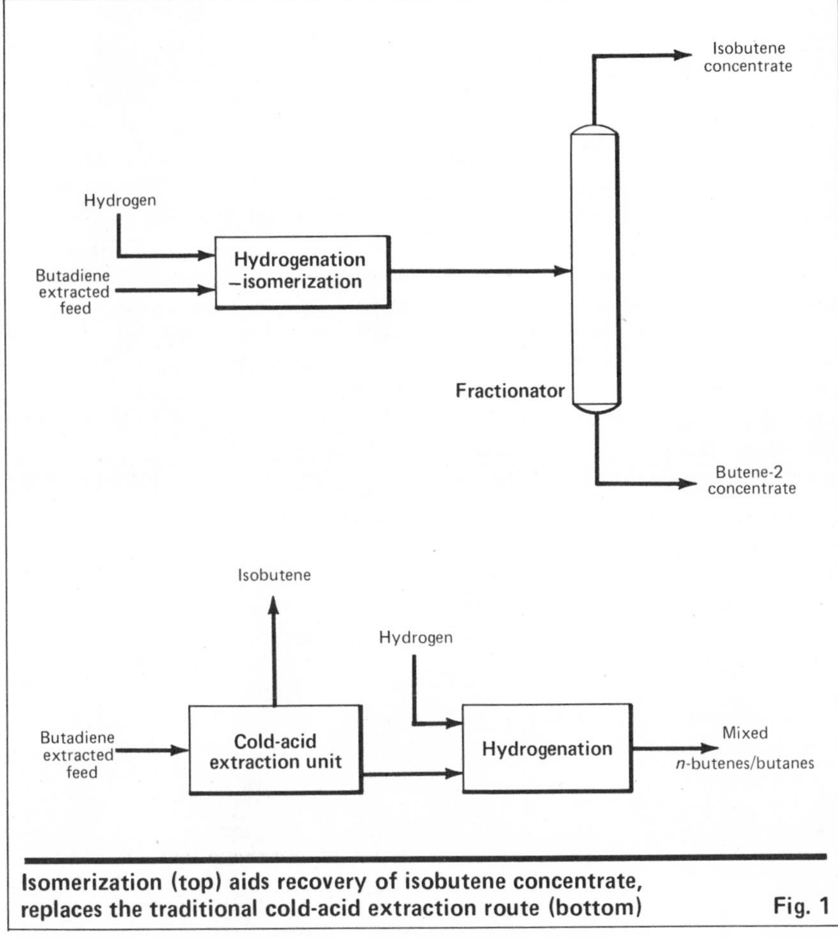

Isomerization (top) aids recovery of isobutene concentrate, replaces the traditional cold-acid extraction route (bottom) Fig. 1

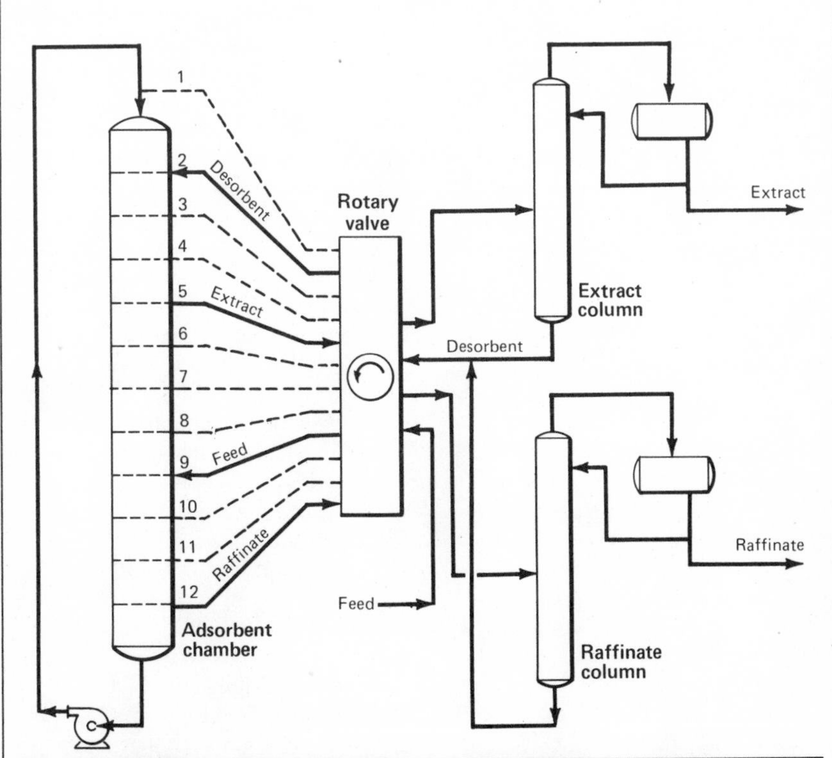

The Sorbutene process is a variation of Sorbex (above), in which a fixed bed is fed by a rotating array of desorbent, extract, feed and raffinate Fig. 2

the butenes business, almost every producer has a different objective. I've diagrammed practically every C_4 user in the world, and no two are alike."

As an example, Guercio cites a firm that once considered butene isomerization just to get more butene-2 needed for a downstream expansion. Most potential users of this technique consider it for removing butene-1 so that isobutene, which boils at nearly the same temperature, can be fractionated from the mixed stream in fairly pure form.

COMPARING METHODS—Isomerization units always follow butadiene extraction steps, so that the feed contains as few dienes and acetylenes as possible. Whether they are liquid or vapor phase, the available processes depend on variations of palladium catalysts, requiring some hydrogen in the feed for best performance. The hydrogen acts to saturate any remaining dienes or acetylenes that might otherwise polymerize over active sites. It also inhibits isobutene oligomerization, and activates and promotes isomerization. (Hydrogen concentration in the feed is less than 1%, so it does not appreciably saturate butenes.)

Processes differ mostly in the way the flowplan is laid out. In making their choice, manufacturers must determine whether the product-isobutene stream must be pure, or if some residual amounts of butene-1 can be tolerated in it.

UOP offers a two-stage version of its basic isomerization route, in which a second isomerization reactor is inserted in the overhead loop of the fractionator following the first reactor. This boosts isobutene purity into the low-90% range.

IFP claims special advantages for its liquid-phase process, the most important of which is a better yield resulting from use of lower temperatures and improved thermodynamics. The firm adds that the process requires less catalyst inventory because the concentration of reactants is higher, and that no bed cooling is necessary.

It is unlikely that Petro-Tex and Phillips Petroleum are installing isomerization capacity simply to obtain more isobutene. Petro-Tex, for one, already gets a substantial amount of isobutene for resale from Oxirane Chemical Co., which produces it as a byproduct of propylene oxide manufacture. Oxirane's process yields large amounts of *tert*-butyl alcohol, which is dehydrated to isobutene. The supply has reportedly been enough to satisfy most of the U.S. need for isobutene over the last few years.

ON PHYSICAL SEPARATION—Pure isobutene can also be obtained from an isobutane-free stream by screening out normal butenes in Linde's Olefinsiv molecular-sieve technique. It works by trapping paraffin C_4's in sieve pores, while rejecting the remaining isobutene fraction. The feedstream must first be lightly hydrogenated to eliminate dienes and acetylenes. These systems use catalysts that closely resemble those employed in isomerization; some shifting of butene-1 to butene-2 takes place.

So far, Linde has had little, if any, commercial experience in separating C_4 olefins. Olefinsiv is an adaptation of the firm's Isosiv technology for separating normal and iso paraffins. And Isosiv is used primarily on streams in the kerosene range.

One feature of Olefinsiv: Since both butene-1 and butene-2 are trapped in the sieve, and since the pretreatment catalyst acts mildly enough to prevent heavy conversion of butene-1 to the butene-2 isomer, processors can add a fractionator to separate the two normal isomers, thereby splitting the butane-free feedstream into all three of its components.

UOP has adopted Sorbutene, a variation of its Sorbex continuous-adsorption process, for the extraction of butene-1 from other C_4 paraffins, isobutene and olefins. The technique has been tested on a pilot scale.

Sorbutene, too, requires pretreatment for removal of dienes and acetylenes. Its major advantage is that it selectively removes butene-1, allowing the rest of the butenes stream to pass through the adsorption bed. Extraction efficiency, according to UOP, is about 92%.

John C. Davis

Caprolactam Without Any Ammonium Sulfate Byproduct

By combining a new process with one developed earlier to make precursor cyclohexanone oxime, caprolactam can be made from cyclohexanone without making unwanted ammonium sulfate.

A. H. DE ROOIJ, DSM
H. A. W. DE VRIES, DSM
A. M. A. HEUNKS, Stamicarbon, BV.

With ammonium sulfate demand in doubt in many areas of the world, output of the material as a byproduct of caprolactam manufacturing may pose troublesome marketing problems. Now, however, by teaming up two Dutch processes, this byproduct can be completely eliminated.

In most conventional routes, about 4.5 tons of the sulfate are made for each ton of caprolactam. About 60% of this is produced in the manufacture of the key precursor, cyclohexanone oxime, and the rest in the conversion of oxime to caprolactam. A method called the hydroxylamine-phosphate-oxime (HPO) process, which avoids sulfate output while making the oxime, has been developed by the Central Laboratory of the Netherlands' DSM and the company's engineering affiliate, Stamicarbon BV. It has already been commercially proven.

Now, DSM and Stamicarbon have gone after the remaining 40% of sulfate output and piloted a method of converting the oxime to caprolactam without any sulfate byproduct. By putting the two processes together, Stamicarbon can offer a totally sulfate-free route to caprolactam.

Like most flowschemes, the new conversion process changes oxime into caprolactam in an oleum medium via the Beckmann rearrangement. In contrast to usual techniques in which subsequent neutralization of the reaction mixture yields ammonium sulfate and caprolactam in separate phases of a two-phase liquid, the new route employs only about half the amount of neutralizing ammonia water to form ammonium bisulfate instead; this bisulfate remains in the same phase as the lactam. Hence, the method is called the bisulfate lactam process.

Further Differences—Production of bisulfate requires some changes in downstream processing. For example, instead of using benzene to extract the caprolactam from the neutralized liquid, the new route employs a commonly available, chlorinated hydrocarbon as the extraction agent. The preferred solvent is an aliphatic hydrocarbon containing one or two carbon atoms.

Also affected are operating and capital costs. Comparing this combined process to one that joins the HPO route with conventional oxime conversion, raw-material requirements (excepting ammonia) are seen to be about 0.3% higher because of the hydrolysis of caprolactam that results from the much lower pH of the neutralization mixture.

Extra utilities are needed as well, mainly for handling and recycling the chlorinated solvent. An additional 1,100 lb of low-pressure steam, 4,000 gal of cooling water (at a temperature gradient of 7°C), and 9,090 kcal of refrigeration are needed per ton of product. Power requirements may also be a few percent higher. However, process-water demand is lowered, by 1,200 lb/ton, and the burden of the sulfate byproduct is, of course, eliminated. Consumables for the combined processes are summarized in the table.

Investment figures are roughly 5% greater than the norm, largely because of the increased size of certain equipment in the oleum production circuit. Furthermore, some additional equipment has to be used for lactam production.

Flowscheme—Oleum and normal-quality cyclohexanone oxime are first fed into an agitated, atmospheric-pressure reactor, where rearrangement to caprolactam takes place. The vessel has special internal geometry to ensure good mixing. The reaction is exothermic, and a heat exchanger is used to maintain the temperature at about 125°C.

The proper choice of oxime-to-oleum mixing ratio is of major importance for both reaction efficiency and product quality. Temperature

Caprolactam-Production Consumables — Combined HPO and Bisulfate Lactam Routes

Materials	Consumption, per short ton of caprolactam
Cyclohexane, lb	2,126
Hydrogen, lb	186
Ammonia, liquid (100%), lb	642
Ammonia, aqueous, lb	500
Oleum, lb	2,730
Caustic soda, lb	202
Chemicals, dollars	3.70
Catalysts, dollars*	11
Process water, lb	5,140
Utilities	
Steam, lb	27,600
Electricity, kWh	430
Boiler feedwater, lb	5,180
Cooling water, gal	387,000
Fuel, kcal	174,000
Refrigeration duty, kcal	44,500

*Estimate depends on noble-metal price.

Originally published March 18, 1974

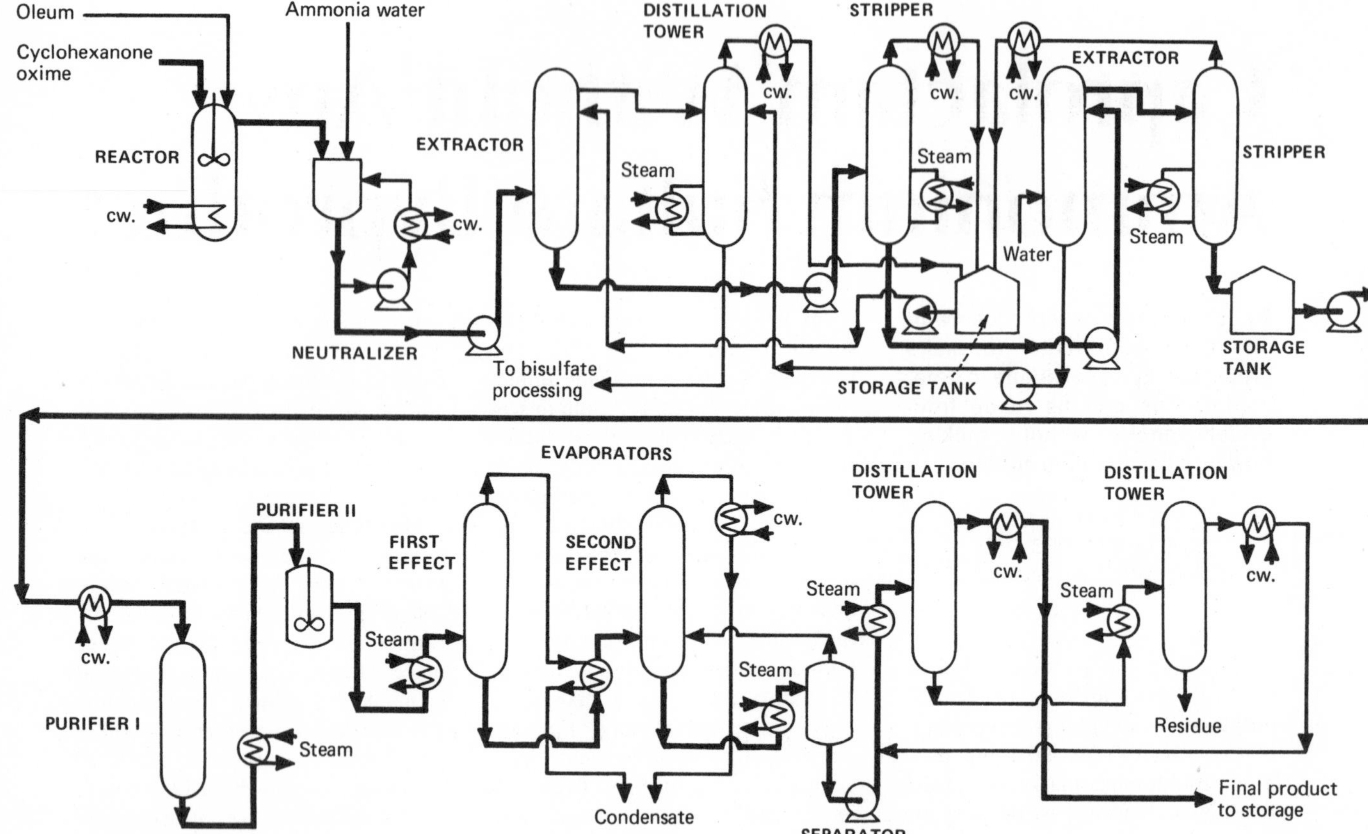

must also be carefully controlled. Under the right conditions, conversion efficiency is at least 98.5%.

The stream leaving the reactor is a rather viscous liquid. It is passed to another atmospheric-pressure vessel, where a 20% aqueous-ammonia solution is fed in equimolar amounts to the ammonium bisulfate that forms as a result. Temperature is held between 30 and 50°C by a cooler.

The caprolactam/bisulfate solution is then sent to an extractor, where it is countercurrently contacted with the chlorinated hydrocarbon. The caprolactam drops out with the heavier-than-water organic phase at a 99% extraction efficiency. This stream is then pumped to an atmospheric-pressure stripper, where the caprolactam level is brought up to about 60% by taking some of the solvent overhead. The overhead stream is condensed and piped to a storage tank for recycle.

The enriched caprolactam solution is pumped to another countercurrent extractor, where it contacts demineralized water. The product goes into the aqueous phase—extraction efficiency is again better than 99%—and leaves the vessel at a concentration of about 30% in the water, along with trace amounts of the solvent.

The residual solvent is taken off as tops in an atmospheric stripper, condensed and passed to the storage tank. The aqueous caprolactam bottoms solution is also sent to storage, prior to being delivered to a conventional caprolactam purification system. The resultant caprolactam product is a high-purity melt with a solidification point of 69°C. The residue and a major part of the condensate produced in this circuit are recycled in the process; the rest of the condensate is discarded.

Utilizing Bisulfate—Meanwhile, the ammonia bisulfate stream from the first extractor, and the impure hydrocarbon phase from the second, are fed to a distillation tower. (This is the only vessel requiring an exotic material; all the others are made of conventional stainless steels.)

From this distillation column, pure solvent goes overhead and is sent to storage for recycle. Bottoms consist of a roughly 50% aqueous ammonium-bisulfate solution plus impurities such as sulfonated products formed in the reactor. The water from this solution is evaporated and the remainder is incinerated to provide a dilute sulfur dioxide gas, which is converted to trioxide over vanadium pentoxide catalyst. The sulfur trioxide is then absorbed in sulfuric acid to make the oleum used in the Beckmann rearrangement.

The bisulfate-to-oleum system is not part of the Dutch knowhow, but it will be provided by other licensors that have expertise in this field, working with the same contractors that Stamicarbon does.

Meet the Authors

A.H. de Rooij is a graduate of Delft Technical University and is head of the department for pioneering research of chemical products at the Central Laboratory of DSM, Geleen, The Netherlands.

H.A.W. de Vries, Ph. D., is a graduate of the Aachen Technical University and is a process engineer in the department for semi-technical development of inorganic products at DSM's Central Laboratory.

A.M.A. Heunks is a professional engineer, a graduate of the Amsterdam Engineering College, and now manager of the caprolactam process engineering group of Stamicarbon.

Caprolactam From Toluene—Without Ammonium Sulfate

A brand-new separation step, meshed into an established process for making caprolactam from toluene, forms no ammonium sulfate byproduct and eliminates troublesome pollutants from discharge waters.

ANDREW HEATH
McGraw-Hill World News
Milan, Italy

Two major difficulties have for some time plagued the conventional synthesis of caprolactam: too much ammonium sulfate byproduct, and discharge during the caprolactam separation step of a water stream carrying several pollutants.

In recent years, a third problem has cropped up for most conventional processes. The cost of benzene, which is commonly needed for making cyclohexane intermediate, has multiplied severalfold, and also availability is low.

A process commercialized a decade ago by Snia Viscosa,* Italy's largest manufacturer of nylon 6, needs no benzene, instead uses toluene as precursor. Lower cost and better availability of toluene (due to increasing output as a byproduct of ethylene cracking and of catalytic reforming) is making the advantages of the Snia synthesis more and more evident. Just last year, an 80,000-metric-ton/yr plant went onstream in Manfredonia, Italy, for Società Chimica Dauna, a 50-50 venture between Snia and ANIC, the chemical arm of Italy's state oil company ENI.

Now, Snia reveals an attractive new wrinkle for the separation step of its caprolactam route. The technique completely eliminates ammonium sulfate formation. Instead of discharging an aqueous waste with impurities generated during the reaction step, it destroys them.

The essential feature: Caprolactam is separated from the sulfuric acid in which it is dissolved during reaction by extraction with an alkylphenol solvent. No ammonia is used, and thus no ammonium sulfate forms. The remaining sulfuric acid is thermally cracked, destroying impurities while forming sulfur dioxide for recycle.

The improvement can be applied to other caprolactam processes as well as its own, says Snia. The firm is already offering the technology for sale and usage in large plants; its 16,000-metric-ton/yr plant at Torviscosa, Italy, will be converted to the new separation scheme shortly.

Forming Caprolactam—There are three main reaction steps in Snia's process: oxidation of toluene to benzoic acid, hydrogenation of benzoic acid to cyclohexane carboxylic acid (CHCA), and nitrosation of CHCA to caprolactam.

Toluene oxidation takes place with air in the presence of a cobalt catalyst. Reactor pressure is 10 atm and temperature is 150-170°C. Most of the reaction heat is recovered as steam.

Unreacted toluene is stripped from the liquid reactor-effluent and recycled. Crude benzoic acid is then purified of reaction side-products in an atmospheric-pressure rectifier. Purity is over 99%, total yield 91%. The heavy fraction from the rectifier can be treated before incineration to recover cobalt catalyst. (This benzoic acid technology can be licensed separately.)

The benzoic acid is then hydrogenated in the presence of a palladium catalyst on an activated-carbon support, to yield CHCA. The reaction occurs at 10 atm and 170°C, in a series of four reactors. A single reactor would be notably larger than the sum of the four, Snia explains. Conversion is 99.8%. Distillation recovers traces of catalyst from the crude CHCA.

Meanwhile, the nitrosylsulfuric acid needed for nitrosation is turned out in a side unit. The reaction of ammonia with air over a platinum catalyst forms nitrogen oxide, which is then absorbed in oleum. The resulting material contains 73% nitrosylsulfuric acid ($NOHSO_4$), 2-3% free SO_3, and the rest sulfuric acid (this excess of sulfuric acid maintains a stable solution).

After blending with oleum, the CHCA is pumped into a multistage reactor. Nitrosylsulfuric acid enters separately into each stage. The nitrosation reaction, in which nitrogen adds to the CHCA ring, takes place at atmospheric pressure and 80°C.

Process Requirements

Basis: Per metric ton of caprolactam

	Conventional process	Modified process
Materials		
Toluene, tons	1.11	1.10
Ammonia, tons	1.31	0.26
Sulfur, tons	1.14	0.05
Hydrogen, normal m³	800	800
Caustic soda, kg	190	40.0
Misc. chemicals, $	11.50	11.50
Catalyst, $	10.80	10.80
Utilities		
Steam, tons	10.6	2.5
Electricity, kWh	1,020	1,020
Cooling water, m³	1,000	1,000
Fuel, million kcal	1.9	16

*Snia Viscosa S.P.A., Via Montebello, 18, Milan, Italy.

Originally published July 22, 1974

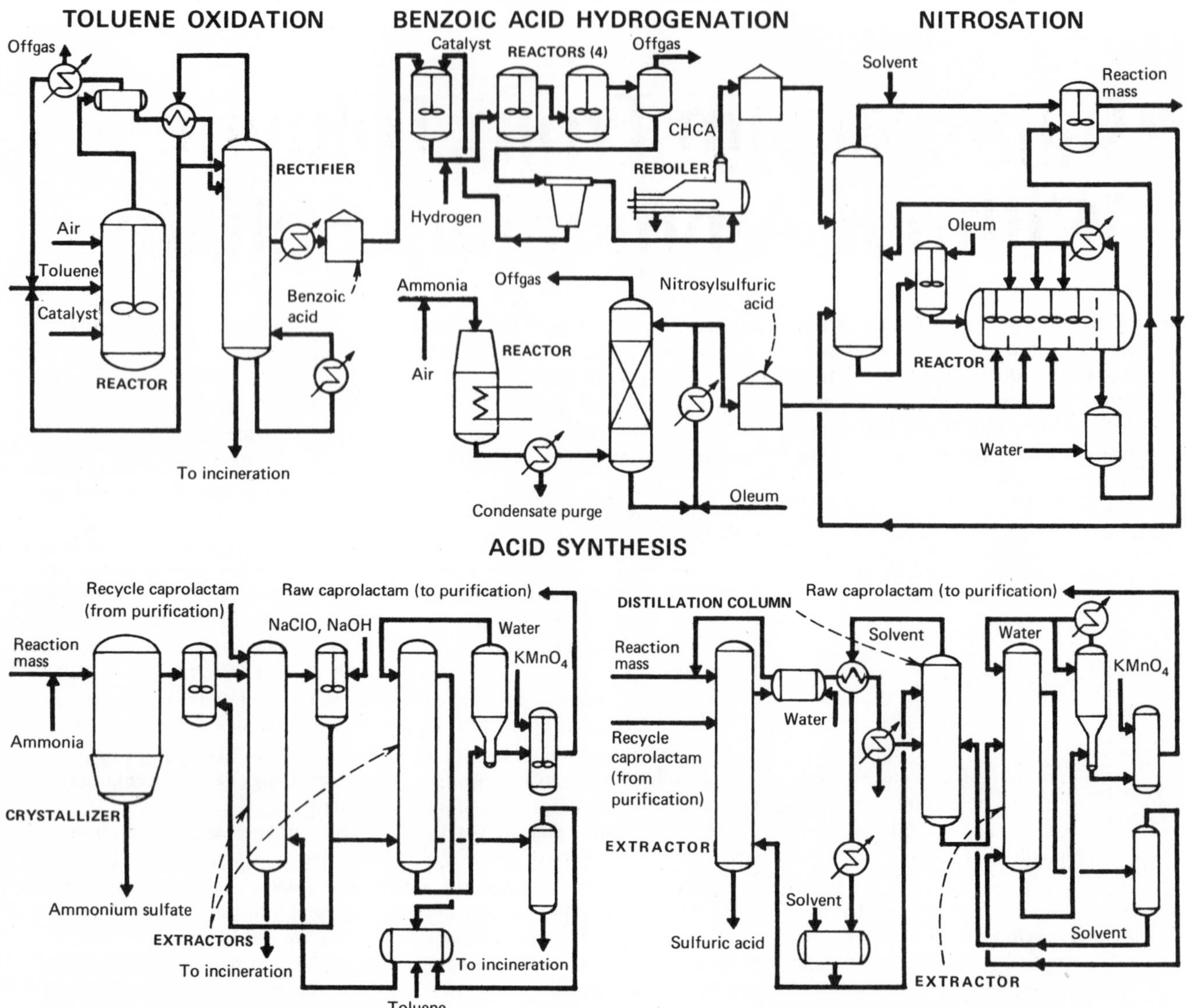

Caprolactam forms as a composite (not just a mixture) with sulfuric acid.

The reactor effluent is diluted with water, and nonconverted CHCA extracted by cyclohexane and recycled.

Old vs. New—The caprolactam/sulfuric acid composite is then ready for separation.

In the basic process, ammonia neutralizes the composite to produce crystals of ammonium sulfate (at a 4.2:1 weight ratio to caprolactam product). Caprolactam oil floating on top of the mother liquor is decanted, and extracted first by toluene and then by water. The water raffinate draws off the impurities formed during nitrosation and is discarded. The crude caprolactam remaining is dehydrated and distilled in a series of thin-film evaporators and rectifiers (not shown in flowsheet), operating under a moderately high vacuum. The final liquid or flake product is fiber-grade.

In the modified product-separation scheme, the acid composite is slightly diluted with water and fed to an extraction unit. An alkylphenol solvent extracts caprolactam from the sulfuric acid, leaving the latter behind with the impurities. The sulfuric acid is mixed with a fuel and thermally cracked. Impurities are destroyed during combustion; the sulfur dioxide thus produced is reused as oleum, for the production of nitrosylsulfuric acid and as reaction medium for the CHCA nitrosation.

After washing, the caprolactam extract is passed through a distillation column for solvent recovery. The caprolactam stream is then fed to another extraction column, where water enters from the top. Caprolactam leaves from the bottom as an aqueous solution 40% strong. Finally, the crude caprolactam is distilled as before.

The new method increases use of fuel, Snia admits, but still holds a significant economic advantage overall. Nevertheless, for those countries where ammonium sulfate remains in demand as a fertilizer, the former processing method might still be selected. #

Meet the Author

Andrew Heath is a correspondent with McGraw-Hill World News in Milan. He holds a B.A. (Econ.) from Britain's Durham University, and has authored several previous Process Technology articles.

Choosing the optimum CO_2-removal system

This article looks closely at the most widely used CO_2 absorption processes, their advantages and disadvantages, and the variables that govern their performance.

Samuel Strelzoff, *Marlboro, Vt.* *

☐ Selecting a suitable CO_2-removal system is very important to the overall economics of producing many gases—e.g., synthesis gases for ammonia, methanol and oxo alcohols; hydrogen-rich gases for refinery hydrotreating; CO_2 itself (for secondary-oil recovery and dry-ice production).

Typical production of many of these gases involves using a hydrocarbon feed—e.g., methane—that undergoes either steam reforming, or total or partial oxidation.

Oxidation reactions, as in CO_2 production, are exothermic:

$$CH_4 + O_2 = 2CO_2 + H_2O + 191.7 \text{ kcal} \quad (1)$$

The reverse is true of steam reforming, as in the production of methanol synthesis gas:

$$CH_4 + H_2O = CO + 3H_2 - 59.1 \text{ kcal} \quad (2)$$

$$CH_4 + CO_2 = 2CO + 2H_2 - 49.3 \text{ kcal} \quad (3)$$

In a typical steam-reforming process—e.g., to make ammonia synthesis gas—a desulfurized hydrocarbon feed reacts with steam at high temperature and pressure over a catalyst that is packed in externally heated tubes suspended in a furnace box.

Reaction products—H_2, CO, CO_2, methane and undecomposed steam—may go to a second reforming stage, in which air is added to supply the nitrogen for ammonia synthesis. The mixture of gases leaving the second reformer, after passing through several purification stages for removing CO, CO_2 and water, contains hydrogen and nitrogen in the correct empirical ratio for ammonia synthesis.

CO is normally eliminated by converting it into CO_2, according to the following reaction:

$$CO + H_2O = CO_2 + H_2 \quad (4)$$

CO conversion—which results in additional CO_2 that must be removed—is also a typical step in the production of methanol and oxo alcohol synthesis gases.

CO_2 is also generated when hydrocarbons heavier than methane, e.g. refinery naphtha, are employed as the feedstock used for steamreforming processes in the production of hydrogen, ammonia, methanol, etc.:

$$C_nH_{2n+2} + \frac{n\text{-}1H_2O}{2} = \frac{2n\text{+}1CH_4}{4} + \frac{n\text{-}1CO_2}{4} \quad (5)$$

The CH_4 in this equation reacts approximately as indicated by reactions (2) and (3).

In all the above-mentioned processes, CO_2 concentration ranges from around 7% to over 30%, depending on the ratio of hydrogen to carbon in the feedstock (Table I). In addition to these variations, processes for hydrogen and synthesis gas must contend with operating pressures ranging from near atmospheric to 500 psig and higher.

Any CO_2 removal scheme must also take into account the possible presence of H_2S, which may be introduced in the feedstock going to partial oxidation processes. The sulfide is eliminated with the CO_2.

Removal by absorption

The removal of CO_2 depends on its so-called acid-gas character; it tends to form carbonic acid in water:

$$CO_2 + H_2O = H_2CO_3 \quad (6)$$

and this carbonic acid undergoes reversible reactions in chemical absorption by solutions of amines:

$$RNH_2 + H_2CO_3 = (RNH_2)_2CO_3 \quad (7)$$

* This article has been extracted from an ammonia-and-hydrogen-plant design manual now being prepared for publication by the author.

Originally published September 15, 1975

CO_2 removal requirements

From ammonia synthesis gas

Feedstock	Generation process	Tons CO_2 per ton of NH_3
Natural gas	steam reforming	1.22
	partial oxidation	1.63
Naphtha	steam reforming	1.42
	partial oxidation	2.09
Heavy fuel oil	partial oxidation	2.49

From crude hydrogen

Feedstock	Generation process	Tons CO_2 per 1,000 scf H_2
Natural gas	steam reforming	14.51
Naphtha	steam reforming	17.90
	partial oxidation	29.32
Bunker C fuel	partial oxidation	32.34

or by solutions of alkaline salts:

$$K_2CO_3 + H_2CO_3 = 2KHCO_3 \quad (8)$$

to form carbonates, which will decompose, on heating, to CO_2 and the amine or salt, so as to drive off absorbed CO_2 and regenerate the absorption solution.

There are a variety of such absorption agents, as well as a number of proprietary additives for increasing the activity of these agents and adapting them to the operating conditions of one or more of the H_2 and syn-gas processes.

In addition, there are physical absorbents, such as water and methanol. And since the acid-gas solutions tend to become corrosive at about 230°F, there are a number of proprietary corrosion inhibitors and processing schemes for reducing regeneration temperature.

The result is a large and complex body of technology from which the chemical engineer is obliged to select an optimum.

Suiting absorbents to the situation

This optimum depends on the primary process variables and the type of absorbent. The primary variables, which are dictated by the gas generation process, are: absorption pressure, CO_2 concentration, the amount of non-CO_2 contaminants present, and the amount of steam generated. Depending on these, the best absorbent may be either physical or chemical as illustrated in Fig. 1 (F/1).

Chemical absorption is characterized by effects quite different from physical absorption. It will remove CO_2 to a lower partial pressure in the treated gas (P_1 vs. P_2 in F/1). However, at higher CO_2 partial pressures (P_3), the physical absorbent will carry more CO_2 (C_4 vs C_3). Thus the slope of the vapor-liquid equilibrium curves for CO_2 in the absorbent is important to the absorbent recirculation rate.

Physical absorption's proportionality with partial pressure can be utilized for solvent regeneration by simple flashing. This improves the economics of CO_2 removal, particularly for large quantities at high operating pressures. For example, when the CO_2 load corresponds to the partial pressures P_3–P_2 in F/1, the useful concentration differential, which determines the solvent circulation rate, is C_4–C_1. At the same partial-pressure difference, the useful concentration difference for chemical absorption is much smaller (C_3–C_2) due to the curvature of the equilibrium line.

In general, chemical solvents are therefore not suitable for flash desorption and have to be regenerated with heat, whereas physical absorbents can also be regenerated completely by stripping with an inert gas, provided that the CO_2 removed from them can be diluted with the stripping gas before it is discharged to the atmosphere.

When the process gas must be pure and the removed acid-gases have to be discharged undiluted, hot regeneration of the solvent is necessary, so chemical absorption is the natural choice. Such regeneration is usually done by reboiling, which not only reduces solubility due to increased temperatures, but also strips the solvent with the reboiler vapors.

When reboiled regeneration is used, the investment for heat exchange between rich and lean solvent can be justified, if the solvent recirculation rate is kept low. The combination of bulk CO_2 removal with flash regeneration, followed by final removal with hot regeneration, sometimes offers advantages.

Variables to contend with

When choosing an optimum absorbent, the first step is to identify the primary variables of the gas generation process, as follows:

- Pressure. The pressure of CO_2 absorption must be that of the process gas.
- Concentration. The concentration of CO_2 in the raw gas is determined by the feedstock and type of reforming reactions, while that of the treated gas is deter-

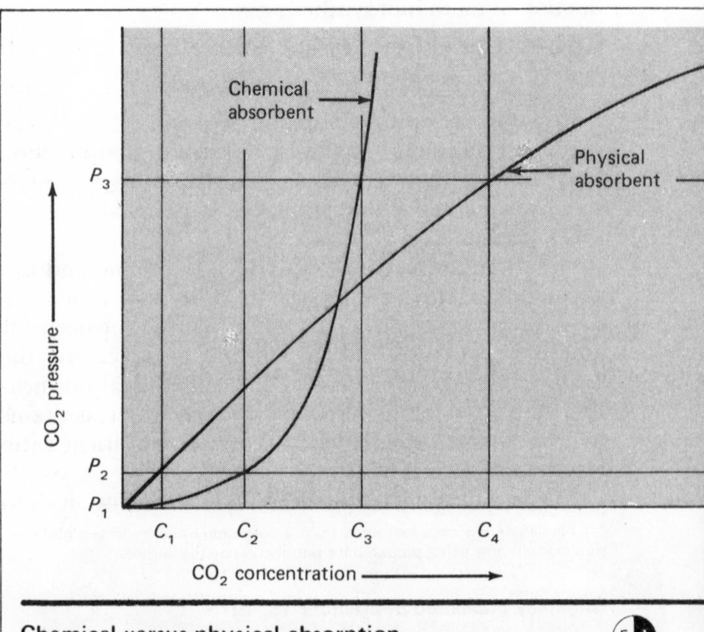

Chemical versus physical absorption 

mined by its use and subsequent processing treatments. Since CO_2 is a poison to synthesis catalysts, its content must be reduced to less than 5-10 ppm for ammonia, methanol and oxo processes. Such low concentrations are costly to obtain by absorption alone, so traces of CO_2 are usually removed by methanation or by nitrogen wash. Thus the cost of removing incremental trace quantities of CO_2 is balanced between methanation and absorption.

■ Poisons. Some absorbing agents are poisoned by compounds in the raw gas, as monoethanolamine (MEA), for example, is poisoned by COS and CS_2.

■ Utilities. Finally, the utilities balance of the overall process affects the amount of low-pressure steam that might be available for regenerating the absorbent and driving the solution-recycle pumps.

One ammonia process, for example, makes use of MEA absorption partly because its steam consumption in the regeneration reboiler balances with steam discharged from backpressure turbines to provide a very efficient energy-conversion process.

With the foregoing conditions prescribed, it is possible to compare absorbents on the basis of a general process scheme (Fig. 2). A heat and material balance drawn around this process scheme makes possible three observations:

1. The diameter of the absorber column is relatively independent of the type of absorbent, because column diameter depends on the superficial velocity of the process gas.

2. The conditions for regenerating absorbent solution are close to 4 psig and 235°F in a regenerator reboiler, because about 4 psi are required to drive vapors through the regenerator tower to the atmosphere, and 235°F is about the maximum for limiting corrosion by the acid-gases.

3. The minimum temperature of lean absorbent and regenerator reflux can be assumed at about 100°F, because of available cooling-water temeratures.

These observations leave five variables to be determined: (1) the height of the absorber, (2) the height of the regenerator, (3) the diameter of the regenerator, (4) the duties of the regenerator reboiler and overhead condenser, and (5) the absorbent recirculation rate. All of these depend on the vapor-liquid equilibrium curves for the absorbent, which are the keys to any absorbent system.

Commercial absorbents

Fig. 3 (F/3) shows various equilibrium isotherms that are the basis of those CO_2 absorption processes most commonly used today. The two groups, chemical and physical, can be clearly differentiated as (1) the highly bent curves for the chemical absorbents (curves e to i) and (2) the fan of almost straight-lined curves for physical absorption, including water (curve a), N-methylpyrrolidone (curve b), the scrubbing agent of the Lurgi-Purisol process (curve c), and methanol (curve d).

Although the curves of (F/3) are isotherms, it should be remembered that temperature is very important to physical absorption. Methanol, for example, makes possible the application of very low temperatures,

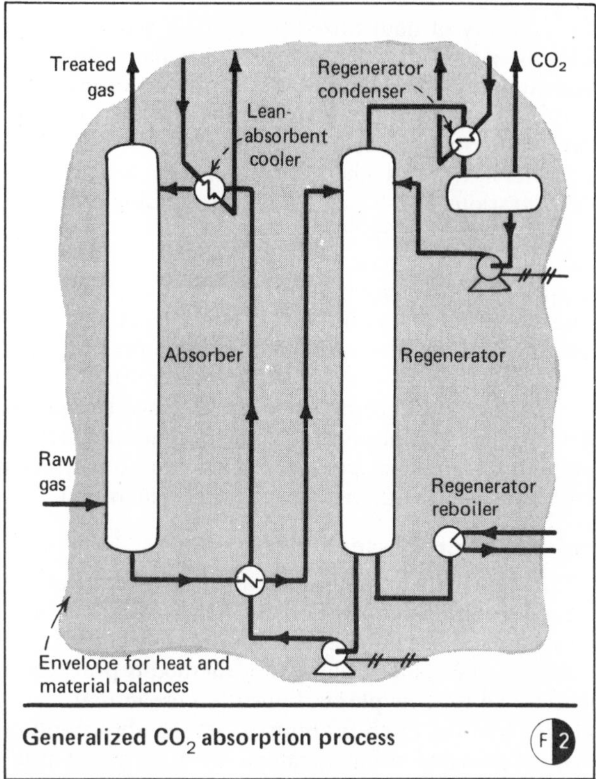

Generalized CO_2 absorption process F2

down to about -70°C. Compared with water at 30°C, it can absorb 10 times as much CO_2 at -15°C, and more than 20 times as much at -20°C. At higher CO_2 partial pressures, these absorption values become practically unlimited as the scrubbing temperatures approach the dewpoint of CO_2.

The range of temperatures is not so great for chemical solvents, because these are used in water solutions, and the concentrations of absorbed CO_2 are limited by

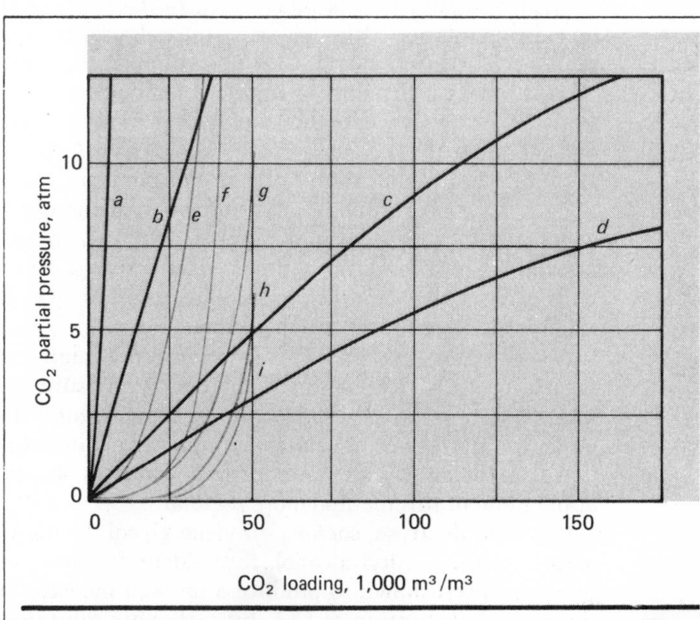

Loading curves for different absorbents 

the solubility of neutralized products in water, as well as by the viscosity, which increases with increasing concentration. Consequently, the concentrations and scrubbing temperatures given for the chemical absorbents' isotherm-equilibrium curves may be considered characteristic. Most chemical absorbents are used in concentrations of 2-3 mol% and at scrubbing temperatures ranging from 30-50°C.

Potassium carbonate and triethanolamine (TEA) are exceptions to this temperature limitation. Although potassium carbonate scrubbing is effective at low temperatures, a high absorption temperature presents heat economies. Increasing the reaction temperature of TEA extends the equilibrium curve—it becomes straighter, like the curve of a physical solvent—so that for a certain range of CO_2 partial pressures, TEA can be used as a scrubbing agent with pressure-release regeneration.

Industrial processes

The most widely used methods of CO_2 removal are water scrubbing, absorption by MEA solutions, and absorption by hot potassium carbonate solutions.

Water scrubbing: The oldest, this method is practically abandoned for new plants, but still in use in many ammonia facilities. High pressure causes the absorption of CO_2; and regeneration is caused by a release in pressure, involving practically no change in temperature. Most of the energy contained in the high-pressure water is recovered by a water turbine (Pelton wheel) coupled to the motor of the high-pressure water pump.

Although this method is simple and inexpensive, it has certain drawbacks. Hydrogen is dissolved along with the CO_2, so that the H_2-loss is appreciable, amounting to 1.5-2.5% when the scrubber is operated at about 200 psig.

Several schemes have been proposed for recovering this hydrogen, but these require additional equipment, and the increased investment often cancels the value of the recovered H_2.

MEA scrubbing: This is widely used for CO_2 removal in existing plants, although its principal use in new plants is for H_2S removal. CO_2 absorption is typically carried out at a little above ambient temperature and up to 600 psig. Regeneration involves pressure release and reboiling the regenerated solution. Operating costs are somewhat higher than water or hot-potassium-carbonate scrubbing, due to the costs of regeneration, which requires close control because of the corrosiveness of the MEA at the reboiler temperatures. Even with close control and pure solutions, stainless steel is usually specified for the reboiler tubes.

Hot potassium carbonate: This solution has a high capacity for CO_2. Regeneration involves a pressure release and a relatively small rise in temperature. In modern installations, the hot potassium carbonate solution is activated by certain additives, which make the application of this method more popular.

Glycerol, dextrose, sucrose, ethylene glycol, levulose, methyl alcohol, ethyl alcohol, formaldehyde, and lactose have been shown to produce a marked increase in the rate of absorption of CO_2 by carbonate solutions. But the only additives known to have found commercial application are held proprietary under such names as Benfield, Giammarco-Vetrocoke, Catacarb, Carsol, etc.

The Benfield process

In this process, the pressurized feed gas containing CO_2 and H_2S enters the absorber tower, where it is scrubbed by potassium carbonate solution at about 240°F. Scrubbed gases containing a residual 5-10 ppm of CO_2 pass from the top of the scrubber to the primary gas process, while the solution is regenerated by reboiling at reduced pressure in a regenerator tower. The regeneration and absorption temperatures are often nearly the same, so that rich-solution-to-lean-solution heat exchangers are eliminated. Typical requirements per 1,000 scf of CO_2 are:

Basis: Natural gas reforming, with 18% CO_2 in the raw gas and 0.05% CO_2 in the treated gas.
Absorber pressure – 400 psig.
Regeneration heat – 87,000-120,000 Btu/ton CO_2
Circulation energy – 1.15 kWh/ton CO_2
Condensing and cooling duty – 100,000-120,000 Btu/ton CO_2

The process claims negligible H_2 losses (less than 0.1%), no vapor carryover to the methanator or other downstream units of the primary gas process, and no poisons for downstream catalysts. Also, a high-purity urea-grade CO_2 (99.5%+ of CO_2)is said to be obtainable.

However, certain parts of the equipment exposed to the hot solution must be made of corrosion-resistant materials. There is a possibility that salt crystals will form if the solution is cooled. And, in some cases, antifoaming agents must be added.

The Giammarco-Vetrocoke process

This process is based on potassium carbonate solution activated by arsenic trioxide (As_2O_3), glycine and various ethanolamines. The solution activated by arsenic compound seems to absorb CO_2 at a rate 10-20 times higher than that of the plain potassium solution (F/4). Given this accelerated absorption rate, the transfer area and thus the volume of the absorption tower may be reduced.

CO_2 absorption takes place at medium temperatures of 50-100°C and at pressures ranging from atmospheric to 75 atm and higher, depending on the raw-gas conditions. Regeneration of the rich solution may take place through either of two methods:

- Reboiling and stripping with steam at atmospheric pressure or higher (F/5).
- Stripping with air at atmospheric temperature (F/6).

The process claims that arsenic trioxide gives the absorbent solutions anticorrosive protection even at high temperatures. However, arsenic trioxide is poisonous and must be handled accordingly. Inventor Giammarco has introduced another activating compound, glycine, which is said to be nearly as good as As_2O_3, and is probably one of the best nonpoisonous activators because of its absorption-velocity constant.

Recent process schemes by Giammarco-Vetrocoke

seem to have reduced the heat requirements. Absorption is carried out at a cooled temperature but without the typical heat exchanger. The raw gas is fed to the absorber at about 220-250°C; and the heat thus added to the bottom of the absorber is carried out in the rich solution without reducing the chemical absorption at the top. The heated rich solution is then flashed and the steam used to regenerate the solution through direct-contact stripping. Consequently, heat contained in the process gas is applied by direct exchange without the need for a conventional reboiler.

Catacarb process

One of the first industrial applications of the Benfield process was made by Chemical Construction Corp. as part of a 180-ton/d ammonia plant for Escambia Chemical Co., Pensacola, Fla. In order to increase its production capacity to over 200 ton/d of ammonia and to ensure the equipment against corrosion, A. G. Eickmeyer, an independent consultant, recommended the addition of certain chemicals to the potassium solution. This suggestion was accepted and be-

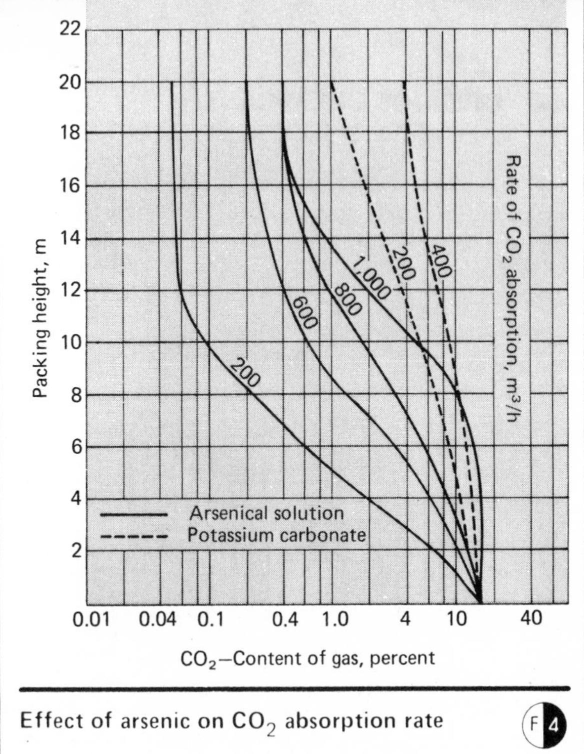

Effect of arsenic on CO_2 absorption rate

F 4

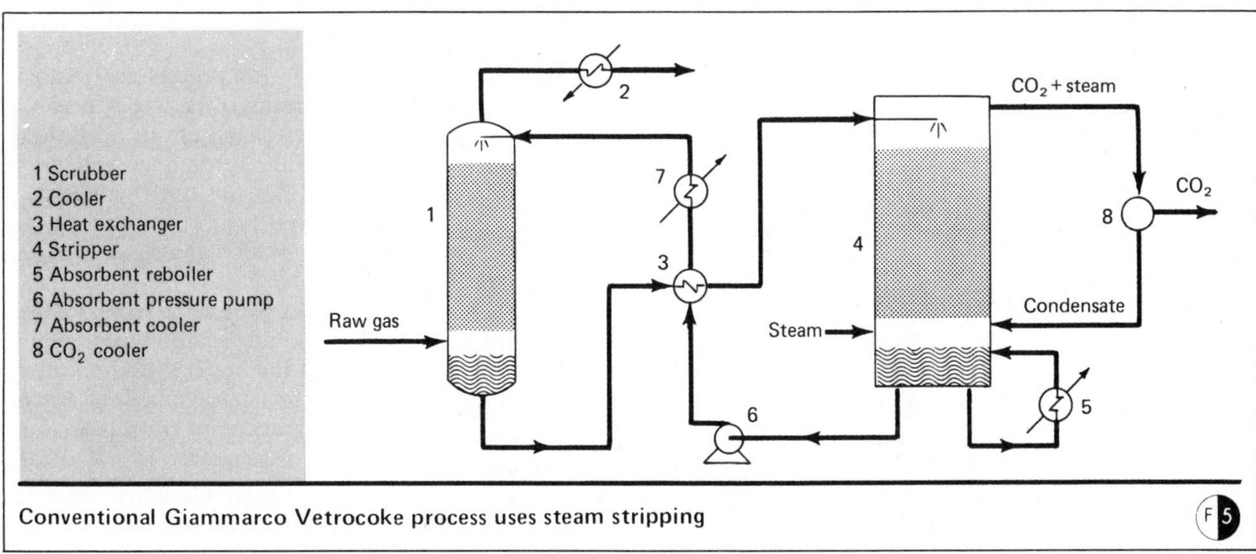

Conventional Giammarco Vetrocoke process uses steam stripping

F 5

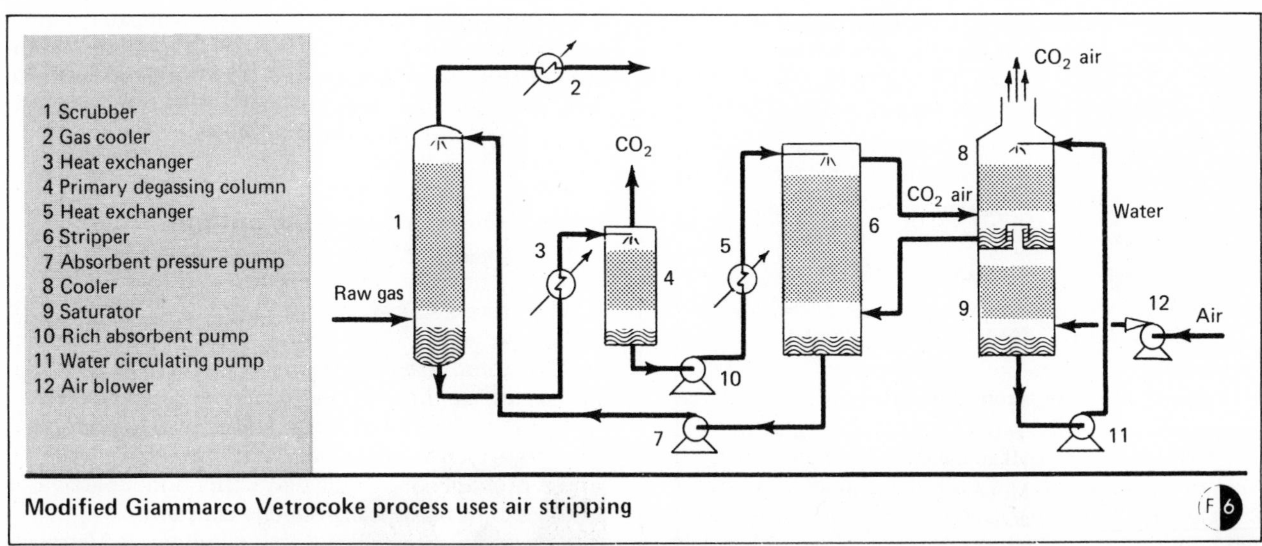

Modified Giammarco Vetrocoke process uses air stripping

F 6

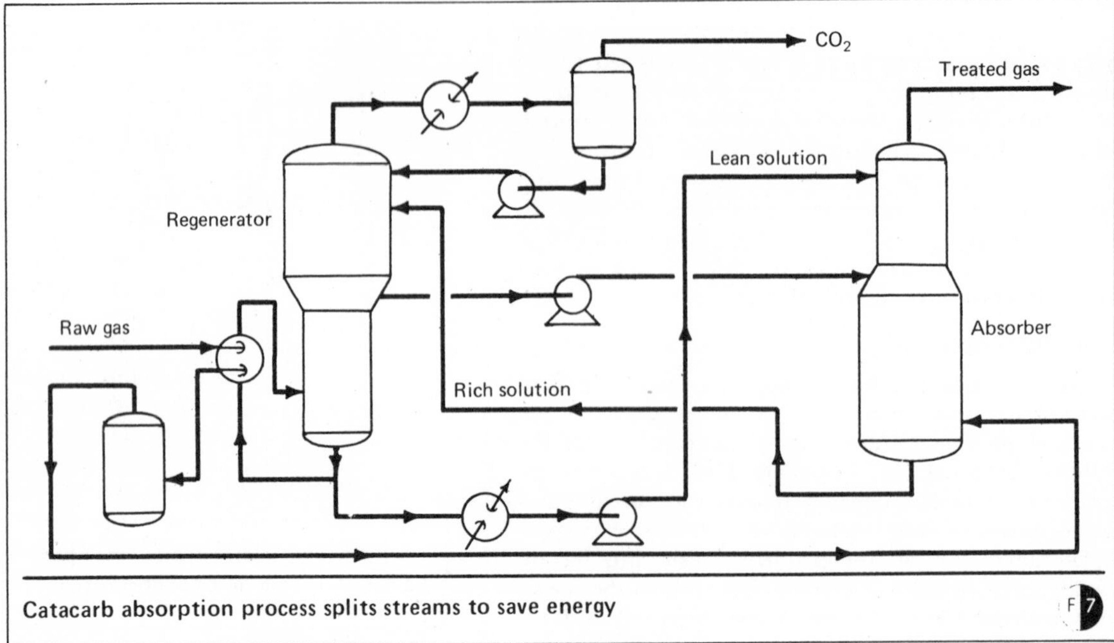

Catacarb absorption process splits streams to save energy F 7

came the successful introduction of the process.

In its present form, this process incorporates a potassium salt solution containing an amine-borate activator, non-toxic-to-nickel catalysts for methanation and reforming, plus a corrosion inhibitor. The activator apparently increases the chemical reaction rate of CO_2 with the potassium solution in both the absorption and the regeneration steps, since the CO_2 is regenerated to lower concentrations than with the straight hot potassium carbonate solution.

Over 40 installations of this process for removing CO_2 and H_2S from ammonia synthesis gas, crude hydrogen gas, sour natural gas, etc., are in operation in plants throughout the world. In ammonia plants, the H_2-N_2-CO_2 mixture leaving the shift converter is scrubbed in packed or trayed absorbers, and the rich solution from the bottom of the absorbers is regenerated by reboiling.

A simple two-stage or split-stream system is usually employed, so that a minor portion of thoroughly regenerated solution is fed to the upper section of the absorber for final cleanup. The heat available in the raw gas from pressure reforming and shift conversion is usually enough to reboil the regenerator (F/ 7). Process temperatures depend on the purity required and the absorption and regeneration pressures, which are ordinarily 300-400 psig and 0-5 psig, respectively.

In summary, the chemical engineer who wants to find an optimum process for CO_2 removal can begin by comparing vapor-liquid curves for CO_2 at the conditions of absorption dictated by his primary process, and then compare the heat requirements for absorbent regeneration with the utility balance.

A list of CO_2-recovery processes

Licensor (process)	Probable absorbent
Allied Chemical (Selexol)	Dimethyl ether of polyethylene glycol plus an alkanolamine
BASF	Alkazid
BASF	Triethanolamine
Benfield	Hot potassium carbonate solution plus an activator
Eickmeyer (Catacarb)	Hot potassium carbonate solution plus an activator
Carbochimique (Carsol)	Hot potassium carbonate solution plus an activator
Giammarco (Vetrocoke)	Potassium arsenite solution plus an organic activator
Linde-Lurgi (Rectisol)	Methanol
Lurgi (Purisol)	N-methyl-2-pyrrolidone
Fluor	Propylene carbonate
Shell (Sulfinol)	Tetrahydrothiophene plus an alkanolamine
Union Carbide (U-CAR)	Monoethanolamine plus an activator

The author

Samuel Strelzoff is a consultant and processing specialist. Prior to this, he served as senior vice-president and advisor to the president of Chemical Construction Corp.—an engineering and construction firm with which he was associated for 27 years. Born and educated in Europe, with a cum laude Master's degree in chemical engineering from the Royal University of Liege, Belgium, he has authored articles on a wide range of engineering subjects for both U.S. and European publications.

Diaminomaleonitrile makers push product's potential

The world's two producers of the specialty chemical see a bright future for it as an intermediate. One firm has just put a semiworks unit onstream, while the other may double its capacity next year.

☐ It's not going to happen overnight, but one day diaminomaleonitrile (DAMN) may make its mark as an intermediate for producing resins, dyes, surfactants, pharmaceuticals, catalysts, herbicides, and numerous other compounds. At least that's the word from the Japanese manufacturers of the tetramer.

As with many specialty chemicals, though, production boosts hinge largely on end-users' attempts to develop applications. Recent actions of the two Tokyo-based diaminomaleonitrile makers—Kyowa Gas Chemical Industry Co. and Nippon Soda Co. (Nisso)—indicate that some new roles for DAMN are already verging on commercialization. In September, for instance, Nippon Soda started up its first semicommercial plant, rated at 20 metric tons/yr, at Mizushima on Honshu Island. And by early spring, Kyowa Gas Chemical may raise capacity of its 10-metric-ton/yr unit at Nakajo, Honshu, to 20 metric tons/yr. Besides this, Nisso is thinking of building a 50–100-metric-ton/mo plant.

At this time, the two firms remain tight-lipped about the reasons for their optimism. But sources say that Nisso, at least, may have received encouragement from its U.S. customers, who are attempting to make various derivatives from DAMN.

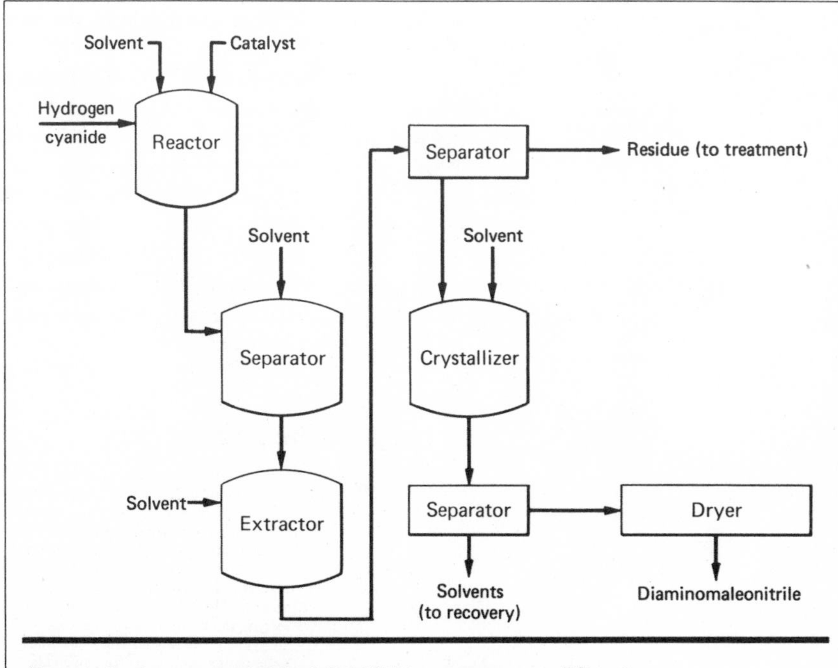

Kyowa looks to up yield of DAMN by reactor modifications

MULTIPLE ROLES—The pale-brown tetramer is a versatile organic intermediate, note Kyowa and Nippon Soda. Its needle-like crystals' functional groups—two amino and two nitrile groups, plus a carbon-carbon double bond—can be reacted to yield a wide variety of nitrogen-containing compounds, such as imidazoles for making pharmaceuticals and synthetic condiments; pyrazines for production of pigments, heat-resistant polymers, and medicines; triazoles for reaction to industrial catalysts and antiviral drugs; imidazolones for resins, dyes and surfactants manufacture; and azoles for making herbicides.

Besides the development work being done in the U.S. by its customers, Nippon Soda is using a portion of its output "for captive consumption," indicating that it, too, is trying to make some of these derivatives. Kyowa, meanwhile, is exporting some product to two unnamed U.S. companies that are thought to be developing DAMN-based new products. Other clients are working on pigments derived from DAMN. If any large-volume applications develop, Kyowa and Nisso are willing to sell the technology to manufacturers interested in making DAMN on their own.

The two companies, 1972 licensees of basic technology developed by Sagami Chemical Research Center (SCRC) of Tokyo, remain as reluctant to disclose process details as they are to talk about progress in applications development. As a Nippon Soda source says: "Regarding technology for known chemicals—say, styrene monomer—many people reveal lots of figures in attempts to sell processes. DAMN is so new that we just cannot do the same."

Actually, a process to synthesize diaminomaleonitrile had been developed as far back as 1950, when Du Pont applied for a patent. But it wasn't until 1969, when SCRC developed its oxidation-of-hydrogen-cyanide route, that the compound became feasible to make commercially. Even then, SCRC's processes (there are two, involving different solvents) required modifications by Kyowa and Nisso to slash the selling price of the laboratory-derived product by about fourfold, down to \$30–\$45/kg (\$13.60–\$20.50/lb).

Originally published December 8, 1975

Nippon Soda can easily up annual plant capacity from 20 to 50 tons.

Until bulk applications pan out, however, the producers say they cannot pare the present pricetag much more: achieving a more-reasonable \$3.30–\$6.60/kg (\$1.50–\$3/lb) price, for example, would demand about a 600-ton/yr output.

UNCHANGED CHEMISTRY—Despite process modifications, the firms use basically the same chemistry in their semicommercial facilities (see flowsheet) as that developed in the laboratory. Specific reactants and catalysts are different, though, says SCRC's Tomio Okada, one of the developers of the technology. A Kyowa Gas manager confirms this, noting that reaction conditions are not drastically changed either.

In the lab route, a cooled solvent-catalyst mixture of dimethylsulfoxide or dimethylformamide and sodium cyanide is reacted with hydrogen cyanide. The solution then is heated at 60–70°C (140–158°F) for 5 h, diluted with water, and the diaminomaleonitrile product, $(CHN)_4$, extracted with ether. Evaporating the ether yields 97%-pure tetramer. To up the product's purity to 99.6%, the crude hydrogen cyanide tetramer is dissolved in water or isopropyl alcohol, heated to 80–90°C (176–194°F), cooled, crystallized and dried.

Not surprisingly, both Nisso and Kyowa have modified the process to up the yield of DAMN over the 70% laboratory mark. For example, Kyowa has increased the yield to 80%. Further changes, especially around the reactor, should boost the number still more, says the firm's senior technical-development staffer, Sho Matsumoto. And although Nippon Soda won't divulge a concrete figure for its yields, a company representative claims that "use of our own catalyst and specific combination of solvents has improved the yield and product quality."

Shota Ushio/Larry J. Ricci

Efficient, nonpolluting ethylbenzene process

A newly developed alkylation catalyst that departs from the traditional Friedel-Crafts types achieves several significant improvements in making ethylbenzene.

F. G. Dwyer, Mobil Research and Development Co.
P. J. Lewis, Badger America, Inc. and F. H. Schneider, Mobil Chemical Co. *

☐ An ethylbenzene process developed through the joint efforts of Mobil Oil Corp. subsidiaries and The Badger Co., Inc. is now ready for commercial use, following a successful demonstration last year in a 40-million-lb/yr plant.

The process uses a solid, non-Friedel-Crafts catalyst to alkylate benzene with ethylene. By contrast, all of today's commercial paths to making ethylbenzene (the feedstock for styrene) employ Friedel-Crafts types, most commonly aluminum chloride in a mixed liquid-gas-phase reaction system.

There is considerable incentive for a process operating over a solid, non-Friedel-Crafts catalyst. Such a process would avoid pollution problems, reduce catalyst consumption, and eliminate the need for highly-corrosion-resistant materials of construction. If the process additionally were vapor phase and run at high temperature, the heat of reaction would be more easily recovered.

The new Mobil-Badger process achieves all these benefits. It is a fixed-bed technique featuring a solid catalyst designated APEB, a proprietary development of Mobil.

HOW IT WORKS—The process flow diagram includes two principal sections: a reaction or alkylation section, and a product recovery section.

After preheating and vaporization, fresh and recycle benzene combines with an alkyl-aromatics recycle stream and with fresh ethylene, and the mix is fed to the alkylation reactor containing APEB catalyst. The alkylation section may have two or more reactors, one of which is being regenerated or on standby while the other is onstream.

The reaction takes place above 700°F, at 200–400 psi, and at high ethylene space-velocities. The ratio of aromatics to ethylene in the feed falls in the range of 5 to 20 mole/mole. Yields and selectivity compare to those of existing commercial plants.

The distillation train in the product recovery system first removes unreacted benzene from the reactor effluent and recycles it back to the alkylation reactor. Since conversion of ethylene is essentially 100%, any unconverted ethylene or light paraffin gases are vented from the system for use as fuel gas, rather than recovered and recycled.

The benzene-free product is then separated into ethylbenzene final product, suitable for styrene manufacture, and a heavier bottoms fraction. The latter is further separated into an alkyl-aromatics stream that recycles to the alkylation section, and a residue. The alkyl-aromatics stream contains all of the nonselective products that recycle to system equilibrium. Residue amounts to less than 2 wt.% on total feed, and like vent gas serves as fuel.

Catalyst regeneration takes place using nitrogen-air mixtures of controlled oxygen content to prevent an excessive temperature rise. Recycle of regeneration gases with air makeup is technically feasible, but requires slight additional capital. Cycle length between regenerations runs several weeks at high ethylene space-velocities, and longer at less-severe operating conditions.

PROCESS DEVELOPMENT—During the initial phase of pilot-plant testing, by Mobil Chemical Co., the product-selectivity effects of both benzene-to-ethylene ratio and operating temperature were evaluated.

As expected from thermodynamic equilibria, diethylbenzene production increases as the benzene-ethylene ratio decreases. At 750°F and 260 psi, diethylbenzene output climbs from 0.78 wt.% to 5.50 wt.% as mole ratio drops from 10.0 to 3.0. A similar trend occurs, though not as well defined, in other byproducts boiling between ethylbenzene and diethylbenzene.

The choice of benzene-ethylene ratio thus depends on the relative economics of recycling benzene or diethylbenzene, as compared with operation with shorter catalyst cycle-times between regenerations as the ratio decreases.

Over an inlet temperature range of 650–875°F, the pilot work has shown a definite reduction in diethylbenzene formation once inlet temperature exceeds 750°F, but the falloff levels out at 850°F. There also appears to be some reduction in byproducts in going from 650°F to 725°F, but little or no change as temperature rises further. In addition, temperature has an important effect on catalyst cycle time, as evidenced by the high aging rate observed while operating at 650–750°F.

Based on these studies, 750–850°F emerges as an optimum range.

Catalyst regenerability has also been shown, one catalyst having

* This article is adapted from a presentation before the First Chemical Congress of the North American Continent, Mexico City, Dec. 1–5, 1975.

Originally published January 5, 1976

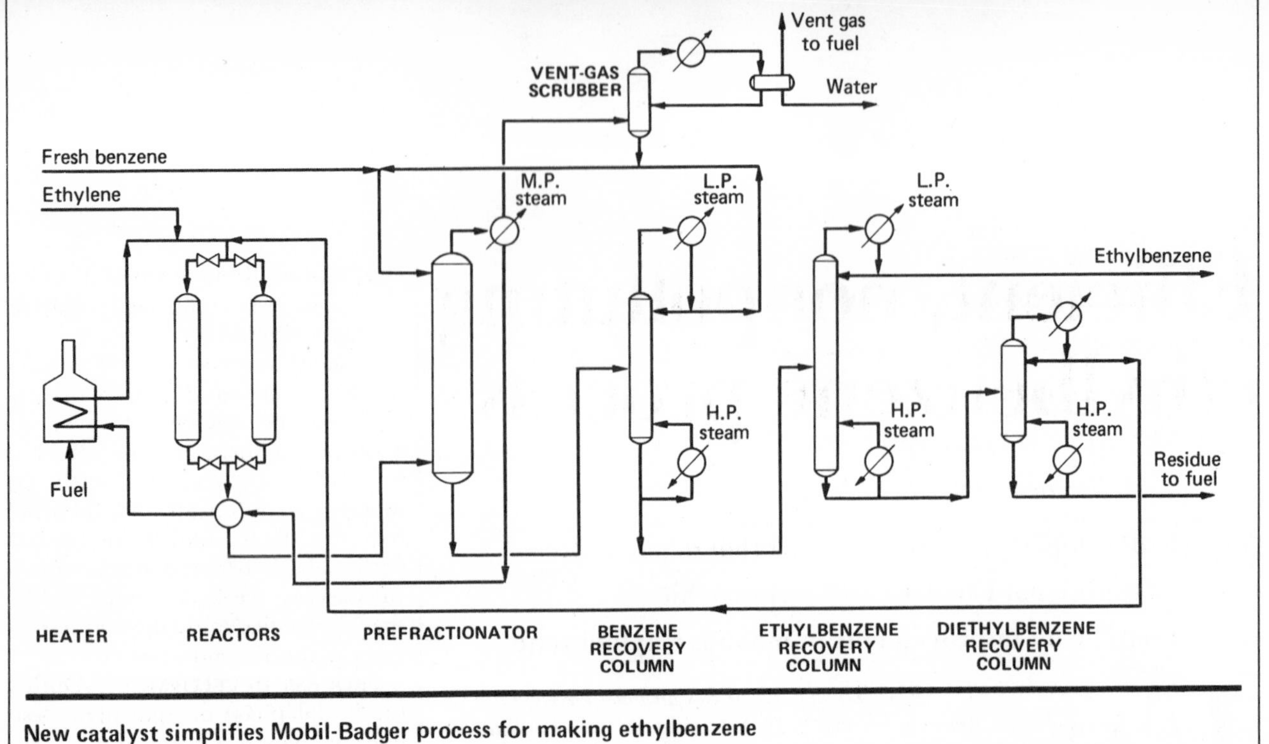

New catalyst simplifies Mobil-Badger process for making ethylbenzene

been regenerated 14 times without loss in activity or selectivity.

The second phase of the pilot-plant work was conducted by Badger. This phase demonstrated a catalyst cycle time of about 2 wk to ethylene breakthrough; product selectivities similar to those obtained earlier; and full recovery of catalyst performance upon regeneration. It also proved that transalkylation of diethylbenzene can be accomplished simultaneously with alkylation; temperatures of 800–850°F thus reduce diethylbenzene recycle.

The ensuing 40-million-lb/yr demonstration plant was installed and operated at Foster Grant Co.'s site at Baton Rouge, La., under supervision of Badger personnel.

Trials showed that minimum overall yields of 98% could be expected commercially. Regenerability of the catalyst was again confirmed, as well as the capability to recycle to steady-state levels the nonselective products boiling between ethylbenzene and diethylbenzene.

MANY PLUSES—The following clearcut advantages are inherent in the Mobil-Badger process:

■ *Energy efficiency*—The high temperature of the vapor-phase alkylation combines with efficient distillation-column design to allow recovery of 95% of the net process-heat input and heat of reaction, as useful low- and medium-pressure steam. A small amount of heat is rejected to cooling water. The onsite process heater requires gas or oil fuel, but vent gas and residue produced in the unit may supply part of this.

■ *Simple design*—This very high level of energy efficiency is attained because the process design is much simpler than for Friedel-Crafts alkylation. The catalytic action is truly heterogeneous—no catalyst leaves the reactor, and thus its effluent can go directly to distillation without catalyst removal or recycle. Reaction and distillation are necessarily highly integrated, but this poses no operational problems. No catalyst preparation or pretreatment is required; reaction occurs immediately upon adding ethylene to benzene at reaction temperature.

The entire plant occupies about two-thirds the onsite plot area of conventional units. There are not nearly so many items of equipment.

■ *Catalyst*—The catalyst is highly productive and has a long service life. It presents no hazards or waste-disposal problems. It is environmentally inert and needs no special handling.

Since the catalyst introduces no corrosive substances, the entire ethylbenzene unit can be fabricated from carbon steel except where equipment design temperatures call for low-chrome steels. Maintenance costs will be correspondingly low relative to existing processes, and should average less than 2% of initial capital investment per year.

■ *Environmentally clean*—No process streams are produced that require treatment. Besides ethylbenzene, the only streams are the light vent gas and heavy residue, used as fuel.

The low plot-area requirements reduce treatment costs for storm-water runoff. Very small cooling-water consumption consequently keeps down the amount of cooling-tower blowdown for treatment.

■ *Economics*—Evaluation of the Mobil-Badger ethylbenzene process with a number of different sets of utility values indicates an operating cost advantage of from 0.3¢-0.5¢/lb of ethylbenzene produced. Initial capital investment for the process will be less than that for competitive processes, whether or not the ultimate cost of pollution abatement is included for these alternatives.

The authors

Frank G. Dwyer is a research associate for Mobil Research and Development Corp., Paulsboro, N.J. He holds a B.S. from Villanova Univ., and an M.S. and Ph.D. from the Univ. of Pennsylvania, both in chemical engineering.

Peter J. Lewis is a process supervisor for Badger America, Inc., Cambridge, Mass. He holds B.S. and M.S. degrees in chemical engineering from Massachusetts Institute of Technology.

Fred H. Schneider is manager of process development for Mobil Chemical Co., Edison, N. J. A registered professional engineer in Texas, he holds a B.S. and M.S. in chemical engineering from Texas A & M Univ.

Revolutionary changes in the availability and price of natural-gas liquids are having a dramatic impact on the economics of petrochemicals, especially of ethylene—the olefin king of the industrialized world.

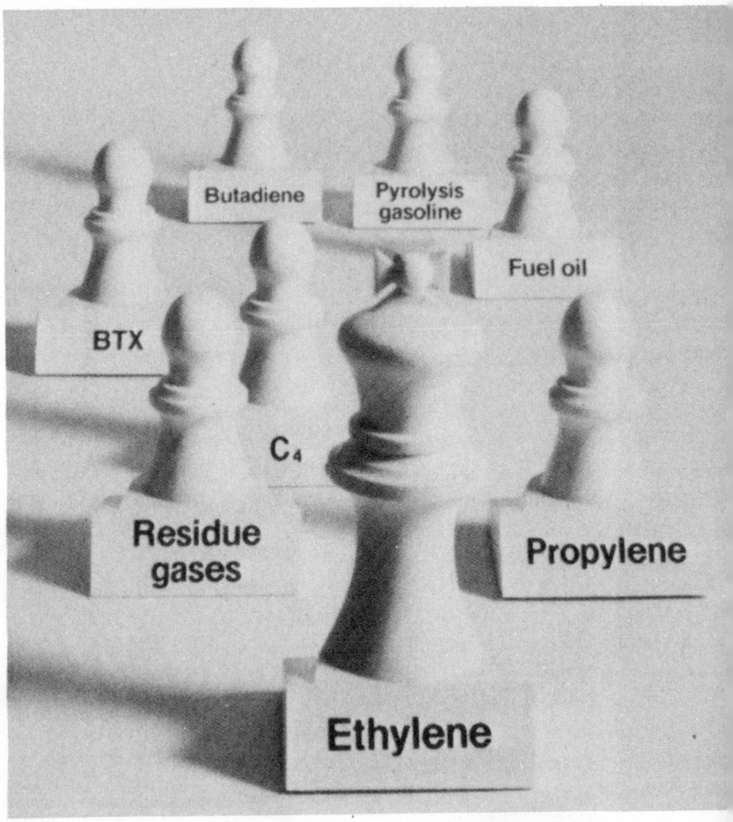

Butadiene
Pyrolysis gasoline
Fuel oil
BTX
C_4
Residue gases
Propylene
Ethylene

Ethylene and its coproducts: The new economics

Theodore B. Baba and James R. Kennedy, Stone & Webster Engineering Corp.

☐ Projections of demand for ethylene and its coproducts in the U.S. during the next decade indicate that a substantial expansion in manufacturing capacity will be necessary. This new capacity will, in all likelihood, be based primarily on heavier feedstocks, such as naphtha or gas oil, because of increasing shortages of ethane and propane, the prevailing U.S. feedstocks.

Reliance on heavier feedstocks will result in more-complex plants and in the production of larger quantities of coproducts, which can pose marketing problems and substantially influence profits.

Market aspects—ethylene and its coproducts

Ethylene—This feedstock has long been heralded as the backbone of the petrochemical industry. In any survey of the largest-volume synthetic chemicals in the industrial world, it is invariably ranked as the number one organic product.

Market studies indicate that the world ethylene demand in 1974 was on the order of 64 billion lb [*16*]. And it has been projected that during the next 10 years this demand will increase to approximately 120 billion lb/yr—a dramatic twofold increase that some observers, nevertheless, consider inadequate.

Originally published January 5, 1976

In the U.S., ethylene demand during this period is projected to increase from approximately 24 billion lb/yr to 48 billion lb/yr [*11*]. This forecast is predicated on an average annual growth rate of slightly less than 7%. It represents a middle-of-the-road estimate that contrasts with the approximately 11% annual growth rate of the past decade. (Considerably higher fuel and raw-material prices, greatly increased capital requirements, inflation in general, and a stable population are expected to be suppressors of future growth.)

To satisfy this continuing demand, the equivalent of 65 new plants, each of a nominal 1-billion-lb/yr capacity, will be needed during this 10-yr period, 24 of which will have to be located in the U.S. It is a certainty that additional ethylene and downstream derivative plants will be built in the less-developed oil-producing countries (particularly in the Middle East and Indonesia), because of their abundant raw materials. The number that will be built will depend on political and economic developments, which will also influence the number of plants that will be constructed in the industrialized world, including the U.S.

Propylene—Together with this substantial rise in demand for ethylene will come an appreciable one for

Large ethylene plant of Shell Nederland Chemie, N.V. at Moerdijk, Holland, is based on gas-oil feed

propylene as a feedstock for chemicals and plastics. At present, slightly more than 20% of the propylene is obtained through pyrolysis. At the end of the decade, over 70% of such propylene may be obtained this way.

Total propylene demand in the U.S. for 1975, for all purposes, has been estimated as 25 billion lb/yr [*17*], with approximately 13 billion lb being used by refiners for alkylate and other purposes. U.S. propylene demand for chemicals and plastics 10 years hence has been projected to range from a low of 19 billion lb/yr to a high of 31 billion lb/yr. In the future, any periodic excess of propylene is expected to find its way to the market via liquefied petroleum gas (LPG), polymer gasoline, and also as a Btu supplement for pipeline gas.

Benzene—Because of the changing raw-material situation, particularly in the U.S., future ethylene plants could become substantial producers of aromatics. This makes worthwhile a look at projected demands for benzene, the major aromatic.

Benzene demand for conversion into chemicals is expected to rise from about 1.5 billion gal in 1974 to about 2.6 billion gal in 1985 [*3*]. In the minds of converters (probably still haunted by the specter of a severe benzene shortage and ever escalating prices) must be the question of from where will all the benzene come.

At present, refiners are the chief suppliers of benzene, which is produced by reformate extraction and toluene hydrodealkylation (the latter being one of toluene's primary outlets). Future olefin plants probably will be the principal suppliers of the higher demand for benzene and other aromatics, with refiners supplying any additional aromatics needed. The future of toluene dealkylation as a source of benzene is less clear and may very well depend upon spot economics.

Butadiene—Another major coproduct of future ethylene plants is expected to be butadiene. In 1974, approximately 3.7 billion lb were produced, with approximately 60% of this made by the dehydrogenation of butane/butene streams. In the future, substantial quantities of butadiene will likely become available from ethylene plants using heavy feed, with the butadiene obtained by extraction.

Excess butadiene has been predicted by several forecasters. The degree of any excess will depend upon what proportion of future ethylene plants will use ethane or ethane-and-propane mixtures as feeds, rather than heavier feeds that will yield substantial quantities of butadiene.

The major outlet for butadiene is rubber for tires, for which the outlook is dim. The non-synthetic-rubber consumption of butadiene accounts for approximately 25% of the total, and this part has been growing at a more rapid rate than consumption by the automotive industry. A balancing of the positive and negative factors in butadiene's future reveals growth projections of

Major derivative feedstocks in the U.S. — **Table I**

	Approximate U.S. production in 1974, billion lb.	Average annual growth (1964-1974), %
Ethylene	23.6	10.6
Benzene	10.9	7.4
Butadiene	3.7	3.9
Propylene	10.0	10.8
Toluene	7.5	7.7
o-Xylene	1.05	12.0
p-Xylene	2.7	25.0

from 4–5%/yr. If a surplus of butadiene does, in fact, materialize, this will furnish the incentive to develop new products based on this major diolefin, such as the Du Pont process for hexamethylenediamine, which is used to produce nylon.

Bases of U.S. organic chemical industry

Questions arise as to possible outlets for these huge quantities of ethylene and coproducts.

Table I provides market data for the major organic feedstocks in the U.S. These feedstocks are among the top 50 chemicals (organic and inorganic) in terms of volume [*5*]. Table II presents data for a limited number of the major derivative (downstream) products of the U.S. organic chemicals industry. These derivatives also rank among the top 50 organic-chemical products.

Major derivative products in the U.S. — **Table II**

	Approximate U.S. production in 1974 billion lb.	Average annual growth 1964-1974, %	Projected average annual growth 1975-1985, %
Acrylonitrile	1.41	9.0	8-10
Adipic acid	1.52	7.1	7.5 - 9.5
Aniline	0.55	12.0	10-11
Cyclohexane	2.34	6.0	7-8
Ethanol (synthetic)	1.9	–0.9	1-2
Ethylene dichloride	7.7	13.4	7-8
Ethylene oxide (all forms)	3.89	6.0	7-9
Isopropanol	1.91	2.4	3
Maleic anhydride	0.28	9.0	6-7
Phenol	2.32	7.7	8-9
Phthalic anhydride	0.98	6.0	5-6
Propylene oxide (all forms)	1.74	12.0	10-11
Styrene	6.0	8.4	7-9
Polyethylene (all forms)	8.81	12.0*	9-11 (all)
Polypropylene	2.25	24.0	11-14
PVC and copolymers	4.85	11.0	7-8
Vinyl acetate	1.4	12.3	10.0
Terephthalic acid (crude, all forms)	3.43	26.0	10-12

*For low density, 16% for high density.

From the foregoing, it is apparent that ethylene, propylene and benzene are the feedstock foundations of the commodity organic-chemical industry (included in this category are plastics and rubber).

Plastics and fibers are ethylene's primary consumer outlets. Polyethylene (all forms) accounts for approximately 40% of ethylene consumption, followed in order by ethylene oxide and ethylene glycol, polyvinyl chloride and styrene—all of which account for over 70% of ethylene consumption.

Propylene similarly finds its way to the consumer primarily as a plastic or a fiber. Polypropylene, acrylonitrile, propylene oxide, isopropanol and cumene (for acetone and phenol) account for over 70% of the propylene consumed by the U.S. organic chemical industry. The pattern is the same for the rest of the industrialized world.

Benzene's three major chemical outlets are styrene, phenol and cyclohexane; consumer usages are the same as for ethylene and propylene.

Outlook for plastics and fibers

A better perspective of the importance of ethylene and its coproducts to the economy can be obtained from an overview of the two previously mentioned consumer outlets, plastics and fibers.

Total U.S. demand for plastics by 1985 will be on the order of 72 billion lb/yr, predicts J.S. Hartman [*11*], an almost threefold increase from 1974's 25 billion lb, of which about 21 billion lb were thermoplastic resins [*6*]. Of this group, polypropylene ranks fourth in terms of production volume, exceeded by polyethylene (all types), polyvinyl chloride and polystyrene.

The large quantities of propylene that will be available from new ethylene plants are expected to create a significant price differential between ethylene and propylene, one that should promote the growth of polypropylene through new-product development. By 1980, a polyolefin market of over 20 billion lb/yr, made up of 10 billion lb/yr of low-density polyethylene and 5 billion lb/yr each of high-density polyethylene and polypropylene, has been predicted by J.M. Jordan [*13*].

Substantial increases in the volumes of other plastics are also expected. Plastics for vehicles are seen rising from about 1.1 billion lb/yr at present to 3 billion lb/yr by 1980, with unsaturated polyesters holding the lion's share [*7*]. Polyurethanes, which have many diverse applications, and for which new ones are continually being announced, have been growing at a rate of about 20%/yr. Some forecasters are projecting a market growth from the present 1.6 billion lb/yr to 3 billion lb/yr by 1980. Isocyanates based upon toluene and benzene, as well as polyols based upon propylene oxide, will be the major beneficiaries.

From the foregoing, it is obvious that the ethylene industry plays a significant role in the economics of the industrialized world. Except for 1974 and 1975, plastics and fibers have declined continuously in price as volume has risen [*15*]. It has been predicted that they will continue to displace their natural competitors on the basis of performance and price, particularly if raw-material cartels come to be established.

With the constantly rising worldwide demand for food, a still larger share of U.S. land can be expected to be diverted from the production of natural fibers. U.S. producers of synthetic fibers foresee demand increasing 60% over the next five years [*8*].

Although 1974 was a poor year for the chemical industry, the production of synthetic fiber amounted to approximately 6.9 billion lb, with polyesters accounting for 2.9 billion lb. At present, as much ethylene glycol goes into producing the rapidly growing polyesters as is used for antifreeze, and there is no doubt that polyesters will be the major future outlet for this chemical. Major hikes in *para*-xylene production can be expected and, in fact, are already underway to meet growing demand for polyesters.

Raw materials in a changing ethylene world

Not too long ago, a prospective ethylene producer was primarily concerned with raising capital for a plant and arranging raw-material and product contracts. Manufacturing costs were on the order of 1.8-2.3¢/lb and the economics were governed by the cost of the raw material and the selling price for the sole product, ethylene. The raw material, ethane, was predictable in price and available. This was when ethane sold for 1¢/lb, and fuel gas was available at 20¢/million Btu.

During the late 1960s and early 1970s, the U.S. ethylene industry was based, as it is today, upon natural-gas liquids, with ethane or ethane-and-propane mixtures as the predominant raw materials. When natural-gas production peaked in the early 1970s, so did the availability of propane and ethane from this source. The economic limits for the application of intensive recovery methods such as the cryogenic extraction of ethane from natural gas, or deep absorption for propane recovery, were reached for the existing price structure. Cryogenic extraction has recently received renewed consideration. (Propane recovery from natural gas is about 75%.) Approximately 15% of the ethane from natural gas is currently recovered in this way.

As gains in natural-gas production narrowed, and propane and ethane prices started to move up, some ethylene producers with diverse chemical operations turned to such raw materials as condensates, naphthas, butanes, and refinery raffinates, because they could use or dispose of the larger quantity of coproducts that were produced. Although U.S. industry has been heading toward liquids cracking, the pace had previously been relatively slow because of an atmosphere of pseudo-stability.

Of course, the bubble burst in October 1973, when the Organization of Petroleum Exporting Countries (OPEC) raised crude prices three- to fourfold. This act rapidly brought the developing deficiencies in supplies of ethane and propane to the surface. The price of naphtha skyrocketed on international markets. The new crude price rippled throughout the world, boosting all unregulated hydrocarbon prices. For the first time, the ethylene industry found itself a competitor for raw materials (for example, with the substitute-natural-gas industry for naphtha and lighter hydrocarbons) that had recently been abundant.

Representative of these changes, ethane has gone from 3¢/gal to 13–16¢/gal, and propane from 5¢/gal to 20–25¢/gal. Intrastate fuel gas now costs about $2/million Btu; naphtha ranges from 26–30¢/gal; imported crude runs from $13–14/bbl, in contrast to about $3/bbl a few years ago. (These prices are as of mid-October 1975, in the Houston area, and represent contract rather than spot prices. Further price increases are still being predicted. Other than for crude, the latest OPEC increase is not fully reflected in these prices; the entire pricing situation is in a state of flux and involved in political controversy.)

Compounding the raw-material problem has been a severe escalation in plant costs, which for new chemical plants in general have soared approximately 60% over the last four years. The impact of the shortage of ethane and propane upon prospective manufacturers of ethylene has been revolutionary. Forced to consider cracking heavier raw materials (such as naphtha and gas oils), they find themselves facing a host of new problems. Not only are capital-investment requirements appreciably higher, but the heavier raw materials produce coproducts in substantial quantities that bear upon economics almost as much as the feed and, at the very least, place an additional burden on the sales department. A prospective ethylene manufacturer may not only have to cope with uncertainty of specific supplies but also with volatile prices. Consequently, a plant must now be capable of handling a variety of raw materials to ensure supply, as well as be in a position to purchase the cheapest satisfactory feed at any time.

The relationship between higher raw-material prices and ethylene-plant economics can be seen in Table III. The raw-material requirement for a 1-billion-lb/yr polymer-grade ethylene plant ranges from 28,000 bbl/stream-day of ethane to 43,000 bbl/stream-day of gas oil, which amounts to approximately $60 million/yr for the former and approximately $170 million/yr for gas oil (the latter amount corresponds to approximately 17¢/lb for ethylene).

Coproduct, feedstock and operations

Table III provides product distribution data for various feedstocks. The corresponding ethylene product basis is 1 billion lb/yr of polymer-grade ethylene (100%) and approximately 500,000 lb/yr of ethane associated with it. The yield structures shown, particularly for the heavier products, are governed by the exact nature of the feedstock, cracking conditions, and how the plant processes various streams. All naphthas and gas oils are not alike [*1, 2*]. The yield structures (which are based on Stone & Webster's extensive commercial and pilot plant experience) are representative for the given feeds and generally accepted product purities. The yield pattern can be varied, to a degree, to suit particular economics.

The table does not reflect the use of any of the streams as fuel, in order to provide a fairly complete material balance picture. The C_2/C_3 acetylenes are hydrogenated.

Some manufacturers recover the contained acetylenes. Depending upon the feed, up to 49 million lb of combined C_2/C_3 acetylenes are formed in cracking,

Yield structure for 1-billion lb/yr ethylene plant — Table III

	Ethane†	Ethane/propane (30/70)	Propane†
Feed	1,244,100	1,744,200	2,111,700
Annual feed, thousand bbl/stream day	28.3	–	35.6
Products			
Propylene‡	–	201,000	325,600
Total C_4 mixture	–	64,800	89,800
Butadiene in C_4 mixture	–	46,800	64,600
Pyrolysis gasoline	–	69,600	102,000
BTX in pyrolysis gasoline	10,300	44,400	68,800
Fuel oil‡	90,700	500	900
Residue gases	152,900	407,800	591,800
Hydrogen for hydrotreating	–	–	1,100

	Kuwait naphtha		Heavy Kuwait naphtha	
Cracking severity	High	Moderate	High	Moderate
Feed	3,125,900	3,707,300	3,388,600	3,962,400
Annual feed, thousand bbl/stream day	37.0	43.9	38.9	45.5
Products				
Propylene‡	419,800	627,000	444,600	649,500
C_4 mixture	248,400	408,000	274,500	450,000
Butadiene in Total C_4 mixture	141,100	171,400	151,800	179,600
Pyrolysis gasoline	740,900	1,101,300	877,200	1,220,300
BTX in pyrolysis gasoline	500,300	412,000	612,200	585,300
Fuel oil **	195,400	117,500	240,600	136,700
Residue gases	520,900	453,000	551,100	505,300
Hydrogen for hydrotreating	5,100	–	–	–

	East Texas gas oil	West Texas light gas oil
Cracking Severity	High	Moderate
Feed	3,951,200	4,362,700
Annual feed, thousand bbl/stream day	40.6	43.7
Products		
Propylene‡	557,400	614,200
Total C_4 mixture	412,500	457,800
Butadiene in C_4 mixture	186,800	202,900
Pyrolysis gasoline	794,200	868,200
BTX in pyrolysis gasoline	409,100	450,900
Fuel oil	729,400	1,006,500
Residue gases	457,200	415,600
Hydrogen for hydrotreating	10,200	–

* Except when noted, quantities are in thousand lb/yr—8,000 hr/yr
† For ethane and propane, 60% and 93% conversions per pass, respectively
‡ Chemical grade
** For the ethane feed case, this stream reflects a mixture of C_3 and heavier hydrocarbons

with the heavier feeds producing the higher quantities. Pyrolysis gasoline, in order to be marketed, must undergo a single stage of hydrotreating to eliminate diolefins. The hydrogen consumed by this hydrotreating operation is reported separately for the various feedstock cases reviewed. The residue gas, as reported in this table, would be the source of this hydrogen and does not reflect this hydrogen consumption. This adjustment is made in the subsequent calculation of transfer price (Table VI). The aromatic residue from ethane extraction is not hydrotreated, because it is used as plant fuel.

Ethane feed

Ethane feed is unique because it produces very few coproducts (Table III). The residue gases are usually burned as fuel, and the aromatic (or fuel oil) residue is crudely separated into a C_3/C_4 LPG fraction and an aromatics concentrate, which are either sold, burned, or absorbed into a refinery operation. Capital and marketing requirements are minimal.

Ethane feed is ideal for the manufacturer solely interested in selling or using ethylene without having to contend with the vagaries of coproduct marketing. So attractive is this feed that some corporate planners have considered importing LNG and separating it into fuel gas and liquids (ethane, propane and butane), selling the former, and using the latter for ethylene production. The refrigerating capability of the LNG could be used in the ethylene plant. The alternative of importing ethane/propane streams has been raised [*9*].

Propane feed

The effect of a feedstock heavier than ethane can begin to be seen in Table III: propane produces a two-and-a-half-fold increase in residual gas, and over a tenfold rise in propylene; twice as much butadiene; and significantly more aromatic pyrolysis gasoline.

The propylene formed is sufficient to support one or two polypropylene lines. For this purpose, it would have to be upgraded to polymer grade, at an incremental transfer price increase of less than 0.3¢/lb. This operation would provide about 1% more propane feedstock for pyrolysis. Although the benzene in the pyrolysis gasoline is insufficient for any major aromatics production, the entire pyrolysis-gasoline stream could be sold as gasoline blending stock, or for benzene-toluene-xylene (BTX) extraction.

The ethane/propane case presented in Table III provides a yield structure intermediate to those of ethane and propane feeds. Although the coproducts are not as extensive as with propane, if the C_3/C_4 and pyrolysis gasoline streams are nevertheless separated and marketed, the equipment and capital investment will obviously approach that for a propane-based plant.

Naphtha feed

Table III shows the variation in feed requirements and coproduct distribution of two naphthas from the same crude, but of a different distillation range, as well as the differences that can be obtained if the severity of cracking is varied. With heavier feedstocks, their origin and detailed analyses are important operating and economic considerations [*1, 2, 10, 12*]. When a range of feedstocks are considered, failure to take into account the physical properties and makeup of the liquid feedstocks in the design and operation of the cracking, waste-heat-recovery, and quenching sections can result in unnecessarily rapid coking of the furnace coils, or fouling of the downstream heat-exchanger equipment, as well as a change in yield structure.

The naphtha feed requirements, on a weight basis, are anywhere from 50% to almost 100% greater than for propane feed. Further, the proportion of coproducts is appreciably higher (and will be so regardless of the severity of cracking), which calls for more heat to the cracking furnaces, and larger downstream processing equipment.

High-severity cracking of Kuwait full-range naphtha compared to propane feed shows propylene about 29% greater, and the C_4 mixture up 280% (Table III). The contained butadiene has risen to 141 million lb. The pyrolysis gasoline has increased seven-fold and contains a substantial amount of aromatics, with a benzene content on the order of 270 million lb/yr. Also, significantly, the heavy ends, in the form of fuel oil, have increased almost 200%.

The other naphtha-feed cases show that the quantity of coproducts can be larger yet, and that the mode of operation is a significant factor in this respect. For both naphthas of Table III, the feedstock required for high-severity cracking is approximately 18% higher than for moderate severity, and this additional feed appears as additional coproducts, with the exception of residue gas and fuel oil (which has decreased). Almost two-thirds of the additional feed appears as pyrolysis gasoline. Further, for each type of naphtha, higher-severity cracking produces additional aromatics in the form of a larger BTX fraction.

High-severity cracking results in a smaller amount of contained C_8 aromatics, particularly xylenes, and correspondingly more toluene and benzene. If benzene were the sole product desired, approximately 500 million lb/yr could be produced by hydrodealkylation of C_7 and C_8 aromatics. The extraction of aromatics from pyrolysis gasoline requires a second stage of hydrotreating in order to eliminate olefins. The required hydrogen could be obtained from the residue gas stream at very little additional cost.

The heavy Kuwait full-range naphtha would be of particular interest to a manufacturer with a need for aromatics, particularly C_8s (such as xylenes), because these make up over 50% of the BTX fraction at moderate-severity cracking conditions. This illustrates the effect of feed distillation-range on coproduct distribution for naphthas originating from the same crude oil.

Some other options are available to the prospective ethylene manufacturer faced with a varying demand for propylene. Although propylene is generally a poor cracking feedstock because of its coking tendencies that shorten onstream time, it has been successfully cracked to ethylene under certain conditions without a significant reduction of onstream time. Less total cracking feed is required with this mode of operation. The desirability of this practice is a matter of economics, because the savings in net naphtha feed must be weighed against the sale of propylene for either LPG or chemicals, or for plant fuel.

Regarding disposal of the C_4 mixture, the couple of available options depend on market circumstances. The C_4 mixture can be totally hydrogenated and recycled to the "crackers" for ethylene production, or it can be sold as LPG. Normally, it would be more profitable to extract and purify the butadiene or to sell the entire stream to a butadiene extraction plant, with the butenes going to a refiner for alkylation or to the LPG market.

Gas oil feed

The yield structure for the cracking of gas oils is given in Table III. The West Texas light gas oil is cracked at moderate severity. The feed quantities are high compared with the preceding naphtha cases, and higher than they would be for moderate cracking of naphtha. The quantity of propylene produced is intermediate between high- and moderate-severity cracking of the full-range naphtha. The C_4 mixture and contained butadiene are similar in magnitude to the moderate-severity naphtha case. The aromatics produced per pound of feed are less than those from the preceding naphthas. The pyrolysis gasoline is similar in quantity to that of high-severity cracking of the full-range naphtha, but the greatest difference is in the pyrolysis fuel oil, which has increased three- to fourfold.

In general, the product outlets required for cracking gas oil or naphtha will be the same, except for the substantial amounts of pyrolysis fuel oil to be disposed of.

The pyrolysis fuel oil, whether produced by naphtha or gas oil cracking, requires special handling. It is a highly aromatic polynuclear material that contains unsaturated gum formers and possibly a small amount of free carbon. It can be blended with paraffinic oils only

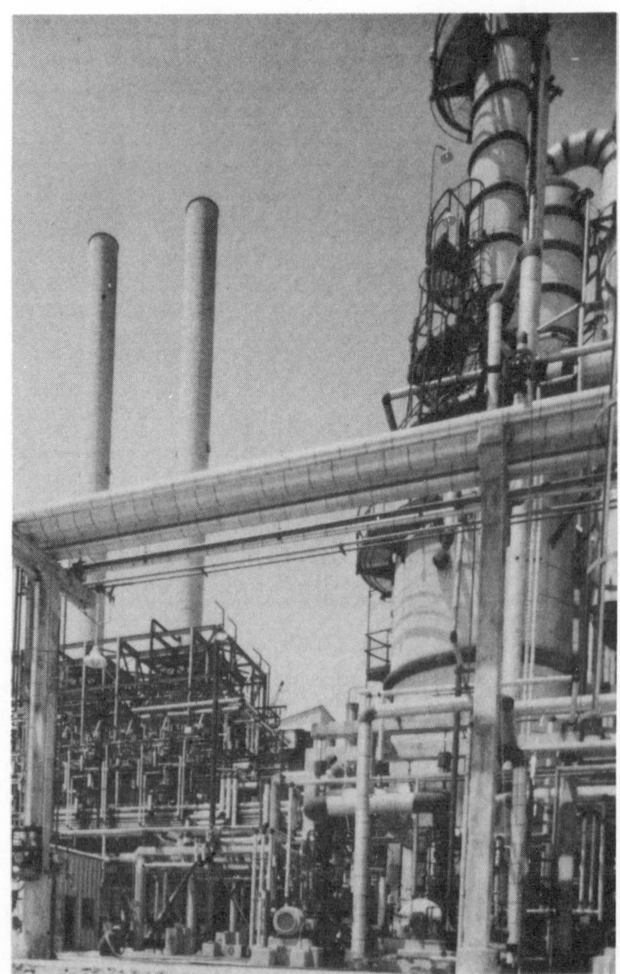

Primary fractionator (lefthand column) maintains thermodynamic efficiency of heavy-feedstock plant

Ethane/propane
Pyrolysis
Direct steam generation
Water quench
Compression and gas treating
Demethanizer
Aromatic residue
Hydrogen
Residue gas
Ethane recycle
Ethylene product
Ethylene tower
Acetylene removal
Deethanizer
Alternate route
C_4 or C_3/C_4 product
Debutanizer
Depropanizer
C_3 product

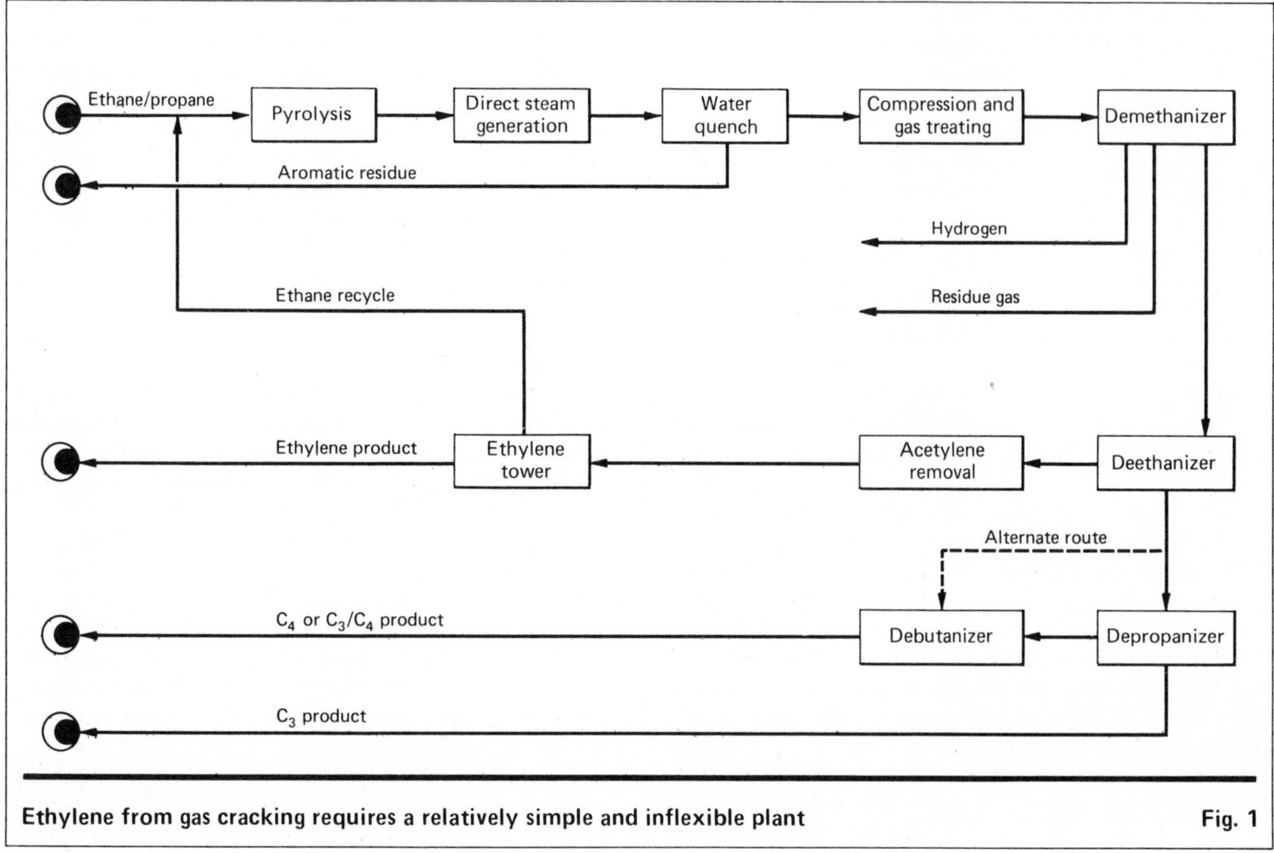

Ethylene from gas cracking requires a relatively simple and inflexible plant **Fig. 1**

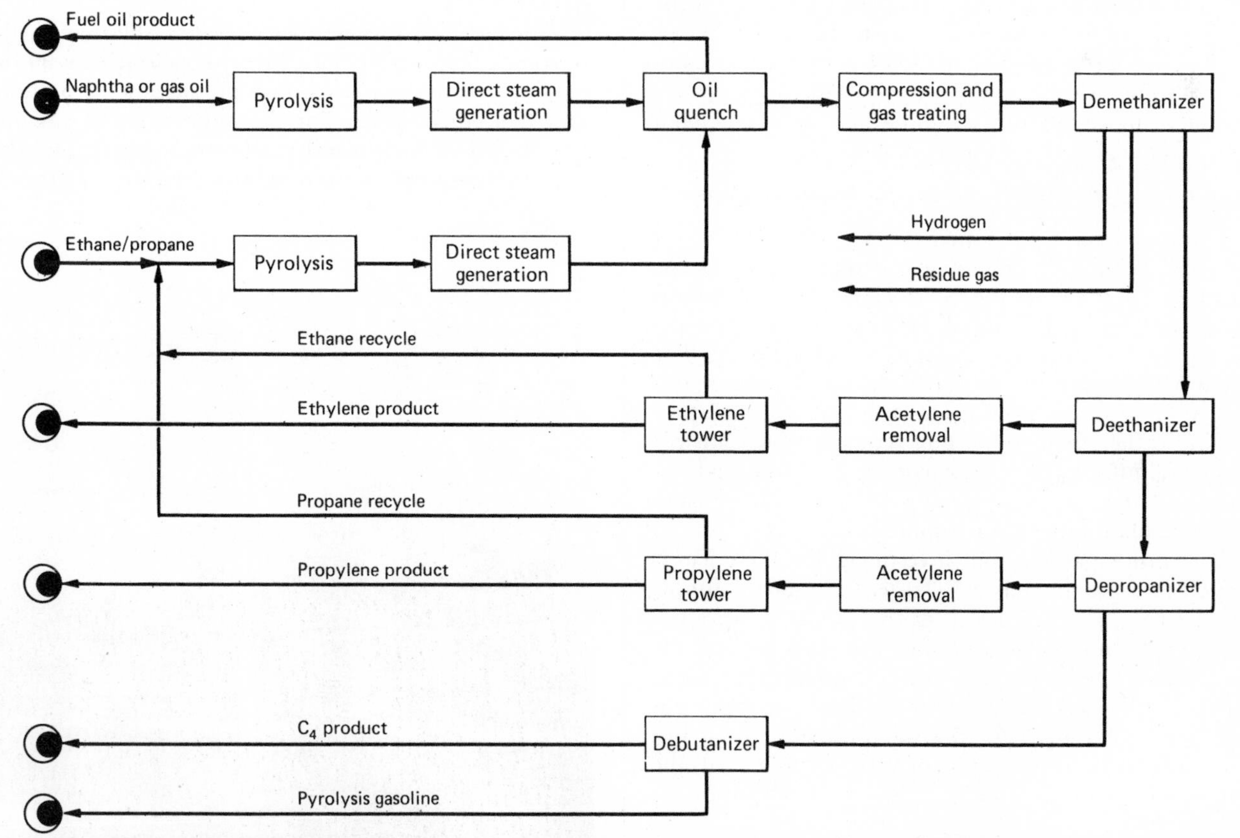

Fuel oil product
Naphtha or gas oil
Pyrolysis
Direct steam generation
Oil quench
Compression and gas treating
Demethanizer
Ethane/propane
Pyrolysis
Direct steam generation
Hydrogen
Residue gas
Ethane recycle
Ethylene product
Ethylene tower
Acetylene removal
Deethanizer
Propane recycle
Propylene product
Propylene tower
Acetylene removal
Depropanizer
C_4 product
Debutanizer
Pyrolysis gasoline

Cracking liquids to produce ethylene calls for a more extensive plant **Fig. 2**

to a limited degree and can present combustion and handling problems. (Nevertheless, it can be burned by a utility or other such large user, or added to a large pool of heating oil, such as is available to refiners. In fact, it has been used to supply much of the pyrolysis furnace fuel required, as well as the entire fuel for the offsite boiler.) These adverse characteristics of pyrolysis fuel oil are reflected in its present market price, which is approximately $1/bbl less than that for Bunker C oil.

The 650°F-plus fraction of the residual oils (referred to as pitch) can also be burned by a utility, or used to produce carbon black feedstock or as a binder pitch for electrodes.

Another outlet seriously being considered for residual fuel oils is as a source of synthesis gas, via partial oxidation—in particular, for the production of ammonia and methanol. It appears that a substantial amount of new ammonia and methanol will probably be based on this feed in the future.

The foregoing discussion has been confined to atmospheric gas oils. Vacuum gas oils can be suitable feedstocks (and indeed have been and are being considered as such), depending upon their characteristics and the economic situation peculiar to a potential manufacturer. Stone & Webster has commercially cracked certain gas oils with end-points above 1,000°F.

In summary, it is obvious that, when cracking a heavy feedstock, coproduct marketing will govern the price of ethylene. This factor would generally eliminate all but major diversified petrochemical producers (or those integrated at least marketwise with a refiner) from considering a heavier feedstock, such as naphtha or gas oil. In fact, the recent construction of ethylene plants has been predominantly by petroleum companies using heavy liquid feedstocks. The future will no doubt bring about further affiliations between major petrochemical producers and refiners.

Feedstock, flexibility and plant investment

Estimated capital requirements for a 1-billion lb/yr ethylene plant in 1975 using different feedstocks are given in Table IV. As the feedstock ranges from ethane to heavier feeds, the capital required increases, particularly for the inside-battery-limit part of the plant. The focus will be upon this section of the plant, because it is the more sensitive measure of technological and processing rather than supporting requirements. Outside-battery-limit capital investment tends to reflect location and derivative supporting facilities, rather than ethylene plant processing, requirements.

The inside-battery-limit investment for West Texas gas oil is approximately 50% larger than for the simple ethane case, a difference in capital alone of approximately $40 million (which, of course, does not include additional investment for outside-battery-limit facilities). The capital requirements for these particular gas oils are higher than for the naphthas discussed, but generally (with the exception of West Texas gas oil) the difference is not an unduly large one, being approximately 10%.

Feedstocks govern the plant

Diagrammed in Fig. 1 and 2 is, respectively, the processing sequence for a gas feedstock, such as ethane or propane, and a liquid feedstock, such as naphtha or gas oil. Obviously, liquids cracking requires a more extensive plant: an additional pyrolysis furnace for cracking coproduct ethane and propane, and an effluent quench exchanger, followed by an oil quench and a primary fractionator for fuel oil separation (in contrast to the simple direct-contact water quench tower of the gas cracking plant). The liquids cracking plant also contains a propylene tower, C_3 acetylene removal unit, and probably (not shown in Fig. 2) a unit for first-stage hydrotreating of pyrolysis gasoline. Cracked fuel oil also appears as a product.

An ethane plant is relatively simple and inflexible. The depropanizer and debutanizer may be eliminated by some manufacturers, and the C_2, C_3 and aromatics residue (all of relatively small quantities) may be combined and burned as fuel, or disposed of.

Among the most expensive components of an ethylene plant are the furnaces [4], and the associated quench boilers and quench towers, including the primary fractionator that, in the heavy-liquids plant, supplements water quenching. Furnace and quench boiler costs will range from approximately 20% to 35% of inside-battery-limit costs for feedstocks ranging from ethane through heavy liquids. An ethane plant normally uses a simple, relatively inexpensive high- or medium-residence-time furnace. A low steam-to-gas feed ratio is used in the furnace, and the relatively nonfouling cracked-gas effluent stream is first cooled in steam-generating heat exchangers and then in a direct-water-quench tower.

This type of furnace becomes less suitable as the feedstock becomes heavier, because yield selectivity decreases, resulting in excessive production of gasoline and fuel constituents. For the naphthas and gas oils, the so-called high-severity, short-residence-time furnaces are required, with a greater quantity of dilution

Battery of short-residence-time, high-temperature furnaces crack gas oil in 1-billion-lb/yr ethylene plant

Approximate capital requirements for 1-billion lb/yr ethylene plant — Table IV*

Feedstock	Inside battery limits	Outside battery limits	Total
	$ million		
Ethane	82	14	96
Propane	92	18.7	110.7
Kuwait full-range naphtha, High-severity cracking	105	26	131
Kuwait naphtha, moderate-severity cracking	107	30	137
East Texas gas oil, high-severity cracking	115	30	145
West Texas gas oil, moderate-severity cracking	122	31	153
Propane and Kuwait naphtha, high-severity cracking	108	†	
Kuwait naphtha and East Texas gas oil, both high-severity cracking	118	†	

1st stage pyrolysis gasoline hydrotreater

Propane	$0.7 million
Naphtha full-range, high-severity cracking	$2.7 million
East Texas gas oil	$3.0 million

*Investments do not have any allowance for satellite units

†Will depend on storage, etc., but will be similar to base cases (Table III)

steam, so as to maximize ethylene yield and minimize coking, and hence to obtain reasonable onstream life between major decokings. These shorter-residence-time furnaces are more expensive, because of higher firebox temperatures and fired duties, and convection-section and quench complexity.

The fouling tendency of the heavier feeds increases through the entire furnace and quench system, particularly as the quantity of fuel oil made increases. It is in these areas that the utmost demands are made on the designers' experience. The cracked gaseous effluent must be kept at a higher temperature, and higher-pressure steam must be generated (on the order of 1,500 psig), to avoid polymer deposition (otherwise, quenching by the direct injection of oil is necessary, depending upon the feed).

As a result of these practices, less heat is recovered from the cracked gases in the effluent heat-exchanger system, and a much larger heat load is imposed on the primary fractionator, which must not only be larger but must also condense and separate oils and tars and maintain the thermodynamic efficiency of the plant by recovering heat at various levels. Special design features should be incorporated into the fractionator to avoid fouling, such as the use of special trays.

The major increase in inside-battery-limit capital expenditures for heavy-liquids feedstocks occurs in the warm end of the plant, just discussed. It is necessary that the compression, purification and refrigeration sections reflect the design requirements for a given feed, but in general these requirements demand more-careful engineering rather than a major increase in capital expenditures.

Bases for economic calculations — Table V

Raw materials*	
Ethane:	4.93¢/lb
Light gas oil:	$12.00/bbl
Naphtha:	$12.60/bbl
Propane:	5.69¢/lb
Catalyst & chemicals for ethylene plant	
Ethane/propane feeds:	$300,000
Naphtha/gas oil feeds:	$500,000
Catalysts, chemicals and inhibitors for hydroheating pyrolysis gasoline	
Propane feed:	$13,000/yr
Naphtha feed:	$58,000/yr
East Texas gas oil:	$62,000/yr
Solvent and chemicals for butadiene extraction unit:	$110,000/yr
Products	
Butadiene, polymer grade:	17¢/lb
C_4 mix product:	10¢/lb
Pyrolysis fuel oil:	$8.50/bbl
Propylene, polymer grade:	8.50¢/lb
Propylene, chemical grade:	8¢/lb
Pyrolysis, gasoline:	$14.28/bbl
Residue gas:	$2/million Btu
Utilities	
Electrictiy:	1.2¢/kW
Treated water:	$0.80/M gal
Purchased water:	$0.40/M gal
Purchased fuel gas:	$2/million Btu
Labor and maintenance	
Operators, including outside battery limits: (8 men/shift for ethane/propane—35 men total, 10 men/shift for naphtha/gas oi'—44 men total)	$21,000/yr/man (including benefits)
Supervision, including outside battery limits: (Ethane/propane plant—3 men, naphtha/gas oil plant—4 men)	$25,000/yr/man (including benefits)
Maintenance:	4% of inside-battery-limit investment
Overhead charges	
Direct:	35% of labor and supervision
General plant:	65% of labor, supervisor, and maintenance
Depreciation, straight line:	10% of inside-battery limits, 5% of outside-battery limits
Insurance and taxes:	1.5% of fixed total capital
Interest on working capital:	10%
Working capital:	10% of fixed total capital

*For transfer price calcualtion, 1% feed loss is allowed.

Flexibility and capital requirements

In recent years, ethylene manufacturers have indicated the greatest interest in a plant sufficiently flexible to produce ethylene from feedstocks ranging from propane through light naphtha, or, more commonly, specified amounts of each. Interest has alternatively focused upon a plant that would use either naphtha or gas oil, or specified amounts of each, as feedstock. By the pairing of such feedstocks, a reasonably broad flexibility can be obtained without inordinate capital and operating penalties.

For the two cases in which inside-battery-limit costs are estimated to demonstrate capital requirements for flexibility (cracking of propane through naphtha, and naphtha through gas oil, Table IV), it is apparent that the capital requirements are set essentially by the heavier feedstock. Going from propane to naphtha as a feedstock involves a substantial increase in capital, whereas going from naphtha to propane does not. For the two cases cited, the additional inside-battery-limit-capital required to provide flexibility (compared to designing a plant solely for the heavier feedstock) is approximately $3 million in each case. Of course, additional investment in outside facilities must be expected (particularly for the propane-and-naphtha combination), primarily for storage.

It is only when processing widely different feedstocks (such as ethane through gas oils, or a good light naphtha through a bad gas oil), while maintaining capacity, that increases in capital requirements become more substantial.

It is not prudent to overgeneralize or oversimplify the matter of flexibility. For an economically viable plant, each feedstock case must be studied in detail to determine the limiting equipment and operating conditions.

Economic analyses

All marketing and engineering analyses must eventually be reduced to dollars and cents, that is, to profits.

No one can rigorously calculate anyone else's production costs and profitability. Each such analysis must be tailored to a certain producer and situation. It must incorporate those items particular to the manufacturer—whether (to name a few) these be accounting practices, plant location, resources, availability of existing facilities, market outlets (open or captive)—and assign transfer prices to the various products, contingent upon the manufacturer's evaluation.

The economic analyses presented here are primarily concerned with determining a representative transfer price for ethylene from several feedstocks, under carefully defined conditions. Additionally, an analysis is presented that evaluates the relative profitability of upgrading crude butadiene to polymer grade, rather than selling the total C_4 stream for its butadiene content.

Despite the limitations stated, the analyses presented should be useful to a prospective manufacturer of ethylene or to an existing or prospective purchaser of ethylene or its coproducts, or both. The purchaser can—by incorporating factors reflecting future escalation in raw-materials, labor and construction costs—develop a reasonable approximation of future ethylene and coproduct manufacturing costs and, hence, selling price.

Transfer price bases

Table VI presents price determinations for the production of 1 billion lb/yr of ethylene from various raw materials. The bases for these determinations can be found in Table IV, which deals with capital-investment requirements, and in Table V, which presents the factors pertaining to operating requirements and amortization of facilities. Applicable items of Table V are also used in Table VIII.

The raw-materials and product prices in Table V are considered representative of new contract prices (rather than spot prices) as of mid-October 1975. For both raw materials and products, a range of prices were obtained that depended on such things as the nature of the contract, and the volume purchased. The prices are generally close to the highest received as fully influenced by the recent OPEC price boost, and federal regulatory changes being considered. The high raw material prices are particularly significant considering the present oversupply of ethylene because of the current recession.

It was difficult to secure several prices from different sources for some of the products, such as for the crude butadiene stream or pyrolysis fuel oil. For the latter, the price of $8.50/bbl may be somewhat low and certainly is not expected to prevail in the future.

The labor rates reflect wages being negotiated for early 1976. Staffing is based on Stone & Webster's experience, with an allowance for the outside-battery-limit facilities, as defined later. This, of course, can be quite variable, depending on company operating philosophy and the extent of offsite facilities.

Overhead charges, although representative, can of course vary from company to company. Although discounted-cash-flow methods are common, straight-line depreciation is used for simplification. Working capital is arbitrarily taken as 10% of total fixed capital investment, which is representative in order of magnitude. Although it is a significant variable, startup cost is not included because it varies and is interpreted differently by companies, as are methods of capitalizing it. Construction interest charges and insurance costs are also not included.

Although the capital investment data in Table IV do not reflect rigorous plant prices, they represent reasonable and consistent approximations for the Gulf Coast during 1975. No significance should be attached to the fact that plant investment figures are not always rounded out to the nearest million. Price fluctuations and escalatory clauses for materials still persist and some deliveries are long term. Of course, even during a period of fairly settled equipment prices, plant costs can vary, and usually will fall within a range dependent upon the purchaser's equipment and plant specifications, and upon plant location and the labor market.

The inside-battery-limit capital requirements include all materials and equipment required for actual processing. The major compressors are steam-driven.

Without doubt, the greatest variable in capital investment can be with regard to outside-battery-limit facilities. To be considered are such factors as location,

Transfer price determinations for production of 1-billion lb/yr ethylene plant — Table VI

	Feedstocks							
	Ethane		Propane		Full-range naphtha, high-severity cracking		Gas oil †	
Total fixed capital, $ million (10% contingency added)	105.6		121.8		147.1		162.8	
	$ thousand, annual	Ethylene, ¢/lb	$ thousand, annual	Ethylene, ¢/lb	$ thousand, annual	Ethylene, ¢/lb	$ thousand, annual	Ethylene, ¢/lb
Raw materials								
Feed	61,952		121,313		157,054		163,943	
Catalyst and chemicals	300		313		558		662	
Total	62,252	6.23	121,626	12.16	157,612	15.76	164,605	16.46
Utilities	3,873	0.39	1,345	0.13	1,972	0.19	2,245	0.22
Labor and maintenance								
Labor and supervision	810		810		1,024		1,024	
Maintenance	3,608		4,048		4,740		5,192	
Total	4,418	0.44	4,862	0.48	5,704	0.57	6,216	0.62
Overhead charges								
Direct	284		284		358		358	
General plant	2,872		3,160		3,747		4,040	
Depreciation	9,790		11,150		13,280		14,630	
Insurance and plant taxes	1,584		1,827		2,207		2,442	
Interest on working capital	1,056		1,218		1,471		1,628	
Total	15,586	1.56	17,639	1.77	21,063	2.11	23,098	2.31
Total of above	86,129	8.61	145,472	14.54	180,411	18.63	190,164	19.61
Coproduct credit (including fuel oil and gasoline)*	0	0	(48,490)	(4.85)	(102,923)	(10.29)	(137,337)	(13.73)
Not manufacturing cost	86,129	8.61	96,982	9.69	83,488	8.34	58,827	5.88
Return of 25% on fixed investment	26,400	2.64	30,450	3.06	36,775	3.68	40,700	4.07
Transfer price	112,529	11.25	127,432	12.75	120,263	12.02	99,527	9.95
*Exported fuel credit as percent of coproduct credit		0	17		7.7		8.3	

† Capital investment and coproduct distributions correspond to East Texas gas oil

Sensivity analyses of transfer-price components — Table VII

	Ethane		Propane		Naphtha, high-severity cracking		Gas oil	
	Annual increment							
Deviation	$ million	Ethylene, ¢/lb	$ million	Ethylene, ¢/lb	$ million	Ethylene, ¢/lb	$ million	Ethylene, ¢/lb
20% of total fixed investment*	8.73	0.87	10.06	1.0	12.01	1.2	13.27	1.33
10% of raw materials cost	6.25	0.63	12.16	1.22	15.76	1.58	16.48	1.65
10% of coproduct credit	0	0	4.86	0.49	10.29	1.03	13.73	1.37

* Variation in working capital ignored

whether a "grass roots" plant or an expansion is involved, whether shipping facilities and docks are required, how much ethylene storage is needed.

The capital investment estimates of Table IV are not intended to represent a "grass roots" plant with indeterminate major outside-battery-limit facilities, but rather a new, or "across the fence," plant with limited and defined offsite facilities.

The outside-battery-limit capital requirements include those for an offsite boiler, cooling towers, a small fuel system, a flare system, tank-truck-loading facilities and railroad spur, water-treatment plant and water storage, piping, pumps, air compressors, fire water-system, fencing, electric substation, API oil-water separator, sewers, communications and lighting, instruments, insulation, painting, three small buildings for an office, a laboratory, a warehouse, and some internal roads. Water is purchased.

Regarding storage, for all cases except as noted, two weeks of feedstock storage, one week of coproduct storage, two days of pyrolysis-fuel-oil storage, no liquid ethylene storage, and 8 hr of propane surge storage are assumed. Ethane and propane are received via pipeline.

The capital requirements include those for all materials and equipment, construction labor, tools, temporary structures and fixed expenses, as well as all costs for engineering, design, supervision of construction, and overhead and profit. The expenses for construction reflect "normal" productivity and a 40-hr working week. Site preparation, land, and pilings are not included.

Transfer price and sensitivity review

As Tables VI and VII indicate, production of ethylene from gas oil results in the lowest transfer price, followed by production from naphtha and ethane. Production from propane bears the highest pricetag. The high transfer price calculated for propane is obviously due to the high cost of the feed. Interestingly, even with a 10% increase in gas oil price, this feed will produce cheaper ethylene than does naphtha.

It has been predicted that, in the future, naphtha in the U.S. may exist in surplus because of reduced, or constant, gasoline demand and greater demand for industrial and residential heating oils [*14*]. This prediction is subject to the contingency that any such excess naphtha will not be soaked up by SNG plants. If this speculation proves correct, naphtha may become a cheaper source of ethylene than gas oil.

The tables reveal the importance of the pricing structure of both raw materials and coproducts. With the exception of the ethane case, no other factor has as great an impact upon transfer price as these two do. As the sensitivity analysis and Table VI show, a 10% variation in raw-materials cost is appoximately equivalent to 70% of all overhead charges. Changes of as much as 20% in fixed investment, although substantial, are still less than the 10% variation in raw materials cost. Over half of the transfer price change resulting from the 20% variation in investment is due to maintaining the return on investment. Noteworthy is the significance of small changes in cost components (e.g., a 1¢/lb decrease in transfer price would represent a $10-million/yr reduction in income, before taxes).

The gas-oil-feed case exhibits the greatest dependence on coproduct credit, with only a 2.73¢/lb differential between raw-material cost and coproduct credit.

The three major components of coproduct credit for both the naphtha and gas-oil-feed cases are propylene, C_4 mixture, and pyrolysis gasoline. To avoid marketing (and, hence, economic) uncertainties and problems, as stated earlier, some corporate planners are considering importing natural-gas liquids or paying the higher price required to increase ethane supplies by deeper cryogenic extraction.

With the exception of the case of ethane feedstock, all feedstock cases result in a self-sufficiency of fuel, and even in an excess that can be sold. (Incidentally, the percent of coproduct credit from the sale of excess fuel, gas or liquid, or both, is reportedly below the transfer price.)

Profitability analysis of upgrading butadiene — **Table VIII**

Total fixed capital, (includes limited outside-battery-limit facilities)	$7,400,000	
	Annual basis, $	
Raw materials		
Feed	–	
Chemicals and solvent	110,000	
Total		110,000
Utilities, total		802,000
Labor and maintenance		
Labor and supervision	231,000	
Maintenance	296,000	
Total		527,000
Overhead charges		
Direct	81,000	
General plant	342,600	
Depreciation (10% of total investment)	740,000	
Insurance and plant taxes	111,000	
Interest on working capital	74,000	
Total		1,348,600
Total of above		2,787,600
Credit for butanes/butenes (5.5¢/lb.)	(5,785,300)	
Net manufacturing cost	(2,997,700)	
Butadiene, polymer grade sales value	(23,460,000)	
Total of above yearly income	(26,457,700)	
Income if total C_4 mixture is sold without separation	(24,843,000)	
Gain in income from separation of C_4 mixture	(1,614,700)	
Return on investment before taxes and royalties	21.8%	

The ethane feed case necessitates the purchase of supplemental fuel, and this accounts for a substantial part of the utility costs for this case. For all cases, only supplemental electricity and purchase of raw water are required.

Recently published spot prices for ethylene place these at 11¼¢–11½¢/lb, with a ½¢/lb increase expected soon. New contract prices for ethylene appear to range from 12½¢/lb to about 15¢/lb, depending on volume and delivery time, with substantial escalating clauses. Table V transfer prices, therefore, appear reasonable, when adjusted for such omitted items as startup costs, and when the temporary recessionary surplus of ethylene is taken into account.

Upgrading butadiene

One of many coproduct options available to the prospective ethylene producer is now considered: the extraction of butadiene from the C_4 mixture, subsequently purified to polymer-grade quality (instead of the sale of total C_4 mixture to a butadiene extractor). The coproduct butanes/butenes could be sold to a refiner for such things as alkylation or LPG.

This example is based on the crude C_4 mixture resulting from the high-severity cracking of the Kuwait full-range naphtha, with a 2% processing loss allowance. Operating requirements are based on licensor data; and the capital investment for a pure-product butadiene stream of 138 million lb/yr is based on estimates by Stone & Webster of the cost of an installation that is integrated with the ethylene plant and is sharing outside-battery-limit facilities.

Factors pertinent to the economic calculations are to be found in Table V, with the exceptionsof an allowance of 2.5 men per shift, and a 10% depreciation (because of limited additional outside-battery-limit requirements).

The profitability analysis shown in Table VIII indicates (for the given price structure) a gain in income, before taxes and royalties, of $1,614,700 per year from such upgrading. (Annualized royalties are quite small.)

An alternate profitability analysis with the butanes/butenes (not shown), credited at a fuel value of $2/million Btu (with the pertinent items in Table VIII kept constant) indicates that such an operation would result in a loss of income of about $41,000.

Table VIII reveals the sensitivity of profit or loss upon price structure. The value of the butane/butene stream is considered reasonable, even though not thoroughly documented. This, as shown in Table VIII, can have a major impact on profit, as it accounts for an almost $6-million annual coproduct credit.

Correspondingly, a 0.5¢/lb variation in the selling price of the total C_4 mixture would increase or decrease income by an additional $1.2 million/yr, which would either make the operation unattractive for external sale, or enhance further the profitability of upgrading the crude butadiene stream. The butadiene extractor buying substantial quantities of olefin-derived butadiene recognizes that the price for the crude C_4 mixture must be sufficiently high to eliminate the incentive for an ethylene producer with substantial quantities of crude butadiene from becoming a competing merchant of polymer-grade butadiene.

Conversely, the ethylene manufacturer that has substantial coproduct butadiene production and that can profitably use or sell the coproduct butane/butene stream resulting from extraction will be strongly inclined to produce polymer-grade butadiene, particularly in rising markets.

Acknowledgments

The authors wish to acknowledge the support and permission to publish this report of the Stone & Webster Engineering Corp., Process Industries Group. Also appreciated are the interest and comments of L. P. Hallee, Manager of Operations, Stone & Webster Ltd., and A. R. Johnson, Vice President and Manager of Business Development. Additionally, the authors wish to recognize the substantial contributions of S. B. Zdonik, Manager of the Process Engineering Dept., and of P. D. Chapman and B. V. Pano of this department.

References

1. Bassler, E. J. others, "Gas Oil Cracking," paper presented at AIChE meeting, Houston, Tex., Mar. 18, 1975.
2. Bassler, E. J., others, How Feedstocks Affect Ethylene, *Hydrocarbon Proc.*, Feb. 1974.
3. Bowman, W. C. and Marbach, M., "Long Term Supply Trends in Benzene," paper presented at AIChE meeting, Houston, Tex., Mar. 18, 1975.
4. Chambers, L. E. and Potter, W. S., Designing Ethylene Furnaces (three-part series), *Hydrocarbon Proc.*, Jan., Mar. and Aug. 1974.
5. *Chem. & Eng. News*, June 2, 1975, p. 32.
6. *Chem. & Eng. News*, May 5, 1975, p 32.
7. *Chem. & Eng. News*, May 20, 1974, p. 10.
8. *Chem. & Eng. News*, June 25, 1975, p. 5.
9. *Chem. & Eng. News*, Mar. 24, 1975, p. 11.
10. Feduska, J. C., others, Gas and Oil Yield Prediction, *Hydrocarbon Proc.*, Nov. 1975.
11. Hartman, J. S., "Olefins Outlook for the Next Decade," paper presented at AIChE meeting, Houston, Tex., Mar. 18, 1975.
12. Haywood, G. L. and Hoggett, P. N., Ethylene Plant Feedstock Flexibility Is Justified, *Oil & Gas J.*, July 28, 1975.
13. Jordan J. M., "Polyolefin Markets—Historical and Future," paper presented at meeting of Soc. of Plastics Engrs., Houston, Tex, Mar. 11–12, 1975.
14. Reid, L., "Energy Resources Development: The Government and the Chemical Industry," speech presented at meeting of Chemical Industry Assn., New York, N.Y., Nov. 25, 1975.
15. Snelling, G. R., "Future Markets for the Plastics Industry," paper presented at meeting of Chemical Industry Assn., New York, N.Y., Oct. 23, 1975.
16. Taylor, E. D., "Ethylene Update," paper presented at meeting of AIChE, Houston, Tex., Mar. 18, 1975.
17. Tobin, H. H., "Propylene," paper presented at meeting of AIChE, Houston, Tex., Mar. 18, 1975.

The authors

Theodore B. Baba is senior staff engineer, Business Development Dept., Process Industries Group, Stone & Webster Engineering Corp. (One Penn Plaza, 250 West 34th St., New York, NY 10001). His activities include corporate planning, market development and new technology. His 30 years of experience has been with both constructors and manufacturers, primarily with new chemical-and-petrochemical projects, and includes research, operations, design, and economic analysis. A member of ACS, AIChE, Chem. Ind. Assn., and Soc. de Chimie Ind., he has a M.Sc.Ch.E. from NYU, and holds patents in the fields of synthetic rubber, vinyl acetate, and acrylonitrile.

James R. Kennedy is a manager in the Process Div. of Stone & Webster Engineering Corp., Process Industries Group, responsible for process engineering activities in New York. He has had 20 years of experience with Stone & Webster in process design. He has been associated with several olefin projects around the world, and most recently was responsible for the process design and construction of two major ethylene plants in the U.S. Holder of a M.Sc.Ch.E. from Massachusetts Institute of Technology, he is a registered engineer in the State of New York and a member of AIChE.

Ethylene diamine route eases pollution worries

Reacting monoethanolamine and ammonia to make ethylene diamine and other polyamines, this catalytic process eliminates waste-treatment problems associated with conventional ethylene-dichloride-feedstock routes. Troublesome byproducts are minimized as well.

Philip M. Kohn, *Associate Editor*

☐ An ethylene diamine (EDA) process that has none of the organic-laden-salt pollution or vinyl-chloride-hazard problems of conventional ethylene-dichloride-feedstock routes has been developed by Leonard Process Co.* (Englewood Cliffs, N.J.), in cooperation with Société Chimique de la Grande Paroisse SA (Paris). (Major uses of EDA are in chelating agents and carbamate fungicides.) In addition, the developers claim that the process can limit less-desirable byproducts, such as piperazines, which are not widely salable (although purified piperazines find use in pharmaceuticals, corrosion inhibitors and insecticides). Yields on raw materials are said to be in excess of 93%, compared with 85 to 90% for ethylene dichloride processes.

* Leonard Process Co., 560 Sylvan Ave., Englewood Cliffs, NJ 07632.

In Leonard's gas-phase route, monoethanolamine (MEA), ammonia and some recycled products react over a catalyst in a hydrogen atmosphere at moderate temperatures and elevated pressure.

Offering a capital-cost saving of around 17% over a conventional ethylene-dichloride-feed plant, the MEA route also boasts a 25 to 50% reduction in utilities expenditures, according to Leonard. And the company feels that the disparity in raw-materials price (approximately 30¢/lb for MEA versus about 11¢/lb for ethylene dichloride) will be more than made up by the route's lack of pollutant effluents—and auxiliary equipment needed to treat them.

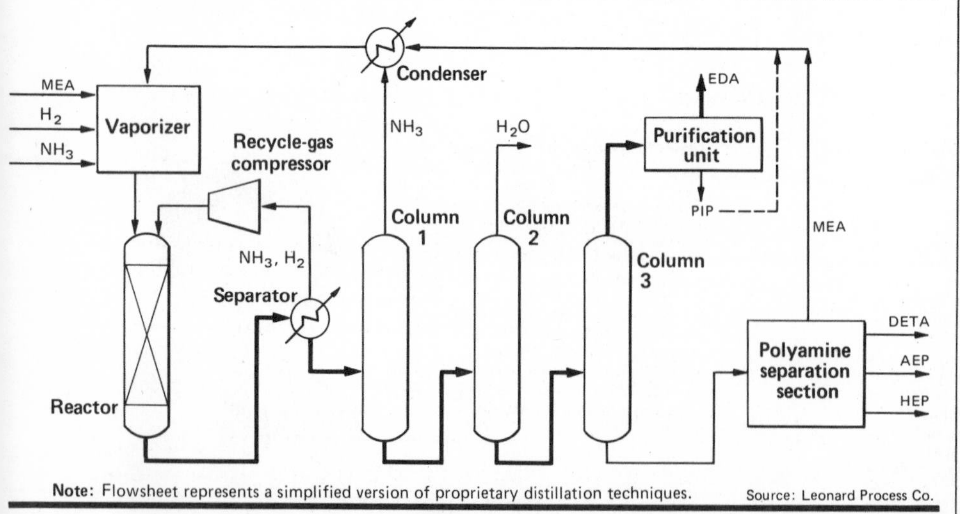

Ethylene diamine process produces neither NaCl nor vinyl chloride.

Leonard is currently negotiating the first commercial licensing of the process.

CATALYST THE KEY—The proprietary amination catalyst makes the process feasible. In ethylene dichloride systems, sodium hydroxide is the catalyst in a liquid-phase reaction that also produces vinyl chloride and HCl as byproducts. In addition, a 10,000-metric-ton/yr plant generates about 30,000 lb/h of a salt solution containing roughly 10 to 12% NaCl. Organics must be removed from this stream, and the salt must be disposed of. With environmental regulations growing ever tighter, this latter task is becoming more difficult to accomplish acceptably.

Numerous attempts have been made through the years to find alternative means to produce EDA, but the various catalyst systems tried either did not work or gave too many less-desirable byproducts in proportion to EDA. Monoethanolamine was not an economical feedstock until it became apparent that the ethylene dichloride route would incur financial penalties associated with tighter pollution- and toxicity-control laws.

(BASF AG, Ludwigshafen, West Germany, has had an EDA-from-MEA process—using a BASF-developed catalyst—in its facilities at Ludwigshafen and at Antwerp, Belgium, for several years. But Leonard indicates that BASF's route produces substantial amounts of piperazine, for which BASF has outlets.)

Finally, in 1976, Grande Paroisse came up with a catalyst that seems to supply an acceptable alternative route to EDA while yielding low quantities of byproducts. Continuous testing has demonstrated a 2,000-h catalyst life, so far, with no dropoff in activity, says Leonard. Ongoing tests will define the ultimate catalyst life-expectancy. The

Originally published March 27, 1978

French firm has applied for worldwide patent protection.

PROCESS DETAILS—In the process (see flowsheet), MEA, ammonia and recycled products are fed to a vaporizer. The gaseous mixture then flows to a fixed-bed reactor in which the amination catalyst is present in pellet form. Reaction takes place in a hydrogen atmosphere, at a temperature not exceeding 300°C and a pressure not over 250 bar. Time of reaction is "on the order of seconds." The amination is mildly exothermic: while the reaction will not run away, according to Leonard, external cooling on the reactor is required to control reaction temperature.

Upon leaving the reactor, the aminated stream flows to a separator (a partial condenser), where the vapor and liquid phases are dissociated.

The vapor phase, consisting almost completely of NH_3 and H_2, recycles through a compressor to the reactor. The liquid phase enters column 1, where any remaining ammonia is removed. The ammonia then passes through a condenser, where it is liquefied, and then returns to the vaporizer. The water distilled off in column 2—containing only trace amounts of organics, according to Leonard—can go to a biotreatment facility.

The bottoms from column 2—containing EDA, DETA, PIP and substituted piperazines, namely aminoethylpiperazine (AEP) and hydroxyethylpiperazine (HEP)—flow into column 3 of the unit's distillation section.

Ethylene diamine and piperazine are stripped off and go to a purification unit. Bottoms from column 3 proceed to the polyamine separation section, which in essence is a batch distillation unit in plants having a capacity of less than 10,000 metric tons/yr. Larger facilities can have continuous stripping operations for separation of the various coproducts and unreacted MEA, which gets recycled to the vaporizer.

RECYCLE FLEXIBILITY—The basic process yields a product mix (by weight) of around 69% EDA, 7% DETA, 14% PIP, and 2% HEP. Leonard notes that output of EDA and DETA can be further maximized—and output of PIP minimized—by recycling the piperazine product stream (shown as a dashed line on the flowsheet). In tests, the route has successfully generated a product mix of 74% EDA, 8% DETA, 4% PIP, 10% AEP and 4% HEP. Ongoing tests will determine the practical limit of the process's flexibility, and requirements of each licensee will determine the degree of flexibility built into an individual process.

ECONOMICS—A key benefit of the monoethanolamine process is a capital investment saving of around 17%, as reflected in the table. Battery-limits plant costs include all processing equipment, plus two-week storage capacity for raw materials and finished products. Excluded are costs of buildings and land. Offsite costs would include such items as waste-treatment facilities and related equipment.

The bulk of the capital investment savings for the MEA process, according to the developer, is due to the Leonard route not needing expensive equipment for separating and processing organics-laden salt streams and vinylchloride gas streams that typically issue from ethylene dichloride processes. In addition, the latter processes require a larger and more-complex reaction system than does the MEA method and, because of their corrosive nature, they need exotic materials of construction.

Even the offsite costs for the ethylene dichloride routes are higher than those for the MEA process, because larger steam boilers and bigger cooling towers are generally called for.

From an operating standpoint, utilities costs should be from 25 to 50% lower than those of an ethylene dichloride process. Actual consumptions per 100 lb of total amine product for the MEA process are: 1,810 lb of steam (at 150 lb/in²), 8,900 gal of cooling water (ΔT = 10°C), and 45 kWh of electricity, as compared with 4,930 lb of steam, 8,330 gal of cooling water and 11 kWh of electricity for an ethylene dichloride process. Here again, according to Leonard, waste-stream cleanup techniques employed in the latter route use substantial quantities of utilities.

Ethylene-diamine cost profile

Process		Ethylene dichloride		Monoethanolamine	
Plant size (metric tons/yr.)		10,000		10,000	
Product mix (metric tons/yr.)		Case 1	Case2	(Recycling PIP)	
	EDA	9,090	2,500	EDA	7,400
	DETA	540	2,500	DETA	800
(triethylene triamine)	TETA	270	2,500	PIP	400
(tetraethylene pentamine)	TEPA	100	2,500	AEP	1,000
				HEP	400
		10,000	10,000		10,000
Investment					
Battery-limits plant cost		$ 9.25 million			$ 7.65 million
Offsite costs		3.00 million			2.50 million
		Total $12.25 million		Total	$10.15 million
Manufacturing costs					
Direct (¢/lb. of total product)					
Raw materials		33.6	37.5		36.2
Utilities*		13.5	8.9		6.3
Labor, supervision, labs and maintenance†		2.8	2.8		2.4
Depreciation, taxes, overhead‡		6.1	6.1		5.0
Indirect					
Sales, shipping, handling		4.0	4.0		4.0
Total delivered cost		60.0	59.3		53.9
(X 10,000 metric tons)		$13.20 X 10^6	$13.05 X 10^6		$11.86 X 10^6
Income from sales¶		$16.84 X 10^6	$19.25 X 10^6		$18.23 X 10^6
Net profit after taxes (at 50%)		$ 1.82 X 10^6	$ 3.10 X 10^6		$ 3.18 X 10^6
Return on battery-limits plant cost		19.7%	33.5%		41.6%
Return on sales		10.8%	16.1%		17.4%

*Steam at $2.50/1,000 lb; water at $0.12/1,000 gal; electricity at $0.015/kWh
† Based on 5% of total capital investment.
‡ Based on 14.5% of battery-limits plant cost.
¶ Selling prices (¢/lb); EDA=75; DETA=90; TETA=90; TEPA=95; AEP=100; HEP=100; PIP=150.

Source: Leonard Process Co.

Fermentation Achieves New High in Flexibility

Highly-versatile fermentation plant turns out not one but a multitude of products, using a single set of processing equipment.

DON L. ISENBERG
Beckman Instruments, Inc.
Microbics Operations

A new fermentation plant that produces a broad spectrum of enzymes and biochemicals—by building up specific process flowsheets from one set of process equipment—has been started up by Beckman Instruments, Microbics Operations, at a $4-million production/laboratory facility in Carlsbad, Calif.

A special fermentation reactor is the key to the plant's unique flexibility. Microbics designed the reactor in collaboration with Chemapec Corp., which manufactured it at its Basel, Switzerland, workshop.

Historically, fermentation has never been considered a versatile unit operation. Conventional equipment is dedicated to a single product (or very few), and integrally tailored for maximum yield of that product from the particular raw materials and biocatalysts.

The Microbics fermentation setup accommodates a large number of low to medium-volume, high-value products. The enzyme dextranase is a typical example (its patented manufacturing sequence is detailed later). Other fermentation flowsheets, all handled by bringing pieces of equipment in and out of service, changing the processing sequence, and adjusting operating parameters, fall into four basic categories:

- Biocatalyst production—enzymes such as glucose-6-phosphate dehydrogenase and glycerol kinase.
- Biomass production—single-cell proteins from petroleum, cellulose or alcohols.
- Metabolite production—alcohols, amino acids, citric acid.
- Substrate transformation—naphthalene to salicylic acid, steroid conversions.

Equipment Setup—Microbics' primary reactor has a working volume of 10,000-l, and is fabricated from 316 stainless steel (as are all other

FERMENTATION SYSTEM (shown above during construction) is built around 10,000-l main reactor; service area (foreground, photo right) includes air compressors and steam generators.

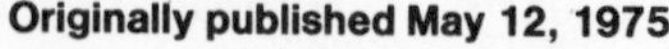

Originally published May 12, 1975

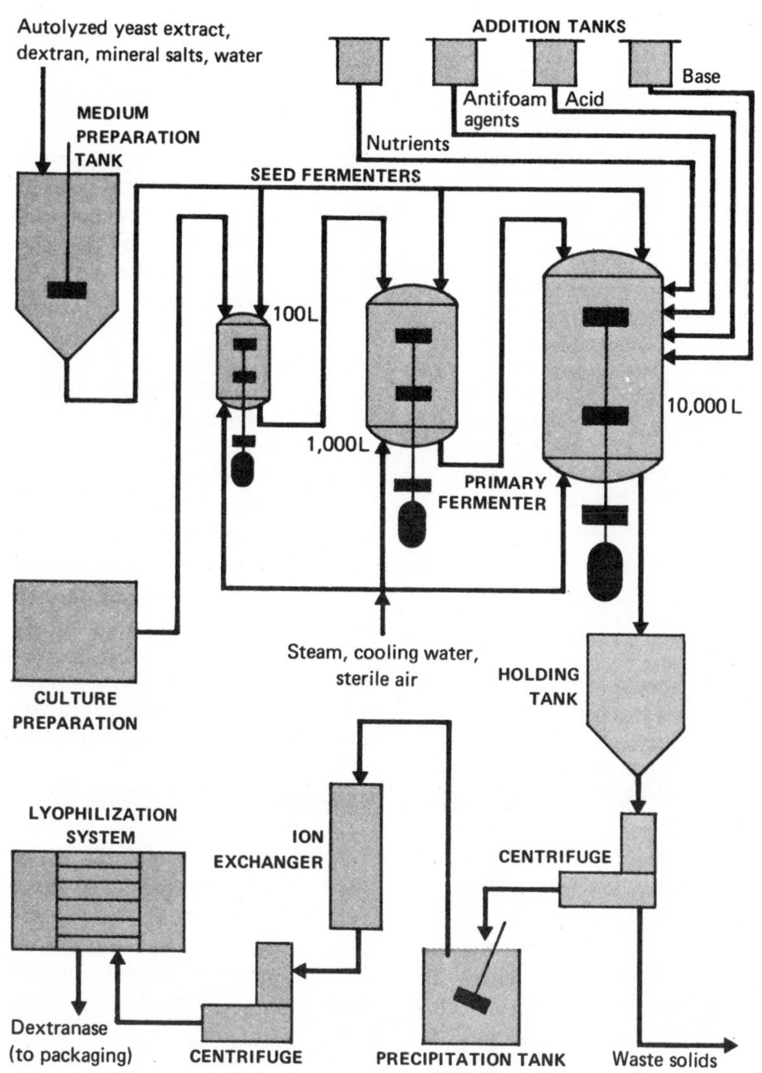

ADDITION VESSELS in foreground feed the two 1,000-l seed fermenters; vessels at rear right feed the main 10,000-l fermentation reactor.

CENTRIFUGE STATION (upper level) has two desludging-type units.

vessels and lines). A bottom-drive shaft with a sterilizable seal provides agitation. The shaft is driven by an external 75-hp motor, equipped with a variable-speed control that provides up to 160 rpm.

Internally, the shaft has five turbine disks, each multiple-drilled to accept up to eight blades. The speed, positioning and configuration of these blades, and their interplay with six peripheral baffles, give a wide range of agitation conditions important for process optimization. Heat-exchange fins are provided for temperature regulation.

Sterile air enters the reactor through a bottom sparger. Air flow can be controlled to provide up to 1 l/m/l of reaction mass at a 2-atm. backpressure. The aeration system includes two 350-cfm oilfree compressors, preceded by dust filters and followed by a glass-wool-pack steam-sterilization system.

Fermentation media are prepared as a concentrate in a 6,000-l tank, then pumped into the fermenter, and diluted prior to sterilization. A group of sterilizable addition vessels, each holding 500 l, provides for either continuous or intermittent charging during the fermentation of acids or bases for pH control, of antifoam agents, and of special nutrients.

The preparation equipment for the primary fermenter also includes a pair of 1,000-l seed (or step-up) fermenters, backed up by two 100-l seed fermenters. Inoculum for the smaller units is transferred aseptically from 30-l fermenters located in a laboratory/pilot plant at the site, using specially designed pressure containers. Inoculum transfer in the major system is through rigid plumbed lines.

Steam boilers provide in excess of 10,000 lb/hr for sterilization and heating. Other services include a 60,000-l chilled-water system, and continuous deionization and reverse-osmosis water-treatment systems.

A 16,000-l holding tank accepts beer leaving the fermenter. The beer then passes to a centrifuge station for removal of cells and other solids. The specific process dictates whether the beer or the cells, or both, undergo further processing in the recovery area.

The primary fermenter can be operated in batch, semicontinuous or

Dextranase: A Product of Many Uses

The leading application for dextranase over the years has been for conversion of naturally occurring dextran (a glucose polymer) to lower-molecular-weight compounds, for use as blood-plasma extenders. Dextranase produced by Microbics via the *Bacillus coagulans* strain is an endodextranase, i.e., its chain-breaking activity is directed internally to the dextran molecule. Exodextranase, by contrast, cleaves small fragments from the ends of the molecule, a far-less-efficient way to destroy it.

A significant new market has been developing for prevention of dental caries and periodontal disease, both believed to result from formation of dental plaque. Since dextran-like polymers formed by streptococcal bacteria in the mouth are important to the integrity of plaque, the otherwise harmless dextranase is important in management of this disease.

At least two other applications are currently under investigation. One relates to utilizing dextran in a final product for treating human and animal iron-deficiency anemias. The other falls in the sugar industry, where formation of dextran-based gums is a problem. Microbics' dextranase, optimized for viscosity reduction, increases process flow by destroying these gums.

continuous modes. To allow continuous operation, the system includes a continuous medium-sterilization capability (a retention coil holds the medium at 143°C for 3 min), and integrated load cells on the addition tanks and the main fermenter. Automatic monitoring and control extends as well to pH, temperature, dissolved oxygen, air flowrate, and backpressure. All probes are steam-sterilizable.

Dextranase Flowsheet—Dextranase has been known and used for many years (see box). Conventional manufacturing starts with fungal sources. A patented Microbics process for dextranase (U.S. 3,787,289) instead utilizes a bacterial source, a specially isolated strain of *Bacillus coagulans*. Yields are significantly higher, fermentation times significantly shorter, and the product requires less purification than with fungal raw material.

The fermentation medium is a combination of autolyzed yeast extract, as an assimilable nitrogen source, and dextran, as an enzyme inducer and carbon and energy source. Both are used at a 1 wt.% concentration in an aqueous medium. Mineral salts are also added, at ¾ wt.%; these are nutrients, and also maintain pH buffering and ionic strength for optimum cellular growth.

Preparation for a production run begins in the laboratory / pilot plant with the inoculation of 50 ml of medium with a working stock of organism (maintained by conventional master / working stock methods). The culture, stepped up through a 1,000-ml stage, then inoculates two 30-l fermenters.

The dual system for these fermenters, and the 100- and 1,000-l step-up fermenters that follow in the main production plant, increases the probability of an optimum inoculum for the 10,000-l stage. This has been a major factor in achieving a zero failure rate.

For the final fermentation, medium is pumped into the 10,000-l vessel as a 10X concentrate. After diluting to required volume with deionized water, steam circulates through the heat-exchange fins until the temperature reaches 90°C. Steam is then injected through the sparger, and the vessel pressure-vents closed to achieve 121°C. This heating process requires about 45 min, starting from ambient temperature, and continues for 30 min at a 121°C sterilization temperature. Circulating chilled water through the fins cools the vessel contents to 32°C, and requires about 30 min.

A slight positive head pressure is maintained throughout the various fermentation stages, to aid aseptic operation. A pH of 7.0, and an aeration rate of 1 volume/h/volume of beer, are also maintained.

Stepup fermentation (which promotes cell growth) takes 20 hr at an agitation rate of 250 rpm. Final fermentation (which uses this cellular medium in forming the enzyme product) in the 10,000-l vessel takes 24 hr at 100 rpm. Yield of dextranase is 1,200 Tsuchiya units*/ml.

At the completion of fermentation, cooling water again circulates through the fins, for 1 h, to reduce the beer temperature to about 10°C. The beer then passes to the holding tank to await scheduling into the centrifuges for cell removal. Two desludging-type units work in tandem to provide a rate of 20 l/min.

In the final recovery area, the clarified beer undergoes conventional enzyme purification. These steps include chemical precipitation followed by ion-exchange separation. The enzyme is then centrifuged out of suspension, passed through a lyophilization stage (a rapid freezing-dehydration used for preservation), and packaged.

Actual purification procedures vary with end-use. Most often, a highly purified product is desired. Overall recovery yield then runs about 40%. The biggest loss occurs during purification (30%), of which 20% is lost during precipitation and 10% during lyophilization. #

*One Tsuchiya unit equals the amount of enzyme that will generate 1 mg of isomaltose reducing units in 30 min.

Meet the Author

Don L. Isenberg is chief scientist of Beckman Instruments, Inc., Microbics Operations, 6200 El Camino Real, Carlsbad, CA 92008. He holds a B.S. in pharmacology from the University of Texas (Austin), an M.S. in microbiology from the University of Houston, and a Ph.D. in microbiology from Louisiana State University (Baton Rouge). He is a member of the Soc. for Industrial Microbiology, American Assn. for the Advancement of Science, and American Soc. of Microbiology.

Molten Salt Process Yields Chlorinated Hydrocarbons

The entire range of C_1 and C_2 chlorinated hydrocarbons can be produced directly from methane and ethane by this safe, pollution-free technique. Chlorine values are obtained from molecular chlorine, hydrogen chloride or even waste chlorocarbon streams.

The world's first commercial plant to oxychlorinate methane—and also the first to use a molten-salt processing medium—is now in the works. The unique facility will introduce Transcat, a technique originated by C-E Lummus* and developed jointly with Armstrong Cork Co.

Scheduled for completion by late 1975, the upcoming plant will turn out 30,000 metric tons/yr of chloromethanes from methane and chlorine feedstocks for Shin-Etsu Chemical Industry Co., at Naoetsu, Japan. The Transcat process, however, is equally applicable for such chlorinated hydrocarbons as vinyl chloride monomer (VCM), trichloroethylene (TCE), perchloroethylene (PCE), and dichloroethane.

*C-E Lummus Engineering, 1515 Broad St., Bloomfield, NJ 07003

Several characteristics distinguish Transcat from other processes:

First, it feeds on low-cost ethane (or an ethane/ethylene mixture) to yield C_2 chlorohydrocarbon products. Countries short on capital but rich in ethane from natural gas can thus upgrade to VCM and other C_2's without investing in an ethylene cracker.

Second, Transcat obtains chlorine values in the form of molecular chlorine, hydrogen chloride or waste chlorocarbons. This allows efficient use of other processes' HCl byproducts, which are very often difficult to sell or consume inhouse. A fluorocarbons plant, for example, integrates with a Transcat chloromethanes plant by supplying the latter with its HCl byproduct, and receiving in turn chloroform and carbon tetrachloride. This chlorine versatility avoids the trouble and expense of burning waste chlorocarbons, too.

Third, Transcat itself discharges no undesirable byproducts or objectionable wastes. All secondary reaction products are consumed.

And fourth, it is a safe-to-operate process. Unlike other routes involving oxychlorination, there is no potentially dangerous contact between molecular oxygen and hydrocarbons. Oxygen is supplied to the hydrocarbons in chemical combination.

A continuously circulating stream of molten inorganic salt is the key to all these features. The salt, circulated by means of a gas lift, catalyzes the chlorination, oxychlorination and

Typical Operating and Cost Data for Transcat Processes

Product	VCM	VCM, TCE, PCE	Chloromethanes[1]	Carbon Tetrachloride
Production rate, million lb/yr	750	500/74/74	250	200
Battery limits capital cost[2], million $	25	22.4	11.9	11.8
Ethane, million lb	452	301	–	–
Methane, million lb	–	–	51.2	28.7
Chlorine, million lb	430	411	212	–
Hydrogen chloride, million lb	–	–	–	190.8
Catalyst and chemicals, $	220,000	200,000	150,000	250,000
Steam (600 psi, 750° F), million lb	2,790	2,410	522	520
Condensate return (300° F), million lb	2,655	2,294	500	500
Cooling water ($\Delta T \approx 23°$ F), million gal	34,050	29,420	7,400	5,850
Electricity, million kWh	112	97	20.1	23.6
Fuel, million Btu	27,000	25,000	23,000	25,000
Operating labor, men/shift	5	5	4	4

[1] Includes 50 million lb methyl chloride, 75 million lb methylene chloride, 100 million lb chloroform and 25 million lb carbon tetrachloride.

[2] U.S. Gulf Coast, 4th quarter 1973 (initial fill of catalyst included).

Originally published June 24, 1974

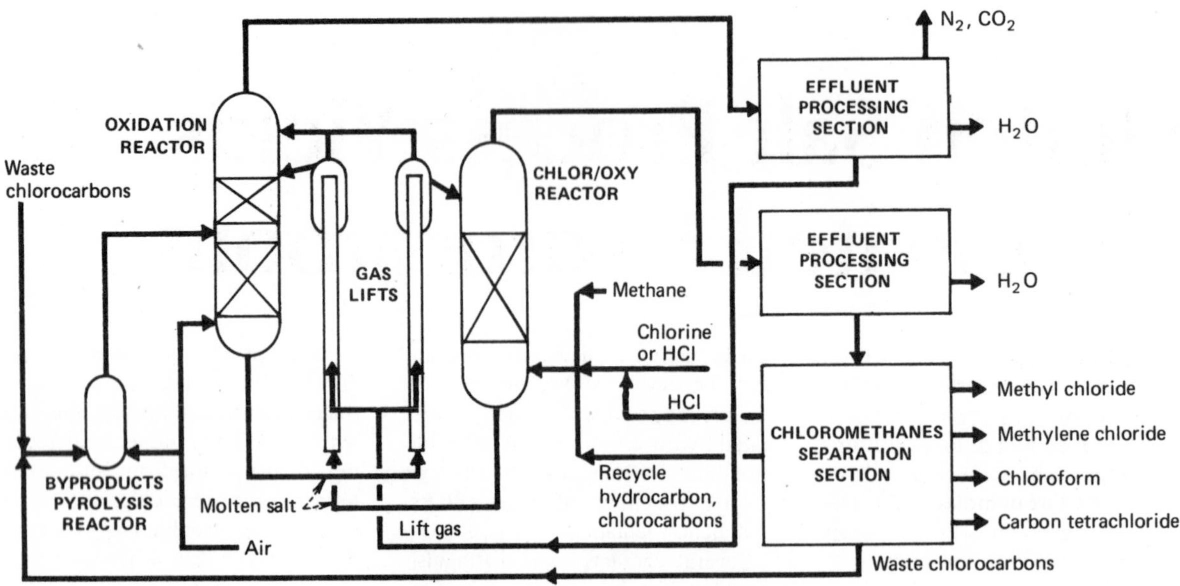

TRANSCAT CHLOROMETHANES PROCESS

dehydrochlorination reactions in Transcat, carries oxygen and chlorine through the system, and transfers heat that is generated.

Byproducts Incineration—Referring to the flowsheet, Transcat first recovers chlorine values from waste chlorocarbons. These include reaction byproducts that cannot be recycled for conversion to desired products, and wastes from other chlorination processes.

An excess of air and the wastes (and sometimes supplementary fuel) are fed to a byproducts pyrolysis reactor. Incineration burns these to hydrogen chloride, chlorine, carbon dioxide and water. Since temperature of the offgases is held below 2,000°F, generation of polluting nitrogen oxides and corrosive attack on vessel materials are low.

By contrast, conventional incinerators for chlorocarbons need a higher offgas temperature of 2,800°F or more, to enhance formation of recoverable HCl and restrict chlorine. The latter would pass unaffected through downstream systems and ultimately vent to atmosphere.

Oxidation Reaction—Gas from the incinerator, which can bear chlorine, is routed to the middle of an oxidation reactor. Entering from the top is a stream of molten salt, consisting primarily of cuprous chloride, cupric chloride and potassium chloride (a melting-point depressant). The salt flows through two packed beds in the vessel. Pyrolysis gases flow upward with air fed in from the bottom.

The cuprous chloride (CuCl) and gases react in three ways. Some CuCl oxidizes to soluble copper oxychloride ($CuO{\cdot}CuCl_2$); some combines with chlorine to form cupric chloride ($CuCl_2$); some reacts with HCl and oxygen to form $CuCl_2$ and water. All of the oxygen later needed for oxychlorination, and some of the chlorine, is thus provided to the salt.

After effluent treatment, gases from the unit (nitrogen, CO_2, steam and some oxygen) are partially vented. The rest return to the system as lift gas, which hydraulically lifts molten salt from the bottom exit of the oxidation reactor to the top entry of the following chlor/oxy unit.

Product Conversion—As in the oxidation reactor, molten salt flows downward in the chlor/oxy reactor through a packed bed. A countercurrent stream of gases consists of feed hydrocarbons, chlorine or HCl, and recycle intermediates.

The conversion of feed to product chlorohydrocarbons—by chlorination, oxychlorination and dehydrochlorination—takes place with the molten salt as catalyst. The salt releases the oxygen and chlorine picked up previously, then exits for circulation back to the oxidation reactor by a second gas lift. Methane feed is used for chloromethanes production; if VCM and/or other C_2 chlorohydrocarbons are desired, the feed is ethane or ethane/ethylene. (The flowsheet, shown for chloromethanes, is similar in all cases.)

Reaction products pass to an effluent processing section for removal of CO_2 and water (slightly basic, weakly saline, copper-free). Distillation finally separates the stream into desired products; HCl, intermediates and unconverted hydrocarbons for recycle; and byproduct residues for burning.

Other Features—Pressures are below 100 psi and temperatures between 700 and 850°F throughout the Transcat reaction sections. To avoid corrosion without resorting to exotic construction materials, an inexpensive ceramic is used to line carbon steel equipment. Packing is ceramic, too, so molten salt does not contact metal at any time.

Turndown capability for a Transcat plant ranges to 25-50%. Product distribution can be varied considerably without major plant modifications, even to the point of revamping a chloromethanes plant to produce chlorinated C_2's. Overall yields are high: 99+ mol % on chlorine; 75-90 mol % on methane to chloromethanes, depending on product distribution; 80 mol % on ethane to VCM; about 90 mol % on ethane when coproducing VCM, trichloroethylene and perchloroethylene.

Typical operating and cost data for Transcat installations are shown in the table.—NRI #

New Route to Hydroquinone

The first hydroquinone plant to employ *p*-diisopropylbenzene (*p*-DIPB) intermediate is now running at more than 20% over design throughput rating.

ALBERT H. OLZINGER
Goodyear Tire & Rubber Co.

Since assuming operation nearly two years ago of a 6-million-lb/yr hydroquinone plant at Bayport, Tex., Goodyear Tire & Rubber Co. has turned the once-troubled unit into an efficient and economical facility.

The first-of-its-kind plant was brought onstream in early 1971 by owner and process developer Signal Chemical Co. The new process, featuring *p*-DIPB intermediate instead of the conventional aniline-to-quinone-to-hydroquinone technique, promised advantages in less expensive raw materials, less environmental concern, and fewer and more-marketable byproducts. After two years of modification and redesign, however, plant productivity had reached only 70-80% of rated capacity, and in January 1973 the plant was shut down. Big Three Industries bought the unit and leased it to Goodyear a few months later.

HYDROQUINONE PLANT includes three processing areas.

Originally published June 9, 1975

Goodyear made several key changes, restarted the plant in late May—and in June set a new production record. The production rate has climbed steadily since, to more than 120% of design, and further boosts to 150% are expected.

From the onset, Goodyear emphasized a continuing program of debottlenecking the plant and improving process efficiency. A pre-startup decision to temporarily suspend manufacture of high-purity photograde hydroquinone provided an immediate production bonus. Numerous refinements in equipment and the process itself have since greatly increased both efficiency and productivity. The keys to efficiency gains (details are proprietary) lie in temperature selections, pH control, and adjustment of purge streams.

Alkylation Section—The fixed-bed alkylation reactor stepping off the process feeds on propylene and recycled cumene. It produces not only *p*-DIPB but also *o*- and *m*-DIPBs and triisopropylbenzenes (TIPBs). The *p*-DIPB is purified and stored for further processing. The *o*- and *m*-DIPBs and the TIPBs are recovered by distillation and sent to a transalkylation reactor. The transalkylator reacts benzene with the materials to yield cumene for recycling back to the alkylator. A silica-alumina catalyst serves in both reactors.

This high-efficiency sequence can be adjusted to use some purchased cumene as feedstock, or to run at near cumene balance by controlling the ratio of *p*-DIPB to materials that can be transalkylated. If the cost of outside cumene runs higher than transalkylated cumene, the purchase of cumene is kept down by maximizing production of transalkylatables. When propylene and benzene prices are disproportionately higher than the cumene price, the yield of transalkylatables is minimized and more cumene bought.

Light and heavy byproducts are burned as fuel, or sold. Light ends come mainly from the propane impurity in the propylene feed.

Oxidation Step—The *p*-DIPB, after addition of caustic, is next oxidized to *p*-diisopropylbenzene dihydroperoxide (DIX). Centrifugation immediately removes water from the stream (necessary for an efficient cleavage reaction later on). The DIX is then purified by crystallization and recovered with a rotary vacuum drum-filter. It is dissolved in acetone, and continuously passed on to the next process stage—there is no in-process storage for DIX.

The filtrate contains large amounts of unreacted *p*-DIPB and monohydroperoxide (MOX), and recycles back to the oxidation reactor. This loop is purified by stripping off a small sidestream of acidic-type oxidation inhibitors.

Cleavage and Finishing Step—Sulfuric acid is added to the DIX/acetone solution in a rearranger, to effect a cleavage reaction that yields hydroquinone and acetone. The acid is neutralized with ammonia to an ammonium sulfate salt recovered as salable byproduct.

Distillation recovers acetone from the reaction mixture. Some of this acetone recycles to the DIX repulping tank; the coproduct portion passes through a finishing column for purification to at least 99.5%.

The acetone-free hydroquinone is dissolved in water and benzene-extracted to remove impurities, prior to concentration by evaporation and crystallization. A centrifuge reduces the hydroquinone slurry to less than 15% moisture; a dryer turns out the final product at less than 1% moisture. The dryer uses a nitrogen rather than oxygen environment, since heat, water and oxygen would combine to degrade hydroquinone. The centrifuge filtrate is reextracted to minimize product impurities.

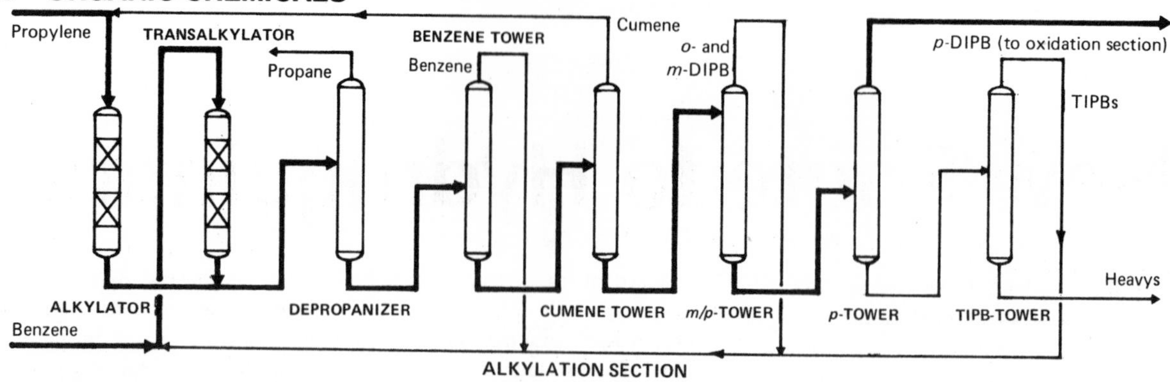

ALKYLATION SECTION

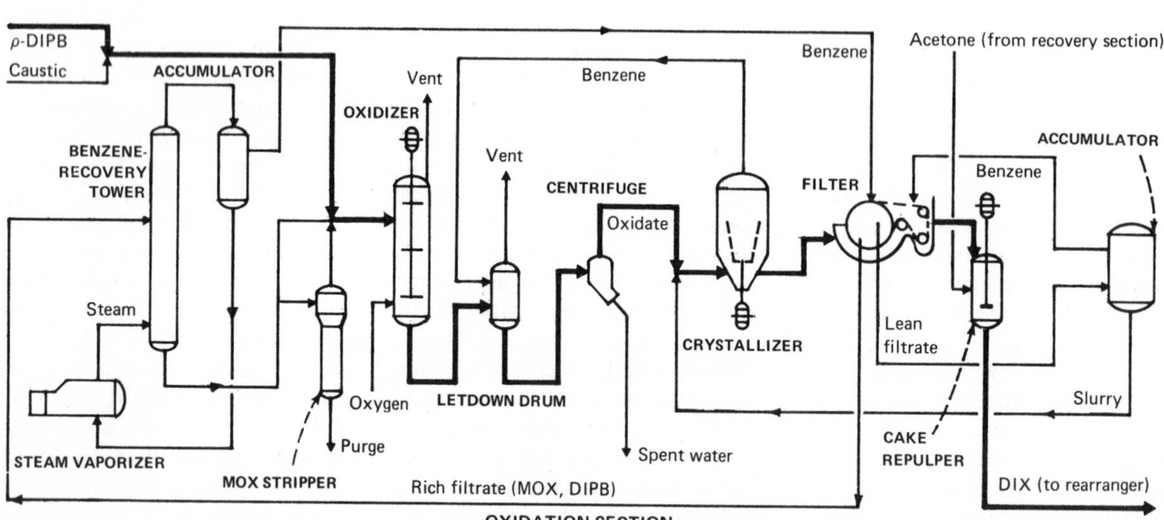

OXIDATION SECTION

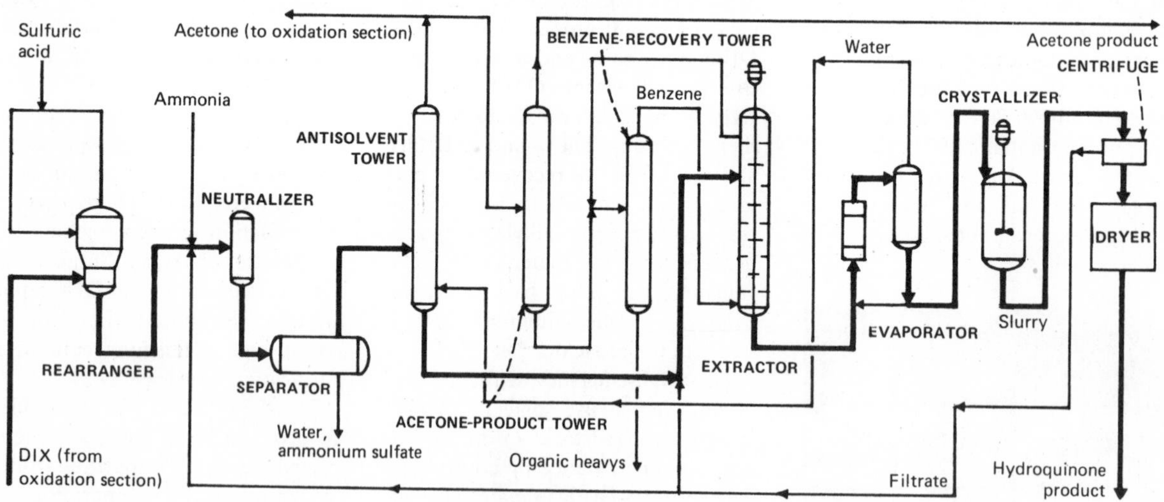

RECOVERY SECTION

Product Flexibility—The bulk of Goodyear's hydroquinone product is a technical grade used for production of rubber chemicals. A second grade serves as a polymerization inhibitor for monomers such as vinyl acetate. Output of photo-grade material is still suspended. Each of the three grades is increasingly free of impurities such as color, ash, water and sulfur.

Other materials can be produced by using process intermediates:

- Both *p*- and *m*-DIPB can be oxidized to their monohydroperoxides (polymerization initiators).
- *m*-DIPB can be converted to resorcinol, using an inhouse process.
- The organic byproduct stream can be used as precursor for high-purity *p*-isopropylphenol.
- DIX can be easily reduced from the hydroperoxide to the dialcohol, for several specialty-chemical uses.
- A stream of 1,3,5-triisopropylbenzene, instead of being transalkylated, can serve as a chemical intermediate. #

Meet the Author

Albert H. Olzinger is manager of plant operations for Goodyear Tire & Rubber Co.'s Bayport plant (Box 669, La Porte, TX 77571), responsible for all technical, developmental and manufacturing functions. He joined Goodyear in 1961 upon graduation from the University of Pittsburgh with a B.S. in chemical engineering. After serving in Akron, Ohio, as a research engineer and section head for bench-scale research, he was transferred to Bayport in 1973 as technical superintendent.

Belgians Tap PA Wastes For Maleic Anhydride

Sizable amounts of maleic anhydride show up in wastes from phthalic anhydride (PA) plants. A Belgian firm explains its new method for recovering this valuable product at relatively low production costs.

It sounds almost too good to be true—an effective pollution control technique that also serves to recover a valuable product at a cost competitive with that of conventional processes. Yet the maleic anhydride (MA) recovery route recently unveiled by Belgium's UCB seems to meet all the above-mentioned requirements.

Essentially, the UCB method makes it easier to deal with wastes resulting from the manufacture of phthalic anhydride (PA) from either *o*-xylene or naphthalene. Effluent gas from PA production contains some PA, which, for environmental reasons cannot be vented to the atmosphere, plus MA, citraconic anhydride and benzoic acid. The wastes are usually scrubbed with water in a packed tower, from which PA and the other impurities emerge as an aqueous acid solution. This usually burns at high temperatures in an incinerator; fuel costs for the operation are considerable.

Instead, the UCB technique concentrates and dehydrates the aqueous acid solution to obtain MA. What is left is an anhydrous liquid residue that burns easily on its own —in fact, in PA plants, it burns much the same as heavy fuel oil, providing much-needed calories.

Low-Cost Anhydride—Recoverable MA is a sizable amount. UCB estimates that the *o*-xylene-to-PA process* produces MA in quantities up to 5% or 6% of the PA obtained. This method now accounts for 80% of total PA output worldwide—i.e., for 2 million metric tons/yr of PA. Therefore, in theory, the UCB development could recover about 100,000 metric tons/yr of MA.

Investment outlay for MA recovery is reasonable. Cost of the process equipment needed is said to be only about 40% higher than that of a conventional incinerator of the kind normally hooked onto the PA scrubber to dispose of the aqueous wastes.

* No figures are available on the amounts of MA obtained via the naphthalene process.

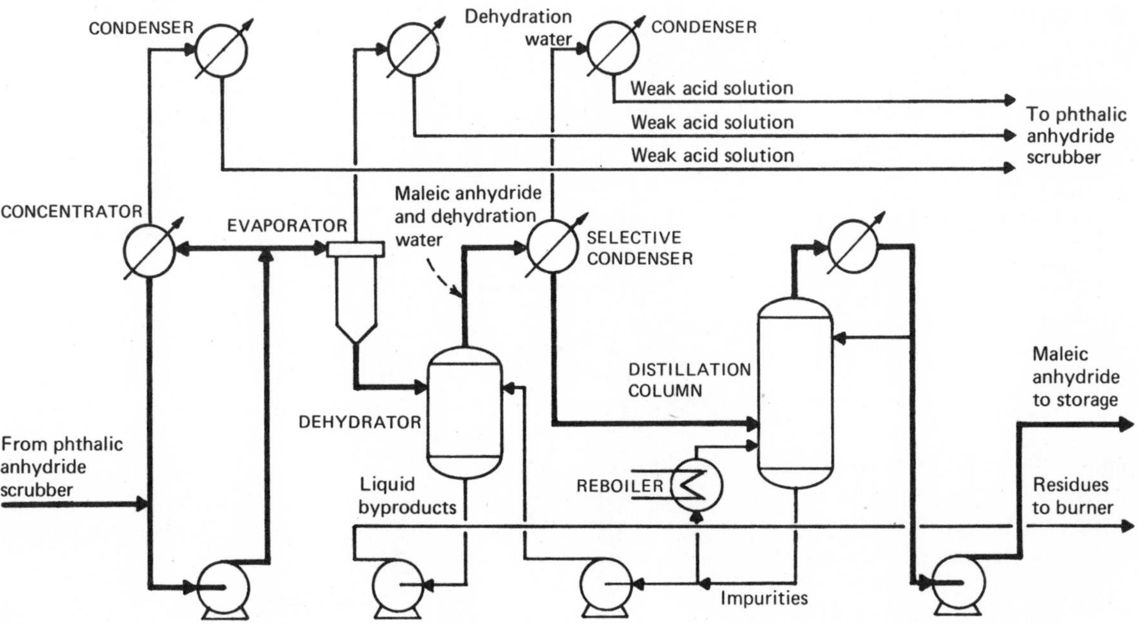

ACID WASTE is a valuable raw material in this streamlined flowscheme for maleic anhydride recovery.

Originally published September 2, 1974

Operating Costs per Metric Ton of Maleic Anhydride

	Process Requirements	U.S. Unit Costs (Dollars)	Cost in U.S. Dollars
Electricity	500 kWh	0.028	14
Steam	14 metric tons	5.00	70
Cooling water	500 m^3	0.008	4
Process water (demineralized)	10 m^3	0.75	7.50
Total			95.50

Production expenses for the MA product are at a bargain-basement level. For instance, there is no charge for raw materials, whereas a chemical company wanting to make MA via an established route—e.g., catalytic oxidation of benzene—would have to start by paying today's inflated prices for benzene feed (from $242 to $264/metric ton in the U.S.), which is needed at the rate of 1.25 tons/ton of MA. UCB concludes that operating costs of its process ($95.50/metric ton of MA) equal only about one-third of raw materials costs of the conventional benzene route.

Process Description — Waste solution from the PA scrubber is kept at a certain concentration of maleic acid (about 150 g/1) in order to prevent crystallization of phthalic acid. In the UCB scheme (see flow diagram), the solution goes first to a concentrator before further concentration takes place in an evaporator. Overhead from these two units is condensed and returned to the PA scrubber as weak acid solutions.

After all solution water is driven off, the acid stream enters into a dehydrator, which converts maleic acid into the anhydride. An overhead condenser condenses only the MA, letting the dehydration water return to the PA scrubber. Purification of the raw MA takes place continuously under vacuum in a distillation column. The concentration of impurities in the column's reboiler is kept constant by recycling part of the bottom discharge to the dehydration reactor.

Although a small amount of citraconic anhydride (about 1.6%) is present in the final MA, this does not disqualify the UCB product for applications such as manufacturing unsaturated polyesters, fumaric acid and maleic esters, all of which account for 70% of MA's industrial uses.

UCB, however, is willing to include a further purification step, which costs less than 5% of the total equipment required for its recovery plant, should it be necessary to produce MA of a quality comparable to that obtained by the conventional benzene route.

Commercial Units—UCB, which developed the technique about 18 mo ago at Havre-Ville, Belgium, says it is best suited to recovery plants producing MA at a minimum rate of 750-1,000 metric tons/yr. The company plans to obtain about 2,000 metric tons/yr of MA when it builds a recovery unit to go with a 45,000-metric-ton/yr PA plant operated by a subsidiary at Ostende, Belgium. Total investment will be about $900,000, excluding cost of buildings, storage and engineering. UCB is willing to license its process.—JAMES SMITH, *McGraw-Hill World News, Brussels.* #

Methanation Routes Ready

Catalytic methanation processes are becoming available to upgrade the Btu value of substitute natural gas (SNG), derived from coal or from heavy oil feedstocks, to pipeline specifications.

NICHOLAS R. IAMMARTINO
Associate Editor

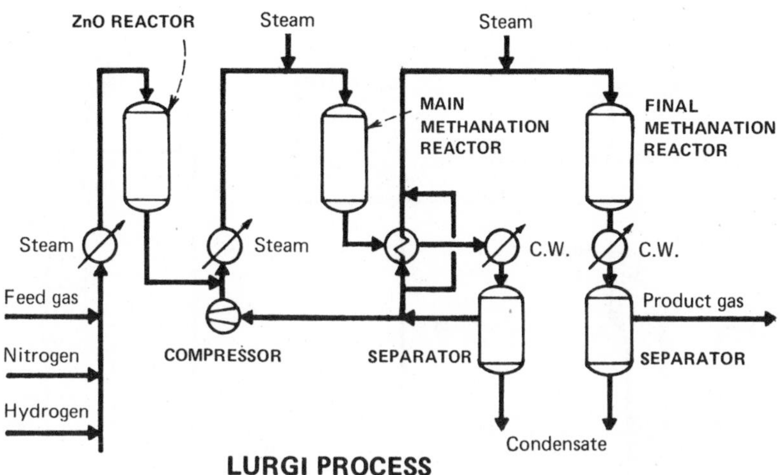

LURGI PROCESS

That's the message heard by some 100 attendees at the American Chemical Soc. (ACS) Symposium on Methanation of Synthesis Gas, sponsored by the Div. of Fuel Chemistry as part of last month's ACS 168th National Meeting in Atlantic City, N.J. Industrial experts gave progress reports on several of the leading process alternatives (and described catalyst formulation and basic methanation mechanisms as well).

Symposium cochairman Len Seglin (Bechtel Associates Professional Corp., New York)* stresses that the gas-recycle methanation routes are ready now for safe scaleup to commercial size. Liquid-phase and steam-moderated approaches, he notes, look promising as second-generation techniques.

Recycle Contenders—Details on the hot-gas-recycle method groomed by Lurgi Mineraloeltechnik GmbH (Frankfurt/Main, Germany) opened up the symposium. A semicommercial plant has been running for over 1½ yr at South African Coal, Oil and Gas Corp.'s coal-gasification facility in Sasolburg, South Africa, reported Lurgi's F.W. Moeller, and another unit is upgrading substitute natural gas from naphtha in an Austrian trial.

The process at both sites (see flow diagram) first passes a heated blend of fresh synthesis gas with H_2 and N_2 over zinc oxide, as a precautionary measure to prevent any chance of sulfur impurities reaching and poisoning the nickel catalyst. (Lurgi's Rectisol purification process is normally adequate as a gas pretreatment method, unless upsets occur.)

Then hydrogen and carbon monoxide in the gas combine exothermically to form methane in a main reactor. Some of the offgas is partly cooled, compressed and recycled; the rest flows through a final methanator and exits as high-Btu product. The hot-gas-recycle stream serves mainly to dissipate heat from the hotter-yet reactor.

Sasol's recent testing established that carbon dioxide in the feed gas brings about more-complete reaction of hydrogen to methane, Moeller said, thus yielding a higher-Btu product gas. Reactor inlet temperature is set between 500-570°F, and outlet temperature between 840-930°F. The influence of total pressure is limited. Resistance of the nickel catalyst to steam deactivation has been good.

Pilot-plant tests for a second system using hot-gas recycle were detailed in a paper by the U.S. Bureau of Mines' (Pittsburgh, Pa.) William Haynes.

Bumines discloses that parallel plates flame-sprayed with Raney nickel catalyst perform significantly better than conventional precipitated-nickel pellets. Production of methane per pound of catalyst is up, residual carbon monoxide in product gas is down, and catalyst life is enhanced. Pressure drop through the catalyst, and thus power for gas circulation, plunges more than 90%. Newer types of both plate and pellet catalysts will be evaluated soon.

Haynes blames catalyst deactivation during some experiments not on the usual culprit, sulfur, but rather on carbon deposits promoted by iron (carried to the catalyst from preceding parts of the system as iron carbonyl). Sintering also damages nickel sites, and nickel carbide formation and residual aluminum on the catalyst are suspected poisons. As feed throughput increases, furthermore, Bumines finds a much faster rate of catalyst deactivation.

W. G. Bair then shifted the discussion to the cold-gas-recycle method being piloted by the Institute of Gas Technology (Chicago, Ill.) as the final step of its Hygas coal-gasification route. A series of fixed-bed adiabatic methanators is used rather than a

*The other cochairman was Richard McClelland, CNG Energy Co., Pittsburgh, Pa.

Originally published October 14, 1974

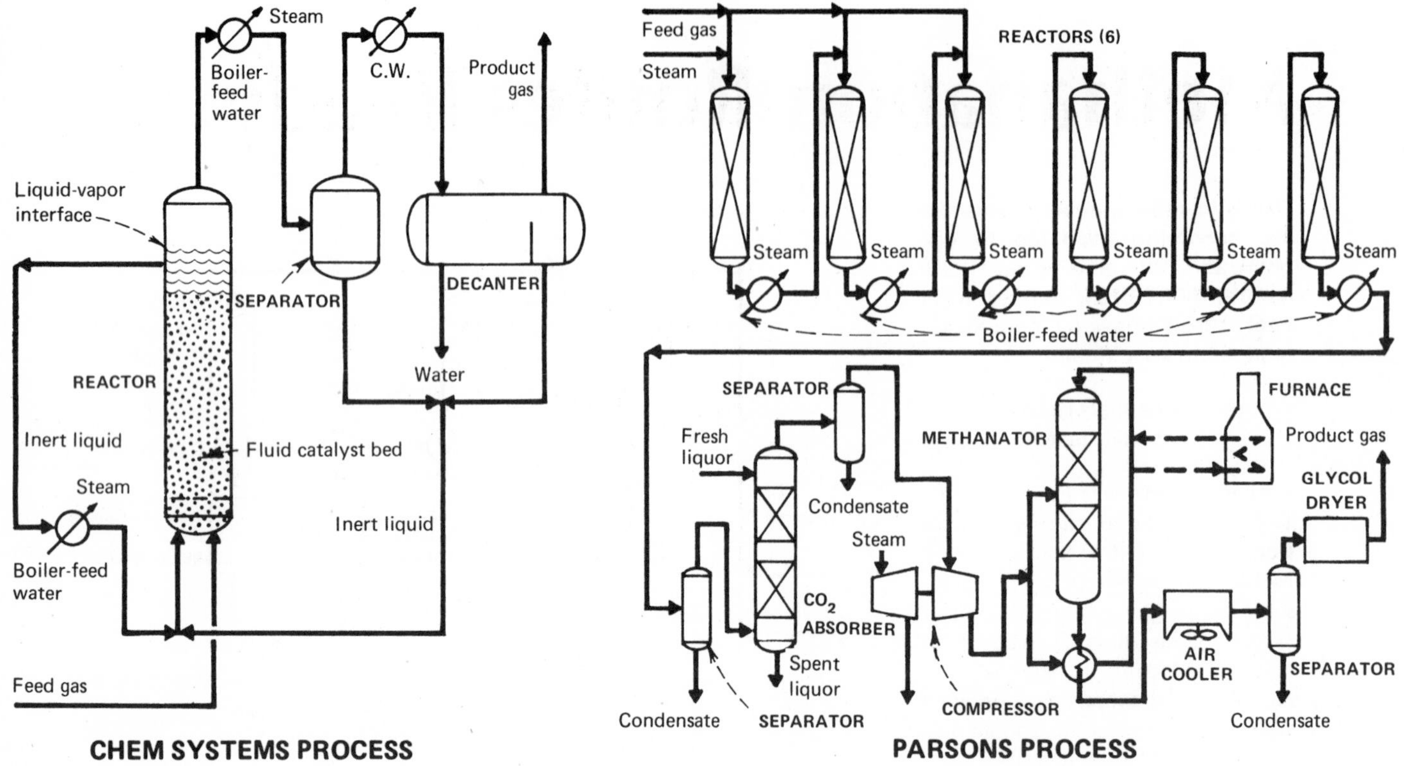

single large reactor, reportedly cutting recycle rate considerably.

Bair pointed out to the ACS audience that maximum outlet temperature for any stage (about 900°F) occurs when all carbon monoxide in the inlet gas reacts. So, temperature can be controlled closely by regulating carbon monoxide content in the inlet gases to each stage, through proportioning of fresh feed, cold recycle gas, and offgas from the previous stage. Inlet temperatures are also set (to 550°F) by regulating the gas blend.

More than 500 h of methanator operation have now been logged over a wide H_2/CO feed ratio, and in all cases Bair reports complete carbon monoxide conversion. No significant catalyst deterioration was noticed.

Liquid-Phase Approach—Development of the Chem Systems, Inc. (Hackensack, N.J.) liquid-phase methanation scheme began on bench scale only 30 mo ago, but by June 1975 a large pilot plant will be running. The skid-mounted design will allow trials at various coal-gasification facilities, points out the firm's David B. Blum.

A circulating stream of inert liquid has the task of heat dissipation (see flow diagram). It flows upward through the reactor cocurrent with the synthesis gas feed, fluidizing the catalyst bed. The liquid mainly picks up sensible heat, but some vaporizes, depending on volatility. Vaporized inert material is condensed from the overhead product gas, both during steam generation and by further cooling, and is then recycled.

Blum says that the methanation reaction goes nearly to completion in a single pass. Further economic studies will be needed, though, to weigh the pros and cons of tacking on a second reactor for final polishing. Catalyst life should be over 1 yr.

Steam-Moderated Method—G. A. White of Ralph M. Parsons Co. (Pasadena, Calif.) then discussed the firm's so-called "RMProcess," now cosponsored by Texaco Inc. and under demonstration at the latter's Montebello, Calif., test plant for SNG by partial oxidation. The dual-purpose technique provides shift-conversion simultaneous with methanation, avoiding the need for separate units to increase the H_2/CO ratio of high-CO synthesis gases.

Desulfurized feed gas at pressures anywhere from atmospheric to over 1,000 psi passes through a series of fixed-bed adiabatic catalytic reactors (see flow diagram). Adding steam to the feed moderates the shift-conversion and methanation reactions by establishing their thermodynamic limits. Combined with heat removal between stages, White says that the steam affords favorable, progressively lower reactor inlet temperatures (typically 1,000°F down to 500°F). No gas is recycled. The steam also prevents carbon formation, said to be a potential trouble in methanation. Final methanation occurs in a cleanup reactor, after absorption of carbon dioxide from the process stream and cooling.

Additional process benefits listed by White include efficient utilization and production of steam; removal of carbon dioxide from the system in a one-stage operation, at a point where gas volume for treatment is minimal. He also reports the ability to methanate at low pressures (for savings in compression duty), and minimum catalyst volume requirement, an upshot of once-through operation at high space velocity. #

n-Paraffins Extraction

Commercially proven technique separates high-purity *n*-paraffins from gasoline, kerosene or gas oils, by adsorption/desorption on molecular sieves.

JOHN GREBBELL
BP Research Center

The British Petroleum pressure-swing process (or MS2 process) has been in commercial operation since 1971 at BP's Ruhr refinery near Dinslaken, West Germany (see photo).

This simple, economical and flexible technique covers the whole carbon-number span from C_5 to C_{25}, and is the preferred process for BP installations and for licensing.

- In the kerosene and gas-oil range, extracted *n*-paraffins have so far served mostly for production of biodegradable detergents and flexible plastics. There is now a rapidly growing interest for use of these *n*-paraffins for single-cell-protein manufacture (as BP Proteins achieved commercially in 1971). The byproducts have much-improved cold properties, and are thus premium middle-distillates for application in cold climates.

FIRST MS2 UNIT turns out 15 million lb/yr of *n*-paraffins product.

Originally published April 14, 1975

- In the gasoline range, the MS2 process is important for increasing octane ratings to allow a reduction in use of lead alkyl antiknocks. It can be used to separate C_5/C_6 product from an isomerization unit, into a high-octane stream, and an *n*-paraffins stream for recycle or petrochemical-feedstock use. Extraction of *n*-paraffins from a C_7/C_{10} stream yields an improved reformer feedstock, and an *n*-paraffins coproduct useful as feed for ethylene production by steam cracking.

Process Mechanisms—The MS2 process is an isothermal, three-stage, cyclic route that operates in the vapor phase.

In the first stage, adsorption, feed hydrocarbon is passed into a molecular-sieve bed at a pressure greater than atmospheric (see flowsheet). The *n*-paraffins are selectively adsorbed to the equilibrium loading for the particular process conditions. The effluent stream coming from the sieve bed during this time is the "denormal" product.

When feeding is completed, the pressure in the sieve bed is reduced, to remove material that is in the sieve voids and adsorbed on the sieve surface. The desorbed *n*-paraffins act as a purging medium and greatly assist the removal of the branched feed components. The total purge-effluent is recycled to feed for maximum recovery of *n*-paraffins.

During the third, desorption, stage, the pressure in the sieve bed is finally reduced to a very low level—*n*-paraffins adsorbed in the sieve cavities are thus removed. The amount desorbed, or the cyclic *n*-paraffin yield, is the difference between equilibrium sieve-loadings at the purge and desorption pressures.

Significant features of the MS2 process include the use of at least two adsorbers to maintain continuous feed-flow. The feed, purge and desorption streams are automatically switched between adsorbers by hydraulically operated valves controlled by a cycle sequencer/timer.

The method of achieving the low pressure required for the purge and desorption stages depends upon the boiling range of the feedstock. With kerosene and heavier material, vacuum (to below 20 mm Hg) is produced by directly condensing the adsorber effluents with a recycling, cooled quench of that material.

Sieve Characteristics—During operation, adsorptive capacity of the 5A-type molecular sieves gradually declines due to buildup of carbonaceous deposits. Periodic regeneration restores the sieves by burning off these deposits with an air/nitrogen mixture.

Though the MS2 process is relatively sulfur-tolerant, certain sulfur compounds accelerate the normal loss of adsorptive capacity. In the gas-oil range, hydrotreated feedstocks containing less than 100 ppm total sulfur are preferred (the Ruhr unit has operated successfully with up to 500 ppm). Feedstocks in the C_5/C_9 range are usually available low in sulfur and olefin content.

Replacing the sieve charge after four to five years onstream is considered the optimum for kerosene/gas-oil units. Economic life for C_5/C_9 feedstocks is estimated at eight years.

Product Quality—Typical feed and *n*-paraffin-product data for operations at Ruhr are shown in Table I. The products are highly satisfactory as chemical feedstocks. They also meet very stringent food-additive tests and have been proven in commercial protein production.

Table II shows the significant decrease in the pour points of the denormal byproducts. This improvement must be evaluated against the decrease in diesel index, but is obviously attractive for cold climates.

In the gasoline range, the MS2 facility at Ruhr has successfully operated on a straight-run C_7/C_{10} benzine. A 95-99%-pure *n*-paraffins stream is typically extracted from a 29% *n*-paraffins feed, providing an especially good steam-cracker feed for ethylene. Denormal product is simultaneously reduced to only 5% *n*-paraffins. This allows a considerable reduction in the severity of catalytic reforming, and this enables significant increases in reformate yield.

The MS2 process can also turn out high-purity normal and denormal streams from a C_5/C_6 cut. This separation is most valuable for improving the octane number of products from an isomerization process.

The *n*-paraffin coproduct could be converted to extinction by continuous recycle to the isomerization reactor. However, it is also a useful petrochemical feedstock.

Economics—Typical economics given below are based on mid-1974 prices in the U.K., for the erected cost (materials and direct labor).

An MS2 extraction unit capable of producing 225 million lb/yr of C_{10}/C_{25} *n*-paraffins (99% pure) requires an investment (excluding molecular-sieve inventory) of $4.7 million. Yearly operating cost is $150,000 for chemicals and sieve replacement. Utilities per pound of *n*-paraffin product are: fuel, 1,900 Btu; electricity, 0.04 kWh; steam, 0.25 lb; and cooling water, 4.3 gal.

A unit capable of producing 440 million lb/yr of 99%-pure denormal C_5/C_6, with 110 million lb/yr of 96%-pure normal C_5/C_6 coproduct, requires an investment (including sieve inventory) of $2.7 million. Yearly operating cost is $13,000 for chemicals and sieve replacement. Utilities per pound of denormal product are: fuel, 800 Btu; electricity, 0.009 kWh; steam, 0.001 lb; and cooling water, 0.01 gal. This MS2 unit stands alone; there would be economic improvement if it were combined with an isomerization unit. #

Meet the Author

John Grebbell is a Project Leader at the British Petroleum Research Center, Sunbury, U.K., in charge of research and development for BP's *n*-paraffins operations. He graduated from the Northern Polytechnic, and is now a fellow of the Royal Institute of Chemistry.

This article is condensed from a presentation before the National Petroleum Refiners Assn. Annual Meeting, San Antonio, Tex., Mar. 23-25.

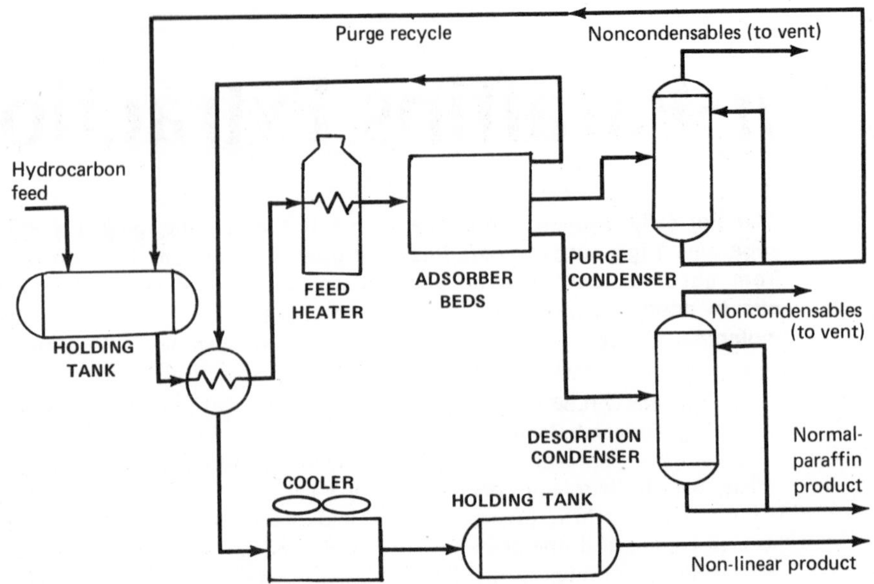

Typical *n*-Paraffin Products From Kerosine and Gas Oil Feedstocks – Table I

Data Source:		Ruhr Unit		
Feed				
Nominal carbon range		$C_{10}-C_{13}$	$C_{14}-C_{17}$	$C_{10}-C_{25}$
IBP	°C	190	250	240
FBP	°C	230	290	350
n-Paraffin content	%wt	27	21	22
Specific gravity 60° F/60° F		0.797	0.827	0.836
Total sulfur	ppm wt	2	30	270
Product				
n-Paraffin content	%wt	99.5	99.3	99.2
Aromatics	ppm wt	20	30	25
Total sulfur	ppm wt	1	2	3
Bromine number	gBr/100g	<0.01	<0.01	<0.01
Pour point	°C	−23	9	18
Food-grade test FDA 121.1146		Pass	Pass	Pass

Properties of Denormal By-Products From Gas Oil Feedstocks – Table II

Nominal Carbon Range		$C_{14}-C_{17}$		$C_{12}-C_{22}$	
Stream		Feed	Denormal	Feed	Denormal
Specific gravity 60° F/60° F		0.822	0.833	0.831	0.844
IBP	°C	239	237	246	250
FBP	°C	296	296	328	327
Cloud point	°C	−17	−46	−7	−40
Pour point	°C	−18	<−57	−9	−45
Aniline point	°C	73.8	68.7	77.1	71.5
Diesel index		67	60	66	58
Cetane number		60	57	63	57
Total sulfur	ppm wt	20	24	135	168
n-Paraffin content	%wt	21.2	3.0	20.9	1.0

Low Air-Feedrate Cuts Phthalic Anhydride Costs

By lowering the ratio of air to ortho-xylene feedstock in the reaction feedstream, this French process uses a smaller than usual compressor. Internal steam production drives the turbine.

JEAN-CLAUDE ZIMMER
Rhône-Poulenc SA

By sharply reducing the amount of air needed to promote oxidation of ortho-xylene to phthalic anhydride, Rhône-Poulenc has come up with a process that enables both lower investment cost and an energy savings.

The investment savings come from the smaller air compressor. While classical vapor-oxidation processes use up to 33 parts of air to each part ortho-xylene feedstock (by weight), this route uses only 22 parts air.

Part of the energy savings is due to the smaller compressor, but the process is also capable of meeting more than its steam requirement through internal heat recovery; about 4 tons of steam per ton of phthalic anhydride are produced. As for electrical consumption, an 80,000-metric-ton/yr plant based on this process uses only about 150 kWh.

The reduced air intake has another indirect energy-saving effect in the tailgas combustor, where no extra energy source is needed.

The process received its first tryout at Rhône-Poulenc's Chauny works in Aisne, France, starting in 1971, hence its name, Chauny 71. The then-existing 40,000-ton/yr Chauny plant was converted partly to the new flowscheme. The firm now is expanding the plant's capacity to 80,000 tons/yr, and the new installations, due for completion in early 1975, will use the new technology.

Two licenses have been sold, one to Resinas Poliesteres, SA, in Spain, and another to Resins Inc. of the Philippines. The Spanish firm is erecting a 30,000-metric-ton/yr unit, which will expand its existing capacity; Resins Inc. started up an 8,000-metric-ton/yr unit last November.

Phthalic anhydride is mainly used to make dioctyl phthalate, which imparts flexibility to a wide range of plastic products such as upholstery, luggage and shoes. Original phthalic anhydride processes were based on naphthalene feedstock, employing fixed- and fluid-bed oxidation over vanadium catalyst.

Ortho-xylene feed began to catch on during the Sixties, in part due to its higher theoretical yield. However, heat of reaction is lower, and energy conservation steps as found in the Chauny 71 route are important. Looking to the future, ortho-xylene's attractiveness as a feedstock will hinge in large part on the price of oil, from which it is obtained. Most naphthalene comes from coal. The

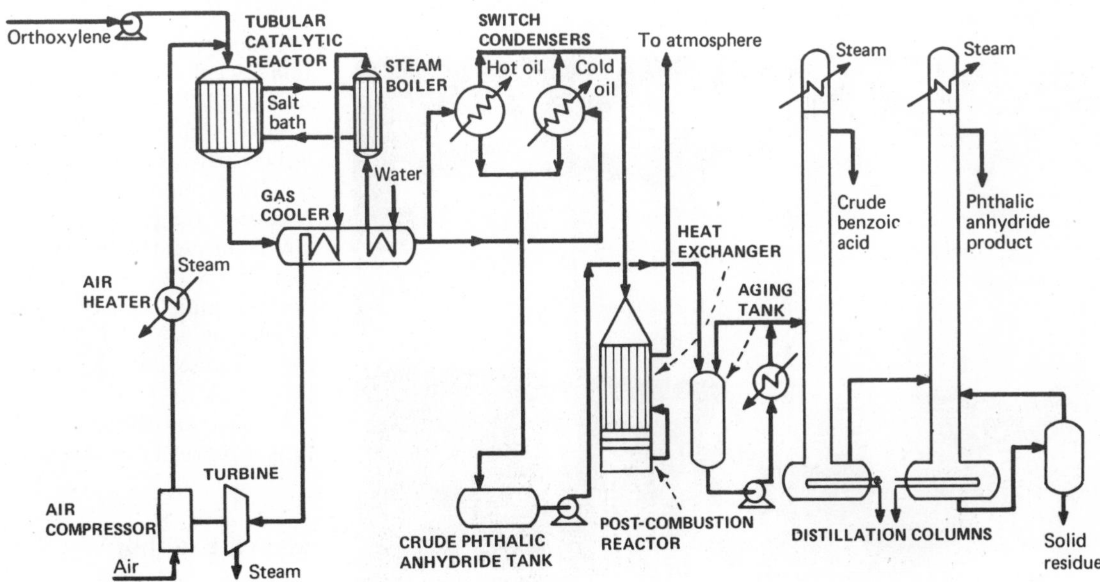

Originally published March 4, 1974

price of finished phthalic anhydride has increased more than orthoxylene feedstock in recent years. This could offset the rising price of oil.

Rhône-Poulenc's researchers are not new to phthalic anhydride. Pechiney-Saint Gobain has manufactured the product since 1939, starting with a fixed-bed air oxidation route. The Rhône-Poulenc group merged its subsidiaries Progil and Pechiney-Saint Gobain to form Rhône-Progil.

Oxidation—In the process, liquid ortho-xylene from storage tanks is mixed directly with hot, filtered air in a vaporizer. The air arrives via the compressor and a steam-heated preheater. Steam to drive the compressor's turbine is produced in the reactor heat-recovery circuit; it leaves the turbine at 160 psi.

The reactor contains vertical tubes through which the reaction mixture passes. There it flows through a fixed bed of vanadium-titanium catalyst. Catalyst life is more than two years. The feed rate over the catalyst is very high—210 g of ortho-xylene per liter of oxidation catalyst each hour. (The initial catalyst load for a 90,000-metric-ton/yr unit is 83 tons.)

The exothermic heat of reaction is removed by liquid salt, which transfers heat to a remote, 700-psi-steam generator. The steam produced here is then superheated by gases leaving the reactor before it goes on to the turbine. The reaction gases also heat the boiler feedwater and in the meantime they are cooled down from 700°F to 320°F.

Recovery—Three switch-condensers alternately cool phthalic anhydride to crystals (that form directly from the gas phase) until a condenser is fully laden, then heat up the product to melt it away. Each condenser consists of a chamber with rows of finned tubes, through which heat-exchange oil passes. The hot and cold heat-transfer oils are respectively heated by steam and cooled by air or water.

The leftover reaction gases, composed mostly of air, leave the switch condensers at 140°F and go to the post-combustion unit for cleanup. The stream is then emitted to the air without any pollution.

The raw stream first enters the heat exchanger portion of the combustor and is preheated against hot, clean gases exiting from the catalytic portion located just below. The catalyst is proprietary. The outgoing gases contain only nitrogen, oxygen, carbon dioxide and water.

Crude phthalic anhydride now goes to distillation for upgrading to a purity of 99.8%. But first it is admitted to an aging tank, where some chemicals (undisclosed) are added and the entire contents are kept boiling under atmospheric pressure. This mixture is continuously tapped and sent to a topping column where benzoic acid and maleic anhydride are removed from the top. Then it goes on to a vacuum distillation column from which the pure product is obtained overhead. Residual tars from this step are recovered as pellets and can be burned as fuel.

Costs—The initial investment for the Chauny 71 is generally 20% less than conventional processes, though an exact price comparison is difficult because the cost of other processes is not clear.

Ordinary steel is used in construction except for the distillation columns, which are made of stainless steel. A plant capable of producing 90,000 metric tons/yr would require an investment of $7.0 million ($1.00=5 francs). (This investment covers the plant within battery limits, and does not take into account such items as initial catalyst loads, offsite storage units for feedstock, power lines, license and engineering fees, taxes, etc.)

Production costs and expenses are broken down in the table. #

Operating Requirements — 90,000-metric-ton/yr plant

Feed stock, per metric ton of phthalic anhydride	
Ortho-xylene, Kg	950
Utilities, per metric ton of phthalic anhydride	
Electricity, kWh	150
Boiler feedwater, m³	4.1
Steam produced (160 psi), metric tons	(3.6)
Labor, men per shift	3
Maintenance, % of battery limits investment per year	3

ALTERATIONS to Rhône-Progil's Chauny works are based on the new process.

Meet the Author

Jean-Claude Zimmer is in charge of licensing for the Chauny 71 process. He holds a Ph.D. in chemical engineering, and before switching to Rhône-Poulenc licensing department, he worked in the firm's research department.

Pressure-swing adsorption: geared for small-volume users

Employing adsorbent carbon to separate air into nitrogen and oxygen at ambient temperatures, this system's simplicity favors its use in small-scale inert-gas generating plants.

Karl Knoblauch, Bergbau-Forschung GmbH

☐ Development work has been under way for some years on new techniques to separate, at ambient temperatures, air into nitrogen and oxygen by means of molecular sieves. The Pressure-Swing Adsorption (PSA) system is a simple, lower-cost method for filling inert-gas generating needs such as purging, inert blanketing, and generating N_2-based controlled atmospheres. Or, the route can even be used to generate oxygen, if needed.

This article is based on a presentation made at the Nitrogen 78 Seminar, sponsored by Petrocarbon Developments Ltd., held at Stratford-upon-Avon, England, in May.

Also, because it avoids the complex equipment—and costs—of conventional separation techniques, PSA is ideally suited to smaller-volume (1,000 m³/h or less) users.

The heart of the process is an adsorbent coke, produced from hard coal, that selectively removes oxygen from a compressed-air supply, leaving nitrogen and argon as the product stream. At the end of the cycle, as selective O_2-adsorption drops off, a vacuum pump draws the oxygen from the bed.

Product gas is essentially free of water and carbon dioxide, and its purity ranges from 95 to 99.9% N_2 plus Ar by volume. In addition, O_2-concentration in the product gas can be controlled to 0.1% by volume.

To date, 14 plants using the Bergbau-Forschung licensed route have been commissioned. System capacities range from 10 to about 1,000 m³/h. Although capital cost varies with nitrogen quality required, PSA plants generally run from about $40,000 for a 10-m³/h unit to approximately $350,000 for a 1,000-m³/h facility.

CONVENTIONAL SEPARATION—Generally, on an industrial scale, nitrogen and oxygen are recovered by cryogenic separation of air: incoming air is liquefied and its components are separated by distillation or rectification, using their differing vapor pressures. Although capital investment and energy consumption costs for cryogenic equipment are high, they do not pose insuperable barriers for large-scale facilities.

For inerting purposes, nitrogen can also be produced by combustion of hydrocarbons—a relatively impure protective gas is obtained. Investment costs for such a plant are relatively low, and energy costs depend largely on the fuel price. Removal of impurities, however, is expensive. For while the proportion of NO_x compounds can be kept low by reducing the air supply for combustion, sulfur compounds must be removed by catalysis, and removal of residual oxygen requires a downstream catalytic purification. Removal of CO_2 and H_2O requires purification by means of additional downstream molecular-sieve adsorbers.

To get around the complexities of conventional systems, a variety of adsorption processes for air separation have been developed. Among their advantages over the classic low-temperature processes: easy readiness for operation, simplicity of process

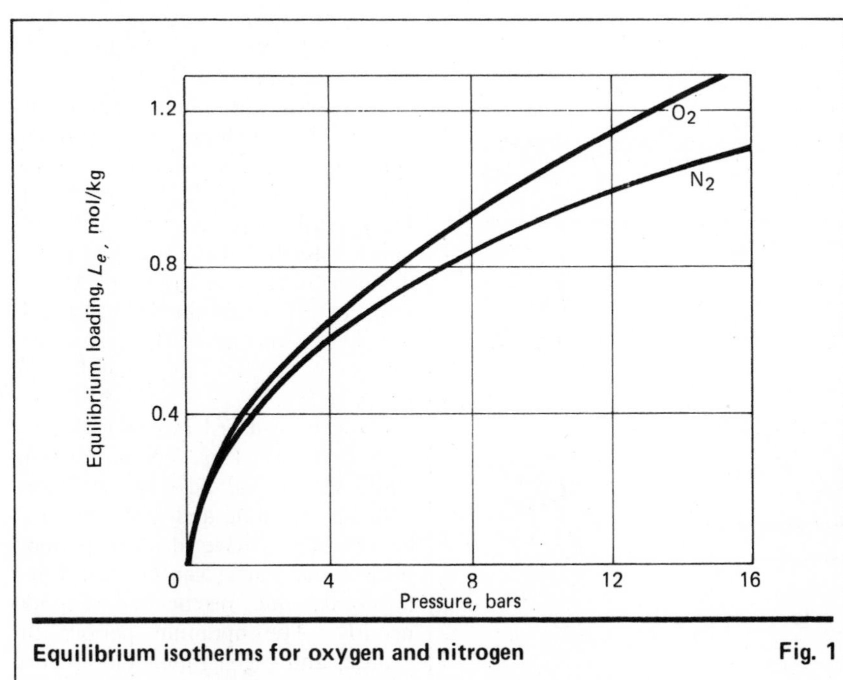

Equilibrium isotherms for oxygen and nitrogen **Fig. 1**

Originally published November 6, 1978

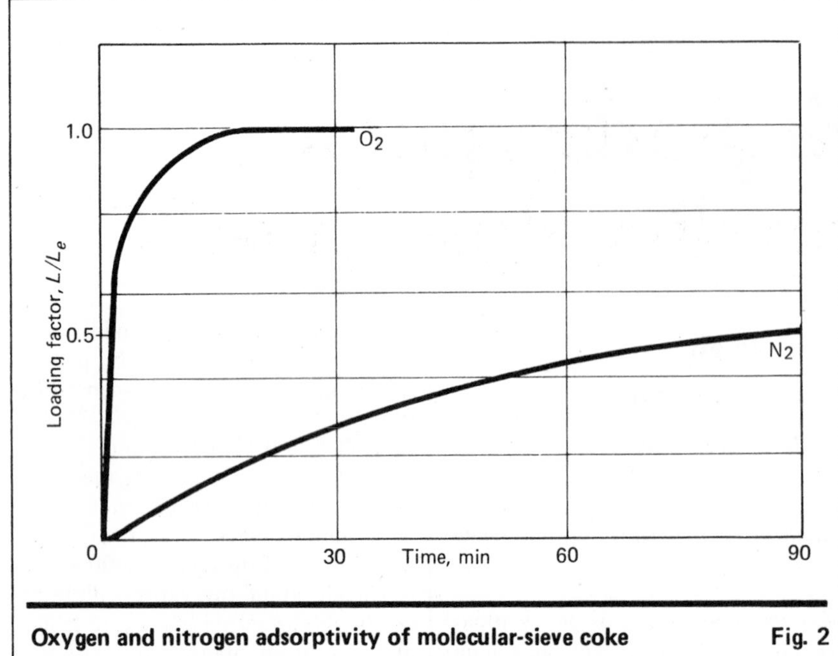

Oxygen and nitrogen adsorptivity of molecular-sieve coke **Fig. 2**

control, operation at ambient temperatures and easy automation. A disadvantage for large-scale plants, however, is their relatively high energy consumption.

CARBON VS. ZEOLITE—There exist two entirely different types of molecular sieve: those using zeolites (hydrated aluminum and calcium—or sodium—silicates) and those using carbon materials for their adsorbent beds. But while the operation of zeolite molecular sieves is typically based on a higher adsorption of the N_2 component, the separation effect of molecular-sieve coke used in the PSA route is obtained by the differing diffusion rates of oxygen and nitrogen molecules within the extremely small pores of the coke.

Among the essential features of molecular-sieve coke are:

■ Under equilibrium conditions, molecular-sieve coke adsorbs oxygen and nitrogen in comparable quantities. On the other hand, the equilibrium loading for zeolite molecular sieves is, at ambient temperatures, considerably higher for N_2 than for O_2.

Fig. 1 shows equilibrium isotherms of N_2 and O_2 for a carbon molecular sieve. The loading increases with increasing pressure. When the pressure is reduced, the individual components are desorbed, and when equilibrium is reached, adsorbent loading corresponding to the respective partial pressure is established.

■ Oxygen diffusion through the narrow gaps of the pore system is at a considerably higher rate than that of nitrogen diffusion. The adsorbed N_2 and O_2 quantities, related to equilibrium load, are plotted on Fig. 2 against time, showing that the rate of oxygen adsorption is considerably faster than the rate of nitrogen adsorption. After a few minutes, the O_2-equilibrium load has reached a value of 80%, whereas N_2-equilibrium loading at the same time is $< 5\%$. The diffusion—determined by means of a volumetric measuring apparatus—amounts to $1.7 \times 10^{-4}\ s^{-1}$ for the oxygen molecules, and $7.0 \times 10^{-6}\ s^{-1}$ for the nitrogen molecules. That is, they differ by a factor of 400.

■ The water vapor and carbon dioxide contained in air are simultaneously adsorbed with the oxygen. But this has only an insignificant influence on the oxygen-adsorption capacity of the molecular-sieve coke.

THE PROCESS—Fig. 3 shows the schematic layout of a plant for producing nitrogen with purity between 97 and 99.9%, by volume. For continuous operation, two adsorbers are required. One adsorber is loaded with air. During this step, oxygen is predominantly adsorbed while nitrogen passes through and leaves the unit. After a short period when the O_2 concentration in the nitrogen-rich product gas reaches a certain value, the loading phase ends.

Oxygen adsorbed by the molecular sieve is desorbed by depressurization, while the second adsorber is loaded with air. Loading and desorption can be adjusted to have identical periods, so that one reactor can be loaded and the other one regenerated simultaneously. The optimum periods for loading and regeneration phases turn out to be 60 s apiece. Desorption pressure is approximately 70 Torr, and the oxygen fraction contains about

N2
Throttle
Adsorber A
Adsorber B
3-way valve
3-way valve
Air
Vacuum pump
Compressor

Adsorbers operate simultaneously, on alternating cycles **Fig. 3**

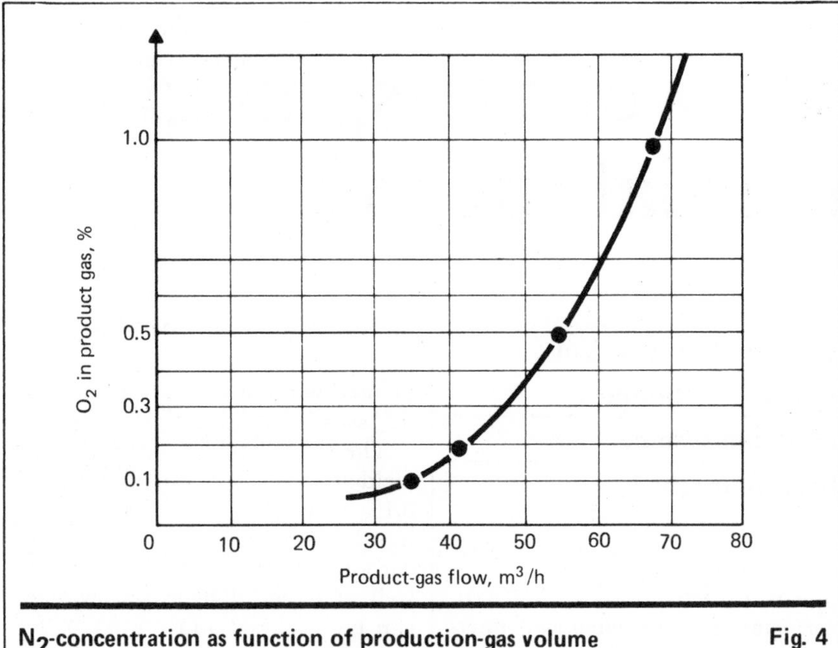

N_2-concentration as function of production-gas volume **Fig. 4**

35% oxygen and 65% nitrogen, as well as CO_2 and H_2O from the air.

Adjustment of the air flow rate through the adsorbers enables infinitely variable control of O_2-concentration in the product down to 0.1%, by volume. This means, of course, that a high degree of purity entails lower product-gas quantities (see Fig. 4).

An increase of product gas can, however, be obtained by increasing operating pressure (as shown in Fig. 5). Operating pressures up to 4 bars turn out to be particularly favorable. Energy consumption for this mode of operation amounts to approximately 0.60 kWh/m³ of nitrogen containing about 0.1% (by volume) O_2. Energy consumption decreases with reduction in purity and amounts to 0.4 kWh/m³ for 99.0% nitrogen.

The only caution is that oil mist—from air compressors, for example—adversely affects the functioning of carbon molecular sieves, as it does in the case of most other adsorbents. Carryover of oil mist, therefore, should be prevented as far as possible.

Nitrogen produced by the PSA system does not contain any impurities other than those in the inlet air. The dewpoint is below −40°C. By means of a downstream catalytic de-oxo unit and the addition of hydrogen, it is even possible to obtain an oxygen-free gas, the dewpoint of which is below 0°C, making the gas well suited for use as a protective agent for welding or annealing operations.

VERSATILE UNIT—The PSA system can also be used to produce oxygen as the main product. For this purpose, the design allows for a maximum quantity of air loading, so that a correspondingly greater amount of O_2 can be adsorbed. The oxygen fraction then recovered by depressurization contains about 50 to 60% O_2. By dividing desorption into three phases, recovering only the middle-phase product and recycling the product of the first and last phases to the main feed, an oxygen purity of 80% can be achieved. Oxygen fractions of higher purity can be obtained, but not in an economically viable way at present. Prospects seem favorable, however, for cost-effective production of higher O_2-concentrations by further development of the technology.

ECONOMICS—The PSA process is competitive in the given small-user market situation when compared with the alternative methods of nitrogen production via combustion or cryogenic techniques.

Systems are available in various standard sizes, including 10, 30, 100, 500 and 1,000 m³/h. Capital costs range from $40,000 to $350,000, as described previously.

Power consumption varies, but generally falls in the range of 0.25 to 0.6 kWh/m³ of product. Lowest overall operating costs are typically attained at pressures of 3 to 4 bars.

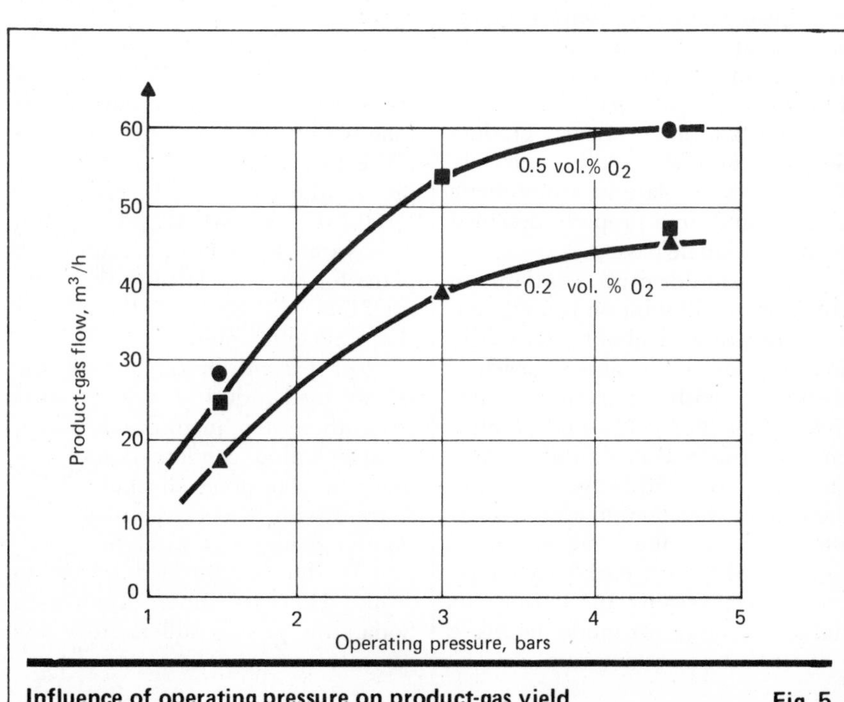

Influence of operating pressure on product-gas yield **Fig. 5**

The author

Karl Knoblauch is department leader for gas-adsorption processes for Bergbau-Forschung GmbH (Franz-Fischer-Weg 61, 4300 Essen 13 (Kray), West Germany), and has been involved in adsorption-process work for over 12 years. He studied chemical engineering at the Universities of Braunschweig and Karlsruhe, holds a Doctor-Ingenieur degree from the University of Aachen and is a member of the Deutsche Gesellschaft für Mineralölwissenschaft and Kohlechemie.

New processes win the spotlight at ACS centennial meeting

The American Chemical Society gathering in New York provided a look at a number of promising techniques, including simplified chemicals-production systems and a municipal-solid-wastes gasification method.

☐ The American Chemical Society celebrated its 100th birthday on April 5th. So, understandably, the group's 171st National Meeting, timed to coincide with the event and held with suitable pomp in New York City, featured ample plaudits for the venerable organization.

But the gathering also served up a wealth of technical information. In fact, more than 1,370 papers vied for the attention of the approximately 10,500 attendees.

Among the highlights of the technical program, one German company and three Japanese firms described chemical-production processes, which they are grooming for commercialization. The routes promise markedly simplified output of: oxamide, cyanoformamide or cyanogen; methylisobutyl ketone; hexamethyleneimine; and trimethyl hydroquinone.

CYANIDE-BASED SYSTEM—At the meeting, Germany's Hoechst AG (Frankfurt/Main) detailed a simple, one-step catalytic process that converts hydrogen cyanide into oxamide, cyanoformamide or cyanogen, depending upon the reaction medium. The firm has successfully operated a more-than-10-ton/mo (of oxamide) pilot plant at Frankfurt for about 6 mo. Hoechst now is thinking of building a commercial plant in Germany, and also is willing to license the technology.

The company notes that oxamide appears attractive for use as a fertilizer, stabilizer or chemical intermediate, while cyanoformamide can serve as an intermediate, and cyanogen already boasts a wide range of applications, such as for organic syntheses and as a fumigant.

The process stems from Hoechst's discovery that the use of aqueous acetic acid or acetonitrile, instead of water alone, significantly speeds some intermediate reactions. In fact, in one key case, the reaction proceeds over 3,000 times faster.

In the flowscheme, as shown on the diagram on the next page, hydrogen cyanide and a moderate stoichiometric excess of oxygen enter near the bottom of a reactor containing a solution of copper nitrate. For oxamide or cyanoformamide output, aqueous acetic acid serves as the solution—while, for cyanogen production, acetonitrile handles the job. Reactions take place at atmospheric pressure and, in a properly designed vessel, in a single pass.

In the oxamide process variant, an 80% acetic-acid solution is used, and a temperature of about 120–185°F. The reaction occurs almost stoichiometrically, with an oxamide selectivity of 99–99.5%. Oxamide continuously forms and precipitates. Crystal sludge collected at the bottom of the reactor then goes to a centrifuge, where almost all of the entrained catalyst solution is recovered for recycle. After washing and drying, the stable, nontoxic oxamide boasts a purity of more than 99%, and contains only 0.02–0.05% copper. It can be used without any further processing, says Hoechst.

The reactions for oxamide also yield about 0.3% carbon dioxide and 0.1% cyanogen. Waste gas, made up mostly of excess oxygen but also the carbon dioxide and cyanogen, as well as traces of hydrogen cyanide, can be scrubbed and vented, or fed to a combustion unit. Overall, the process poses no effluent problems, according to the company.

Using the acetic-acid solution but keeping the reaction temperature below about 115°F spurs the formation of cyanoformamide at 80% yield. Replacing the medium by a 90–95% acetonitrile solution diverts the process to produce cyanogen at more-than-90% yield at 120–185°F. In the latter variant, azeotropic or extractive distillation keeps water concentration at a desired level in the recycling acetonitrile stream.

KETONE PROCESS—Also at the meeting, Tokuyama Soda Co. (Tokuyama City, Japan) highlighted its new process for single-stage reductive condensation of acetone with hydrogen to methylisobutyl ketone (MIBK). The firm has proven the technique in an about-1,100-lb/d (of MIBK) pilot plant. Tokuyama Soda has no immediate plans for commercializing the route itself but is willing to sell the catalyst that stars in the process.

Most present production of the widely used industrial solvent relies on a three-step method: the base-catalyzed aldol condensation of acetone to diacetone alcohol, dehydration with acid of the alcohol to mesityl oxide, and catalytic reduction of that unsaturated ketone to MIBK. However, notes Tokuyama Soda, this process suffers from low

Originally published May 10, 1976

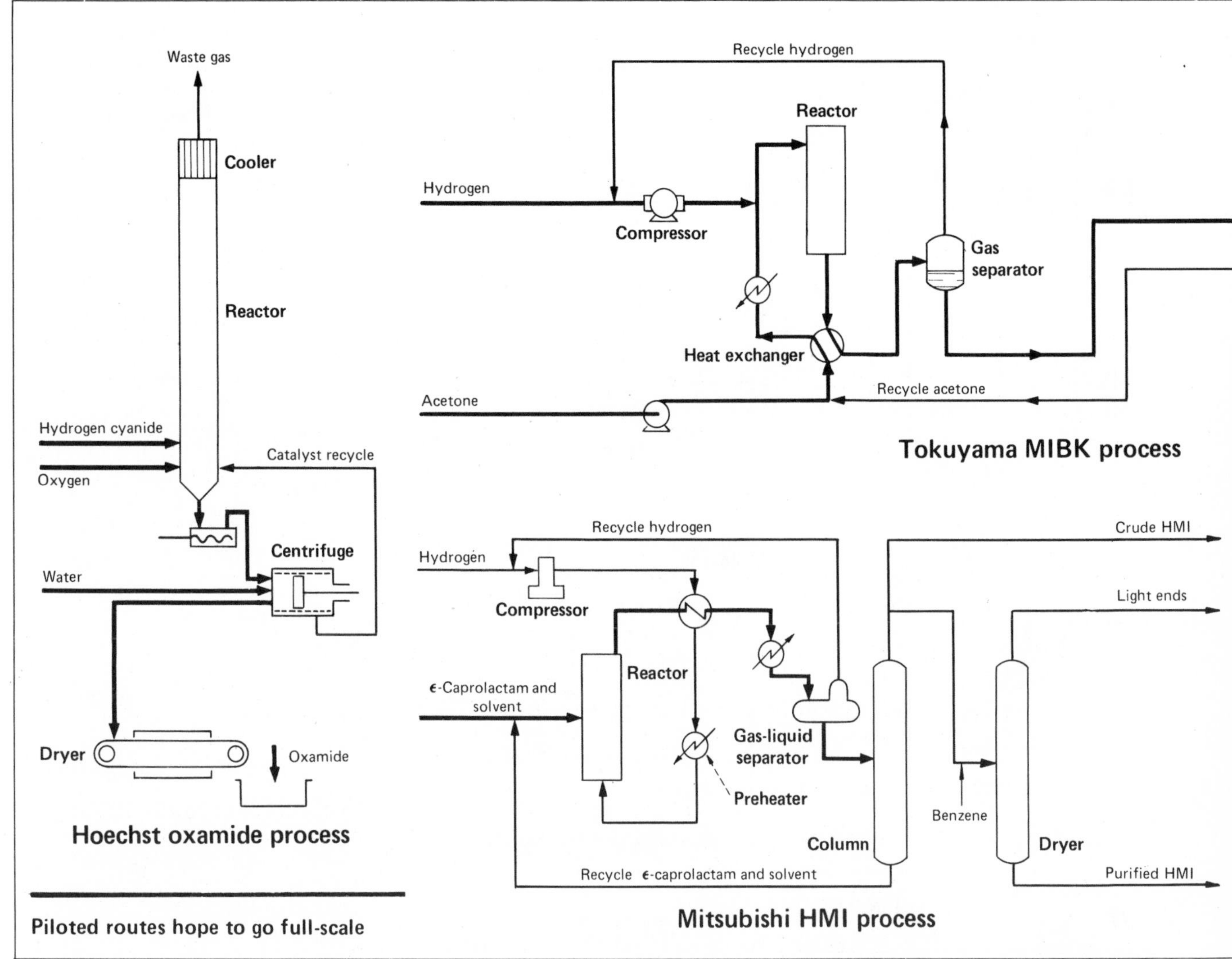

Piloted routes hope to go full-scale

equilibrium conversion of acetone in aldol condensation, as well as from relatively low yield in each reaction stage. In contrast, claims the firm, its route avoids these drawbacks, and promises to considerably reduce plant investment, raw-materials consumption and utilities requirements.

Tokuyama Soda points to its new bifunctional catalyst as the key to its technique. The material consists of zirconium phosphate loaded with fine (average size, less than 30 Å) and uniformly dispersed palladium. The catalyst resists impurities so well that the feed acetone and hydrogen do not require any special purification, says the company. And it adds that the initiator boasts good thermal and mechanical stability, promoting long service life. Pilot-plant performance bears this qut: the catalyst maintained a uniform, high level of conversion and selectivity for more than 3,000 h. And the firm notes that regeneration, if needed, can be easily done.

In the process, acetone and hydrogen (in a molar ratio of 1 to 0.2–0.4) undergo preheating before passing into a fixed-bed reactor operating at 20–50 atm and at about 250–320°F. Conversion of acetone takes place at a rate of approximately 30–40% per pass. And under optimum conditions, selectivity for MIBK exceeds 95 mole-%. Isopropyl alcohol and diisobutyl ketone—but not diacetone alcohol or methylisobutyl carbinol—form as byproducts. This combination of conversion rate and selectivity leads to unparalleled space-time yields, contends Tokuyama Soda, reaching as high as 2,000 g/(l)(h) and allowing significant economies from reduction in the size of the reactor.

A thermal-transfer fluid removes heat of reaction, and the product stream from the reactor cools further by exchange against feed acetone. The product next passes to a separator, where unreacted hydrogen is recovered for recycle. A distillation tower splits off excess acetone, which goes back to the reactor, while leaving crude MIBK as bottoms. Then, another separator eliminates water from the ketone stream, and the MIBK goes through a distillation train for final purification.

As an added advantage of the process, the system does not demand special materials of construction. For instance, stainless steel can serve for the reactor. And Tokuyama Soda states that the process offers the option of turning out mesityl oxide and MIBK, as well as MIBK alone, by controlling reaction conditions.

ABUNDANT FEEDSTOCK—Hexamethyleneimine (HMI) finds use as an intermediate for pharmaceuticals and agricultural chemicals, and in anticorrosive agents for metals, in

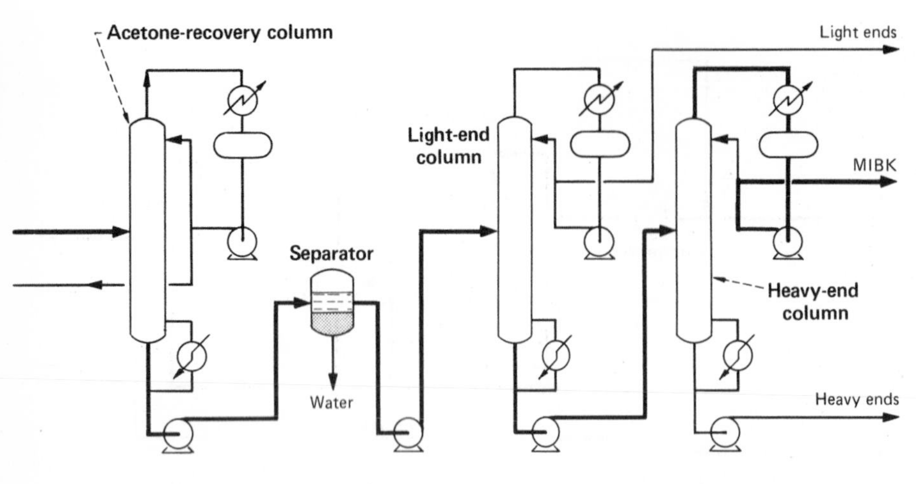

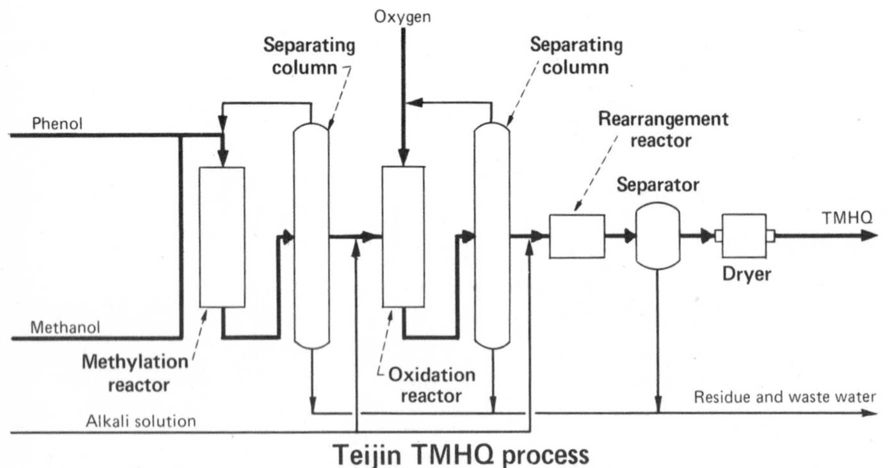

Teijin TMHQ process

softeners for fibers, and in additives for rubber. Availability of the material now depends on its byproduction in the manufacture of hexamethylenediamine. And this limits HMI's prospects. However, a development of Mitsubishi Petrochemical Co. (Tokyo) promises to unfetter the compound.

At the ACS gathering, Mitsubishi Petrochemical detailed a simple hydrogenation process that makes HMI from ε-caprolactam, a material widely available at moderate price because of its use as a feedstock for the production of nylon 6. Heretofore, high catalyst costs, coupled with low yields, have stymied such an approach, says the company. It claims to have overcome these hurdles. The firm has run an about-10-lb/batch pilot plant since the middle of 1974, and wants to license its knowhow.

Its process hinges on the use of a cobalt-and-rhenium catalyst dispersed in a higher-paraffin solvent, such as tridecane, and on the immediate removal of products from the reaction zone. In such circumstances, says the firm, per-pass conversion and selectivity to HMI both run better than 90%.

In the flowscheme, liquid ε-caprolactam joins a preheated, stoichiometric excess of hydrogen in an agitated reactor held at about 20 atm and 390°F. Product stream from the vessel cools by exchange against feed hydrogen and then against thermal-transfer fluid, before passing to a gas-liquid separator. Hydrogen goes overhead for recompression and recycle, while the liquid continuously flows into a distillation column. Unreacted ε-caprolactam, solvent and a small quantity of high-boiling byproducts are removed as bottoms and recycled.

Crude HMI comes off the top of the tower. As is, it suits certain applications, e.g., production of agricultural chemicals and rubber additives, according to Mitsubishi Petrochemical. However, if desired, the HMI can go to a drying column for elimination of water and low-boiling compounds. In this latter case, HMI purity equals or exceeds 99%, with impurities consisting mainly of *n*-hexyl amine, caprolactam and water.

Mitsubishi Petrochemical foresees attractive capital costs for the route because of its simplicity and mild operating conditions.

VITAMIN PRECURSOR—The ACS meeting also saw Teijin Ltd. (Iwakuni, Japan) put the spotlight on its simple technique for making trimethyl hydroquinone (TMHQ), a feedstock for the synthesis of vitamin E, from less-expensive and more-abundant raw materials.

Conventional TMHQ routes start with 2,3,6-trimethyl phenol, convert it to 2,3,6-trimethyl benzoquinone, and react that compound with either isophytol or phytol. However, such systems suffer from the high cost of the trimethyl phenol, and limitations in its supply, because the material stems commercially as a byproduct of cresol manufacture.

Teijin, instead, relies on phenol and methanol as raw materials. And its process involves only three basic steps, and produces no byproducts. The company has operated a pilot plant of about-8-lb/d (of TMHQ) capacity for a few years, and now seeks to license its technology.

In the first—methylation—step of the process, phenol and methanol go into a reactor filled with a magnesium-oxide-containing catalyst. (This catalyst can be regenerated, and gives an overall service life of about 1 yr.) At around 750–930°F, selectivity for 2,4,6-trimethyl phenol (and products convertible to it upon recycle) betters 95%. Trace amounts of unconvertible aromatics, such as toluene, and *meta*-methylated derivatives, such as 2,3,4,6-tetramethyl phenol, also form.

The exit stream from the reactor then passes to a separation column, in which it is fractionated into various components. Convertible compounds are recycled, while residue and wastewater are sent for disposal.

The 2,4,6-trimethyl-phenol fraction next moves to the second—oxidation—stage of the process.

Teijin has devised a new method to oxidize 2,4,6-trimethyl phenol into 4-hydroxy-2,4,6-trimethyl-2,5-cyclohexadiene-1-one, which avoids shortcomings in yields of conventional systems.

The 2,4,6-trimethyl phenol goes into a specially designed reactor, along with molecular oxygen and an aqueous alkali stream, which acts as catalyst. Reaction at low temperature and above-atmospheric pressure yields the product, while ensuring operating safety.

The fluid from this vessel then flows to another column, from which unreacted and not-fully-reacted material goes overhead for recycle, and residue and wastewater remain as bottoms for discard.

The cyclohexadiene-1-one passes into the reactor of the third—rearrangement—part of the process. Rearrangement typically poses problems of TMHQ purity and yield, and catalyst consumption, according to Teijin. However, the company claims to have improved this operation to produce TMHQ of a higher quality, with a better yield and using far less catalyst than ever before possible.

In the reactor, cyclohexadiene-1-one maintained under high pressure and at an elevated temperature forms TMHQ, speeded by the presence of an aqueous alkali solution. The product stream goes to a separator for removal of residue and wastewater, and then to a dryer. TMHQ leaving the dryer meets commercial specifications, says Teijin.

The company adds that its economic data show the process fully competitive with present routes.

Mark D. Rosenzweig

Propylene oxide routes are ready to take off

Companies in Europe and the U.S. are grooming propylene oxide processes that promise to sharply cut the usual output of effluents or coproducts. At least one new route features novel chemistry.

☐ Fundamental changes in the method of manufacture of a major chemical occur rarely, if at all, but propylene oxide production appears headed in that direction.

Such international chemical-process-industries (CPI) heavyweights as Bayer AG (Leverkusen, West Germany) and Solvay et Cie. (Brussels) are among those working on new technology. And some processes are already moving toward commercialization.

Output of propylene oxide—a key intermediate in the production of polyurethanes and polyesters—currently relies on one of two basic chemical approaches: via propylene chlorohydrin or a hydroperoxide. Both methods can pose problems.

The chlorohydrin technique typically reacts propylene with chlorine and water to yield propylene chlorohydrin. This then contacts a lime slurry or caustic soda to form propylene oxide. The method, however, also produces effluent containing chlorine compounds and other contaminants.

In 1969, Oxirane Corp. (Princeton, N.J.)—a joint venture of ARCO Chemical Co. (Philadelphia) and Halcon International, Inc. (New York City)—commercialized the rival, hydroperoxide, approach, with the startup of a plant at Bayport, Tex.

In this route, oxidation of isobutane produces *tert*-butyl hydroperoxide. This epoxidizes propylene to yield propylene oxide; it also generates *tert*-butanol as coproduct. Oxirane has since commercialized a variant that starts with ethylbenzene instead of isobutane, and coproduces styrene monomer. In both versions, about twice as much coproduct as propylene oxide is made.

At the moment, Oxirane and its affiliates are the only ones using the hydroperoxide technique. However, Shell Nederland Chemie NV (The Hague) expects to start up a hydroperoxide-based, 125,000-metric-ton/yr propylene oxide unit at Moerdijk, the Netherlands, by about the middle of next year. Its process will feature a novel catalyst, Shell says, and will yield 325,000 metric tons/yr output of coproduct styrene.

ALTERNATIVE APPROACHES—A number of firms foresee an advantage in coming up with technology that reduces or avoids altogether the effluent or coproduct aspects of proven routes. At least two methods hold promise of such a plus, according to their backers.

One approach makes a peracid, which then epoxidizes propylene. The other first acetoxylates propylene and then cracks the resultant propylene glycol monoacetates.

Two major groups are grooming peracid methods. Bayer and Degussa (Frankfurt/Main) are working together on such a process, as is Propylox S.A. (Brussels), a joint venture of Société Carbochimique S.A. (Brussels) and Interox S.A. (Brussels). Interox itself is an equal partnership of Solvay, and Laporte Industries (Holdings) Ltd. of London.

Chem Systems, Inc. (New York City) and an undisclosed chemical company are developing an acetoxylation/cracking route.

A third option also may surface. One industry source claims that Olin Corp. (Stamford, Conn.) is developing a route in which isopropanol is oxidized to yield a complex hydrogen peroxide stream that then reacts with propylene to form propylene oxide and acetone. The firm will only say that it is working at the mini-pilot-plant stage on many promising alternatives for making propylene oxide.

PERACID PROGRESS—The peracid approach seems most advanced at the

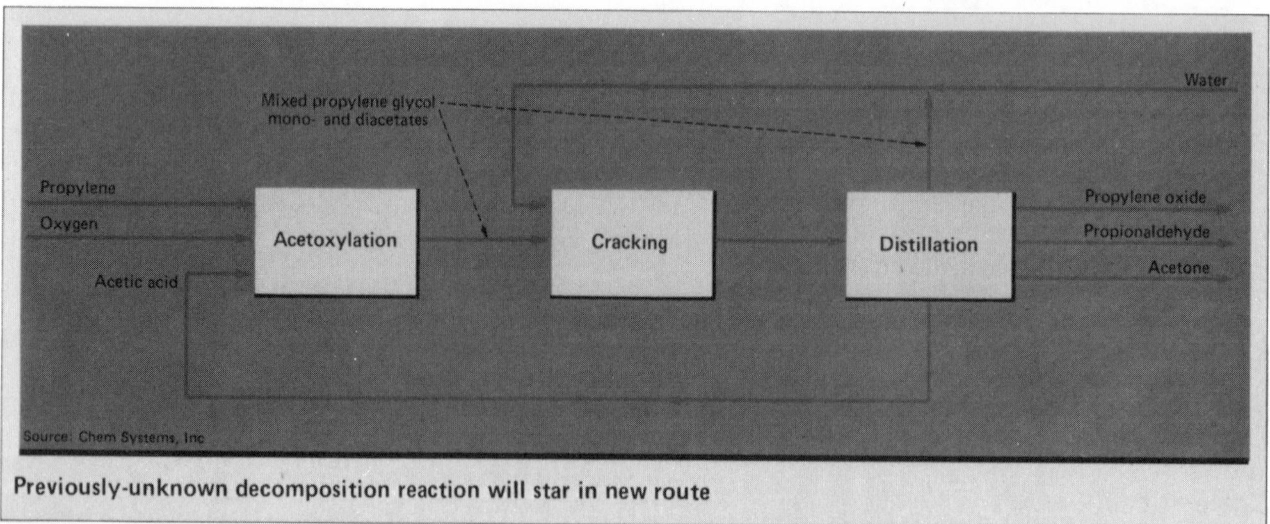

Previously-unknown decomposition reaction will star in new route

Originally published October 24, 1977

moment, with fullscale plants under consideration.

Bayer and Degussa have operated a pilot plant for several years, and now say that planning for a commercial-scale plant has started.

In their route, hydrogen peroxide and an undisclosed organic acid combine to form the corresponding peracid. This then epoxidizes propylene to propylene oxide, and also yields the organic acid, which is recycled. The technique boasts substantially higher yield than chlorohydrin processes, say the firms. Moreover, the route is claimed to be environmentally clean and to create only a negligible amount of byproduct. Dealing with a peracid can pose hazards, but the German team stresses that its route is no riskier than conventional techniques.

Solvay, for its part, notes that its group has already gotten promising results in a pilot plant at Jemeppe-sur-Sambre, Belgium, and is now doing design studies for commercial-size plants.

Propylox's process also is said to start with hydrogen peroxide and an organic acid to yield a peracid for propylene epoxidation. Solvay states that the process should not produce any significant amount of effluents or coproducts. And, the firm adds, the technique will likely consume less energy than do current methods, and safety should not be a problem.

A different peracid process has already proven itself in commercial use in Japan. Daicel Ltd. (Osaka City) has been running a 12,000-ton/yr propylene oxide plant (the smallest in the country) at Otake, Hiroshima Prefecture, since 1969. But its route differs markedly in peracid preparation. And Daicel itself says that the process is feasible only in special circumstances.

The company does not use hydrogen peroxide. Instead, it reacts acetaldehyde with ethyl acetate to make peracetic acid. Epoxidation of propylene then yields propylene oxide as well as a stoichiometric equivalent amount of acetic acid. Daicel welcomes the large output of coproduct, since the firm is a major marketer of acetic acid. The company adds that the route poses no effluent problems.

Daicel and industry observers say that feasibility of the other peracid routes will hinge on the cost of available hydrogen peroxide.

CRACKING—Chem Systems avoids such a worry by relying on different chemistry. In its route, propylene, oxygen and acetic acid react to form propylene glycol monoacetates. Then, in what the company says is a previously unknown reaction, the monoacetates yield propylene oxide.

According to the company, the process boasts high conversion rates and selectivity, and gives only limited output of byproducts.

Chem Systems believes that its route can potentially save as much as 20% in total costs, including capital and operating charges, vs. conventional chlorohydrin and hydroperoxide methods. Also, claims the firm, the process should beat the economics of hydrogen-peroxide-based systems, especially when the cost of making hydrogen peroxide feed for them is included.

The company started work on the process several years ago, and granted an exclusive option in 1976 to an undisclosed international CPI concern. That firm is now working with Chem Systems, and development has advanced to the point that an about-1-ton/mo pilot plant is now being built outside the U.S. This unit should become operational by next spring, and then about 18 months of trials will probably take place.

In the process, as pictured in the box diagram, propylene, oxygen and acetic acid undergo liquid-phase reaction in the presence of an unspecified, proprietary catalyst. The output stream from this reactor contains about 50% propylene glycol monoacetates, and about 50% propylene glycol and propylene glycol diacetate.

The stream then goes to an alkaline cracking step, featuring a heterogeneous catalyst. This step takes advantage of the previously unknown vapor-phase decomposition reaction—propylene glycol monoacetates breaking down into propylene oxide and acetic acid—says Chem Systems, which has filed for patents worldwide on the step. Some propionaldehyde and acetone form as byproducts.

Acetic acid is recycled, as is unreacted diacetate. Propylene oxide yield, based on propylene feed, runs above 80%, says the company, with output of propionaldehyde about twice that of acetone. The modest amount of these byproducts poses no marketing problems, contends Chem Systems.

Mark D. Rosenzweig

Low-pressure oxo process features rhodium catalyst

A new oxo process is now operating commercially in plants for making butyraldehydes and propionaldehyde. The technique features a rhodium catalyst that achieves high product selectivity at low capital and operating costs.

Everard A. V. Brewester, *Davy Powergas Ltd.*

☐ Ten years of research and development work by Union Carbide Corp., Davy Powergas Ltd. and Johnson Matthey & Co. has resulted in the new Low Pressure Oxo process—using a rhodium rather than cobalt catalyst—that makes a complete break with conventional oxo technology.

The production of *n* and *iso*-butyraldehydes by the reaction of propylene with synthesis gas (hydrogen and carbon monoxide) is the most important industrial application of the oxo reaction. While conventional processes attain a product *n*/*iso* ratio of 3:1 or 4:1, the new process achieves in excess of 10:1.

This increase is highly beneficial: First, because the *n* isomer has historically commanded a greater price than the *iso* form (the *n*-butyraldehyde can be hydrogenated to *n*-butanol, a more effective solvent than *iso*-butanol, and can also be converted to 2-ethyl hexanol, an important alcohol for plasticizers). Second, the low *n*/*iso* ratio of conventional processes represents a 20–25% loss of propylene feedstock to the low-value *iso*-butyraldehyde, which often finds use only as fuel or as synthesis-gas feedstock.

The Low Pressure Oxo process also lowers capital and maintenance costs considerably, operating at less than 300-psi pressure instead of the usual 3,000–6,000 psi.

Corrosive media used in recovery of cobalt catalyst from the aldehyde product are avoided, furthermore, which eliminates complex equipment and operating procedures.

COMMERCIAL SUCCESS—The first butyraldehydes plant featuring the new process made its initial product in late January of this year. Located at the Ponce, Puerto Rico, complex of Union Carbide Caribe, Inc., the 300-million-lb/yr unit started up easily and quickly. Excluding external interruptions, the unit suffered just two hours downtime during its first month; 80% load was often achieved within 40 min of introducing feedstock to the reaction section.

Operation since then has remained exceptionally smooth and stable. Catalyst productivity and selectivity have been at least as good as expected. No catalyst deactivation has been observed, and losses of the precious-metal rhodium have been minimal. The unit has overall performed better than design. Davy Powergas was responsible for its basic engineering, while Union Carbide performed the detailed design and construction supervision.

A similar plant is scheduled for startup during the first half of 1979, by licensee Berol Kemi AB, at Stenungsund, Sweden.

Yet-another facility actually preceeded the Ponce Low Pressure Oxo unit. This Union Carbide plant, however, makes propionaldehyde from ethylene and synthesis gas. Since coming onstream in early 1975 at Texas City, Tex., the unit has

Union Carbide plant at Ponce makes 300 million lb/yr of *n*-butyraldehyde

Originally published November 8, 1976

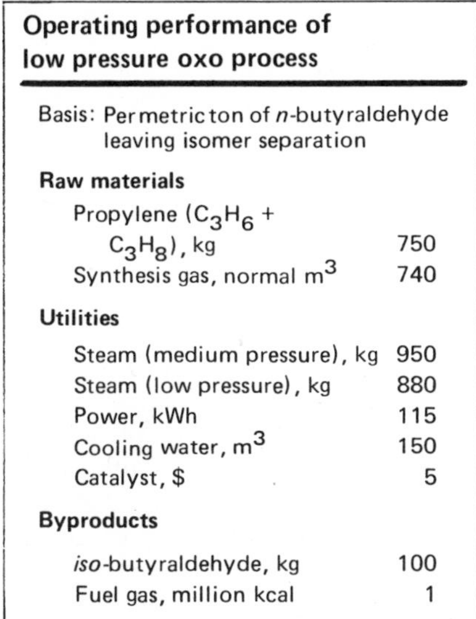

Operating performance of low pressure oxo process

Basis: Per metric ton of *n*-butyraldehyde leaving isomer separation

Raw materials	
Propylene (C_3H_6 + C_3H_8), kg	750
Synthesis gas, normal m^3	740
Utilities	
Steam (medium pressure), kg	950
Steam (low pressure), kg	880
Power, kWh	115
Cooling water, m^3	150
Catalyst, $	5
Byproducts	
iso-butyraldehyde, kg	100
Fuel gas, million kcal	1

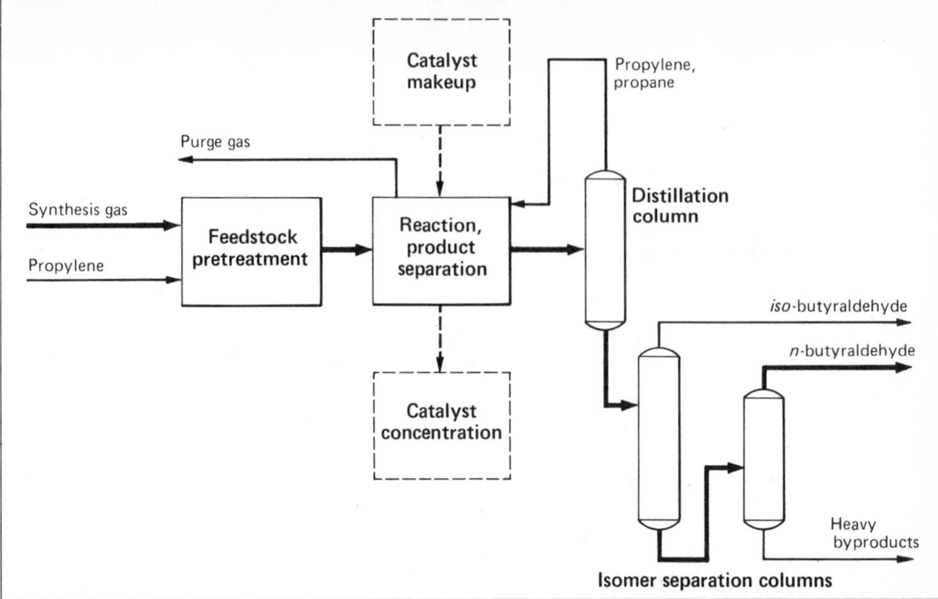

Equipment simplicity makes new oxo process easy and inexpensive to operate

demonstrated a 150-million-lb/yr production capability, about 50% over design.

SIMPLE STEPS—Supplied by Johnson Matthey, the catalyst is an organo-metallic complex of rhodium, in which the metal is bonded by relatively labile linkages to an organic ligand and to carbon monoxide and hydrogen. The catalyst complex is not volatile and does not plate out in the system.

To reduce concentration of sulfur-containing catalyst poisons, the processing sequence (for making *n*-butyraldehyde) starts by purifying the feedstocks propylene and synthesis gas, via proprietary, solid-adsorbent techniques jointly developed by Union Carbide and Davy Powergas. There is no need for synthesis-gas compression or for high-pressure pumps.

The purified feeds pass into one or more reactors, where conversion to aldehydes takes place in a homogeneous liquid phase of catalyst dissolved in reaction products. Operating pressure ranges from 100–350 psi, and temperature is about 100°C (versus the 145–180°C of the conventional oxo process). Reaction conditions are carefully chosen to optimize plant efficiency by balancing *n/iso* ratio (controllable over a span from 8:1 to 16:1), reactant and catalyst concentrations, reactor size, temperature and pressure.

A gas purge from the reaction system removes inerts introduced with the feedstocks, and also removes a small amount of propane formed from propylene by hydrogenation. The purge stream serves as fuel, and constitutes the only continuous liquid or gaseous effluent from the reaction system.

Following reaction, a product separation step yields a relatively pure liquid stream of aldehydes, which is still saturated with dissolved gases, mainly propylene and propane. A product distillation column removes these for recycle.

Mixed aldehydes, containing heavy byproducts formed in the reactor, pass from the base of the column to isomer separation. Here *iso*- and *n*-butyraldehydes are taken as the top products of two conventional distillation columns, respectively; the heavy byproducts exit as a small-volume bottoms stream from the second column, and are incinerated.

Overall, there are only minor feedstock losses, by propylene hydrogenation and heavy-byproducts formation. Production of alcohols (principally butanols) is nil because of the low 100°C reaction temperature, unlike conventional processes in which appreciable amounts do form and thus complicate product recovery and purification.

Catalyst preparation and recovery knowhow are supplied by Johnson Matthey. There is no continuous withdrawal/makeup cycle. Makeup and concentration systems are provided for occasional use only. The former introduces catalyst into the system prior to initial startup; the latter is on hand in order to concentrate catalyst solution before its shipment back to Johnson Matthey for recovery of the precious metal rhodium.

OTHER FEATURES—Rhodium losses during production have been minimal. The manufacturing-cost contribution is in fact lower than the costs incurred using less-expensive cobalt catalysts. Life of the rhodium catalyst is projected at one year or longer.

A Low Pressure Oxo plant also requires fewer operators and less maintenance than a conventional plant using cobalt catalyst. The principal material of construction in the plant is stainless steel.

Typical performance data are shown in the table. These assume a partial-oxidation synthesis gas with 99 vol.% hydrogen plus carbon monoxide, and a 94 vol.% propylene feedstock.

Capital cost for a plant using the new oxo process is approximately 65% that for a unit using cobalt catalyst. The reduced raw materials and utilities consumption and lower capital cost for the Low Pressure Oxo plant result in significantly lower production cost of *n*-butyraldehyde.

The author

Everard A. V. Brewester is a senior engineer in the business development group of Davy Powergas Ltd. (8 Baker St., London W1M 1DA, England). He led the firm's technical effort in the development of the Low Pressure Oxo process (from Union Carbide Corp. research data), and was the lead process engineer for Davy Powergas on the design of both the Texas City and Ponce plants. A member of the Institution of Chemical Engineers, he holds a B.Sc. in chemical engineering from Manchester University.

Section VII
PETROLEUM PROCESSES

Demetallization seeks a heavier refinery role

New and established metals-removal processes may win a wider role in protecting desulfurization catalysts from poisoning from heavier refinery feedstreams.

☐ The need to produce clean fuels from dirty, high-sulfur and high-organometallic crudes is prompting an increasing interest by U.S. refiners in techniques for eliminating metals from the oils.

Sulfur-removal operations are particularly spurring this attention on demetallization: Hydrotreating (fixed-bed hydrodesulfurization) can take out sulfur values from oil, but heavy metals, such as vanadium and nickel, in the crude can poison and deactivate the hydrodesulfurization catalyst. So, to avoid too short a catalyst life, additional steps are mandated to remove these metals.

Refiners are looking at both old and new demetallization methods. Options include pretreating petroleum streams, particularly residual-type feedstocks, by solvent extraction or with scrounging agents to reduce metals content to a level that is tolerable in a hydrotreater. Alternatively, some firms are offering fluid-catalytic-cracking (FCC) systems that can directly handle the streams.

TRADEOFFS—Most crude heavy-metal content now winds up in residual fuel oil. But with residuum selling for about $2/bbl less than raw crude, refiners are not rushing to make any more of the product than they have to. Instead, research and development teams are looking at ways to upgrade the bottom of the barrel, making it suitable as a feedstock for cracking. This is a logical goal because gasoline and middle distillates are the key petroleum sectors in the U.S., and because the heavier, less valuable products, such as coke and heavier industrial fuel oil, could increasingly face inroads from other energy sources, e.g., coal. Juggling crude bottoms is a serious concern for refiners because large capital investments generally are required. The optimum flow-scheme for a given refinery depends upon crude availability and desired product mix—and both of these will usually change with time. So, choosing the most-appropriate demetallization system can play an important role in optimizing refinery operation.

Originally published November 22, 1976

EXTRACTION—Organic solvents are finding a niche in demetallization. For instance, UOP Inc. (Des Plaines, Ill.) and Kerr-McGee Refining Corp. (Oklahoma City, Okla.) both offer solvent-extraction systems for metals removal.

One expert in solvent extraction, Donald Garrett, president of Garrett Energy Research and Development, Inc. (Claremont, Calif.), explains: "There is no reason why solvent extraction couldn't take the metals out of crude oil, because the metals are heavily chelated already."

UOP is using low-molecular-weight paraffinic solvents to demetallize vacuum residuals. Its process, called Demex, splits high-metal-content residuum into a "demetallized oil" (DMO) of relatively low metal level and an asphalt with a high metal content. For example, Kuwait vacuum bottoms can yield about 80% by weight DMO and 20% heavy asphalt, but the asphalt typically would contain 85% of the metals. This translates to a feed containing about 130 ppm nickel and vanadium yielding DMO with under 30 ppm.

The DMO obtained from the process is a desirable feedstock for fixed-bed hydrodesulfurization or hydrocracking, says the company. And it adds that in cases where metal content is sufficiently low, DMO can go to an FCC unit.

Demex is an extension of solvent deasphalting, with emphasis on demetallization. As shown in the flow diagram below, the process consists of four main steps: extraction, asphalt recovery, DMO separation, and solvent recovery.

Solvent passes upward through an extractor, countercurrent to the residuum charge, and dissolves lighter hydrocarbons. This extract phase

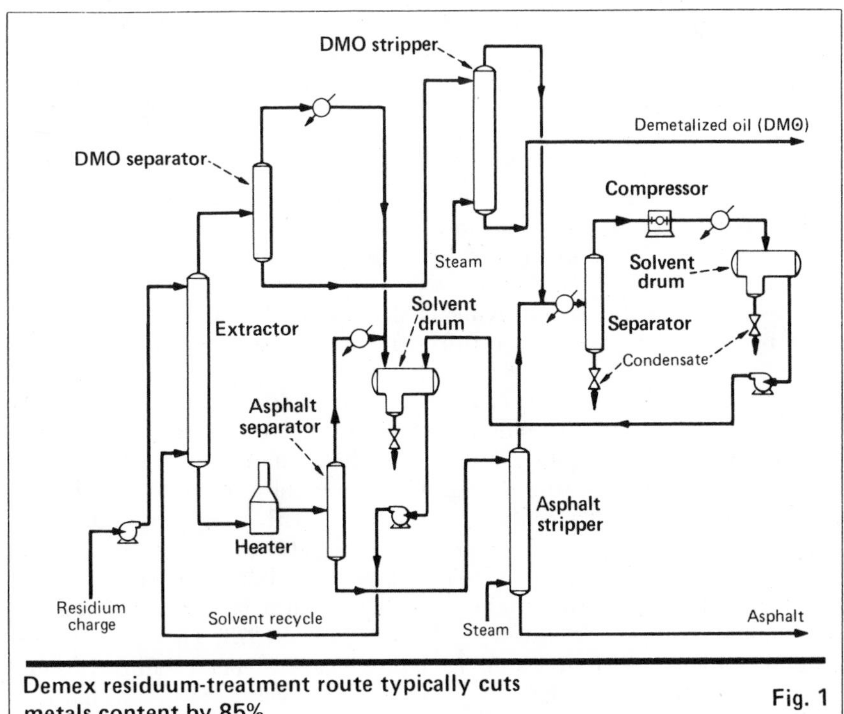

Demex residuum-treatment route typically cuts metals content by 85% **Fig. 1**

then moves to a separator, where the stream is heated. Solvent flashes, and is condensed and recovered. DMO from the bottom of the separator is cleaned of any remaining solvent in a low-pressure wash tower.

Meanwhile, raffinate from the extractor goes to an asphalt separator and then to a stripper, where asphalt is recovered and any entrained solvent and DMO are removed.

Demex already has achieved some commercial success. Three units are now under construction, with an aggregate capacity of 90,000 bbl/stream-d. For instance, Champlin Petroleum Co. chose the route for its Corpus Christi, Tex., refinery. There, the DMO, blended with gas oils, will feed a hydrotreater to make low-sulfur cat-cracker charge. Champlin expects hydrotreater runs of at least one year with the feed.

For its part, Kerr-McGee built and successfully operated a 750-bbl/stream-d unit of its Residuum Oil Supercritical Extraction (ROSE) process at its Wynnewood, Okla., refinery from 1954–1960 for optimizing recovery of asphaltenes and resins. The system now seems promising for demetallization, and the company is running a third-generation pilot plant. Kerr-McGee presently has a 5,000–6,000-bbl/stream-d unit in design.

The company sees an economic plus for the route. Unlike conventional propane deasphalting systems, ROSE employs pentane for extraction (though other solvents are acceptable alternatives in some cases), and this leads to more-economical solvent recovery. A physical separation of phases serves for recovery of solvent, instead of the usual evaporation. So, there are the energy savings from not having to supply the full heat of vaporization.

ROSE typically can remove 70–90% of the metals in residuum. For instance, Arabian light residuum with 75 ppm of vanadium and nickel can give a DMO containing under 10 ppm.

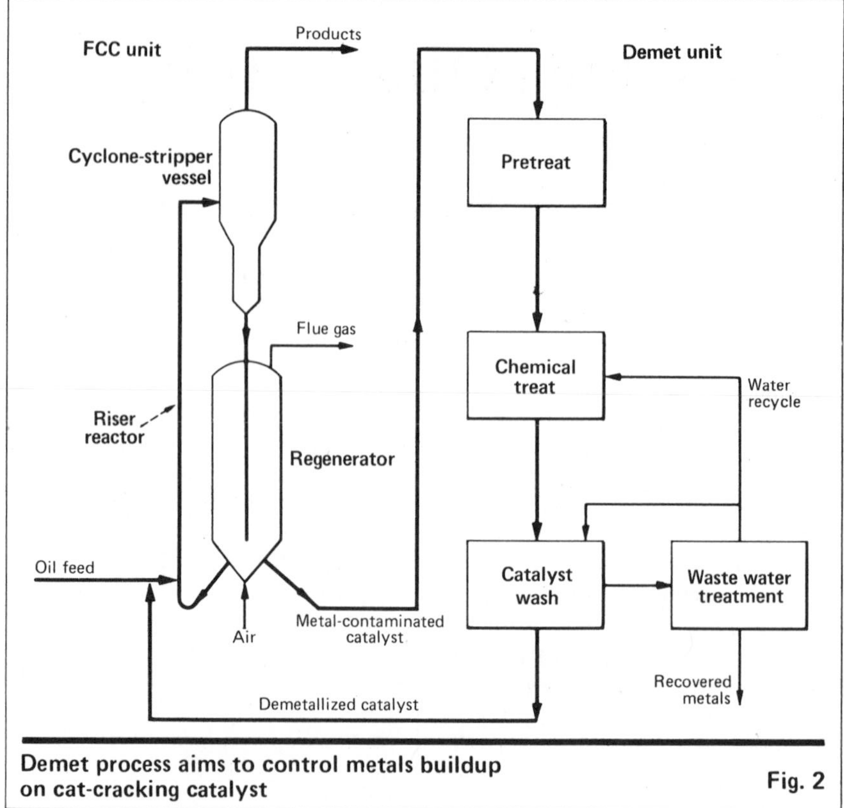

Demet process aims to control metals buildup on cat-cracking catalyst — Fig. 2

SCAVENGING—Removing metals via relatively inexpensive, natural scrounging agents also is drawing attention. Cities Service Research & Development Co. (Cranbury, N.J.) and Shell International Petroleum Co. (The Hague, Holland) have developed processes, and Hydrocarbon Research, Inc. (Trenton, N.J.) also is working in the field.

Cities Service uses an ebullating-bed reactor in its demetallization route, dubbed LC-fining. (An ebullating bed is similar to a fluid bed, except that it uses a liquid as the suspension medium, and larger particles can make up the bed.) The company places the reactor ahead of a hydrotreater when processing a residuum containing large amounts (above 200–300 ppm) of vanadium. The reactor is loaded with an undisclosed natural agent. It costs only 5–10% as much as the cobalt-molybdenum catalyst used in the hydrotreater and so, notes the firm, provides low-cost poison protection.

The company built a 6,000 bbl/stream-d unit at Lake Charles, La., in 1963, which now is offline, being prepared for a summer-1977 demonstration with a solvent-refined-coal stream. However, Petróleos Mexicanos S.A. has operated an 18,000 bbl/stream-d facility at Salamanca, Mex., since 1973. The unit typically provides 90% metal removal from a 250-ppm-average feed, and has converted streams of up to 317 ppm of vanadium and nickel to a product of under 30 ppm.

Shell features an unidentified agent said to have a high affinity for metals and low activity for sulfur. The company employs a moving-bed reactor with special internals to prevent plugging and maintain even distribution of liquid over the full vessel cross-section. Fresh supplies of the scavenger are added and spent material withdrawn at a rate that optimizes the efficiency of vanadium removal. This provides two advantages, according to Shell: First, the demetallized product stream has a constant metal content, which allows easier optimization of the downstream hydrotreater. And second, the unit acts as a filter, protecting the hydrotreater from fouling—a common tendency with certain heavy residual oils.

Shell operated a pilot plant at its Amsterdam laboratory, and gave the technology a full-scale test from March 1973 to August 1974 in a 340-ton/stream-d unit on various streams at the Göteborg, Sweden, refinery of its affiliate Koppartrans AB. For example, an Iranian residuum was reduced from over 160 ppm of vanadium and nickel to under 25 ppm.

A 45,000-bbl/stream-d installation is now under construction at the Seibu Sekiyu K.K. refinery near Onada City, Japan. And a 3,000-bbl/stream-d unit was planned for Shell's Punta Cardon, Venezuela, refinery, but the status of this project now is complicated by the nationalization of the refinery.

Hydrocarbon Research, following tests it conducted for the U.S. Environmental Protection Agency, favors activated bauxite impregnated with a small amount of molybdenum for scavenging. The agent can provide 70–80% metals removal. The company has received demetallization patents: one on metal removal from feedstocks in an ebullating bed, and the other on regeneration of poisoned catalysts.

FCC DEVELOPMENTS—Another approach to the metals problem is to send untreated streams into FCC units, and to remove the metals with the cat-cracking catalyst. Atlantic Richfield Co. (Los Angeles) and the Pullman Kellogg Div. of Pullman Inc. (Houston, Tex.) are championing such a tack.

The Demet III process developed at the ARCO Technical Center in Harvey, Ill., is claimed to be a low-cost catalyst-regeneration system that removes metals from cracking catalysts and thus permits the use of high-metallic crude or residual feedstocks in an FCC unit.

The expense of removing metals that contaminate the catalysts has largely stymied the feeding of residuum to FCC units. But the Demet III process promises to change the situation, hopes the company. Commercialization of the technique is now at the talking stage.

The route involves both chemical and physical treatment. A slipstream of catalyst is withdrawn continuously from the FCC regenerator, processed in a demetallization circuit, and then returned to the FCC unit. According to ARCO, the only fresh catalyst required is makeup for losses of fines. And, the company stresses, the Demet III operation does not damage zeolite catalyst structure, even after repeated metals-removal cycles.

The four-stage demetallization circuit starts with a pretreatment step to activate metals for removal. The metals then are chemically stripped. Next, washing physically separates the metals from the catalyst, which is returned to the FCC unit. Finally, wash effluent is treated to recover metals, particularly the vanadium.

The company is not revealing specifics on Demet III. However, one of the earlier versions (developed for pre-zeolite catalysts) featured an ion-exchange resin for metal removal. Recent pilot programs at Harvey have treated 15% of the zeolite catalyst to maintain catalyst metal content at 1%. On this basis, demetallization costs, subtracting a credit for recovered vanadium, should run about 8¢/bbl.

In contrast the Kellogg process simply runs heavy fractions directly through an FCC unit, and keeps metals level in check by discarding catalyst at a higher-than-usual rate. Kellogg data indicates, for instance, that the route can handle Texas Panhandle atmospheric residuum feed containing 21 ppm of metals, at a catalyst replacement rate of 0.6–0.7 lb/bbl of throughput and at a conversion of about 80% of the feed. Kellogg also has run high-metallic (170 ppm) Iranian Gach Saran atmospheric residuum, at a replacement rate of 1.8–2.0 lb/bbl and a feed conversion of about 47%.

The route has a potential plus in that it is a net exporter of steam.

Phillips Petroleum Co. has operated the Kellogg process for heavy-oil cracking at its Borger, Tex., refinery since 1961. Feed contains 23 ppm of vanadium and nickel, and metals level on the catalyst is maintained at 1.1%. Now, at least five companies are interested in adopting the process, and Kellogg has already signed a letter of intent with one of the firms.

Guy E. Weismantel

Process provides option for nonleaded-gas makers

This process yields a high-octane-number alternative to C_3/C_4 alkylates by dimerizing propylene. Able also to dimerize butylene and co-dimerize C_3/C_4 streams, the technique is cheaper to build and operate than a like-capacity alkylation unit.

***Philip M. Kohn**, Assistant Editor*

☐ Total Petroleum Inc. (Alma, Mich.) is presently constructing the world's first commercial Dimersol unit, to be licensed by France's Institut Français du Pétrole* (IFP). Dimersol is a once-through process that selectively dimerizes propylene or butylene, or co-dimerizes both, to either (1) an isohexene gasoline-component having a blending octane number (research octane plus motor octane, divided by two) of 96, or (2) a heptene or octene oxoalcohol feedstock.

Total Petroleum is in the midst of a two-step expansion of its present fluid-catalytic-cracking (FCC) and alkylation units. The newly expanded alkylation facilities will handle the C_4 stream from the enlarged FCC unit, but are incapable of alkylating all the propylene that will be produced. Since needed propylene-alkylation capacity will not be available until sometime in 1979, Total has decided to utilize IFP's Dimersol technique for the following reasons:

- The process is relatively inexpensive from both a capital and operating standpoint. IFP points out that the process costs about \$900 – 1,200/bbl of product as opposed to an estimated \$1,700 – 2,000/bbl for an alkylation unit. And the licensing firm states that the direct operating cost for a Dimersol plant is about \$0.88/bbl of product as compared with \$1.49/bbl of output for an alkylating facility.
- Dimersol presents a way to convert propylene to gasoline without using relatively scarce, expensive isobutane, the usual alkylation material.
- The technique is an alternative to building expensive, pressurized propylene-storage facilities to hold product that Total cannot handle.
- When Total does complete its planned expansion for alkylation facilities capable of handling the propylene component of its FCC units, the Dimersol process could dimerize purchased propylene, a material that is—and is expected to remain—fairly available in the U.S.

The main feature of the process is a liquid-phase reaction that takes place in the presence of a highly active liquid catalyst dissolved in the reaction mass. Only extremely low concentrations of catalyst are needed because of the degree of contacting allowed by the liquid-phase reaction. In addition, the catalyst, injected continuously into the reactor, is ultimately destroyed at the reactor's outlet, obviating recycle or regeneration of the catalyst. Destruction of the catalyst also eliminates system-performance variations that are due to catalyst aging. The catalyst-destruction aspects of the system are economically feasible because of the minimal level of catalyst present.

Another plus for the system, as IFP notes, is that tests have shown that the all-important octane specification (the blending octane number) for the combined dimerized-C_3/alkylated-C_4 product is higher than it would be if the entire C_3/C_4 olefinic stream from an FCC unit had been alkylated.

Finally, equipment costs are kept low because the process operates at low pressures and temperatures.

One drawback that must be considered, however, is that for the same propylene throughput, Dimersol yields less dimate than an alkylation unit would produce alkylate, using isobutane. That is, you get fewer barrels of product with Dimersol than with an alkylation facility.

THE PROCESS—Dimersol reactions take place through the action of a proprietary catalyst system that comprises a nickel coordination complex and an aluminum-alkyl compound. Composition of the final product depends on the makeup of the feedstock and on the exact nature of the catalytic complex. Dimersol provides a 90 – 97% conversion of C_3 to C_6, by weight, with an isohexene selectivity of 85 – 92% (the remainder of the converted material is C_9 or heavier). (See Table I for the composition of a typical hexene dimate.) IFP says that conversions up to 92% can be accomplished with only one reactor, but that to reach the higher conversion limit a two-reactor system is required.

In the process (see flowsheet), dried C_3 olefin—consisting of approximately 65% propylene, 35% propane—enters the reactor, which is simply a flooded vessel that is back-mixed via a recirculation stream.

A homogeneous liquid catalyst is continuously metered into the recirculation stream via a proportionating pump. The catalyst components are soluble in the reaction mixture, and,

*Institut Français du Pétrole, North American Office, 450 Park Ave., New York, NY 10022.

Originally published May 23, 1977

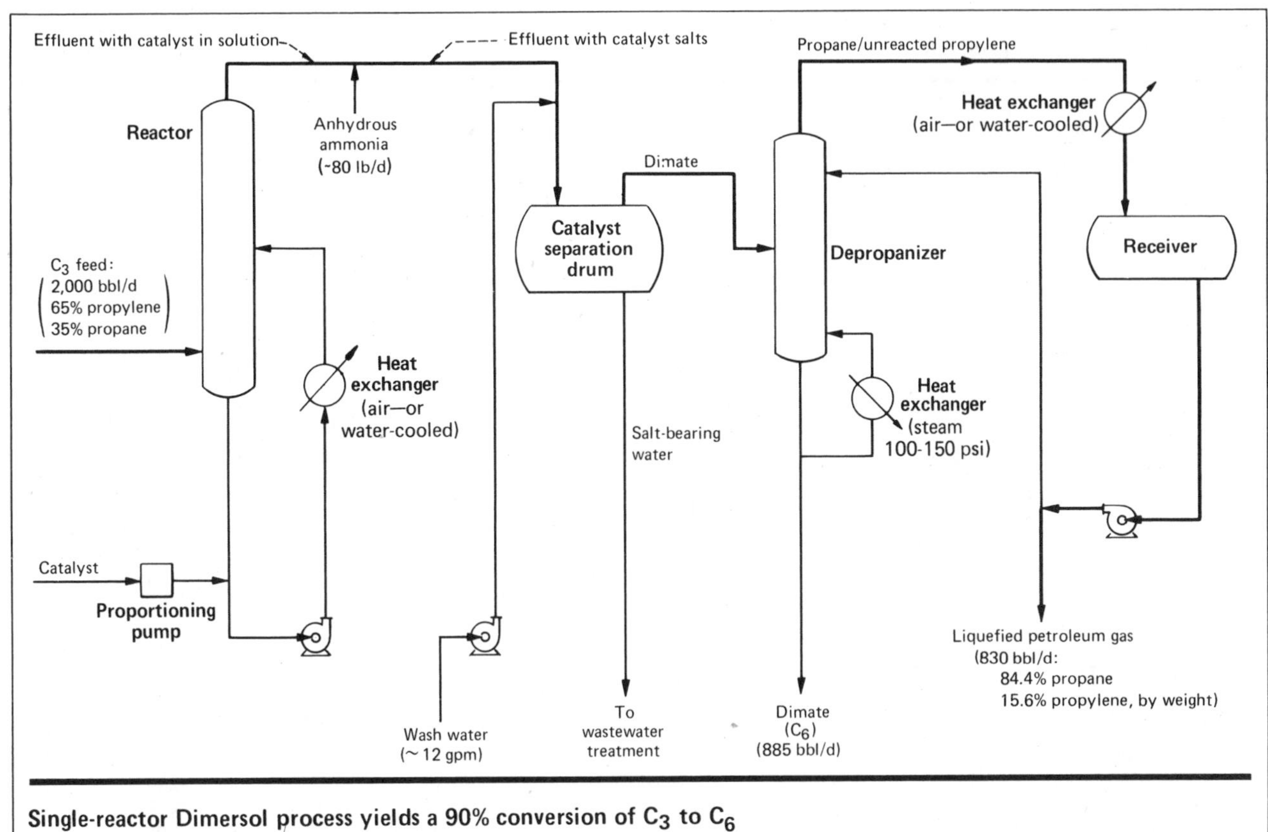

Single-reactor Dimersol process yields a 90% conversion of C_3 to C_6

since the catalyst is very active, only low (ppm) concentrations of the material are needed to maintain favorable reaction rates. The catalyst's solubility enhances mixing and contacting.

No heat is added to the reactor, the dimerization reaction being exothermic. The aforementioned recirculation stream, in addition to performing its mixing function, also cools the reactor by removing the heat of reaction. All dimerization takes place in the liquid phase, and the reactor operates only under sufficient pressure (about 250 psi) to keep the reactants liquid at reaction temperatures to 140°F.

Dimate formed in the reactor leaves from the top of that vessel. Anhydrous ammonia, added to the product stream, stops the reaction by neutralizing the catalyst, forming salts that are insoluble in the dimate. However, because of the small amounts of catalyst involved, these salts remain "lost" in the total volume of dimate, and remain in suspension. Thus, it is also necessary to employ a washing step to remove the salts.

Water is pumped into the salt-bearing dimate stream. This water can be refinery process water, well, city or recovered water from a wastewater treatment facility. The salts preferentially dissolve into the aqueous phase, the aqueous/organic split being made in the catalyst separation drum. The aqueous phase goes to wastewater treatment, while the saltfree organic portion proceeds to the process's fractionation section for separation of the dimate product from the liquefied-petroleum-gas portion of the flowstream.

COMPONENT-SPLITTING—The fractionation section of the process is quite simple, being in essence only a single depropanizer. The unit is of standard design, with heat input effected by medium-pressure (100-150 psi) steam. Propane—which does not dimerize—and unconverted propylene distill off the top of the column, are cooled and collected. Part of this stream is directed back to the depro-

Hexene dimate composition — Table I

Component	weight%	Boiling pt (°F)
4-methylpentene-1	2.00	128.8
2,3-dimethylbutene-1	4.00	132.1
cis-4-methylpentene-2	5.70	133.5
trans-4-methylpentene-2	31.35	137.5
2-methylpentene-1	4.00	143.8
hexene-1	trace	146.3
cis-+*trans*-hexene-3	3.40	151.6
2-methylpentene-2	16.50	153.2
trans-hexene-2	11.30	154.2
cis-hexene-2	4.85	156.2
2,3-dimethylbutene-2	1.90	163.8
nonenes	13	—
heavy ends (C_{12} and higher)	2	—

Hexene plant cost and production parameters **Table II**

Basis: 2,000 bbl/d of C_3 feedstock: 65% propylene, 35% propane
Plant is online 330 d/yr

	90% conversion, by wt. (one reactor)		97% conversion, by wt. (two reactors)	
Feed:	(bbl/d)	(lb/h)	(bbl/d)	(lb/h)
Propylene	1,300	9,900	1,300	9,900
Propane	700	5,185	700	5,185
	2,000	15,085	2,000	15,085
Product:				
Dimate	885	8,910	954	9,605
LPG — Propylene	130	990	39	295
LPG — Propane	700	5,185	700	5,185
	1,715	15,085	1,693	15,085
Estimated capital cost: (Includes royalty and IFP design fee)	$1.35 million		$1.5 million	
Estimated operating-cost factors: (per bbl of dimate)				
Water (gal) (for washing only; assumes air-cooled heat exchangers)	19.5		18.1	
Electricity (kWh)	2.08		1.94	
Steam, for depropanizer (lb)	62.5		58.0	
Catalyst and chemicals ($)	0.36		0.34	
Operating labor (man-h) (Process uses ½ operator per shift, i.e., shares 1 operator)	0.00136		0.00126	
Maintenance ($)	0.03		0.03	

Source: Institut Français du Pétrole

panizer as reflux, while the remainder is gathered as liquefied petroleum gas, its composition dependent on the nature of the original olefin feed and the amount converted.

The dimate, a stabilized C_6 product, comes out of the depropanizer as a bottoms stream and can go directly to the motor-fuel pool.

ADDITIONAL ABILITIES—As stated earlier, the Dimersol process also can produce heptenes and octenes. All that is required is a change in the catalyst system, as well as the availability of a C_4 stream that contains mostly *n*- as opposed to *iso*- compounds. IFP says that Dimersol yields heptene and octene products having limited branching—unlike the many branched hexenes it produces—making the heavier molecules attractive as oxoalcohol feedstock in plasticizer production.

IFP claims that the key to Dimersol's flexibility is the process's homogeneous, soluble catalyst system. For example, if a petrochemical process employs a fixed-bed reactor, the operator is limited to the parameters of that specific catalyst and its load. With a homogeneous soluble system such as is used in Dimersol, however, catalyst concentrations and formulations can be varied on short notice.

Thus, if a Dimersol process operator, producing heptenes, temporarily depletes his supply of butylene feedstock, he can switch over to hexene production. Conversely, with a shortage of propylene, he could shift to making octenes. Moreover, states the licensor, such changes can be handled routinely with no off-spec interim product, because the operator needs only to alter catalyst formulation or injection rate, or both, when the feedstock changes.

FEEDSTOCK REQUIREMENTS—Typically, Dimersol can handle C_3 cuts from cat- or steam-crackers, and C_4 cuts from steam-crackers.

In hexene production, catalytically cracked feeds can generally be charged to the Dimersol process directly, but pyrolysis-plant cuts should first be selectively hydrogenated to eliminate diolefins and acetylenes. These would be removed in any event to permit downstream use of polypropylene, but diolefins and acetylenes left in the Dimersol feedstock would cause excessive catalyst loss. Total acetylenes should not exceed 30 ppm, by weight. In addition, total sulfur and water must be held to no more than 10 and 5 ppm, and 5 and 5 ppm, for catalytically-cracked and steam-cracked feeds, respectively.

Similarly, when processing C_4 streams, isobutylene therein should be limited to 5%, by weight, since it has a tendency to polymerize rather than dimerize.

ECONOMICS—The Dimersol catalyst concept and the relatively straightforward nature of the products result in simplicity of plant design. The reactor is similar in construction to an LPG bullet-type tank. Low-carbon steel is used throughout the process, which IFP says can be fitted to existing equipment.

Table II shown gives capital and operating cost estimates for a hexenes plant having an input of 2,000 bbl/d of C_3 feed. The battery limits of such a plant include facilities shown on the flowsheet.

New additive gives boost to unleaded fuel in Europe

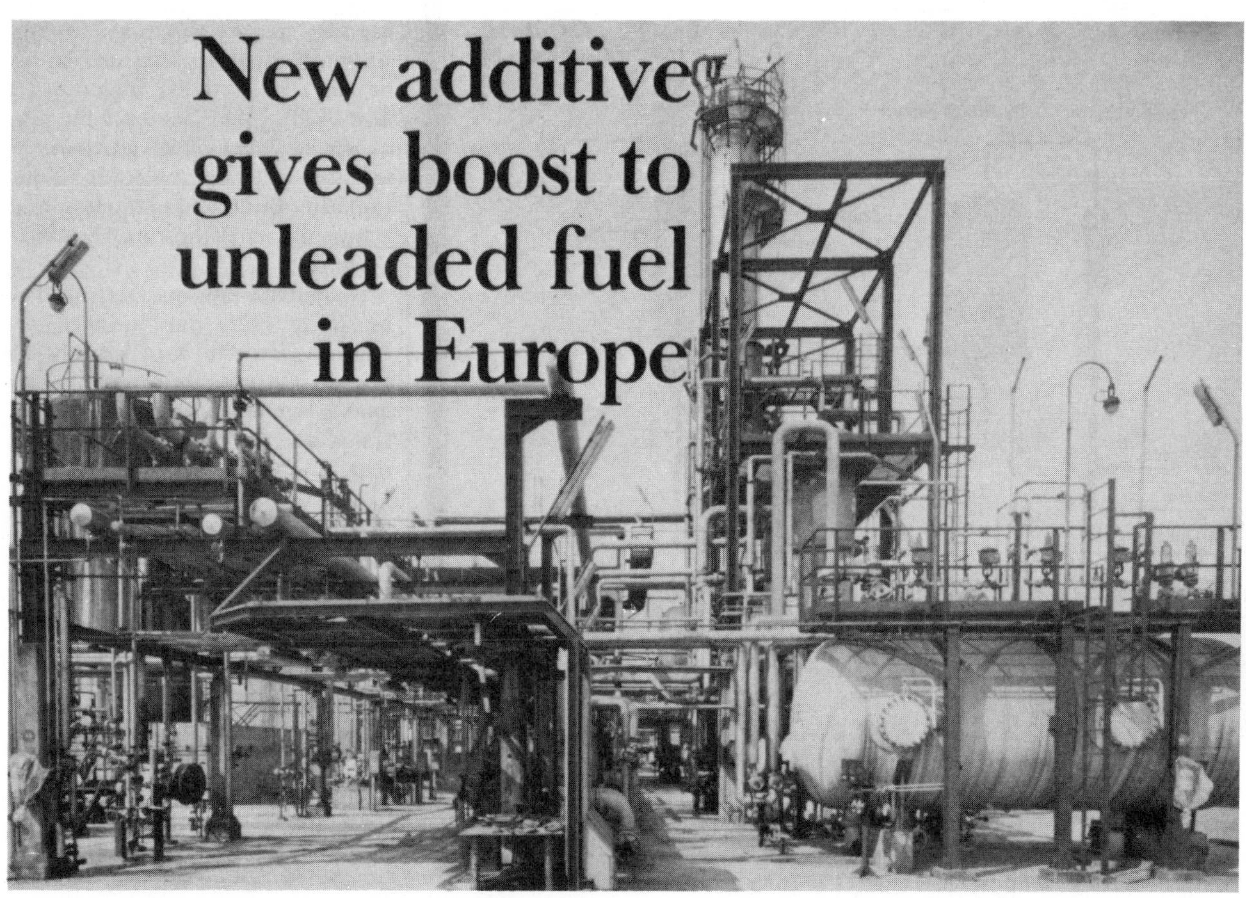

Germany and Italy have become supporters of MTBE, a compound with high-octane blending properties that replaces lead in premium gasolines. Two MTBE routes are for sale, and Germany plans another plant to meet rising demand. U.S. firms are watching the action with interest.

☐ Backed by the enthusiasm of three European firms—Italy's Snamprogetti and Anic, and Germany's Chemische Werke Hüls—the gasoline additive methyl *tert*-butyl ether (MTBE) continues to generate wide interest on the Continent and elsewhere. The Italian firms, which jointly developed an MTBE process that has been in commercial operation since 1973, have announced they are putting their technology on the market. So has Hüls, and all three companies say they are having licensing talks with a number of eager customers in Europe, the U.S. and Japan.

What makes the additive so popular all of a sudden? Possibly a combination of the following factors: excellent properties in gasoline blends, including boosting octane, low water solubility, and a reduction of CO contained in exhaust emissions; ease of fabrication; availability of MTBE raw materials (isobutene and methanol); and attractiveness as a replacement for lead, when compared with other octane-boosting alternatives.

The last item is especially important in Europe because most European countries have imposed severe limits on the lead content of gasoline. And Europeans need increasing amounts of high-octane premium gasoline to fuel their high-compression-ratio automobiles. To produce it, oil refineries would have to make extensive investments in new catalytic reforming capacity, and fluid catalytic cracking and vacuum distillation units—and reforming units consume sizable amounts of virgin-naphtha feed.

Nevertheless, not everybody in Europe is jumping on the MTBE bandwagon. Some experts claim that at a price tag of \$200–\$220/metric ton, the additive is barely competitive with aromatics. And there is an aromatics glut right now in Western Europe; this may quell some of the potential interest in MTBE for as long as one year, say some traders.

Originally published January 31, 1977

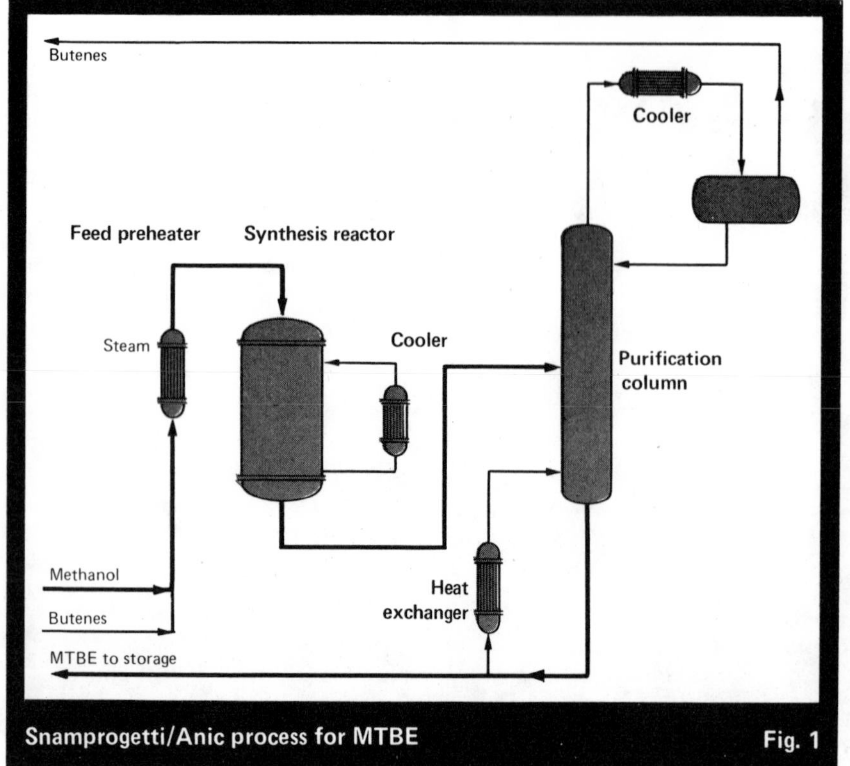

Snamprogetti/Anic process for MTBE Fig. 1

U.S. REACTION—The high cost of MTBE is also a problem in the U.S., where several companies, while closely following the additive's progress in European markets, are not exactly rushing into MTBE production.

Sun Oil Co. (Marcus Hook, Pa.), which started looking at MTBE ten years ago, and conducted extensive automobile tests with it (in 15%-by-volume blends) as early as 1970, says that MTBE is a better gasoline component than aromatics. But the company also thinks that MTBE is a more expensive alternative than alkylation, especially when used at the lower-octane level of the typical U.S. unleaded regular gasoline. A spokesman for Sun Oil adds that MTBE's prospects in the U.S. may brighten if octane requirements go up.

POPULAR IN GERMANY—Few reservations of this kind exist in Germany, where a very low lead ceiling in gasoline (0.15 g/L) was imposed on Jan. 1, 1976. Hüls started up an MTBE plant in time for the deadline, soon increased its capacity to 50,000 metric tons/yr, and now says that, unable to keep up with demand, it is considering building another similar-size facility for completion by 1980 at the latest.

Hüls customers are mainly independent importers, who would have a rough time without MTBE, since gasoline from refineries outside Germany does not meet that country's exceptionally low lead-requirements. The importers buy foreign gasoline before the lead is added, and then make up for the loss of octane by adding MTBE. German refineries are able to meet the new national lead ceiling by virtue of a major investment effort (about $400 million) directed largely at alkylation units.

A Hüls spokesman believes that interest in MTBE will grow in the near future because use of aromatics in gasoline will become "problematic," due to the compounds' suspected carcinogenic effect, and to the fact that available aromatics will probably be more needed as chemical feedstocks.

Hüls adds that it uses basically the same route to MTBE employed by Snamprogetti/Anic, and earlier in the game by Sun Oil. Nevertheless, it refuses to divulge process details until patenting procedures are completed.

SIMPLE TECHNIQUE-The Italian process (see Fig. 1), as used at Anic's 100,000-metric-ton/yr plant at Ravenna, on the Adriatic Coast, is a simple one. Snamprogetti starts from methanol, and isobutene contained in the spent C_4 fractions (from butadiene extraction) of an ethylene cracker. Although mixed refinery butane/butene streams can be used as raw material (as advocated by Sun Oil), the Snamprogetti-Anic choice is particularly advantageous in Europe, where the spent C_4 fraction after butadiene extraction in ehtylene plants is tradtionally used as plant fuel.

(Isobutene raw material need not be pure. Hüls, for instance, uses what in Germany is called raffinate-one—a mixture comprising isobutylene, *n*-butene and butane. The catalyst it uses promotes a selective reaction.)

Methanol and isobutene are fed to a fixed-bed, tube-sheet-type reactor; the catalyst is an acid-based polystyrene/polyvinyl resin obtained from a U.S. firm. Effluent from the reactor contains mainly MTBE and linear butenes, but also unreacted isobutenes, butane, unreacted methanol, and isobutene byproducts.

The mixture is sent to a purification column to separate the C_4's from MTBE. The former are collected as the top stream, along with unreacted isobutene, while the bottom effluent is mainly MTBE (more than 98% by weight) with small amounts of methanol and byproducts.

Snamprogetti engineers claim that the Anic plant was producing MTBE to specifications six hours after it went onstream in 1973. The company suggests that investment costs for a battery-limits plant of 100,000 metric tons/yr in Italy would be $2–$3 million. Labor costs are extremely low—half a worker per shift. And because the MTBE plant can be run downstream of the butadiene plant in a petrochemical complex, raw-material transportation costs would be reduced.

(At Ravenna, there is no ethylene cracker. C_4 and methanol are shipped in from Brindisi, Gela or Porto Marghera, where Anic has other complexes.

THE CASE FOR MTBE—The Italians see MTBE primarily as a new refinery "tool," at a time when most European refineries are under a dual pressure to reduce lead contents to

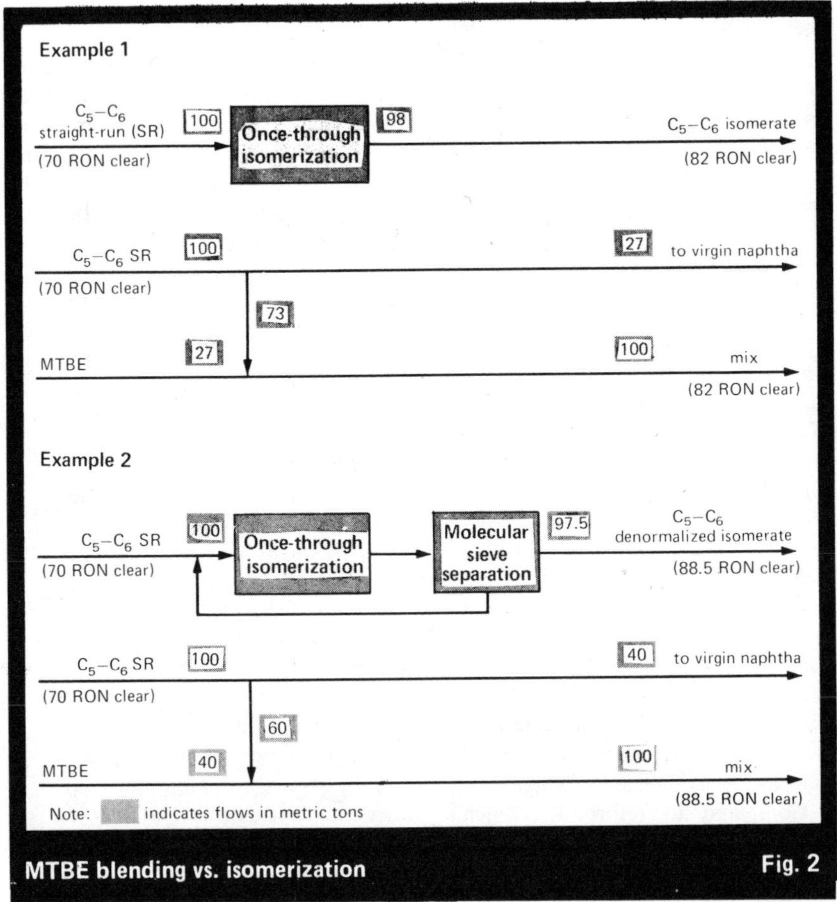

MTBE blending vs. isomerization Fig. 2

impossibly low levels, and release greater amounts of virgin naphtha and aromatics from the gasoline pool to petrochemicals manufacture.

MTBE's high-octane blending values (115–135 RON) allow a conventional topping-and-reforming refinery to maintain gasoline ratings while substantially cutting lead content. Without MTBE, even a modern European refinery would have to boost reforming severity to well over 102 to obtain a 98/99-RON premium gasoline, calculates Snamprogetti.

A rise in reforming severity demands a higher consumption of virgin naphtha, accompanied by a drop in overall reforming capacity. According to Spamprogetti sources, a reduction in premium-gasoline lead content from 0.6 to 0.4 g/l would require a boost of two points (97–99) in reforming severity. This in turn would increase feed consumption by 5%. And the longer reforming times would cut effective reforming capacity by 20–30%.

But a 5–10% blend of MTBE with clear gasoline, Snamprogetti claims, can solve both the excessive severity and reduced-capacity problems. At a constant reforming severity (97), a 5%-MTBE addition allows a refiner to cut the lead content of a 98/99-RON gasoline from 0.6 to 0.4 g/l. And a 10% shot of MTBE allows a lead reduction to 0.15 g/l with little or no increase in reforming severity. Yields and throughput capacity remain the same as for a reforming severity of 97.

A BETTER ALTERNATIVE?—Snamprogetti provides two examples of the way in which MTBE acts as an alternative to isomerization alone, or isomerization plus molecular-sieve separation, in the processing of C_5-C_6 straight-run gasoline (see Fig. 2).

In the first example, 100 metric tons of C_5-C_6 straight-run gasoline pass through an isomerization unit, yielding 98 metric tons of isomerate at 82 RON. However, if 27 metric tons of MTBE are instead blended with 73 of the straight-run gasoline, 27% of the gasoline is released from the gasoline pool to virgin naphtha, and the resulting blended yield is 100 metric tons of fuel at an 82-RON rating.

In the second example, the C_5–C_6 straight-run gasoline passes both through isomerization and molecular-sieve separation. Yield is 97.5 metric tons of denormalized isomerate at an 88.5-RON rating. But a 60/40 blend of the gasoline with MTBE would release 40 metric tons of the straight-run to the virgin naphtha pool, and the yield of the blend would be 100 metric tons at a RON rating of 88.5.

The economic advantages of MTBE blending over isomerization stem from lower investment and operating costs and from lower energy consumption, says Snamprogetti. The blending operation is said to be more flexible than processing via isomerization.

MTBE POTENTIAL—How much MTBE will be produced in coming years depends on worldwide demand. Snamprogetti has pegged production capability at 20% of worldwide naphtha-based ethylene capacity (assuming, of course, that all the isobutene raw material comes from ethylene crackers). Given this premise, Western Europe in 1976 could have produced 3 million metric tons/yr, Japan 1 million metric tons/yr, and the U.S. 600,000. Snamprogetti believes that U.S. capability will increase as the nation installs more naphtha-based ethylene plants.

The major market for MTBE is, of course, premium gasolines. Snamprogetti offers the following comparison between a gasoline/MTBE blend and a typical premium gasoline: no difference in gasoline consumption and horsepower, no corrosion problems (except those normally encountered), no difference in evaporation losses, no effect on paints and elastomers used in commercial vehicles, and no need to modify or adapt automobile systems. In addition, there are some ecological advantages attributed to MTBE blends: reductions of about 20% in carbon monoxide emissions and 70% in polynuclear aromatics. The aromatics are under suspicion as carcinogenic agents.—ANDREW HEATH, *McGraw-Hill World News, Milan;* SILKE MCQUEEN, *McGraw-Hill World News, Bonn;* and RAUL REMIREZ, *Associate Editor.*

Molecular Sieves Star in Gasoline-Upgrading Route

Two commercially proven technologies, working as one to upgrade gasoline octane, exploit molecular sieves both for iso/normal paraffins separation and for light-naphtha isomerization.

NICHOLAS R. IAMMARTINO
Associate Editor

Refiners around the world—all struggling to provide for future gasoline needs in the face of uncertain total demand, lead levels and octane requirements—are finding Union Carbide Corp.'s* total isomerization process (TIP) a flexible and economic answer to their dilemma.

An octane-upgrading route for light naphthas, TIP will make its commercial debut in August in a 6,000-bbl/d unit at Showa Oil's refinery in Kawasaki, Japan. A 4,000-bbl/d facility is under construction at the Shell/Gulf refinery in Cressier, Switzerland, slated for 1976 startup. A third unit is being engineered for an undisclosed European refinery.

TIP closely integrates Carbide's IsoSiv process, for adsorptive separation of branched and cyclic hydrocarbons from straight-chain *n*-paraffins, with Shell International's Hysomer technology, for catalytic isomerization of C_5/C_6 gasoline components (which Carbide offers for license on an exclusive basis). Molecular sieves act as adsorbent and as a component of the dual-function catalyst, respectively. Both processes have been proven on a commercial scale.

Why, Where—Benefits offered by the TIP approach are topped by front-end increases in clear research octane number (RON) and clear motor octane number (MON) of 15-20 (Table I). Isomerization routes that approach this gain by using distillation to separate unreacted *n*-paraffins for recycle consume 60% more energy than TIP (Table II), Carbide says, with virtually no saving in fixed investment. Front-end improvement is especially critical in Europe, where small, high-compression engines demand low-boiling, high-octane blending components, but alternate sources such as light cat-cracked gasoline or alkylate are lacking.

Compared to reforming, which can turn out streams of higher RON, TIP minimizes yield loss to gaseous products and can produce a blending stock of comparable MON. A low-pressure, semiregenerative reformer with bimetallic catalyst, for example, typically yields 4,130 bbl/d of 95-RON / 85-MON clear C_5+ product from a 5,000-bbl/d feed, versus TIP's 4,750-bbl/d of 88.5-RON/85.6-MON clear isomerate. Product sensitivity (i.e., the difference between RON and MON) is also 10 for such a reformer versus only 2.9 for TIP; the relatively high MON for TIP is an important plus for optimum performance of new engines.

A linear-programming model discussed at last month's National Petroleum Refiners Assn. meeting in San Antonio, Tex., identified two prime situations where these TIP ad-

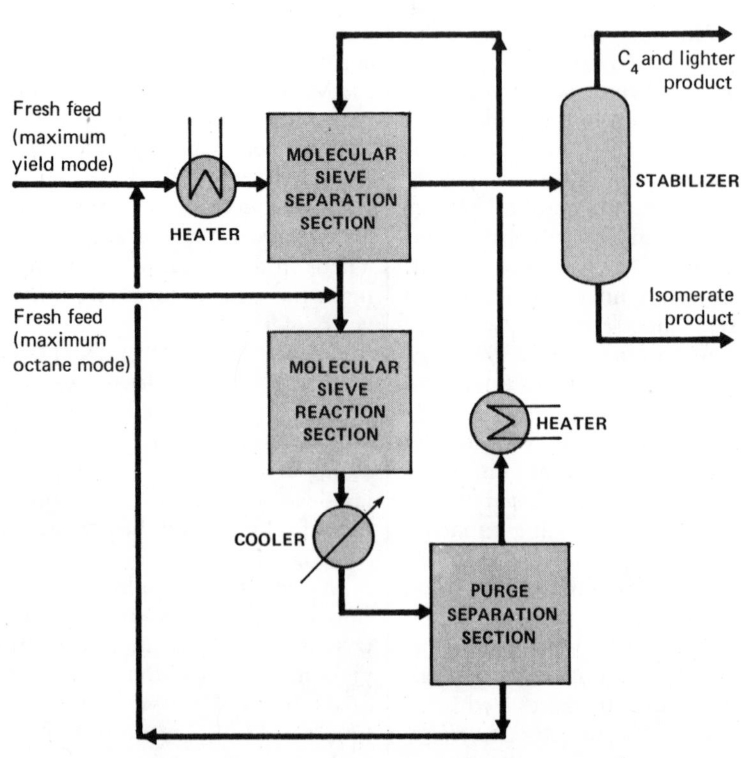

*Union Carbide Corp., Engineering Products & Processes, 270 Park Ave., New York, NY 10017.

Originally published April 28, 1975

vantages can be put to use. First, pointed out Carbide authors M. H. Hainsselin, M. F. Symoniak and G. R. Cann, TIP could replace alkylation if lead remains at its present loading in gasoline, while total output declines and octane level increases. Light ends previously used for alkylation feed could instead be sent to petrochemicals or to other fuel markets.

If lead antiknocks are banned from gasoline, TIP could team up with high-severity reforming to meet interim lead-removal standards during the phaseout, thus postponing major new investment for octane upgrading until the early 1980s.

Other situations expected to favor TIP (but not yet studied) include cases where (1) isobutane for alkylation feed is tight, (2) natural gasoline is an unusually high portion of refinery input, or (3) aromatics are extracted from reformate for petrochemicals.

Two Modes—TIP operates in the vapor phase, at a constant pressure in the 200-500-psi range, and a moderate temperature of 400-700°F. Integration of the Hysomer isomerization and IsoSiv separation sections (see flowsheet) can be in either a maximum octane mode, or a maximum yield mode.

In the maximum yield mode, vaporized fresh feed first enters the separation section, where *n*-paraffins are adsorbed on a type 5A molecular sieve. Other hydrocarbons pass through to a stabilizer column that removes butanes and lighter materials to leave a C_5+ blending stock.

The *n*-paraffins are desorbed from the sieve bed (while a second bed handles incoming feed), using displacement with a purge medium. The *n*-paraffins then pass over a noble-metal-loaded molecular sieve catalyst in the isomerization reactor. A near-equilibrium distribution of *i*- and *n*-paraffins results. The catalyst shows a higher resistance to water or sulfur poisoning than low-temperature halide-activated catalysts.

Reaction products leave the isomerization reactor mixed with purge material. A C_4+ hydrocarbon stream is condensed for recycle to the separation section for recovery of unreacted *n*-paraffins; purge material is reused for desorption.

In the TIP maximum-octane mode, fresh feed passes through the isomerization section before separation. Feed to the IsoSiv section is only the C_4+ hydrocarbon removed from the mixture of purge material and isomerization product. Several factors contribute to the higher-octane end-product. Some monomethylpentanes convert to dimethylbutanes, low-octane heptanes are cracked, and the C_5/C_6 ratio goes up as a few of the feed hexanes are cracked. However, some yield losses result from C_6/C_7 cracking.

How To Choose—The preferred TIP mode depends on feedstock characteristics and relative importance of octane versus yield to the individual refiner.

Feeds with a high heptane or benzene content are best processed in the maximum yield mode to prevent saturation of benzene and cracking of heptane. Stocks with 60+% total pentanes, or less than 25% C_5/C_6 *n*-paraffins, should also be processed for maximum yield. The maximum octane mode is preferred for feeds exceeding 50% *n*-paraffins.

When two or more feedstocks of different characteristics must be handled, capital and operating expenses can be minimized by sending some feeds to the isomerization section, and some to the separation section, of a single TIP unit. #

TIP Performance – Table I

Component, Bbl/D	Feedstock	Isomerate Product
i-Pentane	1,024	2,180
n-Pentane	1,200	54
Cyclopentane	122	87
2,2-Dimethylbutane	33	451
2,3-Dimethylbutane	92	237
2-Methylpentane	713	895
3-Methylpentane	439	560
Naphthenic C_6	800	11
Methylcyclopentane	295	215
Cyclohexane	133	60
Benzene	82	–
C_7+	67	–
Total	5,000	4,750
Feed, Product Qualities		
RON clear	72.2	88.5
MON clear	68.7	85.6
RON at 3 g/gal lead	89.0	99.5
MON at 3 g/gal lead	87.8	98.2
Reid vapor pressure, psi	10.9	12.8

Note: C_4 and lighter byproduct stream consists of 284,000 std. ft³/d C_1-C_3, 195 bbl/d *i*-butane, and 68 bbl/d *n*-butane.

Source: Union Carbide Corp.

Comparative Investment and Utilities Costs – Table II

Feed	TIP[1]	Reformer[2]	Isomerization[1,3] with C_5 Recycle
Rate, bbl/d	5,000	5,000	5,000
C_5+ product, bbl/d	4,750	4,130	4,974
MON clear	85.6	85	84
RON clear	88.5	95	87
Investment, $			
Plant investment[4,5]	3,100,000	3,200,000	3,600,000
Catalyst and chemicals	900,000	450,000	300,000
Total	4,000,000	3,650,000	3,900,000
Utilities, ¢/bbl feed[5]			
Fuel ($1.65/million Btu)	24	47	61
Cooling water (2¢/1,000 gal)	1	1	1
Steam ($2.40/1,000 lb)	10	(8)	–
Power (2¢/kWh)	8	1	8
Total, ¢/bbl feed	43	41	70
¢/bbl C_5+ product	45	50	70

[1] Feed is light straight-run and natural gasoline of Louisiana-Mississippi origin, with 72.2 RON clear and 68.7 MON clear on C_5+ basis.

[2] Low-pressure, semiregenerative reformer, bimetallic catalyst, operating at 95 RON severity. Feed of La.-Miss. origin, 160-350°F ASTM boiling range.

[3] Low-temperature isomerization with deisopentanizer and C_5/C_6 splitter columns.

[4] Plant investment includes paid-up royalty.

[5] Investment and utilities include product stabilizer but exclude feed pretreatment.

Source: Union Carbide Corp.

Resid Conversion Route

Flexicoking, integrating a new coke gasifier with conventional fluid coking, enables refiners to convert vacuum residuum and other heavy feedstocks into easy-to-desulfurize liquids and gases.

TERENCE K. KETT
GERARD C. LAHN
Exxon Research and Engineering Co.

WILLIAM L. SCHUETTE
Exxon Co., U.S.A.

A versatile process called Flexicoking can help meet future energy needs by upgrading heavy residua to a wide range of lighter, readily desulfurizable products, while maintaining a clean plant environment.

Developed by Exxon Research and Engineering Co. (Florham Park, N.J.), the process features coke gasification that converts the usual 15-25% coke byproduct from fluid coking into fuel gas. Overall, about 99% of a typical vacuum residuum ends up as liquid or gaseous product. The remaining 1% leaves as a low-sulfur, salable coke.

This ability to make light, clean products from heavy feeds is becoming more and more important to refiners, as coal, nuclear and other energy sources progressively reduce the demand for heavy fuel oil for electric power generation. Flexicoking is equally attractive, moreover, because yields and processing costs are insensitive to such feed contaminants as organic sulfur, metals and nitrogen. Feedstocks that direct catalytic hydrodesulfurization cannot handle economically can be converted by Flexicoking, and the products desulfurized by conventional processes, allowing refiners to tap new crude sources.

The first commercial Flexicoker is under construction at Toa Oil Co.'s refinery at Kawasaki, Japan. Rated for 22,000 bbl/d, it is scheduled for startup during the first quarter of 1976.

Since April, however, a nominal 750-bbl/d Flexicoker prototype (see photo) has been processing various feeds at the Exxon Co., U.S.A. refinery in Baytown, Tex. Performance has equalled or surpassed expectations—95+% coke gasification versus the design 90+%, for example, and only 0.04 wt.% sulfur in product coke gas versus 0.20% design.

Three Fluidized Beds—The Flexicoking process starts by injecting residuum feed into the coking reactor, where it thermally cracks to a full range of vaporized products, plus coke that deposits on fluidized coke particles already present. The cracked vapors are quenched in a scrubber located above the reactor. Heavy fractions boiling above about 950-975°F condense in the scrubber, and usually recycle back to the reactor. Light fractions flow to a fractionator for separation.

Coke meanwhile passes from the reactor into a fluidized-bed heater, where it partially devolatizes at about 1,150°F to yield a light hydrocarbon gas and residual coke (a very small fraction is removed by elutriation as product coke). One coke stream from the heater circulates back to the reactor to provide heat for the endothermic reaction.

Another coke stream circulates forward to a gasifier vessel. There it reacts at elevated temperature with steam and air, forming a mixture of hydrogen, water, carbon monoxide and carbon dioxide. Gasifier product, or coke gas, next passes through the heater vessel in order to provide some heat for the reactor, then cools down further as it generates steam. Removal of entrained coke dust and sulfur from the gas finally results in a solids-free gas stream, with sulfur content below 0.2 wt.% (fuel-oil-equivalent basis), and a heating value of 100-130 Btu/scf.

Reactor and heater are designed nearly the same as the reactor and burner in a conventional fluid coker. The gasifier has a carbon-steel shell lined internally with refractory material, similar to a catalytic cracking

Flexicoking Yields and Costs for Various Residua

	Iranian Heavy		Bachaquero		West Texas Sour Asphalt	
Feed properties						
Conradson carbon, wt.%	21.4		26.5		34.0	
Vanadium and nickel, ppm	525		1,040		137	
Yields	Wt.%	L.V.%	Wt.%	L.V.%	Wt.%	L.V.%
C_4 and lighter gases	11.8	–	12.6	–	13.5	–
C_5/975°F liquids	61.8	70.5	54.5	63.0	42.6	50.3
Coke	1.2	–	1.5	–	2.0	–
Coke gas (fuel-oil-equiv. basis)	–	15.6	–	21.1	–	30.0
Costs*						
Onsite material and labor, million $	13.4		15.2		17.7	
Throughput-related operating cost, million $/yr	1.03		1.51		2.20	

*Basis: 20,000 bbl/d, Gulf Coast location, 2nd qr 1973.

Originally published December 23, 1974

PROTOTYPE Flexicoker at Baytown has turned in a sparkling performance.

Reactor products to fractionator
Steam
BFW
Cooling water
SCRUBBER
SCRUBBER
Coke fines
Coke gas
Recycle
HEATER
GASIFIER
SULFUR REMOVAL UNIT
Residuum feed
REACTOR
Sulfur
Steam
Air
Steam
BLOWER

regenerator or coker burner, but there are no internal structures.

Process Characteristics—Additional features of Flexicoking confirmed at Baytown are:

- *Environmental protection*—Flexicoking does not emit any gaseous offstreams to the atmosphere. Approximately 50% of the feed nitrogen is converted to harmless elemental nitrogen carried with the coke gas. Liquid products can be cleaned to ultralow sulfur and nitrogen levels by processes such as Exxon/Union Oil Co.'s Go-Fining. Up to 98% of the feed sulfur can be recovered from the product streams as elemental sulfur.
- *Operating stability*—Operation is just as stable as a conventional fluid coker. Circulating coke contains less coarse material than is in a fluid coker, which directionally results in smoother circulation. Control of gasifier temperature and coke gasification level is also smooth. The former deviates no more than ±10°F (after startup), and the latter consistently stays close to 95%.
- *Coke properties*—Circulating Flexicoke is slightly less dense than fluid coke, considerably higher in surface area, and smaller in average particle size. Because of these differences and particularly its lower sulfur content (typically less than 3 wt.%), Flexicoke is more marketable than fluid coke or delayed coke. If desired, coke gasification can even be reduced to turn out more low-sulfur coke for marketing.
- *Coke-gas quality*—Like liquid Flexicoker products, coke gas can be readily cleaned, to essentially zero solids and to sulfur levels below 0.1 wt.%. Reactive nitrogen compounds that would convert to nitrogen oxides during combustion are extremely low in concentration, only about 3 ppm_v or less. No particle-removal units or sootblowers are needed on furnaces or boilers using coke gas. Burning characteristics minimize auxiliary fuel needs.

Where It Fits—Yields and product qualities from Flexicoking will allow refiners to convert more and more of the crude barrel into lower-boiling products in coming years. Product gas-oil can be used initially to yield a high amount of low-sulfur fuel oil. Adding a gas-oil conversion step such as catalytic cracking later on can reduce the fuel-oil yield down to zero, in favor of lighter products.

Since Flexicoking is insensitive to feedstock contaminants, and also handles a wide spectrum of feed types from atmospheric residuum to asphalt, it offers refiners still more flexibility to cope with today's changing feed-supply picture. Yield and economic data for three diverse feeds (see table) illustrate that a substantial 50+% conversion to low-boiling liquids is obtained even for the most difficult asphalt feedstock.

On the other hand, direct hydrodesulfurization for making clean products is not as versatile. This provides little feed conversion, and is impractical for high-metals streams because of catalyst cost.

Because it can process very heavy high-boiling hydrocarbons, Flexicoking can also play a role in exploiting heavy crudes not currently produced in large quantities, such as from the Orinoco tar belt in Venezuela and from the Athabasca tar sands in Canada. High viscosity makes pipelining to refineries expensive and impractical, and conventional refining technology for such heavy crudes is both expensive and pollution-prone anyway.

A Flexicoker at the crude site, however, can convert crude, or residuum from the crude, into a light liquid and coke gas. The former can then be sent to refineries for upgrading, while the latter (and coke product) generates steam at the site for the crude recovery operation. #

Meet the Authors

Terence K. Kett is a senior project engineer in the gas and heavy oil processes div., Exxon Research and Engineering Co. (P.O. Box 101, Florham Park, NJ 07932). He holds M.S. and Ph.D. degrees in chemical engineering from Michigan State Univ.

Gerard C. Lahn, a senior project engineer in the same Exxon division, holds a B.S. in chemical engineering from Pratt Inst., and an M.S. in chemical engineering from the Univ. of Virginia.

William L. Schuette is an engineering associate in the coking and gasification processes section of Exxon's research lab in Baton Rouge, La. He holds a B.S. from the Univ. of Wisconsin, and an M.S. from Louisiana State Univ., both in chemical engineering.

This article is based on a paper presented at AIChE's 67th Annual Meeting earlier this month.

Section VIII
PLASTICS AND ELASTOMERS

Artifical-latex route boasts foamfree stripping technique
Continuous peroxyester route to make its commercial debut
New routes tackle tough plastics-recycling jobs
New catalyst streamlines HD-polyethylene route
LDPE goes low pressure
Water-dispersed polyolefins recovered as microfine powder

Artificial-latex route boasts foamfree stripping technique

Solution-emulsification process disperses bulk elastomeric polymers into water, forming artificial latexes, without the common manufacturing troubles of foaming and coagulation.

☐ A problem-solving approach for making artifical latexes, for a variety of specialty applications, has been selected by Exxon Chemical Co. for a grassroots plant it plans to bring onstream by 1979.

The facility will manufacture several products, including butyl-rubber latex, butyl rubber/hydrocarbon resin latex, and others now moving out of the developmental stage. Exxon Chemical already markets the first two, which are produced exclusive for it by Burke-Palmason Chemical Co. (Pompano Beach, Fla.) in a small plant that has proved the new process.

Artificial latexes are based on polymers, Exxon points out, which cannot be made directly from monomers by emulsion polymerization (the common route to the so-called synthetic latexes, such as styrene-butadiene-rubber latex and vinyl acetate latex). The reason: polymerization catalysts for the precursor polymers of artificial latexes function only in waterfree systems.

The first step in Exxon's route thus dissolves polymer in an appropriate solvent to form a polymer cement. This in turn is dispersed in an aqueous medium. The highlight of the process scheme comes next—a foamfree solvent-stripping technique patented (U.S. 3,503,917 and others) by Oliver W. Burke, Jr., and assigned to Exxon. Finally, concentration of the latex by removal of excess water forms the finished product.

Exxon Chemical's first public discussion of the process took place earlier this year at the 16th Annual Meeting of the International Institute of Synthetic Rubber Producers, Inc. Describing the process for making butyl-rubber latex, before the Rio de Janeiro audience, was Charles P. O'Farrell, research associate in the firm's Elastomers Technology Div.*

LEADOFF STEPS—According to O'Farrell, the initial step of dissolving butyl polymer in solvent reduces the viscosity of the polymer to facilitate its dispersion in water. The boiling point of the solvent, or of the solvent/water azeotrope, should be below the boiling point of pure water. For the toluene solvent used by Exxon, an 80%/20% toluene/water azeotrope boils at 85°C at 1 atm. Hydrocarbons such as benzene, hexane and pentane could also serve but are not as safe as toluene.

Solids content of the resulting polymer cement must be maximized. Within the constraint that viscosity of the non-Newtonian cement not exceed 10,000 cp, approximately 18 wt.% of butyl rubber can be dissolved in toluene.

This viscosity ceiling is imposed to simplify the following process step, dispersion of the cement in the form of droplets into water, using both chemical and mechanical forces.

On the chemicals side, an emulsifying agent (or surfactant) is needed since butyl rubber does not contain enough polar sites to self-emulsify when mixed with water. For butyl rubber, Exxon Chemical uses as emulsifier the sodium salt of the sulfate ester of an alkylphenoxypoly(ethyleneoxy)ethanol, in an amount between 5–10% of the polymer weight. A special combination of theoretical and numerical concepts of emulsion behavior helped pinpoint the material as one having the proper lipophilic and hydrophilic segments (the former are imbedded in the cement droplets; the latter are distributed on the droplet surfaces) to form a stable oil-in-water raw emulsion.

Average size of the droplets must

Stripping section includes (left to right) feed injection device, path length extender and separator.

*Exxon Chemical Co., Elastomers Technology Div., P. O. Box 45, Linden, NJ 07036.

Originally published September 29, 1975

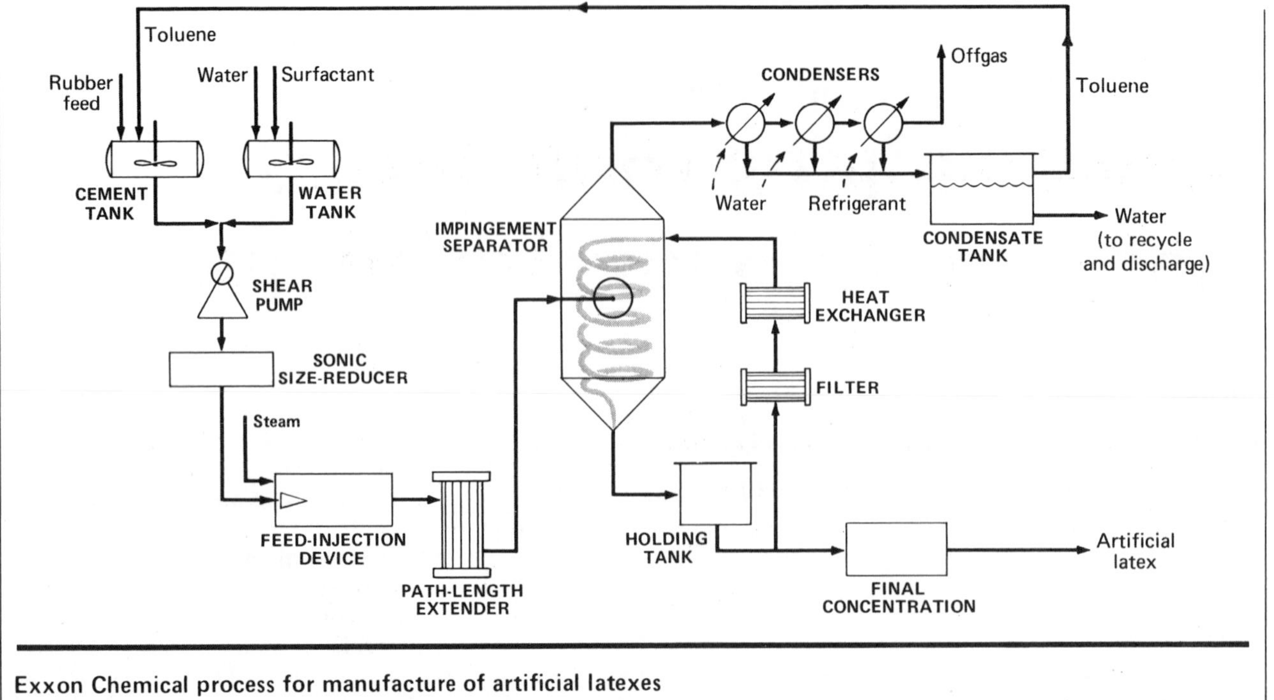

Exxon Chemical process for manufacture of artificial latexes

be below about 10 μ, and preferably below 1 μ. A series of high-shear rotor/stator mixing devices turning at 5,000 rpm does part of the job. The mechanical forces brought into play reduce particle size of the raw emulsion to about 3–5 μ.

The stream passes through a series of sonic size-reduction devices to complete the emulsification step. Pumped in at a pressure of 350 psi, the stream impinges on a vibrating metal reed. Ultrasonic forces generated reduce the cement droplets to about 1 μ (usually about 0.8 μ average size). Emulsion viscosity is about 150 cp at the outlet.

The weight ratio of the cement phase to the water phase usually falls within the 1:1 to 1.5:1 range. More water makes for an easier-to-pump fluid, explains O'Farrell, but boosts energy consumption for its subsequent evaporation. Too little water could lead to an unwanted water-in-oil emulsion rather than an oil-in-water type.

FOAMFREE STRIPPING—The patented stripping section next removes solvent from the emulsion, without the foaming and coagulation problems that beset distillation.

The raw emulsion enters the section through a special feed-injection device that O'Farrell describes as a bayonet-shaped nozzle within a length of pipe. Inlet pressure to the nozzle is relatively low; a vacuum of about 24 in. water is maintained on the outlet side.

The cone of emulsion shooting through the nozzle is contacted and atomized by a coaxial flow of high-pressure steam. Under the influence of vacuum and steam-supplied heat, solvent vaporizes from the atomized latex particles. A plate-type heat exchanger follows the feed-injection device to provide an extended tortuous flowpath and thus adequate residence time (a few milliseconds). But no heating fluid is used in the plate exchanger; heat-input needs are satisfied by adjusting the steam input rate.

The latex particles now comprise only butyl rubber, emulsifier and water. Separation from the steam/toluene gas takes place by impingement of the mixed stream on a moving sheet of liquid latex, which swirls around the inside walls of a conical separator.

Toluene and water vapor are carried overhead, condensed and separate. Toluene and about three-fourths of the water are recycled; the rest of the water is treated and discharged. About 10–20 parts of the liquid latex carried off the bottom of the separator are recycled to form the swirling film, for each part withdrawn as product. A heat exchanger in the recycle loop provides enough heat to keep separator temperature at about 165°F.

Latex withdrawn at about 40 wt.% solids passes through another heating/separation loop for final water removal, to bring solids content up to the final-product level of 61 wt.% minimum.

PRODUCT TYPES—Other typical properties of the butyl-rubber latex, marketed under the name Exxon Butyl Latex 100, include: viscosity, 2,000 cp (5,500 cp maximum); particle-size range, 0.1–0.8 μ (95% limits); pH, 5.5±0.5; emulsifier content, 3.5±0.5 wt.%; and mechanical stability, 100–300% volume increase and less than 0.1 wt.% coagulum when sheared at a speed of 19,000 rpm for 30 min.

Applications fall in the areas of polyolefin adhesives, polypropylene woven-ribbon-fabric coatings, protective coatings, specialty binders, barrier adhesives, and nonwoven binders.

The butyl-rubber/hydrocarbon-resin latex (Butyl Resin Latex MD 602) is made similarly, but with the additional input of resin to the polymer cement. It boasts improved performance for use in polypropylene fabric coatings.

Development latexes now under consideration include EPDM latex, high-strength butyl latex, and latexes based on Exxon's Escorez hydrocarbon resins and Vistanex polyisobutylene polymers.

Nicholas R. Iammartino

Continuous peroxyester route to make its commercial debut

Continuous, solvent-free technique increases processing efficiency, yielding peroxyesters of high purity and at high throughput. The route is considered safer than conventional methods, and costs less, too.

Philip M. Kohn, *Associate Editor*

☐ Ground was broken last month for the first commercial continuous-process peroxyester plant. Slated to cost about $10 million, the plant is being built by Pennwalt Corp.'s Lucidol Div. (Buffalo, N.Y.) at an existing site at Crosby, Tex. The unit, scheduled to start up in the middle of next year, will make four major peroxyesters totalling 10 million lb/yr (twice the next-largest plant capacity), an amount Lucidol believes will satisfy the entire domestic market.

The highly automated plant will be remotely operated, with cameras and microphones, as well as process instrumentation, doing the monitoring. And because the amount of hazardous product present in the unit at any one time is so small, protective barrier walls are unnecessary—a first in peroxide production.

Besides the safety aspects, the firm notes that the process offers other benefits. For example, capital cost is about 25% lower than for a conventional batch-process plant of the same capacity. (Lucidol states, however, that no single-train batch plant has ever been built with a capacity even close to 10 million lb/yr, and the company does not believe such large batch sizes could be handled safely.) In addition, the continuous process is said to provide higher yields and purer product, and results in lower labor, energy and utility costs. And because the process is solvent-free, it in most cases offers peroxyester users the choice of either diluted or undiluted products.

POLYMERIZATION INITIATORS—Peroxyesters are sources of free radicals and are used in plastics manufacturing. The four products made by Lucidol will be: *tert*-butyl peroxybenzoate, *tert*-butyl peroxy-2-ethylhexanoate (commonly known as peroctoate), *tert*-butyl peroxypivalate, and *tert*-butyl peroxyneodecanoate. Sold either as pure liquids or as solutions in odorless mineral spirits or dioctyl phthalate, the peroxyesters find use as initiators in the manufacture of low-density polyethylene, polyvinyl chloride, polystyrene, and polyesters (reinforced plastics). Additionally, they are employed as crosslinking agents in elastomers, and in the making and curing of acrylics ranging from Lucite and Plexiglas to countertops and acrylic paints.

Because the peroxyesters are highly temperature-sensitive and can be violently unstable, conventional batch processing requires use of solvent dilution to minimize explosion hazards. Unfortunately, says Lucidol, the presence of solvents often adversely affects the initiators' efficiency in polymerization reactions.

A BETTER WAY—Lucidol's patented continuous route (U.S. patent 4,075,236) uses two vigorously-mixed reaction zones to produce peroxyesters at a high throughput per unit reactor volume and physical separation units, such as centrifuges and gravity-settling columns, to isolate pure product.

In the process (see flowsheet), reactants—consisting of an appropriate hydroperoxide, an acid chloride and an aqueous alkali-metal hydroxide (NaOH or KOH)—are continuously added to the reactor from storage tanks at controlled flowrates via metering devices. A major portion (about 90 to 95%) of the conversion of the acid chloride takes place at this point, with the temperature of the reaction mixture being maintained within ±1°C of that desired (typically 0° to 40°C), via $CaCl_2$-brine cooling.

The reaction mixture then overflows continuously to the second reactor, where the reaction is completed and the temperature is lowered to about −10° to 20°C, since slight heating occurs due to agitation. The first reactor is equipped with a pH-controller to regulate the pH of the reaction mixture, ensuring complete conversion of the acid chloride to peroxyester, thereby improving product quality and yield.

The cooled reaction mixture continuously overflows into a dilution tank, where water is added. The agitated mass then flows to a centrifuge, where the peroxyester product is separated from an aqueous stream that contains soluble salts, caustics and peroxides. (The product stream—at least 95% pure—can be collected at this point as a pure product.) The peroxyester stream then goes to washtank #1, where it is agitated with a proprietary aqueous wash solution and then fed to another centrifuge. Here the aqueous stream containing impurities—such as unreacted hydroperoxide, undesired byproduct peroxide, or salts—is discarded. These impurities vary depending on the reactants. The process is repeated in washtank #2 and its subsequent centrifuge.

Product from the final centrifuge goes continuously into a packed stripping column, where the peroxyester is stripped of remaining impurities by dry air or another oxygen-containing gas at ambient temperature.

Originally published July 17, 1978

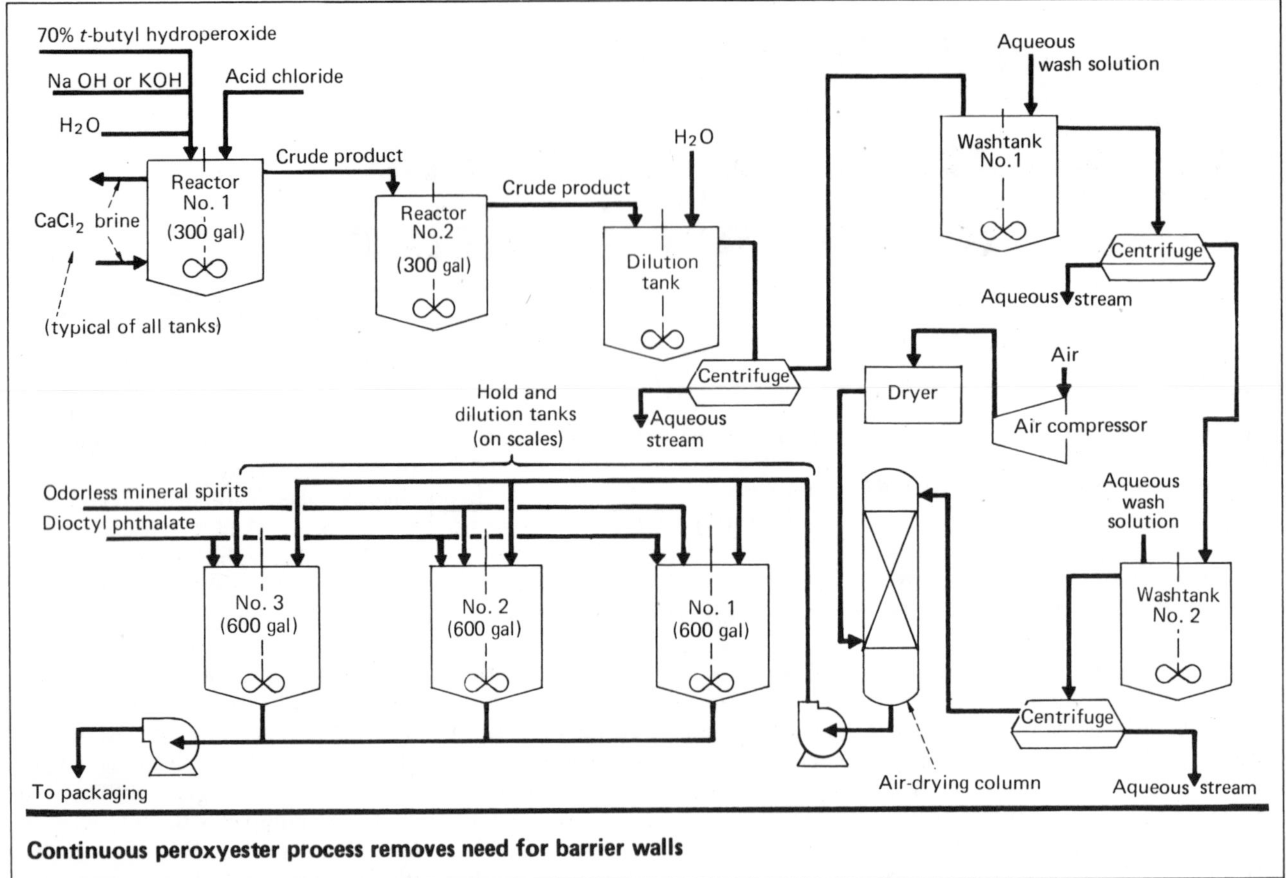

Continuous peroxyester process removes need for barrier walls

Previously, the standard method for drying the product was to slurry magnesium sulfate into the mass, the $MgSO_4$ absorbing moisture. The moisture-laden salt then had to be removed mechanically by filtration. This not only lost production time, but product as well—about one pound of peroxyester per pound of $MgSO_4$ used.

Stripped product goes to three 600-gal hold- and dilution-tanks, which together can store about 2½ h of production. Diluents, for solution makeup, can be added to each tank individually.

Finished product proceeds to a packout section, where three semiautomated packaging lines can each handle about double the operating rate. One line packs the peroxyesters in 1-gal polyethylene containers, while the other two put the peroxyesters in 5-gal packages, the largest allowed by the U.S. Dept. of Transportation.

EFFICIENT PROCESS—Lucidol has extensive experience running all four products through a continuous-process pilot plant, and the firm says the results have been higher yields for all four products than those given by batch methods, varying from 2% more for *tert*-butyl peroxybenzoate to 8.6% for peroctoate.

Besides higher yields, the process turns out product with fewer impurities: <0.1% unreacted hydroperoxides, <0.1% byproduct peroxide, and <100 ppm each of chloride and phosphorus. The firm asserts that these specifications have not been met by any batch methods to date. And the continuous process produces 7 to 10 times more output per reactor volume than do batch processes, where a 1,500-gal reactor is the largest used because of the hazardous materials involved.

COSTS AND CONSUMPTIONS—The project is budgeted for $10 million ±10% (including engineering), with boundary limits incorporating process equipment, storage, process control equipment, and processing and control buildings. Lucidol believes that a batch plant of similar size would cost approximately $13.5 million, but that realistically—because of the hazard—at least two batch plants would be needed to make 10 million lb of product, with a correspondingly higher capital investment.

As for operating costs, more pounds produced with the same manpower obviously provides a lower direct-labor cost per pound. Lucidol projects its labor savings at between 50 to 75%, depending on the product, compared with its own present batch unit.

Energy costs cannot really be finalized until the plant is operational. However, the company notes that the largest single consumer of energy is the refrigeration system. A batch plant must be designed for peak heat loads, whereas the heat load in a continuous plant is a steady value. The new plant will have 150 tons of refrigeration capacity, whereas a batch plant of the same size would require 290 tons.

Lucidol says the continuous process uses water for washes and for dilution much more efficiently than batch routes, and requires only about 40% of the water consumed by conventional processes. This is important, since all of this water is contaminated with organic and inorganic chemicals and must be treated before disposal. Lucidol will use aerated ponds and conventional treatment before getting rid of the water down a 6,600-ft-deep disposal well.

New routes tackle tough plastics-recycling jobs

Companies in the U.S., Germany and Japan are developing specialized technology to permit the recovery of plastics such as polyurethanes and polyvinyl chloride from automotive and other industrial scrap.

☐ Plastics recycling should get a boost from new chemical-engineering technology for treating industrial polymeric scrap that is not amenable to conventional techniques.

Chemical processes promise to break down flexible and rigid polyurethane foams into their basic building blocks, which then can serve to remake the polymers. Other routes aim to recover polyvinyl chloride, either by itself or still compounded with additives. And budding plastics-separation systems may ease the sorting and, thus, recycling of the materials.

Fabricators should welcome the developments. Though only 2–5% of all plastics fed to the average fabricating plant ultimately ends up discarded (because, for instance, grinders and cutters commonly turn clean thermoplastic scrap into powder or pellets that can be remolded), usually 10–30% of the polyurethane and polyvinyl chloride input to some operations goes to waste.

Added to this potential, the increasing use of plastics in automobiles augurs a large possible recovery of polymers from junked-car shredders. Plastics also proliferate in municipal wastes, but economic recycling remains far off (see box appearing on p. 215).

FOCUS ON FLEXIBLE FOAM—Ford Motor Co. (Dearborn, Mich.), General Motors Corp. (Detroit), and Bayer AG (Leverkusen, West Germany) are grooming processes to recover constituents from flexible polyurethane foam.

Such scrap abounds. For instance, the U.S. automotive industry alone gobbles up flexible polyurethanes (mostly for seat cushions) at a better-than-300-million-lb/yr clip, says Jack Milgrom, a plastics expert at Arthur D. Little, Inc. (Cambridge, Mass.). Furniture makers consume huge quantities, too. And because the foam is thermosetting, it can't simply be ground and remolded. Most simply goes to landfill.

Ford's alternative to disposal involves hydrolyzing scrap materials to cause polymer breakdown. First unveiled in early 1972, the decomposition process takes place at 1 atm. or less and 480°F or more, in the presence of steam. The resulting polyols and diamines then can feed a polyurethane production circuit.

So far, Ford has tried out the process only on a laboratory scale, in pound quantities. But the Chemistry Dept. of its Scientific Research Staff is now building a larger "pre-pilot" facility. "We'll go to the pilot-plant stage in one or two years," comments Lee R. Mahoney, the decomposition route's inventor.

In September 1974, Ford entered into an agreement covering the technology with Tenneco Chemicals, Inc.'s (Saddle Brook, N.J.) Foam and Plastics Div. Tenneco has taken chemicals reconstituted by Ford and turned them into new foam. Once Ford has a hydrolyzing setup ready to go, Tenneco will market it, initially to fabricators. (Tenneco itself is said to be working on companion equipment that would allow molders to immediately make new foam on-site from the materials).

Application at auto-shredding plants will come later, Ford predicts. The foam appearing in the light, air-separated fraction of a shredded car contains impurities; more work remains on mechanically separating the polyurethane from other polymers, aluminum, fibers and dirt, and on purifying the product polyols.

SIMILAR SYSTEM—The General

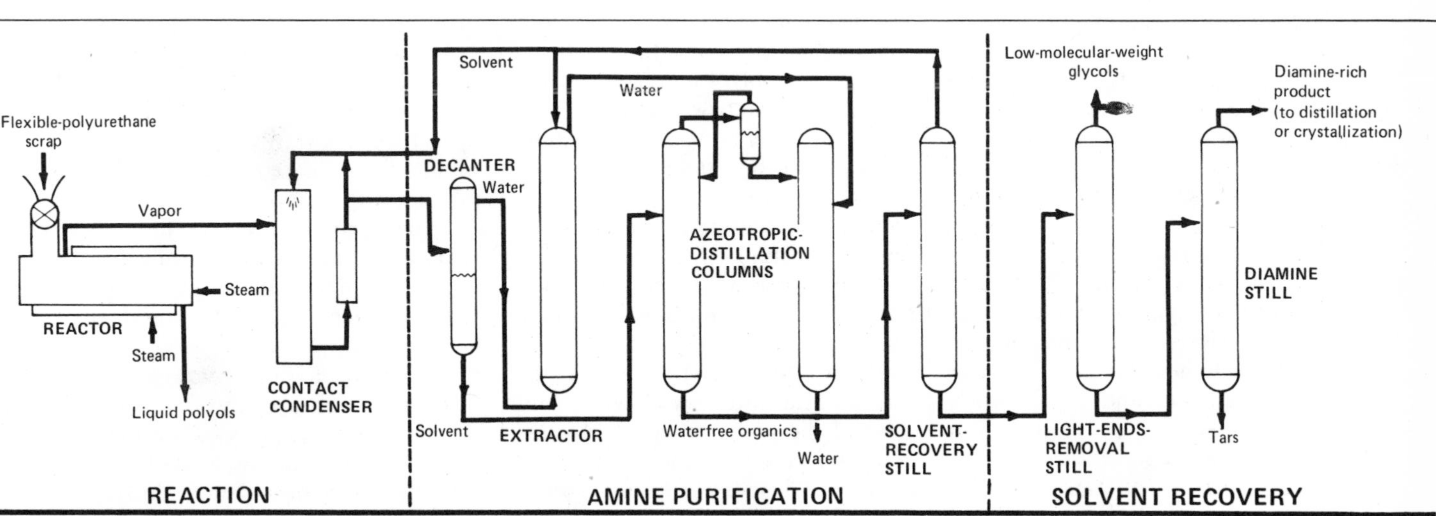

General Motors detailed its breakdown route at September's AIChE meeting

Originally published February 16, 1976

Motors process, announced in October 1974, strikingly resembles Ford's. (Both firms say that patents are pending.) It, too, employs hydrolysis of the flexible foam with steam, at 1 atm and about 600°F.

As shown in the flowsheet on p. 54, the technique divides neatly into three sections. The first, for reaction, recovers polyols directly as a liquid, which is relatively waterfree and ready for reuse after cooling and filtering. Meanwhile vapor from the reactor passes into a spray condenser where it contacts an aniline or benzyl-alcohol solvent. A mixed fluid consisting of steam condensate, solvent and organic products (diamines, glycols and tars) results.

The mixed stream immediately enters a solvent-recovery sequence of decantation, extraction, azeotropic distillation and normal distillation. These steps separate out both solvent and water for recycle, as well as the product organics.

The organics stream continues into the third part of the process. Distillation then yields diamines, the company notes, and glycol-and-tar byproducts, which are suitable for burning as fuel.

According to Philip Weiss, chief of the General Motors Research Laboratories' Polymers Dept. (Warren, Mich.), the firm has successfully proven the process on a laboratory scale, but piloting requires the pinning down of additional parameters. Initial economic projections, at any rate, show a capital investment of $275,000 for a system feeding on about 14 tons/d of polyurethane.

Bayer, for its part, admits to now starting up a pilot plant for its "hydrolysis" project, but will not release any process details. The company is midway through a 3-yr, approximately $460,000 study, funded equally by the German government and the country's Plastics Production Industry Assn. (Verband Kunststofferzeugende) of Frankfurt/Main. Bayer is looking not only at polyurethanes but also at other polycondensation and polyaddition plastics, such as polyesters and polyamines.

DIFFERENT TACK FOR RIGIDS—Upjohn Co. (Kalamazoo, Mich.), meanwhile, has centered its interest on rigid polyurethane foams. Popular for pipe and appliance insulation, these much-higher-molecular-weight foams require a slightly different decomposition approach than that for their flexible kin.

The Upjohn process (U.S. patent 3,738,946) feeds cut-up or pulverized scrap into a nitrogen-purged reactor containing a liquid medium of 90–95% aliphatic diol and 5–10% dialkanolamine. An ester interchange takes place at 365–410°F, yielding a homogeneous liquid that can serve without any further treatment as the polyol ingredient in the preparation of a new foam.

The process works for flexible scrap as well but, notes Upjohn, the liquid polyols generated are incompatible, and typically must be separated before use.

A Japanese licensee is nearing startup of the maiden full-scale plant, notes Upjohn.

RECYCLING VINYL—Just as the automobile industry lies behind most of the polyurethane scrap targeted for recovery, the huge quantities of fabric-backed vinyl waste from car-upholstery manufacture are offering the prime incentives for polyvinyl chloride recycling. Already, Gerwain Chemical Corp. (Frackville, Pa.), Hafner Industries, Inc. (New Haven, Conn.) and Polymer Recovery Corp. (Piqua, Ohio) have devised processes.

All entail chopping the feed scrap, dissolving nonfibrous materials in a solvent, and recovering these materials (after filtering off the fibers) by evaporation and drying.

Gerwain, a joint venture of Horizons Research, Inc. (Cleveland, Ohio) and Ag-Met, Inc. (Hazleton, Pa.), leads the trio in tapping the scrap. The firm started up a commercial-size plant at Frackville in early 1974. Its initial capacity of 1 million lb/yr of recycled polyvinyl chloride will jump to 2.5 million lb/yr by April or May, says the company, which adds that the unit now is running three shifts per day, seven days per week. The PVC product comes out in a compounded form highly suitable for wire coatings.

Horizons Research invented and patented the basic process. However, several major, undisclosed modifications have been made in the route. Besides treating fabric-backed vinyl commercially, Gerwain has tested the process for polyurethane-backed PVC scrap, and for the recov-

Plastics in refuse remain in the dumps

Although recycling of polymers in industrial scrap is advancing, the far more substantial volume of discarded plastics (about 85% of the total each year in the U.S.) that is found in municipal solid wastes remains relatively untouched. And little likelihood exists of recycling such plastics to any significant degree for at least a decade or two.

The big stumbling block: Out of a mountain of municipal refuse, plastics comprise only about 2% on the average, with an upper limit of about 5% or so. Thus, just pulling out a mixed-plastics stream poses a major economic hurdle—and the resultant mixture only suits limited, low-value applications, because of its poor performance characteristics. Splitting the polymers into their different generic types requires additional, costly steps.

Facing these difficulties, many experts advise simply leaving the plastics in the refuse. Since most polymers boast an exceptionally high heating value of 11,000–20,000 Btu/lb (three to four times the average value for combustible municipal waste), they can provide a useful fuel component for the incineration or pyrolysis of the refuse.

Nevertheless, efforts to separate plastics from municipal wastes continue. The U.S. Bureau of Mines, for one, remains active at its College Park, Md., facility. And the Society of the Plastics Industry, Inc. (New York) is sponsoring a comprehensive program at the National Center for Resource Recovery, Inc. (Washington, D.C.). Begun just last year, this project will evaluate techniques to separate a mixed-plastics stream from refuse, and alternatives for sorting the plastics by type.

Trial Mitsui air-flotation unit has run at Mitaka City, Japan, since 1974

ery of polyesters from X-ray- and photographic-film wastes.

Hafner Industries hoped to have a commercial, 50-tons/d (of scrap) plant onstream at Meriden, Conn., by now, but was forced to delay plans when a bond offering fell through. At present, according to company president and process inventor Edwin A. Hafner, proposed ventures involving foreign interests seem particularly promising. Hafner stresses that his method offers in-process control of such factors as color, composition and molecular weight of the recycled plastic. He says that the route can recover PVC either in compound form with plasticizers and stabilizers, or as a simple resin, with a separate output of compounding ingredients, which then can be reused. And, he adds that his method will ultimately accept any mixture of industrial wastes, or organics from municipal refuse, not just specific "clean" scrap plastics.

Polymer Recovery, which expected to have its already built, 18–20-million-lb/yr-of-scrap plant running at Piqua by the end of last year, has suddenly dropped its plans. The firm's president, Wilbur Sussman, says that the unit "is not operating now, and I don't believe it will." Outside sources imply that market, not technical, problems are stymieing the venture.

FOSTERING FLOTATION—Instead of using solvents to separate out a desired polymer, Mitsui Kinzoko Engineering Service Co. (Tokyo) is promoting two flotation units.

Mitsui Kinzoko, a subsidiary of Mitsui Mining & Smelting Co. (Tokyo), explains that plastics generally are not amenable to air flotation in a liquid medium because of the polymers' hydrophobicity—air bubbles attach themselves too easily to the plastics particles and float them all away. But, by using a proprietary pair of wetting agents, the firm states that it can selectively alter a plastic's wetting characteristics, turning the polymer hydrophilic, so that air bubbles can't adhere. Even plastics of identical density can be separated with a high degree of accuracy, claims Mitsui.

The unit (a prototype is pictured at left) can operate continuously and automatically, with no polluting offstreams, emphasizes the company. It recently introduced a commercial version, called the MIPS-600. Rated at about 660 lb/h of granulated plastic feed, the unit can recover PVC from a mixture with other plastics, polypropylene from polyethylene, high-density from low-density polyethylene, polyesters from cellulose triacetate, and numerous plastics from nonplastics.

Mitsui's second flotation separator simply relies on density differences to distinguish among plastics, floating away light materials while allowing heavy ones to sink in the liquid medium. Since introducing the unit in July, the firm has sold three, all for separating agricultural films (polyethylene from polyvinyl fluoride).

The U.S. Bureau of Mines (Washington, D.C.) groomed a similar system a few years ago at its Rolla, Mo., research center. Though this was aimed initially at separating mixed plastic won from municipal wastes, one U.S. firm has taken up the technique for culling plastics insulation from scrap copper wire.

Nicholas R. Iammartino

New Catalyst Streamlines HD-Polyethylene Route

Monomer recycle is eliminated, and nearly all reaction diluent recycled without purification, in this simple and safe suspension process for manufacturing high-density polyethylene. A proprietary titanium catalyst is the key.

HERBERT KREUTER
BERND DIEDRICH
Farbwerke Hoechst AG

The first company in the world to commercialize a Ziegler process for high-density polyethylene (HDPE) was Farbwerke Hoechst AG, whose trendsetting plant at Hoechst, Germany, went onstream in 1955.

Since then, while turning out more than 3 million tons of HDPE, Hoechst has developed a second-generation scheme, one that offers such benefits as a simplified processing sequence, wider flexibility for tailoring product characteristics, safer operation, and more-attractive economics.

HDPE PLANT at Altona, Australia, turns out 40,000 tons/yr of product.

Originally published August 5, 1974

Typical Utility Requirements

	Polymerization	Powder Silo	Granulation	Total
Medium-pressure steam (12 bar abs.), kg	250	–	50	300
Low-pressure steam (4 bar abs.), kg	650	20	150	820
Electricity, kWh	150	30	350	530
Cooling water (25°C), m^3	120	10	50	180
Demineralized water, m^3	0.600	–	0.050	0.650
Nitrogen, normal m^3	30	–	–	30
Transport air (4.5 bar abs.), normal m^3	150	150	100	400
Instrument air (4 bar abs.), normal m^3	225	5	10	240

Basis: Per metric ton of HDPE pellets

Hoechst and its affiliates already use the technique in plants in Germany, India, South Africa and Australia, which produce some 650,000 metric tons/yr. Hoechst will start up another 80,000 tons/yr at Münchmünster, Germany, about mid-1975, and a 50,000-ton/yr plant in Spain in a joint venture with Explosivos Rio Tinto during 1976. The first licensing agreement came in late 1973, when Société Industrielle des Polyolefins modified its existing plant at Le Havre, France, to a 60,000-ton/yr unit using the Hoechst process. A second licensee unit rated for 35,000 tons/yr has just been granted for mainland China, which Friedrich Uhde GmbH will engineer and build.

High-Activity Catalyst—The highlight of Hoechst's technique is a proprietary Ziegler catalyst having enhanced activity and selectivity. It is based on slurry-supported titanium, with aluminum-organic compounds and diluent included as well. All of the ingredients are readily available.

Since HDPE yield is a high 30-100 kg/g titanium, even at low operating pressures below 10 bar, only small amounts of the organic metallic catalyst are needed. This sidesteps the costly and pollution-prone operations otherwise used to purify HDPE of catalyst residues. High catalyst activity allows nearly complete conversion of ethylene monomer, furthermore, without gas recycle.

By slightly modifying catalyst composition, the HDPE molecular-weight distribution can be tailored.

(Regulating the amount of hydrogen present during polymerization controls molecular weight itself; addition of comonomers such as butene-1 or propylene adjusts product density. Overall, HDPE is 0.942-0.965 g/cm^3 in density, with melt-flow index from 0.1-50 g/10 min.)

Reaction Sequence—A two-step catalyst preparation starts off the Hoechst process. In one mixing tank, titanium and aluminum-organic compounds are blended. In a second, diluent is added—this is a safe-to-

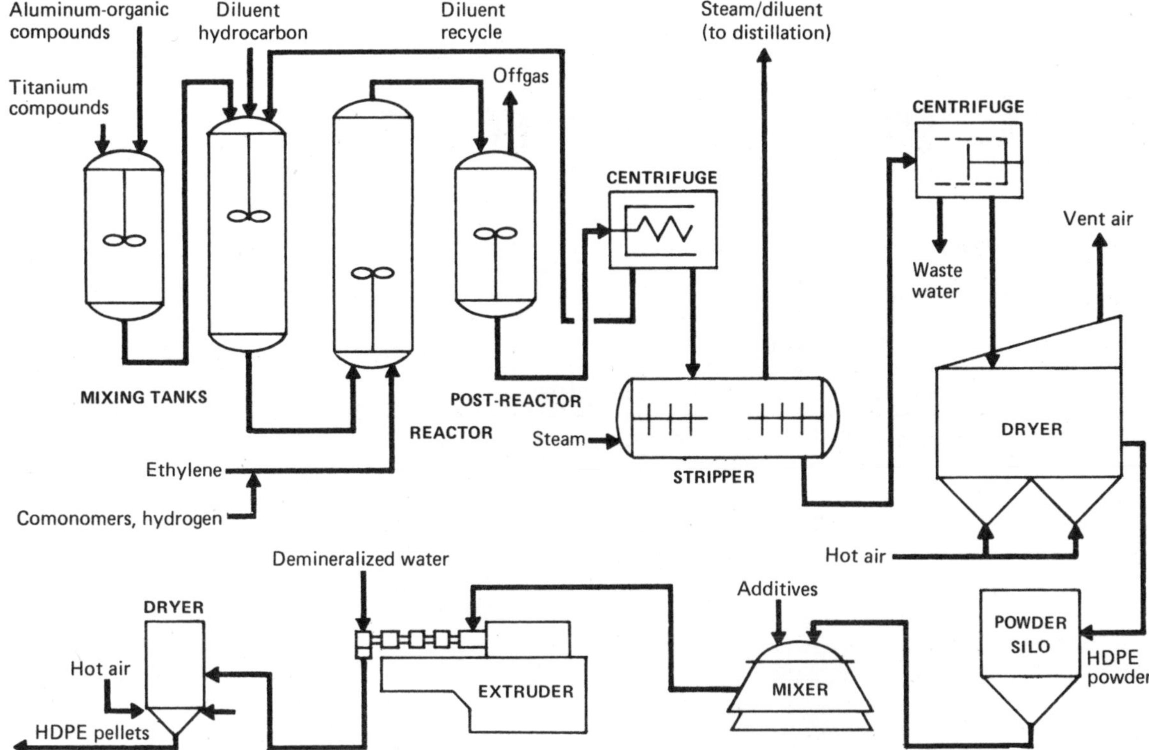

handle hydrocarbon, such as benzene or hydrogenated naphtha, in the 130-160°C boiling range.

The catalyst slurry is then fed into a large reactor, along with polymerization-grade ethylene monomer (99.9+% pure) and comonomers. Temperature is 80-90°C and pressure 8-10 bar. The higher pressure level used in other routes to boost monomer conversion is unnecessary because of the high catalyst activity.

The polymer suspension formed in this primary reactor flows continuously into a post-reactor or receiving tank, sized at about 10% of the primary reactor volume. The small amount of polymerization that occurs here brings the overall synthesis to 98-99% conversion. Mainly, the receiving tank serves to control and smooth out product flow. Tail gases for incineration are also withdrawn from the tank. These consist of non-polymerizable ethane and propane from the monomer, and hydrogen.

Separation and Drying—The HDPE/diluent stream is transferred at about atmospheric pressure to a centrifuge for separation. About 90% of the diluent is withdrawn for immediate recycle to the catalyst-preparation stage, without purification.

The remaining 10% is stripped out by saturated steam at 100°C. A distillation unit recovers this small diluent portion for recycle.

HDPE from the steam stripper is at this point diluent-free, but water-wet. A centrifuge takes out most of the water. Final drying is in a fluidized-bed air dryer. Other processes commonly require inert nitrogen; this is eliminated in the Hoechst method because (1) no combustible diluent remains on the polymer, and (2) polymer particles are an exceptionally coarse 250μ, and thus not susceptible to dust explosions.

Also avoided is alcohol extraction of the HDPE polymer. This step is needed with some other catalyst systems to remove unacceptably high catalyst residues. The extraction spews out a troublesome waste consisting of alcohol, titanium, aluminum and other pollutants.

The dried HDPE powder is then transported by pneumatic conveyor to intermediate storage. The granules are then pelletized, dried and packaged.

Economics—To produce 1 metric ton of HDPE powder, the raw materials consumed in the Hoechst process are: ethylene and comonomers, 1,030 kg; hydrogen, 1 kg; diluent, 25 kg. Catalyst and miscellaneous chemicals cost about $4. Approximately 1,005 kg of powder is needed for each ton of pellets. Utilities consumption is shown in the table.

Capital cost (battery limits) is estimated at about $320/metric ton for a 30,000-ton/yr plant, the smallest feasible size. A typical 80,000-ton/yr plant would run $220/metric ton; the largest single-train plant, 100,000 tons/yr, would be $210/ton.

An added plus for the process is its inherent flexibility. Conventional HDPE plants can be converted to the Hoechst technique without major equipment modification or expense. When market conditions warrant, a Hoechst HDPE plant can be revamped just as easily to turn out polypropylene, on either a short- or long-term basis. #

Meet the Authors

Herbert Kreuter is executive head of the polyolefins group of Farbwerke Hoechst AG, 6230 Frankfurt (Main) 80, Germany. He holds a doctorate in chemistry from Mainz Univ.

Bernd Diedrich heads the polyethylene group in Hoechst's plastics research department, and holds a doctorate in chemistry from West Berlin Univ.

LDPE goes low pressure

In what many observers think is a trend to cut energy and capital costs in the production of low-density polyethylene, a number of producers are exploring low-pressure processes based on new catalyst technology.

Originally published January 2, 1978

☐ For years, one of the ways to cut low-density polyethylene (LDPE) production costs has been to improve conversion rates by raising reactor pressure. But this design philosophy may be undergoing a dramatic change. Some firms are unveiling low-pressure LDPE routes made possible by new catalyst technology. Their motivation: lower energy and capital costs.

Union Carbide Corp. (UCC) has made the biggest splash so far, taking the wraps off a modification of its gas-phase high-density polyethylene (HDPE) process commercialized in the late 1960s. With the aid of a new family of catalysts, and "extensive" modifications to the company's Seadrift, Tex., HDPE plant, UCC has in great secrecy over the past two years made quantities of a new LDPE resin. At least two plastics processors have handled the product as part of UCC's test-marketing program.

UCC is not alone in its efforts. Phillips Petroleum Co. (Bartlesville, Okla.) has been making low-pressure LDPE in commercial quantities since the early 1970s. Lately, the firm has begun an active search for possible licensees. Du Pont of Canada has been producing both low- and high-pressure polymers in the same unit (a petrochemical plant in Sarnia) for an even longer time.

Overseas, Mitsui Petrochemical Industries, Ltd. (Tokyo) says it has been operating a low-pressure LDPE plant for about five years. The 40,000-metric-ton/yr facility is located at Iwakuni, West Honshu.

The list of newcomers includes Dow Chemical Co. (Midland, Mich.), which revealed soon after UCC's announcement that it has been developing its own low-pressure process. Dow, like UCC, has been producing resin for test-marketing over the past ten months, and appears to be almost as close to commercialization.

A number of other U.S. and foreign

polyolefin makers report that they are at least working on the concept. These include West Germany's BASF, which is investigating modifications to its gas-phase polyolefin process, so far commercialized only for polypropylene. The new changes allow HDPE production, with variations in density to lower ranges. A 12,000-metric-ton/yr "semicommercial" plant has been operating for about a year at the firm's Rheinische Olefinwerke GmbH (ROW) joint venture (with Shell) at Wesseling, West Germany. In the U.S., National Petrochemical Corp., a joint venture of National Distillers and Chemical Corp. (New York City) and Owens-Illinois Inc. (Toledo, Ohio), has been doing catalysis work on low-pressure processes but apparently has nothing substantial to report yet.

GAS PHASE—What sets UCC's process apart from the others is its gas-phase fluidized-bed design, which contrasts sharply with conventional liquid-phase tubular or stirred-autoclave reactors operating at much higher pressures.

Reaction pressure in the UCC unit is between 100 and 300 lb/in^2, markedly below the 15,000 to 50,000-lb/in^2 levels of the conventional processes. The fluidized-bed temperature of 100°C is also significantly below the roughly 300°C temperature of high-pressure methods. Conversion rate of monomer to PE is lower in the UCC route, but this is of little consequence to overall economics, since the energy required to repressurize the recirculating unreacted monomer stream is so much lower. (Typical conversion rates for today's most efficient processes range between 30 and 35%; UCC has not disclosed its actual rate.)

Only one compressor is used to feed the fluidized reactor (see flow diagram). Total horsepower is cut from about 13,000 hp to about 2,000 hp. This power requirement is equivalent to that of the booster that feeds the main compressor in high-pressure processes. Overall, energy costs relative to conventional tubular-reactor processes have been reduced 75%, says UCC.

Besides the energy savings, UCC claims important capital-cost advantages. A gas-phase facility, says the firm, would take up only one tenth the area of a tubular-reactor installation. The fluid-bed vessel is vertical and uses much thinner steel. This, coupled with the lack of monomer-stripping equipment and pelletizing machinery, makes for a 50% overall reduction in capital costs. An incidental advantage is that the equipment operates at less-severe conditions, so it usually can be ordered off-the-shelf, resulting in less cost and shorter lead-times.

WEIGHT-REDUCING PLAN—Unspecified comonomer is added along with ethylene to help lower the molecular weight of the polymer. This has been a standard practice of PE producers for years, although it has been difficult to lower resin density much below 0.925 g/cm^3 via comonomer addition without creating undesirable effects on other product properties.

UCC says it can achieve 0.918 to 0.920 density with a polymer meeting the full range of PE applications, including film. The company credits a major part of this achievement to a new family of catalysts, which in day-to-day operation are used selectively to produce whatever resin properties are desired.

Pennwalt Corp.'s Lucidol Div. (Buffalo, N.Y.), the major maker of catalyst-initiators for olefin polymerization, says its field sources report that UCC's new catalysts are not peroxidics—the type made by Lucidol and used universally for free-radical polymerization of ethylene to LDPE.

Peroxidics are, in fact, the cornerstone of Lucidol's business. Last year, the firm announced a new continuous process for making them—a process it will use to expand its production to meet the entire U.S. and Canadian markets.

News of UCC's new process and its nonperoxidic catalysts could conceivably affect Lucidol's long-range market prospects. But the division has no plans to put off its expansion; groundbreaking is set for the first quarter of 1978.

LICENSING EFFORTS—UCC intends to use the low-pressure process to sharply raise its domestic LDPE capacity by 1982. By that time, the firm should have 40% of its total LDPE capacity devoted to gas-phase LDPE. The immediate plan is to build a 300-million-lb/yr unit at Seadrift, Tex., by 1980, and to add a 150-million-lb/yr plant, also at Seadrift, by 1982. A 500-million-lb/yr unit at a still-undecided location elsewhere in the Gulf Coast region is also planned by 1982. Meanwhile, the company is seeking licensees for the process.

UCC's desire to license the technology has raised a lot of questions within the industry. The firm counters these with the argument that it needs more than one gas-phase producer in the market to lure customers. UCC points out that its LDPE differs from conventional resin, and that processors would be too uncomfortable relying solely on one source of supply.

PROPERTIES ARE CRUCIAL—In effect, questions about resin properties, and how the UCC product works at plastics processing plants, account for most of the industry concern. The consensus is that the process will succeed or fail not on the basis of its capital- and energy cost savings but on the acceptability of product properties.

Conventional LDPE comes in pellet form. But UCC's is granular, resembling white laundry detergent. The company says that processors should be able to handle it without major modifications to their own equipment.

The resin reportedly meets all LDPE grades, including injection molding, film, wire and cable, blow molding, pipe, and roto-molding. In the film-grade category, products are offered for both blown-film and slot-cast types. The injection-molding materials have exceptional stress-cracking resistance. According to UCC, pipe grades have a burst strength 25% higher than the best high-pressure product.

INDUSTRY REACTION—UCC's timing has caught at least two U.S. producers and perhaps a half-dozen foreign resin makers—some of these new to LDPE—in the middle of preconstruction expansion plans. Speculation abounds that some may rethink their designs.

In the U.S., Northern Petrochemical Co. (Des Plaines, Ill.) has been thinking of building a 220-million-lb/yr autoclave LDPE unit in Texas, using technology licensed from Britain's Imperial Chemical Industries, Ltd. B.J. Andersen, vice-president of Northern Petrochemical, says that his firm has no plans to change in midstream now, although it has held discussions with UCC. The biggest unknown, Andersen says, is uncertainty about polymer properties. So far, he has not received enough material for a thorough evaluation. The company would be a reluctant licensee anyway, he adds, because this would

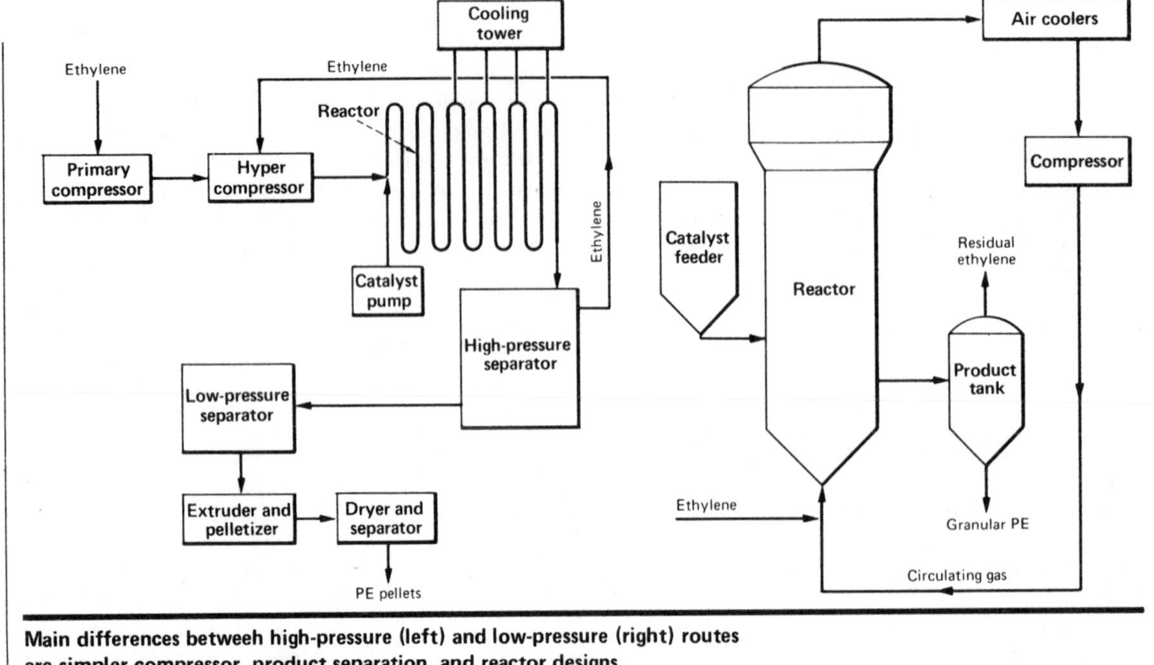

Main differences betweeh high-pressure (left) and low-pressure (right) routes are simpler compressor, product separation, and reactor designs.

mean devoting one fifth of its capacity to a process that makes a seemingly unique product.

Exxon Chemical Co. waited no more than two weeks after the UCC announcement to disclose plans to go ahead with tubular-reactor technology licensed from Dart Industries, Inc., but modified with Exxon's proprietary alterations. The company will build a 300-million-lb/yr unit at Mont Belvieu, Tex.

There is no possibility of switching process design, Exxon says in a prepared statement, although the firm intends to learn more about the UCC process to better assess its potential place in subsequent Exxon expansions. The statement adds, "It is not obvious that [UCC's] new process is fundamentally unique. Also, our initial impression is that the products will be supplements rather than wholesale substitutes for conventional LDPE resins."

OTHER CONTENDERS—BASF is potentially a close competitor of UCC, because it employs a gas-phase process in some polyolefin applications—e.g., polypropylene. However, there are major differences between these two techniques, not the least of which is in the catalyst system. Vessel design is also different: UCC has a fluidized bed supported by a gaseous feedstock admitted through the reactor bottom, whereas BASF stirs its polymer particles to keep them well agitated. Since one process handles PE and the other makes polypropylene, heat duties are much different, as are heat-removal methods.

The ROW joint-venture subsidiary started its first gas-phase polypropylene unit, a 27,000-metric-ton/yr facility, in 1969. A new 50,000-metric-ton/yr unit based on an improved version of the process, yielding a higher percentage of isotactic polymer, is now starting up—also at ROW's Wesseling location. Elsewhere, ICI began construction last summer of a 120,000-metric-ton/yr BASF-developed polypropylene process at Rozenburg, Netherlands. ICI is also building a pilot plant at Teesside, U.K., to study the process further.

Little is known about BASF's ability to modify its process to make LDPE. The company says only that its semicommercial unit at Wesseling may examine variations to the high-density polymer produced there. On the other hand, UCC seems certain that its gas-phase LDPE process can switch to polypropylene production. In fact, the firm has already done this in the pilot plant, according to Heinn F. Tomfohrde, president of the firm's Polyolefins division.

Dow Chemical says that it has worked on and off with low-pressure LDPE technology over the past 15 years. Its new process employs solution chemistry and a new catalyst. The company shies away from gas-phase methods, possibly because this would mean taking a license from UCC. Cost of the new process, according to a Dow spokesman, would not be much different from the rock-bottom cost of the UCC route.

Dow has apparently taken the same development track as UCC—modifying a commercial-scale HDPE unit. A company spokesman claims that Dow can provide commercial quantities of the new low-pressure resin within a few months' notice. The firm expects to begin selling in 1978.

Du Pont of Canada declines to comment on its process, but claims that it is unique. It was developed chiefly so that the Sarnia unit could serve the whole spectrum of the small Canadian market. Last July, Du Pont of Canada more than doubled its capacity, bringing the total to 450 million lb/yr of PE, both low- and high-density.

Phillips Petroleum's process, a variation of its chromium-based-catalyst route for HDPE, has been used to make LDPE of a density as low as 0.925. At the present time, however, 0.940 is more typical of the firm's lower-density polymer, which is mainly used for wire and cable. The company has 15 licensees around the world for the basic HDPE technology, some of which are also privy to LDPE knowhow. However, none is actually making LDPE resin.

Little is known about the Mitsui technique, other than that it is a liquid-phase route that claims energy and capital-cost savings.

John C. Davis

Water-dispersed polyolefins recovered as microfine powder

Low-energy technique turns out small and spherical resin powders that serve especially well for use as textile impregnates, glass-bottle coatings and plastics additives.

☐ The world's first commercial plant to produce microfine powdered polyolefin resin was brought onstream smoothly last March by U.S.I. Chemicals, div. of National Distillers and Chemical Corp. (99 Park Ave., New York, NY 10016).

Located at U.S.I.'s Tuscola, Ill., chemicals complex, the $2.5-million facility can make 10 million lb/yr of the product, tradenamed Microthene F. Offered in high- and low-density polyethylene and in ethylene–vinyl-acetate copolymer, Microthene F consists of extremely fine spherical particles (see photo), averaging about 20μ and having a very narrow size distribution.

A decade-old semiworks unit at U.S.I.'s Cincinnati, Ohio, research laboratory is now on standby for Microthene F product-development tasks. U.S.I. has several patents on the process (the main process patent is U.S. 3,432,483). Recently, inquiries have been received from firms reportedly interested in applying the technology to produce toner chemicals for use in electrostatic copying.

U.S.I. claims significant pluses for its aqueous-dispersion process, compared with competing ways of making thermoplastic powders. High-shear pulverizing of resin pellets, for example, is said to require considerably more energy, as well as to produce larger, irregularly shaped powders with a broader size distribution. Processes that dissolve pellets in solvent and then precipitate a powder, U.S.I. continues, also use more energy, and encounter solvent-removal problems. Even previous processes based on aqueous dispersion are less attractive because of residual dispersant in the product, extra processing steps, and restriction to low-density polyethylene feed.

DISPERSION REACTION—The continuous process highlighted at Tuscola (see flowsheet) starts off by passing polyolefin pellets through a standard extruder. The molten resin mixes with a recirculating stream of surfactant/water, and enters the lower section of an agitated, externally heated reactor.

The surfactant, or dispersing agent, is a block copolymer of ethylene oxide and propylene oxide, containing at least 50 wt.% of the former. Several specific types are used, depending on the polyolefin and on desired product characteristics. For each 100 parts of polymer, 2–30 parts of surfactant are needed, and 0.8–4 parts of water.

Dispersion takes 5–12 min to achieve, and occurs at about 220°C and 250 psi (again, conditions vary with product type).

Slurry withdrawn from the top of the reactor is cooled to about 150°F, and then flows through a control valve for letdown to atmospheric pressure. A rotary vacuum drum fil-

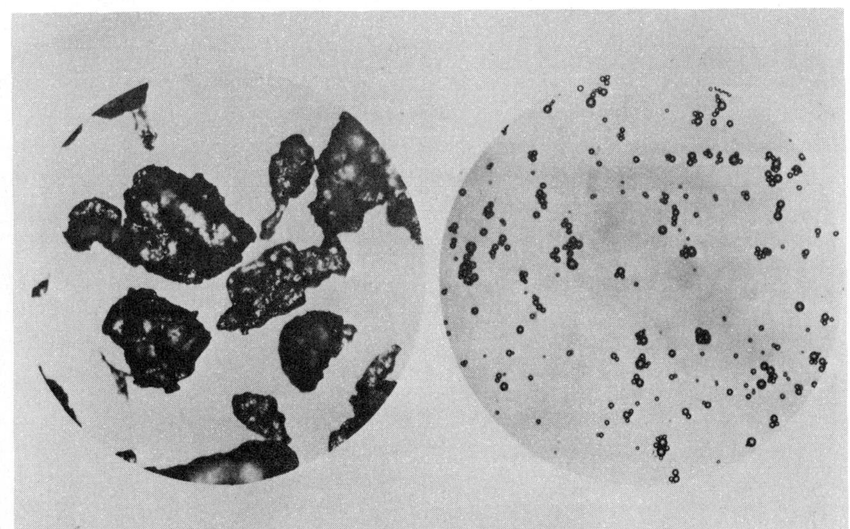

Pulverized powders (left) are large and irregularly shaped; Microthene F (right) is small and spherical.

Originally published August 16, 1976

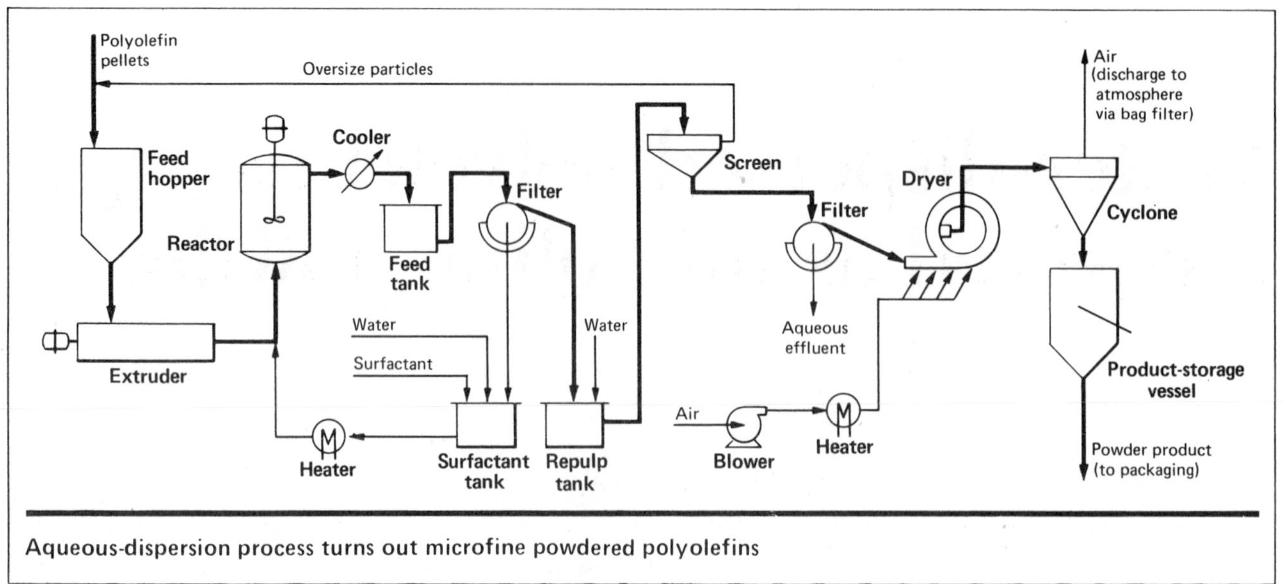

Aqueous-dispersion process turns out microfine powdered polyolefins

ter with string discharge next separates a water/surfactant stream from the polymer solids. The former recycles to the reactor.

PRODUCT FINISHING—In a repulp tank, the polymer solids undergo a simple water wash for removal of residual dispersing agent, down to a typical product level of 0.2–0.3 wt.%. A pair of vibrating-deck screens remove oversize particles for return to the reactor.

The product slurry is again filtered. The separated solids, at about 20% moisture, enter a special dryer where they are entrained in a turbulent circulating-loop flow of hot air. The air enters at about 210°F; flowrate is set for a 110-150°F polymer outlet temperature.

The dried polymer is finally removed from the air flow in a cyclone, and sent onward to storage and packaging.

PROCESS REQUIREMENTS—Typical utilities consumption, per pound of Microthene F product, includes: steam (at 65 psi), 1.8 lb; electricity, 0.50 kWh; cooling water, 12.5 gal; and process water, 0.60 gal. About 10% of the surfactant feed to the reactor is fresh makeup material. Just two operators are needed (one for the process itself and one to oversee packaging), and a part-time supervisor.

There are no corrosive fluids to handle, so 304 stainless steel serves as the major material for equipment, including the reactor, filters, screens and dryer. Tanks and hoppers can be fabricated from 304 stainless steel, aluminum or fiber-glass reinforced plastic.

Only minimal routine maintenance is anticipated—there are no scheduled periodic shutdowns. Switching from one polyolefin product to another, however, can require a brief downtime of eight hours or so to clean equipment.

Air exiting from the product-recovery cyclone is sent to a bag filter for removal of residual fines before discharge to the atmosphere. The water stream from the second filter contains small amounts of surfactant washed off the polymer, so U.S.I. injects it into deep wells. However, the firm is now evaluating reverse-osmosis treatment to recover the surfactant and ease disposal of the stream.—NRI

Filters separate powder from process liquids (left); vibrating screen separates oversized particles for recycle (right).

Section IX
SOLID WASTE PROCESSES

Flash Pyrolysis turns refuse into fuel oil

As an alternative to the troublesome and expensive land disposal of municipal refuse, pyrolysis can harvest liquid fuel as well as glass and metals.

George M. Mallan, Occidental Research Corp.

☐ This October, just 14 months after construction began, a demonstration plant designed by Occidental Research Corp. should be starting up, to turn 200 tons/d of municipal refuse into 7,200–9,600 gal/d of fuel oil. Ferrous metals, aluminum and glass will also be recovered.

The plant, on a 5.3-acre industrial site in El Cajon, Calif., is basing its design on a 4-ton/d pilot unit that has pioneered the Flash Pyrolysis process at Occidental's research center in La Verne, Calif., since 1969.

When fully operational, the facility will be turned over by Occidental to the County of San Diego, which is putting up $2.0 million of the project cost. The U.S. Environmental Protection Agency is providing $4.2 million, and Occidental $8.5 million. Procon Inc. acted as subcontractor on some aspects of engineering design, procurement and construction.

FEED PREPARATION—The first steps in the process involve shredding and classification of refuse into inorganic and organic components. Typically, the raw refuse is just over 50% organic (more than half of that being paper); ferrous metals (7%), glass (7%), and moisture (25%) constitute most of the remainder.

The primary objective is to deliver dry, finely divided, essentially inorganic-free material to the pyrolysis reactor. While the reactor itself is virtually unaffected by inorganics, these materials degrade the quality of the residual char, and increase maintenance costs of the secondary shredder.

In the primary shredding step, a horizontal-shaft hammermill reduces the size of raw refuse (as large as 6 ft) to particles about 3 in. The shredded material passes through a belt-type magnetic separator that extracts the ferrous metal (about 140 lb/ton of refuse) for sale as scrap. The balance goes to storage. This assures uniform flow to the air classifier following, and permits the shredder to operate just 8 h/d while the rest of the plant (except glass recovery) operates 24 h/d.

From storage, the material is fed to a zig-zag air classifier, where a rising column of air catches lighter materials and carries them out the top. The heavy fraction drops out the bottom, for further preprocessing and then transfer to the glass-recovery section (described later).

The lighter fraction, containing mostly organics, is dried (using heat generated in the dryer from burning of pyrolysis offgases) to about 3% moisture. Screening then separates large particles, which go to the secondary shredder, from fine particles, which go to a vibrating air table that further separates light organics from dense metals and glass.

The secondary shredder is a plate attrition mill. The feed passes between counter-rotating disks, and eventually disintegrates into very fine particles (nominally 14 mesh). This material combines with light fines from the air table to form the 1-3-lb/ft^3 organic feed to pyrolysis.

GLASS RECOVERY—Heavy particles from the air classifier, and dense fines from the air table, both pass into the glass-recovery system.

How the properties of pyrolytic fuel oil and No. 6 fuel oil compare

	Pyrolytic fuel oil	No. 6 fuel oil
Carbon, wt. %	57.0	85.7
Hydrogen, wt. %	7.7	10.5
Sulfur, wt. %	0.2	0.7–3.5
Chlorine, wt. %	0.3	–
Ash, wt. %	0.5	0.05
Nitrogen, wt. %	1.1 }	2.0
Oxygen, wt. %	33.2 }	
Heating value, Btu/lb	10,600	18,200
Specific gravity	1.30	0.98
Density, lb/gal	10.85	8.18
Heating value, Btu/gal	114,900	148,840
Pour point, °F	90*	65–85
Flash point, °F	133*	150
Viscosity at 190°F, Ssu	1,150*	340
Pumping temperature, °F	160*	115
Atomization temperature, °F	240*	220

*Assumes oil containing 14% moisture

Originally published July 19, 1976

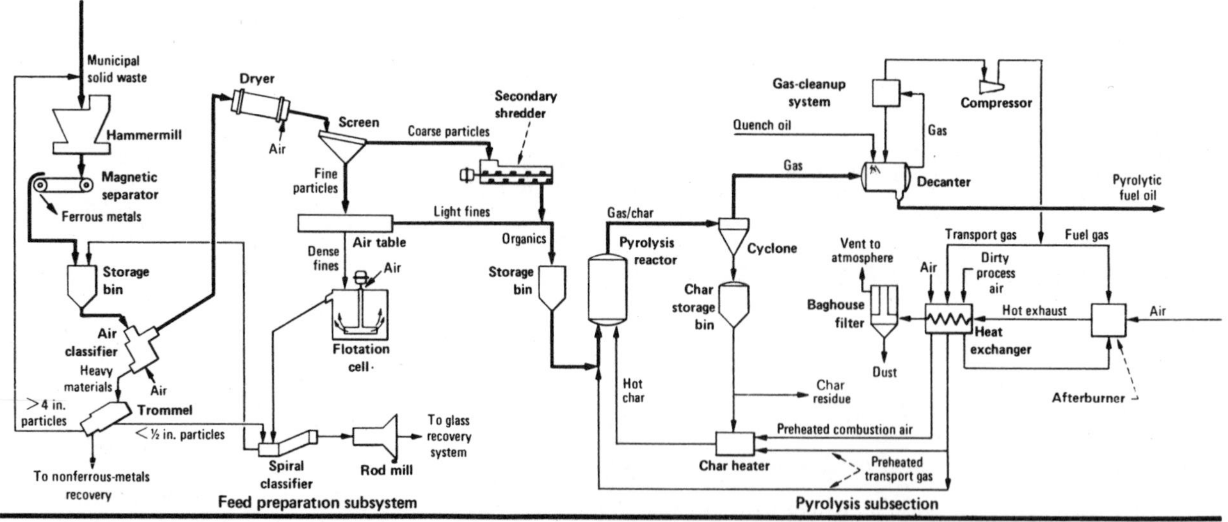

Occidental flash-pyrolysis process includes extensive feed-preparation subsystem to separate organics from other municipal waste, and pyrolysis subsystem to generate liquid fuel oil from the organics

The former however, first undergo additional size classification and reduction in a trommel, spiral classifier, and rod mill.

A series of froth-flotation cells separates the glass product (about 110 lb/ton of refuse) from impurities, using a proprietary set of reagents. By recirculating both float and sink fractions several times, 85–90% of the glass in the glass feedstock is recovered. (Only 75% of the total glass in the raw refuse is recovered, however, due to losses in preceding parts of the process.)

PYROYSIS SUBSYSTEM—A stream of inert carrier gas, meanwhile, charges the organic feedstock into a pneumatic conveying line leading to the bottom of the pyrolysis column, which is a vertical, refractory-lined steel pipe. There the organics break down almost instantaneously into a mixture of vapors (pyrolysis oil, gas, water) and solid char residue. The reaction takes place at 900°F and 15 psi, and is slightly exothermic.

A cyclone removes char from the vapors. In a key process step, part of the char recycles to the reactor as a heat-carrier medium, after external heating to 1,400°F. Excess char is cooled for disposal or sale.

The char-free process vapor is quickly quenched to 175°F by an oil spray, to prevent reactions that would reduce fuel-oil yield. The resulting pyrolytic fuel oil is decanted from the quench oil, and then passed into product storage.

The remaining vapor stream, containing byproduct water and combustible gases, goes through a series of cleanup steps and is compressed. Part serves as an oxygen-free medium for pneumatically transporting both the organic feedstock and the char into and through the pyrolysis reactor. Some burns in an afterburner, along with air from the zigzag classifier and dryer. The combustion products pass to a heat exchanger to preheat combustion air for the char heater, reactor transport gas, and miscellaneous gas streams. The cooled combustion gases are cleaned in a baghouse filter and vented.

PYROLYSIS PRODUCTS—Typical properties of the pyrolysis fuel oil are shown in the table, in contrast to a No. 6 fuel oil. Total product yield is about 40 wt.% of the pyrolyzer feed.

The oil from the plant will be sold to San Diego Gas & Electric, for mixing with residual fuel oil and burning in a pair of steam-generating boilers. The pyrolysis oil will represent about 2% of the daily fuel need. Since the oil contains organic acids, the utility will closely monitor operation for any corrosive effects on construction materials.

The pyrolysis gas, consumed in-process, amounts to about 27 wt.% of the pyrolysis feed and has a heating value of about 550 Btu/ft^3. A typical component breakdown is: moisture, 0.1 mol%; carbon monoxide, 42.0%; carbon dioxide, 27.0%; hydrogen, 10.5%; methylene chloride, 0.1%; methane, 5.9%; ethane, 4.5%; and C_3–C_7 hydrocarbons, 8.9%.

ENVIRONMENTAL ASPECTS—The last pyrolysis product—water at 13%—presents the only potential pollution problem for the Flash Pyrolysis process.

In a large-scale refuse-conversion plant, the water would be condensed from the pyrolysis offgas and sent to secondary sewage treatment or evaporated. Since some of the pyrolysis oil itself is water-soluble, the water carries various organic compounds and has a relatively high chemical oxygen demand.

(In the demonstration plant, the water is removed and sent to a landfill site.)

Wastewater from the glass-recovery operation will be discharged into the sanitary sewer system after removal of settleable solids

All gas streams recirculate within the system, except for those that serve as combustion air in the afterburner, which are made odor-free during combustion and filtered before venting. Residual solid wastes total about 42.6 tons/d, and will initially be sent to landfill (markets are being sought).

In order to fully assess the environmental impact of pyrolysis oil on emissions from its boilers, San Diego Gas & Electric plans a 21-mo analysis program including both laboratory and field tests of numerous flue-gas components.

The author

George M. Mallan is director of the resources-recovery program of Occidental Research Corp., a subsidiary of Occidental Petroleum Corp. (1855 Carrion Road, La Verne, CA 91750). He has worked in the resources recovery field for seven years, and has written a number of articles on the subject. He holds a B.S. and Ph.D. in chemical engineering from the University of Southern California.

Sludge pyrolysis schemes now head for tryouts

One process makes activated carbon and uses it internally to clean up wastes. Another technique mixes trash with sludge to improve the heat value of the pyrolysis gas.

☐ Next month, piloting of a novel sludge-pyrolysis process should begin at Huntington Beach, Calif. And a decision should come imminently on another, different pyrolysis system planned for St. Paul, Minn.

This activity reflects an uplift in the fortunes of that technique for sludge disposal. Richard C. Bailie of the U. of West Virginia (Morgantown, W. Va.), explains, "Communities are turning to pyrolysis because other thermal schemes, like incineration and oxidation, are dirty words. The general public considers these [latter] concepts as being pollution prone, so government officials look at pyrolysis as a process that won't become a political football."

However, Bailie voices caution, ". . . one has to be careful to put pyrolysis in its proper place in the solid-waste-disposal picture." Pyrolysis may carve a niche for itself, but no great shift to the route should occur because it hardly rates as a general panacea. For instance, notes Bailie, "in pyrolysis schemes for municipal sewage you will have to burn all your pyrolysis product just to keep the unit operating."

Stanley L. Katten, president of S. L. Katten & Associates (San Pedro, Calif.), amplifies on the limitations of pyrolysis: "While there is some net energy in raw sewage sludge dewatered to about 20%, pyrolysis of primary sludge is a break-even proposition at best. As sludge is treated, it has even less heating value; and pyrolysis of secondary sludge is a waste of time."

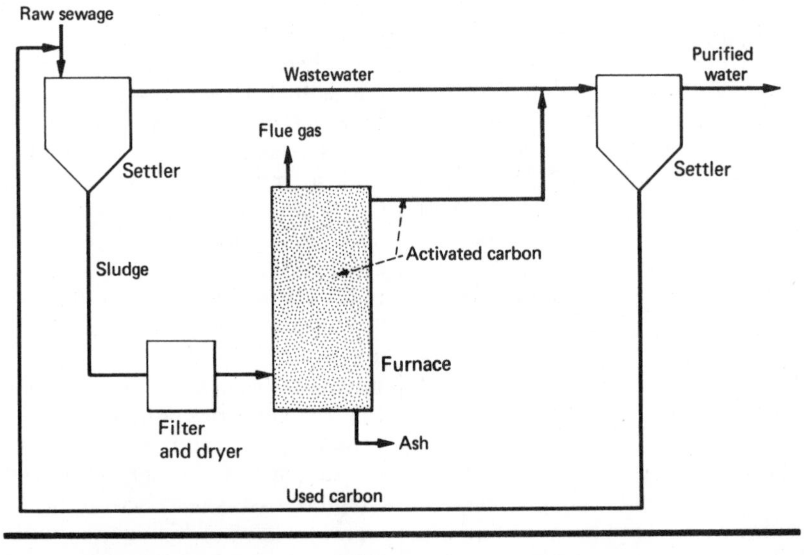

JPL process uses internally produced carbon to clean up sewage Fig. 1

Nevertheless, pyrolyzing sludge to produce a fuel gas and a usable carbon is getting increasing interest. Already two alternative approaches are surfacing. They either feed on sludge alone, or mix it first with municipal refuse before processing.

CALIFORNIA UNIT—Orange County (Calif.) sanitation districts are jointly building a 1-million-gal/d pilot plant at Huntington Beach. The unit will test a technique that uses carbon produced to clean the raw-sewage feedstream. Funded by about $2 million in grants, the facility should start up in January. Trials should last about 6 mo.

The process' developer, Jet Propulsion Laboratory (JPL) of Pasadena, Calif., already has successfully operated a 10,000-gal/d mobile unit at Huntington Beach for more than a year.

According to JPL, the process destroys odors and virtually eliminates sewage solids, while removing heavy metals. In addition, wastewater from the technique should better U.S. Environmental Protection Agency standards for discharge to the ocean—the treated stream should contain less than 10% of the organics in the feed, and boast a biochemical oxygen demand of under 30 mg/l average over a 30-d period. And the route should make enough 300-Btu/std ft^3 gas to cover about 80% of its fuel needs.

Moreover, for a full-scale, 175-million-gal/d facility, the new process should cut capital costs by about 25%, compared with a conventional activated-sludge/water-roughing-filter/solids-incineration setup, says JPL. The savings stem mainly from the elimination of the activated-sludge digesters, and to a lesser extent from replacement of centrifuges by a filter-press.

CARBON RECYCLING—In the process, as shown in the flow diagram at left, raw sewage first goes to a settler. The heavier portion of the stream flows out of the bottom of the vessel to a plate-and-frame filter press, which dewaters the waste to about 40% solids. A flash dryer then brings the solids content up to 90%.

The stream next goes to a calciner—at the Huntington Beach pilot plant, the horizontal, cylindrical furnace will measure 30 in. dia. by 41 ft long, and will accommodate up to 800 lb/h of the wet solids feed. Nat-

Originally published December 8, 1975

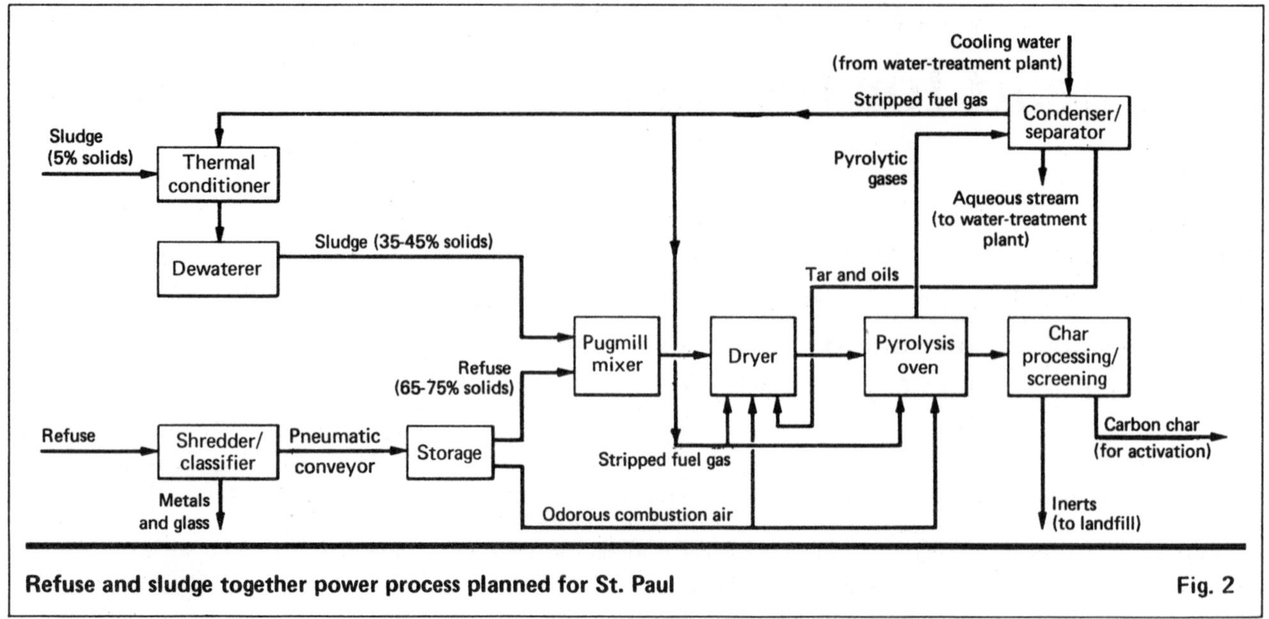

Refuse and sludge together power process planned for St. Paul **Fig. 2**

ural gas, digester gas from a sewage-treatment plant or the pyrolysis gas itself can provide the fuel for externally heating the solids to as high as 1,560°F. Combustion in the absence of oxygen transforms the solids into carbon, which is activated *in situ* by steam. Heavy metals, such as nickel and cadmium, and some of the carbon are removed as a dry, black, odorless ash from the bottom of the calciner and discarded. Pyrolysis gas comes off the top of the reactor.

The activated carbon passes to a second settler, where it joins wastewater from the first settler. The carbon adsorbs contaminants from the water, which then goes through an anthracite/sand polishing filter. Meanwhile, the carbon is recycled to the first settler, where it aids settling. Finally, it returns to the calciner for reactivation, closing the loop.

The system can handle any wastewater impurities removable by activated carbon. Other contaminants, such as chromium, can pose problems. Orange County will get around difficulties with chromium by making sure that it is held to a tolerable level in the feed sewage.

JPL does not want trash in its feedstream. The firm's feeling is that the higher ash content of refuse, which it says is 70–75% after pyrolysis and steam activation, prevents the formation of good activated carbon for water treatment.

MINNESOTA PLANS—In contrast, a $22-million sludge-pyrolysis scheme proposed by the Metropolitan Waste Control Commission of Minneapolis/St. Paul (Minn.) relies on the addition of refuse to the sludge to boost its heat value.

The commission is planning a unit at St. Paul to handle about 180 tons/d of primary and secondary sludge, and 360 tons/d of trash. This amounts to about 6.5% of the solid-waste load expected in 1980 in the region. A hearing on approval of the plans by the governing Metropolitan Council of the Twin Cities will take place about now (in early December). If the unit gets an okay, and construction starts next year, the facility should go onstream in 1979.

The move toward pyrolysis originates from a need for fuel. In the spring of 1974, the commission received notification that its supply of natural gas would be gradually curtailed and finally cut off by the end of 1978. Use of coal would cost about $3 million annually at present prices. Moreover, coal would pose significant storage problems, says the commission, and processing and boiler facilities would total about $12 million more than the pyrolysis plant. Oil was ruled out as too expensive to use.

Primary and secondary sludge alone would not provide enough heating value to support the pyrolysis operation, so the commission has opted for a 1:2 ratio of sludge:trash. This will yield about 10,060 std ft^3/ton of gas containing 620 Btu/std ft^3, as well as some combustible tars and liquids—in all, a major portion of the fuel needed.

MIXING OF WASTES—In the St. Paul flowscheme, detailed above, the sludge is thermally conditioned, and dewatered to about 35–45% solids by plate-and-frame presses and a porous-metal roll-type filter (already proven in the pulp-and-paper industry). The stream next goes to a pugmill mixer where it joins trash fed in by a belt conveyor.

The homogenized sludge/trash mixture from the pugmill then enters a rotary-drum dryer. Moisture content is cut to approximately 10%. The dried mixture goes to a pyrolysis oven, which is a rotary, indirectly fired hermetic unit using pyrolysis gas as fuel. Heating to 1,250°F produces gas and a carbon char.

Gas passes to a venturi scrubber and a tray tower for cleaning. Quenching with water causes tar and oil vapors to condense. Separated from the water, they eventually are used to help fuel the dryer.

The clean pyrolysis gas is piped to a compressor and dryer, where the stream is brought to a pressure of 125 psig and a dewpoint of -40°F.

Meanwhile, solids from the oven are discharged through an airtight chamber to a cooler. Screening then separates inerts, which go to landfill, from carbon, which passes to an activation circuit prior to use in the water-treatment plant.

Guy E. Weismantel

Section X
TAR SANDS, SHALE OIL PROCESSES

Giant oil-sands plant comes onstream

The largest oil-sands extraction and processing plant officially opens this month. Despite some startup woes in the new unit, planners feel that the technology is sound; construction of a third facility is being considered.

Hugh C. McIntyre, McGraw-Hill World News, Toronto

☐ Set to officially come onstream this month is Syncrude Canada Ltd.'s (Edmonton, Alta.) $2.1-billion plant at Mildred Lake, Alta. Situated 30 miles north of Ft. McMurray, in the heart of the Athabasca oil-sands deposit, the facility will produce 129,000 bbl/d of "synthetic crude oil" from 300,000 tons/d of oil-sand fed.

The unit joins an already-working smaller plant—belonging to Great Canadian Oil Sands (GCOS)—in operation since 1968. A third facility—the same size as Syncrude's—has been proposed by Shell Canada Resources Ltd. (Calgary, Alta.). Shell is presently seeking partners to share the estimated $4-billion investment, and the firm expects to make its decision to proceed or not next year. The conceptual design of the plant, to be located near Ft. McKay, Alta., is very similar to that of Syncrude's facility.

Despite startup problems at the Syncrude site and persistent operating problems at the GCOS unit—due, in major part, to the extremely abrasive nature of the oil sands—Canadians feel that oil-sands exploitation is one of the more promising answers to the problem of diminishing supplies of conventional crude oil. And some observers feel that higher oil prices are expected to improve the process economics even further.

SIMILAR TECHNOLOGY—All three plants share the same basic processing route: hot-water extraction of bitumen from the sand; separation of bitumen by solution with naphtha, followed by centrifuging; upgrading the bitumen by coking; subsequent hydrodesulfurization of liquid product from the coker; and reblending into a "synthetic crude" that can be pipelined and refined in existing refineries.

In the Syncrude plant, the four main process areas are (see flowsheet): primary separation, secondary separation, upgrading, and desulfurization. The plant is designed for a feedstock containing 11 to 12.5% bitumen; although some pockets of oil sand contain as little as 7 or 8% bitumen, blending of sands will maintain oil-sand feed of the proper richness.

On entering the primary-separation plant, oil sand from the "dump pocket" (the terminus of the conveyor belts bringing sand from the open-pit mining operation) is mixed with about seven parts (by volume) of hot water, 22% sodium hydroxide, and steam, at a temperature of about 100°C. The mixture is then tumbled in four parallel 18-ft-dia. by 100-ft-long tumblers and pumped over into four primary-separation tanks, also in parallel. Here the bitumen collects on the surface as a wet primary froth and overflows into catchtrays around the periphery of the separation tanks. The middlings are transferred to secondary-separaton tanks, and the sand/water tailings are discarded.

In the secondary-separation tanks, the middlings are aerated, and bitumen again surfaces as froth and is removed, and, again, sand and water are sent to the tailings pond, via an API separator.

The two separation stages remove

Upgrading oil-sand bitumen: Syncrude's inputs and outputs

Inputs	
Oil sand (11 to 12.5% bitumen)	300,000 tons/d
Water	125,000,000 U.S. gal/d
Steam (900 lb/in^2)	88,800,000 lb/d
Electricity	260 MW
Natural gas (for hydrogen-plant feed)	40,000,000 ft^3/d
Outputs	
Upgraded synthetic crude oil	129,000 bbl/d
Elemental sulfur	1,000 tons/d
Coke	200,000 long tons/d
Tailings*	125,000,000 U.S. gal/d

*Tailings consist of sand, clay, water, and about 0.4 to 0.5% (by weight) residual bitumen.

Originally published September 11, 1978

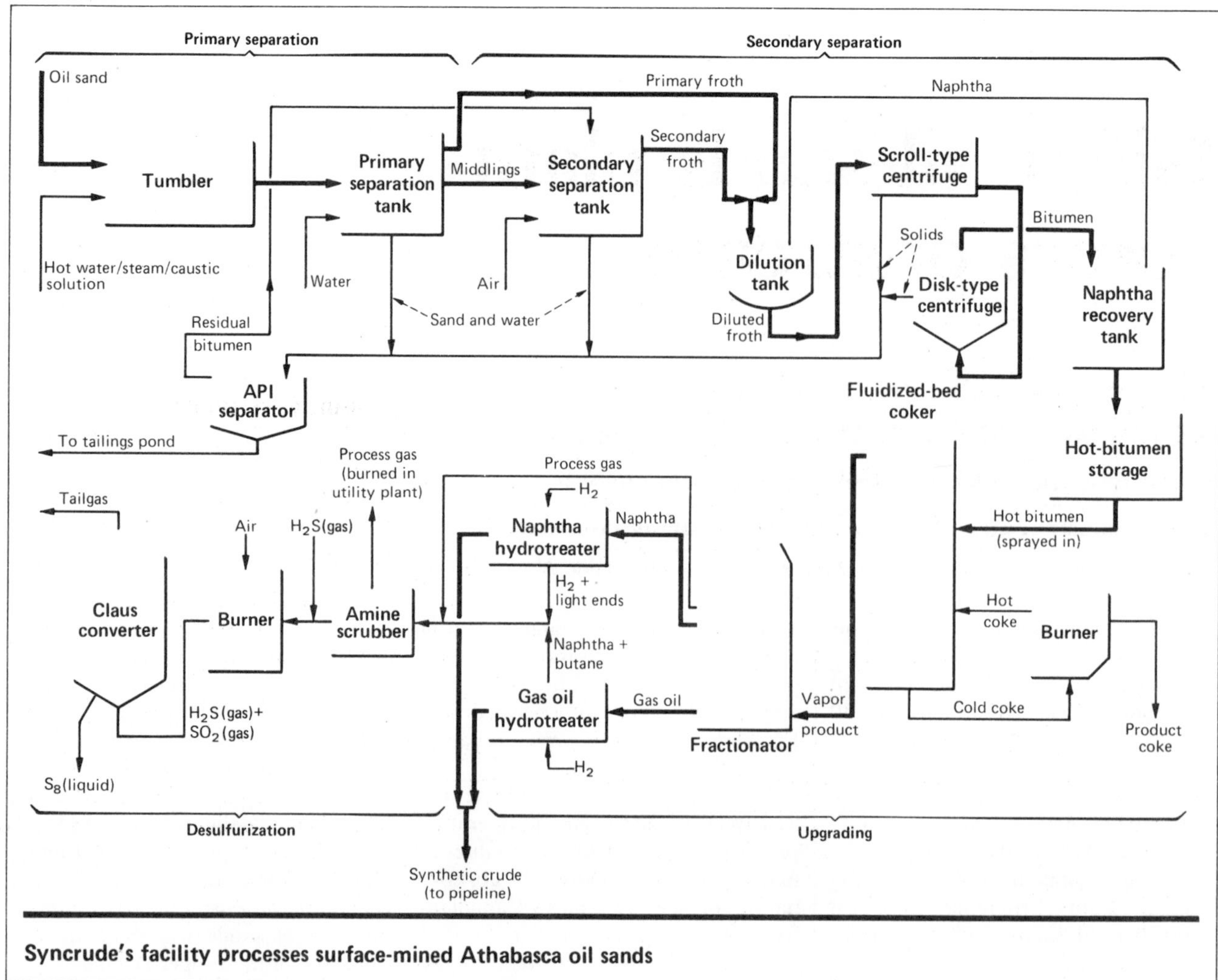

Syncrude's facility processes surface-mined Athabasca oil sands

about 95% of the bitumen contained in the sands. Water from the tailings—after the solids have settled—will eventually be reused for slurrying and washing.

Having been recombined into one stream, the froth goes to dilution, where naphtha from the process is added. The diluted stream then goes to a battery of 22 first-stage, low-speed, scroll-type centrifuges followed by 22 second-stage, high-speed disk centrifuges where remaining sand and water are removed. (Dilution of the froth with naphtha decreases the specific gravity of the bitumen, which is close to 1.00, allowing for more-efficient separation.)

Naphtha is removed from the bitumen by flash distillation, recovered, and recycled to the dilution unit. Pure, hot bitumen goes into interim storage having a capacity of about 2 million bbl, a 17-d capacity that gives great flexibility to separation and upgrading operations. Thus, Syncrude can work the separation and upgrading portions of its process train independently, should one end or the other have to be shut down. In contrast, the GCOS plant—which does not have such an interim storage area—has the two operations very closely coupled.

UPGRADING—The Syncrude plant uses fluid cokers for upgrading the high-gravity (8 to 10 degrees, API) bitumen. The 210-ft-high, 30-ft-dia. cokers are the largest vessels in the process train.

In the units, preheated bitumen is sprayed onto a bed of fluidized coke at 40- to 50-lb/in² pressure and at a reactor temperature of about 510°C (approximately 650°C in the coke burner). The fluidized coke acts as a heat-transfer medium and is replenished by the coking of about 10% of the bitumen feed.

The GCOS plant, on the other hand, uses a delayed coker—that is, a batch-type vessel that holds the reaction mixture in place during coking. As a result, coke buildup must be cut away from the coker walls daily (by using high-pressure water streams), an operation that produces around 2,500 tons of high-sulfur coke per day for GCOS.

While the older GCOS plant is permitted by Alberta environmental authorities to use the produced coke as boiler fuel, the 5%, or higher, sulfur content of such coke precludes this option for the Syncrude facility. Instead, the newer plant uses natural gas and desulfurized fuel gas from the process as its energy source.

Output from Syncrude's cokers is fractionated into four streams: (1) overhead gases, which are desulfurized and used as fuel; (2) a middle, liquid cut of naphtha; (3) a bottom, liquid fraction of gas oil; and (4) combustion products consisting of coke and carbon monoxide, which are burned to raise 900-lb/in² process steam.

In a slightly different arrangement, GCOS separates its liquid product into

Photo: Bechtel Corp.

Syncrude's cokers stand 210 ft tall against the night sky

three streams instead of just two, including a middle distillate having kerosene-like properties. The proposed Shell Canada process, as conceptually designed, will gasify the heavy bottoms of its fractionator output, adding it to the overhead fuel gas, while producing naphtha and gas oil cuts similar to those of Syncrude.

Coker output is fractionated in each variant of the upgrading process only because the sulfur contained in the bitumen is strongly segregated in the gas-oil cut, which cut requires more-severe conditions for hydrodesulfurization than do the other fractions. Thus, Syncrude can hydrotreat its naphtha cut at 500 lb/in^2, while the gas oil must be subjected to pressures of 1,500 lb/in^2 in a nickel-steel reactor having walls 4 to 5 inches thick.

Hydrotreatment saturates the heavy olefinic bonds, stabilizes the product, and removes nitrogen and sulfur compounds. Hydrogen is obtained by steam-reforming natural gas.

Sour process gas in the Syncrude plant is treated in an amine scrubber, the hydrogen sulfide separated there being burned in a Claus converter to produce elemental sulfur.

MIXING—The final stage of synthetic-crude production is the mixing of the various fractions in order to obtain a pipelineable product of 30 to 35 API gravity (Syncrude is aiming for a 33.9-gravity product), having a sulfur content of less than 1%.

In its plant, GCOS stores the three synthetic-crude components separately, mixing them in inline blenders just prior to pipelining the product 300 miles to Edmonton, and thence to the crude's ultimate destination via the Interprovincial Pipe Line. Syncrude, and presumably Shell, will mix naphtha and gas-oil components prior to shipment, and will pump the "syn crude" directly into the line.

DUAL PROCESS LINES—Unlike the GCOS operaton, the Syncrude plant, when in full production, will essentially have two parallel process trains, with crossovers at several crucial points, such as at hot-bitumen storage, centrifugation, and hydrodesulfurization—so that the various units can operate at different rates, and trouble areas may be bypassed.

The first Syncrude train is scheduled to deliver approximately 54,000 bbl/d of product to customers by the end of this year. The second train is scheduled to come onstream this December, seven months ahead of schedule, boosting production to around 100,000 bbl/d. After debottlenecking, full production of 129,000 bbl/d is anticipated, by 1982.

COSTS AND QUESTIONS—As mentioned, the capital cost of the Syncrude facility is $2.1 billion, or about $16,000/daily bbl of output at full capacity. It is estimated that an additional $40 million will be required each year, primarily for continuing development of the mining operatons. Estimated operating cost for the plant is $9.50/bbl at full capacity.

One of the major uncertainties in future technology is the best coking route to follow. Imperial Oil Ltd. (Toronto, Ont.), Exxon's Canadian affiliate, is applying for permission from the Alberta Energy Resources Conservation Board to build a massive upgrading plant for the slightly lighter (10 to 12 API gravity) heavy oil, and will be using the proprietary Flexicoking process developed by Exxon. That route has the advantage of gasifying most of the coke produced in the process. The gas can be desulfurized using existing technology, allowing greater energy recovery from the process. (Syncrude is presently stockpiling the coke it produces, there being no immediate market for it.)

Shell may be able to develop its own coking variant by the time its process line is set up—probably no more than five years from now.

A further option may be the bypassing of coking operations. Hydrocarbon Research, Inc., a subsidiary of Dynalectron Corp. (McLean, Va.), has announced it has successfully demonstrated the economic feasibility of directly converting and upgrading bitumen via the already commercial H-Oil route. In a continuous 30-d pilot-plant run, the firm says it produced 30,000 gal of 31.5-API-gravity, 1.3%-sulfur, residue-free crude oil from Athabasca sands.

Shale Oil—Process Choices

Here's a rundown on some of the routes being groomed for production of shale oil. One of these, the Tosco II process, is already considered proven and is on the verge of commercialization.

With the time ripe for commercial development of the huge oil-shale reserves in the western U.S., engineers are closely scrutinizing a number of process candidates.

All of the alternative technologies share the basic function of retorting oil-shale rock at high temperature, so that the kerogen material it contains will thermally decompose to shale oil and gaseous products. Most do the job in aboveground equipment after the shale is mined and crushed; a few work *in situ*.

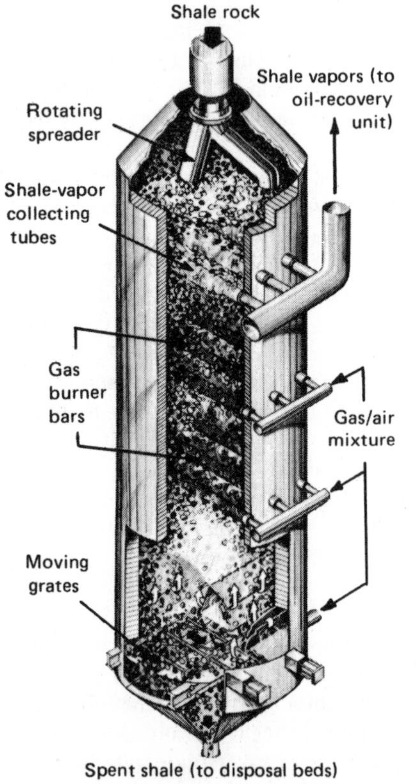

RETORT from Development Engineering uses countercurrent gas/solids flow.

Originally published May 13, 1974

The shale-oil processes described below can be divided into four broad categories. The first three are distinguished by the method used to supply the 600,000-800,000 Btu of heat needed to retort a ton of oil shale:

Solids-to-solids heating—The Oil Shale Co. (Tosco) and Lurgi-Ruhrgas routes.

Gas-to-solids heating with internal gas combustion—U.S. Bureau of Mines, Development Engineering Inc. and Union Oil Co. of California routes.

Gas-to-solids heating with external heat generation—Development Engineering, Union Oil, Petrobras and Institute of Gas Technology routes.

In situ retorting—Occidental Petroleum Corp. route.

Tosco—Industry observers generally agree that the "Tosco II" process is the most advanced. Construction startup by late 1973 is in fact planned for a 50,000-bbl/d plant.

In the process (see diagram on next page), crushed shale is preheated with hot flue gas, and then fed with hot ceramic balls (about ½-in.-dia.) to a horizontal rotating retort. Shale-oil vapors are generated in the retort, and weakened spent shale is pulverized. The vapors are drawn off and fractionated to fuel gas, naphtha, gas oil and resid.

The ceramic balls are meanwhile size-separated from the spent shale by a trommel, a heavy-duty rotating cylinder with many small holes punched in its shell. The balls are reheated (by burning some of the fuel gas produced) and reused.

Tosco has kept final details of the retorting drum confidential (such as ratio of balls to shale, length-to-diameter ratio, rotational speed and inclination angle). Operating temperature, though, is about 900°F, and a pressure slightly above atmospheric prevents admittance of air. The commercial plant slated will have six 11,000-ton/d retorts.

"Our yields are not a strong function of oil-shale grade," says Tosco vice-president Louis Yardumian. All but 5% or so of the carbon in the shale is recovered (the remainder is left on the spent shale as coke). A ton of oil shale typically produces about 252 lb of oil, 31 lb of C_1-C_4 hydrocarbon gases, and 11 lb of C_5, C_5+ gases. Carbon dioxide and monoxide, and hydrogen and hydrogen sulfide are also generated.

Tosco designed its own dilute-phase lift pipe (DPLP) for preheating the oil shale. This is said to offer exceptional efficiency in gas-to-solids heat transfer. Proper residence-time distribution with respect to particle size is the key. Several of the DPLPs are used in series—gas-solids flow is countercurrent between each pair, but cocurrent in any one DPLP. The equipment could prove useful, Tosco says, for other solids preheating, prilling and drying tasks in industry. (Tosco's retort has tested successfully for pyrolysis of scrap tires, coal and municipal wastes.)

Lurgi-Ruhrgas—This second solids-to-solids route (see diagram), developed by Germany's Lurgi and Ruhrgas, features heat carriers of sand, coke or spent shale.

These are mixed with incoming oil shale in a sealed screw-conveyor that acts as the retort. All material discharges into a surge bin. The vapor stream is drawn off to a condenser to produce gas and liquid fractions. Some spent shale is withdrawn from the bin as waste; the remaining solids are routed to the lower end of a lift pipe. Here, they are heated by combustion of the coke on the spent shale (and by burning supplemental fuel if needed). The hot solids are then used once again for retorting.

Bureau of Mines—Dubbed the Gas Combustion process by BuMines, this second-category route to shale oil ran into problems controlling flow of fine shales when last tested back in 1966-67, but is still considered a commercial possibility.

A refractory-lined vertical kiln acts as the retort. Regulated by a special grate discharge-mechanism, crushed shale (¼-3 in) falls sequentially through four functional zones, for preheating, retorting, combustion and cooling. (There are no physical separations between zones.)

Some of the product gas is diverted to the bottom end, and flows upward to pick up heat from the exiting spent shale. About one-third of the way up, more product gas (along with combustion air) is injected. This gas, and coke on the shale in that zone, burn to heat the shale immediately above to the retorting temperature. Shale vapors partially condense at the top of the kiln, for withdrawal as a gas/liquid product. Feed shale is preheated there as the vapors condense.

Development Engineering—A similar technique is used in a process from Development Engineering, the operating arm for a group of 17 firms comprising the Paraho Development Corp. Test runs at 500 tons/d are coming up shortly.

The retort (see diagram) boasts as its key component a patented discharge grate that improves the downward flow of shale over the unit cross-section. It also provides for a careful distribution of incoming gases. The 8½-ft-dia. test retort has multilevel fuel-gas injectors, but uses carbon residue as its primary fuel.

An alternative design will also be studied, in which pyrolysis heat is generated by burning carbon residue in an external vessel, and transferred to raw shale by a circulating stream of product gas. Although the capital expense increases, product gas remains undiluted by nitrogen from the combustion air.

Union Oil—Moving the oil shale upward through a downward flow of gases is the variant technique being pursued by Union Oil. The firm operated an early prototype in the 1950s and has been making undisclosed modifications since.

The same functional zones appear as in the previous two designs, except that here the inlet zone for oil-shale preheating is at the bottom of the vessel rather than at the top. A reciprocating-piston rock pump pushes the shale feed upward; suction blowers draw down the gases. The vessel is shaped like an inverted cone and is open at the top.

Just like Development Engineering, Union Oil is working on an alternative configuration with external heat generation.

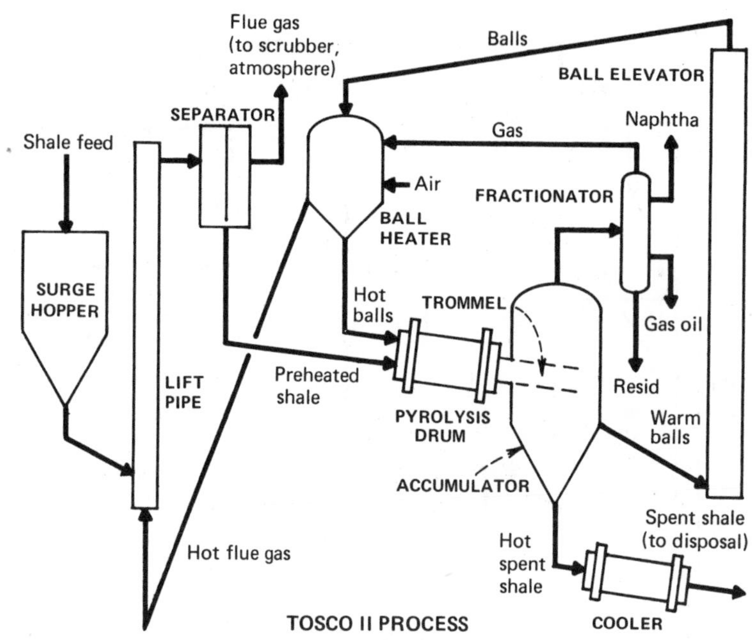

TOSCO II PROCESS

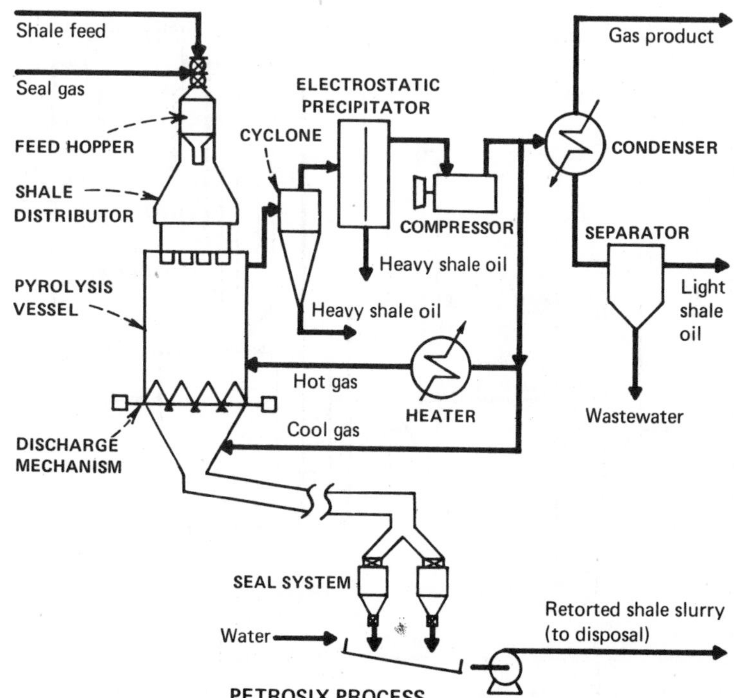

PETROSIX PROCESS

Petrobras—Another gas-to-solids heating method with external heat generation has been running on a semiworks scale (2,200 tons/d) since 1970. Developer and operator is Petrobras, Brazil's national oil company. Called Petrosix, it is said to be similar to the Bureau of Mines process, except that already-heated recycle gas is injected into the shale

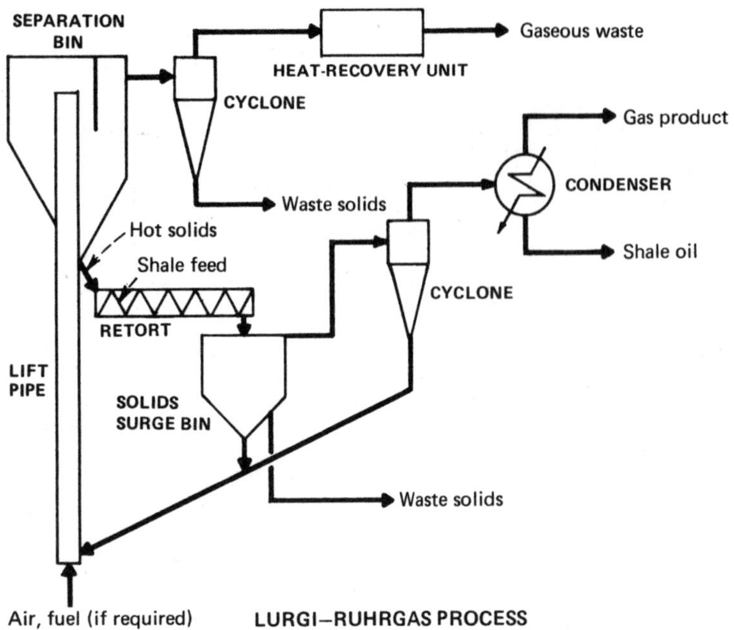

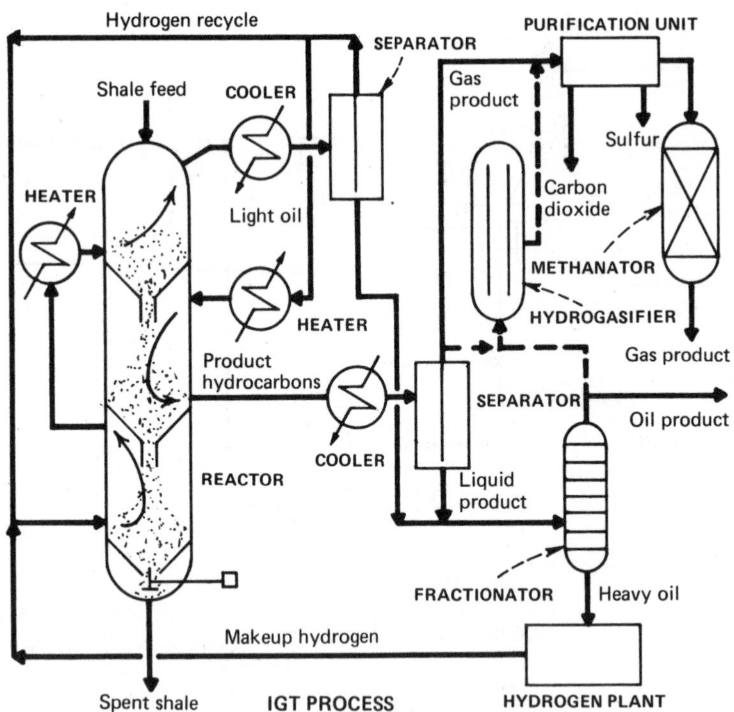

bed rather than fuel gas and air.

The technique (see diagram) handles oil-shale solids up to 15 cm. The external heater operates on fuel gas, shale oil, or the residual carbon on spent shale. An unheated recycle gas is fed into the lower part of the retort to recover sensible heat from the exiting shale, which is pumped in slurry form to disposal.

IGT—Tests at a relatively low 24-tons/d are planned by the end of this year for yet another route using external heat generation, this originated by the Institute of Gas Technology. Its key feature is use of a moderate-pressure hydrogen atmosphere.

A vertical vessel internally divided into three zones is utilized (see diagram). Shale passing downward is prehydrogenated and preheated in the top zone, hydroretorted and/or hydrogasified in the middle zone, and cooled in the bottom zone.

A pair of hydrogen streams are included in the heat-transfer scheme. One picks up some heat from the spent shale and, after additional heating, is used to preheat incoming raw shale. The second is externally heated, and passed through the middle retorting zone in order to hydroretort the shale kerogen. Problems with condensation of shale oils, and plugging of the preheat-zone bed, are reportedly eliminated by having the first stream bypass the middle zone.

By varying reaction temperature, IGT says, the product slate can be varied from mostly liquid (when temperatures are under 1,200°F), to a larger output of gas (over 1,200°F).

Occidental—Though several techniques for *in situ* retorting have been tested over the past years, the one most recently unveiled and gaining the most attention today has been developed by Occidental Petroleum Corp. through its Garrett Research and Development Co. subsidiary.

"In its simplest form, the process consists of three basic steps," says Richard D. Gridley, Garrett's manager of oil-shale research, "a limited amount of conventional mining, blasting of the remaining shale to form the retort, and retorting in place using air and underground combustion."

Air circulates downward through the rock pile. Combustion initiates at the top with the aid of an outside fuel source. The heat released retorts the top shale to yield shale oil, some gas and residual carbon (the latter becomes part of the fuel supply). Offgas is recirculated to control oxygen concentration and burning rate. Oil drains to the bottom of the retort, where it flows through a concrete trough into a sump and then to underground storage.

Gridley says that 70% shale-oil recovery appears feasible, not much below the anticipated yields of aboveground systems.—GEW/NRI#

The authors wish to acknowledge the contributions to this article by numerous individuals and firms, with special thanks going to Thomas A. Hendrickson, Cameron Engineers, Inc., whose presentation at the 7th Oil Shale Symposium provided a particularly large part of this material.

Soviet oil-shale processes offered for U.S. licensing

An audience of potential U.S. licensees was treated last December to operating details and economic data for a pair of Soviet processes for retorting oil shale.

James H. Prescott, Regional Editor, Houston

☐ In a bid against U.S. and German technologies for unlocking the huge energy content of oil-shale reserves in the U.S. and other countries, the Soviet Union is putting up for license two processes it has used itself for decades.

First details of the technologies came out at a Dec. 3, 1975, symposium in Tulsa, Okla. Soviet experts were on hand to explain the process intricacies; and Resource Sciences Corp. (Tulsa), U.S. representative of the U.S.S.R. licensing group V/O Licensintorg and sponsor of the symposium, provided an economic outlook geared to U.S. operations.

The two processes are dubbed the Kiviter and the Galoter routes. The Kiviter process handles coarse oil-shale particles from 1 to 5 in., in a single multifunction retort. The Galoter process accepts fines under 1 in., in an equipment setup including heaters, mixer, and separators, as well as the actual retort.

SOVIET EXPERIENCE—Plants in the U.S.S.R. have commercially processed kukersite, an oil-shale rock found in Estonia and other Baltic areas, for at least 30 years. At first, low-Btu town gas was made for use in Leningrad and Tallin. With natural gas now available in the Baltic, the highly phenolic hydrocarbons in the oil shale are recovered mainly as a raw material for chemicals.

The Baltic oil shale is much richer, at 40–50 gal of oil per ton of rock, than Colorado oil shale, at 25–30 gal/ton. In fact, about three fourths of the 25 million tons/yr of oil shale currently mined in Estonia is burned directly as boiler fuel (like coal) for generating electricity, and only the remaining one fourth is retorted for fuel oil and gas or for chemical feedstocks.

The largest units in service are rated at 250 metric tons/d for the Kiviter process, and at 500 metric tons/d for the Galoter process. A 1,000-metric-ton/d Kiviter retort is under construction, and a complex of four Galoter lines, each 3,330 metric tons/d, is similarly being built, with operation of the first two scheduled for 1978.

KIVITER RETORT—As shown in Fig. 1, the Kiviter retort for coarse oil shale is a downflow, internal-combustion-zone unit, similar to the U.S. Bureau of Mines' gas-combustion retort, the Paraho Development Corp.'s (Grand Junction, Colo.) retort, and Union Oil Co.'s rock-pump retort (except the last is upflow).

A 35-m-high, 11-12-m-I.D. Kiviter unit can process 1,100 tons/d of wet Colorado oil shale (30 gal of oil per ton) and recover 680 bbl/d of shale oil plus 1 million ft^3/d of low-Btu gas. Operating temperature is 800°C; operating pressure is 80 in. water. Thermal efficiency reportedly runs 73% when handling Colorado oil shale.

The Soviet experts recommend a battery of ten 1,100-ton/d units arranged in parallel, with a common oil condensation and recovery section. Battery-limits capital cost of the setup would be $26 million (just under $4,000 per bbl/d of shale-oil product), Resource Sciences figures, but this excludes mining, transportation, process-water supply, spent-shale disposal, engineering, and cost of capital.

Operating costs for labor, utilities and maintenance are estimated to be $2.38/bbl of shale oil. But if the low-Btu gas byproduct (approximately 150 Btu/ft^3) is credited at $1.44/million Btu, its value directly offsets the operating costs.

GALOTER PROCESS—The Galoter process handles oil-shale fines under

Kiviter oil-shale process uses a single multifunction retort — Fig. 1

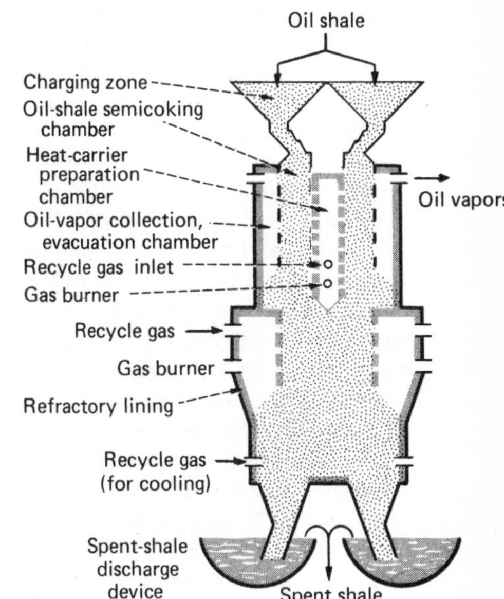

Originally published February 2, 1976

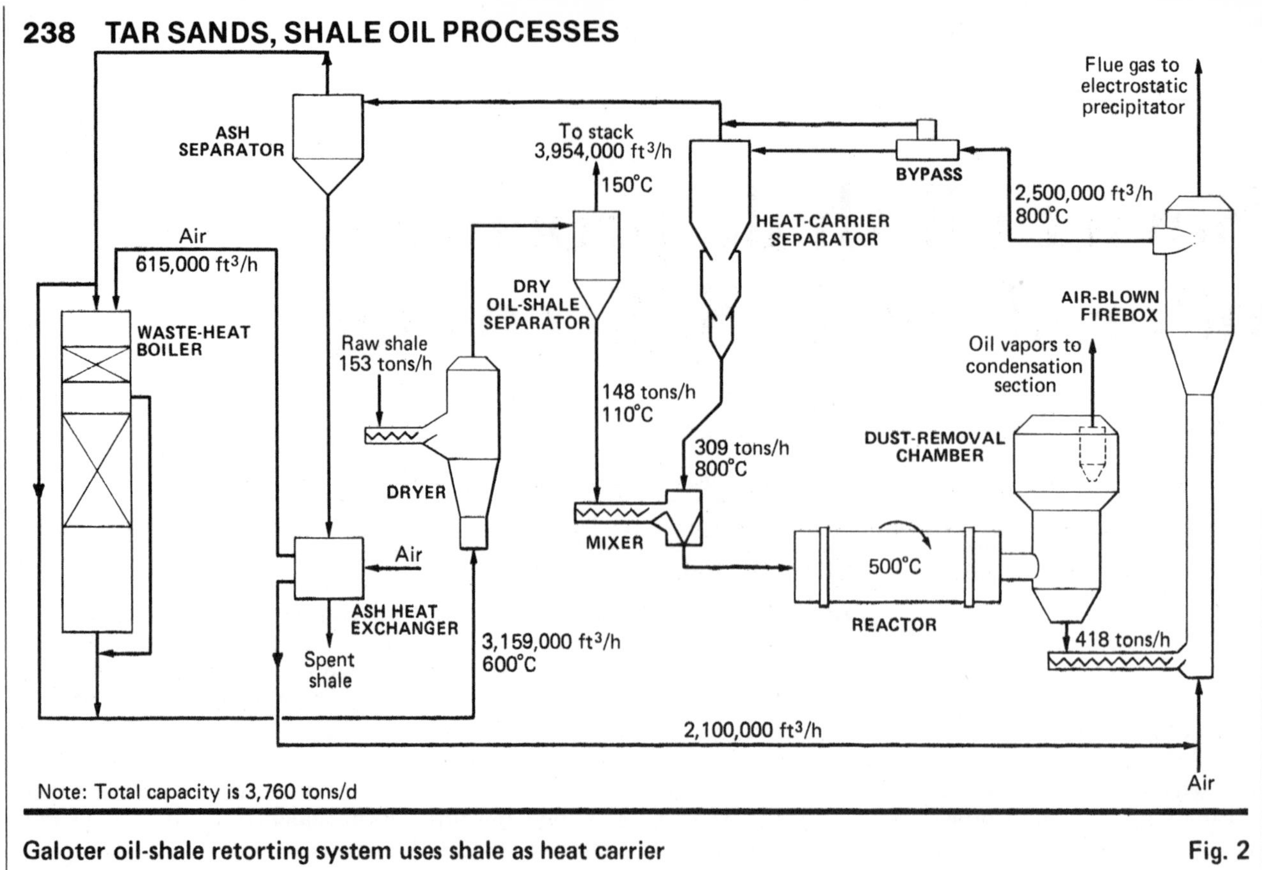

Galoter oil-shale retorting system uses shale as heat carrier **Fig. 2**

1 in., but preferably in the ⅛–¼-in. range. It uses hot oil shale as a heat carrier, and is thus akin to the Lurgi-Ruhrgas process* (which uses spent shale or, for non-shale tasks, hot sand as heat carrier) and The Oil Shale Corp.'s Tosco-II process* (which uses hot ceramic balls).

Selected material flowrates and operating temperatures for the Galoter process are shown in Fig. 2, for a 3,670-ton/d system. The process yields 155 lb of shale oil per ton of Colorado oil shale, together with 800 ft³ of high-Btu gas (1,265 Btu/ft³). Thermal efficiency is 81%.

Capital cost for the recommended Galoter system—three parallel retorting lines feeding a common oil-condensation and -recovery system, with an overall output of 4,880 bbl/d of shale oil—is just under $23 million (excluding offsites). Principal cost component is the reactors (or retorts), at $15 million. Other major outlays include $1.58 million for conveyors, $1.28 million for air blowers, $0.77 million for refractories, and $0.75 million for pumps.

Direct operating costs for the system amount to $2.99/bbl of shale oil product. The 10,000-kW power consumption of the process accounts for $1.48/bbl, half the total. Labor for operation, maintenance, supervision and services adds another $1.02/bbl (including salaries, fringe benefits and labor overhead). Other maintenance costs at $0.38/bbl, and steam and water supply at $0.11/bbl, round out the $2.99/bbl figure.

Resource Sciences points out that more than one third of the overall heat output is in the form of by-product gasoline, high-Btu gas and high-pressure steam (10 lb, 56 lb and 209 lb, respectively, per ton of oil-shale feed). If credit is taken at $10/bbl for gasoline and $1/1,000 lb of steam, a price of $0.72/million Btu for the gas would offset these operating costs. Shale-oil credits remain to account for other costs, including return on investment.

U.S. INDUSTRY COMMENTS—U.S. companies represented at the Tulsa symposium included Exxon, Shell, Union, Tosco and Amoco . . . the number of attendees topped three dozen. But since they hadn't had time to review the data thoroughly, the U.S. experts asked not to be identified with specific comments.

Nevertheless, one of the most frequent remarks was that the Soviet equipment seems pilot-scale by U.S. standards, which aim at single retorts handling 10,000 tons/d or more of oil shale. In fact, U.S. firms consider a 50,000-bbl/d plant as perhaps the smallest commercial size, which would consume about 75,000 tons/d.

In partial reply, Resource Sciences says that the Kiviter retort could actually process three times its rated capacity. But, many of the attendees noted, it would still have a poor specific throughput compared, e.g., to the Paraho retort. Overall, there seems to be less U.S. interest in the Kiviter process than the Galoter.

U.S. experts look favorably on Soviet-claimed run-lengths of 6–8 mo for the Kiviter process and 5–6 wk for the Galoter, and consider the long years of operating experience a plus. Though they don't like the 4–5% carryover of fine particles into the Galoter shale oil, many of the experts believe that better design of the dust-removal chamber will overcome carryover.

The consensus on capital costs is that they would be much higher than the figures presented. Despite efforts to adapt Soviet experience to U.S. conditions, extra costs will likely mount up for air pollution control, waste disposal, water procurement, and the general difficulty of operating in remote U.S. mountain areas.

* The Lurgi-Ruhrgas and Tosco-II processes are detailed in *Chem. Eng.*, May 13, 1974, pp. 66-69. See also *Chem. Eng.* Dec. 8, 1975, p. 81.

Section XI
URANIUM PROCESSING AND ENRICHMENT

Uranium enrichment methods detailed at meeting

The just-held AIChE Annual Meeting revealed that design and construction of nuclear enrichment plants continues in line with mushrooming enriched-uranium demand. Some developers unveiled details about up-and-coming techniques.

Philip M. Kohn, *Assistant Editor*

☐ A three-part symposium at AIChE's 69th Annual Meeting held Nov. 28 to Dec. 2, 1976, at Chicago, Ill., addressed the problem of gearing up to meet world demand for enriched uranium. Attendees reviewed potential reactor-fuel supply problems, but they also heard details about some of the most active work in enrichment technology.

For example, Centar Associates (Fairfield, N.J.) revealed that the centrifuge process for enrichment saves about 25% more energy than was previously reported, using only about 1/13th the power consumed by gaseous diffusion. Centar says that while diffusion technology is mature, the theoretical limits of centrifugation have not been reached.

Spokesmen for firms developing "nozzle" enrichment methods described their processes: For example, Institut für Kernverfahrenstechnik (Karlsruhe, West Germany) detailed its technique that will be used in a 180,000-SWU*/yr demonstration plant in Brazil, while UCOR—Uranium Enrichment Corp. of South Africa (Pretoria)—disclosed particulars about its Helikon process.

CENTRIFUGE ROUTE—Alan M. Fishman, Project Manager for Centar Assoc. outlined the firm's plan for a 3-million-SWU/yr plant using the centrifuge method.

Cylinders containing UF_6 will feed into a high-pressure system, where their contents will be evaporated into a centrifuge cascade (see Fig. 1) at roughly 200°F. The cascades will be grouped into several trains, each train having a capacity of 110,000 SWU/yr. All trains, said Fishman, will conform to one of three standardized designs, yielding product that will assay within ranges surrounding either 1.7%, 2.45% or

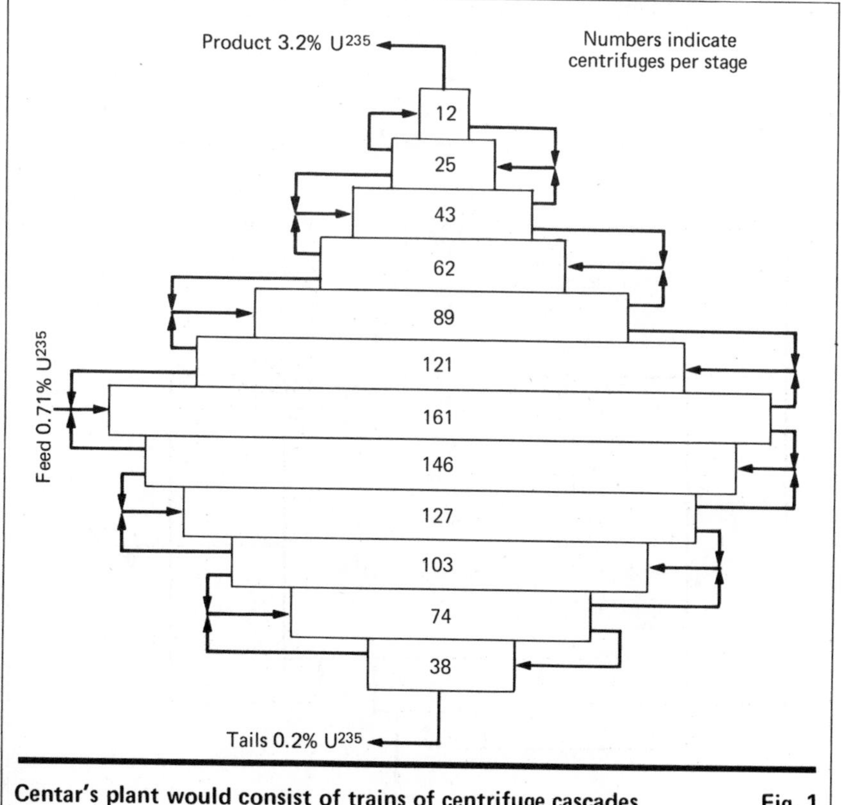

Centar's plant would consist of trains of centrifuge cascades **Fig. 1**

* Since feedstocks and output streams of enrichment plants can be of various U^{235} concentrations, it is impractical to define capacity in quantity-of-material units. Instead, separative work units (SWU) measure the effort spent to separate a given quantity of uranium feed of a given assay into an enriched stream and a depleted stream. For example, 4.3 SWU are required to produce 1 kg of enriched uranium containing 3% U^{235} when the feedstock is 5.5 kg of uranium containing 0.71% U^{235}.

Originally published January 31, 1977

3.2% U^{235}. However, centrifuge flexibilities will permit the cascades to operate over a variable range of assay gradients such that they could meet finished-product specifications that differ from the three design concentrations. Plant product from each train will be cold-trapped out at low pressure, and then transferred into product cylinders as a liquid.

The plant would take in 7.2 million kg of UF_6 feed annually to produce almost 1.3 million kg of enriched uranium. Fishman stated that the total power level required for such a plant is 60 MWe, the cost of which power should amount to \$3–\$4/SWU. However, he pointed out that the centrifuges themselves are the largest cost elements, being approximately 50% of the capital cost and 30% of the operating cost. Thus, said Fishman, the method of procurement of the original-plant centrifuge complement, the method of repair and recycle of failed centrifuges, and the frequency and nature of centrifuge failures, all have major impacts on enrichment costs.

NOZZLE PROCESSES—In the nozzle processes, enrichment is brought about by centrifugal force applied to a gaseous UF_6/H_2 mixture. The heavier isotope, U^{238}, follows an outer flowpath, while lighter U^{235} follows an inner route, resulting in separation and concentration of both.

Erwin Becker, Wilhelm Bier, Klaus Schubert, Rolf Schütte (speaker), and Dieter Seidel of Institut für Kernverfahrenstechnik and Ulrich Sieber of Steinkohlen-Elektrizität AG (Essen, West Germany) authored a paper detailing aspects of German nozzle technology. Schütte noted that the best results for practical application of the nozzle concept were obtained in systems in which a band-shaped jet of UF_6/H_2 was deflected by a curved wall. At the end of the deflection, the flow is split into its lighter and heavier fractions by a knife-edged skimmer blade.

The Institute's process arranges the separation stages into cascades, which produce a net upward transport of the light carrier gas (H_2). To prevent enrichment of the hydrogen in the cascade, light fractions are processed in UF_6-recycle facilities (see Fig. 2), comprising a special separation-nozzle stage backed by a low-temperature freeze-out heat exchanger. The separation nozzle recycles 80–90% of the UF_6 content, while the remaining UF_6 desublimes at temperatures of –30 to –120°C in the heat exchangers.

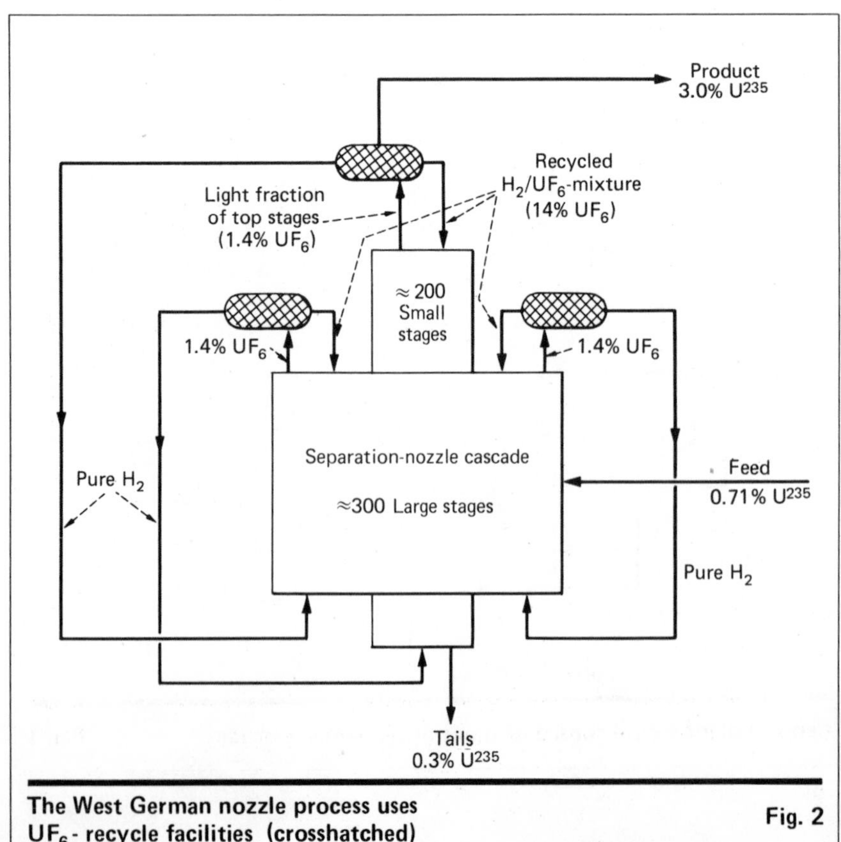

The West German nozzle process uses UF_6- recycle facilities (crosshatched) **Fig. 2**

Schütte noted that developmental work has demonstrated that the nozzle process is both reliable and commercially feasible, and that the basis for actual construction has been established.

SOUTH AFRICAN VERSION—W. L. Grant (speaker), J. J. Wannenburg, and P. C. Haarhoff explained that UCOR's Helikon nozzle variation is quite effective in separating isotopes using small cuts, i.e., highly asymmetrical flows with respect to UF_6 concentrations in the enriched and depleted streams.

In UCOR's process, the feed stream is compressed (the motive force), cooled, and fed into the separating element, where the stream is split into enriched and depleted components. A large number of elements are incorporated into one module, the elements sharing a common axial-flow compressor. The streams from different elements, containing different concentrations of UF_6, are introduced into the compressor in parallel at different points on the inlet circumference. Grant reported that extensive tests have shown that material injected at the inlet travels through the compressor as a narrow band, mixing very little with the other bands that make up the flowstream.

A depleted stream from an element either leaves the module or is introduced into another element of the same module, where it is mixed with an enriched stream from outside that module. Thus, feed material flows in a helical path through each module until it leaves via the most depleted stream of that module. It is this helical motion that gives rise to the term Helikon, Grant noted.

The enriched stream from a given module goes to the corresponding point in the next module. By varying the degree of enrichment over the modules, cascades can comprise modules of fixed size. By extension, the authors pointed out, modules of a given size may be used, within limits, to construct plants of different capacities.

Uranium Processing Update

Several economical methods are available to extract uranium from various ores, and to upgrade recovered material to enrichment-ready form.

NICHOLAS R. IAMMARTINO
Associate Editor

ELECTROLYTIC CELL tested at Mol by Pechiney is rated for 15 kg/h.

The upturn in the uranium demand picture over the past few years has brought with it a corresponding thrust toward development of better processing technology.

Strong-acid leaching routes, for example, can now recover uranium from difficult ores at lower capital and operating costs than can traditional dilute-acid leaching. Developers include the U.K. Atomic Energy Authority (UKAEA) and Pechiney Ugine Kuhlmann (Paris, France).

Electrolytic reduction techniques, one groomed by Japan's Power Reactor and Nuclear Fuel Development Corp. (PNC) in collaboration with Asahi Chemical Industry Corp., and a second by Pechiney, are similarly cost-cutters. These eliminate not one but several of the steps ordinarily used to convert recovered uranium values to uranium hexafluoride .

UKAEA Leaching—The UKAEA strong-acid route, protected by patents in several nations, still seeks its first commercial customer. Testing has been only on a laboratory scale. UKAEA remains optimistic, however—its process excels with the refractory ores that uranium firms will be forced to exploit over the next few years as easier-to-process types dwindle.

Crushing and grinding of mined ore begins the process. After 6N sulfuric acid is sprayed on the ore in a rotary drum to wet the particles to 10-12% liquid, the ore cures in a fixed bed for up to 12 h at 75-95°C. Removal of uranium thus solubilized takes place by percolation of water through the bed, or by reslurrying. Conventional ion exchange or solvent extraction, followed by precipitation and drying, turns out yellow cake (U_3O_8) as a final product suitable for conversion to UF_6.

UKAEA stresses that its technique minimizes crushing and grinding by accepting large 1-2-mm particles (versus a 100μ or less size requirement for dilute-acid leaching). By separating more readily from the pregnant liquor, large particles do away with settling or clarification that would otherwise follow water washing. Static curing is yet another process plus claimed by UKAEA; curing normally takes place under continuous agitation.

UKAEA figures overall capital savings at 20-30% compared with dilute-acid leaching—60% of this from elimination of solid/liquid separation equipment, and the rest split between simplifications in size reduction and curing. Operating cost falls off 30-50¢/lb U_3O_8.

Commercial Already—A 1,200-ton/yr strong-acid-leaching plant, which started up in the Nigerian Sahara in 1970, using a similar Pechiney technology, has been saving some $4/ton ore in producing yellow cake. That 40% reduction brings the average price down to just over $1/lb of U_3O_8. Steps that save still more are reportedly imminent.

Conventional leaching, Pechiney explains, calls for roughly 1,000 l of water per ton of ore. The water carries 50 g/l of acid that actually reacts with the ore, and another 100 g/l that remains free to establish proper chemical potential of the reaction solution. By trimming water addition to just 50 l/ton of ore, free-acid content drops correspondingly to 5 g/l. Acid savings are about two thirds, and water savings even greater.

Pechiney notes that a complicated dry-grinding technique was needed in place of conventional wet-grinding procedures. A rotating cylinder, in which the ore is sprayed with acid, forms the small granules needed for feeding to rotary curing drums. The cylinder avoids turning the ore into an unworkable mudlike mass.

Originally published March 3, 1975

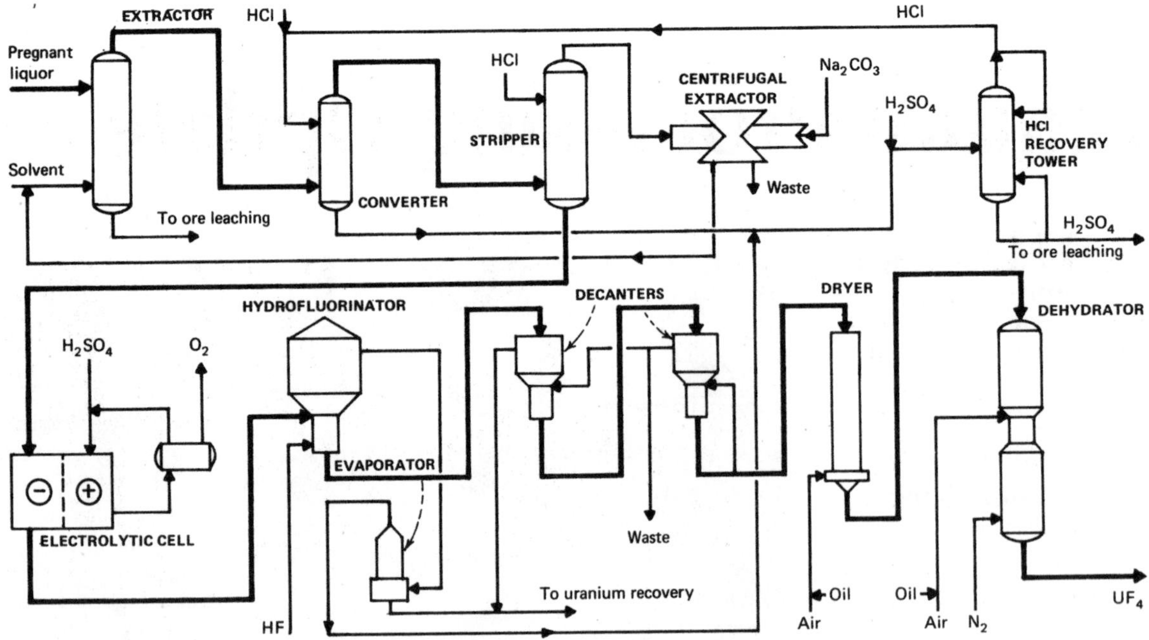

PNC ELECTROLYTIC REDUCTION PROCESS

Since the leaching reactions provide sufficient heat themselves to maintain the Pechiney system at an optimum 60-70°C, external energy sources are eliminated. Also eliminated are oxidizing agents such as manganese dioxide and sodium chlorate, since the acid-soaked granules can react with air for an oxidation effect not possible in a dilute medium. The Pechiney approach can be used for most uranium ores.

Electrolytic Reduction—Japan's socalled PNC technique, tested in a 1-ton/yr pilot plant, skips the production of yellow cake as intermediate from ore to UF_6. The process includes six major steps (see flowsheet): solvent extraction, chloride conversion, electrolytic reduction, hydrofluorination, filtration and dehydration.

A preconcentrated UO_2SO_4 stream from conventional ore leaching (3 g/l uranium) first flows to an extraction tower where tri-*n*-octylamine in a kerosene diluent extracts the uranium. The uranium sulfate complex formed is converted to a chloride complex by 8N hydrochloric acid. The resulting stream for electrolysis contains at least 100 g/l of uranium as uranyl chloride (UO_2Cl_2).

Reduction takes place at room temperature, and features Asahi's cation-exchange membrane and bipolar electrode structure. Reduction of uranyl ion (U^{+6}) to uranous ion (U^{+4}) at the negative electrode yields uranous chloride (UCl_4) at 99.5% conversion efficiency. The membrane prevents reoxidation. Water dissociates at the positive electrode in a sulfuric acid anolyte, evolving oxygen for withdrawal, and hydrogen ion that passes through the membrane to take part in the reduction reaction.

At about 90°C, UCl_4 taken off the cell reacts with hydrofluoric acid to form a hydrated UF_4 precipitate. Separation of the precipitate, drying, and dehydration (at 400°C in inert nitrogen) finish up the process and yield green cake—a UF_4 suitable for further fluorination to UF_6.

PNC estimates that construction and operating expenses are 20-30% lower than conventional technology. Use of low operating temperature in the cell avoids exotic and expensive materials of construction. Current efficiency is a high 70%.

For a 5.6-ton/d UF_4 plant, PNC pegs investment at $4.6 million (1970 dollars, excluding ore preparation), with about half targeted for electrolytic-reduction equipment.

Pechiney Duo—To bypass yellowcake intermediate, Pechiney is grooming not only an electrolytic refining method, but also a wet-chemical approach.

The former has been tested on a semi-industrial 15-kg/h scale, both at both Pechiney's Forez plant in France and the Eurochemic installation in Mol, Belgium. Starting with a uranyl sulfate complex in solvent, at a 15-20% uranium concentration, the electrolysis process first strips off the solvent and puts the uranium material through a further concentration stage. Electrolysis of the concentrated stream takes place in horizontal cells, with mercury cathodes and platinum anodes arranged in a series of three to obtain a reduction efficiency of approximately 99%. The product from electrolysis is precipitated in a hydrofluoric acid medium, and UF_4 granules filtered out.

Advantages cited include simple cell operation (a big plus for service in remote or underdeveloped areas), and electricity usage of only 1.5 kWh/kg of product.

Pechniney's alternative starts with the same uranyl sulfate solution. The scheme consists of adding hydrochloric acid to produce uranyl chloride, stripping off this uranium compound, and finally reacting it in a single vessel with sulfur dioxide, copper and hydrofluoric acid to produce UF_4.

Costs are about the same as for the electrolysis approach. The technique has been tested in a 1-kg/h pilot plant. #

Recovering uranium from wet-process phosphoric acid

Two solvent-extraction processes—one successfully piloted, the other still under investigation—perform a twofold function: conservation of a fuel resource, and removal of a radioactive fertilizer contaminant.

Fred J. Hurst, Wesley D. Arnold and Allen D. Ryon,
*Oak Ridge National Laboratory**

☐ Increasing uranium prices offer the phosphate industry an attractive opportunity to recover uranium as a byproduct of fertilizer manufacture. By removing the material from phosphoric acid, processors would be conserving an important natural resource that could be used to augment the nation's future fuel supply, while removing a radioactive contaminant from fertilizers.

It is estimated that wet-process phosphoric acid produced in 1976 will contain about 3,000 tons of dissolved U_3O_8, and this amount is projected to increase to about 8,000 tons/yr by the end of the century. (Incidentally, sandstone ores currently produce approximately 13,000 tons/yr of uranium, so the amount recoverable from phosphoric acid is significant.)

An independent research organization recently estimated the cost of recovering uranium from phosphoric acid at about \$15/lb. Thus, at present price levels of \$40/lb of U_3O_8, a 1,000-ton/d P_2O_5 plant could yield an annual indicated profit of \$7–8 million (assuming 1 lb of U_3O_8/ton of P_2O_5, a ratio typical of most phosphate-rock deposits).

TWO PROCESSES—Oak Ridge National Laboratory (Oak Ridge, Tenn.) has developed two two-cycle solvent-extraction processes that recover uranium from wet-process phosphoric acid. The DEPA-TOPO, or reductive stripping, process uses the synergistic combination of di(2-ethylhexyl)phosphoric acid (DEPA) and trioctyl phosphine oxide (TOPO) to extract oxidized uranium, while the OPAP, or oxidative stripping, process utilizes a mixture of mono- and dioctylphenyl phosphoric acid (OPAP).

The DEPA-TOPO process has been successfully demonstrated in pilot plant operations at several phosphate plants, and full-scale plants based on the process are presently being designed.

While not as fully developed as the DEPA-TOPO technique, the OPAP process has several advantages over the former. OPAP extracts U^{+4}, the prevailing oxidation state in fresh phosphoric acid, whereas DEPA-TOPO recovers U^{+6}. OPAP is a stronger uranium extractant than is DEPA-TOPO, and is also less expensive (about \$2/gal compared to \$6/gal for the concentrations required). A minor disadvantage of the OPAP system is that the uranium is harder to strip from the solvent, and the stripping itself requires a more-concentrated (54% rather than 32%) acid. Besides, phase separation and crud formation in the extraction steps have been more of a problem with OPAP than with DEPA-TOPO.

TWO CYCLES—Both the DEPA-TOPO and the OPAP processes are very similar: in fact, both can be operated in identical equipment.

The first, or concentration, cycle of each involves three unit operations: (1) pretreatment, to prepare the acid, (2) solvent extraction and uranium stripping, to concentrate the uranium by a factor of about 50, and (3) post-treatment of the acid, to ensure that it will not return to the acid plant contaminated by extractant. The second, or uranium purification, cycle is identical for both processes and consists of extraction, scrubbing and stripping steps.

PRETREATMENT—"Black" phosphoric acid contains considerable organic matter or humus in addition to being supersaturated with gypsum. A significant fraction of this organic matter is soluble or in a colloidal state and cannot be removed by filtration. Some of the material coagulates upon contact with the extractants to form a crud that collects at the organic/aqueous interface, and which will eventually foul the system. Thus, this organic matter should be removed before it enters the extraction system. Pretreatment techniques include oxidation, flocculation-clarification, treatment with activated carbon, and calcination.

Once the organic matter has been removed, the uranium valence is adjusted to the desired state. In the DEPA-TOPO process, $NaClO_3$, H_2O_2 or $Na_2S_2O_8$ oxidizes U^{+4} to U^{+6}; in the OPAP system, U^{+6} is reduced to U^{+4} by adding Fe metal.

The final pretreatment step is acid cooling. Temperature variation controls gypsum precipitation and has a significant effect on uranium extraction and phase separation. An acid

* This article is based on a presentation at the 26th Annual Meeting of the Fertilizer Industry Round Table, held Oct. 26–28, 1976, at Atlanta, Ga.

Originally published January 3, 1977

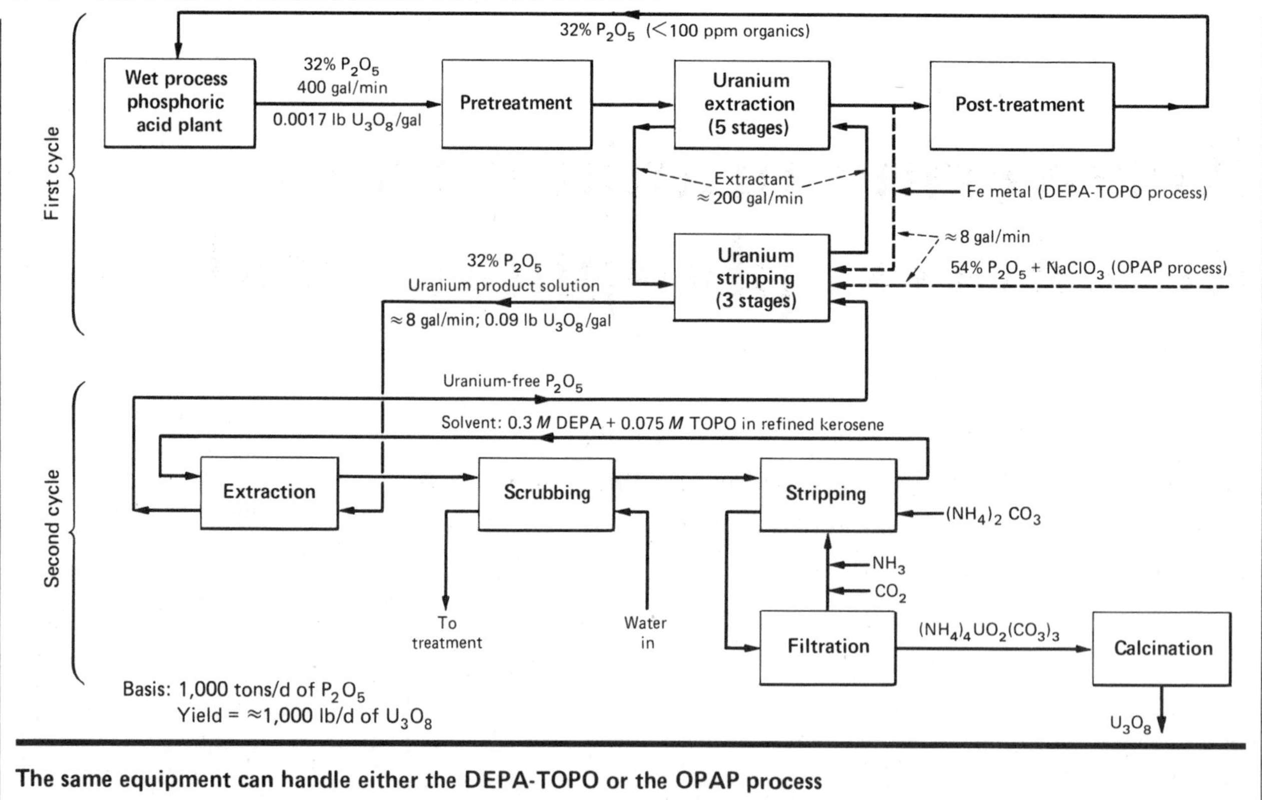

The same equipment can handle either the DEPA-TOPO or the OPAP process

temperature of 40–50°C (104–122°F) is probably optimal.

SOLVENT EXTRACTION—In both processes, pretreated 32% acid is extracted in five stages, with the raffinate proceeding to the post-treatment operations. The solvent used in The DEPA-TOPO process is 0.5-*M* DEPA plus 0.125-*M* TOPO in refined kerosene, while that in the other system is 0.3-*M* OPAP, also in kerosene.

The extract from the DEPA-TOPO extraction is stripped in three stages by 32% P_2O_5 containing Fe metal that reduces U^{+6} back to U^{+4}, enhancing uranium transfer from the DEPA-TOPO to the acid. The acid is obtained from the extraction raffinate stream.

In the OPAP process, a 54% P_2O_5 acid stream—containing $NaClO_3$ as an oxidant that converts U^{+4} back to U^{+6}—strips the uranium in three stages. The concentrated, uranium-free acid-slipstream is obtained from evaporators following the post-treatment section of the first cycle.

Now uranium-rich, the product goes to the second cycle.

POST-TREATMENT—Regardless of how efficiently the solvent-extraction system operates, the uranium-barren acid, or raffinate, will contain entrained solvent. Most of this material must be removed from the acid before it is returned to the acid plant in order to prevent damage to the rubber linings of the acid evaporators and to recover the solvent for reuse. Removal methods include holding tanks, packed columns and air-flotation units.

SECOND CYCLE—For the size plant indicated in the flowsheet, the second cycle is very small, handling an acid feedrate from the first cycle of only about 8 gal/min. Because the phosphoric-acid-solution feed is so highly concentrated in uranium (approx. 0.09 lb/gal), the solvent (0.3-*M* DEPA plus 0.075-*M* TOPO in refined kerosene for both processes) also becomes heavily loaded with uranium, minimizing the extraction of such unwanted impurities as iron and aluminum. The loaded solvent is further cleansed of impurities by scrubbing with water. Uranium is stripped from the solvent by means of an ammonium carbonate solution, precipitating as relatively pure ammonium uranyl tricarbonate. Continuously removed from the stripping system, the precipitate is calcined to U_3O_8. This product is pure enough to go directly to refineries producing UF_6 for enrichment.

POTENTIAL PITFALLS?—Special hazards or problems are minimal in the recovery plant, compared to those already present in the acid plant itself. The major area of concern is probably the fire and personnel-exposure hazard connected with handling kerosene-type solvents. But, since it is necessary to enclose most of the equipment to minimize evaporation losses, this should not pose a significant problem.

Most of the hazardous radionuclides—radium and its daughters—are taken out of the acid when the gypsum is removed in the acid plant and the pretreatment step. The only problem in the recovery plant is from dusting in the calcining and packaging operations. Thus, methods to prevent worker inhalation of dust and its spread to outlying areas must be employed. In addition, health-monitoring procedures must be established according to conditions and limitations specified by the appropriate government agencies.

The authors

Fred J. Hurst is a research chemist in the Chemistry Division of the Oak Ridge National Laboratory, Oak Ridge, TN 37830. He has a B.S. degree in chemistry from Mississippi College, and is a member of ACS and AIME.

Wesley D. Arnold is also a research chemist in the Chemistry Division of Oak Ridge National Laboratory, and he obtained his B.S. in chemistry from the University of Alabama. He is a member of ACS.

Allen D. Ryon is associate head of the Experimental Engineering Section of Oak Ridge National Laboratory. An alumnus of Heidelberg College, he holds a B.S. in chemistry.

Section XII
WASTE GAS TREATMENT/RECOVERY PROCESSES

Cleaning Up Claus Offgas

Flexible, easy-to-operate process teams up reduction and absorption steps to remove virtually all residual sulfur compounds from Claus-plant offgas. Marketable sulfur is the end-result.

Unveiled just two years ago this month, the SCOT (Shell Claus Offgas Treating) technique has quickly won acceptance in the chemical process industries.

The first two commercial installations, both small, skid-mounted units, were put into service at California refineries in June 1973 (see photo). During this second half of 1974, five larger plants are scheduled for startup in the U.S. and Canada, and at least another six will follow in 1975 (see Table I). Headway overseas is just as strong–a minimum of 16 SCOT plants are being tacked on Claus sulfur-recovery units in Japan and Europe.

Researchers at the Royal Dutch/Shell Laboratories in Amsterdam, The Netherlands, originated the fast-moving scheme. Shell Development Co.* handles U.S. and Canadian licensing, in conjunction with a long list of engineering/contracting firms topped (in number of SCOT units) by Ford, Bacon & Davis.

The benefits luring so many Claus operators to the SCOT process are:

- Ability to meet varying emission targets, usually set at 200-500 ppm hydrogen sulfide.
- Use of familiar equipment and process operations.
- Elimination of secondary air and water pollution, and of salt/vanadium wastes for disposal.
- Minimum operator attention.
- Recovery of sulfur as a single, homogeneous product (in the associated Claus plant).
- Flexibility to handle a wide sulfur-input range, with little effect on overall sulfur recovery from changes in feed-gas composition.

* Shell Development Co., Patents and Licensing Division, One Shell Plaza, P. O. Box 2463, Houston, TX 77001.

Catalytic Reduction—The SCOT sequence steps off with a fired heater that raises a mixture of Claus offgas (initially at 140°C or so) and H_2-rich or H_2/CO-rich reducing gas up to reaction temperature. A hydrogen plant or catalytic reformer churns out the reducing gas; or, it can be generated by substoichiometric combustion of a light hydrocarbon (e.g., natural gas) in a direct heater that also serves for preheating.

The offgas/reductant blend then flows to a reactor containing a cobalt/molybdenum, alumina-supported catalyst. At a temperature of 300°C, all sulfur values in the feed gas are converted in this key operation to H_2S. These sulfur values include elemental sulfur, sulfur dioxide, carbonyl sulfide and carbon disulfide.

After passing through a waste-heat boiler for steam generation, product gas from the reduction reactor is quenched with water in a packed tower. The quench water circulates

SCOT Plants, U.S. and Canada — Table I

Process User	Location	Throughput* (Long Tons/D)	Startup	Contractor
Champlin Petroleum Co.	Wilmington, Calif.	15	June 1973	Ford, Bacon & Davis
Douglas Oil Co. of Calif.	Paramount, Calif.	8.8	June 1973	Ford, Bacon & Davis
Murphy Oil Corp.	Meraux, La.	40	Mid 1974	Graff Engineering Co.
U.S. Steel Corp.	Clairton, Pa.	130	Late 1974	M.W. Kellogg Co.
BP Oil Corp.	Marcus Hook, Pa.	160	Late 1974	Ford, Bacon & Davis
Sun Oil Co.	Duncan, Okla.	28	Late 1974	Fish Engineering & Construction Inc.
Marathon Oil Co.	Detroit, Mich.	80	Late 1974	Ford, Bacon & Davis
Shell Oil Co.	Houston, Tex.	325	Early 1975	J. F. Pritchard & Co.
Shell Canada, Ltd.	Waterton, Alta.	2,100	Early 1975	Stearns-Roger Inc.
Southwestern Oil & Refining Co.	Corpus Christi, Tex.	125	Early 1975	Ford, Bacon & Davis
Exxon Co.	Jay, Fla.	100	Early 1975	Ortloff Corp.
Texaco Inc.	Port Arthur, Tex.	235	Late 1975	Ford, Bacon & Davis
Shell Oil Co.	Norco, La.	40	Late 1975	S.I.P. Inc.

*Throughput figures represent capacity of associated sulfur-recovery units.

Process Requirements — Table II

Capital Cost	
Installed equipment, $	820,000
Catalyst and chemicals, initial charge, $	26,000
Utilities	
Steam (50 psi), lb/h	5,700
Cooling water (30°F rise), gpm	860
Electric power, kW	21
Fuel gas, million Btu/h	2.2
Materials	
Catalyst replacement (annualized), $/yr	8,000
Absorbent makeup, $/yr	1,000
Hydrogen, $/yr	6,000
Manpower	
Operators, men/shift	1/6

Basis: 100-long-ton/d Claus unit operating at 94% sulfur recovery; total recovery with SCOT process is 99.85%; Claus offgas at 320°F, 3.5 psi; mid-1973, Gulf Coast costs.

Originally published September 30, 1974

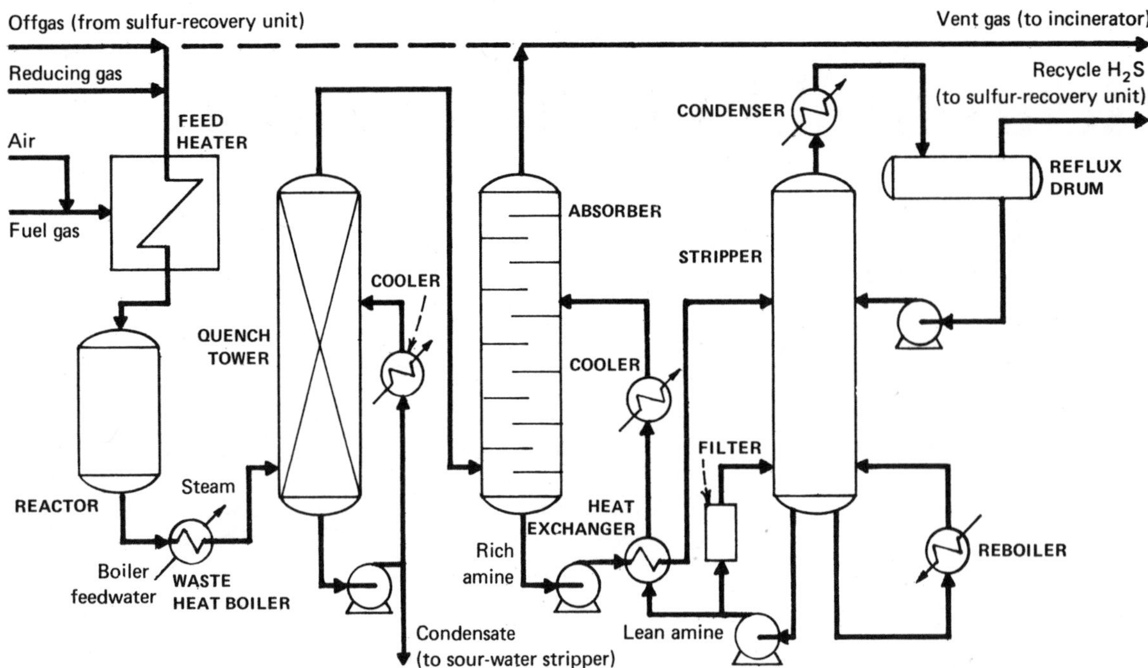

continuously through an air or water cooler. Excess water carried in with the reaction products is withdrawn as condensate. A sour-water stripper removes small traces of H_2S from the condensate before discharge.

PREMIER SCOT PLANT in the U.S. was this skid-mounted unit installed for Douglas Oil Co. in Paramount, Calif.

Amine Absorption—The quenched gas stream, now at about ambient temperature, continues on to the second key step of the process—amine absorption. Unlike conventional high-pressure amine treating, operation is very nearly at atmospheric pressure.

An alkanolamine solvent, usually diisopropanolamine, selectively absorbs nearly all H_2S from the gas stream while coabsorbing only 10-40% of its CO_2 content. Inlet concentrations to the tray tower are normally as much as 3% H_2S and 20% CO_2. The low absorption of the latter in the amine liquor is critical in allowing recycle of H_2S to the Claus sulfur-recovery unit—it avoids buildup of inerts.

Though CO_2 is actually more soluble in the amine than H_2S, Shell explains that H_2S absorption rate is controlled by gas-phase diffusion, while the CO_2 absorption rate is limited by relatively slower liquid-film mass transfer.

The absorber overhead, left with a design amount of H_2S usually between 200-500 ppm, is finally routed to the Claus-plant incinerator for oxidation to SO_2 before atmospheric discharge.

The rich amine liquor meanwhile preheats incoming lean liquor as it exits the absorber, and is then steam-stripped for release of H_2S and CO_2. Steam is removed by condensation, and the remaining H_2S/CO_2 gas is recycled to the front end of the Claus plant for input along with the main acid-gas feed. Typically, total feed increases only 5-7%, and overall H_2S concentration falls off slightly from 90% to 86%. A single sulfur product is thus turned out for marketing from the Claus plant.

Other Features—Nearly all mechanical equipment in a SCOT plant is fabricated from carbon steel, helping to keep capital cost down.

Utilities and materials consumption are also low (see Table II). The cobalt/molybdenum catalyst is estimated to have a lifespan of approximately 3 yr, and is supplied by Shell Chemical Co.

The amine absorbent does not degrade significantly (breakthrough of unreacted SO_2 contaminant during upsets is easily prevented); replacement amine is for mechanical losses.

The SCOT process is suitable, finally, for Claus plants serving not only in refineries and natural-gas-processing plants, but also in such other services as U.S. Steel's desulfurization of coke-oven gases. NRI #

Coke-oven offgas yields fuel, chemical byproducts

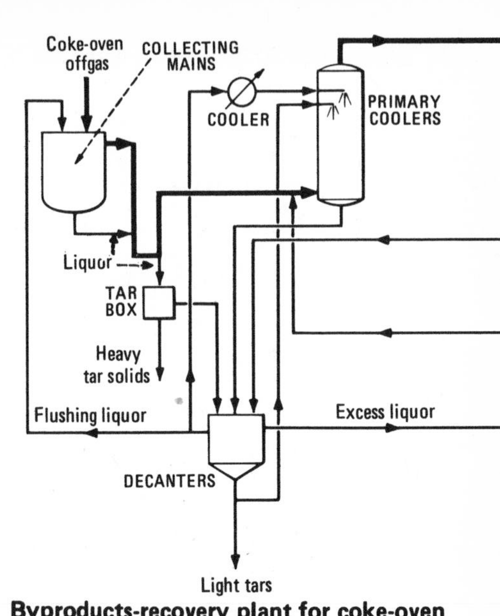

Byproducts-recovery plant for coke-oven

A unique integration of two proven systems for treating coke-oven offgas will highlight a complete byproducts plant starting up soon for Armco Steel.

Louis J. Colaianni, Dravo Corp.

☐ A new pairing of systems for recovering coke-oven byproducts will soon be producing both marketable chemicals and a clean fuel gas, while virtually eliminating the air and water pollution normally associated with coking.

One of the featured systems, developed by Firma Carl Still of Recklinghausen, West Germany, recovers hydrogen sulfide (H_2S) for use as feedstock for sulfuric acid manufacture. The other system, by U.S. Steel Corp. (Pittsburgh, Pa.), recovers ammonia vapor and converts it to an anhydrous product.

The first plant combining the systems should be completed within a few months for Armco Steel Corp., at Middletown, Ohio. Dravo Corp.'s Chemical Plants and Engineering Construction divisions are designing, engineering and constructing the byproducts unit. It will process 84 million ft^3/d of offgas from a new cokemaking facility (a 4,000-ton/d-coke plant also being designed and built by Dravo and Firma Carl Still), and from existing coke ovens.

The Armco byproducts plant also marks the first appearance in North America of the Firma Carl Still desulfurization process (a number of European firms already employ it). The process uses recycled ammonia water for H_2S removal, and thus generates only a clean wastewater.

The U.S. Steel ammonia-recovery technique, called Phosam, serves in a dozen or so installations around the world, but none in conjunction with Firma Carl Still's desulfurization process. Phosam was chosen for the Armco plant because of its unusually high ammonia selectivity, which results in a high-quality anhydrous product needing no further refining.

More-conventional methods, also to be used, will recover other valuable resources, including a light oil containing benzene, toluene and xylenes. The coke-oven gas remaining after treatment will be a clean, 452-Btu/ft^3 fuel stream that Armco will burn in its steelmaking operations. Each cubic foot of the clean gas contains only 0.50 grains of H_2S.

COLLECTION AND COOLING—As shown on the flowsheet, coke-oven gas collected from the ovens is first cooled in the collecting mains by spraying with flushing liquor, a weak ammonia-water solution. This not only cools and condenses various vapors in the gas but also serves as a carrying medium for tars, fine particles and other compounds.

The liquor flows from the collecting mains into tar boxes for separation of heavy tar and solids, and then into decanters for light-tar separation. Some of the flushing liquor recirculates to the collecting mains; another portion passes to an ammonia still (detailed later), and discharges as a clean wastewater after steam-stripping to remove H_2S and ammonia.

The temperature of the noncondensed gas from the collecting mains drops further (from 180 to 95°F) in direct-contact primary coolers, which remove additional tar, some naphthalene, and water vapor. Flushing liquor again serves as coolant. Light tar from the decanters passes into the primary coolers also, to absorb naphthalene from the coke-oven gas.

NAPHTHALENE SCRUBBING—After treatment in three electrostatic precipitators to remove tar mist, the gas enters a naphthalene scrubber—a two-stage spray tower employing a petroleum-fraction wash oil as absorbent. Here, the naphthalene dewpoint of the gas is greatly reduced, but the operating temperature stays above the water dewpoint to minimize corrosion and facilitate the wash-oil stripping step.

The naphthalene-rich wash oil is stripped in a packed tower by 15-psi superheated steam. The lean wash oil recycles to the scrubber after cooling. A portion passes through a reclaimer unit for removal of high-boiling and nonvolatile impurities. The naphthalene-rich vapors flow back into the coke-oven gas stream for absorption in the coolers by tar.

H_2S RECOVERY—Gas exiting the naphthalene scrubber next enters an H_2S scrubber, where ammonia water

Originally published March 29, 1976

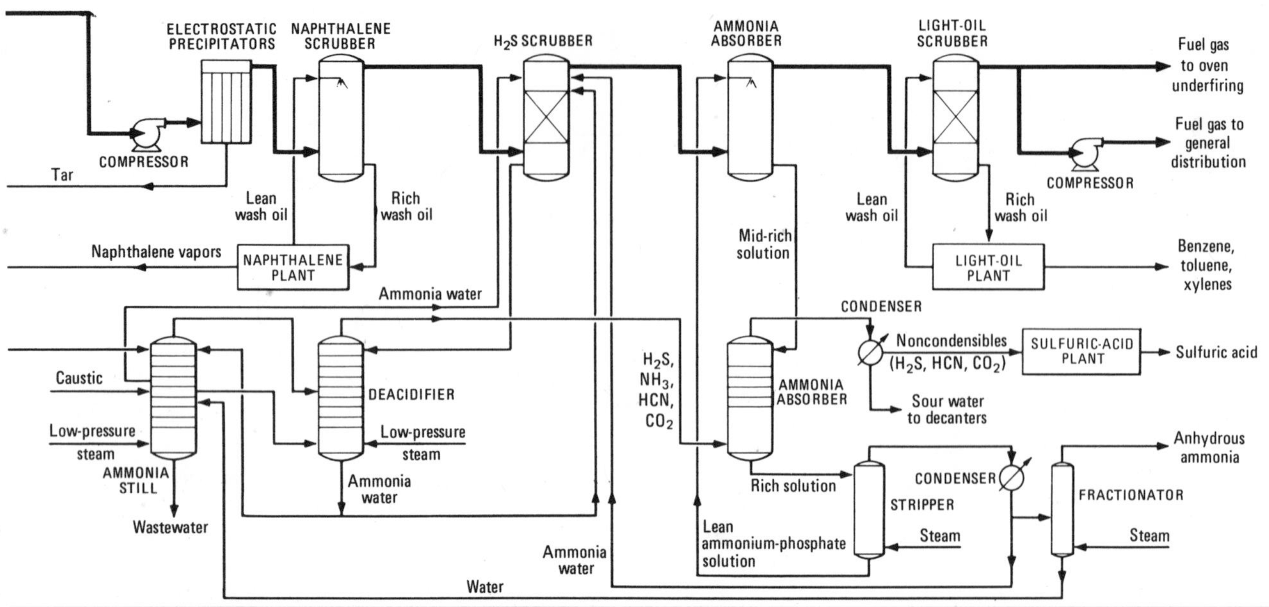

offgas turns out clean fuel gas, ammonia, sulfuric acid, aromatics and tars.

removes not only H_2S but also carbon dioxide and hydrogen cyanide, all as ammonium salts. Salt-rich underflow from the scrubber passes into the deacidifier for steam stripping. The deacidifier turns out a vapor rich in H_2S, ammonia, carbon dioxide and hydrogen cyanide, and also a water stream that primarily recycles to the H_2S scrubber.

The rest of the water enters an ammonia still—consisting of an upper free-ammonia leg and a lower fixed-ammonia leg—along with excess ammonia water from the tar decanters, and ammonia water from a downstream fractionator. Caustic soda added to the middle of the still breaks down fixed salts.

Wastewater from the bottom of the ammonia still goes to disposal after cooling. A water sidestream passes to the H_2S scrubber.

Vapors leaving the top of the still are sent to the middle of the deacidifier, while vapors from the middle of the still enter the lower section of the deacidifier. Thus, the deacidifier needs only a small amount of additional steam. (See table for a listing of process utility requirements.)

The H_2S-NH_3-CO_2-HCN overhead from the deacidifier passes through an absorber for ammonia removal, then through a water condenser, and finally into the combustion chamber of the sulfuric acid plant. Here, HCN decomposes, and H_2S burns to sulfur dioxide. The latter serves as feedstock for sulfuric acid manufacture by the contact process. At design feedrate, the Armco plant will produce about 54 tons/d of 66°Bé acid. The heat liberated during the conversion steps generates high-pressure (600-psi) steam.

AMMONIA PRODUCTION—The H_2S-free coke-oven gas, meanwhile, passes through a second ammonia absorber, for spraying with a lean ammonium phosphate solution. Mid-rich phosphate solution from the bottom of this coke-oven-gas absorber goes to the top of the absorber treating deacidifier vapors.

Rich ammonium phosphate solution exiting the latter absorber passes (via a froth-flotation detarring unit) into the heart of the Phosam circuit. Here, a direct steam-stripper operating at 185 psi removes ammonia from the solution, which recirculates to the absorbers. The stripped ammonia condenses as an aqueous solution containing 10–20% ammonia. A portion of this passes to the H_2S scrubber; the remainder flows to a fractionator operating at 215 psi to produce 25 tons/d of anhydrous ammonia.

LIGHT-OIL REMOVAL—The final treatment for the coke-oven gas is removal of light oil by scrubbing with wash oil (the same hydrocarbon used for naphthalene scrubbing). A steam stripper in the light-oil plant separates 14,500 gal/d of product light oil from wash oil. The latter recirculates to the light-oil scrubber (as in the naphthalene plant, a portion is continuously reclaimed).

The clean gas will serve at Armco's Middletown and Hamilton, Ohio, sites for underfiring of coke ovens, and for general fuel needs.

Armco byproducts plant minimizes utilities use

Steam consumption, lb/h* (400 psi, 600°F)	117,000
Steam export, lb/h (15 psi, 280°F)	25,000
Chilled water, gpm (53°F)	1,667
Cooling water, gpm (85°F)	20,000
Makeup process water, gpm	650
Electricty, kWh	6,000

*Inplant usage is at 400, 250, 165 and 15 psi.
Note: Inplant generation is at 600 psi and at saturated condition, amounting to 7,600 lb/h.

The author

Louis J. Colaianni is a process project engineer in the chemical process department of Dravo Corp.'s Chemical Plants Div. (One Oliver Plaza, Pittsburgh, PA 15222). A member of AIChE, he holds B.S. degrees from Pennsylvania State University in both fuel technology and chemical engineering.

Flue-gas desulfurization produces salable gypsum

Offering high reliability, this process is based on lime but avoids the problems of plugging and scaling common to wet scrubbing systems.

D. R. Kirkby, Davy Powergas Inc.

☐ Unlike conventional lime- and limestone-based systems, the Davy S-H flue-gas-desulfurization process uses a solution instead of a slurry to absorb SO_2. Employing a chemistry that produces a soluble intermediate, this process is able to avoid the problems of wet scrubbing, provide reliability, and produce a gypsum that may be used for wallboard. In the U.S., this system is offered by Davy Powergas Inc., which has an exclusive license from Saarberg-Hölter Umwelttechnik GmbH.

The high reliability of this method has been proved at the 40-MW Saarberg Weiher II power plant (Saarbrücken, West Germany) where 80,300 scfm of tail gas has been treated for over 20,000 h. The system has had an onstream factor (availability) of 96%. An additional unit, which will treat 25% of the flue gas for the Saarberg Weiher III power plant (700 MW), is scheduled for startup this month.

Also, this route produces a gypsum byproduct suitable for sale or for use as landfill. Some of this gypsum has been sold in France to wallboard manufacturers and it is being tested for this application in the U.S.

The process can handle the relatively higher SO_2 concentrations found in chemical plants. A unit for Veba Chemie (Gelsenkirchen, West Germany) went onstream in November 1978. This system treats gas from two Claus units and a sulfuric acid plant. Concentrations of SO_2 may be as high as 6,500 ppm (by volume).

If halogens are in the feed gas, they can be treated. Operating experience has included the removal of HCl and HF as well as SO_2 from the flue gas of a waste product incinerator. Approximately 98% of the HCl and HF, and 90% of the SO_2 were removed, even with variations in SO_2 concentration.

TECHNOLOGY FEATURES—The Davy S-H process is divided into three parts: absorption, oxidation and separation. It is in the chemistry of absorption and oxidation that this route differs the most from conventional scrubbing. When SO_2 is absorbed, a soluble calcium salt is formed, and it is this salt that is later oxidized and converted to gypsum, $CaSO_4 \cdot 2H_2O$.

Conventional lime- and limestone-based processes circulate a slurry and form insoluble calcium sulfite hemihydrate, $CaSO_3 \cdot \frac{1}{2} H_2O$, which causes scaling and erosion. The Davy S-H process employs an aqueous solution of $Ca(OH)_2$ to scrub SO_2 and form the water-soluble intermediate, calcium bisulfite, $Ca(HSO_3)_2$.

Formation of this intermediate is ensured by the use of formic acid as calcium formate, $Ca(COOH)_2$, which buffers the solution and keeps the pH within the proper range. Removal of SO_2 from the flue gas is by chemical reaction:

$$2\,SO_2 + Ca(OH)_2 \longrightarrow Ca(HSO_3)_2$$

$$2\,SO_2 + Ca(COOH)_2 + 2\,H_2O \longrightarrow Ca(HSO_3)_2 + 2\,HCOOH$$

The Davy S-H process can tolerate high chloride levels. The absorption of chlorides actually helps to lower the pH to 4, which is the optimum level for oxidation.

With the addition of air, calcium bisulfite formed during absorption is converted to gypsum in the oxidizer:

$$Ca(HSO_3)_2 + O_2 + 2H_2O \longrightarrow CaSO_4 \cdot 2H_2O + H_2SO_4$$

Economic data

Basis:	
500-MW coal-fired boiler—3% sulfur coal (1.1 million scfm)	
90% SO_2 removal	
U.S. Gulf Coast location, 1978 costs	
Raw-materials and utility requirements:	
Lime, 95 wt% CaO	22,300 lb/h
Formic acid	17.3 lb/h
Makeup water	180–220 gpm
Electric power (includes oxidation air blower)	5.5 MW
Operating personnel	
3 people per shift plus 1 during day shift for supervision. Personnel for routine maintenance included.	
Capital cost	
Total installed cost, $20 million.	
Gypsum Production	
Gypsum (including 20% water)	81,500 lb/h

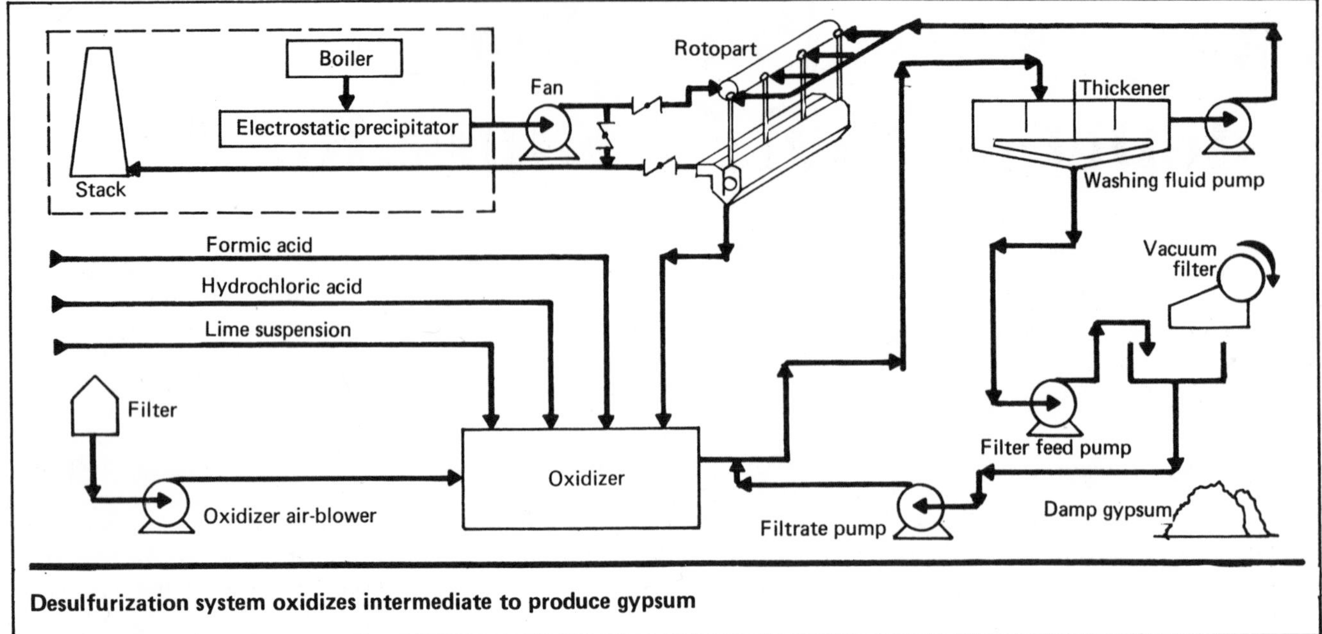

Desulfurization system oxidizes intermediate to produce gypsum

At the outlet section of the oxidizer, calcium hydroxide (as lime and water) is added to replenish the calcium ion in solution. This ion in turn reacts with the sulfuric acid, and neutralizes any HCl present and HCOOH formed during absorption:

$$H_2SO_4 + Ca(OH)_2 \longrightarrow CaSO_4 \cdot 2H_2O$$

$$2HOOOH + Ca(OH)_2 \longrightarrow Ca(COOOH)_2 + 2H_2O$$

$$2HCl + Ca(OH)_2 \longrightarrow CaCl_2 + 2H_2O$$

Lime also raises the pH of the solution to the level required for absorption. And a small amount of formic acid is also added. When gypsum—which is suspended in the resulting solution—is removed, some of the calcium formate is lost. The same is true of calcium chloride, and additions of hydrochloric acid may be necessary if the chloride content of the feed gas is low.

SECOND-GENERATION SCRUBBING—In the Davy S-H process, flue gas with temperatures of up to 400°F can be accepted. Although the system can accept high particulate loadings, pretreatment of the gas for particulate removal is recommended.

Flue gas (see flowsheet) enters an absorption device called the Rotopart. This device has no moving parts and is made of carbon steel with a thermally hardened epoxy lining. The gas goes into a header and then into individual 6-ft-dia. venturi-shaped absorbing ducts.

Absorbent recycled from a thickener downstream also enters the ducts and flows countercurrent to the gas flow. Here, with the absorption of SO_2, calcium bisulfite is formed. The treated gas is then removed from the liquid in the separator section of the Rotopart, and discharged through a stack.

Spent absorbent flows by gravity to the oxidizer, where air is blown through the liquid. The sparged air (approximately 15 psi) oxidizes the calcium bisulfite to gypsum.

As gypsum is formed, it goes into suspension. Additions of formic acid, calcium hydroxide and hydrochloric acid are made. These chemicals restore the absorbent by regenerating its active components and adjusting its pH.

The suspension flows by gravity to a thickener where gypsum crystals are separated from the liquid. This liquid is the regenerated absorbent and it recycles to the Rotopart.

A slurry that is 15% gypsum by weight is pumped from the bottom of the thickener to a vacuum filter. Here, a cake is produced that contains approximately 80% solids by weight (95% of the solids content is gypsum, with less than 0.5% being calcium sulfite). This product is conveyed to storage or shipping as required. The clean filtrate is then returned to the thickener, from where it is also recycled as absorbent.

ECONOMICS—Process economics for a typical installation are provided in the table. Battery limits include lime-receiving and storage facilities, slaking, gas handling, SO_2 absorption and gypsum filtration.

However, costs for the process may vary considerably for individual cases. Gas-handling costs may outweigh those of chemical treatment. Power plants typically have large volumes of gas and low SO_2 concentrations. Chemical plants, such as sulfuric acid units, have higher concentrations but lower volumes.

Also, a credit has not been taken for gypsum production, even though such production may be a special advantage. For conventional wet-scrubbing systems, the capital and operating costs of sludge stabilization and disposal have been estimated at a total $3.20 per ton of coal fired. And sulfite sludge, which is a byproduct of these systems, may be classified as hazardous material, due to its biochemical oxygen demand.

The gypsum produced by the Davy process is chemically stable and, if not used by the gypsum industry, may be stacked in a relatively small area, minimizing load costs.

The author

D. R. Kirkby is Manager of Process Engineering for Davy Powergas Inc. (Houston, Tex.). He is a Registered Professional Engineer in the state of Florida and holds a B.Ch.E. degree from the University of Delaware.

Japan's NO_x Cleanup Routes

There is stiff competition in Japan among new systems that remove the nitrogen oxides in offgases from nitric acid units and from furnaces and boilers in refineries and petrochemical plants.

SHOTA USHIO
McGraw-Hill World News, Tokyo

Prompted by especially tough standards for nitrogen oxides (NO_x) emissions, Japanese engineers have groomed a number of competing techniques to purge the pollutant from offgases.

Dry catalytic routes using ammonia as reducing agent for NO_x have so far taken the lion's share of the treatment-plant market. Hitachi Zosen Co., Sumitomo Chemical Co., Japan Gasoline Co., and a joint venture between Mitsubishi Petrochemical Co. and Hitachi Ltd. are all offering processes.

The chief difference among them is the catalyst, a closely guarded secret. Observers cite as possibilities the oxides of copper, chromium, iron, vanadium, molybdenum, cobalt or nickel, on an activated-alumina carrier. Some of the firms have come up with three or four types.

Operating Costs for Hitachi Zosen NO_x-Abatement System

Item	Annual Cost, Thousand $
Ammonia (45 kg/h)	90
Heavy oil (600 kg/h at 1 kg equal to 10.6 kcal)	300
Water and steam	8.3-9.0
Electric power	333
Catalyst (45 metric tons/yr)	450
Labor and depreciation	1,200-1,300

Basis: 200,000-m^3/h plant, in Japan, operating 8,000 h/yr. Capital cost: $5 million.

Users are opting also for wet absorption approaches to NO_x cleanup, especially for nitric-acid and coke-oven applications. Process developers Mitsubishi Kakoki Kaisha, Ltd., and Nissan Engineering, Ltd., use potassium permanganate solution as the scrubbing liquor. Chiyoda Chemical Engineering & Construction Co., in a hybrid system that takes care of sulfur dioxide as well, scrubs with dilute sulfuric acid containing an iron catalyst.

Catalytic Approach—Hitachi Zosen plans to complete by autumn a 350,000-m^3/h system for an oil refiner, and a 440,000-m^3/h system for a petrochemical firm. At least two smaller plants are in the works.

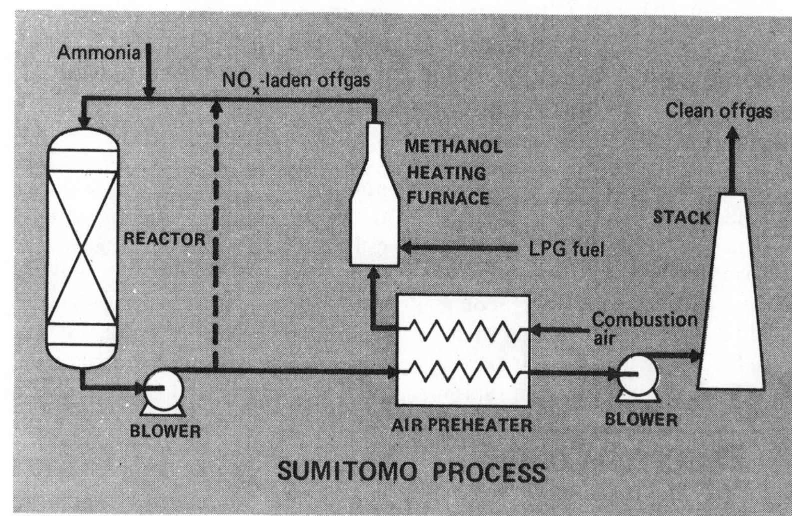

SUMITOMO PROCESS

The process scheme includes only four major elements: gas blower, 250-450°C catalytic reactor, gas-heating furnace, and gas heat exchanger (used when flue gas starts off at a low temperature). To avoid reactor plugging, pretreatment for dust removal may also be required for facilities other than gas- or oil-fired boilers (such as sinter plants, cement kilns, glass-melting furnaces and coal-fired boilers). The catalyst itself, Hitachi Zosen stresses, resists any poisoning by sulfur oxides.

A space velocity of about 10,000 h^{-1} minimizes overall operating costs. Though a higher value decreases catalyst inventory and thus makes for a more compact design, it also raises electric-power consumption and fan capital cost. Ammonia usage is 1.2-2.4 times the quantity of NO_x removed (on an NO basis).

The process can bring NO_x inlet loadings of 350 ppm and higher down to 10 ppm and less on the outlet side. Economic data for annual operation of a 200,000-m^3/h system are shown in the table above.

Already Operating—Sumitomo has proven its catalytic reduction method since early 1974 in Higashi Nihon Methanol Co.'s 800-metric-ton/d methanol plant at Chiba. The system removes 80-90% of the NO_x carried in 200,000 m^3/h of offgas from a heating furnace fired with liquefied petroleum gas (LPG). Reac-

Originally published July 21, 1975

tion temperature is about 250-400°C.

Four more installations are nearly complete, two on boilers and two on ammonia furnaces, bringing total design throughput up to exactly 1 million m^3/h.

The process (see flowsheet) handles clean offgases (e.g., with perhaps 200 ppm NO_x from an LPG furnace) or dirty offgases (400-500 ppm from heavy-oil-fired furnaces). Catalyst life is estimated at two years for the former and one year for the latter.

According to Sumitomo, moreover, its process dovetails with any offgas desulfurization technology. Two additional plants under construction at the company's Chiba works—rated for 250,000 and 300,000 m^3/h of boiler offgas—will use a Wellman-Lord desulfurization unit in conjunction with the NO_x cleanup route.

Capital cost of a typical system: \$270,000-330,000 per 10,000-m^3/h capacity.

Other Catalytic Contenders—Japan Gasoline announced a year ago that it was teaming up its catalyst knowhow with a special parallel-passage reactor groomed by Japan Shell Technology Co. Designed and proven for desulfurization, the reactor avoids plugging from dust in the incoming gas, and also minimizes pressure drop through the catalyst bed. Only 1.0-1.3 moles of ammonia are needed per mole of NO_x, further preventing plugging by minimizing formation of ammonium salts.

A pair of commercial-size units (50,000 and 70,000 m^3/h) are now being constructed for Japanese oil firms, one slated for completion in November and one next spring.

Next summer, the joint Mitsubishi Petrochemical / Hitachi Ltd. venture will commission its maiden unit, a 150,000-m^3/h system at Mitsubishi's Yokkaichi complex. A contract signed last month will bring a huge 500,000-m^3/h unit onstream in October, 1976, for Kawasaki Steel Corp.'s Chiba coke ovens. Kawasaki Steel claims it will be the world's first use of NO_x control technology in such service.

The duo's process is said to feature a reaction temperature of 250-380°C, resistance to sulfur oxides poisoning, and removal efficiency of 90+%.

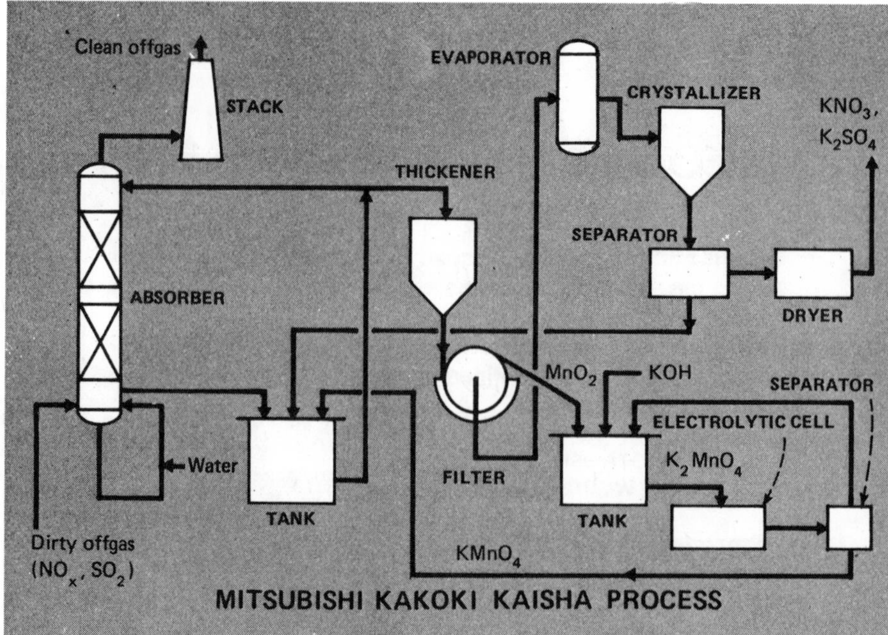

MITSUBISHI KAKOKI KAISHA PROCESS

Wet Routes—Nissan Engineering has been developing not one but four different routes for NO_x cleanup, all using potassium permanganate ($KMnO_4$) as scrubbing liquor. Three plants operating already (100-2,000 m^3/h) use the so-called NE-A route, two for a pigments application and the third for a primary-metals service. The NE-B route is especially suited for nitric acid plants; NE-C handles small volumes of gases containing carbon dioxide; NE-D (still under development) will take on large furnace and boiler applications.

Chiyoda, meanwhile, is grooming its Thoroughbred 102 process for simultaneous NO_x and sulfur dioxide (SO_2) removal. Flue gas is first oxidized with ozone to convert NO to NO_2. Scrubbing with dilute sulfuric acid containing an iron catalyst follows, the acid to absorb SO_2 and the catalyst to reduce NO_x. Status: a 1,000-m^3/h pilot plant is onstream at the firm's Kawasaki research center.

Mitsubishi Kakoki Kaisha has like Nissan opted for $KMnO_4$ scrubbing, and like Chiyoda is working to remove SO_2 along with NO_x. It offers a process developed by Mitsubishi Metal Corp. in conjunction with Nippon Chemical Industrial Co., now operated at a 4,000-m^3/h pilot level.

In the leadoff absorption step (see flowsheet), a slightly alkali solution of $KMnO_4$ and potassium hydroxide (e.g., 1.7% and 0.7%, respectively) reacts at 50°C with NO and NO_2 in the offgas to yield potassium nitrate (KNO_3) and manganese dioxide (MnO_2). If SO_2 is present, potassium sulfate (K_2SO_4) forms also.

The clean gas is vented, and the liquor sent on to the regeneration section. After thickening and filtration, solid MnO_2 is oxidized with potassium hydroxide to form potassium manganate (K_2MnO_4). This is in turn oxidized, electrolytically, back to $KMnO_4$ for recycle. The KNO_3-rich liquor (with K_2SO_4) meanwhile passes from the filter to an evaporator, crystallizer and separator, for recovery as a fertilizer.

Mitsubishi Kakoki Kaisha sees steel mills as a prime application area for the technique. Based on a 200,000-m^3/h plant, designed for 75% removal of NO_x initially at 200 ppm and 100% removal of SO_2 initially at 100 ppm, plant capital cost is estimated at \$5 million. Annual operating expenses include \$1.1 million for fixed costs (labor, maintenance, taxes, depreciation) and \$66,000 for chemicals. Total cost for desulfurizing and denitrifying 1 kl of heavy oil runs \$13.35-15.00, the firm says. #

Meet the Author

Shota Ushio is a correspondent following the chemical process industries for McGraw-Hill World News in Tokyo. He is a graduate of Waseda Univ. and the Univ. of Oregon School of Journalism.

System curbs nitrogen in plant-effluent streams

Designed to control water pollution at a fully integrated nitrogen-fertilizer complex, this route uses waste heat for evaporation and concentration. And, while limiting pollution, the technique saves steam and water and is applicable to other chemical-process-industries segments.

*Allan D. Holiday, Cooperative Farm Chemicals Assn.**

☐ A system of water-pollution control based on evaporation using waste-heat sources is being applied and integrated into plant operation in a way that beneficially affects two major problem areas facing industry today—energy and water conservation.

All too frequently, pollution abatement is attained only through high consumption of energy and other natural resources. Cooperative Farm Chemicals Assn. (CFCA), Lawrence, Kans., has developed a method that keeps effluents within permit limits while reducing steam consumption 29,000 lb/h, cutting overall water requirements by 585 gal/min, slashing treatment-chemical needs by 135 tons/yr, and lowering loadings on various other heat- and waste-removal systems.

Investment cost for the basic concentrator units is approximately $4,000 to $5,000 per million-Btu/h heat-rejection capacity. Power consumption for the concentrators has been measured at about 1.5 kW per million Btu of heat rejection. In addition, operating experience has demonstrated that the system has a high degree of mechanical reliability, due to the simplicity of the concentration system and the basic nature of the process equipment.

And effluent-water control by evaporation could be feasible in many process industries having low-level waste-heat sources, such as stacks, steam vents or heat loads now handled by cooling towers.

PERSPECTIVE—The CFCA plant at Lawrence, Kans., is a fully integrated complex that includes an ammonia plant with a 1,200-ton/d capacity, two urea plants producing up to 890 tons/d, six nitric acid plants having a total capacity of 1,200 tons/d, and two ammonium nitrate plants with a combined capacity of 1,500 tons/d. Urea - ammonium nitrate (UAN) solution also is blended at the site, and urea can be either blended into solution or converted (up to 660 tons/d) into a solid product by means of drum granulators. Most of the ammonia produced at the plant is normally converted to urea or to ammonium nitrate, and final products generated are granular urea, prilled ammonium nitrate, UAN solution, and ammonia.

Early sampling and analytical work found that contaminated water from the older of the two urea plants, the ammonium nitrate plants, and the UAN-solution blending, storage and loading areas was working its way into the plant effluent.

Initially, an effort was made to reduce contamination by collecting the water and concentrating it via natural and induced (from rejected process heat) evaporation. Two 25-million-gal collection ponds were installed, along with a system to pump the water through selected process heat-exchangers in the nitrate area. Runoff water from the more contaminated nitrate-plant areas was directed to the ponds, along with some process wastestreams. Simultaneously, efforts were directed toward recycling contaminated streams to the process, where possible, and toward implementing process modifications that would reduce contaminated-water generation.

During the initial operation of the collection system, the ammonium nitrate-contaminated water from the ponds was used for dilution of the UAN solution. Not surprisingly, the nitrogen content of the plant effluent steadily declined, but it was found that the amount of contaminated water that could be directed to UAN dilution was insufficient to balance input. Thus, the need for additional concentrating capacity became apparent.

A concentrator design was prepared, based on using steam being vented from the ammonium nitrate plant's atmospheric-pressure neutralizer. In principle, the concentrator design is the same as that of a cooling tower, though the operating conditions substantially differ. Ultimately, a second unit was added to further reduce nitrogen contamination of the plant effluent and to add future capacity.

CONCENTRATOR #1—Availability of atmospheric-pressure steam from the nitrate-plant neutralizer served as the basis for design of the concentrator, a cylindrical packed tower with forced-draft airflow (see flowsheet, Fig. 1).

The concentrated solution strength was also a factor in the system design, since the partial pressure of water over the solution decreases with increasing solution strength. A solution strength of 40 wt% ammonium nitrate was selected, being compatible with UAN blending and optimum concentrator operation. At the 40% strength, the vapor pressure above the solution is about 80% that of pure water at the same temperature.

Fairly close agreement between design and actual operating conditions has been experienced on the tower. Concentrator efficiency is about 91%—that is, water evaporated is

*This article is based on a presentation at The Fertilizer Institute's Environmental Symposium, held March 1978, in New Orleans, La.

Originally published August 14, 1978

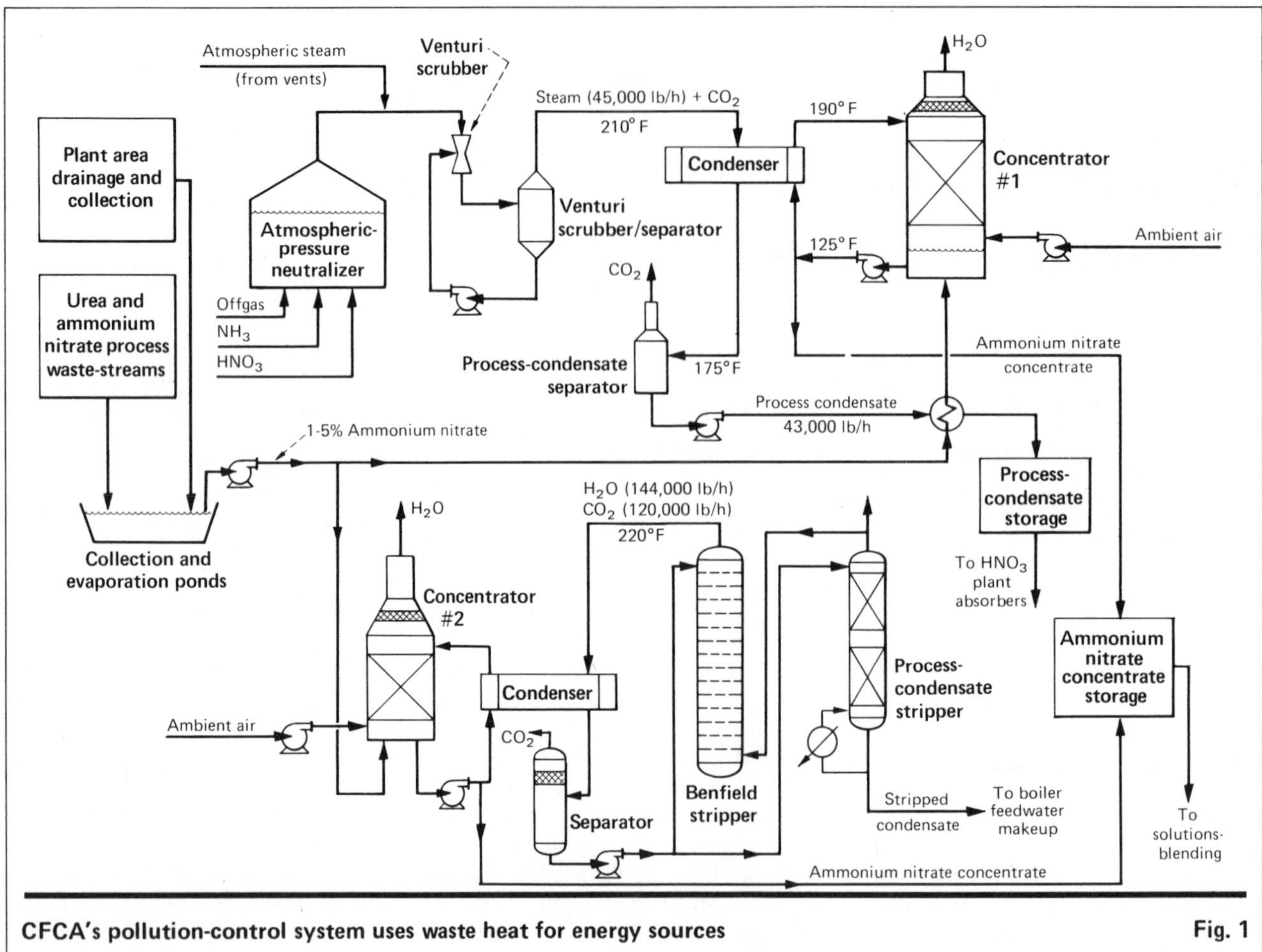

CFCA's pollution-control system uses waste heat for energy sources **Fig. 1**

equal to 91% of the process condensate produced. Also, ammonium nitrate in the vent stack of the concentrator has been measured at < 0.5 lb/h, indicating excellent entrainment control.

Process condensate from the concentrator is pH-controlled at 2.5 and directed to a holding pond before being used as nitric-acid-plant absorber feedwater. The condensate normally contains around 0.5 to 1.0 wt% ammonium nitrate. Use of this condensate as absorber feedwater has reduced the loading on one of the ion-exchange units by 40%, as well as cutting down plant raw-water needs, since demineralized water was previously used as absorber feedwater.

CONCENTRATOR #2—In order to eliminate effluent contamination from the newer urea plant, all of its drainage and runoff was collected and transferred to contaminated-water ponds. Concentrator #2, similar in design to the #1 unit, was then added to handle needed extra evaporative capacity which, in addition to the urea plant, was determined by a water balance to be about 100 to 150 gal/min.

The second concentrator condenses the CO_2-removal-system stripper overhead, which was in part previously cooled by heat exchangers using cooling-tower water on their cold sides. Condensate recovered from the condenser is stripped in the process condensate stripper, along with ammonia-plant process condensate, prior to being used as boiler feedwater in the central plant boilers.

Heat-and-material balances indicate that concentrator #2 is evaporating approximately 230 gal/min of water, as designed. (When the system was sized, it was decided to use the full amount of heat available for evaporation, even though this was about 100 gal/min of evaporative capacity more than was needed. The extra capacity was added for future flexibility.) Also, immediate benefits—including a cut in cooling-tower load and a reduction of hot-lime treating of boiler feedwater—help offset the incremental higher capital cost for the larger unit.

DESIGN FACTORS—Process design conditions and equipment sizing for the two concentrators appear in the table. Conditions were selected for the overall system design by taking into consideration capital and operating costs, as well as the objective of high efficiency in the conversion of heat load to water evaporation. Variables evaluated include: solution strength, solution heating/cooling range, and tower liquid/gas (L/G) ratio.

Enthalpy driving-force diagrams for both the #1 (air - 40 wt% ammonium nitrate solution) and #2 (air - 10 wt% ammonium nitrate solution) concentrators is shown in Fig. 2. On these diagrams are plotted the saturated enthalpy curve for each system, the design operating line for the tower, and the actual operating line(s) for test conditions. The proximity of the operating line to the saturated enthalpy curve is a function of the tower packed-section height (number of contact stages). Within limits, the slope of the operating line(s) can be changed by varying the L/G ratio for tower operation.

Concentrator tower-and-condenser design data

Tower design	Concentrator # 1	Concentrator # 2
Tower dia., ft	12	20
Packed-section depth, ft	18	12
Type packing, stainless Pall rings	3½-in.	3½-in.
Mist eliminator, 4-in.-thick std. weave	10-ft-dia.	17-ft-dia.
Tower heat duty, 10^6 Btu/h	43.4	124.0
Liquid rate, lb/h	890,000	2,274,000
Air rate, std. ft^3/min	28,800	67,850
L/G ratio	6.85	7.3
Liquid temperature in, °F	190	190
Liquid temperature out, °F	125	132
Air temperature in, °F	60	60
Vapor temperature out, °F	170	168
Water evaporated, gal/min	78	232
Process condensate produced, gal/min	86	250
Efficiency,* %	90.7	92.8
Condenser design		
Duty, 10^6 Btu/h	43.4	143.2†
Area, ft^2	4,900	23,470†
Type, per TEMA standards	CHN	CHHN
U, Btu/h/ft^2 · °F	225	147.4
Log mean ΔT, weighted	39.4	41.4
Shellside passes, (parallel)	2	4
Shellside ΔP, lb/in^2	0.50	0.62
Tubeside passes	4	3
Tubeside ΔP, lb/in^2	13	15.7
Design fouling resistance (total)	0.002	0.0027

*Efficiency is water evaporated as a percent of process condensate produced.
†Heat load and rate conditions for heat exchanger were specified nominally 15% greater than the design load for the overall system.

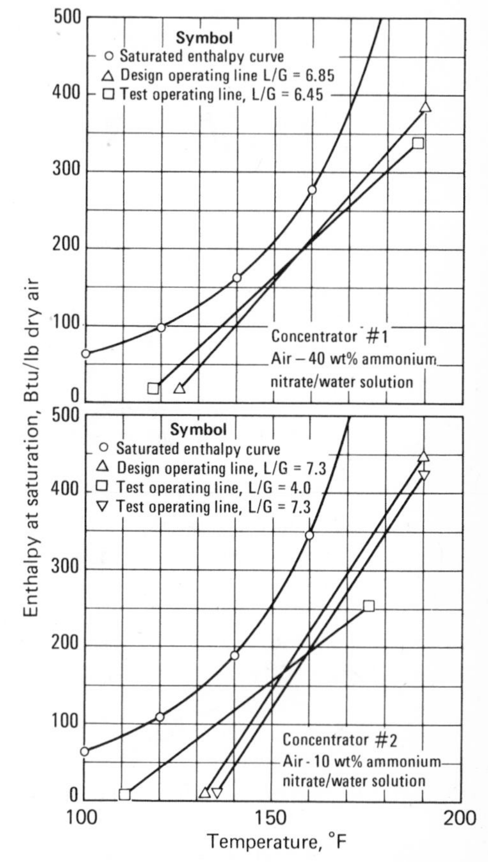

Enthalpy driving-force diagrams Fig. 2

TANGIBLE BENEFITS—In addition to the primary objective of controlling nitrogen contamination of effluent, the following benefits have accrued from using the foregoing system:

- A significant quantity of ammonium nitrate and a lesser amount of urea are now recovered and concentrated into salable product.
- The heat load on three plant cooling towers was reduced by 12% overall, improving process cooling performance of these towers. Load reductions were: nitrate-area tower, by 20 million Btu/h; urea-plant tower, by 28 million Btu/h; ammonia-plant tower, by 40 million Btu/h.
- The ion-exchange load on one unit was dropped by about 40%, from about 200 to approximately 120 gal/min, by using process condensate from the #1 concentrator in acid-plant absorbers instead of demineralized water.
- The load on the hot-lime treater was reduced by 300 gal/min (50% of capacity), cutting the steam requirement by 29,000 lb/h and the need for treatment chemicals (lime, magnesium oxide and soda ash) by 135 tons/yr.
- Overall plant raw-water requirement has been reduced by a nominal 585 gal/min (about 20% of total need).
- Cooling-tower blowdown has been reduced by 44 gal/min, with a corresponding load reduction at a chromate removal system.
- The nitrate-plant neutralizer vent steam is now condensed, reducing this emission to the atmosphere.

COSTS—The economics surrounding this approach to water pollution control are completely dependent on the manner in which the principles are applied and how the system integrates into a given plant.

If a useful condensate can be recovered from a condensing heat load, it could reduce water and treatment requirements. If a cooling tower is unloaded, it can further reduce water needs. If the contaminant being recovered and concentrated is of value, it can help offset the facility cost. Conditions may exist at some plants to go so far as "zero discharge," depending on heat sources available and effluent-water rates.

Investment cost can vary substantially, depending on plant application, capacity, and the materials of construction required. The CFCA facility uses Type 304 stainless steel for wetted parts, and the investment cost of $4,000 to $5,000 per million-Btu/h heat-rejection capacity for the concentrators is applicable to this alloy. Operating costs, too, can vary substantially, depending on how the system is used and the type of peripheral facilities needed. The concentrator system itself, for the process conditions given, uses about one half the electrical power of a typical induced-draft cooling tower of equal capacity.

The author

Allan D. Holiday is technical services manager for Cooperative Farm Chemicals Assn. (P.O. Box 308, Lawrence, KS 66044). Previously employed by Phillips Petroleum Co. and the C. W. Nofsinger Co. in various engineering capacities, he holds 11 U.S. patents on chemical processes, reactor designs and adsorption techniques. He has a B.S.Ch.E from the University of Missouri at Rolla, and is a member of AIChE.

New SO_2-Cleanup Contenders

A double-alkali wet scrubbing system, and a process based on a sodium phosphate scrubbing liquor, are making headway in the race to desulfurize flue gases.

Two of the more recent techniques for scrubbing sulfur dioxide from flue gases are scoring high in early field demonstrations.

At its Parma, Ohio, auto plant, General Motors Corp.* has started up a double-alkali wet scrubber that it developed inhouse. The full-size system is meeting approval by recovering 90+% of the SO_2 in flue gases from a battery of coal-fired boilers; sulfur content of the coal is about 2.5%. High onstream reliability, GM hopes, will be a big system benefit.

Stauffer Chemical Co.'s† phosphate scrubbing process, meanwhile, has chalked up several months of successful operation at Connecticut Light and Power Co.'s Norwalk Harbor station, cleaning a 100-cfm slipstream from a 165-MW boiler. The technique slashes SO_2 emissions to 50 ppm or below, and churns out an elemental sulfur byproduct rather than huge volumes of solid or liquid waste for disposal.

Development Program—Though its Parma system is full commercial size (handling 150,000 cfm from four boilers), it is still considered experimental by GM. "Design and construction cost was over $3 million," says Ernest S. Starkman, vice-president in charge of GM's Environmental Activities Staff, "and operation will add about $10 to the cost of a ton of coal, including writeoff." Considerable cost reduction will be needed before widespread application by GM or others.

This month, GM plans to begin a one-year intensive evaluation of design and operating characteristics in order to optimize the process. GM is funding the project itself, but the U.S. Environmental Protection Agency will have a hand in overseeing the results.

Whether the double-alkali system will be appropriate for large utility power-plants is still a question mark. The 40-MW-total Parma boilers are considerably smaller than most central power stations, and the system there was tailored for the flyash loadings, excess air and load fluctuations typical of industrial boilers.

Process Details—The GM technique starts off by contacting SO_2-laden flue gas (350-600°F, flyash already removed) with a countercurrent stream of 110-120°F, 0.1-molar caustic soda. A Flexitray scrubber is used, points out GM staff engineer Robert J. Phillips, not only for its good removal efficiency at the low operating pH of 5, but also for its wide turndown ratio and ability to take out any flyash lingering in the gas.

The SO_2 and caustic soda combine through a series of reactions to form sodium sulfate (Na_2SO_4), sodium sulfite (Na_2SO_3) and some sodium bisulfite ($NaHSO_3$). About 75% of the exit liquor is recycled, and the rest bled off. Clean gas vents to atmosphere.

The liquor withdrawn is routed to the first of two mix-tanks. Here calcium carbonate ($CaCO_3$) converts any $NaHSO_3$ to sodium and calcium sulfites. In the following mix-tank, lime (CaOH) is added. This starts reacting with the sodium compounds to yield calcium sulfate ($CaSO_4$) and calcium sulfite ($CaSO_3$). Both precipitate out of solution.

The reaction finishes off in a reactor-clarifier. Solids from the bottom of this 60-ft-dia. unit are dewatered in a vacuum filter to about 50% moisture. The cake is high in $CaSO_4$ (gypsum), but also contains $CaSO_3$, flyash and small amounts of sodium as contaminants, so is not suitable for wallboard. It is inert, however, and is sent to landfill. (GM and others are working on ways to utilize such gypsum values.)

Overflow liquor from the clarifier,

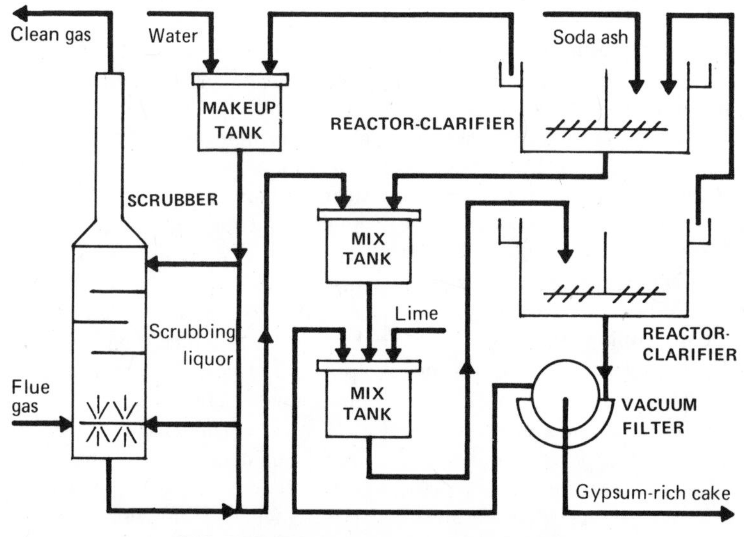

GM DOUBLE-ALKALI PROCESS

*Environmental Activities Staff, General Motors Corp., Technical Center, Warren, MI 48090.

† Stauffer Chemical Co., Licensing Dept., Westport, CT 06880.

Originally published July 8, 1974

containing 500-600 ppm of calcium ion, and saturation amounts of sulfate, sulfite and hydroxide ions, is pumped to a second reactor-clarifier. In a softening step, soda ash (Na_2CO_3) knocks out calcium as a calcium carbonate precipitate (for recycle back to the first mix-tank). The regenerated caustic liquor that is recycled to the scrubber carries only 250 ppm calcium. Calcium reaction with sulfur in the flue gases is thus minimized, avoiding precipitates that would plug and scale the scrubbing vessel.

Versatile Offering—The phosphate process developed by Stauffer has had onstream reliabilities of 98-99+% during its recent field runs at Connecticut Light and Power. Abatement performance is equally impressive: tests with a 2,000-ppm SO_2 inlet concentration that simulated burning a 4%-sulfur oil showed an outlet level of only 20 ppm; a 2,850-ppm influent equivalent to 3.5%-sulfur coal was slashed to 35 ppm. (In both cases, extra SO_2 was added to normal flue gases from the oil-fired boiler.)

According to J. D. Sheehan, Stauffer's manager of research services and technical manager of the abatement project, the firm would like to install and operate a 25-50 MW system (50,000-100,000 cfm) on a coal-fired power plant within the next two years.

In April, Stauffer announced that marketing and design of the system for utility customers will be handled by Chemical Construction Corp. (New York). If a utility is reluctant to run its own system, Stauffer will do the job via an SO_2 abatement service dubbed Powerclaus. This includes day-to-day operation as well as supplying scrubbing liquor and marketing byproducts. (Stauffer already provides a similar service to petroleum refiners, for sulfur recovery and sulfuric acid regeneration.)

For a 1,000-MW power plant that burns 3.5%-sulfur coal, Stauffer estimates battery-limits capital cost at $19.4 million, and total fixed investment at $32.2 million (July 1973, U. S. Gulf Coast basis). Total annual operating cost is pegged at just under $3 million. This translates to 1.46 mills/kW, about 7-8% of the cost of electricity delivered to a home.

Annual production of sulfur is 84,220 long tons. Also extracted are 10,900 short tons of sodium sulfate, suitable for use by detergent makers.

The process can handle nonutility SO_2 streams as well. It was in fact originated and tested by Stauffer in 1971 to clean up emissions from a sulfur recovery plant. Stauffer is evaluating the process now for such applications.

Phosphate Absorbent—In the Stauffer process, raw flue gas must first be treated to bring dust loadings down to 0.04-0.05 grains/ft^3, and humidified to adiabatic saturation temperature (about 130-135°F for coal boilers, 160-165°F for oil boilers). Preconditioning can be by a venturi scrubber, an electrostatic precipitator and spray nozzle, or other new or existing equipment, depending on the installation.

The gas is then passed upward through a packed absorber. A dilute 10% solution of sodium phosphate (a stable, inexpensive and widely available inorganic salt) flows down, at a ratio of 30 gal/1,000 ft^3 when burning 3.5%-sulfur coal. Operating far from saturation of the solution helps to sidestep problems from scaling, plugging and corrosion.

The salt solution acts as a buffer, stabilizing pH at about 4. This allows SO_2 to absorb readily in the water to form sulfurous acid (H_2SO_3), and the acid to dissociate to bisulfite and hydrogen ions. The 130-160°F, sulfur-rich liquor passes next to a stirred-tank reactor. Overall, a Claus-type reaction (a more-complex version than normal, in that it occurs in the liquid rather than gaseous phase) combines SO_2 with H_2S to yield a sulfur precipitate and water.

The sulfur slurry (2-3% corresponds to 2,800-3,000 ppm SO_2) is then separated. In the pilot installation, a continuous vacuum filter is used; commercial systems might use a centrifuge or other filtration device. Regenerated liquor is recycled to the absorber. The sulfur is melted in an autoclave to 99.9% purity.

One-third of the sulfur is available as principal byproduct. The rest serves to generate H_2S by reaction with a hydrocarbon. Operation with natural gas is well established, reports Sheehan; all the steps needed to generate H_2S with oil or coal have at least been proven individually.

The sodium sulfate byproduct results from the small amount of SO_3 in the flue gas, and from some H_2S that reaches the absorber. A crystallizer separates the substance from a liquid slipstream taken off the sulfur separator. At Norwalk Harbor, only about 1% of the sulfur is purged as Na_2SO_4. Formation is kept down by the acidic absorption conditions, and by the antioxidant effect of sodium thiosulfate (a small amount forms as intermediate during the complex reactions of the process).—NRI #

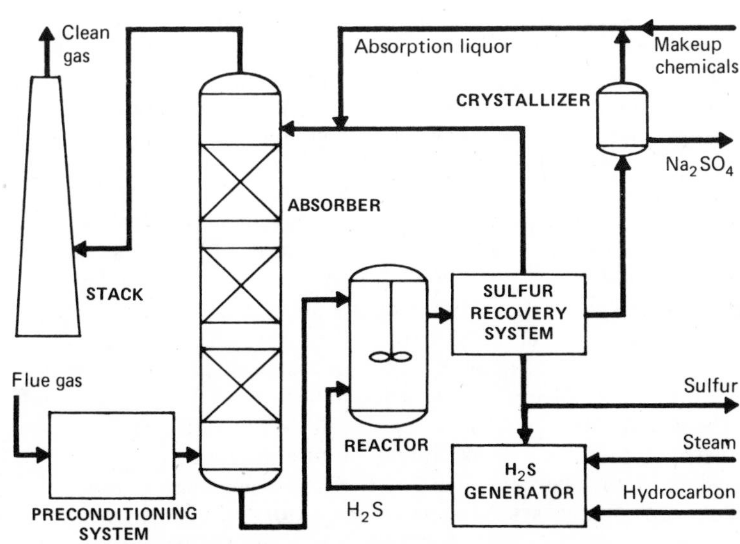

STAUFFER PHOSPHATE PROCESS

Limestone + magnesium: A new SO_2 control team

As the price of lime rises along with energy costs, limestone is looking better for wet alkali scrubbing. And a touch of magnesium is said to considerably improve the desulfurization capability of limestone.

☐ There is a new wrinkle in the desulfurization of flue gas with a high sulfur content (over 1,000 ppm SO_2): Instead of a lime slurry for scrubbing, try limestone enhanced with a magnesium compound—e.g., magnesium oxide. This, say experts, results in important savings in operating costs, but imposes no loss in SO_2-removal efficiency.

The new system will soon receive its commercial baptism. Using technology developed by Pullman Inc.'s Pullman Kellogg Div. (Houston, Tex.), Associated Electric Cooperative Inc. is building the first U.S. large-scale magnesium-promoted limestone-scrubbing unit, at Huntsville, Mo. The facility will remove more than 90% of the SO_2 from 6.2 million lb/h of flue gas containing 5,500 ppm of SO_2. Power-plant capacity is 675 MW.

The U.S. Environmental Protection Agency (EPA) hopes that the Huntsville plant will confirm some of its own observations regarding magnesium enhancement. Recent tests at Shawnee, Ky.—the center of EPA's research on alkali wet-scrubbing—indicate that certain chemicals (including magnesium ones) promote SO_2 removal in flue-gas desulfurization (FGD) systems when dissolved in circulating limestone slurries.

Harlan N. Head, project manager for Bechtel Corp. (San Francisco), which operates the Shawnee scrubbing unit, went one step farther at a recent (June 25-30) meeting of the Air Pollution Control Assn., held in Houston. According to Head, one of the most important variables affecting SO_2 removal in FGD systems based on scrubbers, spray towers and turbulent contact absorbers, is effective magnesium-ion concentration.

Originally published September 11, 1978

PROS AND CONS—EPA's prototype testing program at Shawnee aims to optimize lime and limestone FGD systems in several aspects—e.g., sludge disposal; hardware and process reliability; forced oxidation; process economics; and SO_2-removal efficiency. EPA also wants to overcome some of the systems' basic shortcomings (scale formation, in particular). In the case of limestone, a major drawback is poor mass-transfer rates that can limit SO_2-removal rates.

Says Cliff Wang, assistant project manager at Shawnee: "Lime is about one hundred times more soluble than limestone, so there is enough alkali for scrubbing. But limestone doesn't dissolve easily. Addition of magnesium increases sulfite concentration in the scrubbing liquor, and this improves SO_2 removal."

Tests at Shawnee seem to confirm the fact. In one experiment, SO_2 removal went from 65% (no magnesium) to 96% with the addition of 5% by weight $MgSO_4$.

To be sure, magnesium addition is not without problems, says John Williams, EPA's lime/limestone contract officer. He warns, "One must take the necessary precautions to control pH and eliminate chlorine, which can form magnesium chloride—not a scrubbing species."

According to Wang, "The scrubbing system can tolerate 0.1 to 0.2% chlorides, but chloride ions tie up magnesium and calcium, and increase scaling potential. Most of this potential is due to gypsum; the liquor can get supersaturated in calcium sulfate. But this should not be a problem if you stay below gypsum's incipient scaling level—i.e., 135% supersaturation."

SOME ECONOMIC PLUSES—Joseph C. Yarze, senior process manager for Kellogg at Hackensack, N.J., says: "It would be difficult to meet EPA standards with limestone . . . unless this system were promoted. For unpromoted systems, the lower reactivity of limestone translates into a need for more-expensive absorbers."

Magnesium-promoted limestone so increases efficiency, according to Kellogg, that scrubbers actually operate at lower horsepower requirements. Adds Yarze: "A detailed investment cost analysis comparing lime and promoted limestone has not been developed specifically for this design, but it is clear that the annual reagent-cost advantage gained with use of promoted limestone more than offsets the marginally higher investment requirement."

Lime, of course, is much more expensive than limestone. Depending on location, the price of lime used in FGD studies has varied from \$35 to \$45/ton, whereas limestone's price has been \$6 to \$10/ton. Murray Wells, Radian Corp.'s (Austin, Tex.) assistant vice-president for research and development, notes that many scrubbing units now in operation use lime. "But," he says, "as the price of natural gas needed for calcination increases, the margin between lime and limestone will get even wider, and limestone will become even more attractive."

This is the case for a plant the size of Associated Electric's. Annual cost of reagent, based on 300 days' operation, is about \$3.39 million with promoted limestone, vs. \$10.12 million with unpromoted lime. The saving: \$6.73 million. Says Yarze; "This difference can be expected to increase in the future as the cost of lime goes up to reflect rising energy costs."

OXIDATION IS IMPORTANT—Economy may not be the sole allure of

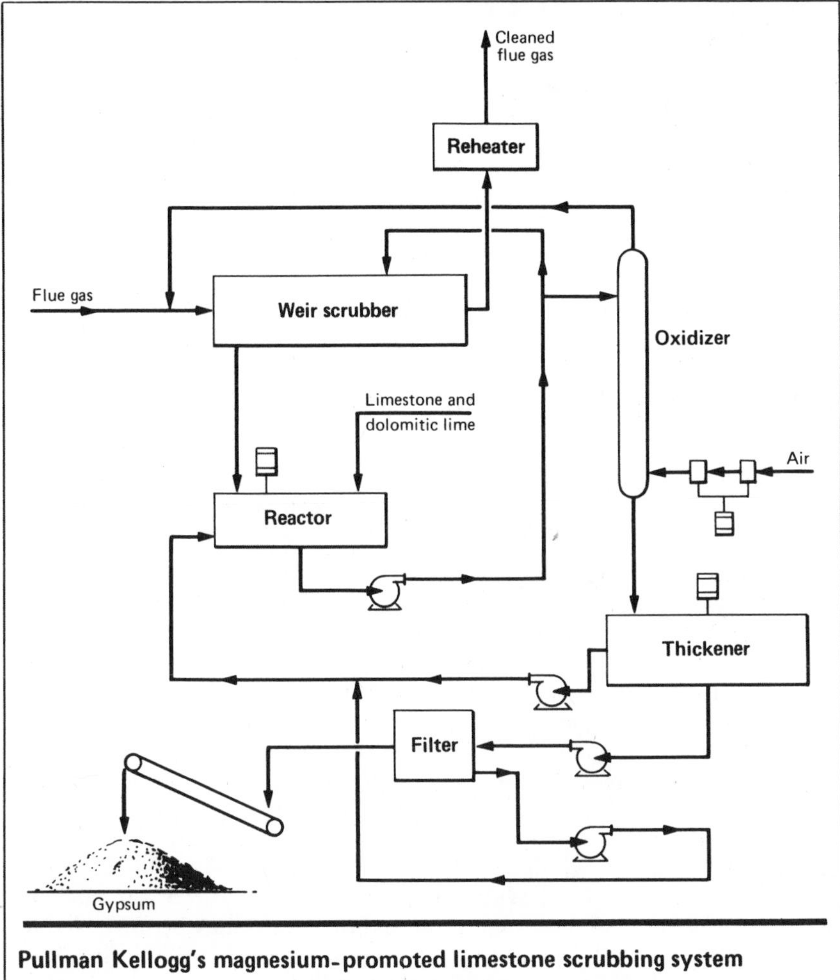

Pullman Kellogg's magnesium-promoted limestone scrubbing system

magnesium-enhanced limestone. According to a spokesman for a large supplier of pollution-control equipment, "Use of magnesium oxide brings such additional benefits as the conversion of sulfite to sulfate in the scrubbing liquor. This is a conversion that promotes growth of bigger crystals and a thicker sludge."

Oxidation of calcium sulfite to sulfate is becoming increasingly important to EPA because disposal of the former, which contains more water of hydration, is difficult without fixation i.e., combination with cement-like compounds to obtain landfill material.

Notes an EPA researcher: "Forced oxidation from *ite* to *ate* allows dewatering. Calcium sulfate sludge is more amenable to disposal, eliminating the need for expensive fixation schemes."

To this end, EPA is looking at additives that promote oxidation—e.g., benzoic and adipic acids. Preliminary work indicates that the latter is, according to EPA, "a little cheaper, not affected by chlorine, and provides the necessary ions for sulfur removal, even in subsaturated operation."

SETUP AT HUNTSVILLE—In addition to the saving that stems from the lime-limestone price differential, the Associated Electric plant will economize by maximizing recovery of magnesium from soluble salts, which minimizes the requirement for makeup magnesium reagent.

The Huntsville facility, which will use dolomitic lime as a source of magnesium makeup, will recover reagent via slurry thickening to a 30%-solids sludge, filtration to obtain a 55%-solids cake, and subsequent washing to reclaim magnesium. Notes Yarze: "According to Kellogg's evaluation, the recovery scheme reduces dolomitic-lime requirements to only about 2% of the limestone feed.

Kellogg is installing four horizontal scrubbing modules at the Huntsville power plant. The units, with no moving parts or internal trays to collect scale, have an unlimited turndown ratio. For Associated Electric, this means that EPA regulations can be met, even with three of the four scrubbers in service and fourteen of the sixteen slurry pumps in operation.

Yarze points out that the scrubbers effectively remove a high percentage of residual particulates. He emphasizes that "it is possible to trap 90% of the incoming particulates with four stages in service. The electrostatic precipitator/scrubber combination can be optimized as a particulate-removal system." Yarze claims that a precipitator of only 95 to 97% efficiency is "satisfactory" to meet EPA regulations, when used in conjunction with the Kellogg scrubbers.

THE RIGHT MINERAL?—The bright prospects for magnesium-enhanced limestone in FGD systems may be good news for Texas Architectural Aggregate, San Saba, Tex. The company owns a sizable (800 million tons) deposit of perdazite north of Van Horn, Tex. The mineral, a mixture that seems made to order for limestone-based FGD, is 60% calcite—$CaCO_3$—and 40% brucite—$Mg(OH)_2$.

Researchers at Texas Christian University (Arlington), Radian Corp., and the Tennessee Valley Authority station at Muscle Shoals, Ala., have confirmed this assumption. Says Joseph Williams, president of Texas Architectural, "An obvious advantage is that you don't have to calcine. The brucite is held mechanically in the calcite, and can be mechanically isolated. Preliminary tests show that perdazite requirements could run up to 10% of the limestone feed."

Phillip Lowell, a consultant on FGD, based in Austin, Tex., agrees. "Perdazite," he notes, "represents an excellent potential magnesium source for wet-limestone scrubbing because it gives high magnesium utilization."

In addition to these qualities, perdazite may be useful in NO_x control. An unidentified company has been testing the mineral as a curb for NO_x emissions and reports good results.

Guy E. Weismantel

Removal of SO_2 from industrial waste gases

Combustion of sulfur-bearing fuels accounts for more than 75% of SO_2 emissions in the U.S. These can be controlled to a degree by burning low-sulfur fuels or by pretreating the fuel to lower its sulfur content. Currently the most practical control involves scrubbing the gas to remove the sulfur dioxide.

Norman Kaplan and Michael A. Maxwell,
Industrial Environmental Research Laboratory—RTP, U.S. Environmental Protection Agency

☐ President Carter's plan to reduce this country's dependence on foreign oil stresses the conversion of oil- and gas-burning facilities to coal. This plan projects a massive need for scrubbers, since industrial coal converters will consume coal at a level of 60,000 MW equivalent by 1985. This compares with just 8,900 MW without the Carter plan [*1*].

WHY CONTROL SULFUR OXIDES?

Sulfur oxides have become among the more dangerous air pollutants to human health, and are clearly the most harmful to vegetation and to some building materials. Numerous epidemiological studies plainly show an association between air pollution, as measured by SO_2 concentration, and health effects of varying degrees of severity.

Mortality and morbidity can be correlated to ambient levels of SO_2, as evidenced during air-pollution episodes in New York City, Donora, Pa., and London, U.K. Adverse health effects do occur when 24-hour-average levels of SO_2 exceed 0.11 ppm for 3 to 4 days.

Sulfur dioxide is a highly toxic gas. For comparison, carbon monoxide, another gas with a reputation for high toxicity, causes adverse health effects only when concentrations exceed 9 ppm [*2*].

Synergism between SO_2 and Total Suspended Particulate matter (TSP) causes a combination of these pollutants to create more health problems than the sum of their individual effects would indicate. This has an obvious implication in the control of combustion sources that emit both SO_2 and particulate matter. In fact, today's primary air-quality standards for SO_2 and TSP were derived from population studies involving both pollutants [*3*].

More recent evaluations confirm that the two pollutants in association relate directly to health effects. However, their presence might only indicate the existence of other hazardous pollutants, such as sulfates. Evidence now shows that these other compounds may, in fact, be primarily responsible for the health effects originally attributed to SO_2 and TSP.

Atmospheric sulfates, especially sulfuric acid aerosol, are produced from SO_2 emissions through several chemical reactions in the atmosphere.

The presence of both SO_2 and TSP probably does indicate the presence of sulfates, although the converse does not necessarily hold true. It is possible to have high sulfate levels and low SO_2 or TSP, because sulfates may be windborne for considerable distances. For instance, sulfur dioxide emissions from rural sources (ore smelters or power plants out in the country) may be converted to sulfates and transported to areas where SO_2 emissions and atmospheric levels of SO_2 are low.

Development of a separate regulatory program for sulfates is an ultimate goal [*4*], but little information to

This article has been reviewed by the U.S. Environmental Protection Agency and approved for publication. Mention of tradenames or commercial products does not signify endorsement or recommendation for use.

Originally published October 17, 1977

Applicable regulations

As mandated by the 1970 Clean Air Amendments, the EPA administrator established six so-called "criteria pollutants" (particular matter, sulfur oxides, photochemical oxidants, carbon monoxide, hydrocarbons, and oxides of nitrogen) for which primary and secondary national ambient air-quality standards (NAAQS) were established. Primary standards were designed to protect public health, while secondary standards were prescribed to protect the public welfare:

NAAQS for sulfur oxides	Primary	Secondary
Annual mean	80 $\mu g/m^3$ (0.03 ppm)	—
Maximum 24-h concentration*	365 $\mu g/m^3$ (0.14 ppm)	—
3-h concentration*	—	1,300 $\mu g/m^3$ (0.5 ppm)

*not to be exceeded more than once a year.

Summary of federal standards of performance limiting sulfur oxide emissions from new stationary sources **Table I**

Source category	Affected facilities	Maximum emissions
Fossil-fueled steam generators	Coal-fired boilers, oil-fired boilers (most utility and some industrial boilers)	Solid fuel: 2.2 g $SO_2/10^6$ cal (1.2 lb $SO_2/10^6$ Btu) Liquid fuel: 1.4 g $SO_2/10^6$ cal (0.80 lb $SO_2/10^6$ Btu)
Sulfuric acid plants	Process equipment	2 kg $SO_2/10^6$ cal (4 lb SO_2/ton of H_2SO_4) 0.075 kg acid mist/metric ton of H_2SO_4 (0.15 lb acid mist/ton H_2SO_4)
Petroleum refineries	Refinery process equipment, including waste-heat boilers and fuel-gas combustion devices	Fuel gas limited to 230 mg/dry Std. m^3 of H_2S (0.10 grains/dry Std. ft^3) maximum
Primary copper smelters	Roaster, smelting furnace, copper converter	0.065% SO_2 by volume
Primary zinc smelters	Roaster, sintering machine	0.065% SO_2 by volume
Primary lead smelters	Sintering machine, sintering-machine discharge end, dross reverberatory furnace, electric smelting furnace and converter	0.065% SO_2 by volume
Petroleum-refinery sulfur-recovery plants	Claus plants	0.025% SO_2 by volume (oxidation or reduction plus incineration) or 0.030% SO_2 by volume of reduced sulfur compounds, and 0.0010% by volume of H_2S (reduction only)

establish standards and control programs for sulfates now exists. A major research program is underway to find the sources and determine the transport of sulfates in the environment, to study chemical composition of sulfur compounds in the air, and to pin down specific health effects of these pollutants (see box). While research proceeds, regulatory programs to control SO_2 emissions appear to be the only practical means, however incomplete, of reducing levels of potentially hazardous sulfate aerosols in the air.

To meet the primary standards, the Clean Air Act amendments require each state to adopt (and submit to the Administrator) a State Implementation Plan (SIP) to provide for implementation, maintenance, and enforcement of the primary standard as soon as practicable, but not later than three years from the date of approval of the SIP. Requirements of SIP to implement, maintain and enforce the secondary standard must specify "a reasonable time at which such secondary standard will be attained."

New Source Performance Standard (NSPS) set by the Environmental Protection Agency with respect to sulfur oxides for a number of industrial source categories are provided in the Code of Federal Regulations (40 CFR, Part 60). Table I summarizes these NSPS, including also the NSPS for petroleum refinery, sulfur-recovery plants proposed by the *Federal Register* (41 Fr 43866) on October 4, 1976.

Between 1970 and 1975, more than 97% of sulfur oxide pollution in the U.S. resulted from fuel combustion in stationary burners and from industrial processes. The electric-utility industry caused about 60% of all SO_x pollution in that period; another 37% was split, half and half, between other stationary-source combustion and industrial processing; the remaining 3% resulted from miscellaneous sources, including combustion for transportation and solid waste.

About 60% of all the SO_x emissions from the industrial-process category (over 11% of the total) comes from metal refining and processing. Most of the remainder of this category comes from chemicals manufacturing, petroleum refining, and mineral-products processing. The chemical industry in 1975, as a case in point, accounted for 1.0 million tons of SO_x emissions, or about 19.2% of the industrial-process output (which, in turn, was 15.8% of total U.S. discharge of SO_x that year).

UTILITY INDUSTRY SO_2 CONTROL

Well over 50 flue-gas-desulfurization (FGD) processes have been invented; many have undergone testing in small-scale laboratory or pilot-plant operations. But relatively few have served to date for SO_2 control of full-scale utility boilers in this country. In Japan, a few more systems have gone on to full operational status. For most purposes, a full-scale utility boiler is defined as one with a steaming rate equivalent to 100 MW of electric-generating capacity, with heat input of approximately 2.5×10^6 cal (10,000 Btu) per kWh generated.

In the United States, there are currently six types of FGD systems in use for controlling full-scale utility boilers. Table II summarizes these systems by type and extent of application. If alkaline-fly-ash scrubbing is considered a variation of lime-slurry scrubbing (since much of the alkalinity in the ash is due to CaO), the Table shows that 90% of all utility FGD operating capacity involves lime- or limestone-slurry scrubbing processes. These are "throwaway" processes, as opposed to "salable product" processes, since the SO_2 is removed in a form that must be discarded. Sodium carbonate scrubbing, also a throwaway process, has only limited application, since it produces a liquid waste-stream containing the sulfur in a soluble form (Na_2SO_4 and Na_2SO_3). (Nevada Power Co. uses this technology.)

The magnesium oxide and Wellman-Lord processes, on the other hand, produce salable sulfur products, such as elemental sulfur, sulfuric acid, or liquid sulfur dioxide (see Table II).

Utility flue-gas desulfurization processes* **Table II**

Process	No. of boilers	Total controlled generating capacity, MW
Lime-slurry scrubbing	9	2,702
Limestone-slurry scrubbing	6	2,037
Alkaline-fly-ash scrubbing	2	720
Sodium carbonate scrubbing	3	375
Magnesium oxide scrubbing	1	120
Wellman-Lord/ Allied Chemical	1	115
Total	22	6,069

*"Summary Report—Flue Gas Desulfurization Systems," Jan., Feb., Mar., 1977, PEDCo Environmental, Inc.

Lime/limestone-slurry scrubbing

In concept, lime- or limestone-slurry scrubbing processes are very simple. In practice, however, the chemistry and the system design for a full-scale operation can become more complex than seems evident at first glance.

These systems use a slurry of lime or limestone in water to absorb SO_2 from power-plant flue gas in a gas/liquid scrubber. The slurry generally ranges from 5% to 15% solids. Various types of scrubbers or gas/liquid contact devices are employed commercially: spray towers, venturis, marble-bed scrubbers (packed beds of glass spheres), and turbulent-contact absorbers (lightweight hollow plastic spheres—Ping Pong balls—held between restraining grids in a countercurrent scrubbing tower). These scrubbers usually operate with a liquid-to-gas (L/G) ratio of 6 to 15 L/normal m^3 (40 to 100 gal/1,000 actual ft^3).

Fig. 1, a schematic of the 10-MW prototype limestone system at the TVA/EPA test facility at TVA's Shawnee steam plant, shows a turbulent-contact absorber with an open chevron-type mist eliminator. Flue gas flows up through the tower and contacts slurry which sprays downward over the packing, countercurrent to the gas stream. The gas/liquid counterflow keeps the spheres in turbulent motion, which improves gas/liquid contact.

The overall absorption reaction taking place in the scrubber and the hold-tanks for a limestone-slurry system produces hydrated calcium sulfite:

$$CaCO_3 + SO_2 + \tfrac{1}{2}H_2O \rightarrow CaSO_3 \cdot \tfrac{1}{2}H_2O + CO_2 \quad (1)$$

With a lime-slurry system, the overall reaction is similar but yields no CO_2:

$$CaO + SO_2 + \tfrac{1}{2}H_2O \rightarrow CaSO_3 \cdot \tfrac{1}{2}H_2O \quad (2)$$

(The actual reactant in Eq. 2 is $Ca(OH)_2$, since CaO is slaked in the slurrying process.)

In practice, some of the absorbed SO_2 is oxidized by oxygen which is also absorbed from the flue gas. This shows up in the slurry as either gypsum ($CaSO_4 \cdot 2H_2O$) or as a calcium sulfite/sulfate mixed crystal [$Ca(SO_3)_x(SO_4)_y \cdot zH_2O$]. Slurry is recycled around the scrubber to obtain the high liquid-to-gas ratios required.

Lime- and limestone-slurry scrubbing systems can be engineered for almost any desired level of SO_2 removal. Commercial utility systems are generally designed for 80% to 95% removal; however, some systems have at times achieved more than 99% removal. The higher removal rate is not costly: investment savings realized in designing for 80% rather than 90% SO_2 removal amount to only about 3.2% to 4.5% [*5*].

Alkaline-fly-ash scrubbing

Western-U.S. low-sulfur coals appear particularly suitable to SO_2 control by scrubbing the waste gases with a slurry of the alkaline fly ash which results from the combustion process. There are two ways to add the alkaline ash to the system: (1) collecting the fly ash in an electrostatic precipitator upstream of the scrubber and then slurrying the dry fly ash with water so that it can be pumped into the scrubber circuit, and (2) scrubbing the fly ash directly from the flue gas by the circulating slurry of fly ash and water.

Most Western coals have a low sulfur content (less than 1.0%). They also usually have a low heating value and consequently require close control to hold their

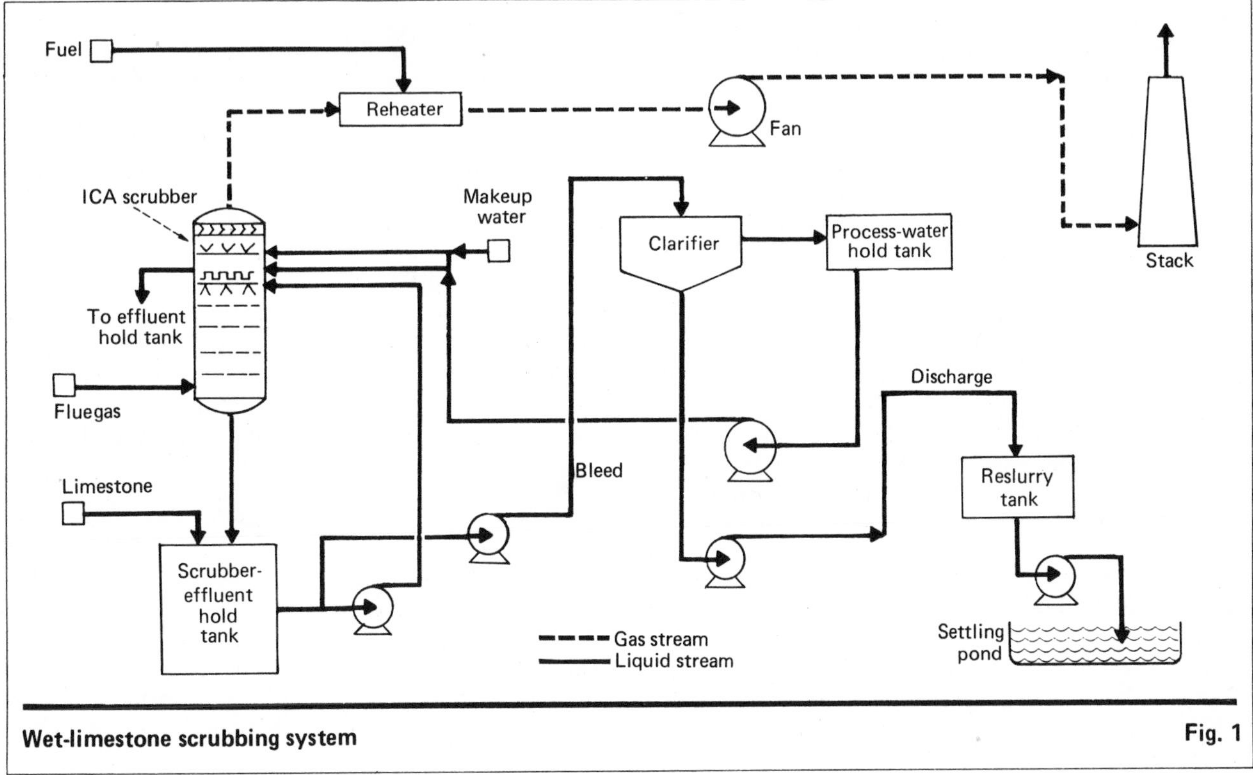

Wet-limestone scrubbing system

Fig. 1

combustion emissions within federal limits for New Source Performance Standards (NSPS), defined as mass emissions per unit heat input. While typical Eastern bituminous coals have heating values of about 22,000 cal/g (12,000 Btu/lb), lignite coals may have heating values as low as 12,000 cal/g (6,800 Btu/lb). The bituminous coal could contain as much as 0.7% sulfur and still meet the NSPS limitation of 2.2 g/10^6 cal, without controls. But the lignite would have to have no more than about 0.4% sulfur.

Generally, the coals best suited to this method of control are the Western lignites and sub-bituminous coals. Certain of these (from North Dakota, Montana, Wyoming) have large amounts of alkaline metal oxides in the ash (e.g., CaO, MgO).

Sodium carbonate scrubbing

This method of controlling SO_2 involves scrubbing the flue gas with a solution of sodium carbonate and bicarbonate, to produce a mixture of sodium sulfite and sulfate by these reactions:

$$Na_2CO_3 + SO_2 \rightarrow Na_2SO_3 + CO_2 \quad (3)$$

$$2\ NaHCO_3 + SO_2 + H_2O \rightarrow Na_2SO_3 + 2\ H_2O + 2\ CO_2 \quad (4)$$

$$Na_2SO_3 + \tfrac{1}{2}O_2 \rightarrow Na_2SO_4 \quad (5)$$

The sodium carbonate does not have to be pure. Nevada Power uses trona salt, a naturally occurring mineral containing 60% $NaHCO_3$, 20% NaCl, 10% sulfates, and 10% insolubles [*6*].

This process has definite limitations in large-scale utility applications, since it requires a relatively cheap source of sodium carbonate or bicarbonate and an ability to dispose of large volumes of waste salt solution. Nevada Power is located near trona deposits—in an area where natural evaporation rates far exceed rainfall and where available land is relatively abundant. The company processes its liquid waste in solar evaporation ponds and deposits the crystallized waste salts back at the mine.

Magnesium oxide scrubbing

The magnesium oxide process differs from any of the previously described processes in that it is a "regenerable" or "salable product" process rather than a "throwaway" operation. In other words, it does not produce waste material—SO_2 removed from the flue gas is concentrated and used to make marketable H_2SO_4 or elemental sulfur.

Employing a slurry of MgO—or $Mg(OH)_2$—to absorb SO_2 from flue gas in a scrubber, this process yields magnesium sulfite and sulfate. When dried and calcined, the mixed sulfite/sulfate produces a concentrated stream (10% to 15%) of SO_2 and regenerates MgO for recycle to the scrubber. Carbon added to the calcining step reduces any $MgSO_4$ to MgO and SO_2.

In commercial applications, the scrubbing and drying steps would normally take place at the power plant. The regeneration, and the sulfur or H_2SO_4 production steps, might be performed at a conventional sulfuric acid plant. Alternatively, a central processing plant could produce sulfur from mixed magnesium sulfite/sulfate brought in from other desulfurization locations.

Wellman-Lord system

The Wellman-Lord process (sold by Davy Powergas, Inc.) is also a regenerable, or salable-product, system.

When coupled with other processing steps, it can make salable liquid SO_2, H_2SO_4, or elemental sulfur.

The W-L process employs a solution of Na_2SO_3 to absorb SO_2 from waste gases in a scrubber or absorber, converting the sulfite to bisulfite:

$$Na_2SO_3 + SO_2 + H_2O \rightleftharpoons 2\ NaHSO_3 \quad (6)$$

Thermal decomposition of the bisulfite in an evaporative crystallizer can regenerate sodium sulfite for reuse as the absorbent:

$$2\ NaHSO_3 \xrightarrow{\Delta} Na_2SO_3 + SO_2 + H_2O \quad (7)$$

The evaporative crystallizer produces a mixture of steam and SO_2 and a slurry containing sodium sulfite/sulfate plus some undecomposed $NaHSO_3$ in solution. As water condenses from the steam/SO_2 mixture, it leaves a wet SO_2-enriched gas stream to undergo further processing for recovery of salable sulfur values.

NON-UTILITY STATIONARY-SOURCE COMBUSTION

Non-utility stationary-source combustion includes residential, commercial, and industrial heaters as well as industrial boilers to make steam for processing needs and, in some cases, to generate electricity for internal plant use [7].

It is difficult to classify emission sources on a common basis according to flue-gas flowrate, because at least six factors appear to influence these rates: (1) stoichiometric firing rate, (2) excess air, (3) leakage, (4) corrections for stack temperature, (5) boiler-load factor, and (6) fan-flow safety margin [8]. As a comparison, every megawatt of electric-generation capacity requires treatment of about 3,750 normal m^3/h (2,200 Std. ft^3).

Most of the flue-gas-desulfurization systems applicable to utility boilers can also control non-utility combustion sources. However, some important differences exist in these applications. For one thing, utility boilers generally release flue gas at about 150°C, while other combustion sources may discharge flue gas in the 200 to 260°C range. This higher-temperature flue gas can serve more effectively for stack-gas reheat.

It also generally holds true that non-utility combustion applications operate with higher excess-air rates than utilities do. As a result, the SO_2 will be more dilute and the oxygen more concentrated in the flue gas. This affects those wet-scrubbing systems, such as dual-alkali and Wellman-Lord, which respond adversely to high oxidation rates (i.e., oxidation of the sulfur species collected in solution: sulfites to sulfates).

Control equipment for non-utility systems, smaller than utility systems, lends itself better to shop fabrication than to field erection. For small-size applications, "package" or skid-mounted systems are available.

Frequently, industrial users of FGD systems will have excellent outlets, or uses, for the sulfur compounds they remove. For example, the pulp-and-paper industry can regenerate sodium sulfite scrubbing solutions in its manufacturing operation. The chemical industry often has captive markets for sulfuric acid. The petroleum industry has sources of H_2S that it can use to make sulfur from the SO_2 recovered in a regenerable FGD system.

Preliminary results of a continuing survey of SO_2-control systems for non-utility combustion and process sources [9] lists 27 known industrial-boiler FGD applications in the U.S., including 17 operational systems (Table III). More recent estimates indicate that the actual number and capacity of systems described in this survey may actually amount to two or three times these figures.

Sodium-alkali scrubbing

Most industrial boilers use a sodium-alkali scrubbing process—basically as described for sodium carbonate scrubbing in the utility-boiler section, and using sodium carbonate, bicarbonate, or hydroxide—which converts sulfur in the flue gas to sulfite/bisulfite and sulfate (Fig. 2).

Disposal practices for spent sodium-scrubbing liquors include: Consume in pulp/paper manufacturing; discharge to evaporation ponds; treat (mainly by air-oxidation) and discharge to city sewer system; and treat and discharge to rivers.

Many sodium-alkali users want to regenerate scrubber liquor by treating the spent liquor with calcium hydroxide. This would actually give them dual-alkali systems with the attendant advantage of eliminating a liquid-waste stream.

Dual alkali

Table III shows that dual-alkali systems have become the second most-prevalent type for SO_2 control of industrial boilers. They may become the first choice as more sodium alkali systems are converted to dual-alkali in the face of new regulations that may limit disposal of liquid wastes containing large amounts of dissolved salts (Fig. 2).

Dual-alkali processes, like lime/limestone-slurry scrubbing, are throwaway systems. In the operation as a whole, lime is consumed to produce a wet solid waste (mainly calcium sulfite/sulfate) just as in lime-slurry

The Ralph M. Parsons Co.

Twin-designed H_2SO_4 plant uses double catalysis/double absorption process **Fig. 2**

U. S. industrial-boiler SO_2-control systems — Table III

Control system	No. of systems	Total equivalent capacity, MW*	Status
Sodium alkali scrubbing	12	783	Operational
	2	173	Under construction
	1	20	Not operating
Dual alkali	4	110	Operational
	2	200	Under construction
	1	12	Planned
	1	10	Not operating
Lime/limestone scrubbing	1	20	Operational
Wellman-Lord	1	100	Planned
Water scrubbing	1	1	Not operating
Citrate process	1	50	Under construction
Total	27	1,479	

*1 MW ≈ 4,500 kg/h (10,000 lb/h) of steam production.

scrubbing. Also, dual-alkali systems require a small amount of sodium-alkali makeup.

A solution of sodium sulfite/bisulfite and sulfate in a scrubber will absorb SO_2 from the flue gas or other waste gas. Only the sulfite is active in absorbing SO_2, forming bisulfite as in the Wellman-Lord system:

$$\textit{Absorption: } SO_3^{=} + SO_2 + H_2O \rightarrow 2\ HSO_3^{-} \qquad (8)$$

The bisulfite-rich liquor, treated with lime in a reaction tank, regenerates active alkali for recycle to the scrubber.

$$\textit{Regeneration: } 2\ HSO_3^{-} + Ca(OH)_2 \rightarrow SO_3^{=} + CaSO_3\downarrow + 2\ H_2O \qquad (9)$$

$$SO_4^{=} + Ca(OH)_2 \rightarrow 2\ OH^{-} + CaSO_4\downarrow \qquad (10)$$

The sulfate and sulfite precipitate as a hydrated mixed crystal, or as a gypsum phase ($CaSO_4 \cdot 2\ H_2O$) plus a hydrated mixed crystal, depending upon the concentration of dissolved species. Also, depending upon solution concentrations, the mixed crystal is predominantly calcium sulfite, with up to about 25% calcium sulfate coprecipitated. Sulfite/bisulfite oxidation by oxygen in the flue gas produces sulfate in the system.

Lime/limestone slurry

The only lime-scrubbing system listed in Table III is the A. B. Bahco system installed at Rickenbacker Air Force Base. It controls SO_2 and particulate emissions from seven industrial boilers, equivalent to about 3 MW each.

The Bahco system, developed in Sweden, is sold in the U.S. by Research Cottrell, Inc., through a licensing agreement. Of particular interest for industrial boiler-size applications, it is sold as a package unit—i.e., offered in several standard sizes up to about 50 MW equivalent, with most of the unit preengineered and constructed before installation. The Bahco system consists of a two-stage vertical scrubber with a cyclone-type mist eliminator. The two stages operate in series, using flue-gas flow to entrain or inspirate the scrubber slurry. Bahco systems can use either lime or limestone slurry.

The boiler system at Rickenbacker burns 3.6%-sulfur coal with about 160% excess air. The scrubber is preceded by a mechanical dust collector, which is about 70% efficient. Lime was used in the scrubber initially but now limestone serves effectively. Underflow waste solid from a thickener discharges into a five-acre (0.02 km^2) Hypalon-lined disposal pond located about 120 meters from the scrubber. Clear water from the pond returns to the process. The system started operation in February 1976 and generally achieves about 90% SO_2 removal.

Citrate process

The single application of the citrate process shown in Table III is a 50-MW-equivalent system being installed by St. Joe Minerals Corp. at its G. F. Weaton Power Station under a project cofunded by St. Joe, U.S. Bureau of Mines, and EPA. The station burns 2.5%-sulfur coal and supplies power for St. Joe's zinc-smelting operation, and also sells power to the local grid. The process, being installed by United Engineers and Morrison-Knudsen, is designed to obtain 90% SO_2 removal. It should start operation in September 1978.

SO_2 is absorbed by a solution containing sodium sulfite, bisulfite, sulfate, thiosulfate, and polythionate. Citrate ion is present as a pH buffer, tying up hydrogen ions as un-ionized citric acid. The spent solution reacts with H_2S to precipitate elemental sulfur. The sulfur is concentrated by oil flotation or froth flotation, and the concentrated sulfur slurry then passes into a melter where the molten sulfur is drawn off. Approximately two thirds of the sulfur is used to produce H_2S by reaction with methane or water gas ($CO + H_2$), for recycle to the process. Regenerated solution must be processed to remove sulfates formed by oxidation in the absorbent loop. The sodium sulfate purge is reported to be 1 to 2% of the recovered sulfur [*10,11*].

METAL SMELTING

The metal-smelting processes—mainly copper, lead, and zinc—are second only to combustion processes in emissions of SO_2 in the U.S. Copper, lead, and zinc occur naturally as sulfides. To produce the metal, the sulfide ores are concentrated and then roasted or sintered to convert them to the metal oxides, with liberation of SO_2. The impure metal is then recovered by reduction of the oxides in electric reverbatory furnaces and flash-smelting systems.

Copper sulfide concentrates are normally roasted in either multiple-hearth or fluidized-bed roasters to remove the sulfur. About half of the smelters eliminate the roasting step and remove the sulfur as SO_2 by blowing the ore with air during the smelting operation. Lead sulfide, converted to the oxide in a sintering machine, is then reduced to the metal in a blast furnace. Fluidized-bed or multiple-hearth roasting, occasionally followed by sintering, removes sulfur from zinc ores. Metallic zinc is produced from the roasted ore by the horizontal or vertical retort process, or by the electrolytic process for high-purity zinc.

In the U.S., only two processes are used to control SO_2 emissions from copper, lead, and zinc smelters: (1) manufacture of byproduct sulfuric acid in single- and double-absorption systems, and (2) DMA N,N-dimethylaniline) process is the other (used at only two locations in the U.S.). In Japan, wet-limestone scrubbing and magnesium oxide scrubbing are also used. In British Columbia, the Canadians have used ammonium bisulfate scrubbing.

Smelter gases are classified as "weak" or "strong" SO_2 gases if SO_2 concentration runs, respectively, below or above 3.5 to 4%. These gases generally range in temperature from 600 to 1,000°C, considerably hotter than gases emitted from combustion of fossil fuels. As a first step in any processing for SO_2 removal, these gases must be conditioned by injecting water to cool and humidify them and to remove residual particulate matter. Storage and shipping considerations call for acid strengths of 93% and greater. Since the acid strength depends upon the SO_2 and moisture concentration of the processed gas, SO_2 concentration levels below 3.5 to 4% would require other, more expensive, means of gas cooling/drying (water balance) to produce concentrated sulfuric acid.

In addition, since the conversion of SO_2 to SO_3 is an exothermic reaction, sulfuric acid manufacture from smelter gases containing over 4% SO_2 is autogenous (requires no external heat). However, weak SO_2 streams do require heat to raise the temperature of the gases to the desired temperature (425 to 460°C) for catalytic conversion.

In the U.S., all of the strong (and none of the weak) SO_2 smelter gases are processed to control SO_2, although New Source Performance Standards (NSPS) will require control of both on any new plants. The preferred method of controlling SO_2 from new smelting plants is to design the plants initially to produce only strong SO_2 gas streams that can be controlled with byproduct sulfuric acid plants. Smelter-gas SO_2 concentration depends upon air leakage into the system and upon whether or not a large amount of fuel is burned with the amount of air required to sustain the roasting or smelting operation. Conversion of sulfides to oxides is exothermic, and so, with proper design, can be autogenous. In copper smelting, reverberatory furnaces that burn large amounts of fuel over the charge, produce gases containing 0.5 to 1.5% SO_2. The newer electric furnaces and flash-smelting systems, on the other hand, produce gases containing 4 to 6%, and 10 to 12%, SO_2 respectively.

Byproduct sulfuric acid manufacturing

The manufacture of sulfuric acid from smelter gas streams involves gas-conditioning, drying, catalytic conversion, and absorption [*12*].

Gas-conditioning may require a scrubbing tower, a cooler, and an electrostatic mist-precipitator, or other equivalent equipment combinations. To prevent acid contamination, minimize catalyst fouling, and reduce corrosion, this step removes particulate matter and cools the gas down to about 55°C.

The strong-SO_2 gases are dried in a tower using 93% sulfuric acid to absorb the moisture. In sulfuric acid plants, using weak-SO_2 gas feed, refrigeration may be necessary for part of the drying process.

Catalytic-converter feed-gases are preheated to 425 to 460°C by heat exchange with gases in the converter. The conversion reaction ($SO_2 + 1/2 O_2 \rightarrow SO_3$) is exothermic. The gas stream then enters a three- or four-stage catalytic converter that uses a vanadium pentoxide catalyst to convert the SO_2 to SO_3. A staged reactor permits the reaction to take place at successively lower reaction temperatures, taking advantage of the trade-off between kinetics and equilibrium. High temperature increases reaction rate, but at the expense of conversion efficiency.

Gas leaving the final converter stage is cooled in a heat exchanger and fed to an absorber where concentrated sulfuric acid absorbs the SO_3. The gas stream leaving the absorber—containing small amounts of unabsorbed SO_3, unconverted SO_2, nitrogen, and oxygen—vents to the atmosphere.

In double-absorption (or double-contact, double-absorption) systems, the gas stream from the second or third stage of the reactor is cooled and passed through an absorber that uses concentrated acid to absorb SO_3. The gas, minus the absorbed SO_3, is then reheated to converter temperature and reintroduced into the third or fourth reactor stage. This double absorption allows a greater conversion of SO_2 to SO_3.

Typically, single-absorption sulfuric acid plants achieve an overall conversion efficiency of 98%, and the double-absorption plants can achieve in excess of 99.7% conversion (emissions of less than 250 ppm SO_2) [*13*]. Accordingly, the single-absorption plants generally do not meet NSPS, but the double-absorption plants do.

Dimethylaniline/xylidine process

In this process, the aromatic amines, DMA (N,N-dimethylaniline) or xylidine, absorb SO_2 from the gas stream. Xylidine is used in aqueous solution, DMA in an anhydrous form [*12*]. (U.S. systems use DMA as the absorbent.) The American Smelting and Refining Co. (ASARCO) developed the DMA/xylidine process in the late 1940s.

CHEMICAL INDUSTRY

Sulfuric acid plants

Sulfuric acid plants represent by far the major emitters of SO_2 in the chemical industry. All commercial sulfuric acid is made by either the lead-chamber process or the contact process. However, the contact process accounts for almost all of the acid produced in the U.S. and therefore is the only one considered here [*14*]. Categorized by raw material, 68% of the acid is produced in sulfur-burning plants; 18.5% by processing spent acid, refinery sludge, or hydrogen sulfide; and 13.5% by processing metal-smelter gases.

The identified control methods for U.S. sulfuric acid plants include double absorption, Wellman-Lord, and wet-limestone scrubbing, all of which were discussed previously [*9*]. The preferred method of SO_2 control with new, large contact sulfuric acid plants is the use of double-absorption (or double-contact, double absorption)—an integral part of the plant. Other SO_2 control methods include ammonia scrubbing and molecular-sieve absorption.

Wood-pulp processes

Wood is processed to produce cellulose pulp for paper, wallboard, particle-board, and plywood by dissolving the lignin that binds the cellulose fibres together [*14*]. The principal processes used in chemical pulping are the kraft, sulfite, neutral sulfite semichemical (NSSC), and soda.

The kraft process uses an aqueous solution of sodium sulfide and sodium hydroxide in a pressurized digester, to dissolve the lignin from wood chips. This process accounts for about 60 to 65% of all the pulp produced in the U.S. The spent cooking liquor, called black liquor, is concentrated and burned in a recovery furnace to produce heat and dispose of the organic matter. By controlling the amount of excess air, Na_2CO_3 and Na_2S are collected in the bottom of the furnace as a molten smelt. This is reused after treatment with slaked lime to recausticize it. Calcium carbonate is calcined to lime for reuse.

The kraft process emits a very characteristic odor, due to the presence of such sulfur compounds as H_2S, methyl mercaptan, and dimethyl sulfide. These compounds come from the various black-liquor concentration steps and from the recovery furnace. The various reduced-sulfur compounds can be incinerated to SO_2 (sometimes in the lime kiln) to cut down on the stronger odors that characterize reduced sulfur compounds.

Most of the remainder of the pulp is produced by the acid sulfite and NSSC processes, which emit mainly SO_2. The acid sulfite process uses a sulfurous-acid-based solution buffered with sodium, magnesium, calcium, or ammonium ion. The NSSC process uses sodium sulfite and bicarbonate solution. The SO_2 scrubbed from waste gases from these processes can be used as a makeup chemical for the system.

FMC Corp.

Soda ash scrubbing system is effective in reducing high SO_2 concentrations **Fig. 3**

PETROLEUM REFINING, CLAUS PLANTS

Federal NSPS for SO_2 control in petroleum refineries, unlike other SO_2-control regulations, can limit the concentration of H_2S to about 160 ppm in the fuel gas burned, as a means of controlling SO_2 emissions, since most of the SO_2 emissions result from burning or incinerating gases containing hydrogen sulfide and other reduced-sulfur compounds. It is usually more economical in such cases to control SO_2 emissions by desulfurizing the fuel—treating it with an amine to absorb H_2S from the refinery gas. Subsequent regeneration of amine scrubbing liquid produces a concentrated H_2S stream, from which a Claus process can make elemental sulfur.

The Claus process, now a well-established technology, is primarily used to convert H_2S or mixtures of SO_2 and H_2S to elemental sulfur, as follows:

$$H_2S + 3/2\ O_2 \rightarrow SO_2 + H_2O \qquad (11)$$

$$SO_2 + 2H_2S \rightarrow 3S + 2H_2O \qquad (12)$$

$$\text{(overall)} \quad 2H_2S + O_2 \rightarrow 2S + 2H_2O \qquad (13)$$

Claus plants can recover sulfur from sour natural gas, coke-oven gas, and other gases containing H_2S, in addition to refinery fuel gases.

In refineries, H_2S escapes from the barometric condensers on vacuum-distillation columns, from hydrocracking units, and from hydrogenation units. Regeneration of catalyst from catalytic-cracker units, combustion of spent acid sludges, and residues from various chemical treatment units produce the SO_2 emissions. Emissions of H_2S and SO_2 can be simultaneously treated in Claus plants to recover elemental sulfur, if the mole ratio of H_2S to SO_2 is greater than 2:1.

There are three principal approaches to tail-gas treatment in Claus systems [*15*]:

1. Continuation of the Claus reaction at lower temperatures on solid catalysts or in liquid media.
2. Catalytic hydrogenation of SO_2, COS, CS_2, and CH_3SH to H_2S, which is recovered by absorption.
3. Incineration of tail gas to convert all sulfur compounds to SO_2 followed by treatment through an SO_2-control system.

In the first category are the Sulfreen and the IFP-1500 processes. The Sulfreen process continues the Claus reaction in a fixed-bed reactor on a carbon or alumina catalyst at reduced temperature. The sulfur is retained on the catalyst and must be removed during regeneration by a hot inert gas. In the IFP-1500 process [*16*] the Claus reaction is extended at 120–130°C in a packed tower where Claus tail-gas contacts polyethylene glycol solvent containing a proprietary carboxylic acid catalyst. Sulfur is drawn off from the bottom of the tower as a liquid through a seal leg.

The Beavon and Shell SCOT (Shell Claus Offgas Treating) processes in the second category, are expected to make the largest impact in achieving the proposed NSPS for Claus plants. These processes use a cobalt molybdate catalyst to promote hydrolysis of CS_2 and COS, and provide hydrogenation of SO_2 to H_2S [*15*]. The SCOT process concentrates H_2S by absorbing it in an alkanolamine solution and then recycles it to the Claus plant. The Beavon process uses the Stretford process to oxidize the H_2S directly to elemental sulfur, while the SCOT process simply recycles the recovered H_2S to the Claus plant.

The third category of Claus tail-gas processes can use any acceptable SO_2-control process, after incineration of the tail gas, to convert all gaseous sulfur compounds to SO_2. These controls include those previously described, but certainly favor processes, such as Wellman-Lord, which produce a concentrated SO_2 stream. The SO_2 can then be recycled to the Claus plant to increase the yield of sulfur.

MINERALS PROCESSING

No Federal NSPS limits exist for SO_2 emissions from minerals-processing plants. However, some local regulations do require SO_2 control in certain areas.

Some of the industrial operations that emit sulfur oxides include: portland-cement kiln operations, fiberglass manufacturing, minerals- and chemicals-production from sulfate or sulfide ores, clay kilns, and production of expanded aggregate. In some mineral-process operations, sulfur in those minerals containing sulfates is reduced to SO_2 by a fossil fuel or coke that is burned in direct contact with the mineral. The fossil fuel itself also produces some of the SO_2 during combustion.

Some of the SO_2-control processes used in this type of operation include water scrubbing (with or without alkali addition for pH control, lime scrubbing, and dual-alkali scrubbing. In some cases, the mineral dust emitted from the process as particulate matter is itself alkaline and, when scrubbed from the process waste-gas in a wet scrubber, aids in control of SO_2.

An example of a mineral-process control system is the dual-alkali (or double-alkali) system installed at FMC's Modesto, Cal. (Fig. 3), chemical plant controlling emissions from two (barium and strontium) sulfate-reduction kilns [*17*]. The plant produces barium and strontium chemicals such as oxides, chlorides, carbonates, and nitrates by reducing the sulfates to oxides in the kilns, followed by further chemical processing. Charged with ore and coke, the kilns produce a high-temperature off-gas containing 4,000 to 8,000 ppm SO_2 at a maximum rate of about 26,000 normal m^3/h. This control system reportedly has achieved greater than 99% SO_2 removal, with greater than 95% operability since December 1971.

References

1. *Electrical Week*, June 13, 1977, pp. 1–3.
2. Princiotta, F. T., and Kaplan, N., "Control of Sulfur Oxide Pollution from Power Plants," Oct. 1972.
3. Environmental Quality—1976, Seventh Annual Report of the Council on Environmental Quality, Sept. 1976.
4. U.S. Environmental Protection Agency, Position Paper: "Regulation of Atmospheric Sulfates," Research Triangle Park, N.C., Office of Air Quality Planning and Standards, EPA-450/2-75-007 (PB-245-760), Sept. 1975.
5. Slack, A. V., and Hollinden, G. A., "Sulfur Dioxide Removal from Waste Gases," 2nd ed., Noyes Data Corp., Park Ridge, N.J., 1975, p. 137.
6. Gerstle, R. W., and Isaacs, G. A., "Survey of Flue Gas Desulfurization Systems, Reid Gardner Station, Nevada Power Co.," EPA Report No. EPA-650/2-75-057 (PB-246-852/AS), Oct. 1975.
7. Putnam, A. A., et al., "Evaluation of National Boiler Inventory," EPA Report No. EPA-600/2-75-067 (PB-248-100/AS), Oct. 1975.
8. Choi, P. S. K., et al., "SO_2 Reduction in Non-Utility Combustion Sources—Technical and Economic Comparison of Alternatives," EPA Report No. EPA-600/2-75-073 (PB-248-051/AS), Oct. 1975.
9. PEDCo Environmental, Inc., "Survey Report on SO_2 Control Systems for Non-Utility Combustion and Process Sources," EPA Contract No. 68-02-2603, Task No. 4, May 1977.
10. Ponder, W. H., and Christman, R. C., "The Current Status of Flue Gas Desulfurization Technology," presented at the 68th Annual American Pollution Control Assn., Boston, Mass., June 15–20, 1975.
11. Nissen, W. I., et al., "Citrate Process for Flue Gas Desulfurization—A Status Report," EPA Report No. EPA-600/2-76-136b (PB-262-722/AS), May 1976, pp. 843–864.
12. Mathews, M. C., et al., "SO_2 Control Processes for Non-Ferrous Smelters," EPA Report No. EPA-600/2-76-003 (PB-251-409), Jan. 1976.
13. Mandelik, B. G., and Turner, W., Selective oxidation in sulfuric and nitric acid plants: current practices, *Chem. Eng.*, Apr. 25, 1977, pp. 123–130.
14. U.S. Environmental Protection Agency, "Compilation of Air Pollutant Emission Factors," AP-42, Feb. 1976, pp. 5.17-1 to 5.17-8.
15. Semrau, K., Controlling the Industrial Process Sources of Sulfur Oxides, in "Sulfur Removal and Recovery from Industrial Processes, "Advances in Chemistry Series 139, American Chemical Soc., Washington, D.C., 1975, pp. 1–22.
16. Barthel, Y., et al., Sulfur Recovery in Oil Refineries Using IFP Process, in "Sulfur Removal and Recovery from Industrial Processes," Advances in Chemistry Series 139, American Chemical Soc., Washington, D.C., 1975, pp. 100–110.
17. "Capabilities Statement Sulfur Dioxide Control Systems," FMC Corp., Technical Report 100, Mar. 1976.

The authors

Michael A. Maxwell is EPA's Chief of Industrial Environment Research Laboratory, Emission/Effluent Technology Branch, Research Triangle Park, NC 27711. His primary responsibilities include the directing of nonregenerable flue gas desulfurization programs. Before, he was discharged from the Air Force with the rank of Captain after having served four years in the Air Force Systems Command's Research and Technology Div. He holds a B.S. degree in chemistry from The Citadel, and attended graduate school at the University of Georgia. He is a member of the American Chemical Soc.

Norman Kaplan is project officer in EPA's Utilities and Industrial Power Div. of Industrial Environment Research Laboratory, Research Triangle Park, NC 27711, where he has been involved for 5 years mostly with the development and demonstration of dual-alkali technology, and in full-scale industrial-boiler systems. He has served also as project officer in the General Motors industrial-test program. Before, he worked for General Electric Co. in the production of resins and with Allied Chemical Corp. as unit supervisor. He holds a B.S. in chemical engineering from the City University of New York.

Section XIII
WASTEWATER TREATMENT/RECOVERY PROCESSES

New extraction process wins acetic acid from waste streams

The route's low operating and capital costs promise to free previously uneconomic-to-recover acid from chemical-plant, refinery and other effluents.

☐ A special solvent system is opening up new possibilities for retrieving acetic acid from a wide variety of waste waters.

For many such streams, notes Hydroscience Environmental Systems (Knoxville, Tenn.), the acid appears at such a low concentration that the tab for recovery by conventional extraction processes can outweigh the acid's value. However, the firm now is grooming a route, featuring the new solvent, that largely overcomes this barrier. Hydroscience claims that its technique can economically win acid from about three-quarters of these low-acid-level effluents.

Acetylation and esterification processes expectedly turn out a wastewater containing acetic acid, but the acid also crops up in offstreams elsewhere, such as those from other chemical syntheses, high-pressure petroleum-refining operations (e.g., cracking and reforming), and wood pulping. In many cases, these units can spew out a sizable amount of the acid. For instance, Hydroscience points out, one pulp-and-paper producer is losing about 10 million lb/yr as part of a foul condensate stream from a mill's chemical-recovery circuit.

Conventional extraction systems, using a solvent such as ether, can handle effluents containing as little as approximately 10% acetic acid, says Hydroscience. But below that concentration, the company adds, problems arise: the acid recovery demands a large amount of solvent, adding to inventory and process equipment size, and also boosting energy requirements to strip the raffinate and fractionate the solvent from the acid.

Hydroscience lessens these difficulties by using trioctylphosphine oxide (TOPO) as the key component of the solvent stream. The company calls the chemical an excellent extractant of low (down to as little as 0.5%) concentrations of acid in aqueous solutions.

Patented for such applications by Dow Chemical Co. (Hydroscience's parent firm), TOPO boasts good chemical stability, high boiling point, and low solubility in water. However, the chemical also features a high melting point, and a density close to that of water. This pair of properties forced the Knoxville company to find another solvent component—to act as a carrier for TOPO to avoid solidification problems, and to ease the separation of the acetic acid from the oxide. Hydroscience ultimately opted for a proprietary, kerosene-like, aliphatic hydrocarbon.

FOUR-PART FLOWSCHEME—In the process, as shown on the diagram below, wastewater contacts the solvent system in a conventional, countercurrent multistage extractor. Acetic acid and some water go into the organic, extract phase. (The composition of the extract depends on such parameters as the ratio of solvent to wastewater, and the temperature, pH and acetic acid concentration of the waste stream.) Raffinate passes to a sewer. Then, single-stage flash distillation removes a large part of the water from the organic stream. This water, which contains a small amount of acid, condenses and returns to the extractor.

The partially dehydrated extract passes to another distillation column. TOPO, the solvent carrier and some acetic acid leave as bottoms in this tower, and recycle to a solvent makeup tank. Meanwhile, the remaining bulk of acid and any residual water go overhead from the still to another fractionator. This unit drives off the water. Crude glacial acetic acid comes off as product, which may require some further refining, depending upon desired end-use. Trace solvent exits through the bottom of the tower.

For waste waters containing 0.5–10.0% acetic acid, the new process boasts unrivaled costs, says Hydroscience. As detailed in the table on p. 60, recovery cost of acid from a 200-gal/min, 3% stream runs almost one-third less using the TOPO-solvent system compared with using diethyl ether.

The economic advantage stems

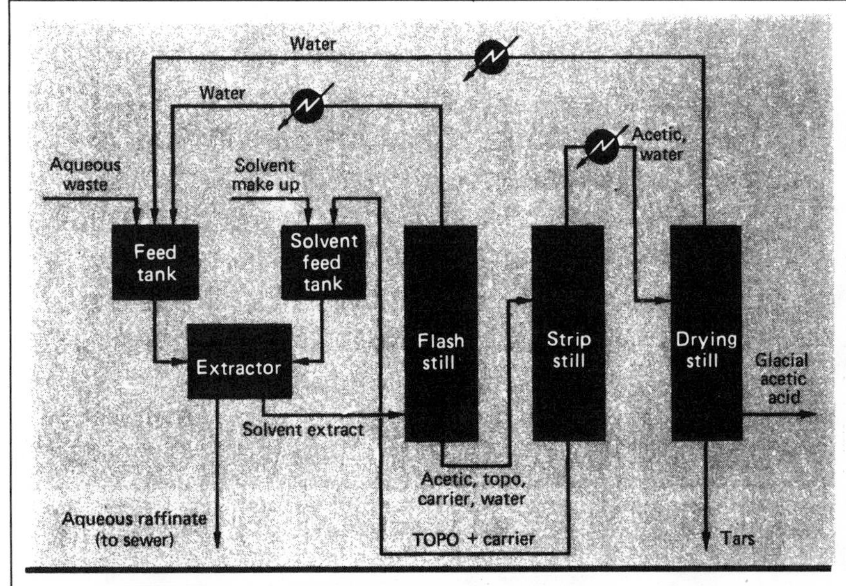

Recovery system cuts equipment size by featuring a lower-boiling solvent

Originally published March 15, 1976

New route slashes both capital and operating costs

	Solvent extraction system (for treating 200 gal/min of 3%-acetic-acid wastewater)	
	TOPO	Ether
Capital costs, $	2 million	3.2 million
Operating costs, $/lb acetic acid		
Utilities ($3/$10^6$ Btu)	0.006	0.027
Solvent losses	0.040	0.030
Labor	0.010	0.010
Maintenance (5% of capital)	0.004	0.007
Depreciation (10% of capital)	0.008	0.013
Total acetic-acid-recovery cost	0.068	0.089

Source: Hydroscience Environmental Systems

from both operating- and capital-cost savings. Cuts in steam consumption accrue because the solvent's low solubility in water obviates raffinate stripping, and its high boiling point allows acetic acid, rather than the much larger volume of solvent, to go overhead in the fractionation train. This overhead load runs considerably less than for conventional systems. So, the route can get by with smaller process equipment. This translates to cutting capital costs by as much as 40%.

The bill for solvent losses, however, can add up to the same or more for the new flow scheme. Actual volume losses wind up less than for conventional processes, says Hydroscience, but the carrier/TOPO system carries a selling price of approximately $5/gal, versus $1.20/gal for alternative solvents.

The company now is running a miniplant at Knoxville, currently seeking to establish design parameters for a solvent-extraction system to treat a pharmaceutical-plant effluent containing 5% acetic acid. If all goes well, the firm expects a full-scale unit to start up in 1978. Other prospects are brewing, too. Hydroscience adds that work is pending on circuits to handle pulp-and-paper, petrochemical and coal-processing waste streams.

Philip M. Kohn

Choosing a process for chloride removal

The presence of traces of organic chlorides in wastewater now looms as a major problem in many segments of the chemical process industries. Here is a summary of techniques for removing these compounds.

M. F. Nathan, *Crawford & Russell, Inc.*

☐ Many desirable properties—such as inflammability, biological inertness, useful vapor pressures, and solvent power—have for years made organic chlorides attractive both to chemical processors and domestic consumers for a wide range of uses varying from chemical intermediates, industrial solvents and dry-cleaning fluids to aerosol sprays and an anesthetic. In these uses, the organic chlorides have been generally considered inert and relatively harmless.

Examples of such chlorides: chlorobenzene, an early precursor for production of phenol; vinylidene chloride, a monomer for the production of plastic film; trichloroethylene, a vaporous degreasing agent for fabricated metal products; perchloroethylene, a dry-cleaning solvent for clothing; trichlorofluoromethane, an aerosol; chloroform, an anesthetic; and polychlorinated biphenyls (PCBs) as transformer fluid and condenser dielectric.

However, organic chlorides are no longer considered harmless. Various governmental regulations now require that they be kept out of the environment. Vinyl chloride monomer in air is now under stringent regulation by OSHA and EPA. Chlorofluorocarbons have recently been banned for use as propellants. OSHA is reported to be reducing exposure to chloroform, carbon tetrachloride and trichloroethylene. PCBs are banned. Further, the chlorination of water as a hygienic practice has come under scrutiny and probably will be curtailed since chlorination converts hydrocarbons in the water to carcinogenic chlorinated organics.

Consequently, removal of organic chlorides from water has become an important problem. This problem is compounded by existence of some of the same properties that in the past have made organic chlorides so desirable. Such chlorides are considered to be resistant to biodegradation and in some cases are completely nonbiodegradable.

Also, although the distribution of organic chlorides between water and hydrocarbons is favorable for extraction, such extraction results in a wastewater that contains solvent, and thus requires further treatment. Although organic chlorides can be adsorbed on activated carbon and polymeric resins, these adsorbents are too expensive to be used once and discarded, so that regeneration is necessary. In the case of carbon, the cost of first dewatering and then regenerating it in a multitiered reducing-atmosphere furnace, followed by an oxidizer to destroy the material, becomes expensive. In the case of polymeric resin, regeneration involves an expensive distillation system for recovering a regeneration solvent and rejecting the water that ends up in the regeneration streams.

Finally, although the high activity coefficients of many organic chlorides make steam stripping possible, these same high activities mean that tray efficiencies will be low or that the heights of packed towers will be excessive.

Match the process to the chloride

On the other hand, mutant bacteria have been developed to destroy many organic chlorides; and when

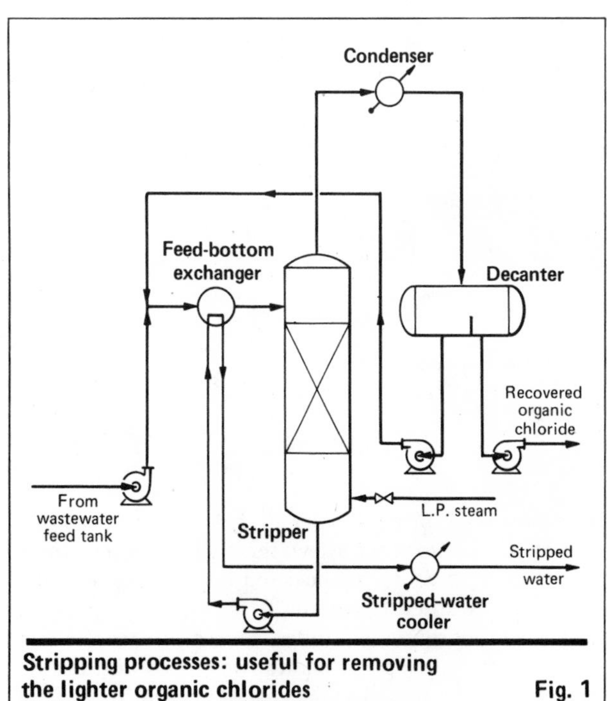

Stripping processes: useful for removing the lighter organic chlorides **Fig. 1**

Originally published January 30, 1978

These properties help identify the process for removing organic chlorides from wastewater **Table I**

Organic chloride	Formula	Mol. weight	Vapor pressure mm Hg	Vapor pressure @ °C	Organic solubility in water, weight %	Water solubility in organics, weight %
Methyl chloride (Chloromethane)	CH_3Cl	50.49	4028	25	0.46 @ 20°C	
Methylene chloride (Dichloromethane)	CH_2Cl_2	84.94	420	25	1.96 20°C	0.167 25°C
Chloroform (Trichloromethane)	$CH\ Cl_3$	119.38	197.4	25	0.8 @ 20°C	0.932 @ 25°C
Carbon tetrachloride (Tetrachloromethane)	CCl_4	153.84	115.25	25	0.16 @ 25°C	0.0116 @ 25°C
Ethyl chloride (Chloroethane)	CH_3CH_2Cl	64.92	1199	25	.574 @ 20°C	
Vinyl chloride (Chloroethylene)	$CH_2{=}CHCl$	62.5	2660	25	0.68 @ 20°C	0.9 @ 20°C
Ethylidene chloride (1,1, Dichloroethane)	CH_3CHCl_2	98.97	225	25	0.537 @ 30°C	0.115 @ 30°C
Ethylene dichloride (1,2 Dichloroethane)	CH_2ClCH_2Cl	98.97	82	25		0.187 @ 25°C
1,1,1 Trichloroethane (Methyl chloroform)	$CCl_3{-}CH_3$	133.42	125	25		0.0339 @ 20°C
Trichloroethylene (Trichloroethene)	$CCl_2{=}CHCl$	131.4	74.3	25	0.11 @ 25°C	
Tetrachloroethylene (Perchloroethylene, Ethylene tetrachloride)	$Cl_2C{=}CCl_2$	165.83	19	25	0.015 @ 25°C	0.0105 @ 25°C
Ethylene chlorohydrin (2-Chloroethanol)	$ClCH_2CH_2OH$	80.52	8	25	miscible with water	
Propyl chloride (1-Chloropropane)	$CH_3CH_2CH_2Cl$	78.54	335	25	0.27 @ 20°C	
Isopropyl chloride (2-Chloropropane)	$CH_3CHClCH_3$	78.54	505	25	0.31 H_2O @ 20°C	0.33
Allyl chloride (2-Chloro-1 propene)	$CH_2{=}CHCH_2Cl$	76.53	360	25	0.1	0.1
Propylene dichloride (1,2-Dichloropropane)	$CH_3CHClCH_2Cl$	112.99	50	25	0.27 @ 20°C	0.06 @ 20°C
Chloroacetone (1-Chloro-2-propanone)	$ClCH_2COCH_3$	92.53	20	30	9	
1,3-Dichloro-2-propanol (Glycerol dichlorohydrin)	$CH_2ClCHOHCH_2Cl$	128.99	1	28	9.9 @ 19°C	
n-Butyl chloride (1-Chlorobutane)	$CH_3CH_2CH_2CH_2Cl$	92.57	105	25	0.07	
sec-Butyl chloride (2-Chlorobutane)	$CH_3CH_2CHClCH_3$	92.57	155	25	0.1 @ 25°C	
tert-Butyl chloride (2-Chloro-2-methyl propane)	$(CH_3)_3CCl$	92.57	300	25	sparingly soluble in water	
Isobutyl chloride (1-Chloro-2-methyl propane)	$(CH_3)_2CHCH_2Cl$	92.57	142	25	insoluble in water	
Butylidene chloride (1,2-Dichlorobutane)	$CH_3CH_2CHCl_2$	127.02	760	113.5	practically insoluble in water	
Chlorobenzene (Benzene chloride)	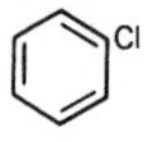	112.56	760	131.7	0.0512 @ 25°C	0.049 @ 30°C
o-Dichlorobenzene	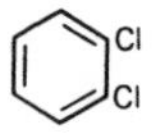	147.01	1.282 760	25 180.5	0.0145 @ 25°C	
m-Dichlorobenzene	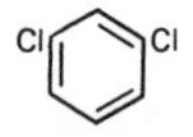	147.01	760	173	0.0123 @ 25°C	
p-Dichlorobenzene	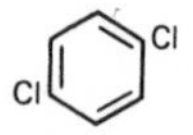	147.01	0.4 760	25 174	practically insoluble in water	
1,2,3 Trichlorobenzene	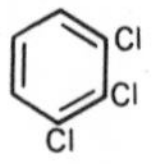	181.46	1	40	insoluble	

the chlorides represent a small fraction of the total organic content of the wastewater, such bacteria can be added daily to a biochemical oxidation system to destroy the unwanted compounds.

Also, some organic chlorides, such as chlorophenols, exhibit a solubility great enough to justify distillation and solvent loss, as well as the cost of preventing solvent from entering the environment. In the case of low concentrations, a nonregenerable carbon bed can be used at reasonable operating cost; or the regeneration of carbon or a polymeric resin can be accomplished with caustic (for chlorinated phenols, etc.). Finally, the one theoretical tray that can be achieved in a flash drum is often sufficient to attain adequate stripping of organic chlorides.

The key, therefore, to solving the problem of chloride removal lies in identifying the chlorides in a wastewater with their properties, like vapor pressures and solubilities. Once the compounds and their properties are identified, they can usually be matched to a suitable re-

These properties help identify the process for removing organic chlorides from wastewater **Table I cont'd**

Organic chloride	Formula	Mol. weight	Vapor pressure mm Hg	Vapor pressure @ °C	Organic solubility in water, weight %	Water solubility in organics, weight %
o-Chlorophenol	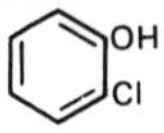 [structure: OH, Cl]	128.56	2.25	25	2.77	
m-Chlorophenol	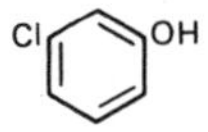 [structure: Cl, OH]	128.56	1	40	2.53 @ 20°C	17.7 @ 23.1°C
p-Chlorophenol	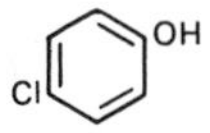 [structure: Cl, OH]	128.56	1	50	2.63 @ 20°C	15.98
2,4-Dichlorophenol	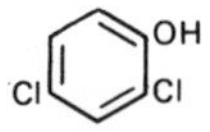 [structure: OH, Cl, Cl]	163.01	1	60	0.45 @ 20°C	
2,4,6 Trichlorophenol (Dowicide 25)	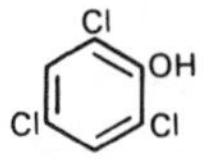 [structure: Cl, OH, Cl, Cl]	197.46	1	77	<0.1	
Phosgene (Carbonic dichloride)	$COCl_2$	98.92	1418	25	slightly sol. in water; hydrolizes	
Aldrin	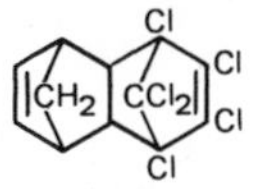 [structure: CH_2, CCl_2, Cl]	364.93	760	104	insoluble in water	
Chlordane	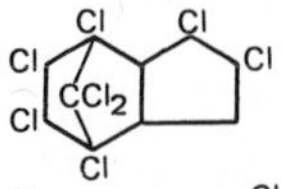 [structure: CCl_2, Cl]	409.8		155 (m.p.)	insoluble in water	
3,3' Dichlorobenzidine	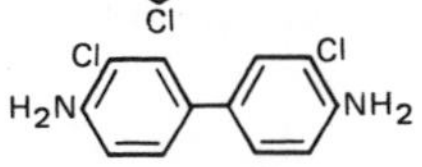 [structure: H_2N, Cl, Cl, NH_2]	253.13		132-33	almost insoluble in water	
Dieldrin	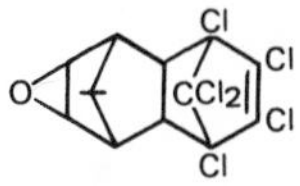 [structure: O, CCl_2, Cl]	380.93	1.8×10^{-7}	25	<0.1 ppm	
Endrin	 [structure: O, Cl_2]	380.93	2×10^{-7}	25	<0.1 ppm	
Heptachlor	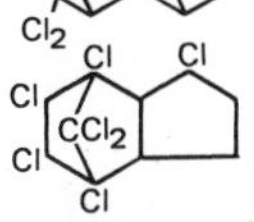 [structure: CCl_2, Cl]	373.35	4×10^{-4}	25		
Lindane (γ isomer)	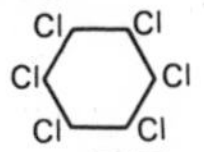 [structure: Cl]	290.85	9.4×10^{-6}	20	insoluble	
p,p'-DDD (TDE)	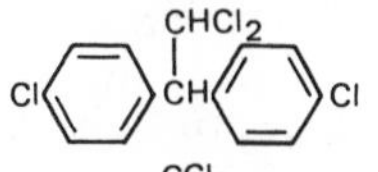 [structure: Cl, $CHCl_2$, CH, Cl]	320.05		109-10 (m.p.)		
DDE	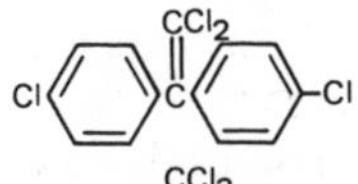 [structure: Cl, CCl_2, C, Cl]					
DDT	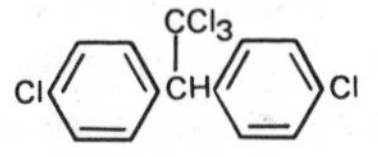 [structure: Cl, CCl_3, CH, Cl]	354.5	1.5×10^{-7}	20	practically insoluble	
Toxaphene (Chlorinated camphene—	Mixture of at least 175 poly-chloro-derivatives: $C_{10}H_{10}Cl_8$			65–90 (m.p.)	practically insoluble	

moval process. A list of potential organic chlorides and their properties is shown in Table I. Some processes for their removal are stripping, adsorption, biological oxidation, and extraction.

Stripping

A typical stripper system operating at atmospheric pressure is shown in Fig. 1. Wastewater containing organic chlorides is preheated against bottoms and fed to the top of a packed stripper tower. Low-pressure steam is introduced at the bottom of the tower. The mixture of steam and vaporized organic chlorides leaving the top is condensed and separated in a decanter, from which the water is returned to the feedstream while the organic layer is removed for disposal. After passing in heat exchange with the feedstream, the tower bottoms can be further cooled.

This system can also be operated under vacuum. Although vacuum operation eliminates the need for a feed-bottoms exchanger, it incurs the added costs of the

These data are typical of processes to strip ethylene dichloride (EDC) from wastewater **Table II**

Flowrate of wastewater stream	10 gpm
Temperature of wastewater stream	90° F
EDC Content of wastewater stream	0.87 weight %
Recovery of EDC	99.9%
Average stripping-tower temperature	221° F
Solubility of EDC in Water at 221° F	1.95 lbs/100 lbs water
Solubility of water in EDC at 221 °F	0.99 lbs/100 lbs EDC
Vapor pressure of wastewater at 221° F	25 psia
Tower design data	
Diameter	28 in.
Packing	20 ft of 3/4 in. Raschig rings
Low-pressure steam flow	1,565 lb/h
Bottoms pressure	2 psig
Bottoms temperature	225° F
Overhead pressure	1.5 psig
Overhead temperature	216° F
$K=\frac{y}{x}$ for hydrocarbon	408
Activity coefficient for hydrocarbon in water	269
Overhead vapor	865 lb/h of water vapor plus 43.4 lb/h EDC
Theoretical trays	less than 2

equipment and utilities for maintaining the vacuum. Also, the quantity of stripped water is increased with vacuum, since the vacuum steam-jet gets contaminated.

Finally, low tray-efficiencies often make it desirable to substitute a flash drum for a stripping tower.

In calculating a stripper, the tower temperature can be assumed equal to the boiling point of water at the tower pressure. Normally, the pressure differential over the tower is small, and an average equilibrium constant at an average tower temperature can be assumed for the organics. The presence of the hydrocarbon can be ignored in performing the heat balance to obtain the rate of vapor traffic in the tower. Vapor loads can be assumed constant, except for the top tray. The combined feed is generally at a temperature lower than the tower temperature, with consequent condensation of some of the stripping steam on the top tray.

As an alternative to ignoring the hydrocarbon in calculating the vapor load, a correction in top-tray vapor can be made on the basis that most of the hydrocarbon entering the tower is vaporized on the top tray. Thus the top-tray vapor can be condensed to supply the heat required to vaporize all the hydrocarbon, and the hydrocarbon vapor can then be added to the water vapor to obtain the total top-tray load.

Hydraulics for the stripping tower are the same as for other distillation systems; i.e., conventional packed-tower correlations or tray calculations are used to set pressure drop as a function of tower traffic.

In addition to the liquid and vapor loads for sizing tower diameter, the equilibrium constant for the organic chloride is needed to calculate the number of theoretical trays.

Equilibrium constant: When a water phase and a hydrocarbon phase coexist at the same temperature, the total pressure is equal to the sum of the vapor pressures of the two pure liquids. There is, however, a small amount of hydrocarbon in solution in the water phase, as well as water in the hydrocarbon phase; and the fugacity of the hydrocarbon in the water phase must equal its fugacity in the hydrocarbon phase.

In terms of activity coefficients, this fugacity relationship is shown as

$$(\gamma_{hc}P_{ohc}x_{hc})_{w\phi} = (\gamma_{hc}P_{ohc}x_{hc})_{hc\phi}$$

where γ_{hc} is the activity coefficient of the hydrocarbon, x_{hc} is the mol fraction of the organic chloride, and P_{oho} is the vapor pressure of the pure hydrocarbon. The subscripts $w\phi$ and $hc\phi$ refer to water phase and hydrocarbon phase, respectively.

As the mol fraction of organic chloride in the hydrocarbon phase approaches 1.0, the activity coefficient approaches 1.0. Thus the activity coefficient of the organic chloride in water can be determined from the saturation solubility data for the two phases:

$$(\gamma_{hc})_{w\phi} = (x_{hc})_{hc\phi}/(x_{hc})_{w\phi}$$

This activity coefficient can be assumed to apply from the saturation solubility down to zero solubility of the organic chloride in the water phase at a given temperature. The equilibrium constant for the organic chloride, $K_{hc} = y_{hc}/x_{hc}$, can be obtained from the vapor-phase/water-phase relationship:

$$(\gamma_{hc}P_{ohc}x_{hc})_{w\phi} = y_{hc}\pi$$

where π is the total pressure, and y_{hc} is the mol fraction in the vapor phase. Then:

$$K_{hc} = y_{hc}/x_{hc} = \gamma_{hc}P_{ohc}/\pi$$

The pressure at any point in the tower determines the temperature at that point. (It should be noted that the solubility of other hydrocarbons in the wastewater has the effect of reducing an activity coefficient calculated from pure hydrocarbon-water data.)

Theoretical trays: Assuming that the amount of organic chlorides in the stripped water has been set to meet some standard, there are infinite combinations of theoretical trays versus stripping steam that will remove the hydrocarbon from the water. The choice should be based on both economics and the variations expected in day-to-day operation. Also, there are practical limitations on the number of theoretical trays, since tray efficiencies for stripping organic chlorides are generally low, with correspondingly high packing height for a theoretical tray.

The stripped-water composition plus the amount of stripping steam can be used to determine the tower's bottom-tray composition through heat and material balances. This composition plus the equilibrium constant can then be used to calculate the concentration of organic chlorides in the vapor leaving the bottom tray; and a material balance on the organic chlorides entering and leaving the bottom tray then determines the amount of organic chlorides entering that tray in the liquid from the tray above. Since the volume of that liquid is determined by heat and material balances, the quantity of contained organic chlorides sets the composition, which with the equilibrium constant can be used

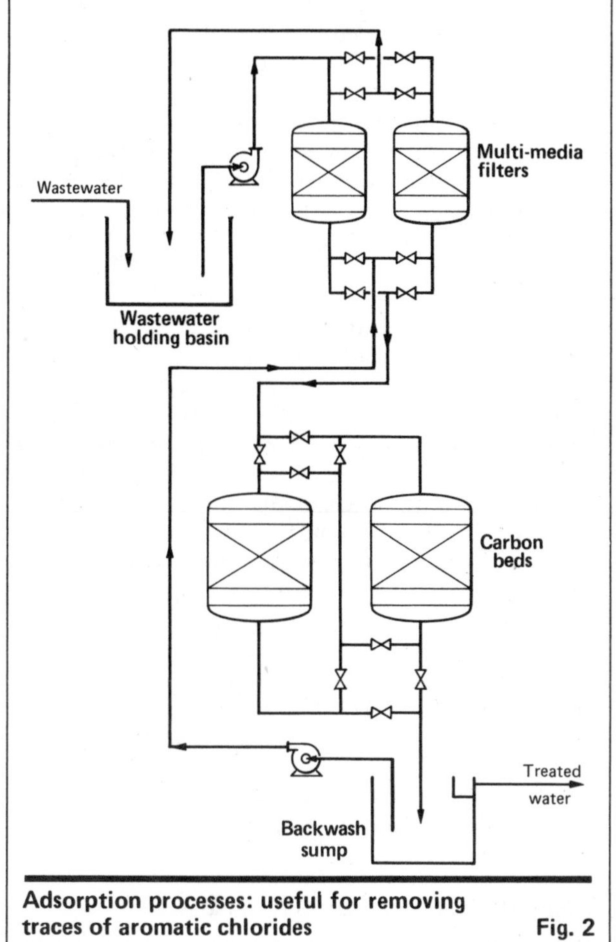

Adsorption processes: useful for removing traces of aromatic chlorides Fig. 2

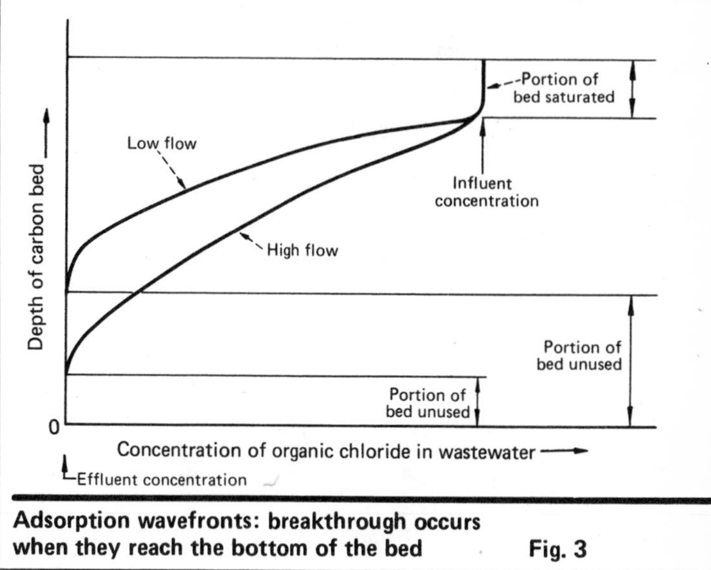

Adsorption wavefronts: breakthrough occurs when they reach the bottom of the bed Fig. 3

to calculate the organic chlorides leaving the second tray from the bottom.

This tray-to-tray calculation, starting from the bottom, is continued up the tower, until the calculated concentration of organic chlorides in the liquid equals or exceeds the concentration in the incoming feed. If such calculations show that the concentration of organic chlorides in the liquid from a tray above is the same as that leaving the tray just calculated, a pinch-point has been reached, and the calculations must be repeated using more stripping steam.

Once a tower has been calculated in this manner, day-to-day variations should be assumed and checked against the design. It may be desirable to design for more steam and fewer theoretical trays, rather than have an unstable situation, even though optimum economics may not be achieved.

Table II summarizes the design data for a system that will remove 99.9% of 0.87 wt % ethylene dichloride in a 10-gpm stream of wastewater occurring at 90°F. A feed/bottoms exchanger was not used in this example; its use would cut steam requirements substantially.

Adsorption on activated carbon

Activated-carbon adsorption of organics dissolved in water proceeds slowly, so that the plot of organic concentration versus length of passage through an adsorbent bed shows a gradual curve, or long wave-front (Fig. 3). At any given time, the first layer of carbon is saturated, while the last layer remains unused until the wave reaches that layer and "breakthrough" occurs.

In a typical carbon-adsorption flowsheet (Fig. 2), wastewater is first pumped downward over one of two parallel multi-media filter beds to remove suspended particles and undissolved hydrocarbons. From the filters, the wastewater passes through two carbon beds functioning in series. Each of these beds should be sized so that it alone can remove the organic chlorides to the desired level. Then when breakthrough occurs from the first bed, that bed can be bypassed and removed from service while its carbon is replaced, and subsequently returned to service downstream of the other bed. Maintaining two beds in series permits better carbon usage, since each bed can be saturated before it is removed from service.

A sump and pump, which are provided for backwashing the filter, should also be sized to permit periodic backwashing of the carbon beds, since they tend to accumulate suspended solids over a period of time, despite presence of the filters. The backwash stream is preferably returned to the wastewater holding basin, so that its solids may settle out. Otherwise, a separate filter must be provided to filter particles out of the backwash water.

Adsorption of chlorinated organics from water occurs in surface pores of the granulated activated carbon. These pores must thus be large enough to permit the organic chlorides to enter. Commercially available activated carbons, made from a variety of base materials, typically have surface-pore areas from 500 to 1,400 m^2/g [*3*]. In addition to its effect on pore size and surface area, the choice of base material also affects the hardness of the carbon granules and thus their resistance to abrasion. Abrasion resistance is particularly important when the carbon is to be transported, as a slurry for example, in a regeneration process.

For any given carbon, the organic compounds differ in their tendency to be adsorbed, according to polarity, molecular structure and molecular weight [*4*]. Highly

Adsorption isotherms for toxic organic chlorides on activated carbon **Table III**

Compound	pH	Wastewater concentration: Initial microgram/1	Wastewater concentration: Effluent microgram/1	% reduction	Carbon's adsorptive capacity*
Aldrin	7.0	48	<1.0	99+	30
Dieldrin	7.0	19	0.05	99+	15
Endrin	7.0	62	0.05	99+	100
DDT	7.0	41	0.1	99+	11
DDD	7.0	56	0.1	99+	130
DDE	7.0	38	<1.0	99+	9.4
Toxaphene	7.0	155	<1.0	99+	42
Arochlor 1242 (PCB)	7.0	45	<0.5	99+	25
Arochlor 1254 (PCB)	7.0	49	0.5	99+	7.2

*Mg of toxic chemicals adsorbed/g carbon

polar substances are generally more soluble and do not adsorb easily. Aromatic rings, on the other hand, adsorb well, and the addition of chlorine to such molecules increases this tendency. Higher molecular weights are also conducive to adsorption, so long as the molecules are smaller than the pores. Large molecules have a limited surface-area capacity [*4*].

Suppliers of carbon will usually recommend that laboratory isotherms be run, and column tests carried out, for a given wastewater. In such tests [*5*], it is important that the carbon be ground, in order to allow the test bed to reach equilibrium within a reasonable time. The column tests will permit estimates of contact time, carbon usage rates, and requirements for pretreatment. These tests thus permit sizing the bed in terms of time onstream before being bypassed for carbon renewal.

Contact times can vary considerably, from as little as 20 minutes to several hours and more. Flowrates over carbon generally range from 2 to 10 gpm/ft^2. Beds as long as 40 ft have been reported [*5*] and even longer beds can be used. A flow of 2 gpm/ft^2 in a 40-ft bed results in a contact time of 150 min.

The data in Table III [*6*] are based on isotherms for various pesticides and polychlorinated biphenyls, showing 99+% removal and 7 to 130 milligrams of toxic chemicals adsorbed per gram of carbon. Based on these data for DDE, and assuming a contact time of 60 min., calculations indicate the carbon would have to be changed once in 15 years. Actually, it would probably have to be changed more frequently. But even so, this is essentially a nonregenerative system.

Pressure drop through the bed—a function of hydraulic loading, bed depth and particle size—is typically 0.5 to 2.0 in. of water per foot at a flow of 2 gpm/ft^2, and 3.5 to 10.0 in. of water per foot at a flow of 10 gpm/ft^2. Carbon suppliers generally have data on pressure drop per foot of bed, and will make such data available.

The EPA has published a process design manual for carbon adsorption [*2*], indicating that carbon-contacting systems can be upflow or downflow, series or parallel, single- or multi-stage, pressure or gravity downflow, packed-bed or expanded-bed upflow. The advantages and disadvantages of each type, and how they should be related to a specific problem, are outlined.

Depending on how the carbon beds are managed, they can be nonregenerated or regenerated onsite, or regenerated under contract. If they are regenerated onsite, multiple-hearth reducing-atmosphere furnaces or rotating kilns are required [*29*]. While carbon can adsorb most organic chlorides from water, its major use will probably be for wastewaters or relatively small volumes of waste containing large molecules of low vapor pressure. The carbon for such systems is replaced as often as several times a year; i.e. the systems are not regenerable onsite.

Since carbon is abrasive, long-radius bonds are recommended for slurry service; and the piping should be designed for multiple cleanouts and flushing connections. The EPA design manual recommends ball valves and plug valves for slurry service, and suggests controlling flow by throttling the water supply rather than throttling the slurry.

Finally, industrial users have an alternative to owning a carbon adsorption unit. At least one of the carbon suppliers will design and build such a unit tailored to the need, using a modular system. The unit is leased and operated by the carbon supplier, who also makes guarantees, does analytical work, carries out maintenance, and removes and regenerates carbon offsite for a monthly fee [*8*].

Adsorption on polymeric resins

The flowsheet for polymeric adsorption will be similar to that for carbon-bed adsorption, with a multimedia filter ahead of the adsorption beds. However, regeneration with a solvent requires a distillation system to separate the solvent from the adsorbed organic chloride as well as from water brought into the regeneration system during rinse cycles. Thus one tower is needed to distill out the water, and a second tower to separate solvent from the organic chloride.

Pressure drops through polymeric adsorption beds range from about 8 in. of water per foot of bed at 2 gpm/ft^2 to about 75 in. of water per foot of bed at 10 gpm/ft^2. Data for the adsorption of chlorinated phenols [*11*] are shown in Table IV, along with information on the effects of sodium chloride in the wastewater [*12*].

When a bed is ready for regeneration, excess water is drained to a level that will keep the bed wet. Then, solvent is passed at a low flowrate (e.g., 1 to 4 bed-volumes/h) through the bed. A rinse with washwater is recommended before the bed is put back onstream. The regeneration liquids are then separated in the distillation system.

Although the polymeric resins are more expensive per pound than activated carbon and generally have a lower adsorptive capacity, they also have a higher density—about 44 lb/ft^3 compared with 30 lb/ft^3—so that the volumetric capacity is more nearly equivalent.

A new carbonaceous adsorption material has been developed that can be regenerated with steam after removing low levels of halogenated hydrocarbons from drinking water. Such steam regeneration, however, would apply only to light materials, such as chloroform.

Suppliers of resin will aid in engineering the system and in performing test work required. The patent situation and the need to license must be investigated by the user.

Biological oxidation

For organic chlorides to be successfully removed in a biological oxidation system, the microbial population of that system must contain components capable of withstanding and of degrading the toxic organic chlorides. The rate at which these organisms develop, in relation to other organisms and to the total amount of food available, will determine their relative concentration in the biomass. On the other hand, mutant bacteria for degrading specific organic chlorides can be added daily to a conventional activated-sludge system in order to enable the system to handle the organic chlorides.

A sufficient concentration of mutant bacteria must be maintained to destroy the organic chlorides. The amount of such bacteria required for any given waste system must be determined via laboratory studies.

Mutant bacteria capable of degrading many of the aromatic organic chlorides have been developed by commercial laboratories (Tables V, VI) [*13*]. As more chlorine is attached to the benzene carbon rings, the resultant compounds become more difficult to degrade [*14*]. However, 100% ring disruption has been achieved for mono-, di-, and even tri- chlorinated benzene; and once the ring has been broken, the chlorinated material is biodegradable. Table VI shows that chlorinated phenols can be destroyed.

Extraction

Wastewater is fed to the top of an extraction tower (Fig. 4), while solvent enters the bottom and flows countercurrent to the downward-flowing water. Rich solvent from the top of the extraction tower passes to a distillation tower, where the organic chloride is separated, and the regenerated solvent then returns to the extraction tower.

Raffinate water from the bottom of the extraction tower meanwhile passes to a steam stripper, where most dissolved solvent and unextracted chlorides are vaporized overhead along with water, and both are condensed. Solvent and water are then separated in a decanter, with the solvent from the decanter returning to the main solvent system, while the decanted water is recycled to the feed to the stripper.

Either a polar or an aromatic solvent can extract organic chlorides from water. Polar solvents, such as *n*-butyl acetate, have much higher distribution coefficients than aromatic solvents, such as benzene and toluene; and higher distribution-coefficients result in a lower solvent-to-water ratio for a given extraction, with consequent lower energy costs for separating the extracted organic chloride in the distillation column.

Whether aromatic or polar, however, some of the solvent is dissolved in the raffinate water, which is why the stripper is provided (Fig. 4). This stripper also removes part of the organic chloride not extracted from the water, and thereby serves to counterbalance the

Adsorption of chlorinated phenols on a polymeric resin — Table IV*

Flowrate: 0.5 gpm per cubic foot (15 min. contact time)

Material	Solubility in water, ppm	Solute in influent, ppm	Solute adsorbed, % at 0 leakage
m-Chlorophenol	26,000	350	5.45
2,4-Dichlorophenol	4,500	430	11.6
2,4,6-Trichlorophenol	900	510	27.2
m-Chlorophenol (13% NaCl)		350	7

*Source: Ref. 8

Times required to biodegrade organic chlorides — Table V*

Organic chloride	Concentration mg/l	Ring disruption, % Parent	Ring disruption, % Mutant	Time in hours Parent	Time in hours Mutant
Monochlorobenzene	200	100	100	58	14
o-Dichlorobenzene	200	100	100	72	20
m-Dichlorobenzene	200	100	100	96	28
p-Dichlorobenzene	200	100	100	92	25
1,2,3-Trichlorobenzene	200	87	100	120	43
1,2,4-Trichlorobenzene	200	92	100	120	46
1,3,5-Trichlorobenzene	200	78	100	120	50
1,2,3,4-Tetrachlorobenzene	200	33	74	120	120
1,2,3,5-Tetrachlorobenzene	200	30	80	120	120
Hexachlorobenzene	200	0	0	120	120

* Biodegradation by mutant Pseudomonas species at 30°C
Source: Ref. 9

Times required to biodegrade phenol and chlorophenols — Table VI*

Compound	Concentration mg/l	Ring disruption,% Parent	Ring disruption,% Mutant	Time in hours Parent	Time in hours Mutant
Phenol	200	100	100	25	8
o-Chlorophenol	200	100	100	52	26
m-Chlorophenol	200	100	100	72	28
p-Chlorophenol	200	100	100	96	33
2,4-Dichlorophenol	200	100	100	96	34
2,5-Dichlorophenol	200	60	100	120	38
2,3,5-Trichlorophenol	200	100	100	100	52
2,4,6-Trichlorophenol	200	100	100	120	50
Pentachlorophenol	200	7	26	120	120

* Biodegradation by mutant Pseudomonas species at 30°C
Source: Ref. 9

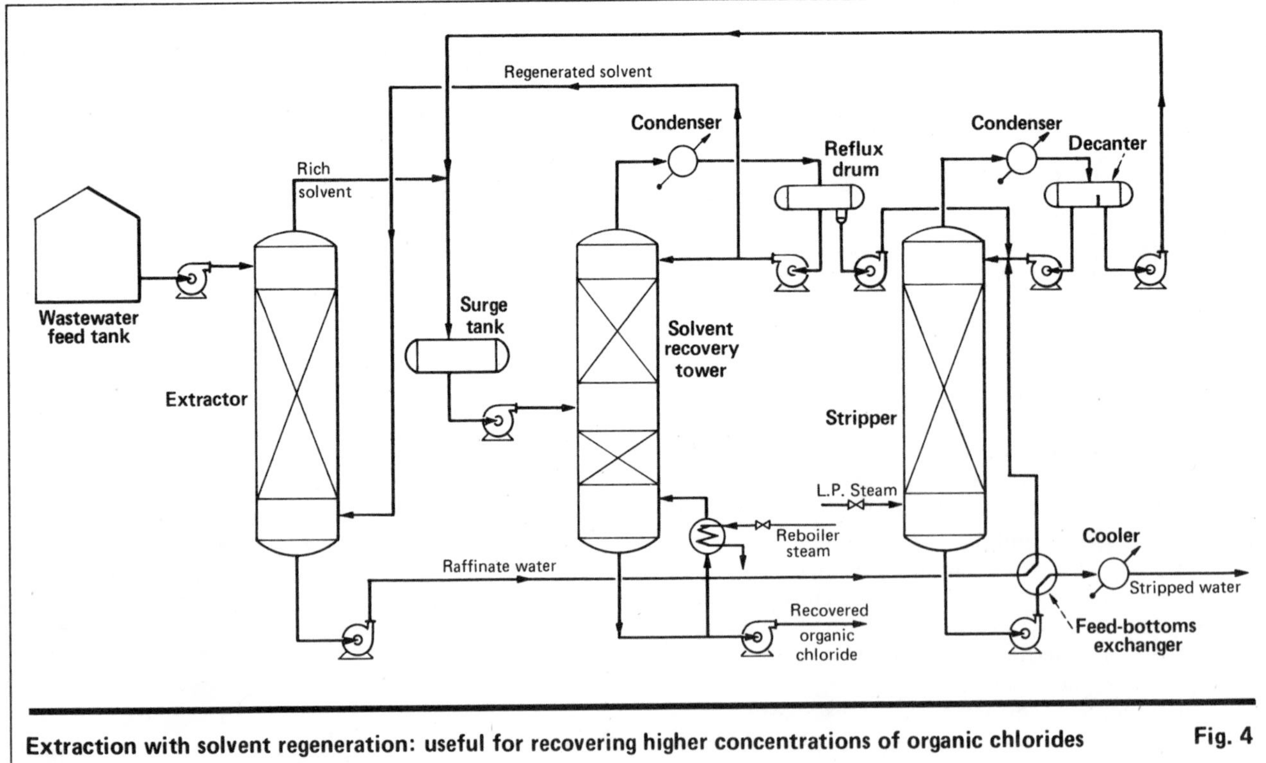

Extraction with solvent regeneration: useful for recovering higher concentrations of organic chlorides — Fig. 4

difference in distribution coefficients between polar and aromatic solvents. Although the polar solvents leave less organic chloride in the raffinate water, they are more water-soluble than the aromatic solvents, and the result is about the same total of organic-chloride-plus-hydrocarbon solvent removed by the stripper. Thus there is not much difference between polar and aromatic solvents for the flowsheet shown in Fig. 4.

A dual-extraction process has also been suggested to minimize energy requirements [*15*]. However, that approach requires two extraction towers and two distillation towers.

J. P. Earhart, others [*11*] have presented data on distribution coefficients, based on experimental data from the literature, together with solvent boiling points and solubility of solvent in water. The distribution coefficient is the weight-fraction of solute in solvent per weight-fraction of solute in the water.

Corrosion

Organic chlorides are not considered corrosive at normal temperatures, so carbon steel is permitted. However, dilute HCl is formed to the extent that organic chlorides are hydrolyzed in water solutions, and Monel is recommended up to temperatures of about 250°F and for cases where no O_2 is present.

If the HCl is neutralized, the resulting sodium chloride presents a potential for stress corrosion. Thus, 316L stainless steel has been used under process conditions where salts are not naturally concentrated, or up to 140 ppm sodium chloride at temperatures not more than 140°F. Above 140°F, and in situations where the sodium chloride is more concentrated, Carpenter 20 or titanium are often specified. In any given situation, however, a metallurgist should be consulted to make sure that a seemingly innocuous component doesn't alter the corrosiveness of the system.

References

1. U.S. Environmental Protection Agency, "Process Design Manual for Carbon Adsorption," Oct., 1973.
2. Calgon Corp., Pittsburgh, Pa., "Basic Concepts of Adsorption on Activated Carbon," Brochure 27-18.
3. Bernardin, F. E., Jr., Selecting and specifying activated carbon adsorption systems, *Chem. Eng.,* Oct. 18, 1976, p. 77.
4. Calgon Corp., Pittsburgh, Pa., Bulletin 20-52.
5. Hager D. G., Waste Water Treatment via Activated Carbon, *Chem. Eng. Prog.,* Oct., 1976, p. 57.
6. Calgon Corp., Pittsburgh, Pa., "Calgon Adsorption Service," Brochure 2702B.
7. Kennedy, D. C., Treatment of Effluent From Manufacture of Chlorinated Pesticides with a Synthetic Polymeric Adsorbent, Amberlite XAD-4, *Enviro. Sci. Tech.,* Feb., 1973, p. 138.
8. Rohm & Haas, Philadelphia, Pa., "Amberlite XAD-4," Technical Bulletin, 1976.
9. Polybac Corp., New York, N.Y., "Polybac," Technical Bulletin, 1973.
10. Zanitsch, Roger H., and Lynch, Richard T., "Selecting a thermal regeneration system for activated carbon," *Chem. Eng.,* Jan. 2, 1977, p. 95.
11. Earhart, J. P., others, Recovery of Organic Pollutants Via Solvent Extraction, *Chem. Eng. Prog.,* May, 1977, p. 67.

The author

M. F. Nathan is manager of environmental engineering for Crawford & Russell, Stamford, CN 06904, where he has worked in the capacity of a process design engineer and manager since 1970. Prior to joining Crawford & Russell, he was manager of the Process Development Div. of M. W. Kellogg, with broad experience in refining and petro-chemical process design. He holds a B.S. in chemical engineering from The Pennsylvania State University, as well as an M.S. and Ph.D. in chemical engineering from the University of Illinois.

Wastewater cleanup processes tackle inorganic pollutants

A trio of techniques described at the just-held AIChE National Meeting not only solve pollution problems for the chemical process industries, but at the same time recover metals, metal-containing substances or fertilizers.

Nicholas R. Iammartino, Associate Editor

☐ Turning pollution into an asset was one of the more-prominent themes at AIChE's 82nd National Meeting, held Aug. 29 to Sept. 1 at Atlantic City, N.J.

Attendees heard, for example, the success story of FMC Corp.'s new treatment process for zinc hydroxide sludge generated during rayon-fiber manufacture. The technique recovers zinc as a concentrated and purified zinc sulfate stream suitable for recycle.

E. I. du Pont de Nemours & Co., further, detailed the commercial-scale testing of its extended-surface electrolysis method for removing trace heavy metals from wastewaters. The metals are first electrodeposited on specially designed electrodes, and then acid-leached to yield a concentrated metal-ion solution either for reuse in the original process or for metal recovery.

And, Mallinckrodt, Inc., described how it has solved a major disposal problem for an ammonium sulfate solution containing aniline—it now purifies and concentrates the liquor for sale as a fertilizer.

SLUDGE PROBLEM—FMC's Fibers Div.* had been accumulating zinc hydroxide in lagoons at its Front Royal, Va., rayon plant since 1949 . . . amounting to 80 million lb of zinc by the time its recovery process went onstream early last year. In making rayon, explained FMC coauthors John H. Cosgrove (speaker), Lee B. Bowen and John H. Mallinson, zinc sulfate serves, in the acid-spinning bath, to control regeneration of dissolved cellulose to fiber form as well as to increase fiber orientation. Fiber-wash liquid containing the bath chemicals is treated with lime, neutralizing the acid and converting zinc sulfate to zinc hydroxide precipitate. Clarification removes the precipitate along with other solids, and the solids are pumped to the storage lagoons.

Before turning to its current treatment approach, FMC investigated 22 alternatives on a laboratory, pilot-plant and/or commercial scale; all fell short, Cosgrove said, because of either poor economics or a high impurity level in the output zinc-sulfate solution (which prevents reuse).

WINNING PROCESS—The successful technique (see Fig. 1) starts with sludge containing 2–6% solids (design value is 3%). The zinc portion is a slimy hydrated oxide hard to dewater; calcium sulfate, cellulose organics, and small quantities of calcium oleate, iron and other metals are also present.

By heating the sludge stream to 280–300°F, the amorphous zinc hydroxide particles convert to more-filterable crystals. Steam flashed from already-converted sludge preheats the cold feed, and steam sparging further raises the stream temperature. About 42% of the required heat energy input is recovered via the flash steam, Cosgrove told the AIChE audience.

Centrifugal pumps next transfer the stream to a filter press, having 98 1.5-×-2.0-m plates made from ductile iron in a recessed chamber configuration. Solids content of the cake runs about 35%; such other dewatering equipment as centrifuges, rotary vacuum filters and belt filters were not able to match that performance, especially over the entire 2–6% sludge solids range. Inlet temperature to the filter press is held at 212°F to keep hemihydrate calcium sulfate crystals (converted during the previous step from the original gypsum form) from blinding the polypropylene-felt filter cloths. With relatively clean cloths, dumping the cake takes just 20 min. Typically, the cake solids contain 34% zinc, 0.5% iron and 8% calcium sulfate.

Digestion with sulfuric acid generates a 25–30% zinc sulfate solution. Since about 9–10% of the zinc values in the digester feed is zinc sulfide (rather than zinc hydroxide), hydrogen peroxide is added to assist in sulfide conversion to sulfate. The unit operates on a batch basis to minimize undissolved zinc but can operate continuously as well.

A second filter press then removes calcium sulfate, organics and other impurities from the zinc sulfate stream. The filtrate still carries 2,000–6,000 ppm iron, however, which must be eliminated before reuse to prevent rayon-fiber contamination.

So, hydrogen peroxide is added to oxidize ferrous iron to the ferric

* Plans for sale of FMC's Fibers Div. to a new firm called Avtex Fibers, Inc., were recently announced—*Chem. Eng.*, July 19, p. 71.

Originally published September 13, 1976

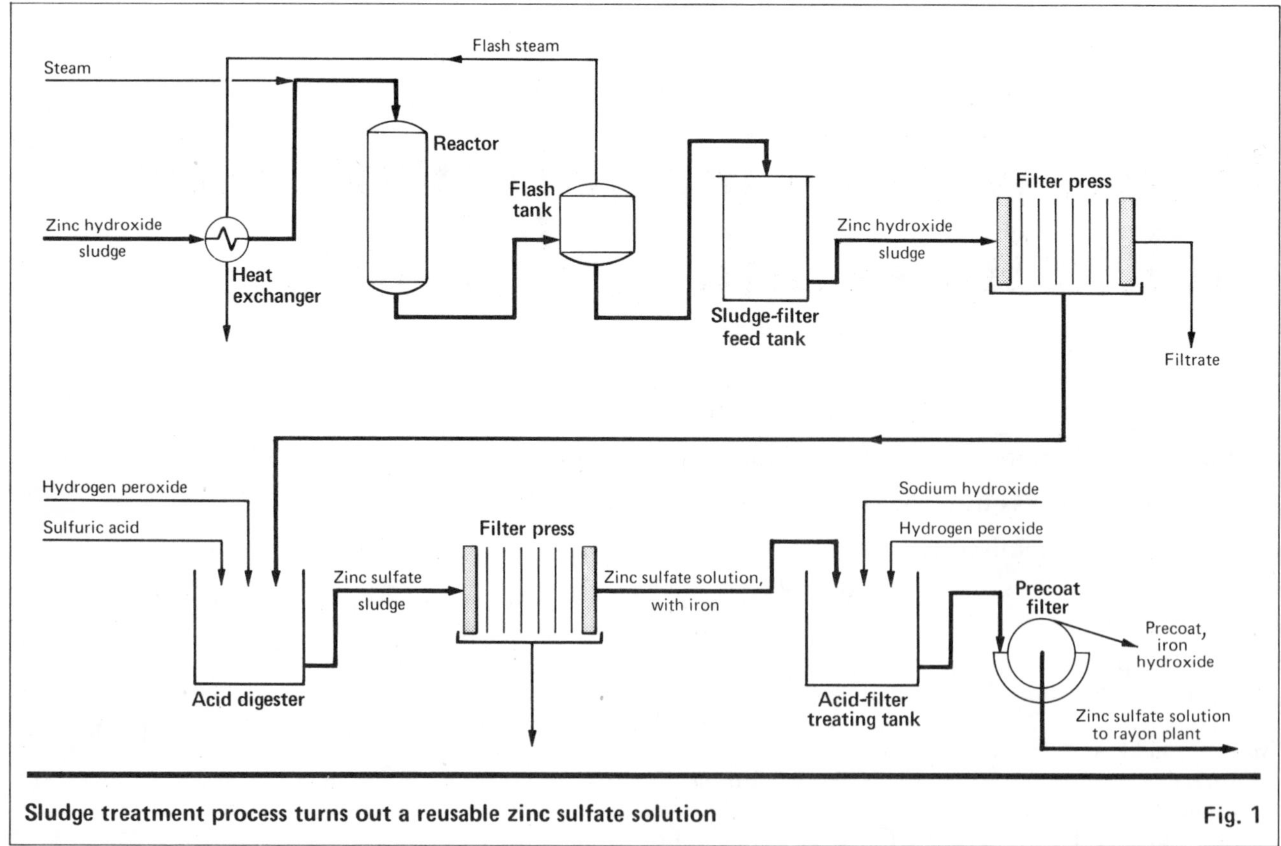

Sludge treatment process turns out a reusable zinc sulfate solution **Fig. 1**

state, and caustic soda is added to increase solution pH to about 4.5, to precipitate hydrated iron oxide (a pH over 5 would cause zinc losses by precipitating zinc hydroxide as well). A rotary vacuum precoat filter measuring 8 ft dia. by 8 ft long finally removes the iron, leaving a concentrated and purified (typical iron content: 60 ppm) zinc sulfate stream for return to the rayon plant.

OTHER FEATURES—Raw-materials and utilities data listed by Cosgrove, Bowen and Mallinson are shown in the table. Design yield factor for the Front Royal plant is 80% recovery of the zinc contained in the sludge cake.

Process requirements for zinc recovery from sludge

Basis: Per lb of zinc recovered

Chemicals	
H_2SO_4 (100%), lb	2.25
NaOH (100%), lb	0.18
H_2O_2 (100%), lb	0.10-0.14
Precoat, lb	0.043
Utilities	
Steam, lb	19.6
Filtered water, gal	4.94
Soft water, gal	0.56
Cooling water, gal	4.46
Power, kWh	0.98

Source: FMC Corp.

Materials of construction for equipment preceding the first, sludge press were not subject to corrosion, but did have abrasion to contend with (especially pumps and control valves) and were selected for high abrasion resistance.

"The digester is literally a maelstrom of corrosive materials," including zinc sulfate, sulfuric acid, hydrogen sulfide, hydrogen peroxide and oleic acid (converted from calcium oleate), Cosgrove continued. The original rubber-covered agitator in the digester lasted less than a year; replacement with one made from Carpenter 20 alloy steel or fiberglass-reinforced plastic (FRP) is now being considered. The digester itself is fabricated from Atlac 382 FRP; the tanks and piping that follow are also FRP or FRP-lined (principally Atlac 382). These components have "all performed well" since startup.

ELECTROLYSIS IS READY—The Du Pont extended-surface electrolysis (ESE) technique has proved itself in testing with a 100-gpm commercial-scale unit, reported J. M. Williams (speaker) and M. C. Olson (both with Du Pont's Engineering Dept. in Wilmington, Del.). ESE is now "a mature technology and a serious alternative for the removal of heavy metals from wastewater."

But though electrolysis has often been studied by researchers, because of its intrinsic specificity for contaminants without addition of treatment chemicals, the authors admit that no commercial installations of any system, including ESE, are yet operating.

CELL DESCRIPTION—Because of the inefficiency of conventional flat-plate electrodes, Du Pont directed its initial efforts to finding porous, corrosion-resistant, high-surface-area, flow-through types.

"Basically, ESE cells utilize a sandwich-like construction, with a thick (about 1 cm) 'fluffy' cathode, a porous separator layer, a thin (0.25–1.0 mm) sheet anode, and another separator layer," reported Williams at Atlantic City. The assembly is either repeated several times in a planar configuration, or is rolled into a spiral. Small lab cells and pi-

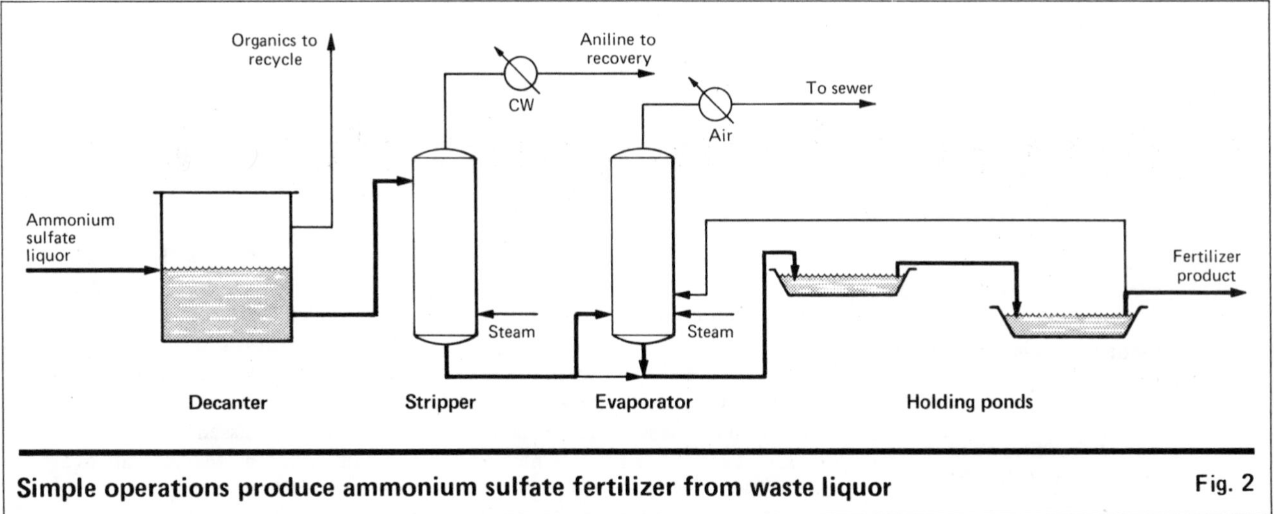

Simple operations produce ammonium sulfate fertilizer from waste liquor Fig. 2

lot cells used the spiral design, but the 100-gpm unit is planar for ease of fabrication and lower cost.

The cathodes are multiple layers of knitted or woven stainless-steel mesh, having a 2–5-mil filament diameter. The structure yields a specific surface area of 40 cm^2/cm^3, and a void fraction greater than 0.9, Williams noted. The anodes are catalyst-coated titanium. Cathode and anode compartments are also stainless steel and titanium, respectively; the cell body is FRP.

The electrodes are arranged so that the process fluid passes upward, parallel to the electrodes and at a right angle to the direction of charge transfer. With a fluid velocity of about 10 cm/s, initial pressure drop through the cell does not exceed 2 psi/ft of active cell length.

The 100-gpm prototype has a solid-state, 12-V, 2,000-A current source. It is designed to remove 80% of the entering contaminant, such as copper, silver, mercury, gold, manganese, antimony or lead. A typical operating sequence would deposit heavy metals on the cathodes during a 4–24-h onstream period (depending on waste composition), and follow with a 30-min leaching operation that would remove the metal for subsequent recovery (e.g., by another electrolysis procedure) or reuse.

TEST RESULTS—The data reported at the AIChE meeting covered a battery of tests that employed the 100-gpm ESE unit to treat a simulated wastewater containing 10–100 mg/l of copper in an acid solution (3.0 pH).

Conclusions reached by Williams and Olson included:

- The optimum applied current is about twice the amount theoretically needed to remove all the copper in one pass through the cell. Greater excesses have little effect and may actually reduce cell performance by blinding the cathodes with gas.
- The apparent mass-transfer coefficient depends very strongly on fluid velocity, implying that long, thin cells are more economical (to a point) than short, fat ones.
- Pressure drop across the cell, as copper accumulates, rises more slowly than anticipated, and presents no operating problems.
- Copper removal efficiency increases with onstream time, because the real surface area of the cathode increases as copper is deposited.

The leaching step, which used a hydrogen peroxide/sulfuric acid mixture, proceeded at a satisfactory rate and with acceptable decomposition of the peroxide (about 25%), the authors added. A backflush-rinsing step (used after leaching to remove leachate and any suspended solids that were filtered from the wastewater during electrolysis) was also rapid, taking just 2–5 min.

FERTILIZER RECOVERY—An ammonium sulfate waste stream generated by Mallinckrodt's Raleigh, N.C., *p*-aminophenol plant—the stream has been stored in ponds since the plant's 1970 startup—is now being processed for sale as fertilizer, reported the firm's S. Noble Robinson. The simple treatment basically serves two purposes: removal of toxic aniline contaminant, and concentration to marketable strength.

The waste stream, discharged by centrifuges that turn out the main *p*-aminophenol product, first passes to a decanter for removal of most aniline and other organics (see Fig. 2). The solution, now containing about 17% ammonium sulfate and 3% aniline, is steam-stripped to raise the former to about 28% while reducing the latter to a trace level. Further concentration takes place by evaporation, which yields a saturated, 41%, solution of ammonium sulfate. This is kept in a pair of holding ponds; liquor withdrawals from the second are sold to farmers mostly within a 50-mi radius. The fertilizer carries about 40 ppm aniline and 1% dissolved organics; though the product falls below the state's usual minimum value for nitrogen content, tests by Mallinckrodt showed it acceptable for certain food and ornamental grass crops (as well as nontoxic), so North Carolina approved sale of the liquor in April 1975.

Processing cost amounts to $220/equivalent ton of nitrogen. This includes depreciation, general operating expenses and steam—the outlay for steam, in fact, is by far the largest single item, 75% of the total. Though selling prices cannot recoup the total processing cost, Robinson points out that "Mallinckrodt's objective has been to *dispose* of the liquor via marketing as a fertilizer, and to this end we have been willing to sell it for prices down to zero dollars per ton."

Mercury Cleanup Routes—I

Special activated carbons and ion-exchange resins remove mercury from wastewaters in a trio of pollution-control systems newly commercialized in Europe.

MARK D. ROSENZWEIG
Regional Editor, Europe

As chlor-alkali producers and other users of mercury know only too well, stringent wastewater standards for mercury pollution have been taking effect worldwide over the past few years.

Some firms have shuttered older facilities, and others have revamped basic processes, to comply with ceilings usually set at a few ppb. Most firms, however, are paring mercury pollution by turning to one or more of a fast-expanding arsenal of cleanup techniques.

Three routes detailed here have already chalked up a year or two of operating success—activated-carbon methods by Billingsfors Bruks AB (S-660 11, Billingsfors, Sweden) and the Organization for Applied Scientific Research in the Netherlands, TNO (Industrial Liason Dept., P.O. Box 215, Schoemakerstraat 97, Delft, The Netherlands), and an ion-exchange approach by Akzo Zout Chemie Nederland B.V. (P.O. Box 25, Hengelo (O), The Netherlands). Other contenders are onstream or ready for marketing in the U.S. and Japan.

Special Adsorbent—Billingsfors Bruks' "BMS process" is intended as a final step in treating wastewaters from chlor-alkali units. Operating at ambient temperature and pressure, it follows up normal pretreatment schemes such as filtration or precipitation, cutting effluent content at least to 10 ppb mercury.

Two 90,000-metric-ton/yr chlorine plants in Sweden use the process. Elektrokemiska AB commissioned its unit at Bohus in late 1972; KemaNord AB came 100% onstream inside a month at Stenungsund last summer.

The latter system (see flowsheet) sends raw water (rain, makeup, etc.) first through a sand filter to remove clays, rust and other solids. Chlorine-plant effluent, already cleaned from 5 mg/l to only 0.1 mg/l by KemaNord's own proprietary treatment based on hydrazine and electrical-potential reduction, joins this stream in a pH adjustment / redox control tank. A 2-m^3 capacity handles 8-10 m^3/h total flow. Tank fabrication is of glass-fiber-reinforced polyester, lined with polyvinyl chloride.

In the tank, a 50-60% spent sulfuric acid readily available from the chlorine-plant drying system reduces pH of the wastewaters from 10-11 to 4-5. Simultaneously, chlorine oxidizes any mercury present as metal to the ionic state, which is affected by the later adsorption. At Stenungsund, the 10-50-ppm free chlorine carried by the spent sulfuric acid does the job, so a separate chlorine feed is not required. Redox potential is adjusted to −0.9 V.

Chlorine poisons the BMS adsorbent, so the stream next passes through a bed of activated carbon to remove any residual free chlorine. KemaNord's rubber-lined steel vessel holds about 2.5 m^3 of a 0.5-1.0-mm-dia. material, which needs backwashing periodically but replacement only annually.

The special BMS-process adsorbent finally takes out the mercury to wrap up the treatment. An activated carbon similar in size to the chlorine-removal type, BMS adsorbent is made selective to ionic mercury by a concentration of sulfur compounds on its surface.

A proprietary salt solution can reactivate the material once or twice, says Billingsfors Bruks. On this basis, and with 0.1-0.5 mg/l feed mercury, lifespan is about 1 yr. The solution recycles through the pH adjustment / redox control tank. Spent BMS adsorbent is distilled to recover mercury, leaving a safe carbon residue for disposal.

Washwaters passed through either the sand filter or the activated-carbon beds pick up some mercury as they remove deposited solids, so they too are best distilled.

For an 8-10-m^3/h plant, capital outlay runs approximately $120,000. No extra labor is required, and energy consumption is small, as only a mixer and several pumps must be driven. BMS material costs $9,500-$12,000 per bed change, bringing overall operating charges to around $0.25/m^3 of feed. Mercury credits do not balance this, but do cover distillation expenses.

The BMS process contends with upset conditions, adds Billingsfors Bruks. If the pretreatment circuit fails and 5-mg/l wastewater reaches the system, an outlet of about 0.1 mg/l can be expected, but at the expense of adsorbent life.

Versatile Performer—TNO's cleanup route similarly uses a specially tailored activated carbon to adsorb mercury, but was developed for organic mercury compounds that pesticide and agricultural-disinfectant producers typically contend with. A number of such firms in The Netherlands now apply the process. Additionally, the organization says its process is suitable for removing inorganic and metallic mercury from wastewaters as well, and is looking into service for sludges and waste solids.

Typical inlet loadings run 25-50 mg/m^3 mercury, and outlet loadings 10 mg/l or less. A coagulation/flocculation step starts off the process; this removes suspended solids and colloids to prevent plugging of the following series of two adsorbent beds.

Originally published January 20, 1975

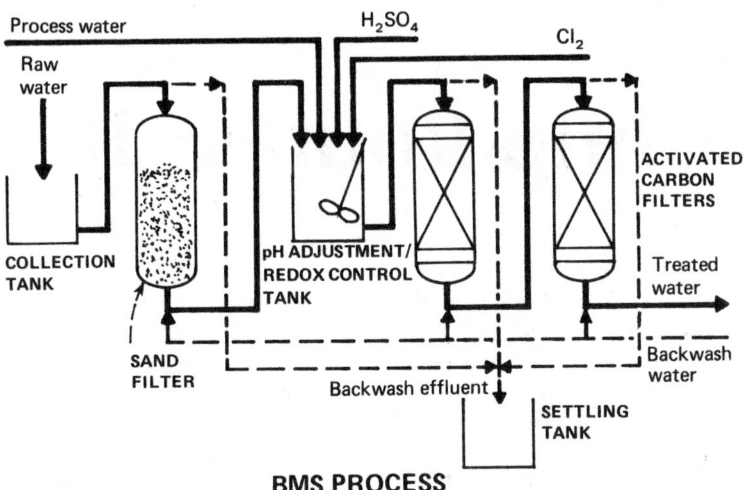

BMS PROCESS

Linear velocity is about 2 m/h through these beds. When mercury reaches about 10 wt.% in the first bed and breakthrough occurs, the second bed takes over and treats the mercury-laden, 5-50°C influent. The spent bed is renewed, and restored to service as second-stage adsorber until breakthrough next occurs.

Ion-Exchange Approach—Mercury effluent less than 5 ppb is guaranteed by Akzo for its ion-exchange-based Imac TMR process. The firm is aiming mainly at chlor-alkali plants, but is considering other applications also. Five installations will be running by mid-year: two of Akzo's at Delfzijl and Rotterdam in Holland, one for Elektrochemie Ibbenburen at Ibbenburen, West Germany, and two for Norsk Hydro at Porsgrunn, Norway. All are rated for 10-15 m^3/h of wastewater.

Operations are at ambient conditions. Wastewater first flows into an oxidation reactor together with a slight excess of sodium hypochlorite or chlorine (see flowsheet). Metallic mercury oxidizes to ionic mercury, and pH is controlled to about 3 in order to keep iron in solution and prevent problems in subsequent filtration from iron hydroxide.

This sand or cloth filter not only removes solids that would plug the ion-exchange beds, but also retains and oxidizes mercury droplets that might have passed through the oxidation reactor. Mercury-free solids are periodically discarded.

The liquid is next dechlorinated, to prevent contamination of the mercury-removing resin, in a rubber-lined steel column containing a special activated carbon. The liquid then passes to a pair of similarly fabricated vessels containing the Imac TMR ion-exchange resin.

Akzo describes the mercury-selective resin as a polymeric mercaptan in which thiol groups are attached to a chemically inert and mechanically strong matrix (a macroporous styrene / divinyl-benzene copolymer). It does not contain any nitrogen compounds, avoiding formation of explosive nitrogen trichloride. The resin boasts high thermal stability, and maintains its performance under wide fluctuations of pH (1-14), temperature, and concentration of sodium, chlorate and sulfate ions.

Each of the ion-exchange vessels handles a flowrate of 10 volumes wastewater per volume resin per hour. Cycle life depends on feed mercury level. For a 10-ppm feed, Akzo says life runs about 1,000 h; for a 1-ppm feed, it stretches to 6 mo. At the expense of cycle life, the resin takes care of any loading to safeguard against any disruptions in upstream pretreatment systems.

Resin can be regenerated up to five times. For chlor-alkali plants Akzo recommends a hydrochloric acid solution, widely available because of its use for pH adjustment in the electrolysis brine circuit. After stripping mercury from the resin, the acid can in fact be used for brine-circuit pH control, thus returning the mercury to the electrolysis stream.

For non-chlor-alkali services, a proprietary regenerant is offered.

When the resin is finally spent, its residual mercury content amounts to only 100 g/m^3, and will not leach out into rainwater or brine when discarded. Though the spent resin no longer satisfies process needs, it retains an affinity for mercury and will remove it from groundwaters.

Battery-limits investment runs about $300,000 (1975 basis) for a 10-m^3/h unit. This includes initial resin and activated-carbon charges; replacements will cost about $27,000. Operating costs, after credit for recovered mercury, work out to approximately $0.15/$m^3$ for treating 7-ppm influent to 5-ppb effluent. #

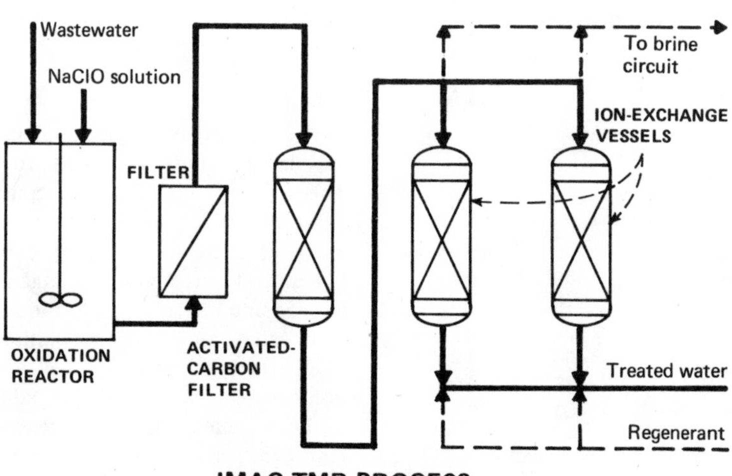

IMAC TMR PROCESS

Mercury Cleanup Routes—II

Here are three more new techniques for removing mercury from waste streams, all of which are already on the job commercially.

NICHOLAS R. IAMMARTINO
Associate Editor

Processes for preventing mercury pollution have emerged recently not only in Europe (see pp. 288 and 289), but also in Japan, and the U.S.

Nippon Electric Co., Georgia-Pacific Corp. and FMC Corp.* are all successfully operating new mercury-cleanup systems on a full-scale basis. Nippon's batch or continuous scheme purges mercury and a dozen other heavy metals from such various effluents as offgas-scrubbing liquor from municipal-waste incineration, and wastewater from electroplating of metals. The other two processes handle wastes from chlor-alkali plants.

Magnetic Ferrites—The so-called "Re-elixirization" technique by Nippon Electric is an offshoot of its expertise in manufacturing ferrites for telecommunications. For water-pollution control, Nippon came up with the idea of exploiting heavy metals in wastewater as a ferrite precursor.

The metal-bearing water first mixes with a divalent ferrous salt—usually ferrous sulfate, a widely available byproduct of iron and titanium dioxide manufacture. Two moles or more of salt are typically needed per mole of metal. With divalent iron ion (Fe^{++}) now co-existing with nonferrous-metal ion (M^{++}) in aqueous solution, the stream is neutralized with alkali, forming a dark green hydroxide mixture according to the reaction:

$$x\,M^{++} + Fe^{++}_{(3-x)} + 6(OH)^- \rightarrow M_xFe_{(3-x)}(OH)_6$$

Oxidation with air follows, during which redissolution and complex formation take place, yielding a black ferrite by the reaction:

$$M_xFe_{(3-x)}(OH)_6 + \tfrac{1}{2}O_2 \rightarrow M_xFe_{(3-x)}O_4 + 3\,H_2O$$

A magnetic separation removes the insoluble, ferromagnetic ferrite from solution, either for disposal or for possible application in microwave transmission products, cleanup of phosphates, algae, bacteria and viruses from water, and as ferrofluid.

The neutral aqueous liquid remaining is discarded. Performance data for two commercial units now onstream show mercury reductions from 6 mg/l inlet to 0.005 mg/l outlet, and 7.4 mg/l to 0.001 mg/l. The first, operated by the Osaka Prefectural government, treats 4.5 metric tons/h of liquor from a scrubbing tower that cleans stack gas generated during incineration of municipal refuse; the second handles 1 metric ton/h of wastewater at Nippon Electric's Central Research Laboratory in Kawasaki. Reduction of other heavy metals at the Osaka facility is shown in the table. Metals removed also include nickel, bismuth and tin.

Re-elixirization Performance Osaka Unit

Metal	Concentration, mg/l Influent	Effluent
Mercury	6	0.005
Arsenic	0.7	<0.01
Chromium (III)	25	0.01
Chromium (VI)	0.5	Not detected
Lead	480	0.05
Cadmium	15	<0.01
Iron	3,500	0.04
Zinc	650	0.5
Copper	23	0.08
Manganese	60	0.5

Re-elixirization has several advantages over conventional alkali precipitation of heavy metals as metal hydroxides, says Nippon. Ferrite precipitates do not redissolve as do simple metal hydroxides, and are easy to separate from liquid because of their larger particle size and ferromagnetic behavior. Operating a Re-elixirization system involves only a single precipitation for all heavy metals present, rather than a series of hydroxide precipitations tailored for each particular metal.

Extraction Approach—FMC's mercury-cleanup route was groomed by engineers at the firm's 175-ton/d chlor-alkali plant in Squamish, British Columbia, Can., and has served there since early 1971. Though available for licensing, the patented process (U.S. 3,691,037) is still looking for its second user.

Operations at Squamish, nevertheless, have been impressive. Mercury removal from treated streams tops 99%, FMC reports. Discharged wastewater averages less than 3 ppb mercury, and wet solids sent to landfill only 20-30 ppm. Total mercury discharge at Squamish amounts to about 0.14 lb/d in liquid and 0.07 lb/d in solids.

The system first collects all mercury-laden liquids in a concrete tank. These liquids include (1) a continuous stream of purge water from each electrolysis cell's tail box (where sodium/mercury amalgam is washed and cooled), (2) caustic-filter washwater, and (3) any spills or other streams collected in cell-room drain sewers. Sodium sulfide added to the collected liquids in a reaction tank converts soluble mercury com-

*NEC Environmental Engineering, Ltd., NEC Bldg., 33-1, Shiba-Gochome, Minato-ku, Tokyo 108, Japan.

Georgia-Pacific Corp., Bellingham Div., 300 Laurel St., Bellingham, WA 98225.

FMC Corp., Industrial Chemicals Div., 633 Third Ave., New York, N.Y. 10017.

Originally published February 3, 1975

MERCURY-RECOVERY SYSTEM by Georgia-Pacific treats wastes from 200-ton/d chlor-alkali plant.

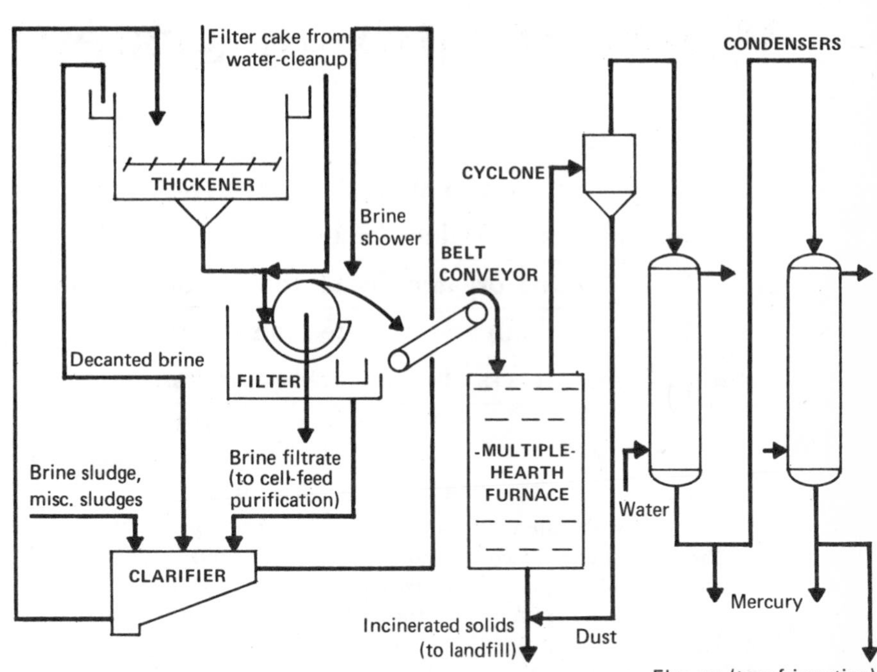

GEORGIA-PACIFIC SLUDGE-TREATMENT PROCESS

pounds to insoluble mercury sulfide (a standard procedure). A rotary vacuum precoat filter separates the dilute slurry into a mercury filter cake, and a clean filtrate for discharge.

Filter-cake processing is the heart of FMC's scheme. The cake is mixed with sludge from the brine pretreatment clarifier. This sludge contains mercury carried out of the electrolysis cell by recycling depleted brine, as well as calcium and magnesium compounds purged from the fresh salt input.

The mixture passes into an agitated vessel, along with chlorine gas and caustic (together forming sodium hypochlorite), and some brine. At a pH between 9 and 12, mercury salts and elemental mercury in the sludge mixture convert to brine-soluble mercury compounds, according to the general reactions:

$$HgS + 4(OCl)^- \rightarrow HgCl_4^= + SO_4^-$$

$$Hg + 3\,Cl^- + ClO^= + H_2O \rightarrow HgCl_4^= + 2(OH)^-$$

A rotary vacuum precoat filter again separates the stream. Brine laden with extracted mercury recycles to the electrolysis cell, while remaining sludge is mixed with sand and sent to landfill.

Roasting Method—Working partially under a grant from the U.S. Environmental Protection Agency's Office of Research & Development (Project 12040 HDU), Georgia-Pacific designed and built mercury-cleanup facilities for its Bellingham, Wash., chlor-alkali plant. Construction was completed in January of last year, at a cost of \$145,000 for water treatment (100 gpm) and \$365,000 for sludge treatment (8 tons/d, dry basis). Operating costs run \$0.50/1,000 gal, and \$35/ton, respectively.

The water system is a conventional sulfide precipitation. Liquid wastes with dissolved mercury are first carefully adjusted to 5-6 pH with spent sulfuric acid. A 1-3-ppm excess of sodium sulfide next added reacts with dissolved mercury compounds to form solid mercury sulfide in a dilute solution (0.02 wt.%).

A precoated pressure filter removes these solids for further processing. Discharge water contains approximately 10-125 ppb mercury (averaging 50 ppb), from a feed of 0.3-6 ppm, for a removal rate of 87-99% (averaging 96.8%). A carbon or resin polishing could cut effluent loading still more, the firm notes.

Georgia-Pacific's more-innovative technique is its roasting scheme for sludge treatment (see flowsheet). Sludge from the brine purification circuit and other plant sources, typically only 5-10% suspended solids, collects in a clarifier and then passes through a thickener to a rotary vacuum filter. Mercury-laden solids from water purification also enter the filter, which dewaters the combined input to 50-60% solids.

A gas-fired, multiple-hearth furnace fed with this dewatered stream produces an output clinker two or three times lower in mercury than output solids from chemical processes, Georgia-Pacific found. Operating at 1,350-1,400°F, the furnace removes 99.8% of the feed mercury, leaving a 0.5 to 0.8-ppm-mercury clinker for landfill disposal.

Still-lower clinker levels could be achieved by an acid pretreatment Georgia-Pacific discovered for the furnace input (U.S. 3,814,685). The approach is not used yet at Bellingham, the firm says, but could be incorporated later on if increased sludge throughput or higher-than-expected clinker mercury occurred.

Overhead gas from the furnace, finally, is cooled with water to condense and recover mercury. #

Water treatment system cuts organics

Activated carbon, oxygen and gamma radiation unite in this hybrid process to reduce bacteria and viruses by as much as 99% and organics by up to 98%. Said to cost less to build, maintain and operate than competing techniques, the process also cuts energy needs.

Philip M. Kohn, *Assistant Editor*

☐ The Radox Wastewater Treatment Process is an advanced technique that uses a combination of methods, including aeration, oxygenation, carbon adsorption and irradiation. The system gets its first commercial tryout this month, when it is scheduled to begin handling the 125,000-gal/d effluent from the plant of an East Coast meat-packing concern.

Designed and built by International Purification Systems, Inc. (Atlanta, Ga.), and marketed by International Purification Systems,* Radox allows users to comply with 1983 EPA requirements, according to the marketer. The system is said to cut organics by 98% or better, while deactivating almost 100% of viruses, bacteria, odor and color.

Consisting basically of two components, an oxygenation system and a carbon-column irradiator, Radox requires less than 50% of the land needed by a conventional activated-sludge system of comparable capacity. In addition, the activated carbon is continuously regenerated in place by gamma radiation-induced oxidation of the organics adsorbed onto the carbon, saving downtime, carbon loss and regenerative costs. Energy needs are only about 50 to 60% of those of conventional systems.

Wesmark says that the process, reflecting the above advantages, offers a capital investment saving of as much as 40 to 50% over other competitive processes, while operating and maintenance costs will be only about 40% of those for comparable techniques.

*International Purification Systems, 3355 Lenox Rd., Atlanta, Ga. 30326

THE PROCESS—Radox emulates natural water-purification processes by simulating three basic operations: sedimentation, oxygen ingestion and solar irradiation. Radox's holding tank, oxygenation system and carbon-column irradiator, respectively, perform these functions.

In the Radox process (see flowsheet), wastewater enters the holding tank after first undergoing primary treatment for removal of solids. The holding tank is a treated concrete vessel of 285,000-gal capacity. Its purposes are to provide shock control by eliminating surges, to homogenize the feed to Radox's downstream components, to facilitate sedimentation of suspended pollutants, and to regulate flow.

Heavier suspended solids in the holding tank tend to gravitate to the bottom, where they are exposed to ambient air introduced into the holding tank via sparge pipes. Exposure to the air causes a gradual oxidation of the organics, reducing their density. The now-lighter pollutants rise to a higher level in the tank, to be replaced by other, heavier suspended solids. Sludge does not accumulate in the bottom of the holding tank, according to Wesmark, because the oxidation process is continuously "floating" the at-least-partially-oxidized pollutants. Normal retention time in the holding tank is approximately 48 hours, during which time a partial reduction in the overall organic content occurs. The presence of air in the tank retards anaerobic bacterial growth, thus minimizing odor generation.

The supernatant liquor—the holding tank's surface layer—is continuously pumped over to the next phase of the Radox system, the oxygenation process.

THE ONE-TWO PUNCH—The supernatant enters the oxygenation system where air is injected into the flowstream, raising the dissolved-oxygen level to 20 to 25 mg/L. This high dissolved-oxygen level is needed in the system to enable reduction of the organics to carbon dioxide and water. It is also essential to the proper and efficient performance in the carbon-column irradiator. There is no holdup in the oxygenation system, which accommodates a full flow of roughly 75 gal/min, depending on the operating parameters of the system. Oxygenation of the system by ambient air occurs at atmospheric pressure; the system uses a proprietary technique for the actual introduction of the air.

The oxygenated stream then enters the irradiator—a stainless-steel tank containing approximately 12,000 lb of activated carbon. Pencils of cobalt, located inside the carbon column in a geometric configuration, provide a homogeneous radiation field throughout the tank. The calculated residence time for the organics is about 10 hours.

Wastewater flows upward through the carbon column, with the organic components adsorbing onto the carbon. The activated carbon, as in conventional schemes, removes the organics, odor and color, while the gamma radiation from the energy source induces oxidation of the organics, dynamically regenerating the carbon. The radiation is also the agent that inactivates the viruses and bacteria.

THE ENERGY SOURCE—The irradiation source generally accounts for about 20 to 25% of a plant's capital costs, but this percentage can vary significantly depending on such factors as organics-loading and cobalt dosage.

The half-life of cobalt is 63 months, and the replacement rate is approximately 10% per year. Cobalt replace-

Originally published August 15, 1975

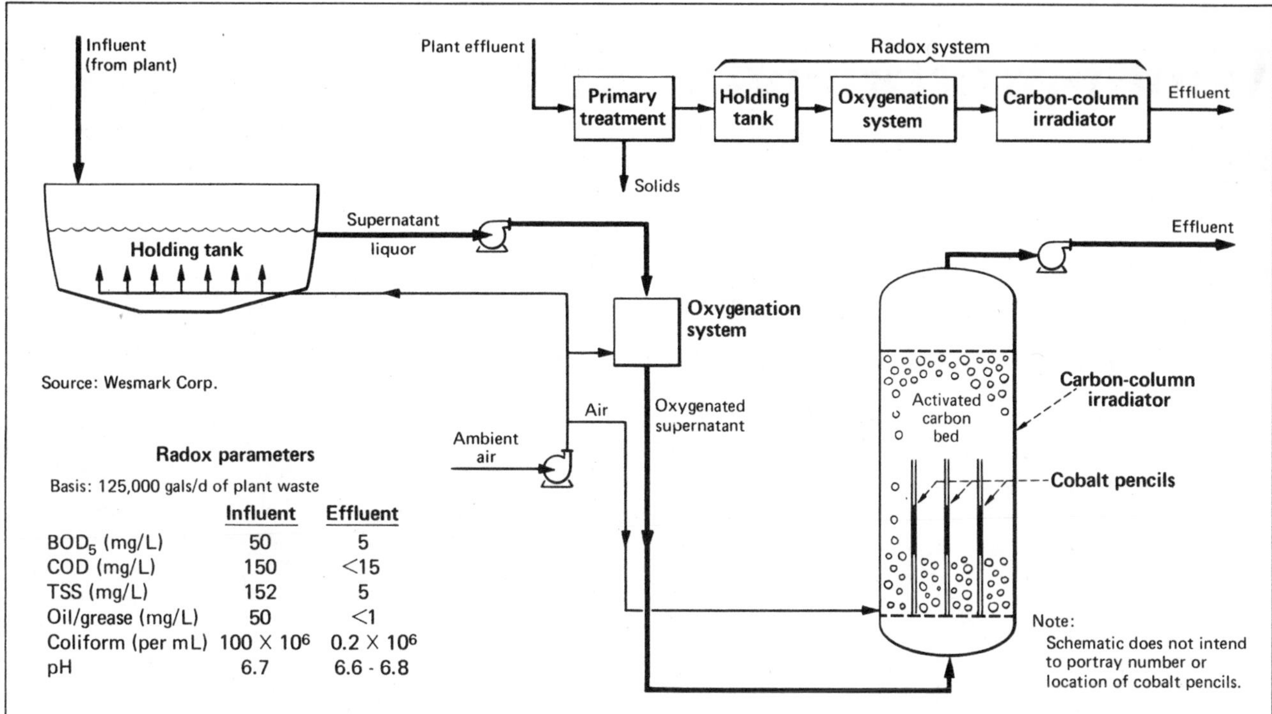

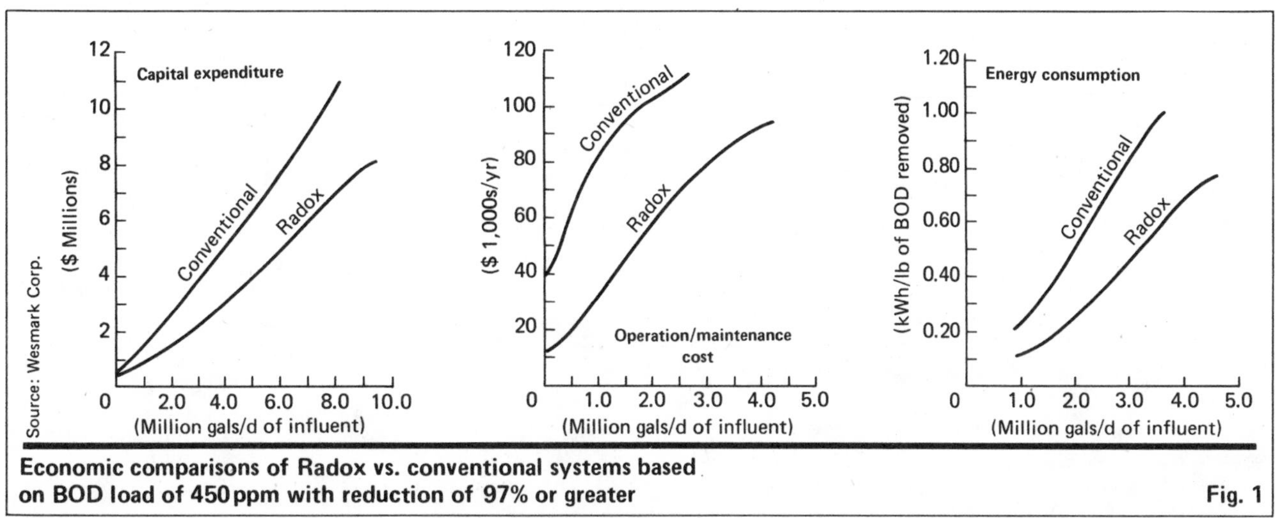

Economic comparisons of Radox vs. conventional systems based on BOD load of 450ppm with reduction of 97% or greater **Fig. 1**

ment is best handled contractually, typically by the supplier of the radioactive material.

Radox's oxygenation system and carbon-column irradiator are housed in a reinforced concrete structure that acts as a radiation shield. International Purification Systems and Wesmark state that there is no detectable radiation outside the concrete structure, and no trace of radioactivity in Radox's effluent, since gamma radiation leaves no residual activity.

EASY TO OPERATE—Wesmark states that the labor needed to operate Radox is one semi-skilled operator, part-time. Since organics are burned off in the irradiator, there is no sludge buildup and no need for regeneration. Non-oxidizable inorganics are typically removed from the influent stream by conventional means, such as flocculation or filtration, prior to the holding tank.

ECONOMICS—One of Radox's strong points is its economy of operation. Because the process' radiation dosage is concentrated just on the organics, rather than on the entire wastewater stream, the system's operating efficiency is reported to be substantially higher than competitive processes while costs are lower.

Above are graphs comparing Radox to comparable conventional treatment systems using activated sludge or carbon adsorption. Wesmark notes that as the volume handled by Radox increases, the process becomes more and more effective. In applications where volume is less than 300,000 gal/d, Radox capital costs are about 10 to 15% less expensive than other processes, while for systems handling 500,000 gal/d and up, Radox is typically 20 to 35% less costly than competing techniques. And Wesmark states that few competing processes are capable of matching the high reductions in organics and bacteria that typify Radox.

Photo-processing facility achieves zero discharge

Recycling process and washwater streams, this route discharges no industrial sewage, allows for silver recovery and reagent regeneration, and cuts costs.

Philip M. Kohn, Associate Editor

☐ What is claimed to be the first zero-discharge water-treatment system in the photographic industry has gone into operation. In August, PCA International, Inc., dedicated—at its Matthews, N.C., headquarters*—a technology/laboratory facility that houses a system that recovers $600,000 worth of photographic silver in a year, reduces water-treatment operating costs by 66%, saves $300,000/yr in chemicals costs, and extracts enough ammonia to make fertilizer for the extensive lawns on the 57-acre site.

The system, designed in-house, recycles 75% of its photographic solutions plus 90% of its washwater, and discharges no effluent, thereby reducing the load on the community's limited waste-treatment resources and meeting federal zero-discharge goals ahead of schedule.

Introduced in three phases, the system cost about $1 million, including research and some capacity expansion. The third phase, which enabled the actual achievement of zero discharge, cost about $425,000. In full operation since August, the route will have an annual operating cost of about $25,000, most of which is for electric power.

BACKGROUND—PCA, billing itself "the world's largest portrait photographer," operates its photo-processing facility around the clock, five days a week. In an average week, the company processes 30 miles of film and 180 miles of 10-in.-wide paper for prints, the equivalent of almost one and one-half million 8 x 10-inch prints. Proposed U.S. Environmental Protection Agency effluent guidelines for industry—expected to be finalized in December 1979—call for a discharge of no more than 4 gal of wastewater per ft^2 of sensitized material processed. When tested in August 1977, PCA had an output of 0.129 gal/ft^2.

Edward Schiller, PCA's director of product quality, and the person responsible for overseeing the operation of the waste-treatment system, says that the company originally had an effluent of 20,000 gal/d with production only about 5% of the present rate. Effluent output had peaked at about 40,000 gal/d, when PCA began implementing its treatment system. Output was curtailed to about 18,000 gal/d by August of this year, at which time the final phase of treatment was completed, eliminating effluent entirely. Schiller points out that curtailment of wastestreams was effected while the facility's production rate was increasing exponentially.

THREE-PHASE PROGRAM—A major force in motivating PCA to improve its waste-treatment program has been, of course, the federal regulations mandating zero discharge by 1986. The company has come in ahead of the deadline by instituting a three-phase program, beginning in 1972.

In Phase I, PCA began regenerating 100% of the bleach-fix chemicals used in developing and printing portrait paper. Installation of a reverse-osmosis unit made possible the recycling of 85% of the washwater used in portrait-paper processing machines. In addition, all of the film-bleach solutions and half of the film-fix solutions were recycled. Remaining process solutions and washwater from film-developing and paper-processing systems went through an ozone-treatment process and an aeration pond for chemical-oxygen-demand removal before being discharged into the municipal sewer system. An additional 12,000 gal/d of cooling water—primarily from in-plant compressors—was also discharged.

*PCA International, Inc., 801 Crestdale Ave., Matthews, NC 28105

PCA's reverse-osmosis units remove dissolved solids from washwater

Originally published December 4, 1978

Phase II saw introduction of a proprietary metals-removal route that eliminated 95% of heavy metals, such as silver, cadmium, lead and iron. Regeneration of paper developers was begun, as well as recycling of half of the paper final-bath solution, and addition of more cooling water to the recycle loop.

In Phase III, a second reverse-osmosis unit was added, permitting recycling of 90% of film-processing washwater. All cooling water was added to the closed-loop recycle system, and an evaporation step was put in for solids concentration and recycling of all 18,000 gal/d of wastewater generated.

THREE-PART SYSTEM—The zero-discharge system can be divided into three steps: chemical mixing, chemical recovery, and wastewater treatment.

The beginning and end of the cycle is the chemical-mixing step (see flowsheet). Here, all the process solutions are prepared. A separate tank is used for each solution, and preparation is computer-controlled. A premeasured package of chemicals is mixed into a set amount of water, pumped into a storage tank, and metered to the various processing stages.

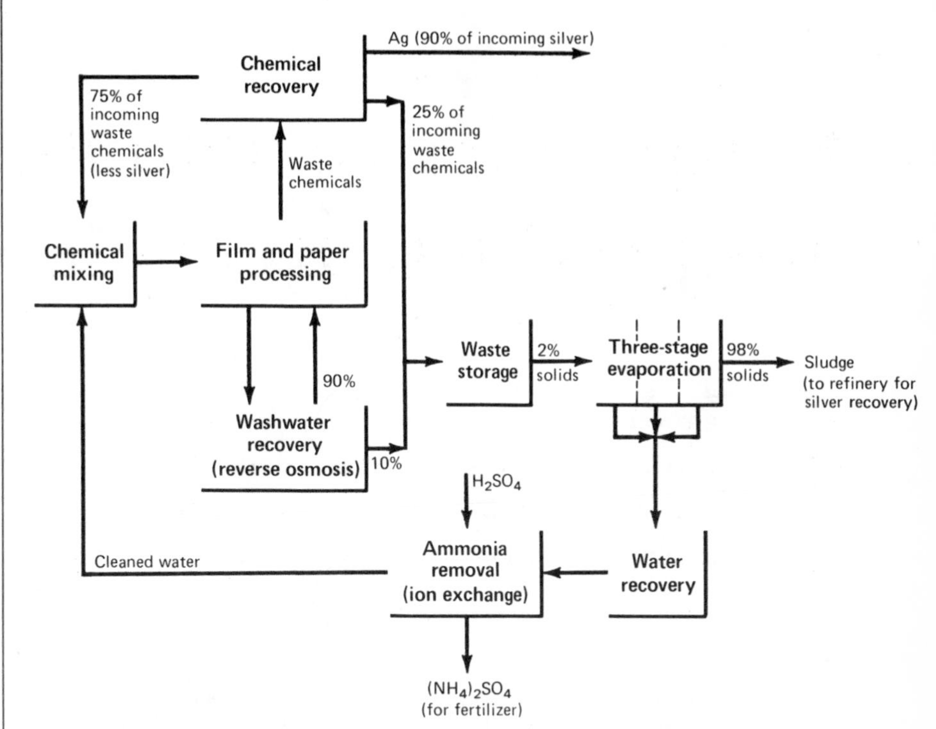

Simplified flowsheet of PCA's treatment route emphasizes "looping" characteristics

The next step, chemical recovery, is an integral part of the processing operation. Seven out of a total of eleven processing solutions are either partially or totally recycled. The remaining materials are treated in the wastestream. Materials to be recycled are collected. Unwanted byproducts and impurities (picked up during processing)—as well as about 90% of the recoverable silver—are removed either through filtration, ion-exchange, reverse-osmosis, or electrolytic-recovery techniques, depending on the materials in question. The purified process solution is them pumped back to chemical mixing, where it is analyzed and adjusted, if necessary, before reuse.

Washwater streams from processing operations are collected and passed through two 32-gal/min reverse-osmosis units: one treats water from the film-processing section, the other, from the paper-processing line. The reverse-osmosis units remove dissolved solids from 90% of the washwater, concentrating them into the remaining 10% of the original volume. Cleaned water is reused, while the pollutant-laden stream goes to the final step, waste treatment.

Effluent reaching the evaporating stages contains about 2% solids—low, but still too high for reuse. A three-stage evaporation system concentrates the solids: the first stage takes off 90% of the water; the second, the next 5%; and the final, the rest, concentrating solids to about 98%. Having a toothpaste-like consistency, these solids are sent to an outside refinery, where more (practically all of the other 10% of the total) photographic silver is recovered.

Water from the evaporators still contains some ammonia (picked up from the processing operations), however, and must be treated before being recycled. The ammonia, removed in an ion-exchange unit that uses sulfuric acid as its regenerating agent, reacts with the sulfuric, forming ammonium sulfate. After removal by backflushing from the ion-exchange columns, the ammonium sulfate is applied as fertilizer to PCA's lawns.

The clean water is then returned to the chemical-mixing step, completing the cycle.

Chemical analyses are performed on all processing solutions once each shift, and on wastewater streams every three hours, to make sure that all are within specifications.

SIZABLE SAVINGS—PCA's Schiller says that it is difficult to assess the actual capital cost for the treatment system alone, since it was developed gradually, and in conjunction with other research and capacity-enlarging efforts. A good estimate is approximately $1 million, however. Phase III, instrumental in reaching zero-discharge, cost approximately $425,000 to effect. The heaviest portion of the capital expenditure went for the evaporation system.

The firm used to spend approximately $65,000/yr for waste-treatment prior to implementation of the present system, most going for chemicals used in treatment processes. Now, according to Schiller, costs have dropped to about $25,000/yr, about 80% of which is for electric power.

Because 75% of the processing solutions are now reused, PCA estimates that photo-processing-chemical costs have been cut by about $300,000/yr over previous levels.

An added bonus is that the route allows the company to recover 100% of the silver it uses. For 1978, recovered value should be about $600,000; that figure should rise to approximately $700,000 next year at silver prices of around $5/troy oz.

Wastewater reclamation system ups productivity, cuts water use

Combining reverse osmosis and solar evaporation, this system reclaims plating wastewater, reducing water consumption, minimizing product waste, and meeting the 1983 EPA goal of zero discharge.

Jay E. Warnke, Kenneth G. Thomas and Samuel C. Creason,
*Beckman Instruments, Inc.**

☐ A plating-wastewater reclamation system, designed by Abcor, Inc. (Wilmington, Mass.), in collaboration with Beckman Instruments, Inc., is now in operation at the latter's Porterville, Calif., facility. Having a reverse osmosis (RO) unit at its heart, the system also features a solar evaporation pond for disposal of removed contaminants.

Since its installation, the system has reduced water consumption by about 70%, lowered the cost of shop rejects to less than a third of its previous amount, and achieved the 1983 EPA goal of zero discharge.

The Porterville facility produces printed-circuit boards for process and scientific instruments. Processing about 14,000 ft^2 of material per month, the plant uses conventional subtractive processes for copper-, tin/lead-, and gold plating.

Prior to installation of the reclamation system, the plant used about 1.5 million gal/mo of water, and the rate was increasing. Two streams fed the plant: Porterville city water and recirculated, contaminated rinse water. Additional city water could not be used without extensive modifications to the Porterville city water system. And, increasing the use of recirculated rinse would only increase the 10% circuit-board reject rate, which was due primarily to contaminants depositing out on the boards.

Thus, the plan of attack was to: (1) cut overall water usage, (2) improve the quality of the recirculated rinse water, and (3) minimize or eliminate waste discharge.

WATER CONSERVATION—As a first step, rinse-tank controllers were installed on the plating line, to sense the conductivity of reclaimed water in the rinse tanks and to admit fresh water if the conductivity rose above a prespecified amount. By eliminating manual water-flow control, use of the sensors cut water consumption to 750,000 gal/mo.

Having slashed water usage to a tolerable level, the following system (see flowsheet) was devised to improve the water quality and cut effluent discharge.

WATER TREATMENT—Wastewater from the plating shop splits into two streams. About 2,500–3,000 gal/d flow directly to the solar pond with no attempt at reclamation. About two thirds of this stream is previously-reclaimed rinse water; the remainder is contaminated city water. No reclamation is attempted, because the stream consists mainly of rinses containing tin, lead and palladium salts, pumice, silicates or methylene-chloride, which are either detrimental to the RO membrane, or unaffected by it.

The other stream, about 12,000–15,000 gal/d, proceeds to the pretreatment operation and then to the reverse-osmosis unit. In addition, approximately 90% of the contaminated concentrate from the reverse-osmosis system is returned to pretreatment for recycling, while the remaining 10% goes to the pond.

PRETREATMENT—For the RO membranes to operate at good efficiency for any length of time, the

Reclamation-system costs:

Capital	
Prefiltration system, reverse-osmosis pump and membranes, polishing unit	$60,000
Storage tanks, enclosure for reverse-osmosis unit, plumbing and electrical wiring	60,000
Pond construction	40,000
Rinse-tank controllers, pH and conductivity monitors	15,000
	175,000
Operating (per month)	
Materials, including resins, chemicals, filters	1,000
System maintenance (including full-burden costs, such as depreciation, overhead, etc.)	1,000
	2,000

* This article is based on a presentation given at the 31st Annual Purdue Industrial Waste Conference, held May, 1976, at W. Lafayette, Ind.

Originally published March 28, 1977

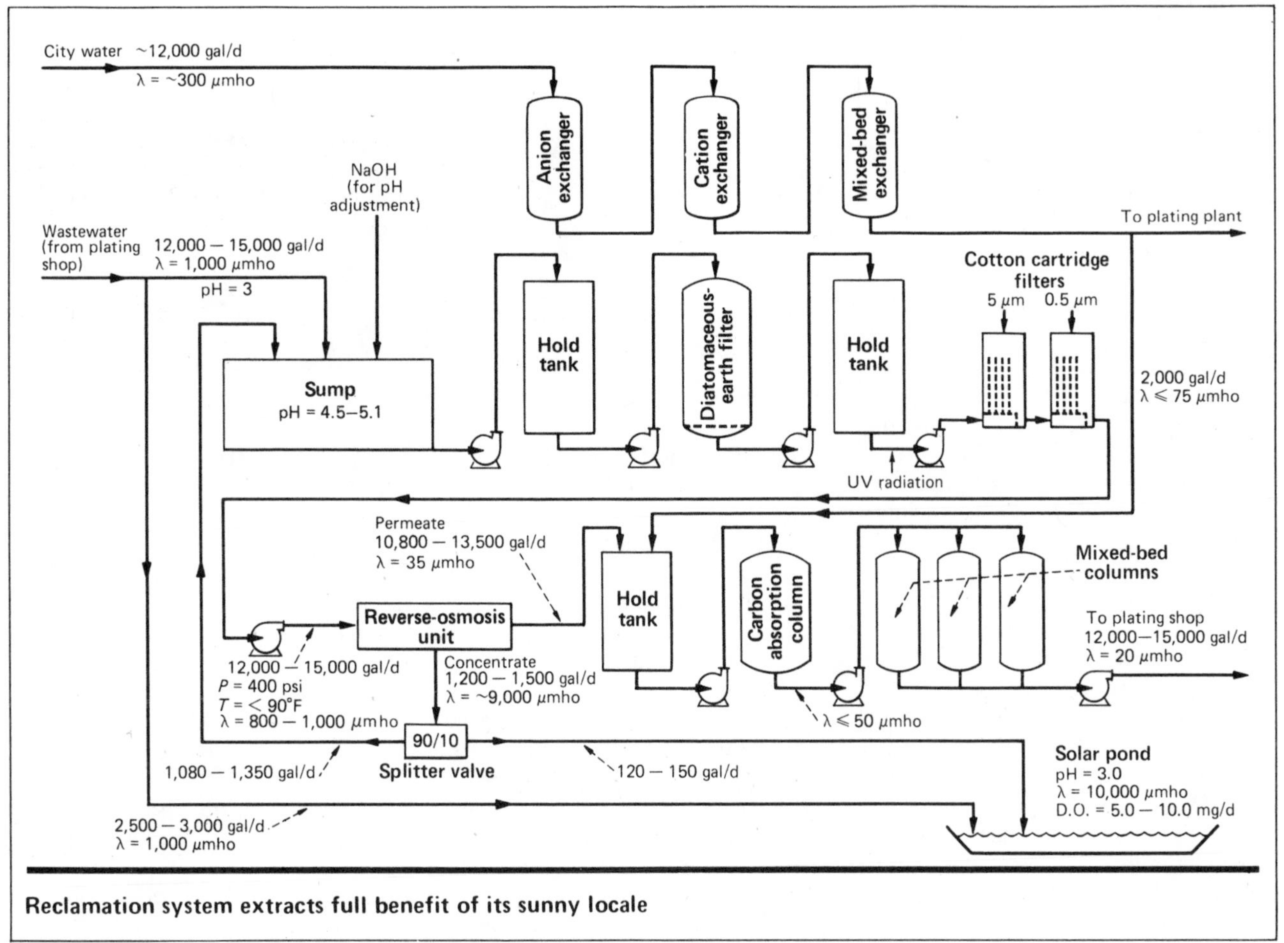

Reclamation system extracts full benefit of its sunny locale

feedstock's pH must be adjusted to 4.5–5.1, and the feedstock must be filtered so that the membrane will not clog. Because the average pH of the used rinse water is near 3.2, pH is monitored in a collection sump and at the RO unit's inlet. Adjustment is carried out by addition of caustic soda in the sump, and is controlled by an automatic system.

The feed then passes through a diatomaceous-earth filter containing approximately 14 lb total of three different grades of filter material and operating at flowrates to 96 gal/min. The filter area is about 48 ft^2, and under present conditions the filter must be repacked about once a month. After the filter is repacked, untreated water is circulated through it in the forward direction and sent back to the pretreatment holding tank for reprocessing. If this is not done, the material that initially passes through the repacked diatomaceous earth will clog the subsequent cotton filters.

The 5-μm and 0.5-μm filters consist of cotton-wound cartridges. The coarser filter comprises six 10-in. cartridges that must be changed every two weeks, while the finer unit has twelve 10-in. cartridges that can last about three months before needing to be changed.

Because of the importance of prefiltration, filter operation is monitored by a pressure-drop-measuring device. Too high a pressure drop through the filter indicates the system needs attention.

Precaution must also be taken against an unlikely problem source—algae. Although the wastewater from the plating shop is acidic and contains a relatively large amount of copper, algal growth occasionally occurs. Flushing the pretreatment system with hypochlorite solution every 2–3 months and constantly sterilizing the feedstock by ultraviolet radiation at a point just prior to the 5-μm cotton filter is very effective in controlling unwanted flora.

REVERSE OSMOSIS—The reverse-osmosis unit was manufactured by Abcor and will process 25 gal/min of waste at a design efficiency of 90%. It consists of a pair of 8-in. polyamide membranes in series, which feed a pair of 4-in. cellulose triacetate membranes in parallel. The feedstock is supplied to the membranes at 400 psi and must be maintained at a temperature below 90°F, which in practice has not been a problem, even with occasional ambient temperatures of 110°F at Porterville.

The lifespan of the membranes is estimated at about 3 yr, and a spare of each type is kept on hand. One membrane per week is removed from service for maintenance. Typically, the membrane is flushed with a circulating rinse of a solution of 2% ethylenediaminetetraacetic acid and 0.1% of a cleansing agent, and then rinsed with demineralized water. If the unit shows signs of bacterial growth, it is soaked in a 1% formaldehyde solution prior to flushing.

The conductivity (λ) of the feed to the RO membranes is about 800–1,000 μmho. However, the characteristics of the membranes are such that the conductivity of the permeate is not necessarily lowest when the feedstock's conductivity is lowest. For this reason, the concentrate is normally recycled to the pretreatment system, which procedure has lowered the conductivity of the permeate to around 35 μmho from about 100 μmho. The concentrate is often as high as 9,000 μmho. However, if the feed to the RO membranes rises above 1,100 μmho, more than the standard 10% of the concentrate is sent to the solar pond.

After treatment in the RO unit, the reclaimed water is held in storage. City water that has passed through anion, cation and mixed-bed columns is added at this point as makeup. A portion of the treated city water is also sent on to the plating shop for purposes other than plating-rinse. The reclaimed water and makeup water are then passed through a carbon absorption polishing-filter and mixed-bed columns.

POLISHING—While over 99% of the heavy metals contained in the original feedstock are removed by the RO membranes, the permeate must be given a final polish to eliminate virtually all of the remaining metals and particulates before being sent to the plating shop.

The carbon absorption column contains 18 ft^3 of granular-carbon filter material and 400 lb of ⅛- × 1/16-in. gravel, and operates at flowrates up to 110 gal/min. Approximately 3,000 gal of RO permeate are flushed through the unit, the filtrate typically having a conductivity of 50 μmho. The filter is equipped with a self-contained flushing unit having an automatic backwash feature; backwash is performed about once every six months and requires about a half hour.

Three mixed-bed columns are situated in parallel, each containing about 5 ft^3 of resin, which must be regenerated about once a month.

WASTE DISPOSAL—The solar pond is the reservoir for all the heavy-metal-laden wastes, as well as for any waste streams that are not reclaimable. In cross-section, the pond consists of a 10-mil polyethylene liner laid directly over compacted subgrade. Above this is an 8-in. gravel fill, topped by a 20-mil PVC liner and 12 in. of uncompacted earth. The pond is 8 ft deep and has a ¾-acre area. The gravel zone between the two liners is monitored for leaks by a perforated pipe leading to a manhole that is checked weekly for water.

Highly concentrated rinse streams are sent directly to a solar pond.

For the Porterville area, the average annual rainfall is 11 in., while the average annual evaporation is 60 in. The maximum allowable discharge to the pond under the terms of the applicable EPA permit is 6,000 gal/d; present maximum discharge is only 3,000 gal/d. It has been estimated that the pond will not overflow even under conditions of maximum allowable discharge and greatest annual rainfall (which was 22 in., in 1969).

The pH, conductivity, dissolved oxygen (D.O.) content and depth of the pond are measured weekly: the pH is about 3.0, the conductivity is around 10,000 μmho, and the dissolved oxygen content fluctuates between 5.0 and 10.0 mg/l. The maximum depth observed has been 5 ft. It has been estimated that about 50 years will pass before the heavy-metal-laden sludge will have to be removed from the bottom of the pond. The sludge will then be processed for metals recovery.

ECONOMICS—The costs for the reclamation system can be found in the table. In addition, an extra $100,000 was spent to upgrade the plating facility itself, but a like expenditure might not be necessary at another location.

Since the installation of the system along with the plating-operation upgrading, the reject rate of circuit boards has dropped to 2.5% from 10%, and the scrap rate (unsalvageable rejects) has gone from 3% to less than 0.5%. The portion of the reduction in scrap rate attributable to the water reclamation system represents a yearly saving of about $17,500. (The facility processes over $1.5-million worth of circuit-boards per year.) Hence, without consideration of the costs of water or of sewage discharge that would otherwise be needed, a ten-year writeoff is anticipated.

Further, the total operating cost, including materials and labor, is offset by elimination of a monthly carbon treatment of the plating tanks for removal of organic growth that no longer flourishes.

The reclamation system has also made possible denser circuitry on the boards with higher quality than previously achieved, with no net increase in cost.

Likewise, city water usage has decreased to 0.35 million gal/mo, and EPA discharge requirements are satisfied. It should be noted, however, that in non-predominantly-sunny locations, discharged wastes would have to be treated by using evaporators rather than solar ponds, and the resulting reclamation system most likely would not be cost effective.

The authors

Jay E. Warnke is quality assurance manager and manufacturing engineering manager for Beckman Instruments, Inc. (167 W. Poplar Ave., Porterville, CA 93257). Holder of a B.A. in social science from California State University at Long Beach, he is a member of the California Circuits Assn.

Kenneth G. Thomas has the title of process manufacturing engineer at Beckman's Porterville facility. He obtained a B.S. in chemical engineering at Rocky Mountain College and holds the rank of Commander, U.S. Navy (Ret.).

Samuel C. Creason is principal chemist in Beckman's Process Instruments Div. Holder of a bachelor's degree from Chico State College and a master's degree from Sacramento State College, both in chemistry, he obtained a Ph.D. in analytical chemistry from Northwestern University. He is a member of the American Chemical Soc., Instrument Soc. of America, American Soc. for Testing and Materials, and the Technical Assn. of the Pulp and Paper Industry.

Finns make saltpeter from wastewater

Being groomed for commercial use, this liquid-liquid extraction process selectively removes nitrates dissolved in wastewater and yields potassium nitrate, which must now be imported into Finland. Treatment costs should be more than offset by the sale of this product.

☐ A promising technique for removing soluble nitrates from wastewaters is now completing successful pilot plant tests in Finland. The liquid-liquid extraction process, developed at Kemira Oy's research laboratories in Oulu, is said to have distinct economic advantages—under conditions prevailing in Finland—over established nitrate-removal methods such as ion exchange, and chemical or biological decomposition of nitrates.

Indeed, the new process is especially valuable to the Finns because it recovers nitrate in the form of KNO_3, which Finland imports at a price of $220/metric ton (CIF) for use as a specialty, chlorine-free fertilizer and in black-powder manufacturing. According to calculations provided by Kemira Oy, KNO_3 recovery results in operating-cost surpluses that can go toward amortizing the investment cost of a commercial unit in 12 years. Under the right circumstances, the return on investment can be bigger.

The other plus of the Finnish development is an environmental one. About 85% of the nitrates in the concentrated wastewaters treated (nitrate content: 10–20 g/l) wind up as potassium nitrate product, so a source of pollution that may become the target of future regulatory laws is thereby largely eliminated.

Although Kemira Oy intends to commercialize its process, it has yet to reveal any definite plans. The company has been operating a 0.5-m^3/h pilot plant that treats acid residues from a cellulose nitrate plant, and has applied for patents in 20 countries; Belgium, East Germany and Austria have already granted such protection.

HOW MUCH REMOVAL?—Compared to biological and ion-exchange methods, which are known to reach nitrate-removal efficiencies of 99+%, the Finnish development leaves a fair amount of nitrates in the wastewater. But Kemira Oy seems unconcerned about this. The company does point out that it is possible to increase process efficiency to more than 95%, although this means an increase in operating costs.

At any rate, the issue of nitrate-removal capability is not a crucial one yet in Finland and in many other industrialized nations. The U.S., for instance, has set no limits on the nitrate content of industrial effluents; current restraints apply only to drinking water, which must not exceed a nitrates concentration of 10 mg/l.

When nations do get around to placing limits on nitrates, the rulings will affect a variety of industries that use nitric acid, such as electronics, fertilizers, photochemicals and explosives.

AMINE IN KEROSENE—Although Kemira Oy says it can change process conditions to accommodate a variety of nitrate-removal appli-

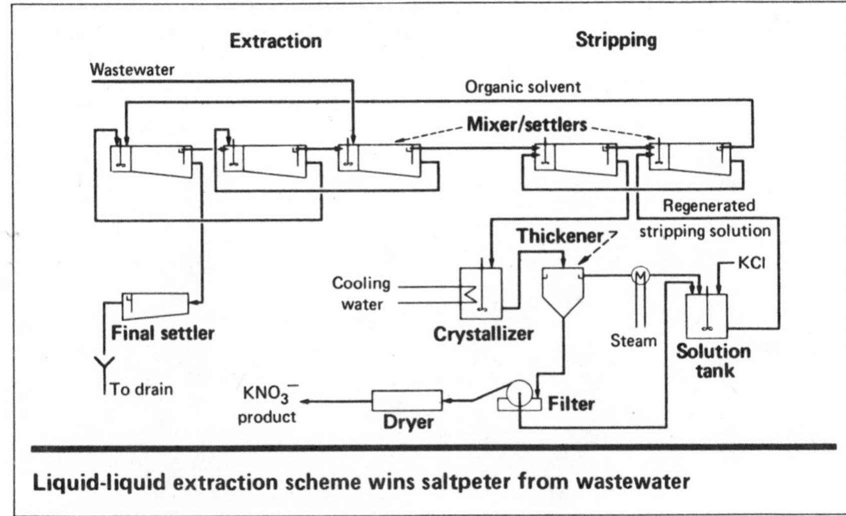

Liquid-liquid extraction scheme wins saltpeter from wastewater

Originally published July 5, 1976

Annual operating costs (U.S. dollars)

Raw materials (KCl)	$79,000
Solvent losses (kerosene and amine)	5,675
Salaries	38,500
Utilities	
electricity (100 kW)	12,000
steam (80 kg/h)	4,000
cooling water (20 m^3/h)	1,500
Maintenance	12,750
Quality control	4,500
Miscellaneous	6,250
Total	$164,175

cations, it has initially focused on the explosives industry. A large-scale unit tied to a commercial cellulose nitrate plant is being considered.

The Finnish extraction/stripping technique is based on the following chemistry:

$$(R'R''NH_2Cl)_{org} + (NO_3^-)_{aq} \underset{\text{stripping}}{\overset{\text{extraction}}{\rightleftharpoons}} (R'R''NH_2NO_3)_{org} + (Cl^-)_{aq}$$

$R'R''NH_2$ is a heavy, water-insoluble, liquid secondary amine, which is dissolved in kerosene to form a 5–10% by volume solution. During stripping, nitrate is removed from the organic phase by a salt solution containing about 15% by weight of KNO_3 and 23% by weight of KCl. The chloride content of the salt solution is sufficiently high to drive the reaction toward the left, and as this happens, the KNO_3 content of the stripping solution increases while its KCl content decreases correspondingly.

High-purity (99+% after washing) potassium nitrate crystallizes when the stripping solution is cooled—the amount being proportionate to the stripped nitrate. After KCl is added to compensate for crystallized KNO_3, the stripping solution is reused.

Timo Mattila, director of the Oulu research laboratory's process development dept., says the process is very selective to nitrate ion. Sulfate ion present in the wastewater does not move in great amounts into the organic phase, and therefore does not wind up in the potassium nitrate. Only a few ppm of organic solvent remain in the final product.

PROCESS DESCRIPTION—Extraction takes place in a series of mixer-settlers (see diagram) and wastewater under treatment and the organic solvent run countercurrently from stage to stage. "Nitrate-free" wastes go through a final settler before entering the drainage system or undergoing further treatment, while nitrate-loaded organic solution from the first extraction unit passes on to other mixer-settler units for stripping.

As the organic phase runs countercurrently to the KNO_3/KCl aqueous solution, the latter exchanges its chloride ions for nitrate ones. The regenerated organic phase goes back to extraction, while the stripping solution enters into the crystallizer, where potassium nitrate is obtained when the solution is cooled from 25°C to 20°C. The KNO_3 crystals are separated in a thickener, passed through a filter and dried. Meanwhile, KCl is added in a separate tank to which the overflow from the thickener and the mother liquor from the wash filter are sent. This compensates for the removal of KNO_3 crystals and restores the stripping solution.

Process equipment for the Kemira Oy technique must be able to resist severe corrosion. The best material for mixer-settlers is reinforced plastic, although for bigger units, concrete or steel vessels coated with plastic are recommended. Heat exchangers for the stripping liquid should be lined with Teflon; plastics are also indicated for pumps and piping.

ECONOMIC PICTURE—On the basis of pilot plant tests, Kemira Oy has worked out figures for a plant able to treat the effluent from a 1,100-metric-ton/yr cellulose nitrate facility. The plant would handle 15 m^3/h of wastewater for a total operating period of 4,800 h/yr (based on 200 days).

Based on current exchange rates (1 U.S. dollar = 4 Finnmarks), capital investment for the entire extraction plant is about $775,000. This includes a building to house the facility, which is considered necessary in Finland.

Operating costs are given in the table; cost of treatment is about 3.5¢/m^3. If income from the sale of KNO_3 is taken into account (about $228,000/yr), there is a profit of about $64,000/yr, which could help amortize the treatment plant in 12 yr. Profits can be bigger for larger plants, or for those in more temperate climates (where no building would be needed).—RAUL REMIREZ.

Swedes Recover Zinc From Rayon Wastes

A new technique based on solvent extraction economically recovers zinc that can be reused in the spinning baths. The developer claims important advantages over existing zinc recovery methods.

A new process, billed as the world's first commercial solvent-extraction method to recover zinc from rayon-plant wastewaters, went onstream last month at the works of Svenska Rayon AB in Valberg, Sweden. The route reduces zinc concentration in effluents to levels lower than those economically possible with competitive techniques. It also returns to the rayon-processing circuit a value of metal that considerably surpasses the cleanup system's operating costs, claims the developer, MX-Processer AB of Mölndal.

Many rayon producers can find use for such a recovery technique. Sizable quantities of zinc turn up in plant wastewaters after this metal is used in acidic spinning baths to help regulate the speed of coagulation of cellulose. Until recently, the Valberg works alone had been discarding about 1 metric ton/d of zinc through its effluent. Environmental concern and the rising price of zinc are good reasons to cut this loss.

MX-Processer's technique, dubbed the Valberg for the location of a $700,000-plus unit built by Svenska Rayon, is notably simple. Plant wastes are mixed with a kerosene, stream containing an organic acid that reacts with zinc to form an organo-metallic compound.* Further on in the process, the zinc is recovered from this compound with sulfuric acid, and the resulting zinc sulfate solution goes back to the rayon-plant bath.

The Valberg route is said to be able to pare the level of zinc in waste streams to less than 4 ppm—a performance unrivaled by other recovery processes, says the Mölndal firm. In addition, the developer, which is interested in engineering other units and licensing the technology in partnership with Svenska Rayon, claims that the method can handle any other acidic wastestreams containing zinc.

Better Than the Competition—The Valberg process compares well with existing zinc-recovery routes. Neutralization with caustic soda, for instance, causes the zinc present in rayon-plant wastes to precipitate as a hydrophilic zinc hydroxide sludge, from which it is hard to remove the water. Addition of water is undesirable. (Water is unwanted because it upsets the material balance in the rayon process.)

Lime neutralization precipitates calcium sulfate along with the zinc hydroxide, and it is very difficult to separate the mixture, the company notes. There is a two-step neutralization scheme that uses lime first, then caustic soda or soda ash to precipitate zinc hydroxide; but this method is unable to prevent some contamination of the zinc precipitate with calcium sulfate.

Still another technique comprises an initial neutralization step followed by reaction with hydrogen sulfide to precipitate zinc as a sulfide. However, this introduces the danger of secondary pollution by hydrogen sulfide, which can be worse than that caused by zinc.

Electrochemical recovery of the metal is theoretically possible, but MX-Processer notes that no successful operation of this kind has yet been reported.

Some rayon plants use ion-exchange units. But that approach has its drawbacks, too, according to MX-Processer. Since the exchanger's ion selectivity is often poor, large quantities of sodium are absorbed along with zinc. This reduces the resin's zinc absorption efficiency, and increases the consumption of sulfuric acid needed for elutriation. And, of course, sodium must be separated before the elutriant can go back to the rayon spinning bath.

High Extraction—The new Swedish process can handle very concentrated wastes, although those with more than 0.5 g/l of zinc must be diluted before treatment. If the waste stream has an inordinately high solids content, a filtra-

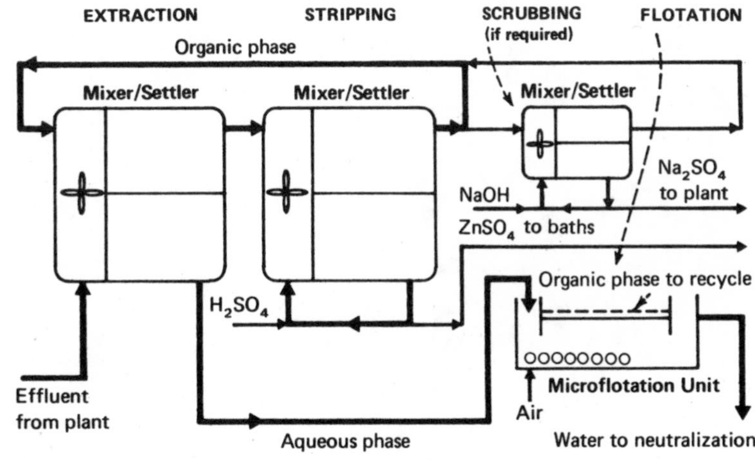

SWEDISH PROCESS includes scrubber for removing iron impurities.

* Thus, the Valberg process is an instance of solvent extraction that entails chemical interaction between the solute and the solvent stream.

Originally published April 28, 1975

tion pretreatment may be necessary in order to prevent clogging in the system's mixer/settlers. At Valberg, about 3,500 l/min of 1.8-pH effluent containing 0.2 g/l of zinc go directly into a conventional mixer/settler made of glass-fiber-reinforced polyester. There, the waste comes into contact with an organic phase—kerosene with 20-25% of di (2-ethylhexyl) phosphoric acid (dioctyl phosphoric acid). Zinc reacts with the latter at ambient temperatures and pressures to form an organo-metallic compound. About 95% of the zinc is extracted into the organic phase in one step; however, two countercurrent steps are used to recover all but 4 ppm or less of the metal. Size of the mixer/settler is 55 m^3; the settling area occupies 50 m^3.

The ratio of the organic to the aqueous phase is about 2:1, which means that at Valberg 6,500 l/min of the kerosene solution enters the mixer/settler.

Raffinate from this unit goes to a microflotation vessel, where any entrained organic phase is recovered for recycle. Organics in the effluent amount to 10 ppm or less—i.e., their level of solubility in the aqueous phase. Wastewater, which is still acidic, undergoes neutralization and is then discarded.

Stripping—The organic phase passes to another mixer/settler. There, it is joined by an aqueous phase consisting of a 6M solution of sulfuric acid mixed with some zinc sulfate recycled from the settler portion of the vessel. The ratio of organic to aqueous is again about 2:1. The acid solution takes zinc from the organo-metallic compound to form an aqueous zinc sulfate solution also containing excess acid.

This solution has about the same ratio of metal to acid as that of a rayon spinning bath. The only difference is that zinc concentration (about 80 g/l) is high in comparison with that of a bath (0.4 — 0.5 g/l), mainly because of the high-acid-concentration required for stripping. Nevertheless, the stream from this second mixer/settler can be recycled directly to the rayon-plant bath; if fed at low rates, it is suitably diluted by the liquids in the bath.

Although this does not happen at Valberg, iron impurities can build up in the organic phase. In this case, a sidestream of it is scrubbed with alkali prior to its return to the first mixer/settler. Size of the scrubber, typically about 1 m^3, would depend on the iron content.

Scrubbing leaves a raw sodium sulfate, which can be handled by the sodium sulfate recovery unit available in many rayon plants. If scrubbing is required, and the rayon facility lacks a recovery unit, the sodium sulfate can go into the neutralization circuit.

Economic Considerations—In assessing costs, MX-Processer notes that sulfuric acid added for stripping would be used anyway as makeup in the spinning baths. The Valberg process merely uses it first before sending it on to the baths. Its consumption, therefore, is not an additional expense. Makeup for organic losses, as noted previously, is only 10 ppm on the aqueous throughput. Power consumption at Valberg runs to about 100 kw.

MX-Processer claims an average operating cost of \$115,000/yr. Zinc recovered annually is worth about \$350,000. Metal losses at Valberg have been cut from 1 metric ton/d to 50-100 kg/d.—MARK ROSENZWEIG, *European Editor.*

Section XIV
WOOD, PULP AND PAPER

Bagasse Now Becomes a Newsprint Source, Too

Making paper from bagasse is nothing new. But a process developed in Mexico, and about to make its debut in Peru, can convert this sugar-cane waste into the special-quality stock that is needed for printing newspapers.

After years of research and active promotion, a technique for making newsprint out of sugar-cane bagasse is ready to go commercial. The first newsprint plant to use the Cusi process—named for its inventor, Mexican engineer Dante Cusi—will be built at Trupal, Peru. Scheduled to go onstream in 1977, the $50-million facility will produce 112,000 metric tons/yr of finished newsprint, claimed to be competitive in price and quality with the one made from wood pulp.

Proven technology for making paper from sugar-cane bagasse has been available for years, and it flourishes today in cane-growing areas such as the Philippine Islands, Cuba, Taiwan and South America. But the processing of bagasse into paper of the special qualities required for newsprint—e.g., proper ink-absorption, resistance to tearing—has taken longer to reach the commercial stage. In addition to Cusi, only the U.S. firm W. R. Grace & Co. seems to have developed a marketable bagasse-newsprint system.

No figures comparing the Grace and Cusi methods are available. But Mexican engineers say the two were tested side by side during the U.S. firm's trial of its newsprint technique in Paramonga, Peru, in 1972. Apparently as a result of the outcome, Peru's state-owned agency Induperu chose the Cusi method for the Trupal newsprint plant.

In comparison with traditional bagasse-pulping methods—for making regular paper—the Cusi process comes out the winner. According to extensive tests performed under the supervision of Mexican government agencies and private firms, it yields nearly twice as much pulp per unit weight of bagasse, while using less steam and caustic soda.

Tailored for Bagasse—The design of the Peruvian plant, which will be built by Bufete Industrial,* Mexico's foremost private engineering and construction firm, (with help from the U.S.'s Rust Engineering Co.) strives to deal effectively with the highly heterogeneous nature of sugar-cane bagasse.

"Traditional methods, based on wood, try to adapt the raw material to the technology available," says Cusi's assistant Julio Amador. "Wood, after preliminary stripping procedures, is a highly homogeneous material. But bagasse, though

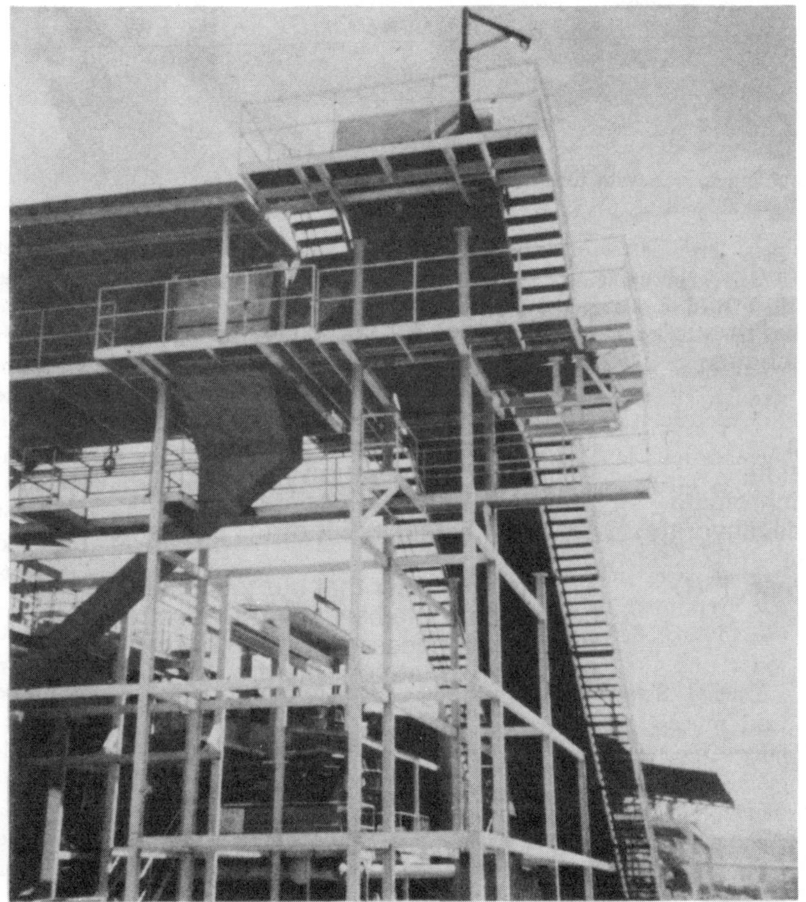

CONVEYOR BELT has a magnet to remove metallic particles from bagasse.

* Patents covering the Cusi process are owned by Cusi's engineering company, Procesos y Tecnicas Industriales. Bufete Industrial holds licensing rights for North and South America, and the British group Simon-Carves for the Eastern Hemisphere.

Originally published November 25, 1974

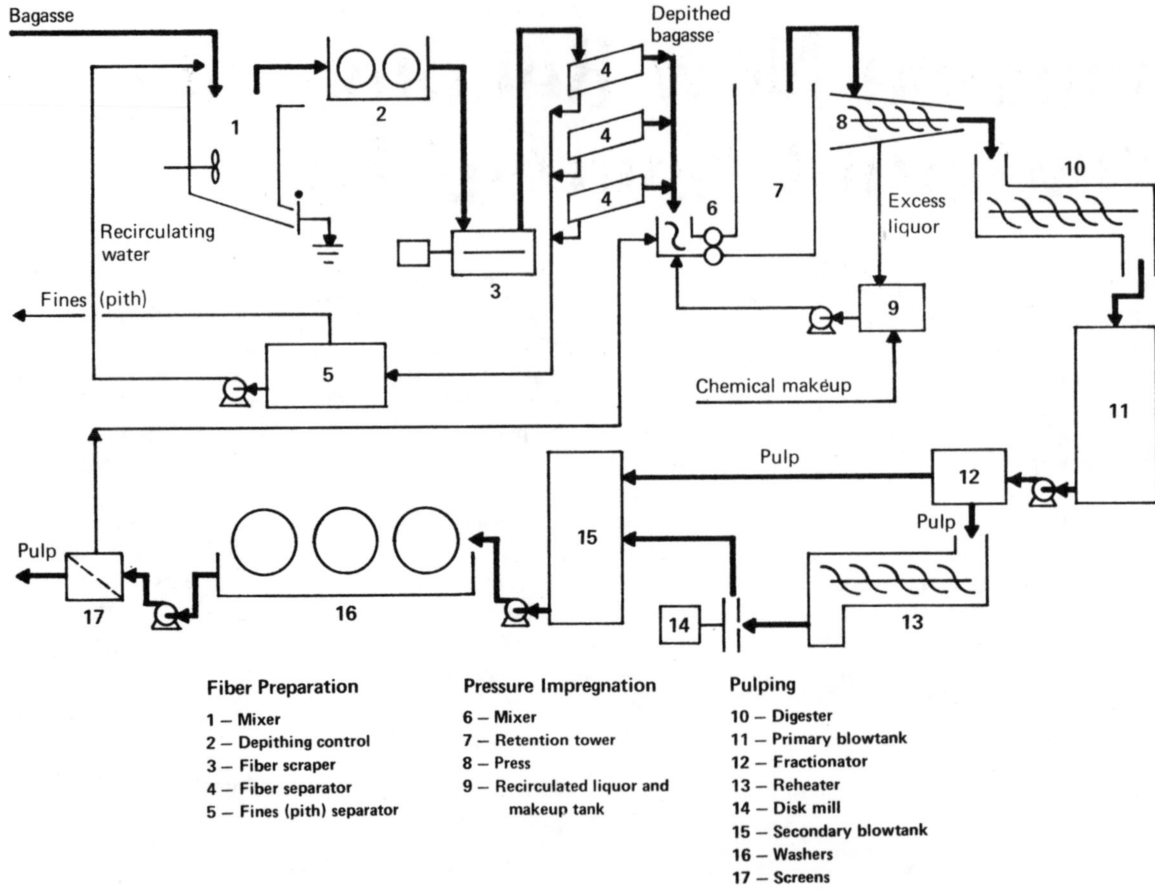

CUSI DESIGN makes use of special equipment to deal with the varying quality of bagasse fibers.

chemically homogeneous, is a highly heterogeneous mixture of plant tissues that vary from the rigid rind supporting the cane stalk, to the large pithy tissues, built like honeycombs, that store the sugar juice in the green stalk. The transitions from one tissue to another are extremely gradual."

Mexican engineers say that conventional bagasse-pulping squanders almost three-fourths of the raw material. It usually begins with a harsh depithing operation — in which, ideally, the useless, softer tissues are separated from the rest by scraping—that rejects 40-50% of incoming bagasse. Accepted material then cooks in a chemical digester at high temperatures (170°C) in the presence of a caustic soda solution (12-14%, and up to 20% NaOH).

This strong cooking further reduces yield, says Cusi, because the physical properties of bagasse tissues, even after depithing, vary widely. He explains: "If porous, light tissues are cooked along with dense ones, the former will wind up overcooked, while the latter will remain practically raw."

Cusi estimates that conventional methods lose 44% of the depithed bagasse in the digester. Therefore, final yield of pulp is only about 30% by weight of bagasse (some bagasse-paper mills in Mexico obtain even lower yields because they use 15% of the incoming bagasse as fuel).

Crucial Steps—Success of the Cusi process (see flow diagram) hinges on two carefully designed operations that take place before pulping: fiber preparation (depithing) and pressure impregnation.

All of the raw material is subjected to depithing (Steps 2-5 in diagram), which involves a highly selective scraper—designed by Bufete Industrial—for cleaning and scraping bagasse fibers in a water solution. Since the Cusi technique employs only mild cooking, the fines rejected by the separator (25% of incoming bagasse) are sufficient to provide fuel for the process.

The next operation, pressure impregnation (Steps 6-9 in diagram), aims to equalize the absorbency of different bagasse tissues. Depithed material undergoes chemical treatment under pressure in a mixer, and then it is allowed to soak for 30-60 min in retention towers. The chemical bath used is a solution whose composition varies according to desired GE-brightness in the end-product. For instance, for newsprint of 60°GE, Cusi uses a 7.5-8% NaOH, 12% Cl_2 solution (the chlorine is provided by sodium or calcium hypochlorite).

Between the retention towers and the digester, the material is drained and pressed; this breaks down porous tissues, consolidates the mix and eliminates excess liquor. Cooking in caustic soda takes place in the digester at mild—e.g., 155-165°C—temperatures.

The strongest tissues emerge only partially cooked; they are separated in a subsequent fractionator, and ground in a disk mill. After grinding, they go back into the pulp mixture to provide strength and tear resistance.

The pulp is bleached, washed, screened and sent to paper machines in the same manner as wood pulp. Cusi figures show a 52.5% yield of pulp by weight of raw material.

A Cusi facility can produce various grades of paper by varying the chemical treatment in the digester. For instance, the San Cristóbal paper mill in Mexico has been using the Cusi process for several years to produce $60-million/yr worth of bagasse writing-paper, tissues, pulp and other products. The Cusi bagasse-newsprint has been tested successfully by newspapers in Mexico, West Germany and Canada.

Acceptance Grows—Bufete Industrial engineers are optimistic about the method's future, for newsprint and other paper. "Bagasse will become a major factor in the development of a competitive paper industry in many of the sugar-cane-growing countries," says a company spokesman.

"The first [newsprint] contract was the hardest because we were promoting a new process," he adds. "If we had already built 15 such plants it would have been easier. But negotiations are now advanced for projects of this kind in Ecuador, Bolivia, Brazil, Venezuela, and, of course, Mexico."

At Cusi's engineering company, Procesos y Técnicas Industriales, a spokesman says a contract with the Mexican government is "imminent."—TIMOTHY BERRY, *McGraw-Hill World News, Mexico.* #

ClO_2 Generator Cuts Byproduct Sulfuric Acid

A boon to kraft pulpmills seeking ways to reduce the amount of sulfur emissions, this bleach-plant process makes no unwanted sulfuric acid.

EDWARD S. ATKINSON
Hooker Chemical Corp.

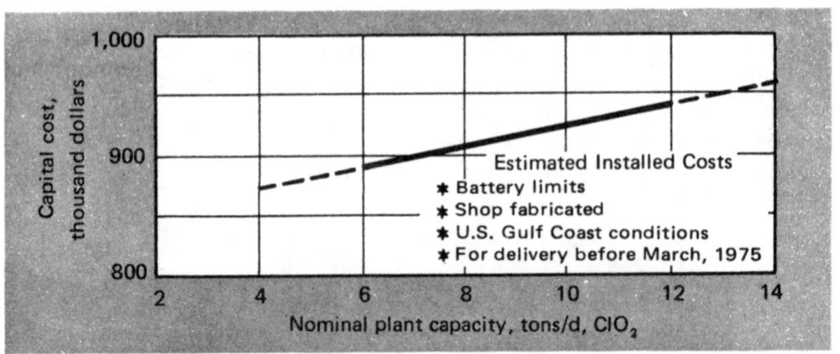

Tighter restrictions on sulfur emissions at kraft pulpmills have had an adverse effect on the mills' ability to reuse byproduct spent acid from bleach-plant chlorine dioxide generators. The acid, which contains sodium sulfate and sulfuric acid, can no longer be fully utilized in the pulp-liquor recovery process as generally practiced, because steps taken to limit total reduced sulfur emissions elsewhere in the mill will be adversely affected.

One way out of the dilemma is the SVP* process developed by Hooker Chemical Co. It produces the least sulfur byproduct of any bleaching route, while making chlorine dioxide at minimum cost. The key is a unique chlorine dioxide generator that yields product in concentrations up to 36% at reduced pressure. In conventional atmospheric-pressure routes, such concentrations would be explosive. But by reducing pressure, diluting the gas with steam, lowering residence time, and installing proper control interlocks, Hooker has commercialized this technique as a very stable one.

Since the process' introduction four years ago, eight units have been installed and five others are on order. In October 1973, Hooker announced an agreement to sell the technology to V/O Prommashimport for a new mill to be located at Ust-Ilimsk, U.S.S.R. In order to produce the ClO_2 necessary for bleaching, the dual-train unit will have a huge 22.7-metric ton/d capacity.

No Acid—The SVP process is unique in that while it produces byproduct sodium sulfate, it produces no sulfuric acid. The advantage is that the sulfate is easily utilized in the pulpmill recovery circuit, whereas any acid, if produced, would have no sodium value to offer to the recovery circuit and would only pose a net sulfur load. The SVP route reduces sulfur output by 33-65%, if not completely, compared with other processes.

Moreover, the economics of conventional processes is very dependent on the successful utilization of the spent acid, which is valued at 4.5 to 8.5¢/lb of ClO_2 produced. But mills are now faced with the need to throw away some of this credit and send the spent acid to the sewer rather than allow all the sulfur to get into the air via the recovery furnace. The cost of chlorine dioxide increases, and pollution is merely spread around. Reducing the production of ClO_2 is not an attractive alternative; this would go against the industry trend to use more of the bleaching agent.

How It Works—The reactions that produce chlorine dioxide in the SVP route are the same as those of the conventional method in which sodium chlorate is reduced by sodium chloride in the presence of sulfuric

* SVP, a registered trademark of Hooker Chemical Corp., stands for Single-Vessel Process; ClO_2 generation, evaporation and crystallization take place in a single vessel.

Chemicals and Utilities*
Table I

Products, lb/d	
ClO_2	16,000
Cl_2	9,920
Na_2SO_4	36,800
Chemicals Consumed, lb/d	
$NaClO_3$	26,080
NaCl	16,000
H_2SO_4	25,440
Utilities Consumed	
Process water, gal/min	
For ClO_2 absorption (50°F)	170
For Cl_2 absorption (75°F)	250
Cooling water, gal/min	
At 80°F	350
For the chiller condenser	400
Steam, lb/h	
For reboiler (60 psi)	5,000
For ejectors (150 psi)	1,500
For chiller (150 psi)	5,000
Electricity, kWh	3,400

*Figures are typical of an 8-ton/d unit. In evaluating total cost, add a royalty of 1.3¢/lb of ClO_2 produced.

Originally published February 4, 1974

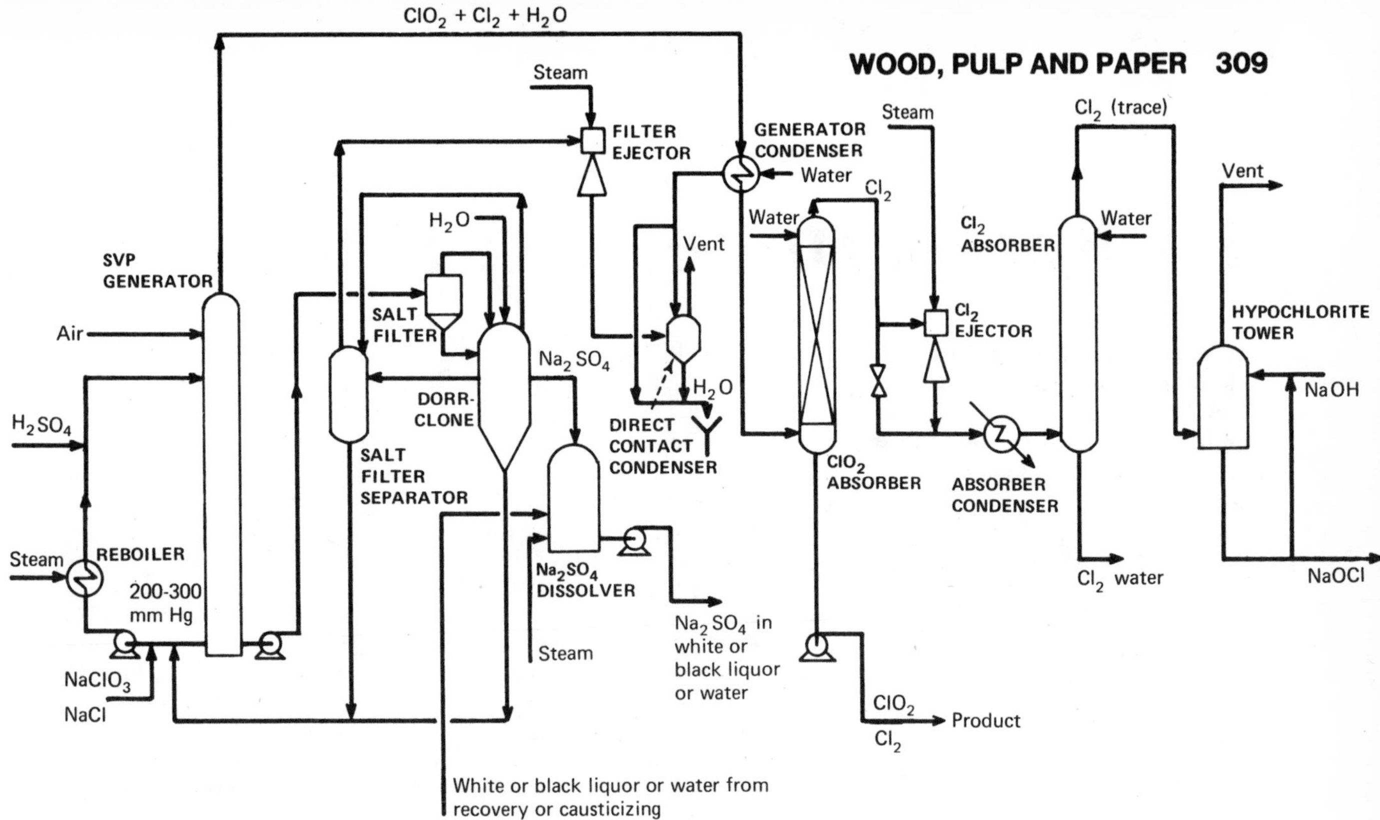

acid. However, hydrochloric acid can be substituted in the SVP route (this would eliminate the sulfur output).

The SVP system can also use spent acid from conventional processes. Four of the units already onstream operate in this way by working alongside the older plants. In this manner, most of the benefits of the SVP route are passed to the conventional process; the capital cost of increased ClO_2 capacity is reduced; and the overall byproduct-sulfur output can be decreased to 25-47% of that of other processes.

In operation, a slurry of sodium sulfate crystals in the reaction solution is tapped from the bottom of the generator, mixed with sodium chlorate and sodium chloride feed, reboiled, and mixed further with sulfuric acid. This stream is returned to the top of the generator.

In the generator, steam and the gaseous reaction products (chlorine and chlorine dioxide) are flashed off. These gases, along with a small amount of dilution air that is admitted near the top of the generator, flow through the condenser, where the steam is condensed and the gases are cooled prior to entering the packed absorber. Keeping the temperature low minimizes possible absorption and storage problems.

The gases leaving the condenser contain about 36% ClO_2, 25% Cl_2, 10% air and 29% water vapor. Most of the chlorine dioxide and part of the chlorine are absorbed in 40-55°F chilled water, which when combined with condenser condensate results in a solution containing 8 g/l ClO_2 and 1.6 g/l Cl_2.

These concentrations fall well within the explosive range for chlorine dioxide at atmospheric conditions. A steam syphon in the SVP system keeps the pressure low at 200-300 mm Hg to prevent spontaneous detonation. This, plus other measures

Comparative Sulfur Byproduct
Table II

Process	Byproduct Sulfur, as Equivalent Saltcake, Lb/Lb ClO_2
SO_2 (Mathieson)	3.66
MeoH (Solvay)	4.08
R-2 (Hooker)	6.94
SVP	2.30
Cascaded Systems	
SO_2/SVP	1.75
R-2/SVP	2.30
MeoH/SVP	1.81

already mentioned, keeps the mixture stable. The relatively few instances of chlorine dioxide decomposition that occur are of little consequence; in fact, these are largely undetectable except for a gradual rise in temperature and the absence of product and color in the absorption-tower solution.

Operation at reduced pressure and the use of highly corrosion resistant metals and plastics add considerably to the cost, but units in the 7-ton/d range can be installed for less than $900,000. Titanium and polyester are used extensively.

Costs—Chemical costs and credits of the SVP process are listed in Table I, along with chemical and utility requirements of a typical 8-ton/d unit. The net materials cost is 11.3¢/lb of ClO_2, which compares well with the 15.2¢/lb cost of the conventional processes (taking all byproduct credits into account). The comparative reduction in byproduct sulfur is detailed in Table II. Capital costs for a typical shop-fabricated plant are estimated in Fig. 1.

Meet the Author

Edward S. Atkinson is manager, special projects for pulpmill systems at Hooker Chemical Corp., Niagara Falls, N.Y. He holds a B.S.Ch.E. from Tulane University and has worked for St. Regis Paper Co. and Buckeye Cellulose Corp.

Fir-Bark Conversion Route

A unique solvent-extraction technique has been groomed to turn out several marketable products from Douglas fir bark, until now a troublesome and valueless waste.

ROBERT D. GOOD
Blaw-Knox Chemical Plants Div.,
Dravo Corp.

FRANK S. TROCINO
Bohemia Inc.

The 14 million tons of bark removed from trees each year has long been an expensive nuisance to the wood-products industry, mostly in terms of the actual stripping operation required. But recently, disposal of this waste material has become even more of a problem. The once-accepted method of burning is under strong attack because of pollution concerns. Sales of bark as fireplace fuel and as mulch have not been nearly enough to dispose of mounting accumulations.

A new technique has now been developed to put Douglas fir bark to use. Design of a $3.2-million processing plant that will extract such marketable products as wax, cork and resin extender from bark has been completed. Construction of the facility for Bohemia Inc. (Eugene, Ore.), a wood-products company that will own and operate the plant, is expected to be finished by July.

Bohemia collaborated with the Blaw-Knox Chemical Plants Div., Dravo Corp. (Pittsburgh, Pa.) in developing the new solvent-extraction process. A pilot plant was operated for about four years at Bohemia's research and development laboratory in Eugene. In March 1973, Bohemia awarded an engineering, procurement and construction contract to Blaw-Knox for the plant going up.

This full-scale installation will initially convert about 2,000 tons (dry basis) of bark per month into a high-quality vegetable wax, a thermosetting-resin extender, and a phenol substitute. It is estimated that 70-80 tons of wax will be turned out each month, 1,700-1,800 tons of extender, and 200 tons of phenol substitute.

Ultimate monthly production is tentatively projected to include 600-750 tons of cork, and 500-600 tons of bast fiber, a material suitable for strengthening plastics. (There will be a corresponding reduction in extender output when cork and bast fiber are produced.)

Vegetable wax from the process is a hard, light-brown material that can be used in polishes, carbon paper and many types of cosmetics. Like cork, it is an item that to date has not been manufactured in the U.S. Douglas fir bark contains a higher content of such wax than most tree species.

Special Extractor—The basic Bohemia process is an adaptation of one already used to extract meal and vegetable oil from soybeans. Blaw-Knox refined the system specifically to extract wax from bark. These refinements are headed by use of the Blaw-Knox Rotocel, a continuous countercurrent extractor for solids.

Bark is first cut and ground (to a size predominantly between 7 and 80 mesh), and dried in a rotary steam-tube dryer. Particles smaller than 100 mesh are immediately screened out and sold as a phenol substitute. This material contains so little wax that extraction would be uneconomic.

The fraction of larger bark particles is slurried with miscella—a mixture of solvent and freshly extracted wax—and fed to the Rotocel. This unit consists of a rotor within a vaportight circular tank. The rotor is divided into sector-shaped cells, and turns slowly within the tank. The solvent is a mixed one of both paraffinic and aromatic compounds.

The ground bark forms beds in the cells through which liquid can percolate by gravity. As each cell moves around the circular path from bark inlet to outlet, the bark is flooded by successive washes of miscella, which are gradually approaching fresh solvent in composition. After a final spray, with fresh solvent, the extracted bark is permitted to drain

DESOLVENTIZER is shown in this view of Bohemia's new plant (taken during construction).

Originally published May 27, 1974

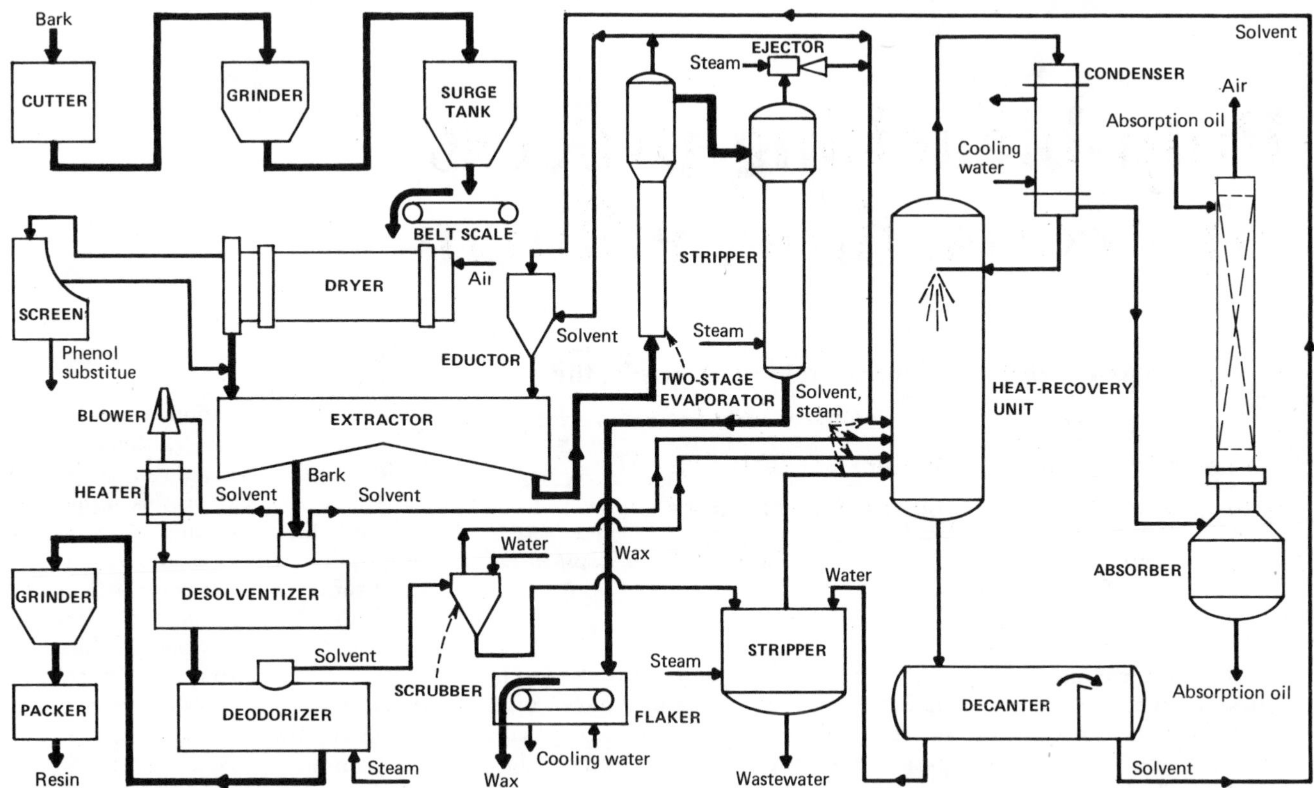

completely by gravity, Still solvent-wet, it is then discharged through hinged doors at the bottom of the cells.

Drained liquid collects under the rotor in separate compartments from each cell. Stage pumps route this solvent/wax mixture from cell to cell, toward the bark inlet. The stream becomes richer and richer in wax, and is finally withdrawn and routed to the solvent/wax recovery section.

Bark Treatment—A second refinement of the Bohemia process by Blaw-Knox is the desolventizer-deodorizer system that treats the extracted bark.

Upon discharge from the Rotocel, the spent bark passes to a horizontal vapor-desolventizer. Here, a special cage mechanism conveys and showers the bark through superheated solvent vapors. The bark moves through the unit continuously, remaining in the superheated vapors for 3-4 min. Temperature is kept under 250°F. About 99% of the solvent evaporates from the bark particles.

A recycle of solvent vapors supplies the necessary vaporization heat. Cooled vapor withdrawn from the desolventizer by a blower is reheated in a steam-heated tubular exchanger (temperature rise is 120-140°F), and returned to the vessel.

Desolventized bark then goes to a deodorizer for final solvent removal. The deodorizer is another horizontal vessel containing a conveying and lifting device, which is designed to assure close contact between bark and stripping steam. Pressure within the vessel can be adjusted from about 15-in-Hg vacuum to atmospheric. Under vacuum, steam does not condense on the bark but acts totally as a stripping medium; at atmospheric pressure, some of the steam condenses. This allows control of product moisture.

The deodorized bark is fed to a grinder for size reduction (smaller than 100 mesh), and bagged as a powdered extender. Alternatively, some of the deodorized bark can be size-separated to cork and bast-fiber fractions, by a complex series of mechanical screenings.

Wax/Solvent Recovery—The miscella from the Rotocel, containing both wax and solvent, is meanwhile pumped to a two-stage evaporator.

The first stage evaporates the miscella to about 10% wax. Solvent released at this point is used to maintain heat in the Rotocel, or is sent onwards for solvent recovery. The second-stage evaporation takes place under vacuum, and concentrates the wax up to about 92%.

The remaining solvent is removed from the wax by vacuum steam-stripping. The wax is solidified on a continuous-belt flaker, and packed.

Solvent vapors collected from the evaporators, steam stripper and other points in the plant are routed along with steam through a spray-type liquid/vapor contactor for heat recovery. Both solvent and water condense in an overhead exchanger, and pass down through the contactor to a decanter. Solvent overflows out for recycle; water is drained, steam-stripped to remove traces of solvent, and discharged.

Air carried through the system by the vapors is treated in a mineral-oil absorber for solvent recovery, and released to atmosphere.

Utilities consumption is estimated for the Bohemia plant as: electricity, 750 hp; cooling water, 650 gpm; steam, 11,500 lb/h at 225 psi; fuel, 4,000 lb/h wood waste (wet basis). #

Meet the Authors

Robert D. Good is a consultant for extraction, food and biochemical technology for Dravo Corp.'s Blaw-Knox Chemical Plants Div. (One Oliver Plaza, Pittsburgh, PA 15222). A 27-yr veteran with the firm, he holds a B.S. in chemical engineering from Carnegie-Mellon University.

Frank S. Trocino has been technical director for Bohemia Inc. (P.O. Box 1819, Eugene, OR 97401) since joining the firm six years ago—specializing in development of useful products from wood-processing wastes. He holds a B.S. in chemical engineering from the University of Iowa.

Pulp-bleaching process cuts costs, time, effluent

Using the current displacement-bleaching principle, this compact system features only two bleaching towers, rather than several. It also offers utilities-consumption savings over conventional bleaching techniques.

Philip M. Kohn, Assistant Editor

☐ The first commercial displacement-pulp-bleaching installation in North America is now online at the Temple-Eastex, Inc., pulp-and-paper mill in Evadale, Tex., and a second facility has started up at the Plymouth, N.C., mill of Weyerhaeuser, Inc. Designed to bleach hardwood or softwood furnish, the process was developed by the Kamyr organization and supplied here on a turnkey basis by U.S.-based Kamyr, Inc.* The units' design capacity is 500 a.d.† tons/d of 85+ G.E.-brightness** pulp, with a total bleaching time of about 115 min.

The principle of displacement bleaching is that pulp can be bleached with reduced retention time by passing high-strength chemicals through a bed of the pulp, rather than by simply mixing the chemicals with the pulp and then letting diffusion rates determine the bleaching time, as is the case in conventional bleaching systems. And, if fresh chemicals can replace the preceding ones without mixing, washing the pulp between bleaching stages becomes unnecessary.

To accomplish its goal, Kamyr employs several diffusers in a single tower to achieve desired bleaching sequences. Doing this also eliminates transfer points, as well as exposure of the pulp to the atmosphere during the bleaching process.

Advantages of the displacement process over conventional techniques are: (1) lower capital cost, since only two towers—rather than several—are required, and interstage washing and transferring are not necessary; (2) space requirements are diminished—only 6,000 ft^2 are needed for the complete bleach plant; (3) effluent volume and water usage are cut by about 80%; and (4) the total cost of steam and power is almost halved.

SIMPLE FLOWSHEET—The system is straightforward, comprising a conventional hot-chlorination stage with a drum-washer, followed by an *E-D-E-D-W* (caustic soda *E*xtraction - chlorine *D*ioxide bleach - caustic soda *E*xtraction - chlorine *D*ioxide bleach - hot water *W*ash) displacement tower.

In the process, (see flowsheet), brown stock from the screening operation passes over the decker and is pumped from the decker chest. Chlorine is added to the stock and mixed in via two static mixers followed by a final mixer. In the latter, retention time is 2½ min. The feed then enters the chlorination tower, which has a retention time of 25 min at design tonnage. Of fiberglass construction, the tower is 12 ft in dia. × 65 ft, 8 in. high.

In the chlorination tower, the chlorine reacts with the lignin embedded in the fiber walls of the pulp. Complete removal of this lignin is necessary for the production of high-brightness pulp. Reactions involve substitution of chlorine for hydrogen in the 5 and 6 positions of lignin's benzene ring, partial demethylation, and at least a partial depolymerization of the aryl alkyl ether bonds.

The chlorinated pulp then goes to a chlorination washer having a tile vat and a stainless steel drum. Caustic soda for the first extraction stage in the displacement tower is added at a repulper following the washer; the washed pulp advances first to a steam mixer and then into the displacement tower via a high-density pump.

In the extraction stages, the sodium hydroxide removes further quantities of the partially depolymerized chlorinated lignin as sodium salt. The chlorine-dioxide stages further demethylate the lignin and break down its aromatic ring.

The pulp proceeds upward through the tower, progressing through the various stages, until it emerges from the tower and is pumped to bleached-pulp storage. Retention time in the displacement tower is about 90 min. The tower has an inside diameter of 14 ft, 8 in., and an overall height of 78 ft.

AUXILIARY EQUIPMENT—There are only three stock pumps associated with the bleach plant: (1) a low-density pump that transfers pulp from the brown-stock decker chest to the chlorination tower; (2) a high-density unit that transports chlorinated pulp from the steam mixer to the displacement tower; and (3) a second high-density pump that carries

* Kamyr, Inc., Ridge Center, Glens Falls, N.Y. 12801.

† Pulp is marketed in "air-dry tons," defined as containing a theoretical 10% moisture. Thus, an a.d. ton contains 1,800 lb of actual moisture-free pulp.

** G.E. brightness is defined as the percent reflectance of light of 457-mμ (nm) wavelength, based on a scale on which MgO gives a reflectance of 100%.

Originally published February 28, 1977

Displacement-bleaching plant consumes these resources — Table I

Basis: 414 a.d. tons/d of bleached pulp production

Item	Consumption
Connected horsepower	2,050
Consumed horsepower	1,665
Horsepower/a.d. ton	3.7
Steam, lb/a.d. ton	853
Water, gal/a.d. ton	4,669

bleached pulp from the displacement tower to the bleached-pulp storage area.

Four 8-ft-dia. filtrate tanks are associated with the displacement tower, and are designed for a retention time of 3 min at the nominal production rate. The tanks are of a "stacked" design, with one set for the extraction stages and one set for the dioxide stages.

The design and the operating firms feel that computer control and automatic testing and sampling of the bleaching process are very desirable. In line with this, chlorine addition is controlled by an automatic system, with a sensor located between the second static mixer and the final mixer. The bleaching system is likewise computerized, and a continuous sampling system monitors the pH and conductivity of all filtrates from the displacement tower.

CORROSIVE SERVICE—To provide corrosion-resistance, the tower itself is lined with acid brick, while diffusers and internals in the upper section of the tower are 100% titanium. Distribution nozzles and the lower section of the tower's central shaft are made of Hastelloy C. There are about 75,000 lb of titanium and 31,000 lb of Hastelloy C in the structure.

The four filtrate tanks are made of fiberglass, and all of the filtrate pumps are of titanium. Displacement-tower filtrate piping is Kynar-lined carbon steel.

ASSESSMENT—The performance of the displacement bleach process is encouraging. Indications are that the cost of steam and power would be cut in half compared with a conventional bleaching plant. And the displacement plant yields an effluent volume (see Table II) and a water usage about five times smaller than that of typical processes. The firm notes, however, that preliminary data suggest that other effluent values—e.g., pH, BOD, COD, etc.—are in the same ballpark for both conventional and displacement plants.

Effluent characteristics of bleaching plant — Table II

Basis: 414 a.d. tons/d of bleached pulp production

Item	Value
Volume, gal/a.d. ton	4,883
Temperature, °F	180
pH	4.5
TOC, lb/a.d. ton	21.4
COD, lb/a.d. ton	179.0
BOD_5, lb/a.d. ton	17.2

In the operation of the facility, the following points can be made based on experience gained so far:

- Uniformity of stock entering the displacement tower is important; stock flow and pulp consistency should be constant.
- Although flowrate control of ±10 gal/min was targeted, fully-acceptable operation has been obtained with flowrate accuracies of ±50 gal/min.
- The requirement for skilled maintenance labor is somewhat higher than that for conventional bleaching plants.

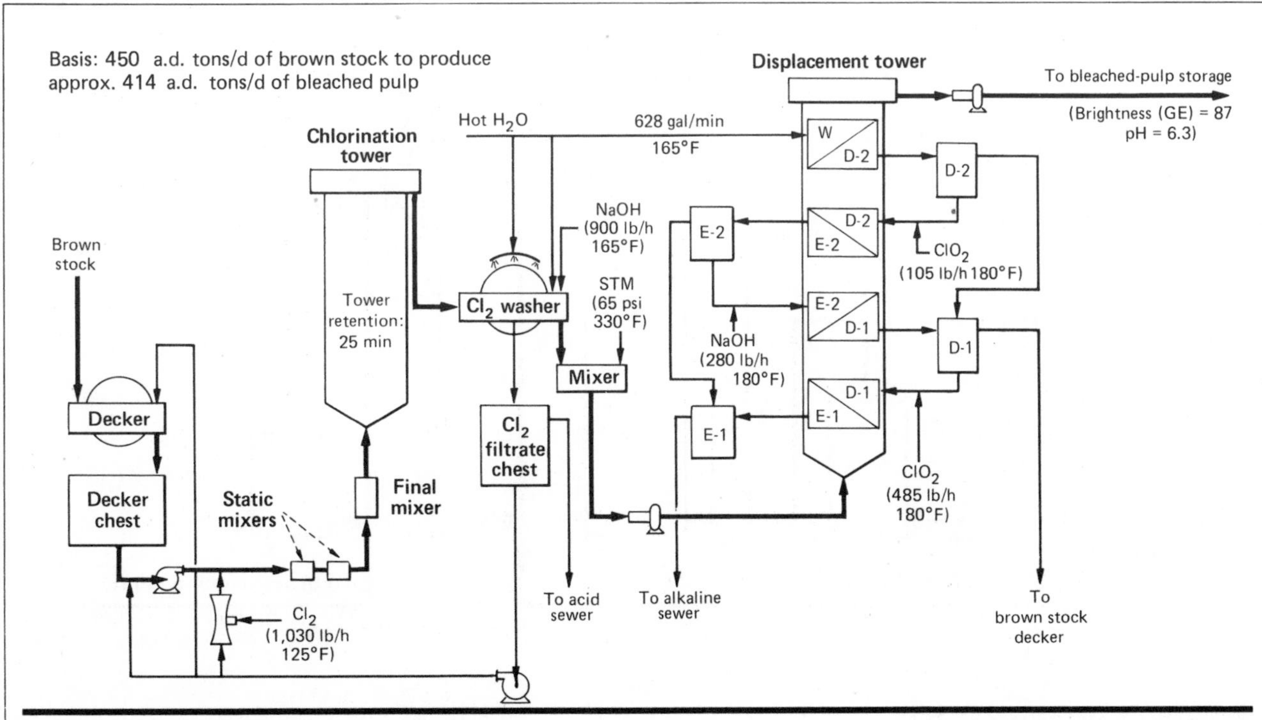

Kamyr's system includes four bleaching stages in one tower

Pulping/Bleaching Scheme Boosts Yields, Cuts Waste

An intermediate treatment of pulp with oxygen and alkali, snuggled between conventional pulping and bleaching, handles part of the task of each. The result is better yields and less need for cleanup of aqueous effluents.

Sometime next year, an innovative pulping/bleaching process groomed by International Paper Co. (IP) will make its debut at the firm's Ticonderoga, N.Y., mill. The recently patented technique (U.S. 3,832,276) exploits oxygen and alkali to remove residual lignin from cellulose.

The Ticonderoga unit is only "a step in the development of a commercial system," but nevertheless full-commercial size, handling all of the kraft pulp made at the mill. Since the process combines functions of both pulping and bleaching, the unit will be installed between the existing equipment for these operations.

The process was originated at IP's Corporate Research and Development Center at Sterling Forest, N.Y., and demonstrated on a pilot-plant scale at one of the firm's Southern mills during the past two years.

How It Works—Feedstock for the new IP process is a low-consistency pulp, preferably 3-4 wt.% but possibly up to 10%. Alkali brings the dilute slurry up to about 12 pH. Sodium hydroxide is typical, but other alkalis such as ammonia or sodium carbonate are suitable.

A high-shear mixer evenly distributes oxygen into the stream, at a rate of about 0.2-0.8 wt.% for softwood and 0.2-0.4 wt.% for hardwood. Undissolved oxygen bubbles must be avoided, IP stresses, so a vent tank is also included to remove any bubbles exceeding 1/16-in. dia. or so. This minimizes agglomeration and channeling as the pulp passes through a following bleaching tower, and prevents the uneven bleaching this would cause.

Optionally, oxygen may be injected into the recycling alkali, rather than directly into the mixing chamber. Before entering the vent tank, the stream may also pass through a small and inexpensive high-pressure vessel, where it is momentarily subjected to pressures of 2-10 atm to help delignification.

Bleaching takes place at 90-110°C as the mix rises through a 40-300-ft-high tower—this height corresponding to a bottom pressure of about 17-135 psi. Pressure gradually declines as the mix nears the top (other pressure-reduction techniques may also

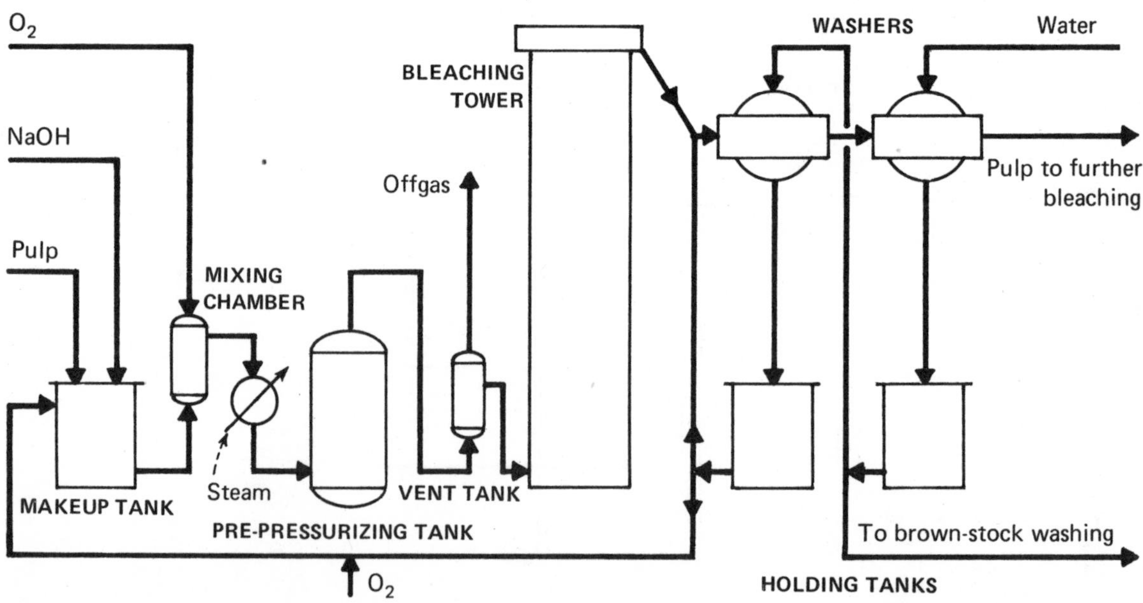

Originally published March 31, 1975

be chosen). Residence time in the tower generally falls between 5 min and 2 hr, depending on pressure and required degree of bleaching.

Effluent from the tower is washed first by recycling alkali, then by water. The washed pulp then moves to succeeding stages of bleaching such as chlorine dioxide treatment.

Many Pluses—Unlike other routes using oxygen for pulping and bleaching, the IP process needs neither high-consistency (20-30%) feed pulps nor extensive high-pressure operations to achieve delignification. Capital costs for large high-pressure systems and for dewatering are thus eliminated.

IP's approach also minimizes unwanted depolymerization of cellulose. Such chemicals as magnesium carbonate, otherwise needed to control degradation, are not employed, saving both purchase and waste-treatment costs. Somewhat gentler kraft-pulping conditions enhance this low-degradation benefit, as does bleaching at declining rather than constant pressure. Less consumption of bleaching agents offers raw material economies, just as does elimination of magnesium carbonate, and also lowers the amount of chlorides and color bodies carried to waste-treatment with the bleaching effluent. #

Salt-recovery process allows reuse of pulp-bleaching effluent

A new technique for recovering salt from pulpmill white liquor makes recycling of bleaching effluent feasible, and thus avoids the usual effluent treatment and discharge.

Conrad F. Cornell, *Erco Envirotech Ltd.*

☐ The last step toward achieving the closed-cycle bleached kraft pulpmill has now been taken with the development of Erco Envirotech Ltd.'s new SRP (salt recovery process) system.

The concept of the closed-cycle mill calls for the elimination of contaminated aqueous discharges through internal recycling. Contaminant buildups that would result from recycling, however, must be avoided. The SRP system meets this challenge by removing sodium chloride contaminant introduced into the pulpmill circuit through reuse of bleaching effluent. At the same time it makes the purified salt available for use in regeneration of bleaching chemicals.

A 250,000-ton/yr mill now under construction for Great Lakes Paper Co. at Thunder Bay, Ontario, will feature the first commercial-size SRP system. Mill startup is scheduled for mid-1976. By teaming up the SRP system with other, standard mill practices (see box below), the new facility will be the world's first closed-cycle bleached kraft mill.

The concepts for the closed-cycle mill and for the SRP system were originated by W. H. Rapson and D. W. Reeve in their research work at the University of Toronto. Erco Envirotech continued the development work at its Salt Lake City facility on a pilot-plant evaporator-crystallizer employing commercial white liquor supplied by several kraft pulpmills.

Continuous runs on the pilot plant from November 1972 to April 1975 confirmed the early research work, and also provided design data for heat-transfer coefficients, scaling potential, crystal sizes and liquid-solids separation rates. Additional studies determined that type 304 stainless steel, Monel 400 and Inconel 600

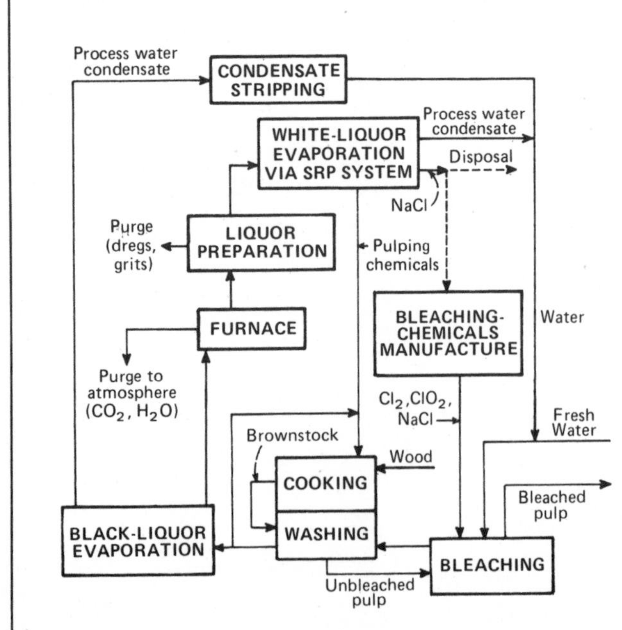

The closed-cycle pulpmill

A closed-cycle bleached kraft pulpmill using the SRP system in conjunction with standard industry steps to minimize aqueous discharges is shown in the diagram. Complete countercurrent washing in the bleach plant reduces water usage from approximately 25,000 gal/ton of pulp to only 4,000 gal/ton. Effluent from the leadoff chlorination stage of the bleaching sequence (which features a 70% substitution of chlorine dioxide for chlorine, to cut the salt load in half as well as to enhance pulp quality) dilutes the feed of high-consistency unbleached pulp. The remaining bleachery effluent—waters carrying salt generated during bleaching—serve for washing brown stock from the cooking or digestion step, or for other washing tasks in the unbleached mill.

Heavy metal ions and silica contaminants are removed in the dregs and grits solid discharges during recausticizing. Contaminated condensates are steam-stripped to remove volatile materials (which are burned), and the condensates are then recycled for reuse in the pulpmill.

The SRP system for white-liquor treatment completes the setup. It separates salt for reuse in generating bleaching chemicals, as it concentrates the alkali pulping chemicals.

Originally published November 10, 1975

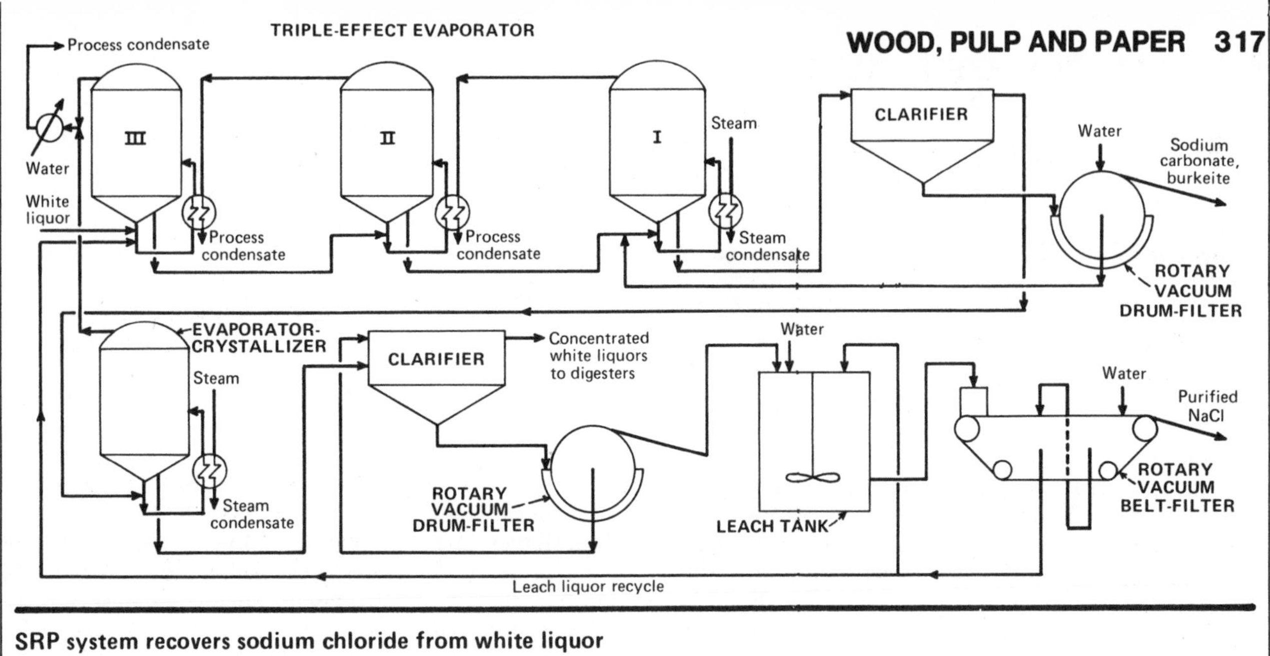

SRP system recovers sodium chloride from white liquor

could be used as materials of construction depending on severity of conditions in various system sections.

FIRST STAGE—The SRP system feeds on white liquor taken from the liquor preparation (recausticizing) section of the pulpmill. The white liquor initially contains approximately 2–3 wt.% of sodium chloride, and 10–11 wt.% of sodium hydroxide (NaOH) plus sodium sulfide (Na_2S).

A triple-effect set of evaporators employing backward feed concentrates the liquor to about 26 wt.% NaOH + Na_2S. All three evaporators are forced-circulation units with external heat exchangers. Final liquor temperature runs 250–260°F.

At 26 wt.% NaOH + Na_2S, salt concentration will be near saturation. Sodium carbonate (Na_2CO_3) and sodium sulfate (Na_2SO_4) present in the white liquor will crystallize as anhydrous Na_2CO_3 and the double salt burkeite ($Na_2CO_3 \cdot 2Na_2SO_4$). A gravity sedimentation unit (clarifier) removes the crystals, leaving a clear liquor that overflows to the second stage of the system.

The underflow from the clarifier passes into a rotary vacuum drum-filter for crystal dewatering. Most of the crystals recycle to the recausticizing section, though a small amount is sent to the recovery furnace to reduce sulfates to Na_2S. The filtrate is recycled to the first evaporator.

SECOND STAGE—An evaporator-crystallizer further concentrates the clarifier overflow from stage one to about 36 wt.% NaOH + Na_2S. Most of the sodium chloride crystallizes out of solution.

The crystal slurry enters a gravity sedimentation unit that clarifies the stream to yield a concentrated white liquor. It contains relatively small amounts of NaCl, $NaCO_3$ and Na_2SO_4, but all of the active alkali for pulping. After dilution with recycled process streams, such as weak black liquor or condensate, the white liquor flows to the digester for use in pulping.

Underflow from this clarifier is pumped to a rotary vacuum drum-filter for dewatering of the salt crystals. Filtrate recycles to the clarifier. The dewatered crystals pass to an agitated leach tank to dissolve the residual amount of Na_2CO_3 and burkeite crystals present with a small amount of water.

Slurry from the leach tank is fed to a horizontal vacuum belt-filter, where a countercurrent washing procedure turns out a 99+%-pure NaCl cake. The material can serve for manufacture of chlorine dioxide for pulp bleaching. Part of the filtrate from the belt filter returns to the leach tank, while the remainder passes to the third effect of the stage-one evaporation system.

SRP ADVANTAGES—Benefits of the system start with a considerable reduction in raw-water pretreatment costs, due to the reduction in raw-water volume achieved by recycle. The expense normally incurred to provide primary and secondary treatment of effluents is also eliminated, since only uncontaminated cooling waters and cleaned condensate waters will be discharged to receiving waters.

Fibers, organic solids and chemicals previously lost in the effluent are recovered through the recycling technique. Countercurrent washing in the bleachery, and reuse of bleachery effluent similarly retain heat values normally lost. In fact, the heat savings with the closed-cycle system are more than the SRP system steam requirements.

The salt recovery process has the added advantage that it can be shut down for several days if necessary without disturbing mill operation. The pulpmill can continue to run at full production and to recycle effluent internally. The salt contamination level builds up only slowly in the mill circuit, and quickly drops to a minimum as soon as the SRP system comes back online.

The system can be applied to any existing or new alkali pulpmill, and, furthermore, to any bleaching sequence now in commercial operation. It is flexible enough to meet the demands of a wide range of pulping and bleaching conditions without the need to change existing plant facilities.

The author

Conrad F. Cornell is general manager for Erco Envirotech Ltd. (P.O. Box 300, Salt Lake City, UT 84110). He holds a B.S.Ch.E. degree from Northwestern University, and is a member of both the Technical Assn. of the Pulp and Paper Industry and the Canadian Pulp and Paper Assn. He has been associated with liquid-solids separation and the pulp and paper industry for 22 years.

How to get water out of lignite, wood and peat

A high-temperature, high-pressure beneficiation process removes moisture from these materials to yield a high-Btu product at relatively low cost.

☐ Within the next few months, SRI International (Menlo Park, Calif.) expects to start offering licenses for a process (Chem. Eng., Feb. 27, p. 62) that upgrades moisture-laden fuels into high-Btu products. Devised by Edward Koppelman, an inventor from Encino, Calif., the method takes lignite, peat or wood, and reduces moisture content while raising the heating value from about 7,000 Btu/lb to around 12,000 Btu/lb.

SRI has been developing the process with funds supplied by Koppelman, who says he holds patents on about 20 inventions. Based on pilot-plant runs, SRI estimates a commercial plant fed 1 million tons/yr of lignite (costing \$6/ton), would produce a 12,000-Btu/lb fuel at a cost of 60.2¢/million Btu. Koppelman says this figure increases to 66.5¢, after including amortization of the plant, which is estimated to cost \$16.3 million. He adds that this figure does not include finance charges, and further notes that the 66.5¢ compares with about \$1.10 to \$1.25/million Btu for coal of the same heating value (bought for about \$30/ton f.o.b. mine).

PROCESS STEPS—In Koppelman's scheme, lignite is crushed, slurried and fed into a reactor under a pressure of 1,500 lb/in.^{2} (see figure). Temperature is kept at 1,000°F and residence time is 10 min. The process is said to be economical because moisture driven off from the coal is not converted into steam, and because the internal structure of the lignite is sealed to prevent reabsorption of moisture.

"We don't make steam," says Koppelman. "We couldn't afford it—it would use too much energy."

Feed entering the reactor is 50–60% solids by weight. Slurry water comes from the product-dewatering process. In the reactor, water—including moisture from the lignite—is filtered out by a proprietary method. Pilot-plant tests have reduced moisture content of lignite from around 30% to under 5%.

Gases driven off in the reactor include methane, carbon dioxide, hydrogen and nitrogen. The mix varies, depending upon feed makeup. Tests with lignite have produced gases that average 400 to 500 Btu/std ft^3.

The gas goes through two recovery turbines for generation of electricity. Its pressure drops from 1,500 lb/in^2 to about 300 lb/in^2. Then, it is fired in a third turbine to produce more electricity. Exhaust gases from the third turbine, plus incoming air, are mixed with fresh gas from the process to fire the reactor furnace.

The process yields, in addition to beneficiated lignite, small amounts of condensible organics (propane and heavier). SRI's pilot runs show that for the 1-million-ton/yr plant, about 16,000 fuel-oil-equivalent bbl/yr of condensible hydrocarbons, and 3,100 tons/yr of organic liquids, can be recovered.

"The process is 90 to 92% efficient," says Koppelman, "and it provides all its own energy once it gets started."

After processing, the lignite is discharged at 600°F through lock-hoppers to a cooling system. In the pilot plant, a continuous belt is used. This is cooled indirectly with water at 150 to 160°F.

SEALED PRODUCT–The lignite coming from the process is claimed to be sealed against moisture reabsorption. Water, says Koppelman, is unable to penetrate into the product's inner structure.

Tests on lignite are being done in a pilot plant with a continuous input of 180 to 400 lb/h. The lignite is sized at 1 in. or less, but material up to 1½ in. has been processed as well. The

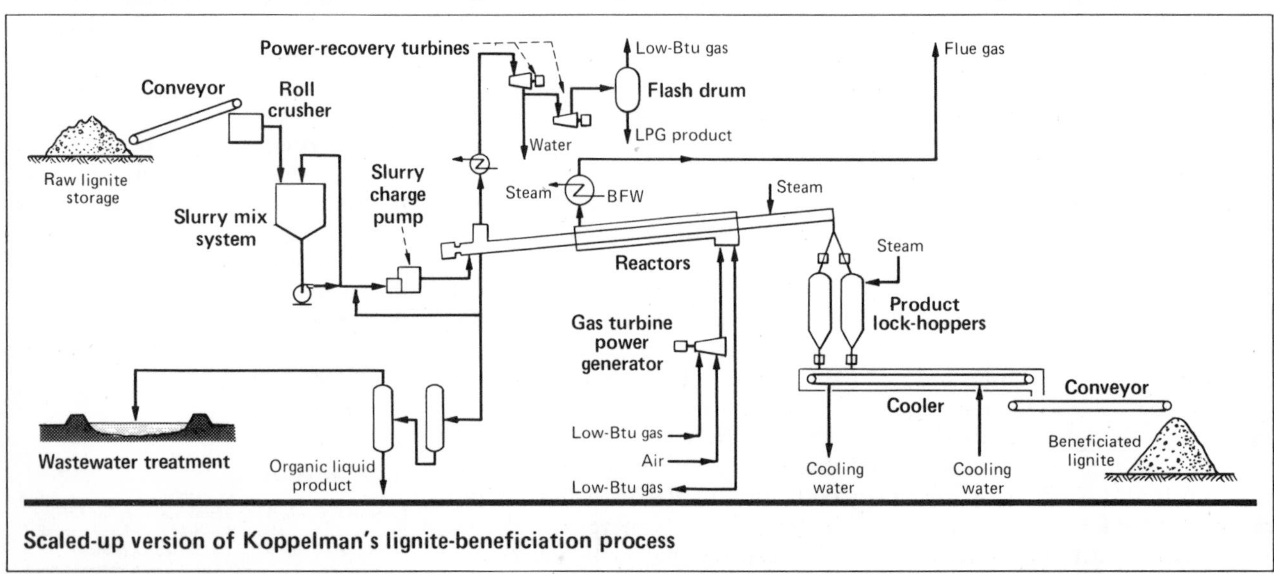

Scaled-up version of Koppelman's lignite-beneficiation process

Originally published March 27, 1978

reactor measures 6 in. dia. by about 12 ft long and is inclined slightly from the horizontal. Koppelman declines to say how material moves through the reactor, but states the reactor is stationary and has no belt conveyor.

A 50,000-ton/yr (input) commercial reactor would measure 15 in. dia. and 40 ft long. Koppelman envisions a commercial plant consisting of a number of modules. The plant "would have four or five reactors in parallel, heated by one furnace and working from one feed system that would divide into four of five trains," he says. Each reactor would be rated at an output of fuel equal to 30,000 tons/yr.

PEAT TESTING—SRI started up another pilot plant earlier this month to test peat and wood. Licensing of the process is being delayed until this plant's results come in, because Koppelman wants to offer licensees a choice of using lignite, peat or wood. Economic data from the plant are expected in about two months.

The facility produces about 75 lb/h of peat or 125 to 150 lb/h of wood, both with a heating value of 13,000 Btu/lb. Wood and peat of this quality have already been produced in the laboratory. Koppelman feels the process would be ideal for preparing low-moisture fuels for a gasification plant.

"The most economical way to get fuel into high-pressure gasifiers is to slurry it," he says, "but this gives excess water that results in inefficiency. This is because a lot more oxygen is needed for gasification and more heat is needed to vaporize the water. With our process, we could put bone-dry feed into the gasifier, so that the proper amounts of steam and oxygen could be added for optimum efficiency."

Interest in the process has been expressed by Land O' Lakes, Inc. (Minneapolis, Minn.). The firm is planning to build a Wellman gasifier demonstration plant at Perham, Minn. Costs are being shared by the company and the U. S. Dept. of Energy, with construction scheduled to start this summer. A company spokesman said Land O' Lakes is "impressed" with Koppelman's process and hopes to get output from SRI's pilot plant to run through its own plant for a day, or possibly more. Using feed from Koppelman's process, says the company, would really cut costs: The firm has done an economic analysis and claims the figures look good, even if they should be off by as much as 30%. Further, the company is located close to peat deposits in North Dakota.

SULFUR ADVANTAGE—Koppelman says that peat generally contains only about 0.14% sulfur. He adds that lignite has around 0.54 to 1% S, and processed lignite, closer to 0.52%.

Not only is peat low in sulfur, it is plentiful in the U.S. Known deposits are 120 billion tons and contain about 1,440 quads of energy, according to testimony delivered to a House subcommittee hearing last Sept. 29 by Ezekail Clark, then assistant director for gasification with the former Energy Research and Development Administration's Office of Fossil Energy. Clark added that the major obstacle in using peat for energy has been its high moisture content of 90 to 95%. Koppelman says his process has given a product with a moisture content as low as 1½%—way below what Clark calls a "usable level of about 35%."

Gerald Parkinson

Section XV
MISCELLANEOUS

Non-Caustic Refining of Edible Oils and Fatty Acids

This steam-refining alternative increases yield and cuts pollution. It is now being applied to crude palm oil from Malaysia, which is gaining more favor because of shortages of other edible oils.

FRANK E. SULLIVAN
Frank E. Sullivan Co.

Palm oil has a natural versatility among edible oils because of its glyceride structure. It has good non-foaming properties and oxidative stability that makes it highly suitable for use in commercial deep frying. The plantations of Malaysia are now producing high-quality palm oil in sufficient quantities to permit its use in the production of frying fats, as well as margarine and shortenings.

Because of the demand for quality frying oils to supply the potato-processing industry in the northwestern U. S., a completely new 250-ton/d refinery to process Malaysian palm oil was started up late in 1973 by Palmco Inc., a joint venture of Mitsubishi International and Koppel (Long Beach, Calif.). The plant was engineered and designed by the Frank E. Sullivan Co.

The $3.5-million refinery, in Portland, Ore., is the first edible-oil refinery to eliminate the caustic refining step and to process crude palm oil by steam-refining/deodorization to produce a high-quality finished product. The technique cuts pollution and increases yield.

Better Yield—Comparing the new process to the conventional caustic-refining system for this type of oil, the yield-loss advantages are obvious. Caustic refining of crude palm oil is difficult, especially certain grades. Published refining losses usually range from a low of 1.7 times free-fatty-acid (FFA) to over 2.2 times FFA. On a crude oil of 4.0% FFA with an average refining factor of 2.0, the loss would be 8%, or a refined yield of 92 lb for every 100 lb of crude oil purchased. The refined oil would then have to be bleached and steam-deodorized.

The large amount of soapstock produced from this 8% refining loss becomes a difficult byproduct that must be processed further. It is usually acidulated with sulfuric acid to make a low-grade acid oil. The effluent from the acidulation process is very high in biochemical oxygen demand and must be neutralized and treated before discharge into the sewer system.

By comparison, when the free-fatty-acids are removed by steam refining, they are collected as an overhead fraction. The condensed distillate from palm oil usually analyzes to 90-94% as high-grade, light-color, distilled free-fatty-acid, and it is continuously collected and stored in fiber glass tanks prior to loading in tank cars or trucks. The small amount passing through the fatty-acid condenser is collected in the cooling-tower skim basin. The fatty-acid layer that forms is continuously skimmed, heated to separate entrained water, and pumped to a skim-basin oil storage tank, where it is collected along with any floor-drain or truck-wash oil; it is periodically shipped out as animal-feed additive, similar to low-grade acid oil.

Since there is no acidulation of soapstock to result in acid waste water, there are no problems caused by sewer disposal of effluents. In addition, 5% more oil is recovered.

Past Efforts—Deacidification of certain high-FFA crude oils using steam distillation as a primary refining step has been carried out in Europe for many years as a replacement for the basic caustic refining method. In practice, it had been found difficult to reduce the free-fatty-acid content of the oils below 0.1-0.2%. As such, the steam distillation was usually carried out to reduce the FFA to the 0.5-0.6% range, and then the refining was completed by the basic caustic refining process. In order to produce a finished edible oil, the caustic refining step still had to be followed by conventional bleaching and steam deodorization.

While much of the fundamental technical information on steam distillation has been available for some time, most of the common oils used over the years for salad oil, shortening or margarine have not been amenable to steam stripping for fatty-acid removal.

Until the 1940s, cottonseed was the principal source of crude oil for shortening or salad oil in the U.S. This dark crude oil could not be steam refined due to color problems.

ONE-MAN control is possible from this control panel.

Originally published April 15, 1974

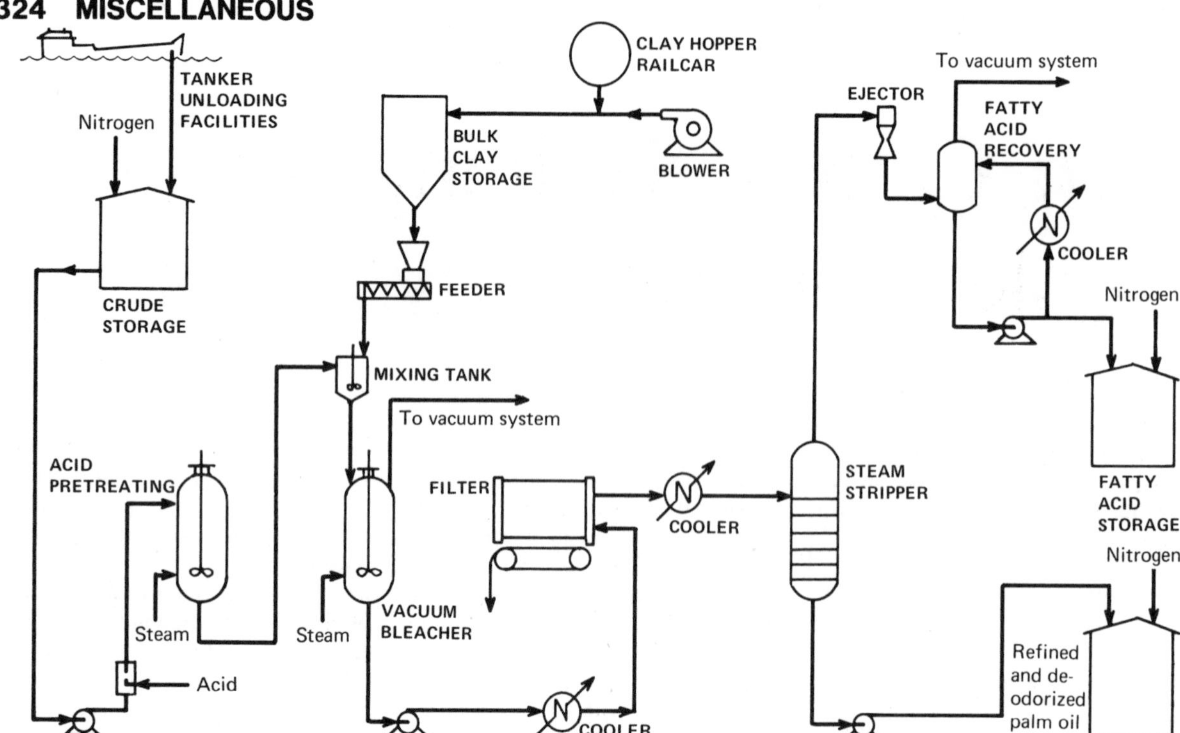

It was absolutely necessary to caustic-refine the oils for fatty-acid removal as well as removal of color bodies.

Cottonseed oil was followed by soybean oil, which is also difficult to steam refine. The principal non-oil constituents of this oil are phospholipids. These contaminants must be thoroughly removed from the oil prior to high-temperature steam deodorization. Phospholipids vary in their composition, primarily due to growing conditions of the basic soybean.

Caustic refining is the only way to remove phospholipids; nevertheless, certain types, for example nonhydratable phospholipids, require special acid pretreating before caustic refining, followed by a good bleaching with activated clay to produce an acceptable product for deodorization.

Recent shortages in other domestic edible oils have accelerated interest in steam distillation or steam-refining/deodorization to produce a finished palm oil. Certain other oils also lend themselves to steam refining, namely the lauric acid oils such as coconut or babassu, and a variety of animal fats.

Palmco Operation—The crude palm oil is received dockside in 15,000-metric-ton shipments arriving directly from Malaysia every two months. It is stored in agitated, insulated tanks under inert-gas blanketing to protect it from oxidation or degradation.

Processing of the crude oil to a finished product is a continuous operation that eliminates storage at any intermediate stage. Because there are no batch operations, there always seems to be little activity at the plant, even though it is turning out product at a rate of 22,000 lb/h.

As seen in the flow diagram, the oil is pretreated with an inorganic acid for reaction with the many trace metals commonly present. The pretreated oil then flows continuously to a vacuum-bleaching vessel, where it is contacted with acid-activated clay. During the clay contact stage, certain color bodies and trace metals precipitated in the previous step are adsorbed. The oil then flows continuously to filtration for separation of the clay, then to the steam-refining/deodorization stage, and finally into finished-oil storage.

The actual deacidification or fatty-acid removal is accomplished in equipment similar to that of conventional steam-deodorization, with certain modifications. The deaerated oil, after heating to 170°C by steam, is steam stripped at 270°C, using Dowtherm as the heating medium.

As the oil progressively passes from tray to tray under 4-6 mm Hg absolute pressure, the free-fatty-acids are removed and the oil is completely deodorized. The cooled, finished oil is pumped from the last tray to storage, and from there to rail-car or tank-truck loading.

Again, all finished-oil storage tanks are agitated, insulated and blanketed with inert gas. The entire processing from crude-oil storage to loading of finished oil is conducted either under an inert-gas atmosphere (nitrogen) or a high vacuum to protect the oil in all stages of processing. The plant is highly automated and easily operated by one person.

Another interesting feature of this plant is the automation of the bulk clay system. Clay from Mississippi arrives in sealed railroad hopper-cars in quantities of approximately 30 tons. The clay is fluidized by low-pressure air, 5-10 lb/ft, and transported in this condition to the bulk-clay storage silo. The silo holds 80 tons, so two cars can service the plant. This procedure eliminates the costly, and also dusty, procedure of unloading 50-lb bags of clay from box cars by hand. The clay is instead fed into the vacuum bleaching-vessel by a worm screw. Except for connecting the pipes to the hopper cars, the entire clay-handling system is controlled from the panel board. #

Meet the Author

Frank E. Sullivan is president of the company that bears his name, which specializes in fluid processing systems. He holds a B.Ch.E. from Pratt Institute and an M.Ch.E. from Stevens Institute. He has taught at several eastern U.S. universities.

Beneficiated-ilmenite process recycles HCl leach liquor

Hydrochloric acid (HCl) leaching technique beneficiates ilmenite ore to a feedstock suitable for the environmentally clean chloride route to titanium dioxide—via a closed-loop system that avoids creating pollution woes of its own.

Nicholas R. Iammartino, *Associate Editor*

☐ From a dozens-long list of methods proposed for upgrading ilmenite ore, Benilite Corp. of America's* "BCA Cyclic Process" has emerged as the definite frontrunner in worldwide acceptance.

Its maiden commercial plant started up just over a year ago; two units are coming online this year; and three will follow during 1978–79 (see Table I). License options have also been granted for three more plants, which would add 180,000 metric tons/yr of product capacity to the 355,000 metric tons/yr accounted for by the first six units.

The BCA Cyclic Process offers titanium dioxide (TiO_2) makers a beneficiated ilmenite ore, tradenamed Benilite, having a high 92–96% TiO_2 content and a low 1–2% iron impurity. Benilite converts to TiO_2 pigment by the environmentally clean chloride process, which otherwise calls for the increasingly scarce rutile ore.

But the firm shuns dubbing its product by the oft-heard "synthetic rutile." President S. T. Weng explains that the Benilite product has some significantly different properties than natural rutile—the high porosity and reactivity, for example, lend themselves to a micronized, buff-colored form termed Hitox, which can substitute for pigment where white color is not needed.

As further evidence of the product's flexibility, the firm recently announced Benilite-S, developed (though not yet commercialized) as a feedstock for TiO_2 manufacture by the sulfate process. The advantage over unprocessed ilmenite ore: Sulfuric acid consumption plunges from about 3½ tons per ton of pigment to only 0.7–1.0 tons, with a corresponding drop in waste-acid volume for disposal. Exploiting Benilite-S, Weng stresses, can substantially reduce the equipment size of New TiO_2 plants, bringing capital cost savings that could justify small local installations.

* Benilite Corp. of America, 233 Broadway, New York, NY 10007.

SIMPLE FLOWSHEET—The patented process (U.S. 3,825,419) starts off (see figure) by reducing some of the ferric iron impurity in the ilmenite ore into ferrous iron, which reacts more readily with HCl during subsequent leaching. There is no need to grind the ore, or to reduce its iron values all the way to metal.

Reduction takes place in a rotary kiln, where ilmenite and 3–6 wt.% reductant are heated, countercurrently, to a relatively low 1,600°F. The reductant can be solid or liquid, such as coal, coke, char or heavy fuel oil. Depending on ilmenite grade, the reduced ore will carry 80–95% of its total iron as ferrous iron.

Six commercial plants will make Benilite by 1979 — **Table I**

Plant owner	Location	Capacity (metric tons/yr)	Startup date
Taiwan Alkali Co.	Kaohsiung, Taiwan	30,000*	First phase—March 1975; Second phase—March 1976
Malaysian Titanium Corp.	Ipoh, Malaysia	65,000	Mid-1976
Kerr-McGee Chemical Corp.	Mobile, Ala.	100,000	Oct. 1976
Sakai Chemical Industry Co.	Sakai, Japan	30,000	1979
Kerala Minerals & Metals, Ltd.	Kerala, India	30,000	1978
Indian Rare Earths Ltd.	Orissa, India	100,000	1978

*In addition, a 5,000-metric-ton/yr Hitox plant is operating.

Originally published May 24, 1976

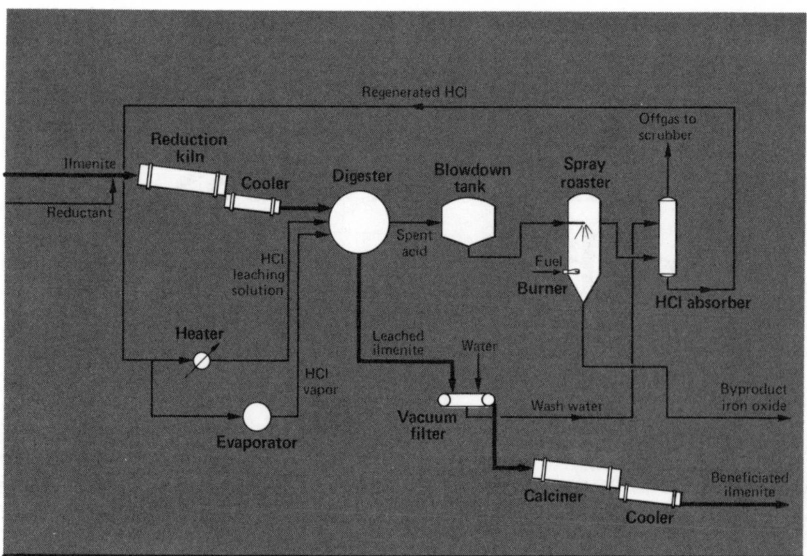

BCA Cyclic Process beneficiates ilmenite without generating troublesome waste streams

A brand-new twist in the reduction step is addition of about 2 wt.% sulfur to certain ilmenites to catalyze the reaction, and ultimately allow greater recovery of TiO_2 from ore.

Upon leaving the kiln, the reduced ilmenite is quickly cooled to 200°F to avoid reoxidation, and passed into a spherical rotary digester. Leaching with a regenerated 18–20% HCl solution takes place usually in two stages, each lasting about 4 h. The 290°F, 35-psi digester rotates continuously at 1 rpm.

Injecting live 18–20%-HCl vapor into the digester provides the necessary heat—a key innovation that sidesteps the leach-liquor dilution problem of earlier systems using steam for heating, and thus allows use of the weak HCl solution for the actual leaching.

After each stage of leaching, solids settle in the digester, as spent acid decants for transfer to regeneration.

The leached ore passes from the digester onto a continuous-belt vacuum filter for water washing and filtration. The solids then go to another rotary kiln for calcination at 1,600°F, to remove free and combined water, and emerge as the final beneficiated-ilmenite product.

The regeneration loop, meanwhile, employs a conventional spray-roasting technique. Iron chlorides, and most of the other metallic chlorides in the spent acid, thermally decompose into HCl, iron oxides, and other metallic oxides. Wash water from the belt filter absorbs the liberated HCl to form the 18–20%-strong leaching solution.

A 90%-pure iron oxide stream comes off as a valuable byproduct, typically sold in powder or pellet form to steel or cement mills.

Since all spent acid and wash waters recycle, there are no routine wastes (except for spillages, blowdowns, floor washings, etc., which are neutralized and discarded).

The firm hasn't yet revealed the process modifications needed to make Benilite-S, but does point out that a key aspect of the technology is avoiding changes in the crystal structure of the ilmenite that would make the ore insoluble in sulfuric acid. Patent-pending modifications in the reduction-leaching steps, also still under wraps, overcome leaching troubles experienced with fine particles in ilmenite, yielding a lighter-colored beneficiated product more suitable for making Hitox.

SIMPLE EQUIPMENT—Standard equipment fits readily into the BCA Cyclic Process, Weng says. The spherical digester must, however, be modified somewhat from versions used for pulping and other mineral processes.

Materials of construction are also not exotic. Though small quantities of special titanium alloys are used in the regeneration section, and graphite is used for the acid evaporator, carbon steel vessels elsewhere are simply lined with acidproof brick or rubber as needed.

Complementing flexibility in product type, equipment/operating parameters such as acid/ilmenite ratio and leaching time can be tailored to suit practically all ore grades, Weng says (though the economics may vary).

ECONOMICS—Battery-limits capital cost for a plant turning out 100,000 metric tons/yr of Benilite, based on 1975 costs and a U.S. Gulf Coast location, totals approximately $32 million. This includes engineering, equipment, construction, waste-disposal system and control house, but excludes cost of land. The design ilmenite feed is 54% TiO_2, and Benilite product is 94% TiO_2. Operating data for this typical plant are shown in Table II.

Typical Benilite plant has these operating needs — **Table II**

Basis: 100,000 metric tons/yr Benilite (94% TiO_2) production, along with 60,000-70,000-metric-tons/yr byproduct iron oxide, using 54%-TiO_2 ilmenite feedstock

Item	Consumption
Raw materials and utilities, per metric ton of product	
Ilmenite, metric tons	1.83
Makeup HCl (31.5%), metric tons	0.15
Heavy No. 6 fuel oil, metric tons*	0.54
Steam (100 psi, sat.), metric tons†	1.25
Electricity, kWh	300
Cooling water, thousand gal.	3.0
Direct labor, men/shift	12-15
Maintenance labor, total men	12
Maintenance materials, % of capital investment/yr	3

*Including use both as reductant and fuel.
†Fuel oil for steam generation included in fuel-oil figure.

Tobacco supplement seeks to catch fire in Britain

A relatively straightforward process yields a material that may win a prime role in milder—low tar, low nicotine—cigarettes. Impending fullscale marketing in the U.K. may show how smokers accept supplement-containing cigarettes.

Mark D. Rosenzweig, Managing Editor—News

☐ A cellulose-based tobacco supplement that's been 20 years in development will soon get its big chance to prove itself commercially. In July, smokers in the U.K. should start seeing three cigarette brands featuring the material Cytrel, from Celanese Corp. (New York City).

The debut will come only about three months after Britain's Independent Scientific Committee on Smoking and Health—chaired by R.B. Hunter of the University of Birmingham—gave its okay for controlled, fullscale marketing of cigarettes containing the supplement. But two major U.K. cigarette makers—Gallaher, Ltd. (London) and Carreras Rothmans Ltd. (Basildon, Essex)—have worked closely over the last few years with Celanese to ready the cigarettes.

Gallaher will introduce Silk Cut King Size with 25% Cytrel and Silk Cut Extra Mild, containing 40% supplement, while Carreras Rothmans will bring out Peer Special probably featuring about 25% Cytrel.

ADVANTAGES—Cytrel consists of about 20% modified cellulose, 40% inorganic fillers, and 30% combustion modifiers. Humectants (moisture retainers) and minor modifiers, e.g., colorants, make up the remainder.

The product boasts a number of performance advantages. Prime among these, the supplement's total lack of nicotine, and low tar-delivery (only one third to one seventh that of all-tobacco cigarettes), result in a proportionate reduction of these materials in the smoke of Cytrel/tobacco blends. For instance, Gallaher says that it achieves a more-than-20% reduction in tar-delivery in its 25%-Cytrel Silk Cut, compared to the all-tobacco version.

Also, the smoke from Cytrel contains fewer components than tobacco smoke, and no new ones. And certain components appear at greatly reduced levels, says Celanese.

Moreover, the supplement provides similar performance to tobacco for key cigarette parameters such as smoldering rate (the number of puffs in a cigarette) and ash formation. The material tastes bland on its own, so it barely affects cigarette taste. Cytrel resembles tobacco in density, and even in color. It currently comes in two shades—similar to the hues of burley and flue-cured tobaccos. And, once in cigarettes, says Celanese, Cytrel is indistinguishable from tobacco. In addition, the supplement can be cut and handled on conventional tobacco-processing and cigarette-making machinery at normal efficiencies.

Cytrel's allure goes beyond performance, adds Celanese. Output can take place continuously, yielding a uniform, reproducible product. In contrast, a seasonal crop like tobacco mandates the keeping of large inventories, and may vary in quantity and quality from harvest to harvest. And while natural leaf requires aging, Cytrel does not.

Price also favors the supplement. At the moment, Cytrel's tab runs somewhat less than that for most tobaccos, says Celanese, and the firm feels that in the long run it is in a better position to control costs than tobacco growers. For one thing, hoped-for boosts in Cytrel output promise economies of scale. And the product depends on relatively abundant raw materials, widely used elsewhere.

THE PRODUCT—Cytrel's modified cellulose component, essentially a food-grade carboxymethylcellulose, serves as the primary combustible material, and also as a key film-former during production. It plays a major role in controlling tar-delivery and smoke "taste."

The fillers, such as calcium carbonate, and combustion modifiers, e.g., perlite, markedly influence film density during manufacture, and smoldering rate and ash formation during smoking. Glycerol and other humectants plasticize the supplement during processing, and prevent the product from becoming brittle and friable. Colorants, such as caramel, provide the desired tobacco-like shade for Cytrel.

Supplement output relies on merchant raw materials. However, Celanese notes that feeds must meet its specifications, including strict limits on such contaminants as trace metals. And the company stresses that it rigorously checks the material during processing, as well as the end product. Testing extends to actual cigarettes manufactured, so as to check smoke chemistry and total performance.

PRODUCTION—Celanese started thinking about supplements in 1957, and set up a pilot plant at Charlotte, N.C., in 1968. The company brought onstream a full-sized, 9-million-lb/yr unit at Cumberland, Md., during the

Originally published June 20, 1977

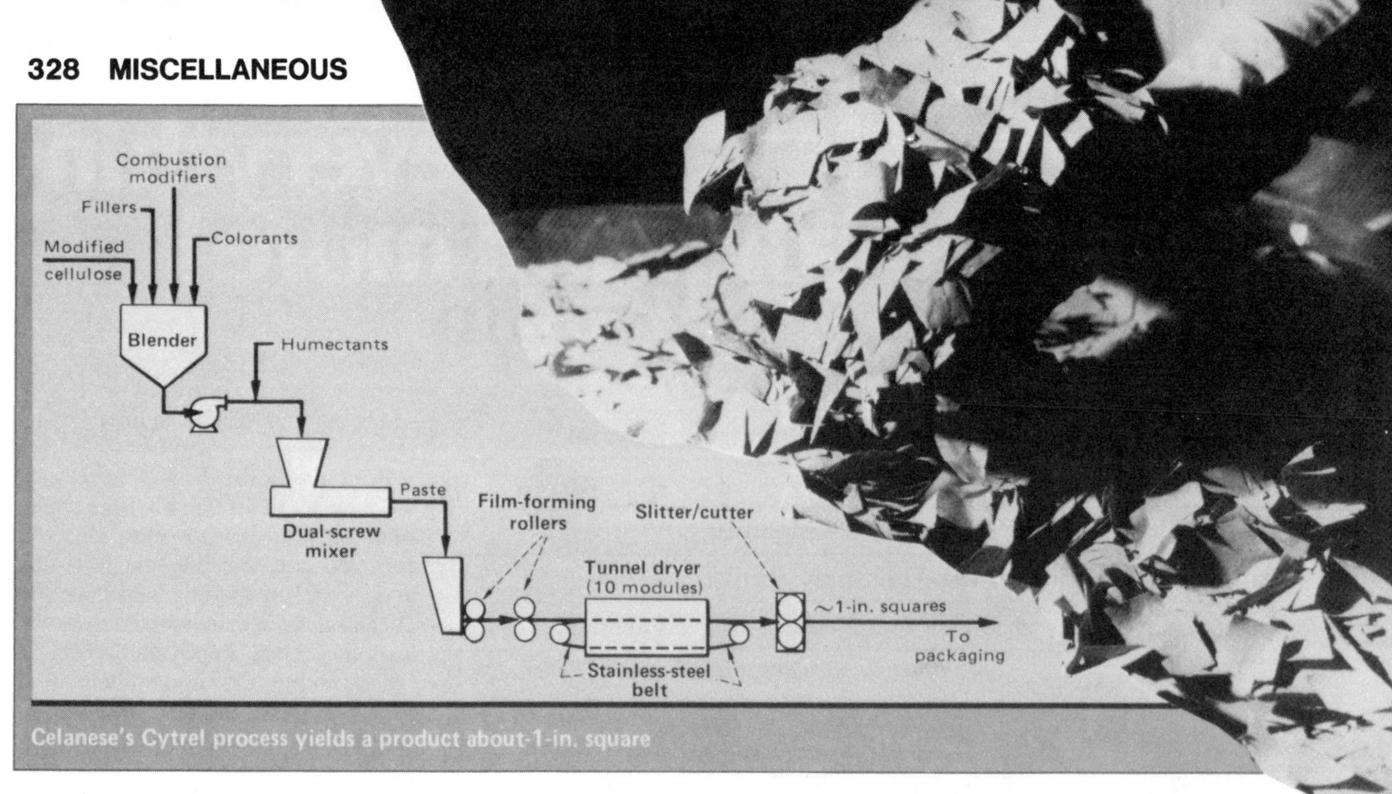

Celanese's Cytrel process yields a product about-1-in. square

summer of 1974, and more than doubled capacity there to 20 million lb/yr in late 1975. Operation of one of two process lines at the plant takes care of current demand, but Celanese says that it can readily and quickly expand the plant's rating.

At Cumberland, dry ingredients—the modified cellulose, fillers, combustion modifiers, and colorants—are blended (see flowsheet) in one building and then piped to the main Cytrel structure. There, they join a liquid stream containing humectants, and move into a dual-screw continuous mixer (with stainless-steel internals), which yields a paste.

The paste gets spread onto a stainless-steel conveyor belt about 5 ft wide. The belt extends for 240 ft, and can travel at up to 240 ft/min.

For its last 200 ft, the belt passes through a tunnel dryer. The dryer consists of ten modules, each with its own steam-heated coils, circulating fan, and temperature control. All but the last module are held at the same temperature, while the temperature in the final drying zone varies so as to control product moisture.

A 7-mil-thick sheet of Cytrel exits the dryer, and moves onto a slitter/cutter, specially designed by Celanese. Featuring self-sharpening blades and the flexibility to produce different product shapes and sizes, the unit can handle sheet material that is difficult to cut by conventional means.

The slitter/cutter conventionally turns out about-1-in squares of Cytrel, which then are baled in 400-lb cases.

All process steps take place at atmospheric pressure and all, save drying, at ambient temperature. And Celanese notes that the overall process requires relatively modest investment, and only a few operators per line per shift.

BRITAIN: THE KEY—The Hunter Committee gave a go-ahead to tobacco supplements after four years' study, but only to cigarettes with agreed-upon levels of specific materials.

The action of the Hunter Committee, which is widely regarded as the world's most-authoritative body studying tobacco supplements, may spur the prospect of marketing approvals in other countries. However, cigarette makers undoubtedly are waiting to gage the reaction of U.K. smokers to the supplement-containing brands—particularly since companies in West Germany and Switzerland have offered cigarettes with Cytrel for some time, but with inconclusive consumer acceptance.

In December 1974, following the granting of an interim, 2-yr approval by the German Ministry of Health, Martin Brinkmann AG (Hamburg)—a corporate relative of Britain's Carreras Rothmans Ltd.—launched Peer Leicht with 20% Cytrel, and B.A.T. Cigaretten-Fabriken GmbH (Hamburg) came out with Leichte Klasse containing 25% supplement. In late 1976, the Ministry extended its approval for another two years. Both brands remain on the market, but neither rates as a particularly strong seller.

Industry sources contend, however, that German regulations work against the supplement. German advertising strictures prohibit claims about the benefits of using a low-tar supplement, while the nation's food law requires the listing of any nontobacco ingredient. Seeing "something chemical" on the label probably puts some smokers off, notes B.A.T. Celanese adds that sales of the brands reflect the particular situation in Germany, rather than cigarette performance.

To emphasize this, Celanese cites more-encouraging results in Switzerland, where advertising is less restricted. Following clearance by the National Health Dept., Laurens Rothmans AG (Geneva) introduced 20%-Cytrel Peer Special in early 1975. The brand now takes more than 2% of Swiss cigarette sales—a respectable, if not spectacular, showing.

U.K. results should be more telling. For instance, Gallaher is already running advertisements stressing its adoption of Cytrel. And Silk Cut packs will clearly proclaim their supplement content. Carreras Rothmans will also promote Peer Special's use of Cytrel, as well as the acceptance, limited as it is so far, of brands marketed by its related firms.

Hydroprocessing for white oils

This catalytic hydrogenation process can be tailored to fit almost any product slate of medicinal- and technical-grade white oils. It offers a number of significant advantages compared with the conventional acid-treating procedure.

John B. Gilbert *and* ***Christopher Olavesen,*** *Imperial Oil Enterprises Ltd.*
Clinton H. Holder, *Exxon Research and Engineering Co.*
Jacques Lecomte, *Esso SAF*

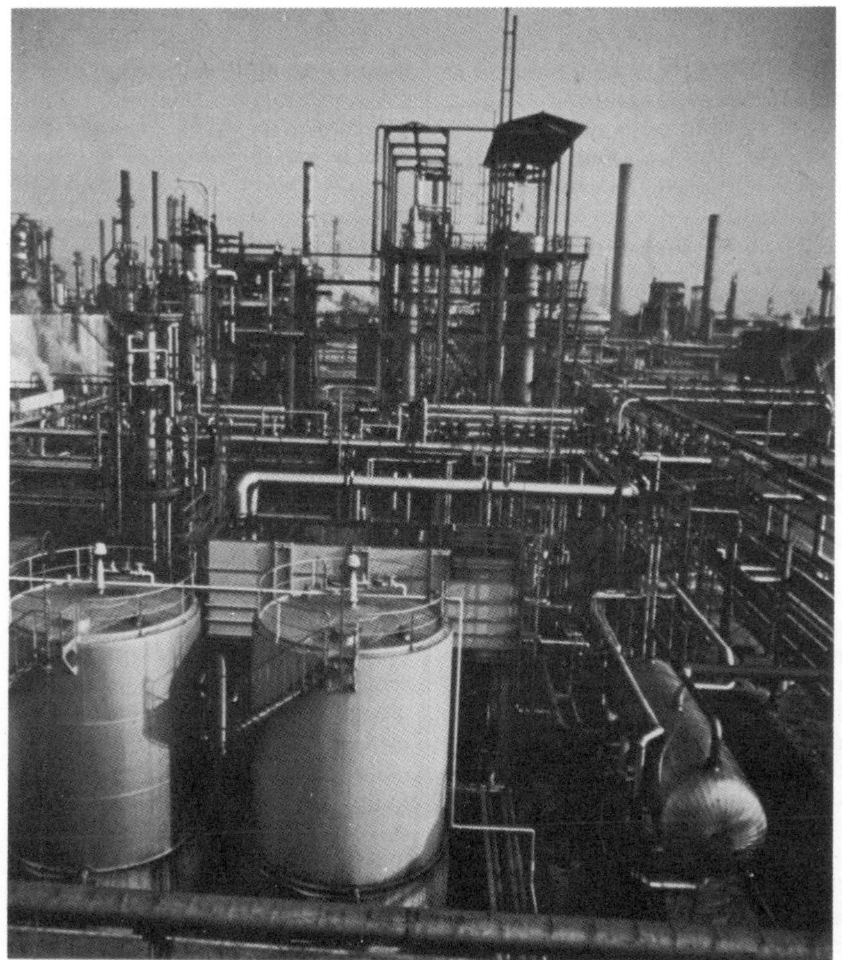

White oils plant at Port Jerome came onstream quickly and smoothly.

Originally published September 15, 1975

☐ A two-stage catalytic hydrogenation process for manufacturing white oils, developed by Exxon Research and Engineering Co., has performed up to or better than design specifications since its October 1973 debut at the Port Jerome, France, refinery of Esso SAF.

The new process accepts various feedstocks derived from either naphthenic or paraffinic crude oils. It can produce both medicinal- (or food-) and technical-grade white oils, over an exceptionally broad range of product viscosities.

A proprietary hydrogenation catalyst used in the second stage enables the purity requirements of medicinal-grade product to be met at less-severe operating conditions than with existing commercial catalysts. The white oils are free from aromatic hydrocarbons, particularly polynuclear aromatics, and are colorless, odorless and tasteless.

The first, hydrotreating, stage can also stand on its own—it can prepare white-oil charge stock for conventional acid-finishing, or for specialty applications such as making rubber/plastics extender oils and transformer oils.

HYDROGENATION BENEFITS—In acid-treating procedures, a suitable lube-oil distillate or raffinate generally undergoes four steps: multiple treating with oleum or gaseous sulfur trioxide, neutralization of the oil layer with alkali, extraction of oil-soluble sulfonates, and clay treatment to remove trace impurities.

Disadvantages of the acid process—particularly low white-oil yields, high yields of acid sludge, and disposal problems for both the acid sludge and spent clay—have become more acute with rising feedstock costs and environmental pressures. Also, the economics of the

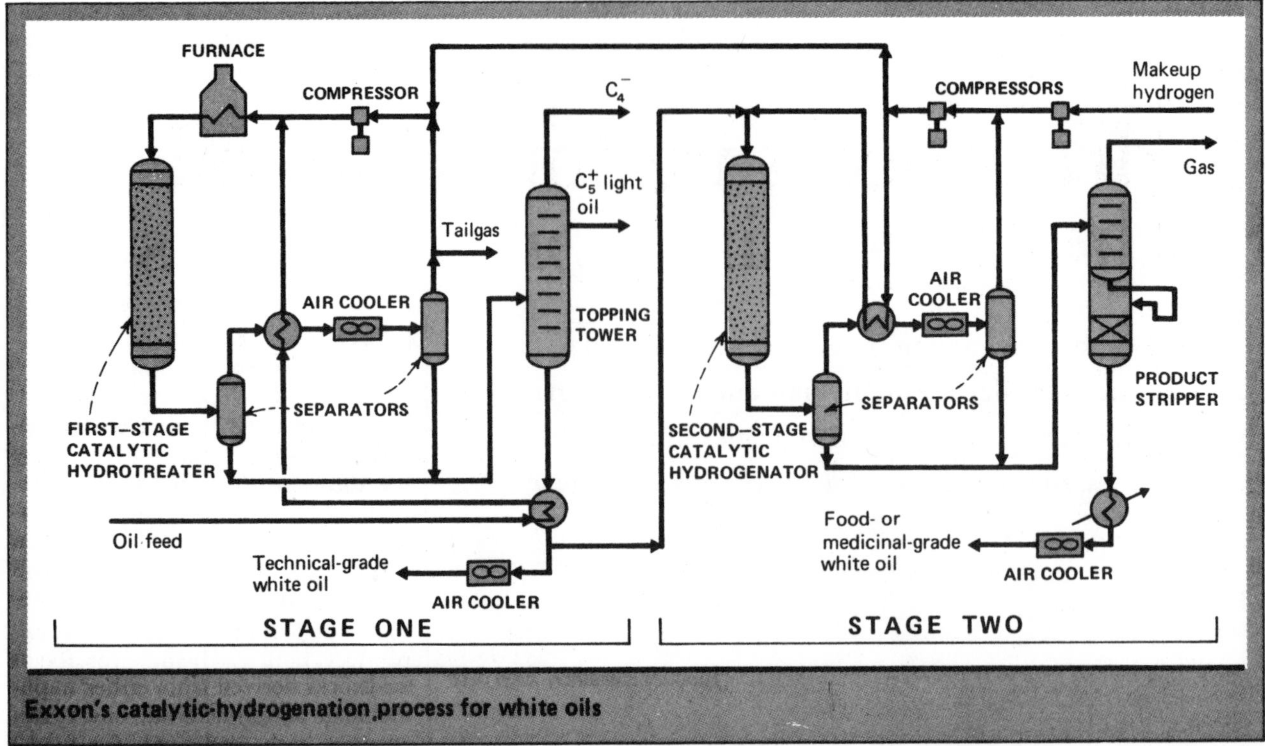

Exxon's catalytic-hydrogenation process for white oils

acid-treating routes have been hurt in recent years by fluctuations in market demand and price for the sulfonates byproducts (caused by the introduction of synthetic sulfonates). Moreover, high-viscosity grades of white oils are difficult to process from paraffinic crudes.

Catalytic hydrogenation overcomes these troubles. It offers higher white-oil yields, a wider viscosity range for products, the flexibility to handle less-severely-refined feedstocks, and freedom from environmental pollution problems.

TWO STAGES—In the first half of the new Exxon technique, feedstock is desulfurized by catalytic hydrotreatment over a conventional metal sulfide catalyst. Feeds of low sulfur content are necessary to ensure an economic operating life for the deep-hydrogenation catalyst that serves in the second stage; this catalyst would react readily with any sulfur compounds present and lose its activity.

The oil feedstock and a hydrogen-rich treat gas are thus contacted at elevated pressure with the metal sulfide catalyst. Reaction temperature and flow are adjusted to reduce the sulfur content of the oil to a very low value, typically less than a few ppm. Hydrogenation also occurs under these conditions, greatly reducing aromatics content (though not to final-product level).

Desulfurized oil is next separated from the treat gas, and stripped to remove dissolved gases and to adjust the oil viscosity and flash point.

The liquid then passes to the second-stage deep hydrogenation, again combined with a hydrogen treat-gas. The mixed stream contacts the Exxon catalyst at elevated pressure. After separating treat gas from liquid on the outlet side, the treat gas flows back to the first stage of the process, while the liquid undergoes a mild stripping and is sent to white-oil product storage.

PROCESS DEVELOPMENT—The catalyst for the first stage was chosen on the basis of pilot-plant screening tests of a number of cobalt-molybdenum, nickel-molybdenum and nickel - tungsten oxide catalysts, as the one showing maximum activity for aromatics hydrogenation, along with adequate hydrodesulfurization.

Extensive pilot-plant studies at the Imperial Oil laboratories have defined the effects of the major operating variables—temperature, pressure, and liquid space velocity—on the sulfur and aromatics content of various white oils. Hydrodesulfurization, for example, is a strong function of temperature, while aromatics saturation is more sensitive to pressure and space velocity.

Overall, the hydrotreating stage must be carefully designed to process a selected range of feedstocks so as to give adequate sulfur removal and aromatics saturation, but not cause excessive yield or specific-gravity losses, or adverse changes in other product characteristics such as pour and cloud points.

With suitable operating conditions, technical-grade white oils can be made directly by hydrotreating and topping, without second-stage processing. The partially hydrogenated stream, moreover, is a high-quality feedstock for acid-treating. For existing acid-treating plants, installation of a hydrotreating stage to upgrade charge stocks could be an economical first step toward phasing out acid-treating operations in favor of catalytic hydrogenation.

For the second-stage reaction, catalysts containing Group VIII metals, such as platinum, palladium and nickel, have the hydrogenation activity needed to achieve the deep levels of aromatic-bond saturation. But these catalysts work only at relatively high hydrogen partial-pressures, and low liquid space-velocities.

Inhouse research by Exxon, however, has revealed a novel technique for producing a nickel-based catalyst with much higher hydrogenation activity at less severe operating conditions. Subsequent work in conjunction with the Ketjen Catalyst Dept. of Akzo Chemie NV (Amsterdam, The Netherlands) has scaled up the method through the pilot plant to commercial size, for supplying the Port Jerome plant.

This catalyst, designated EK, has a reactor density of about 44 lb/ft^3. It turns out medicinal-grade white oils at 5–10 times the space velocity of conventional types, when compared at the same pressure and at the optimum temperature for each.

Due to the high selectivity of the EK catalyst, only minor changes in white-oil physical properties (such as viscosity and specific gravity) occur in the second stage. Yields after final stripping are nearly 100%; the step mainly removes dissolved gases.

PRODUCT QUALITY—The table shows typical feed and product inspections for light and heavy medicinal-grade white oils produced at Port Jerome. Feedstocks are solvent-treated raffinates from naphthenic distillates. Both products pass the appropriate U. S. National Formulary and U. S. Pharmacopoeia tests for acidity, sulfur compounds and solid paraffins. The products readily meet stringent tests for carbonizable substances and for polynuclear aromatics (even the tough British Pharmacopoeia test for the former).

The excellent yield and quality obtained for white oils, particularly for heavy product, illustrate the extremely high aromatics conversion and selectivity levels inherent in Exxon's route. Conventional oleum treating of the stocks would give only 75% and 68% yields of light and heavy products, respectively.

Operating on a solvent-treated naphthenic distillate for production of technical-grade white oils, the Port Jerome unit converts 95% of the feed to ongrade product.

ECONOMICS—The following investment and operating cost estimates illustrate the economics of the process for a plant making 10 million gal/yr of medicinal/food-grade white oils and 5 million gal/yr of technical-grade.

Capital investment for a Gulf Coast location in 1974 was $5.33 million. Inclusion of investment for the first catalyst charge brings this to $5.60 million. This figure includes all equipment and direct labor costs to construct onsite facilities, specifically major process equipment and associated piping, instrumentation, insulation, foundations, structures, freight to site, and sales taxes on materials. Excluded are field-labor overheads and burden, contractors' engineering, contractors' fees, escalation, contingencies, offsites and utilities investment.

A source of hydrogen at 95% purity and 250 psi is assumed available. But the plant could also be designed to operate on catalytic-reformer tail-gas containing as little as 60 mole% hydrogen, for example.

The estimated average yield on feed for the above product slate is 90 vol.%, and chemical hydrogen consumption 375 std. ft^3/bbl. These figures are typical, but will vary with feedstock and product slate. Average capacity of the plant in the estimate is 1,280 bbl/d.

Utilities and fuel costs total 3.2¢/gal of product. Some 2.01¢ represents steam consumption (380 lb/bbl feed, at $2/1,000 lb). Other components are electric power, 0.52¢ (7.8 kWh/bbl feed, at 2.5¢/kWh); fuel, 0.63¢ (86,000 Btu absorbed/bbl feed, at $10/fuel-oil-equivalent bbl); and cooling water, 0.01¢ (354 gal/bbl feed, at 1.3¢/1,000 gal).

Catalyst makeup costs are 1.1¢/gal. EK catalyst life in this particular case is 1.5 yr. In a grassroots plant, this could be longer or shorter, depending on the economic balance between reactor investment and plant downtime for catalyst change.

Commercial production of hydrogenated medicinal-grade white oils

	Light white oil		Heavy white oil	
	Feed	Product	Feed	Product
Yield, vol.%	100	95	100	89
Viscosity at 100°F, cSt	16.3	15.8	92.8	77.0
Viscosity at 100°F, SSU	82.6	80.7	430	357
Specific gravity, 60/60°F	0.861	0.851	0.880	0.867
Sulfur, wt.%	0.19	Nil	0.15	Nil
Aromatics, wt.%	16	Nil	18	Nil
Flash point, COC, °F	338	342	442	453
Pour point, ASTM, °F	-22	-20	+5	+10
Color, Saybolt	–	+35	-30	+35
Odor	–	None	–	None
Taste	–	None	–	None
U. S. Pharmacopoeia polynuclear aromatics (by UV absorption)	–	0.01	–	0.01
British Pharmacopoeia carbonizable substances (Lovibond color units, red/yellow)	–	0.3/1.5	–	0.5/1.5

Note: Data obtained from Port Jerome unit, operating on solvent raffinate feed from naphthenic distillate.

The authors

John B. Gilbert is a research chemist with Imperial Oil Enterprises Ltd., Research Dept. (P. O. Box 3022, Sarnia, Ont. N7T 7M1, Can.). He is involved in the development of hydrotreating processes and catalysts for manufacture of white oils and specialty products. A fellow of the Chemical Institute of Canada, he earned a B.S.C. and Ph.D. at the University of Hull, Yorkshire, England, and has done postdoctoral work at the University of Alberta, Edmonton, Alta., Can.

Christopher Olavesen is a chemist with Imperial Oil Enterprises at Sarnia, also concerned with development of hydrotreating processes. He received a B.S.C. and Ph.D. in chemistry from the University of Birmingham, England.

Clinton H. Holder is a senior research associate with Exxon Research and Engineering Co., Florham Park, N.J. He is responsible for helping to develop processes for lubes, waxes and specialty products. He joined Exxon in 1939, after earning a Ph.D. in physical chemistry from McGill University, Montreal, Que., Can.

Jacques Lecomte, a senior research chemist with Esso SAF, Centre de Recherches, at Mont-St.-Aignan (near Rouen, France), is involved with both research and sales engineering in industrial oils, including white oils, insulating oils and compressor oils. He was graduated from Ecole Nationale Supérieure de Chimie Industrielle in Rouen in 1958. He belongs to the European Technical Working Group on White Oils, and is a member of the Wax and Paraffin Committee at Bureau de Normalisation du Pétrole.

Award Winning Processes

1973—

Proteins from hydrocarbons
New catalytic route to acrylamide
Gas-phase, high-density polyethylene process

1975—

Complete combustion of CO in cracking process
Integrated use of oxygen in pulp and papermaking
Shortened route to pure nickel
Arc-plasma dissociation of zircon

1977—

Low-pressure oxo process yields a better product mix
Nitrogen trifluoride by direct sythesis
Winning more from heavy oils
Better path to ethylbenzene
Carbon monoxide from lean gases

BP PROTEINS LTD.

Proteins From Hydrocarbons

Top honors in the 1973 Kirkpatrick Chemical Engineering Achievement Award go to BP Proteins Ltd. for more than a decade of innovative research and engineering that has led to two commercial processes for growing high-protein-content yeast on normal paraffins.

The accomplishments of BP Proteins Ltd. (London) that have led to its winning the 22nd biennial Kirkpatrick Chemical Engineering Achievement Award are marked not only by innovative engineering but by outstanding social significance. The company, a subsidiary of British Petroleum Co. Ltd., has the added distinction of being the first non-U.S. firm to be honored by the panel of educators given the demanding task of selecting the most important chemical-engineering feat commercialized during the past two years.

For more than a decade, British Petroleum and its subsidiaries have worked to develop two separate processes for cultivating a high-protein-content yeast on normal paraffins. The result—industrial-scale production of proteins from hydrocarbons—is especially noteworthy in view of the world food situation.

A large segment of mankind is presently suffering from protein-deficient diets, and traditional methods of producing food are falling behind rather than catching up with protein demand. For example, estimates point to a worldwide protein shortage on the order of 20 million tons/yr. by 2000.

Originally published November 26, 1973

BP DEMONSTRATION UNIT at Grangemouth, Scotland, produces proteins from pure normal paraffins.

BP's Record

BP is clearly one of the leaders in preventing a diet disaster. Its product, a yeast tradenamed Toprina, is presently being used in animal feeds, thus freeing supplies of traditional protein sources such as soya bean and fish meal for human consumption. Assuming public acceptance, the company hopes that eventually its yeast will be used as food for people.

The parent company established BP Proteins in 1970 to commercialize yeast-production technology developed within the British Petroleum Group. After exhaustive toxicological and nutritional testing, BP set up a demonstration unit capable of making 4,000 metric tons/yr. of Toprina at Grangemouth, Scotland, in 1971.

The following year a unit based on the second BP process and rated at 16,000 metric tons/yr. went onstream at Lavera, France. Then, in February of this year, the company engaged Foster Wheeler to construct a 100,000-metric-ton/yr. plant in Sardinia. This is expected to be commissioned early in 1975 for Italprotein, a subsidiary of British Petroleum that is associated with ANIC, the Italian state chemical enterprise.

Process Problems

Yeasts have been nurtured for decades on carbohydrate-containing wastestreams such as those from sugar refining and sulfite pulping. However, the amount of biomass that can be harvested is limited—first by the volumes of wastes available and second by the economic restrictions of collecting them. In practice, this has meant a ceiling on yeast output of 10,000 to 15,000 metric tons/yr. for each unit.

Use of hydrocarbons avoids the shortcomings of wastestream feedstocks and still gives yeast of similar composition and quality. The hydrocarbon route, however, does involve formidable obstacles that had to be overcome by BP.

Carbohydrate-containing substrates provide carbon, hydrogen and oxygen, which are necessary for cell growth, in an aqueous solution to which other essential elements can be easily added. In contrast, a hydrocarbon medium furnishes only carbon and hydrogen in a form that is practically insoluble in water. Thus, large quan-

PROTEIN PLANT at Lavera, France, uses alternate manufacturing process based on gas-oil fraction.

The Man Behind the Process

If one man can be singled out as the guiding light behind the BP protein-from-hydrocarbons processes, it would be Dr. Alfred Champagnat.

Educated as both a mechanical engineer (at Aix-en-Provence) and a chemist (at Strasbourg), Champagnat began his career in 1934 as a research engineer at the predecessor company to the present Société Française des Pétroles BP. He was promoted to director of research in 1941.

By the late 1950s, his interest in microbiology led him to seriously consider commercial protein production from hydrocarbons. In May 1959, Champagnat proposed a research plan at a meeting of l'Association Nationale de la Recherche Technique. He implemented the plan that year at BP, leading to development of the gas-oil process at Lavera a few years later.

In 1964 a new research group was formed within BP—Société Internationale de Recherche BP. It was natural that Champagnat be named its director, a post he held until his retirement in 1968.

Champagnat's pioneering protein work has already brought him a number of honors: In 1963, he was made a Chevalier de la Legion d'Honneur by France; in 1968, the Societe de Chimie Industrielle awarded him and two of his coworkers—Charles Vernet and Bernard Laine—the prize commemorating its 50th anniversary; in 1971, the British Institute of Petroleum bestowed its Redwood Medal; and just last month, Champagnat was awarded the Gairn Gold Medal of the Society of Engineers in the United Kingdom.

tities of atmospheric air must be blown into the process, and intensive mixing is needed so that the yeast makes contact with both water-soluble and insoluble materials. And these are not the only problems.

Metabolic conversion of hydrocarbons proceeds via oxidation to yield cellular proteins, carbohydrates and lipids, while carbon dioxide, water and heat are liberated. The reaction is similar to carbohydrate conversion except for two differences that greatly complicate large-scale operations. Hydrocarbon assimilation requires about 2.5 times as much oxygen, and more than twice as much heat must be removed from the process.

Work Begins

In 1959, Dr. Alfred Champagnat, head of research at Société Française Pétroles BP decided that the promise of alleviating the protein shortage through hydrocarbon-based yeast production demanded an attack on the bioengineering problems involved.

Work was started that year at the French company's complex at Lavera, near Marseilles. Researchers focused on cultivating biomass on the waxy normal-paraffins present in a middle-distillate fraction of crude. They reasoned that this approach to growing yeast would have the bonus of being able to return the remaining fraction, dewaxed, to the refinery.

As work progressed, BP decided to also explore an alternate scheme: Start with pure normal paraffins instead of a mixed fraction requiring selective consumption by yeast. This would eliminate the byproduct petroleum stream, which might be desirable in some cases. Work on this front was placed with a group at BP Grangemouth, Scotland, near Edinburgh.

Both teams were faced with the need to promote and control fermentation in a four-phase system: two liquids (aqueous nutrients and hydrocarbons), a gas (air for oxygen supply), and a solid (yeast). Before designing a suitable fermenter, optimum levels had to be determined for a host of key parameters—composition of the aqueous medium, hydrocarbon concentration, degree of aeration and agitation, pH, temperature, and residence time.

The much greater stoichiometric requirement for oxygen in hydrocarbon assimilation compared to that of carbohydrates meant that BP had to design a fermenter with correspondingly higher oxygen-transfer rates. The company has developed units with transfer rates up to 15 kg./(cu.m.hr.), about three times those of previous reactors for carbohydrates. In addition, fermenter sizing was an important consideration because only about 30% of the oxygen is actually consumed by the yeast (the process requires 30 tons of air per ton of yeast).

Equipment Design

The Lavera team settled on an airlift design in which incoming air provided both aeration and agitation, while the group at Grangemouth opted for a mechanically agitated, baffled vessel. To save development time, the company decided to have each group concentrate on its own fermenter choice, rather than evaluate both types. (BP notes, however, that the type of vessel is not specific to one process or the other.)

Then there was the major problem of cooling. For a 100,000-metric-ton/yr. facility, the heat output is equivalent to that of a 115-Mw. power plant. The traditional method of cooling fermenters with a water jacket was out of the question, and an external heat exchanger was obviously necessary.

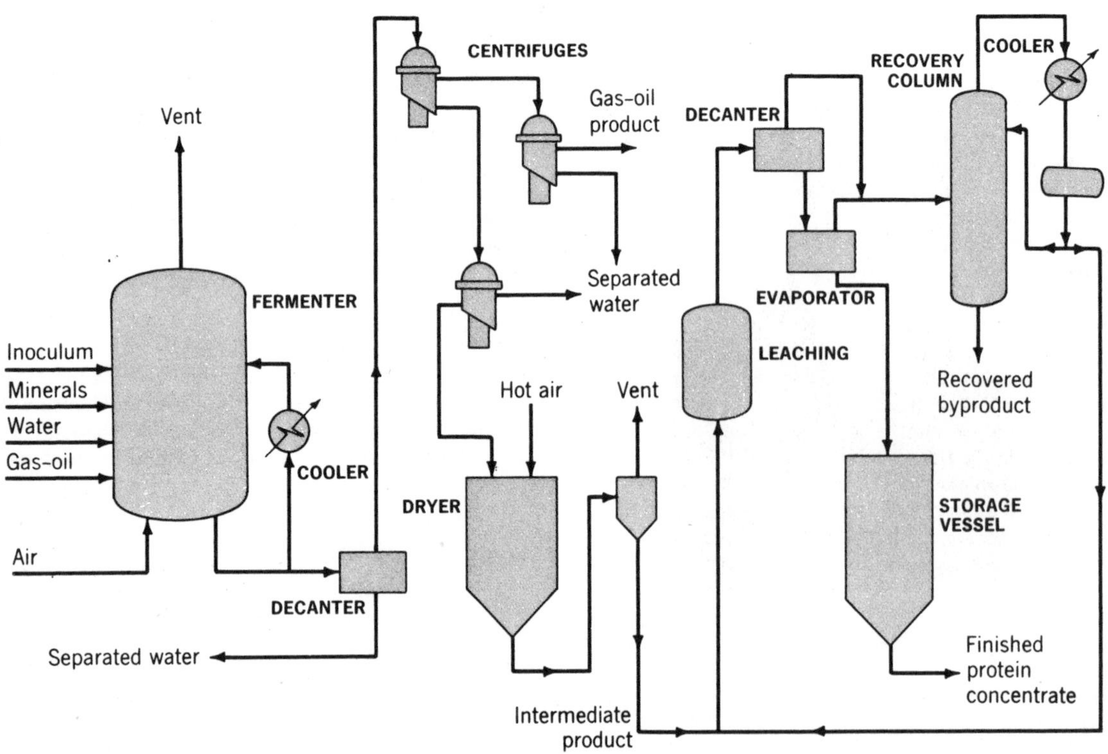

GAS-OIL PROCESS includes novel solvent-extraction circuit for removing residual hydrocarbons from product.

Considerable experimental work went into establishing the heat-transfer coefficient for the aerated broth from the fermenters. Moreover, because of the corrosiveness of the hot broth and the chosen coolant, special care had to be taken in selecting materials of construction for the exchanger. The company also discovered that preliminary deaeration of the broth was essential to improve pumping efficiency.

Special Problems

Each process posed specific problems. Though both routes consume only normal paraffins, in the gas-oil process developed at Lavera, paraffins comprise 10% or more of the hydrocarbon feed. Consequently, there is some unmetabolized gas oil remaining in the yeast, even after centrifugation. Through a novel countercurrent solvent-extraction circuit, residual hydrocarbons are reduced to 0.08% in the product. The difficulty of this step can be appreciated by considering that the particles being separated are roughly 1 to 5 microns in diameter.

BP is closemouthed about extraction details. It will only say that the solvent system is made up of widely available chemicals and that it had to construct special solids-leaching, contacting and separation equipment. The company also notes that precise control in the prior fermentation and centrifugation steps is essential to provide yeast cells having the optimum physical properties for the separation.

The Grangemouth group found itself in a different dilemma. Unlike the gas-oil process, in which selection of proper operating conditions allows the desired yeast to predominate, pure normal paraffins are such a rich substrate for fermentation that there is no easy way to control which microorganisms thrive in the fermenter. The solution was to sterilize all feed streams, and then inject the desired strain of yeast into the vessel.

Liquid-stream sterilization is handled by conventional heat treatment, although the piping design was complicated by the need to have total sterilization during startup and changes in feedrates. The air stream, too, must be treated. BP concluded that available equipment would not be economical because of the high flowrate and extreme sterility required. So, the company developed its own air-filtration system.

Details of this system are guarded. But BP will say that air required in large plants can be treated to give a bacteria count of not more than 1/1,000 hr. of filter operation. (The company may make this knowhow available.) The air system is designed so that there is a measurable pressure drop between the compressor and the fermenter. The discharge temperature from the compressor helps reduce the sterilization load on the filter.

Commercializing the Processes

Even with all of the engineering achievements during process development, and startup of two pilot plants (the

Product-Testing Program

The product of the BP Protein award-winning processes is a cream-colored, odorless, tasteless yeast known by the tradename Toprina. It is being used as a constituent in animal feeds, having passed a number of stringent toxicological and nutritional tests.

Specific characteristics of the yeast depend to some extent on how it was grown—whether via the refined-normal-paraffins route or the gas-oil process. In general, though, Toprina contains at least 60% crude protein—more than other yeasts now used in animal feeds—and 8% or less moisture. The remainder consists mainly of lipids, ash, calcium and phosphorus. Its amino acid content is similar to that of soya bean and fish meal, except for a slightly lower methionine level, which is typical of yeasts. Toprina is richer in lysine than either soya bean or fish meal; animal-feed additives are often lacking in this essential amino acid. Overall, biological availability of the amino acids in the new yeast is at least 92%.

For any products used in foodstuffs, toxicity is naturally of paramount consideration. It is in the area of toxicological testing that BP is due major credit for pioneering work that it sponsored to assess and assure the safety of Toprina.

When the company started working on hydrocarbon-based protein production in the early 1960s, there was no internationally recognized plan for testing such materials. To devise one, BP joined with an independent research institute. The soundness of the program developed is shown by the fact that in 1970 the Protein Advisory Group (a unit that makes recommendations for checking the suitability of materials for human consumption under auspices of the United Nations) adopted a toxicity-testing protocol for novel sources of protein that is essentially the same as the one pioneered by BP.

Toprina's safety was evaluated by the Central Institute for Nutrition and Food Research, an independent, state-supported organization based in Zeist, The Netherlands. Testing—which started in 1964 and is still continuing—involves six-week, three-month, and up to two-year terms on laboratory animals.

In the test programs, particular attention has been paid to tumor formation, since hydrocarbons may contain cancer-causing polynuclear aromatics. BP feedstocks are controlled so that the polynuclear aromatic level is less than one part-per-billion and the yeast has no detectable level of these carcinogens. While the international protocol requires examination of three generations of rats, BP has already checked 13 generations of rats and 23 generations of quail without finding ill effects of any kind.

Nutritional value of the yeast also has been the subject of extensive research. The company commissioned another independent Dutch organization, the Institute for Agricultural Research Into Biochemical Products, Wageningen, to evaluate Toprina as high-level constituent of animal feeds.

Nutrient experiments began in October 1965. They concentrated on single-stomach animals—poultry and pigs, for example—because such livestock depend on a more-limited range of protein sources than ruminants. The work has now been extended to a wider spectrum of animals.

One animal group was given a feed containing Toprina at or above the level expected in commercial practice. A control group was fed a conventional mixture based on fish meal or soya bean. The program evaluated parameters such as egg production, fertility and hatchability of eggs, rate of growth, efficiency of feed conversion, number of young born, etc. Results, now extending through the fifth generation of poultry and pigs, show that animals thrive on the yeast. This has been confirmed during the last three years in trials run by British and French feed manufacturers.

Backed by such a long and exhaustive testing program, Toprina-containing animal feeds have been approved for use in Denmark, West Germany, The Netherlands, Italy, The United Kingdom, Belgium and South Africa. BP believes that Toprina is the only hydrocarbon-derived yeast that has been approved for marketing in all these countries.

gas-oil process at Lavera in 1963 and the pure normal-paraffin system at Grangemouth in 1965), considerably more work was required to groom the processes for commercialization. For instance, correlations had to be determined for predicting yeast production from fermenter geometry and operating conditions, and for forecasting power requirements. Other scaleup factors had to be determined and, of course, the yeast had to be thoroughly tested for safety and nutritional value.

In the gas-oil process, the middle-distillate fraction, containing at least 10% normal paraffins up to C_{18}, is metered directly into the fermenters. Other starting materials are water, water-soluble nutrients (potassium, magnesium, iron and zinc cations, along with sulfate and phosphate anions) and other compounds essential for proper cell growth. Ammonia is added to supply necessary nitrogen values and to control pH. Altogether, hydrocarbons account for 5 to 20% of the liquid feed.

Temperature is maintained at about 86 F. by the external heat exchanger. Carbon dioxide and excess air are vented directly from the fermenter.

The product stream from the heat exchanger flows to a decanter, in which a large portion of the water is separated for reuse. The remaining froth is deaerated to a "cream" and sent to the centrifugation circuit. Here, the bulk of the gas-oil and some water are removed (and further separated in another centrifuge) and then most of the remaining water is taken out by a second centrifuge. This leaves a slurry of yeast cells containing about 15% solids, plus traces of unmetabolized gas-oil.

Product is converted to a powder, either in a single drying stage or by evaporation followed by drying. The powder is fed to the solvent extraction system, in which leaching removes almost all of the gas-oil and the greater part of the cellular lipids. The protein-rich solvent phase is separated by decantation and sent to an evaporator. Solvent from the overhead is distilled and recycled, leaving marketable lipids in the bottoms fraction. Mean-

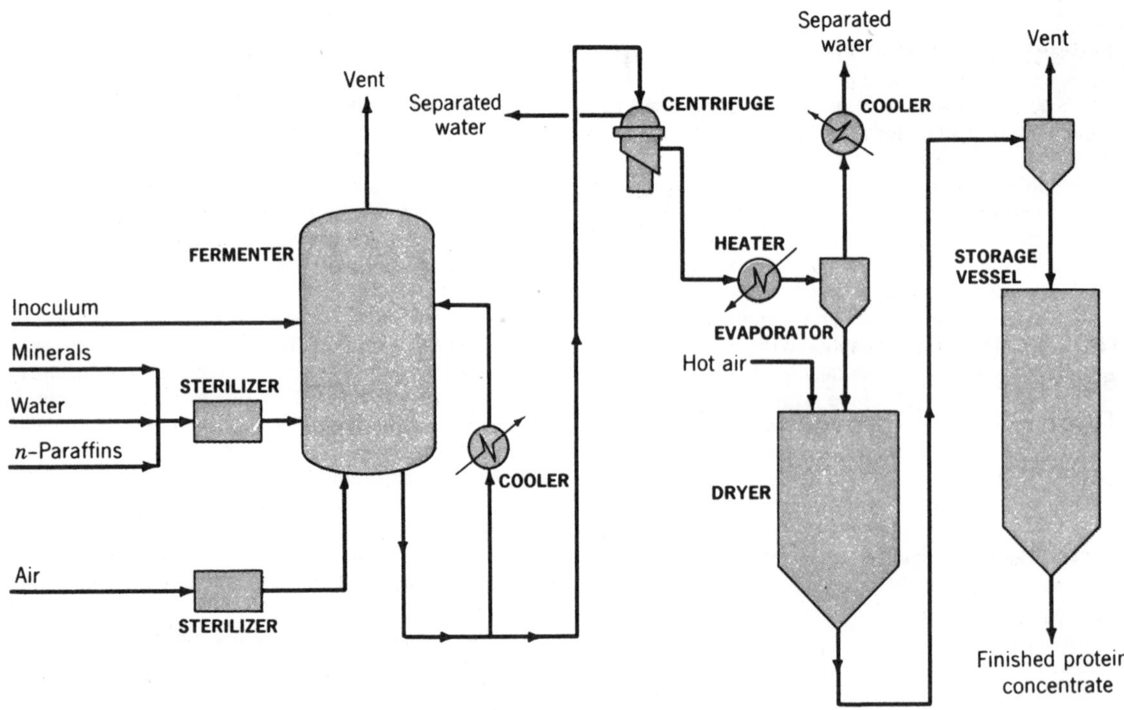

NORMAL-PARAFFIN ROUTE requires special provisions for sterilizing all feed streams by heat treatment.

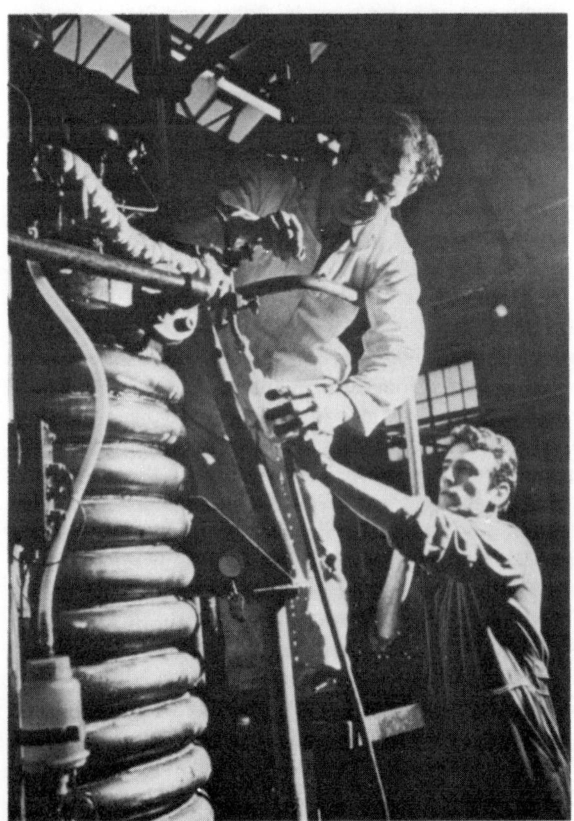

WORKERS at the Grangemouth, Scotland, pilot plant remove sterile samples from the fermentation step.

while, product yeast from the evaporator is sent to storage prior to packaging.

Paraffins Route

In the pure-normal-paraffins route, hydrocarbon feed is prepared by a proprietary molecular-sieve process and pumped to a sterilization unit where it joins the other process liquids. Microorganisms in the mixture are killed by steam injection. Heat is recovered by passing the feed to the sterilization unit through an exchanger.

Sterilized broth enters a baffled, mechanically agitated fermenter. Treated air is supplied through the special filtration system so that the vessel operates under aseptic conditions except for the yeast that is injected separately. Temperature is maintained at about 86 F.

Carbon dioxide produced during fermentation, and excess air, are vented after passing through another sterile filter, which assures that there is no back-contamination. The yeast-containing product stream is taken from the external cooling circuit and decanted to remove most of the water. The resulting "cream" is fed to a multiple-effect evaporator, which provides heat economy by interstage vapor decompression. Condensate is reused. Next, the yeast moves through a dryer prior to storage.

BP says the choice between the two processes largely depends on local circumstances. Product quality is comparable. Costs are in the same range because the saving with gas-oil feedstock is generally counterbalanced by the added capital charges for purifying the feed. ■

THE DOW CHEMICAL COMPANY

New Catalytic Route to Acrylamide

Direct hydrolysis of acrylonitrile to acrylamide via a new catalyst system greatly improves yields and product purity, eliminates byproducts, and cuts both fixed and operating costs.

Highlights of New Acrylamide Process

- 98.5% yield based on acrylonitrile
- 99.5 + % acrylamide purity
- No byproducts to dispose or market
- 35 to 45% savings in direct fixed-capital costs
- 25% variable cost advantage over next best method
- Process suitable for hydrolyzing most organic nitriles to the corresponding amides

PLANT based on Dow's new acrylamide process went onstream in 1971 and already has been expanded 50%.

Originally published November 26, 1973

The benefits of Dow Chemical's new catalytic route to acrylamide sound like a chemical engineer's dream. Compared to other commercial processes, the new method produces higher yields and product purity, eliminates troublesome byproducts, and reduces both fixed and variable costs.

The award-winning achievement began with a planned search for a catalyst that would directly convert acrylonitrile and water to acrylamide. Laboratory work started in 1967; pilot-scale tests in 1969; and commercial production in 1971. Pilot-plant data were used primarily to optimize running conditions in the production unit, which was designed by scaling up laboratory data 200,000 times.

Other-Process Faults

Acrylamide has been an important chemical intermediate since the early 1950s, when researchers discovered that high-molecular-weight polyacrylamides were excellent flocculants. These are now widely used in the paper, sugar, coal and minerals industries, as well as in wastewater treatment. But despite the success of polyacrylamides, previously developed processes for producing monomer have serious faults.

In conventional acrylamide plants, approximately equal molar quantities of acrylonitrile, water and sulfuric acid are reacted to produce acrylamide sulfate. This step is not particularly difficult. Most of the problems arise in isolating acrylamide from the salt. The usual separation method involves neutralizing sulfuric acid with ammonia. Insoluble ammonium sulfate is removed by filtration, and acrylamide is crystallized from the filtrate.

Dow has followed a different path for all of its acrylamide production. A number of years ago, the company developed an ion-exclusion technique in which acrylamide sulfate is diluted with water and passed through a column containing cation-exchange resin—a sulfonated copolymer of styrene and divinylbenzene. The product of the ion-exclusion process is a water solution of acrylamide.

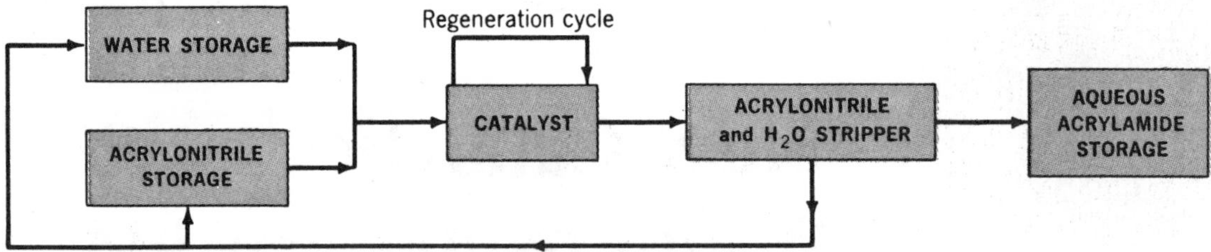

CATALYTIC ROUTE to acrylamide eliminates troublesome byproducts and increases product yields.

Both of the conventional methods require relatively high capital investment and give yields of 80 to 90% based on acrylonitrile. In each case, there is a byproduct that cannot be easily sold and thus presents a disposal problem. (Acid neutralization results in about 2.2 lb. of ammonium sulfate/lb. of acrylamide, while ion exclusion gives approximately 1.6 lb. of sulfuric acid.) Further, the dry crystal obtained from acid neutralization presents a potential hazard during packaging, shipping, and conversion to end-products. Acrylamide affects the central nervous system, and studies on animals indicate it can be absorbed through unbroken skin.

Search for Ideal Method

In the search for an improved process, Dow workers felt from the beginning that the various versions of the sulfuric acid catalyzed route had too many problems to be fertile ground for research. Since acrylamide is a highly reactive monomer with low vapor pressure, purification by either distillation or crystallization would likely lead to the same troubles found in existing systems.

Dow decided that the ideal process should yield a high-purity monomer without the capital and technical problems of conventional schemes. In other words–direct conversion of acrylonitrile and water by contact with a noncontaminating catalyst. A literature search uncovered several possible techniques based on both heterogeneous and homogeneous catalysts. All were tried, but none had high enough reaction rates and selectivity to be commercially attractive.

Copper-Chromium Combination

After months of testing transition-metal and rare-earth compounds as catalysts, all the evidence pointed toward copper-containing materials as the best generic class of compounds. Nevertheless, there were still a number of side-reactions in the hydrolysis step that had to be eliminated.

Continuing analytical work, catalyst synthesis and process engineering eventually led to a copper-chromium combination that produced the desired goal–high yield plus a rapid reaction rate. Dow points out that its new process is one of the few examples of a commercial system based on a fixed-bed heterogeneous catalyst in combination with liquid-phase reactants.

In the commercial plant, a solid copper-chromite catalyst converts acrylonitrile and water directly to acrylamide with essentially no byproduct formation. Product from the catalyst bed travels through a stripper that removes unreacted materials. The process is continuous and operates at moderate temperatures and pressures–less than 100 C. and 50 psig.

Catalyst Details

Catalyst compositions developed by Dow are covered in six issued and two allowed U.S. patents, as well as numerous foreign applications. U.S. Patent 3,597,481, assigned to Dow by Ben A. Tefertiller and Clarence E. Habermann, covers combinations of 10 to 90% by weight of oxides of copper, silver, zinc or cadmium with 10 to 90% by weight of oxides of chromium or molybdenum. These are prepared by coprecipitation from soluble salts.

A later patent (U.S. 3,642,894) refers to copper oxide or copper chromite catalysts that are reduced with hydrogen, maintained under an inert atmosphere, and placed in a heated reactor (70 to 85 C.) with water and a nitrile.

Spent catalyst can be regenerated without substantial degradation. According to the Dow patents, the catalyst can be oxidized with hydrogen peroxide at room temperature, or with air at elevated temperatures. In either case, oxidation is followed by hydrogen reduction.

Problems Met and Solved

Even though the new Dow process is outwardly simple, many problems were resolved during its development. The company used a taskforce approach involving research, analytical, pilot-plant and production groups. They relied on a kinetic model of the reaction parameters that was coupled with computer analyses, detailed analytical work and fundamental process-engineering techniques.

Aside from the major effort on catalyst investigation, Dow developed the required operating conditions, inhibitors and reactor designs to prevent reactor plugging and achieve high product quality. Catalyst regeneration, which involves highly exothermic reactions, was a particular problem that was overcome. Finally, safety was made an overriding concern because of the dangerous nature of acrylonitrile and acrylamide.

The commercial success of the process is demonstrated by the fact that Dow has already expanded the original unit by 50%. Also, the company is planning a grass-roots facility in Europe and is studying proposals for use of the technology in Japan. ■

UNION CARBIDE CORPORATION,
CHEMICALS AND PLASTICS DIVISION

Gas-Phase, High-Density Polyethylene Process

By using raw-material ethylene as the fluidizing gas and heat removal agent, and product polyethylene as the bed material, Union Carbide's new process for high-density polyethylene eliminates the need for a carrier solvent and trims both capital and operating costs.

Classic simplicity is the hallmark of Union Carbide's fluidized-bed, gas-phase process for making high-density polyethylene.

This impression comes through most strongly at the focal point of the flowsheet: the reactor where ethylene is polymerized. In the company's solvent-free approach, the ethylene is not only the main reactant monomer, but also the fluidizing gas and the medium for removing heat of reaction. Moreover, the product—polyethylene—also serves as the fluidized-bed material.

TWIN FLUID-BED reactors dominate new polyethylene resin plant located in Stenungsund, Sweden.

This strikingly simple economy of function is echoed throughout the process. For example, there is no need for carrier-solvent circuits such as are used in other processes, or equipment for drying the product and removing catalyst from it.

The payoff shows up in both capital and operating costs. Carbide estimates that the required investment can be as much as 15% lower than for a conventional high-density-polyethylene plant; operating costs can drop 10%. And the absence of a solvent gives the process a definite environmental advantage.

Three commercial plants have already demonstrated the viability of the flowsheet. A 30,000-metric-ton/yr. plant of Unifos Kemi A.B., came onstream at Stenungsund, Sweden, in December 1971. Carbide's Australian affiliate has operated a similar unit at Altona since November 1972. And recently, the U.S. company elevated to fully commercial status a 13,600-ton/yr. unit at its Seadrift, Tex., complex, which had started up as the prototype in 1968.

Additional plants are now being designed or constructed. Among them are an 80,000-metric-ton/yr. unit for Chemopetrol at Zaluzi, Czechoslovakia, and a second Carbide facility at Seadrift, this one rated at 70,000 tons/yr. The latter is due for completion late in 1974.

A Profile of the Process

A detailed process description appeared in CHEMICAL ENGINEERING's Sept. 18, 1972, issue, pp. 104, 105. The processing sequence (see diagram) begins with purified ethylene and powdered catalyst, which are fed continuously into the reactor. Bed material (i.e., polyethylene) is disengaged—usually by gravity—in a widened overhead section of the reactor vessel and allowed to fall downward. The gas goes to external air or water coolers, and then is compressed and recycled to the reactor. Mean-

Originally published November 26, 1973

while, polyethylene powder is removed intermittently from the vessel at a rate that maintains the fluidized bed at an approximately constant level. The product can be marketed as is, or pelletized.

Originally, Carbide had also included cyclone separators in the process to remove fines from the gas after it left the disengagement section of the reactor. But more-recent designs have omitted the cyclones and their attendant pressure-drop penalty, because engineers found that the recycle system could tolerate relatively large amounts of entrained polymer.

Conversion per pass through the reactor is about 2 to 3%. Overall conversion (weight of polymer produced per unit weight of monomer fed to the reactor) can be as high as 99% if unreacted monomer is recycled from the product discharge tank. Average residence time for particles in the bed is about 3 to 5 hr., during which time they grow to an average size of 500 microns dia.

Reaction pressure is normally about 300 psig. Density of the product is controlled by adding alpha-olefin comonomers, such as propylene or 1-butene, to the ethylene. To obtain a given average molecular weight, the operators adjust the reaction temperature (around 80 to 105 C.) and/or the concentration of a chain-transfer agent, which terminates the growth of one molecule and initiates that of another. Molecular-weight distribution is controlled by the specific grade of catalyst and by adjusting reaction conditions.

Carbide spent some ten years in developing this technology. The process successively graduated into bigger and bigger reactors: first, 4 in. dia.; then 24 in.; and finally 8 ft. for the prototype at Seadrift. In a sense, scaleup has not ended–Carbide's second Seadrift unit will employ a 14.5-ft. vessel.

Catalysis Is the Key

Of all the process considerations, catalysis probably got the most attention during the ten-year gestation period.

To make the process economically attractive, it was necessary to avoid a separate step for removing catalyst from the product. This meant the polymer could carry away only minute traces of catalyst. To meet this requirement, the amount of catalyst in the reactor system had to be small, which dictated the need for a catalyst with very high activity.

Carbide's engineers soon found that the main cause of low activity with the then-available catalyst grades was the presence of impurities–those in the catalyst itself and those otherwise introduced into the polymerization system. The answer was a two-pronged effort aimed at cleaning up catalyst raw materials and removing impurities from the incoming ethylene.

The process uses a family of catalysts containing chromium. Carbide makes them by depositing chromium compounds and optional modifying agents on dehydrated silica or other supports. Activity of these materials is so great (i.e., the amount of catalyst present in the reactor system is so minimal) that the polyethylene product generally contains less than 1 ppm. of metallic chromium, a "completely innocuous concentration."

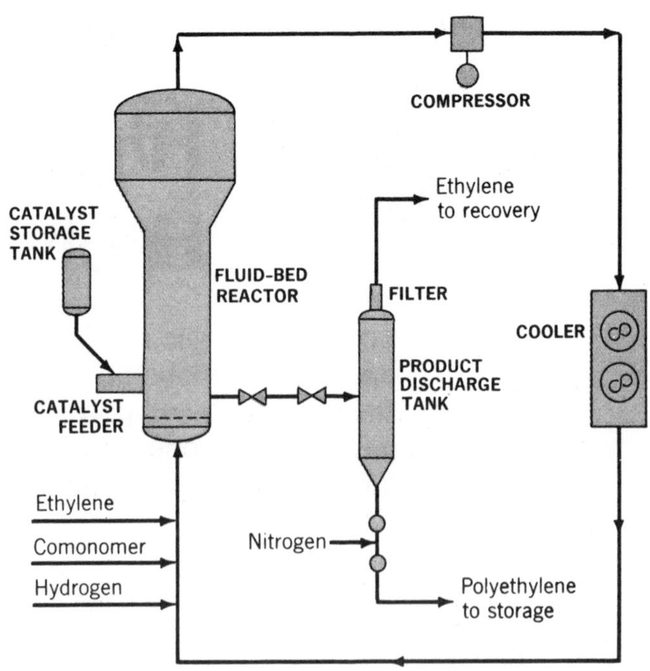

SIMPLICITY is the hallmark of Union Carbide's gas-phase process for high-density polyethylene.

A related problem surmounted by the company's engineers entailed catalyst feeding: how to introduce solid catalyst into a reactor under 300 psig. pressure, and how to assure rapid and uniform distribution throughout the vessel. Carbide came up with feeders that operate successfully over a wide range of feedrates.

Designing for Good Distribution

Another achievement was designing an effective gas-distributor plate for the bottom of the reactor. The stream of incoming ethylene and comonomers enters the reactor through this plate, which is located under the fluidized bed.

The plate has to withstand fouling by the reactive fines that are entrained in the recycled ethylene. Perhaps more important, however, is the need to be sure of uniform gas distribution over the cross-section of the bed. This is vital for efficient heat transfer. The polymerization reaction is exothermic, and the normal operating temperature is only about 25 to 30° C. below the softening point of the polyethylene that constitutes the bed. Without properly distributed gas flow, local hot spots would form in the reactor, creating chunks of polymer.

Carbide's distributor-plate design forestalls the problem of hot spots. Uniform gas flow and active fluidization assure good radial and axial mixing of the solids within the bed, enabling it to behave nearly isothermally. Usually the upper 90 to 95% of the bed is at constant temperature, with virtually all of the temperature gradient showing up immediately above the plate. ■

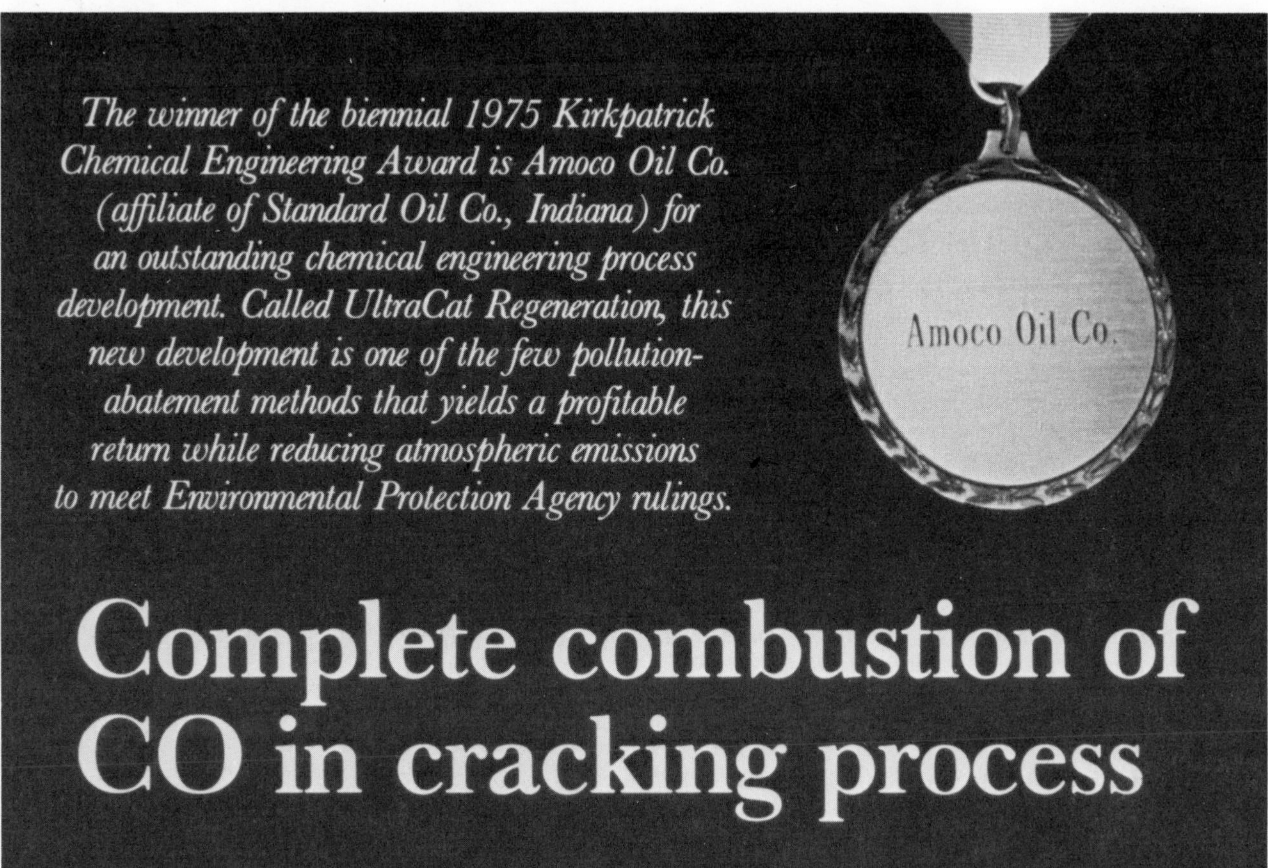

The winner of the biennial 1975 Kirkpatrick Chemical Engineering Award is Amoco Oil Co. (affiliate of Standard Oil Co., Indiana) for an outstanding chemical engineering process development. Called UltraCat Regeneration, this new development is one of the few pollution-abatement methods that yields a profitable return while reducing atmospheric emissions to meet Environmental Protection Agency rulings.
Amoco Oil Co.
Complete combustion of CO in cracking process

☐ Worldwide research efforts in some 200 units having the fluid-catalyst-cracking (FCC) process probably reflect the greatest chemical engineering activity ever applied to a petroleum-refining process.

Historically, the process has used excessive energy. Of the energy available from the combustion of coke deposited in the catalyst cracking step, about one-third (equivalent to 2% of feedstock) is lost as CO in the regenerator flue gas, unless the unit is equipped with an expensive downstream CO boiler.

Another deficiency has been the incomplete removal of carbon from the catalyst during the catalyst regeneration step, which results in a subsequent loss of catalyst activity in the reactor. With the development of *UltraCat Regeneration** by Amoco Oil Co. (affiliate of Standard Oil Co., Indiana), these deficiencies have been overcome, with resulting large gains in energy conservation, improved liquid-product yields, reduced CO emission to the atmosphere, and improved unit reliability.

What UltraCat regeneration does

With conventional regeneration, large quantities of CO are generated in the dense bed in the regenerator vessel. The CO-containing flue gas then passes through a dilute-phase zone and through internal cyclone separators before being discharged to the atmosphere. In the past, when sufficient oxygen was present to sustain substantial CO burning above the bed, the dilute-phase temperature would skyrocket and damage catalyst and equipment.

With UltraCat regeneration, complete combustion of CO in the regenerator vessel is obtained by means of special designs and control techniques, which evolved through the use of steady-state and transient-response simulations of the regenerator and reactor, and from studies conducted on commercial FCC units. This led to a definition of operating conditions and equipment details for controlling the CO combustion rate, and for transferring heat between catalyst and gas in both the dense and dilute phases, without exceeding permissible temperatures for catalyst or equipment. The low CO emission level also meets Environmental Protection Agency requirements.

On the reactor side, the proper kinetic environment was defined to obtain full benefits from the additional heat evolved, and from the very active catalyst produced in the regenerator. These studies revealed that complete CO combustion in the regenerator—when combined with proper reactor design—will produce higher yields of gasoline at the expense of less-valuable coke and gas. This makes possible further significant advances in FCC process technology, beyond those achieved by the introduction of riser cracking and molecular-sieve catalysts.

UltraCat regeneration is now fully proven and has

* U.S. Patent 3,909,392, issued Sept. 30, 1975. Counterpart patents also issued in a number of other countries.

Originally published November 24, 1975

been used commercially both by Amoco Oil Co. and by licensees of Standard Oil Co. (Indiana). The technology has been used routinely on seven FCC process units and has been tested on six other units; additional conversions are planned.

The fluid-catalytic-cracking process

The petroleum refining industry uses the fluid-catalytic-cracking process to convert the middle third of crude oil mostly to gasoline and other light products. In the U.S. alone, over 4 million bbl/d of this stock (called gas oil) is processed. The uncracked portions are used for fuel oils. Much coke is formed on the catalyst as byproduct.

In the fluid-cracking unit, the catalyst is circulated between the reactor and regenerator vessels at a weight rate of 4–10 times the feedstock-charge rate. This circulation of coked catalyst to the regenerator makes possible frequent removal of the contaminant coke from the catalyst surfaces, and serves to transfer the heat of combustion to the reactor to raise the charge temperature from the 300–750°F range to 900–1,050°; it also makes it possible to meet the endothermic-reaction heat requirements.

The spent catalyst is returned from the reactor to the regenerator in a slip-stream of air, while the main air supply enters through dispersion devices that afford maximum catalyst-gas contacting in the regenerator bed. The temperature of the bed is conventionally 1,100–1,250°F, but the dilute-phase zone between the bed and the catalyst-recovery cyclones may run somewhat hotter.

Since World War II, the petroleum industry has devoted enormous technical attention to reaction rates, yield optimization, and effects of variables on feedstock conversion. Regenerator technology has also received much attention, as have vessel geometry, optimum catalyst inventory, mechanical-mixing devices, coke burning rates and residual-coke levels. Despite these efforts, basic problems in regeneration existed until 1973 (when UltraCat regeneration was developed) that were impeding progress toward high clean-air standards and improved production.

Unburned CO—Large residual concentrations of CO in regenerator flue gases was a problem since the inception of the catalytic process. The chemical reactions of regeneration—aside from oxidation of hydrogen and sulfur in coke—are:

$$C + O_2 \rightarrow CO_2$$
$$2C + O_2 \rightarrow 2CO$$
$$2CO + O_2 \rightarrow 2CO_2$$

Surprisingly, under the wide variety of catalysts and operating conditions encountered over the years, the CO in the flue-gas Orsat analysis has averaged about 7–10%, representing a ratio of approximately 1.3–0.8 CO_2/CO. Typically, this ratio was about 1.0 during the past decade. Such variables as higher bed levels of catalyst, higher temperatures, and more free oxygen, promoted deeper catalyst regeneration but had little effect on the ratio of CO_2/CO leaving the bed.

CO afterburning—This refers to combustion of CO in the dilute-catalyst phase, in the cyclones or in the flue-

Cat crackers at Amoco's Whiting, Ind., plant where UltraCat regeneration was discovered

gas line. In these zones, the highly exothermic combustion of CO causes large temperature responses. Because of this, burning any appreciable amount of CO has been avoided. Moreover, high-alloy metallurgy, special ceramic liners, and carefully designed automatic cut-in cooling sprays of water or steam, have been used to prevent damage by CO afterburning. Nevertheless, afterburning has caused millions of dollars of damage in plant equipment, with regenerator cyclones having been especially vulnerable.

In an effort to prevent the loss of the atmosphere-vented CO fuel value, the petroleum industry had started in the mid-1950s to install special external recovery equipment. This consisted of long flue-gas ducts leading from the elevated vessel structure to furnaces located at grade, where the CO was burned in the presence of large supplementary fuel supplies to ensure against flameout. The hot gases were passed through boilers to recover the heat.

Unfortunately, such equipment (which are called CO boilers) is expensive, which made uneconomical their installation except at large units. Although the energy crisis has made CO boilers more economically attractive, the majority of fluid-cracking installations are still not equipped with them, and CO continues to be vented to the atmosphere.

Incomplete regeneration—The regeneration of coke from spent catalyst was also incomplete. Between 1945 and 1965, the coke content on regenerated catalyst was

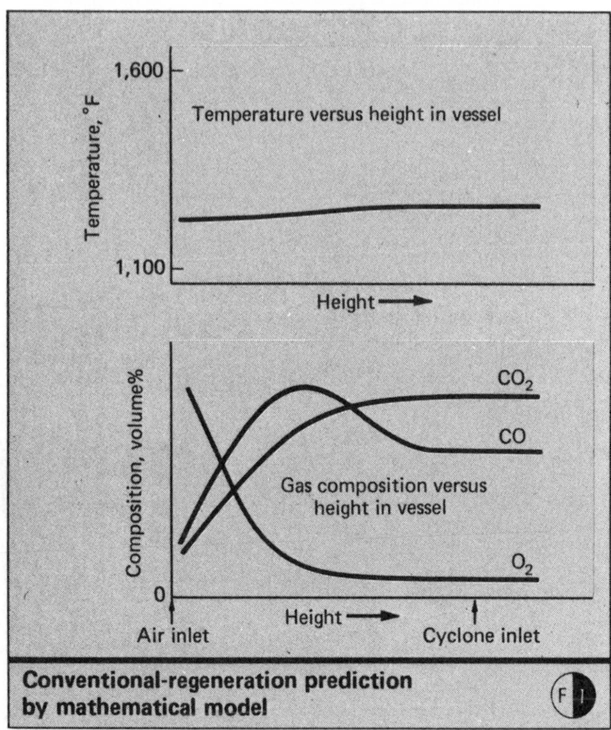

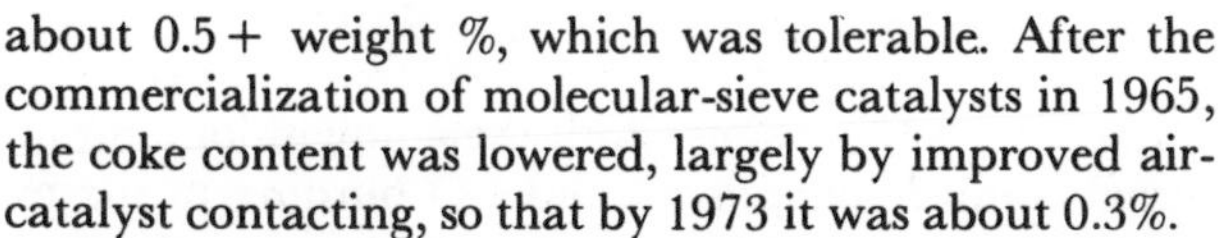
Conventional-regeneration prediction by mathematical model F 1

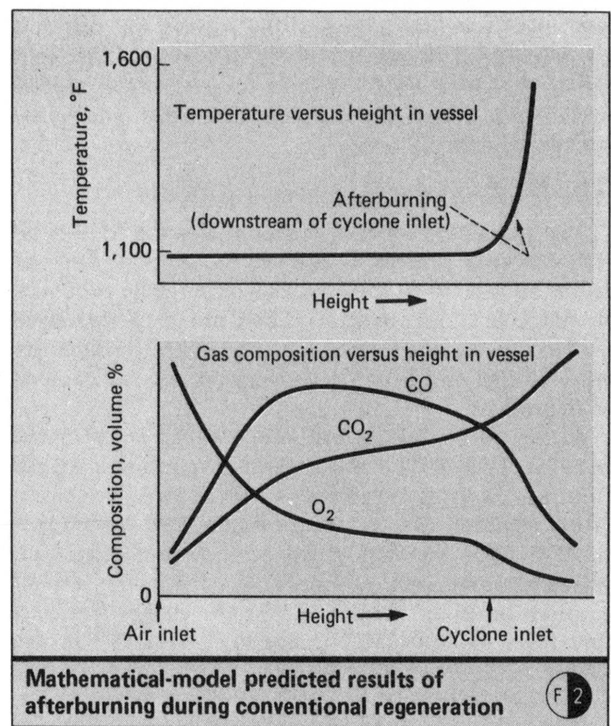

Mathematical-model predicted results of afterburning during conventional regeneration F 2

about 0.5+ weight %, which was tolerable. After the commercialization of molecular-sieve catalysts in 1965, the coke content was lowered, largely by improved air-catalyst contacting, so that by 1973 it was about 0.3%.

This level, however, is now considered unacceptable. Molecular-sieve catalysts require regeneration to below 0.1% coke content to realize full benefits; but the danger of afterburning prevents the use of more-severe regeneration conditions and the use of more air to raise the oxygen content. While the high extra capital and operating costs could be borne, the risk of overheating the regenerator and forcing the unit offstream was deemed too great.

Inadequate heat supply—In the early years of fluid cracking, heat availability in the regenerator exceeded the needs of the reactor, because of high coke yields. Refiners therefore installed waste-heat boilers, which removed the extra heat from the catalyst bed by the recycle of a stream of hot catalyst.

Later the situation was reversed by a trend that moved away from mixed-bed toward transfer-line cracking. This caused less catalyst residence time in the oil, with resulting lower contents of coke on the catalyst. Thus, acceptable regenerator temperatures were obtained without coolers.

When molecular-sieve catalysts were introduced, the trend to transfer-line cracking with little or no bed cracking was accelerated. At this point, excessively low regenerator temperatures frequently arose. Some refiners combated this by installing preheat furnaces and supplied them with expensive fuels.

Emissions of CO—By the early 1970s, the presence of large amounts of CO in the flue gases entering the atmosphere was considered seriously objectionable. Various states promulgated regulations requiring that current practices be amended. The industry had no way to do this, except by installing expensive external equipment trains for combustion and heat recovery.

Innovations in UltraCat regeneration

Each of the above problems was solved by the development of the UltraCat regeneration process, which burns the CO in the regenerator completely. UltraCat also recovers the extra energy within the process, and involves full regeneration of the spent catalyst. The regeneration air rates have been held constant for constant-feed conversion. The following innovations were accomplished:

Heat flow—Amoco's initial use of complete CO burning permitted much of the combustion to occur in the dilute-phase zone. Since the burning there occurs readily, the problem was one of heat-flow control. Extreme temperatures and fast temperature transients were unacceptable. Therefore, lifting of extra catalyst from the dense phase to the dilute phase by use of a steam-driven catalyst fountain was considered a basic step. Such catalyst fountains were designed and installed in several units. After the catalyst is lifted and picks up heat, it returns directly to the bed, or is collected in the cyclones and then returned. The heat supplied to the bed is transmitted to and used in the reactor.

It was recognized that, during startup or during transitions caused by equipment failures, the heat-flow control by the catalyst fountain and other systems would be imperfect. A backup system was needed for rapid use by the operator or automatic controllers. Existing cooling systems were hazardous if used rapidly, or if used on the heat-removal scale contemplated, because they could cause direct impingement on the metal, thereby inducing temperature shock that would

UltraCat regeneration in Salt Lake City, Utah, combined with latest technology in riser cracking

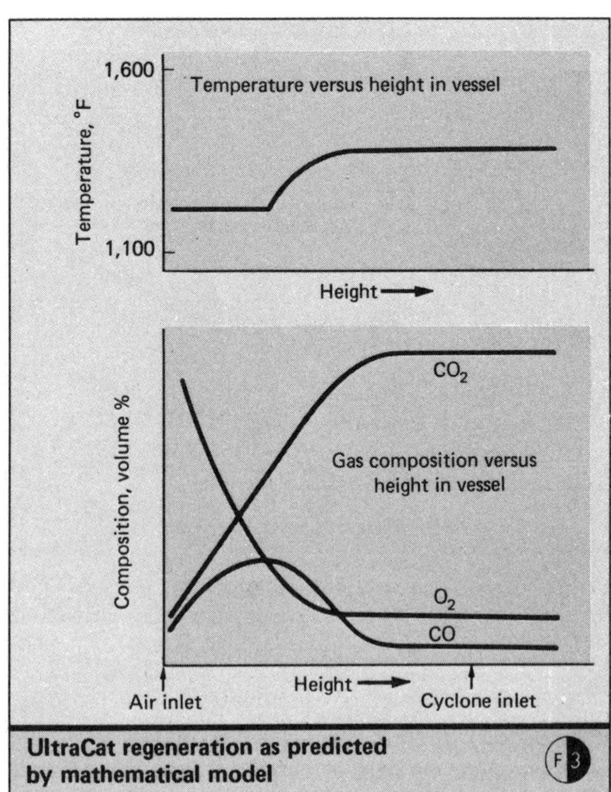

UltraCat regeneration as predicted by mathematical model

open up critical welds, as well as causing other damage.

Since the solution was to develop better cooling technology, thorough distribution of temperature-sensing elements was provided throughout the regenerator, with special controls associated with some of the elements. New water and water/steam cooling systems were installed, which allowed all needed cooling to be applied quickly without a single instance of runaway temperature or thermal shock.

Autoignition of CO—When oxygen is gradually raised, sudden ignition and rapid burning occur at some point. Moreover, when the temperature rises, the large fuel inventories represented by the residual coke on the catalyst can burn rapidly and create further, fast temperature rises. Even the new emergency cooling facilities were not expected to cope with these effects.

Such problems were met by a series of new proprietary procedures, which permit the normal operation to be shifted to fully burn the CO—including much burning in the dilute phase—without temperature shock. The temperature and heat-flow transients are adjusted to the desired extent with minimal use of water or steam coolants. Both the autoignition and internal combustion provide full CO burning in the regenerator.

Coke reduction—Combustion of CO would normally require large increases in air-rate flow. By taking advantage of the extra heat released, and returning it to the reactor, the coke yield was reduced. And by using the riser-reactor environment under proper conditions, the net yield was improved. Because of this, no increase in air-rate flow to the regenerator was needed to burn all the CO, and the requirement for additional air-blower capacity was averted.

Mathematical models—Development of the UltraCat regeneration process had to be made without the aid of pilot plants, because the existing ones could not be suitably adapted, and new ones would have required long delays for design and construction. However, strong predictive capabilities were needed. After initial plant trials, the available mathematical models were greatly expanded to meet the needs.

To improve the understanding of the dynamic behavior and control of regeneration, and to investigate the conditions needed for the burning of CO, a large real-time kinetic model of the regenerator was developed. This operates on a hybrid analog-digital computer, which solves the necessary partial differential equations, and simulates the performance of the regenerator in regard to both time and position in the vessel. The model predicts the behavior of carbon on the regenerated catalyst, catalyst circulation rate, and the temperature profiles with regard to height in the vessel, as well as gas-phase composition. The model can be used to study the effects of air rate on CO burning,

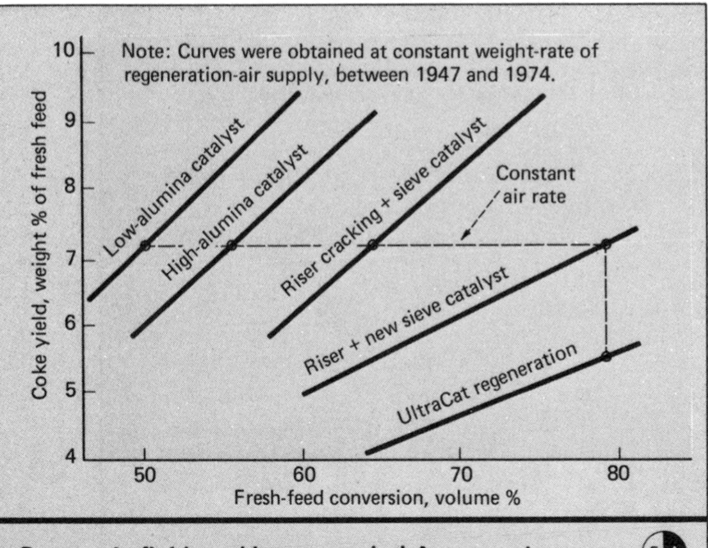

Progress in fluid cracking at a typical Amoco unit F/4

coke on spent catalyst, coolant streams, reactor temperature, feed preheat-temperature, feed rate, catalyst entrainment, regenerator geometry, and overall regenerator performance.

The training of operating supervisors was facilitated by the use of cathode-ray-tube traces of variable responses, as new operating conditions were dialed to the input. This hybrid-computer model also provided for exploration of most boundaries between acceptable and unacceptable operating inputs. Such work was accompanied by major mechanical engineering studies of allowable stresses in regenerator equipment.

F/1 through F/3 show some hybrid-model outputs that depict various stages of regenerator operation. The first two, taken from the cathode tube with a hard-copy unit, show the instantaneous variation during conventional regeneration and afterburning. The last pair shows the transition to, and full use of, UltraCat regeneration; the first curve of this last pair is a computer strip-chart print, whereas the second is another cathode-tube copy. A digital computer model was used to cover practically all remaining responses and to achieve maximum accuracy.

Spent-catalyst injection and air distribution—The spent-catalyst injectors, air distributors and other regenerator internals must be able to provide for regeneration of catalyst particles, if air-catalyst mixing is not thorough. Even the catalyst particles entrained in the dilute-phase zone must be well regenerated. The intense conditions of the process can cause particles that enter this zone with excessive coke contents to undergo ignition and rapid burning. This causes the particles to exceed ambient temperatures and to deactivate. Plant experiments and computer runs have indicated the need to avoid this. Therefore, new, superior spent-catalyst distributors have been installed. Catalyst activity is thus fully maintained.

Modified heat-flow control—As studies proceeded, it was found that, in some cases, the catalyst fountain for recovering heat in the dilute-phase zone was not necessary with a modified design of the spent-catalyst distributor, as well as with other regenerator improvements. Some of the CO burning that had earlier occurred in the upper zone was transferred into the dense bed, where the temperature rise is smaller and acceptable.

Avoidance of flameout—The transition to conventional regenerator operation, in preparation for unit shutdown, is easily carried out—although an abrupt flameout of the CO is unacceptable. The exact conditions at which a flameout occurs are not reliably predicted by theory-based models. Therefore, correlations were developed to predict conditions for stable combustion.

It was also found that the existing systems for sampling and analyzing the flue gases were not sufficiently reliable because of the particulates and condensate that were present. Therefore, new highly sensitive and reliable systems were designed.

Results of new technology

Reduced emissions—The concentration of CO in flue-gas emissions to the atmosphere runs 7% or more at units not having CO boilers. Since most facilities are not so equipped, the total CO emission is substantial. This can be eliminated almost completely by applying UltraCat regeneration. Concentrations of 0.1% (1,000 ppm) are easily achieved with the use of 1% oxygen in the flue gas. During tests to determine ultimate capabilities, concentrations down to 8 ppm were achieved.

By 1973, a number of states had issued regulations requiring fluid-catalyst units to use direct-flame afterburners for CO cleanup. But because Amoco has informed EPA and many state agencies of the UltraCat system, federal regulations for new units will permit CO combustion within the regenerator. CO emission will be limited to 500 ppm, which the process can meet. Texas has approved the process and adopted an emission limit of 1,500 ppm for existing units. Illinois has also approved the process for existing units.

Improved operating stability—The conventional regeneration of molecular-sieve catalysts involves a balance between afterburn and incomplete regeneration, in which regenerator conditions are relatively unstable and can shift into afterburn of CO, with consequent damage to equipment. The danger of afterburning is eliminated by full CO combustion by means of UltraCat regeneration. Operating conditions can be substantially shifted without encountering problems, and with greater ease of control.

Improved product yields—F/4 shows coke-yield reductions by the new process and compares them with advances of past years in fluid-catalytic cracking. The tables provide further detail on the operating and yield changes between conventional, modern operations on molecular-sieve catalysts and the ones observed on UltraCat regeneration.

With Amoco's new process, coke yields decline about 30%, representing almost 2 weight % of fresh gas oil feed. The reactor temperature and the required heat supply to the reactor remain essentially unchanged. However, full CO combustion increases the heat of

Advantages obtained by means of UltraCat regeneration

Operating conditions and results	Units in which UltraCat regeneration was installed					
	A		B		C	
	Before	After	Before	After	Before	After
Operating conditions						
Fresh feed, thousand bbl/d	15.0	16.1	33.3	30.1	49.3	49.3
Weight % of carbon on regenerated catalyst	0.17	0.03	0.30	0.05	0.29	0.03
Reactor temperature, °F	920	948	974	965	965	965
Feed preheat temperature, °F	605	608	526	503	687	687
Throughput ratio	1.22	1.13	1.14	1.09	1.36	1.16
Catalyst circulation, % decline	–	25.6	–	37.5	–	39.0
CO boiler	Yes		No		No	
Results						
Conversion, volume % of fresh feed	75.0	75.0	79.7	79.5	63.8	63.8
Coke, weight % of fresh feed	6.8	5.0	7.3	5.8	6.2	4.4
Flue gas, dry-volume %						
Oxygen	1.9	4.1	0.2	0.7	0.2	1.8
Carbon dioxide	9.9	14.2	10.3	17.3	10.3	16.3
Carbon monoxide	9.3*	0.0*	11.8	0.1	10.2	0.0
Product octane number						
Research, clear	88.9	92.0	93.5	93.2	92.2	92.5
Research, + 3 cc tetraethyl lead†	96.9	98.0	99.6	96.6	96.6	96.7
Motor, clear	79.5	80.2	81.5	81.2	78.1	78.0
Motor, + 3 cc tetraethyl lead†	86.4	85.7	87.5	88.0	82.6	82.7

*Before the CO boiler.
†Standard dosage for fuel testing.

combustion of the coke from about 12,000 to 17,000 Btu/lb, which allows the coke-yield reduction while maintaining about constant heat release. Only about 10% of the heat increment from the added CO combustion is required for the higher flue-gas temperature. The heat loss to coolants is negligible, although the magnitude of the additional heat release is large; at a typical Amoco unit, it exceeds 100 million Btu/h.

UltraCat regeneration yields (constant conversion)

Product	Unit yield change, volume %		
	A*	B*	C*
Ethane and lighter products, weight %	+0.5	−0.3	0.0
Propane	−0.2	−0.3	+0.2
Propene	+1.0	−0.1	−0.8
Iso + *n*-butane	−1.2	+0.2	−1.6
Butylene	+1.5	+0.4	0.0
Gasoline	+0.7	+1.6	+4.7
Cycle oil	0.0	+0.4	0.0
Coke, weight %	−1.8	−1.5	−1.8

*These units were built between 1944 and 1959, with widely different design features. One design is now pure transfer-line cracking; the others are predominantly transfer-line cracking. The units represent a spectrum of modern cracking operations, with many different types of charge stock.

By means of optimized control of reactor operating variables—combined with the proper reactor geometry, including transfer-line cracking—the reduced coke yield means gains in the most valuable products. Such reduction of coke on the catalyst—from a typical value of about 0.3% to about 0.05%—constitutes full regeneration, which is important in maintaining the conversion constant. Without such regeneration, there would be a shift of gasoline to heavy fuels. As mentioned, the air-blower rate has been held constant despite full CO combustion. This avoids large investment and operating expenses for additional air-blower capacity.

Improved profitability—The profit improvement is about 10¢/bbl of gas oil feed. The gasoline volume gain is also considerable. If applied to all U.S. fluid cracking, it would amount to about 0.5% of U.S. gasoline consumption. While the benefits of the process are restricted where a CO boiler is already in use, or where the existing unit design is difficult to modify, the process nevertheless has wide applicability.

UltraCat regeneration has been applied regularly to seven units, including those of licensees; additional installations are planned. Investment costs are usually moderate and are only a small fraction of the cost of a CO boiler. When a sensible-heat boiler can be justified, the investment comparison becomes closer, but still favors the process by a substantial margin. Moreover, the CO boiler does not provide the yield advantages that UltraCat regeneration offers.

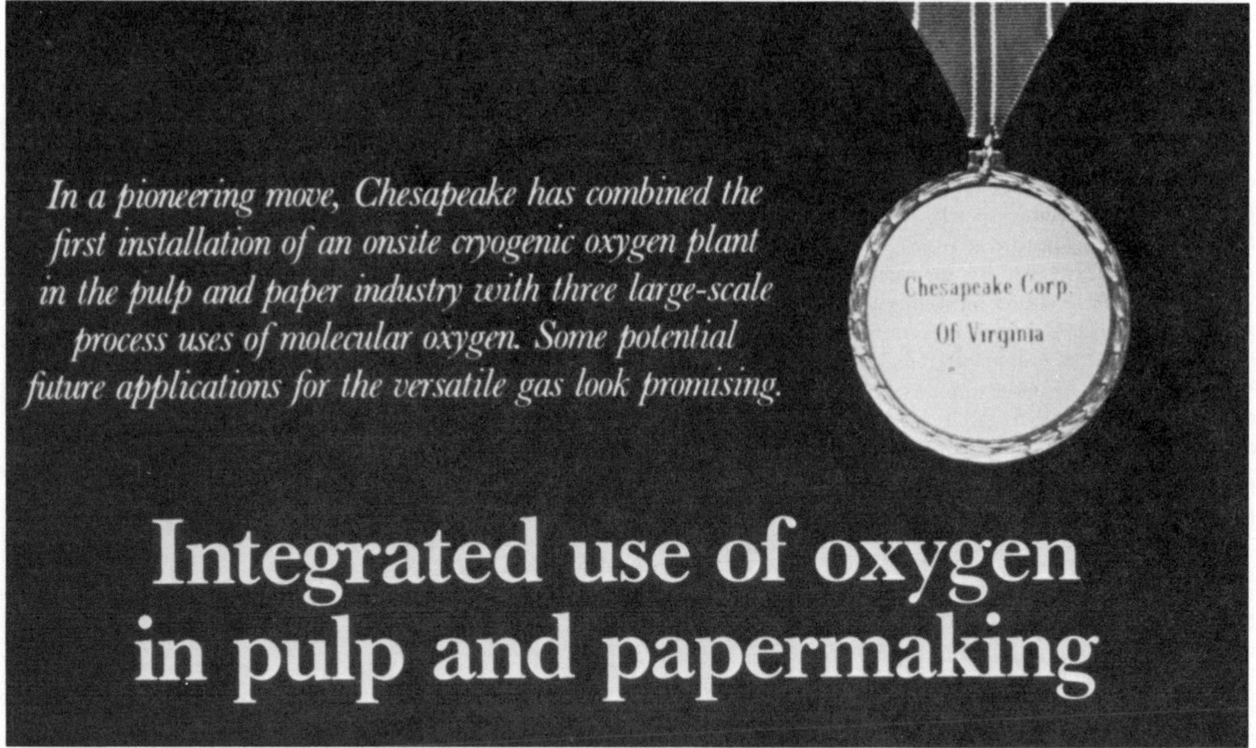

In a pioneering move, Chesapeake has combined the first installation of an onsite cryogenic oxygen plant in the pulp and paper industry with three large-scale process uses of molecular oxygen. Some potential future applications for the versatile gas look promising.

Integrated use of oxygen in pulp and papermaking

☐ The post–World War II breakthrough that linked onsite tonnage oxygen plants to oxygen processing of steel may have a counterpart in the pulp and paper industry. At West Point, Va., the Chesapeake Corp. of Virginia has pioneered the use of high-purity (99.5%) oxygen for:

- Bleaching high-quality, hardwood market pulp in the first commercial venture in the world to use the pulp bleaching sequence: chlorine-dioxide/chlorine, oxygen, chlorine dioxide.

Aerial view of bleach plant, including chemical makeup and pulp storage

- Treating wastewater in the first application of its type in the pulp and paper industry.
- Oxidizing black liquor in the first such system to use oxygen as the oxidant, pipeline-sparged into the weak black liquor.

By integrating these achievements with the onsite production of 100 tons/d of oxygen, Chesapeake demonstrated outstanding group effort in chemical engineering to win a 1975 Merit Award in Chemical Engineering Achievement. And these accomplishments have been wholly or partly emulated by at least five other companies.

Looking to the future, the company has already completed plans for using oxygen in its lime kiln to reduce odor emissions. Furthermore, it is in a position to explore and commercialize oxygen pulping and low-cost production of ozone for improved power factor, for bleaching, for bactericidal use and for tertiary wastewater treatment.

Prior to adopting oxygen for processing, Chesapeake Corp. of Virginia had reached the point where timber in its procurement area was 45% pine and 55% hardwood; the company found it necessary to develop a product that would use much more of the available hardwood. Chesapeake's choice of product and process faced such requirements as: compatibility with present manufacturing and marketing positions (domestic and foreign); a reasonable capital return; optimum use of forest resources; no increase in ground water withdrawal; and release of less than presently allowable air and water pollutants.

The product and process that proved best to satisfy most needs was hardwood market pulp bleached by a high-purity, molecular oxygen process. While such a product—based on liquid oxygen bought at \$30-\$50/ton—was competitive with pulp bleached by conventional systems, an opportunity was seen to apply oxygen to other usages if the gas were available at \$10/ton, as from an onsite separation plant. Virginia law requires adding secondary-waste-treatment facilities when a plant expands its production. And oxygen treatment of wastewater seemed the way to go, if \$10/ton gas were available to make feasible black-liquor oxidation with pure oxygen.

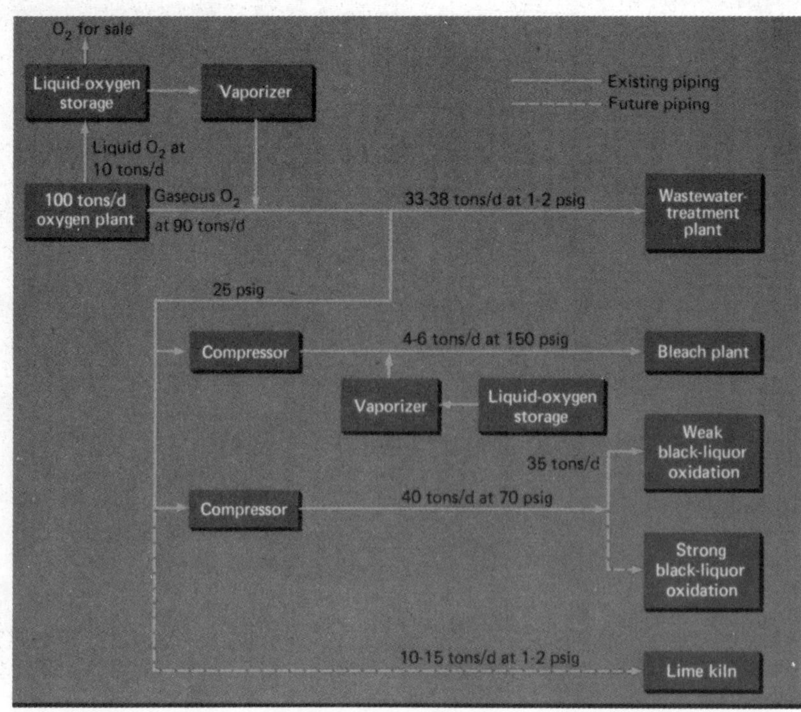

High-purity oxygen-distribution system for processes in pulp and paper mills

The oxygen bleaching process

When Chesapeake began to consider oxygen bleaching, it investigated a softwoods process used in South Africa. However, the five-stage sequence of acid pretreatment, oxygen, chlorine dioxide, caustic extraction and chlorine dioxide did not satisfy cost and quality criteria.

Subsequently, Chesapeake worked with Mo och Domsjö AB and Suncds AB in Sweden, Canadian Industries Ltd. (Montreal) and Chemtics Ltd. (Vancouver) to confirm development by those firms in bleaching hardwood with oxygen and to extend their results to Chesapeake pulp. Out of this effort came the decision to use a three-stage sequence (chlorine dioxide/chlorine, oxygen, chlorine dioxide) to produce 88+ brightness market pulp.

The industry was skeptical of Chesapeake's decision, since conventional sequences to achieve this quality require from five to seven stages. Furthermore, oxygen reacts equally with lignin and cellulose. However, by leading off with a chlorination stage, the Chesapeake sequence produces chlorinated lignins that the oxygen attacks preferentially in the second stage. In the first stage, unbleached pulp at 3–4% consistency (% dry fiber) mixes with an aqueous mixture of chlorine dioxide and chlorine. The chlorine dioxide moderates and inhibits the normally drastic reactions of chlorine with the pulp at the high (125°F) temperature in that stage.

Within 10 min, the pulp passes through two series-connected towers and is washed and dewatered by pressing to a 32 to 35% consistency. A 7% solution of sodium hydroxide is mixed with the pulp before feeding it into the pressurized oxygen reactor, which is 10 ft dia. by 40 ft high.

A device at the top of the reactor fluffs the pulp to enhance oxygen transfer. A rotating rake moves pulp from the bed in the upper section of the reactor to the lower section for dilution to 4% consistency and discharge. Oxygen enters the reactor just above the rake at a rate that will maintain a 90-psi operating pressure. Retention time is 30 to 45 min.

Following another washing step, the stock at about 15% consistency undergoes a 4–5-h retention in the final chlorine dioxide stage of the bleaching sequence.

Improved operations and product

Chesapeake's oxygen bleach plant has produced a daily average of 275 tons of pulp and has operated at or above its 330-ton/d maximum design capacity, occasionally reaching 400 tons/d. Both capital and operating costs are less than for conventional plants. With fewer stages, energy is conserved.

High water-recycle and countercurrent washing greatly reduce fresh water demand. And effluent BOD_5 loading and color decrease by 30% and 65%, respectively. Pulp quality is high, with less than two points of brightness reversion compared with a four- to five-point figure for conventionally bleached fiber. Oxygen also breaks up small hard clumps of fiber (shives) so that they are undetectable by the end of the bleaching sequence. The pulp's strength is comparable to others. But beating, whereby fiber surface area is increased to improve hydrogen bonding in a sheet of paper, is easier to carry out with the oxygen-bleached pulp.

Chesapeake's wastewater treatment facility, based on Union Carbide's Unox oxygen process, was designed following an eight-month cooperative pilot-plant study by Chesapeake, Union Carbide and Hydroscience, Inc., a wastewater consulting group. Sized for 16.3 million gal/d of wastewater and operating at 12.5 million gal/d, it is reducing BOD_5 from the 350–400 mg/l level to 20–30 mg/l (total), with an oxygen utilization efficiency of 90%.

The black-liquor oxidation system, based on bench and pilot-scale studies by the company's research department, is reducing the level of sodium sulfide from 7.5 g/l to 0.1 g/l at a 1,950-gpm flowrate. Oxygen utilization efficiency runs in the 75–85% range. Higher efficiency will be sought.

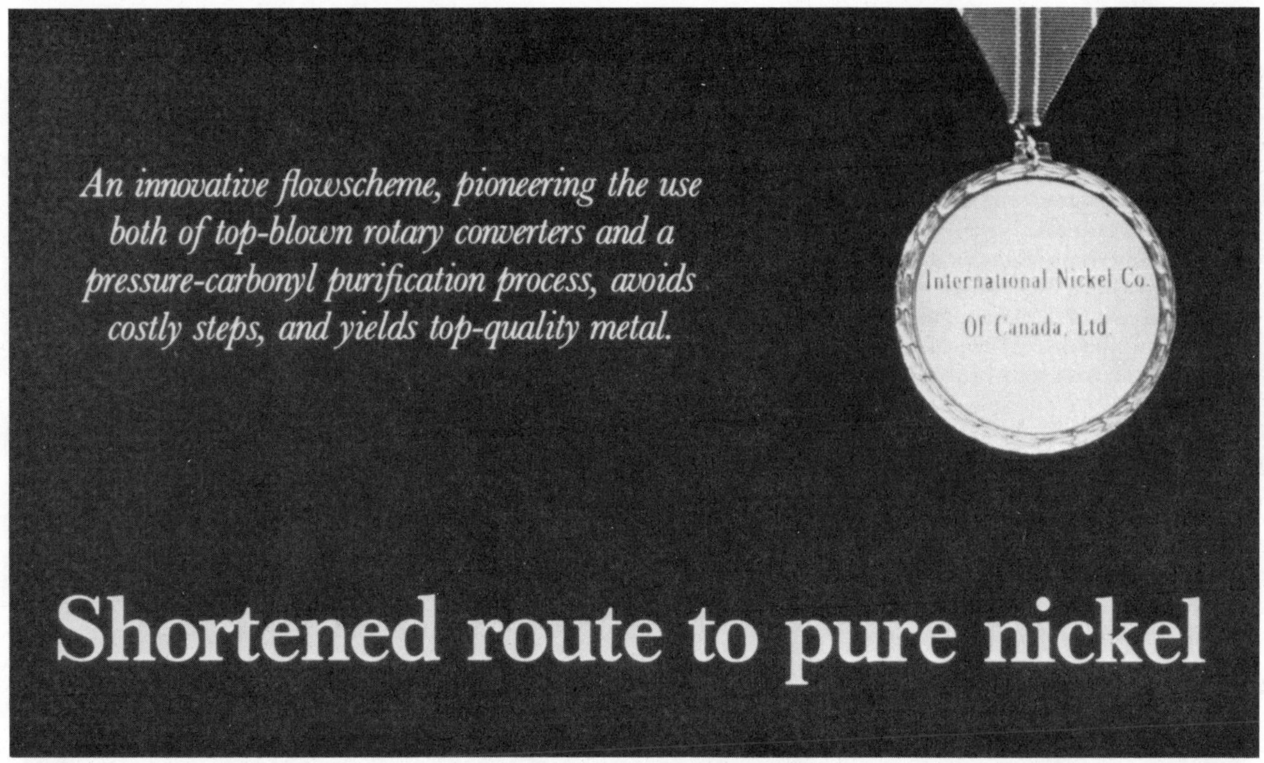

An innovative flowscheme, pioneering the use both of top-blown rotary converters and a pressure-carbonyl purification process, avoids costly steps, and yields top-quality metal.
International Nickel Co.
Of Canada, Ltd
Shortened route to pure nickel

☐ Producing high-purity nickel from a wide variety of nickel-sulfide materials hallmarks the merit-award-winning technology used by International Nickel Co. of Canada, Ltd. (Toronto) at its 125-million-lb-of-nickel/yr Sudbury, Ont., refinery. Culmination of over a decade of development when it started up in 1973, the highly automated Sudbury plant features the first use of top-blown rotary converters (TBRCs) for nickel processing, and also the initial application of the Inco Pressure Carbonyl (IPC) refining process.

Using TBRCs, nickel sulfides that range from crude ores to residues can be transformed into nickel-rich alloy. This avoids the intermediate, oxide-formation step of conventional schemes, and converts the sulfide inputs into the finely divided metallic-alloy feed required by the IPC purification process. It, in turn, refines the alloy without the need for the usual copper-nickel separation stage, to yield nickel containing a level of metallic impurities that is generally less than that detectable within the sensitivity limits of analytical methods.

Avoiding oxide—For three quarters of a century, nickel researchers strived to directly convert sulfides to metal, as is done in copper processing. However, the molten nickel sulfide invariably oxidized to solid, refractory nickel oxide, which plugged tuyeres (air nozzles) and produced hard-to-handle dry slag.

The chemistry for making the metal via a sulfide-oxide reaction had been theorized:

$$Ni_3S_2 + 4NiO \rightarrow 7Ni + 2SO_2$$

but it was not until 1959 that International Nickel Co. of Canada demonstrated a method for promoting the reaction while controlling solid oxide formation. Its technique centered on surface blowing with oxygen of molten nickel sulfide contained in a rotating vessel. The surface blowing avoided tuyere plugging and provided the high (2,900°F) temperature for the reaction, while the rotation enhanced contact between the reactants and spurred an even heating of the furnace.

After refinement of the system, two 50-ton units were built. Each of these TBRCs weighs about 200 tons, and can spin at up to 40 rpm around a tilted axis. A retractable lance provides oxygen and natural gas for melting the charge, or high-pressure oxygen for metal production. The converters have wide flexibility in controlling temperature, mixing, and furnace atmosphere.

Processing sulfides—A crude nickel-sulfide feed, containing about 15% sulfur, is charged to the rotating furnace, and melted to about 2,500°F by the combustion of oxygen and natural gas. Oxygen blowing desulfurizes the melt to about 4% by sulfur dioxide generation—and raises the temperature to more than 2,900°F, limiting the formation of solid nickel oxide. Petroleum coke reduces any excess dissolved oxygen.

Offgas from the furnace undergoes evaporative cooling via sonic nozzles. Next, it passes to a precipitator, from which high-value fines are recovered. Meanwhile, alloy melt is transferred by bottom-teaming ladling to a granulator.

Granulation in high-pressure water jets effectively quenches the material. Also, the cooling preserves the melt's uniform sulfur dispersion—which is necessary for the catalysis of the subsequent carbonylation reaction—and granulation provides a high-surface-area solid, which promotes gas-liquid contact.

Novel refining route—The alloy granules, now typically 65–75% metallic nickel, are purified by treating them

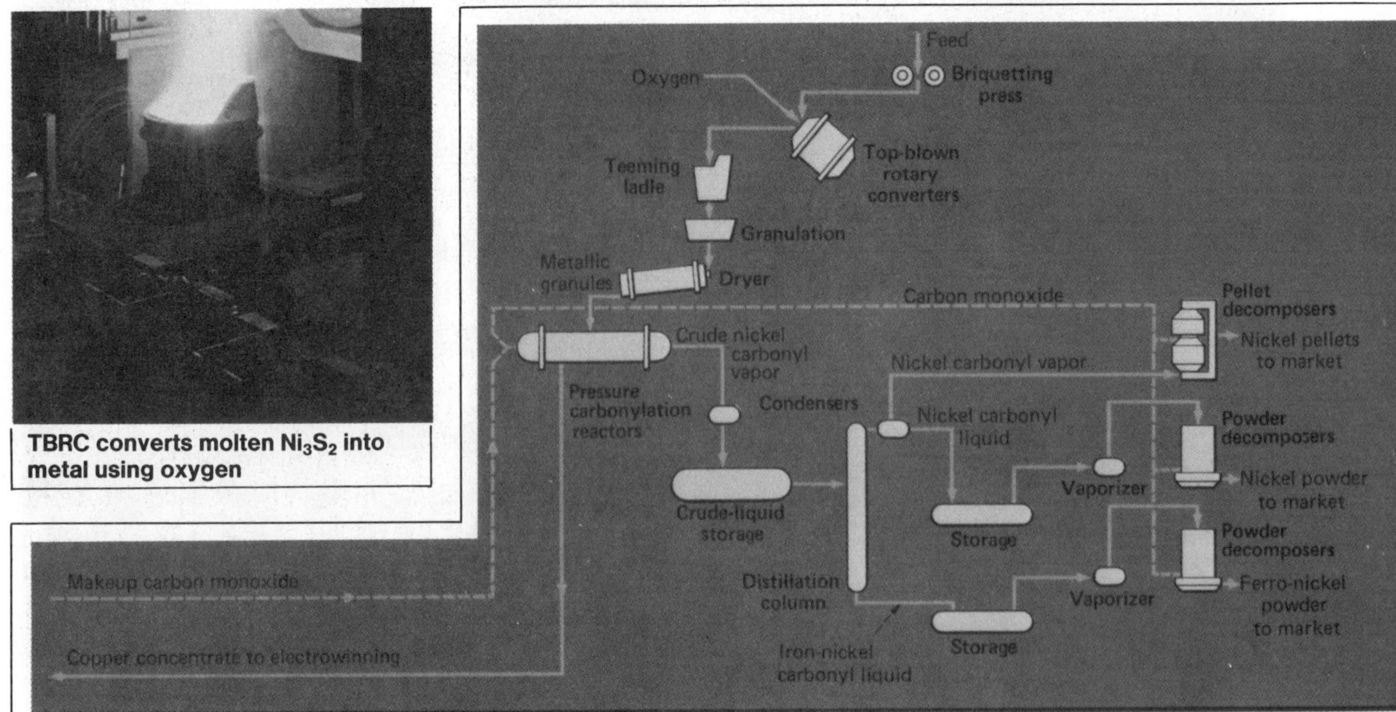

TBRC converts molten Ni_3S_2 into metal using oxygen

Manufacture of nickel powder and pellets, as well as of ferro-nickel powder by means of top-blown rotary converters

with carbon monoxide, so that straightforward but highly specific carbonyl reactions take place:

$$Ni + 4CO \rightarrow Ni(CO)_4$$
$$Fe + 5CO \rightarrow Fe(CO)_5$$

The carbonyls are then broken down to yield metal, obviating conventional, expensive techniques for removing copper from the nickel.

Increasing the pressure favors carbonyl formation and stability, and high-pressure operation boosts the process' tolerance of copper. Reactions occur at 1,000 psig and 350°F. Under such conditions, even when the copper-to-nickel ratio in the IPC feed runs 1-or-more to 3.5, extraction of nickel as carbonyl exceeds 96%.

Removal of iron as carbonyl can be controlled to between 20–50% of the input. Iron concentration in the alloy rarely goes beyond 4%; the level of iron carbonyl at most reaches about 2%. Cobalt in the feed can form a carbonyl, too, but the compound is unstable under the conditions at which the process is run. So, output is negligible, as is production of other carbonyls.

Massive vessels—At Sudbury, three reactors comprise the heart of the IPC circuit. Each weighs over 250 tons, and can take alloy-granule batches of up to 150 tons. The vessels rotate to enhance gas-solid contacting, while internal water-cooling coils remove over 16 million Btu/h of reaction heat from each unit. Filters within the vessels prevent dust contamination. And the reactors feature complex, rotary gas- and water-seals, which were the outgrowth of extensive pilot-plant work. Three five-stage compressors deliver the 1,000-psi carbon monoxide to the units, while two, 2,000-ton refrigeration units provide glycol for condensing the produced carbonyls, and other cooling duties.

The condensed product stream, containing mutually soluble nickel- and iron carbonyl, is piped to an accumulator. Then it goes into a controlled fractionation system, which required the development of activity coefficients and other data on the binary system.

Overhead vapor from the distillation column is split to make two different nickel products. A portion is condensed and then revaporized before thermal decomposition into nickel powder; the rest is thermally decomposed directly to yield nickel pellets. Both products consistently contain less than the 20-ppm iron-content limit specified for the overhead vapor from the fractionator. This purity is particularly noteworthy in the case of pellets, considering that the bulk of their contamination actually stems from abrasion of the carbon-steel equipment in which they are handled.

The mixed carbonyl bottoms, about 70:30 iron:nickel, are vaporized and thermally broken down to yield a marketable ferro-nickel powder. Carbon monoxide formed in the three decomposition operations is recycled.

Safety precautions—Toxicity of the carbonyl compounds mandates close control over process operations. A total of 52 samplers spread over all working areas in the plant continuously monitor the atmosphere in parts per billion for carbonyls and parts per million for carbon monoxide. Moreover, all products are purged of such contaminants, as well as all process equipment prior to its maintenance and inspection. All purge gases are then incinerated, with the nickel- and iron-oxide products of carbonyl combustion being captured in a baghouse. The clean gas, consisting of carbon monoxide, nitrogen and water vapor, finally is released to the atmosphere.

By careful, controlled injection of fine zircon sand into a plasma zone created by uniquely designed electrodes, zirconium dioxide is formed in uniform round crystallites. The product is unlike any achieved with other processing techniques, and is opening up new avenues for applications of zirconium dioxide.

Arc-plasma dissociation of zircon

☐ Plasma-reaction engineering has always been an important topic among chemical engineers, but few have achieved anything close to a viable commercial application for it.

A bright exception is the technology developed by the interdisciplinary research team assembled at Ionarc Smelters Ltd (Bow, N.H.), which since the late 1960s has delved into such applications as direct chlorination of precious metal ores, production of pure metals from oxides, preparation of zirconium dioxide and columbium pentoxide, production of nickel from laterite ores, and a direct reduction route to aluminum.

Choosing targets

While these and other possibilities seemed feasible at first, most could not be undertaken by a company as small and resource-limited as Ionarc. Management decided that whatever the project, its initial scale would have to be limited to 500,000 to 1 million lb/yr of product having a price range of 50¢/lb and higher. The plasma furnaces would have to be relatively small—35–1,000 kW—to be in the realm of present technology and still satisfy a commercial market. Larger-scale projects, such as iron ore reduction and direct chlorination of minerals, could wait for later.

After a couple of false starts, the choice fell to zirconium dioxide. The material could be won from a readily available resource, zircon sand ($ZnO_2 \cdot SiO_4$), and the market appeared broad-based and loaded with untapped potential. Ionarc's product offered unique properties, particularly the uniformity of its round, 0.1–0.2-micron crystallites. Applications include: ceramic tile coloring, refractories, chemicals and abrasives. Strong inroads have already been made into the first of these.

Ionarc is now getting ready for a twofold scaleup (to 10 million lb/yr) for a plant scheduled for completion by early 1977. This facility will use parallel 350-kW reactors similar to the one now operating at the company's headquarters at Bow, N.H.

Problems to overcome

The process consists of a simple one-step dissociation of zircon into zirconium dioxide and silica, followed by separation in a caustic leach, and centrifugation.

The zircon feed is merely screened to a particle-size range of 20–200 microns prior to admittance to the plasma reactor. It must be metered carefully and steadily into the reactor to ensure maintenance of proper operating conditions in the plasma zone. Right away, the feed meets the 13,000°K plasma jet that is created by electrical arcing between carbon electrodes. Within a split second, it dissociates and falls through the plasma, then continues falling down through the reactor while it cools and reforms into separate zirconium dioxide and silica species.

Though simple in concept, some enormous engineering problems had to be solved to make the process practical. Among these:

- Ensuring efficient heating of particles through proper electrode design.
- Design of a continuous and uniform zircon-feed system.
- Prevention of silica buildup over injection ports, as well as of zirconium dioxide on cool reactor walls.
- Development of suitable methods for separating the product from byproduct silica.

New electrode design

To increase electrical efficiency over that of previous designs, Ionarc engineers developed a unique electrode

Originally published November 24, 1975

Plasma furnace showing door at the bottom

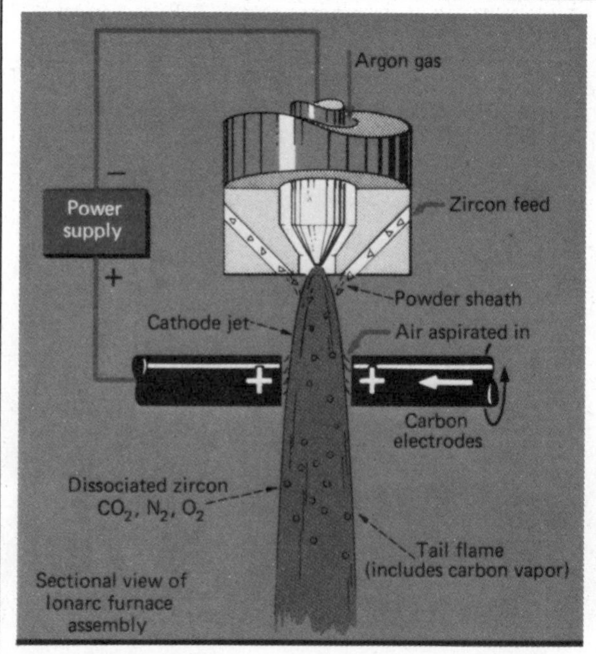

Sectional view of Ionarc furnace assembly

that allows reactants to fall through the hottest portion of the plasma, while at the same time ensuring that the particular electrode geometry is strictly maintained. These constraints enable maximum electrical efficiency and prevent the processing of off-specification product.

Ionarc came up with a design (see diagram) in which feed enters through ports surrounding a vertical, tipped cathode, and then falls past three pairs of symmetrical, horizontal anodes. During operation, the anodes are continuously fed inward to compensate for vaporization of the tips.

A chief advantage of this design over others is that the reaction stream passes directly through the current path between the cathode and the anode, thus maximizing electrical efficiency. About 30–50% of the energy in the Ionarc system ends up as sensible heat in the particles; all other previous devices netted only about 5–15% efficiency.

But heating efficiency does not depend on electrode design alone. Another major variable is the uniformity of zircon feedrate, which is necessary to prevent oscillations in plasma pressure and heat extracted from the arc. Fluctuations in pressure are especially troublesome because they cause feed powder to be periodically blown away from the plasma, thus magnifying the inefficiency.

This problem was overcome with a proprietary feeding system. Ionarc has stated that this feeder starts with a screw operating at approximately 150 rpm, followed by a vibrating baffle to smooth out variations caused by the screw itself. Then the feed passes through a labyrinth and onto a spinning disk that distributes all the feeder powder into the six feedports. Carrier gas is introduced at this particular point to transport powder through the ports.

Unwanted deposits

Even the most well-designed injection system would not have worked unless ways to avoid silica deposition on the nozzle could be found. The problem was attacked on four fronts: first, feedstock specifications were tightened to limit the amount of fines; second, the particles were heat-treated prior to feeding to eliminate fracturing as they entered the hot zone; third, injection angles, velocities and carrier-gas-to-powder flow ratio were optimized. These measures still only enabled 1½–3 h of continuous running. So, as a fourth measure, a cleaning technique was developed in which the furnace is shut down, cleaned, and put back onstream within a minute. (The reactor can be started up instantaneously, without warmup.)

Zirconium dioxide forms unwanted deposits over reactor walls. Since the material is an extremely good insulator, the wall's cooling system can be easily stymied. Eventually, the zirconia glazing gets so heavy that it falls under its own weight.

This problem was overcome by operating the furnace under a slightly negative pressure and developing an air-distributor ring that produces velocity patterns that keep product from building up.

Product recovery

Furnace product is transported in buckets to a leach reactor and mixed with 50% sodium hydroxide. The batch reaction takes about 1 h. Then, hot mother liquor goes to a centrifuge to extract silicate. The product is dried in a kiln. Centrifugation has been chosen over filtering because of the high viscosity of the silicate solution, which would require tremendous dilutions before succumbing to this type of treatment. As it now stands, the silicate byproduct is suitable for sale.

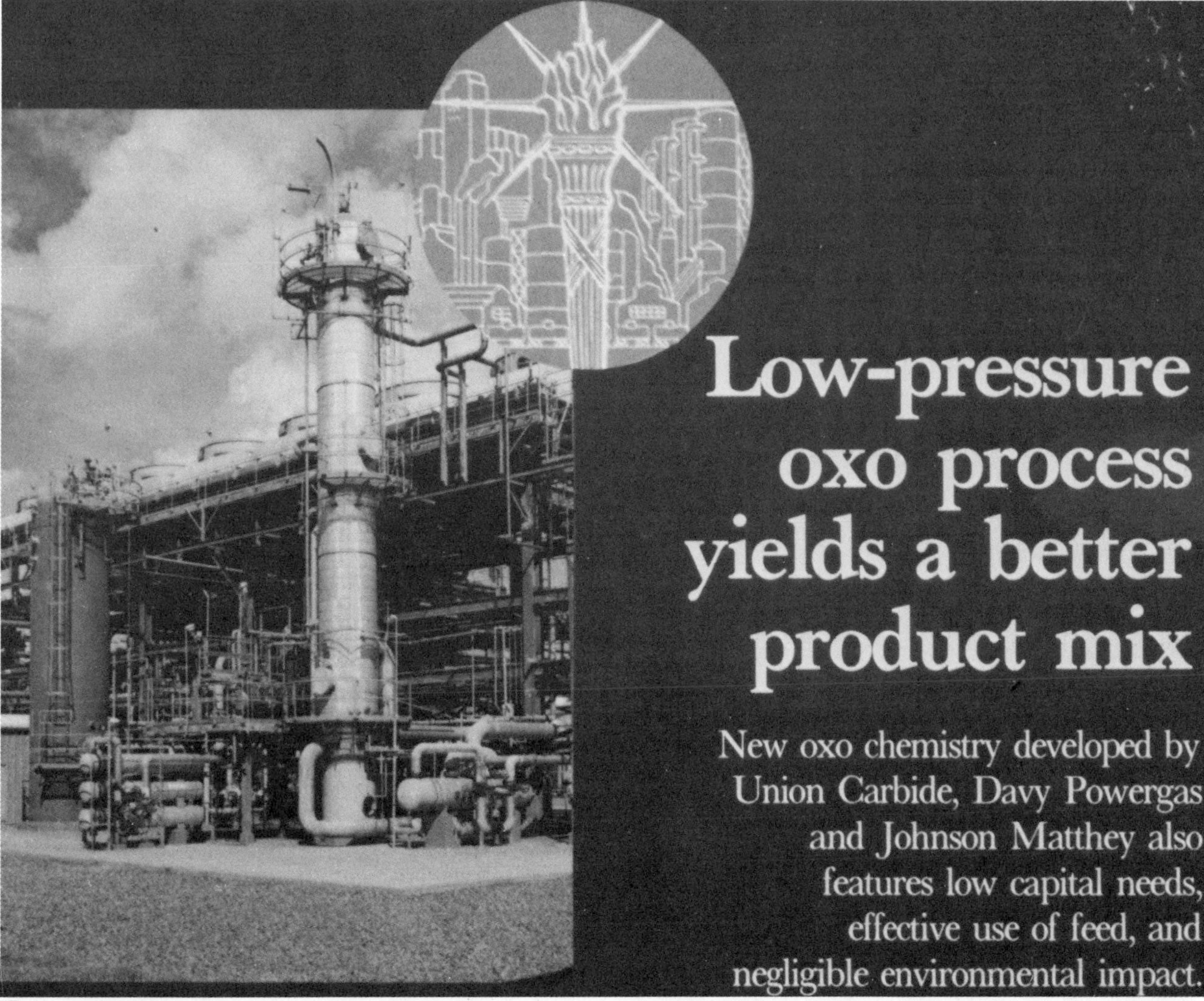

Low-pressure oxo process yields a better product mix

New oxo chemistry developed by Union Carbide, Davy Powergas and Johnson Matthey also features low capital needs, effective use of feed, and negligible environmental impact.

☐ A switch from the conventional cobalt-based to a new rhodium-based catalyst has greatly enhanced the oxo process. For developing and commercially implementing the technology that underlies this switch, Union Carbide Corp., Davy Powergas Ltd. and Johnson Matthey and Co. Ltd. (U.K.) win the 1977 Kirkpatrick Chemical Engineering Achievement Award.

Called the Low Pressure Oxo (LPO) process, the improved method is embodied in a 300-million-lb/yr propylene-to-butyraldehyde plant that Union Carbide started up at Ponce, Puerto Rico, early in 1976. Like any other oxo plant, it makes aldehyde by a hydroformylation reaction of olefin with a mixture of carbon monoxide and hydrogen (i.e., water gas, or synthesis gas):

$$CH_3CH{=}CH_2 + CO + H_2 \longrightarrow \underset{(n\text{-butyraldehyde})}{CH_3CH_2CH_2CH{=}O} + \underset{(\text{isobutyraldehyde})}{CH_3CHCH_3CH{=}O}$$

But the use of a triphenylphosphine-modified rhodium catalyst, instead of the conventional cobalt hydrocarbonyl or the sometimes-employed tributylphosphine-modified-cobalt version, leads to a host of technical and economic attractions.

The most important of these is the product mix from the plant. Manufacture of butyraldehyde from propylene, the main commercial application of oxo chemistry, has for many years been plagued by an unfortunately low ratio of valuable *n*-butyraldehyde to less-desirable isobutyraldehyde in the product mix. When homogeneous cobalt hydrocarbonyl catalyst is used, this ratio has been 3:1 to 4:1, according to Union Carbide. The LPO process, by contrast, allows ratios of 10:1 or greater. And it does so without requiring the relatively high temperature needed when cobalt is modified with tributylphosphine to improve its isomer ratio.

Hand in hand with the high yield of *n*-butyraldehyde in the new process is a very low incidence of the other side reactions prevalent with the cobalt-catalyzed process. These include: hydrogenation of propylene to propane; hydrogenation of butyraldehyde to butanol; reactions of the butanols to yield esters and acetals; and formation of heavy aldehyde-condensation products. A key reason is that the LPO process operates at 80 to 120°C, whereas conventional cobalt-catalyzed manufacture requires 140 to 180°C, and the tributylphosphine-modified-cobalt approach needs 180 to 200°C.

More significant than the temperature difference, however, is the low pressure at which the LPO process operates: only 200 to 400 psi, in contrast with 4,000 to 5,000 psi and 800 to 1,500 psi for the two cobalt-based

Originally published December 5, 1977

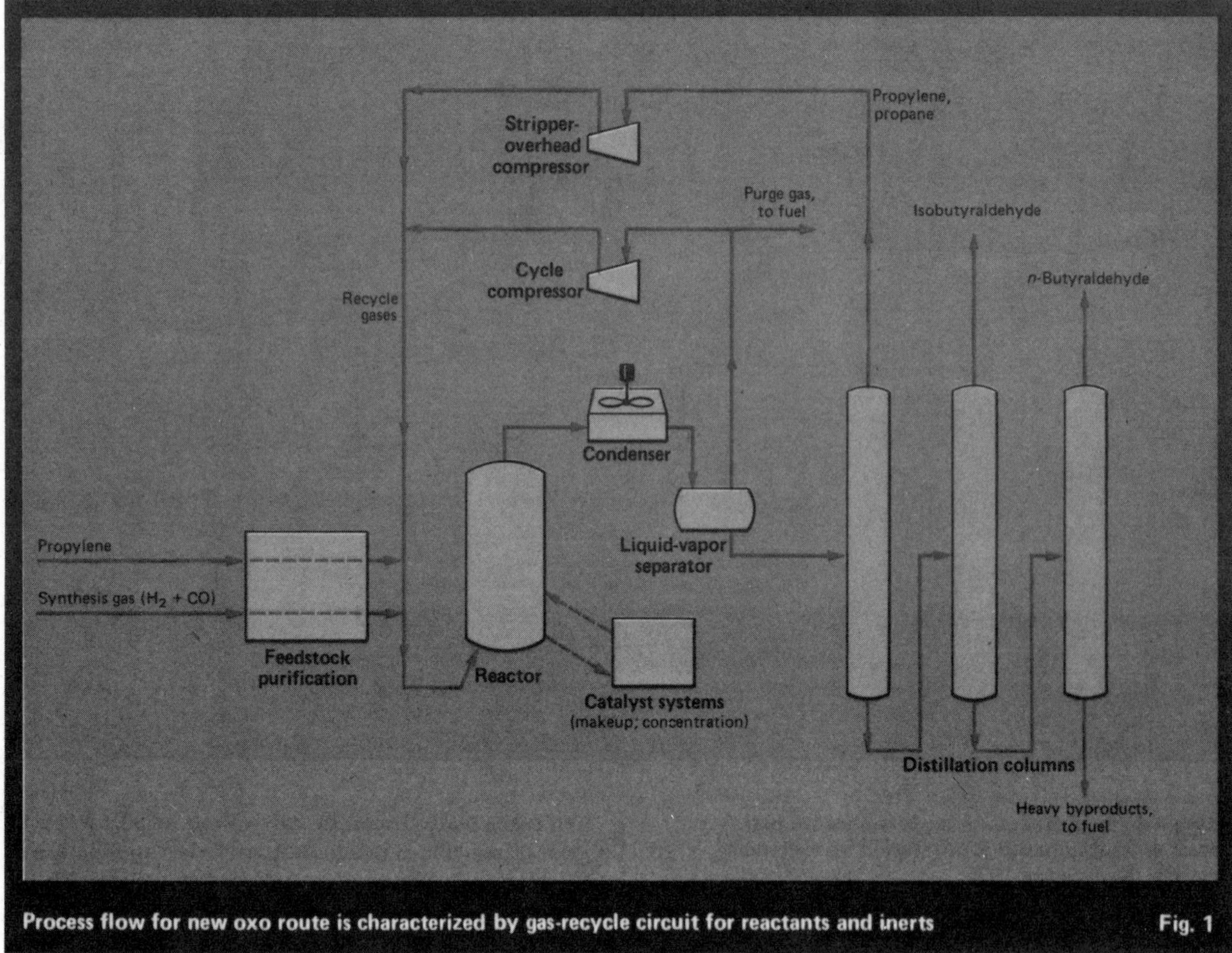

Process flow for new oxo route is characterized by gas-recycle circuit for reactants and inerts Fig. 1

versions. The resulting saving in capital, operating and maintenance costs is obvious. For instance, there is no need to compress the synthesis gas. This, together with the fact that only minimal distillation is needed in the process, helps keep energy requirements likewise low.

Environmental impact is negligible, points out Union Carbide. The process has only two effluent streams: a liquid containing the minor output of heavy byproducts, and a gas stream. Both can be used as fuel.

The triphenylphosphine-modified rhodium catalyst not only brings about all these advantages; there are also some other attractions that directly center upon the catalyst system itself. The LPO flowsheet was designed to minimize in-process rhodium inventory and losses, and there is no need for a catalyst-recycle system. If the catalyst is accidentally poisoned or deactivated, the rhodium can be recovered with little loss, using techniques developed by Johnson Matthey.

In the conventional oxo process, by contrast, the cobalt hydrocarbonyl is generated at high pressure, either outside or inside the hydroformylation reactor. Because the catalyst is volatile, and stable only in the presence of carbon monoxide, it must be decomposed before the reaction product is recovered. The cobalt-catalyst recovery cycle is traditionally cumbersome and costly.

Process flow

The flow diagram shows the LPO process to be both elegant and simple (Fig. 1). The synthesis-gas and propylene streams are first purified via proprietary solid-adsorbent techniques developed by Union Carbide and Davy Powergas, to remove catalyst poisons such as hydrogen sulfide or carbonyl sulfide. Then they join with recycle gas, and the combined stream enters the base of the hydroformylation reactor through a distributor. Here, the rhodium-based catalyst is present in a homogeneous liquid phase, dissolved together with free triphenylphosphine in a mixture of butyraldehyde and heavy byproducts from aldehyde trimerization.

The discovery that these aldehyde-condensation products are good solvents for the catalyst was important to the development of the commercial process. It allows the unit to operate without need for other solvents, and forestalls the need for a catalyst-recycle step.

Because the catalyst is very active, only a low rhodium concentration of several hundred parts-per-million is needed. The triphenylphosphine level is kept much higher—typically, several percent by weight. This favors production of the *n*- isomer, as explained below.

The gaseous reactants pass from small bubbles (with high interfacial area) into the liquid phase, where reac-

Butyraldehyde plant employing low-pressure oxo process at Ponce, Puerto Rico, started up extremely smoothly during early 1976. Figs. 2 and 3

tion takes place. It can be carried out at 80 to 120°C. Heat of reaction is taken out partly via vaporization of aldehydes into the overhead gas stream, and partly by circulating a coolant through coils inside the reactor.

Reactor effluent is removed in the gaseous phase, passing through entrainment separators that minimize rhodium loss. This effluent then goes to a condenser where aldehydes and byproducts drop out; this mixture is removed in a separator. The liquid stream from the separator contains appreciable amounts of dissolved gases, mainly propylene and propane. A product stripping column distills these out. The liquid stream from this stripper goes through two distillation columns in series that remove *iso*- and *n*-butyraldehyde as overhead products, respectively. The only liquid effluent from the overall operation is the small byproduct stream that leaves the bottom of the second column; this stream can be used as fuel.

The propylene-rich overhead product from the stripping column is compressed and recycled into the gas stream to the reactor. Meanwhile, the overhead gas from the liquid-vapor separator likewise is compressed and recycled, after a portion of it is bled off in a purge stream that controls the level of propane and other inert gases (e.g., methane introduced with the feedstock) in the gas circuit. By regulating the flow of this purge gas, the operators can control the reaction pressure very accurately. The purge stream is the only gaseous effluent from the operation; and, like the liquid effluent, it can be used as fuel.

A key feature of the process is its use of the gas-recycle system, as described above. It makes possible the notably simple process flowscheme, because: (1) reactor byproducts can be easily removed (in the gas purge-

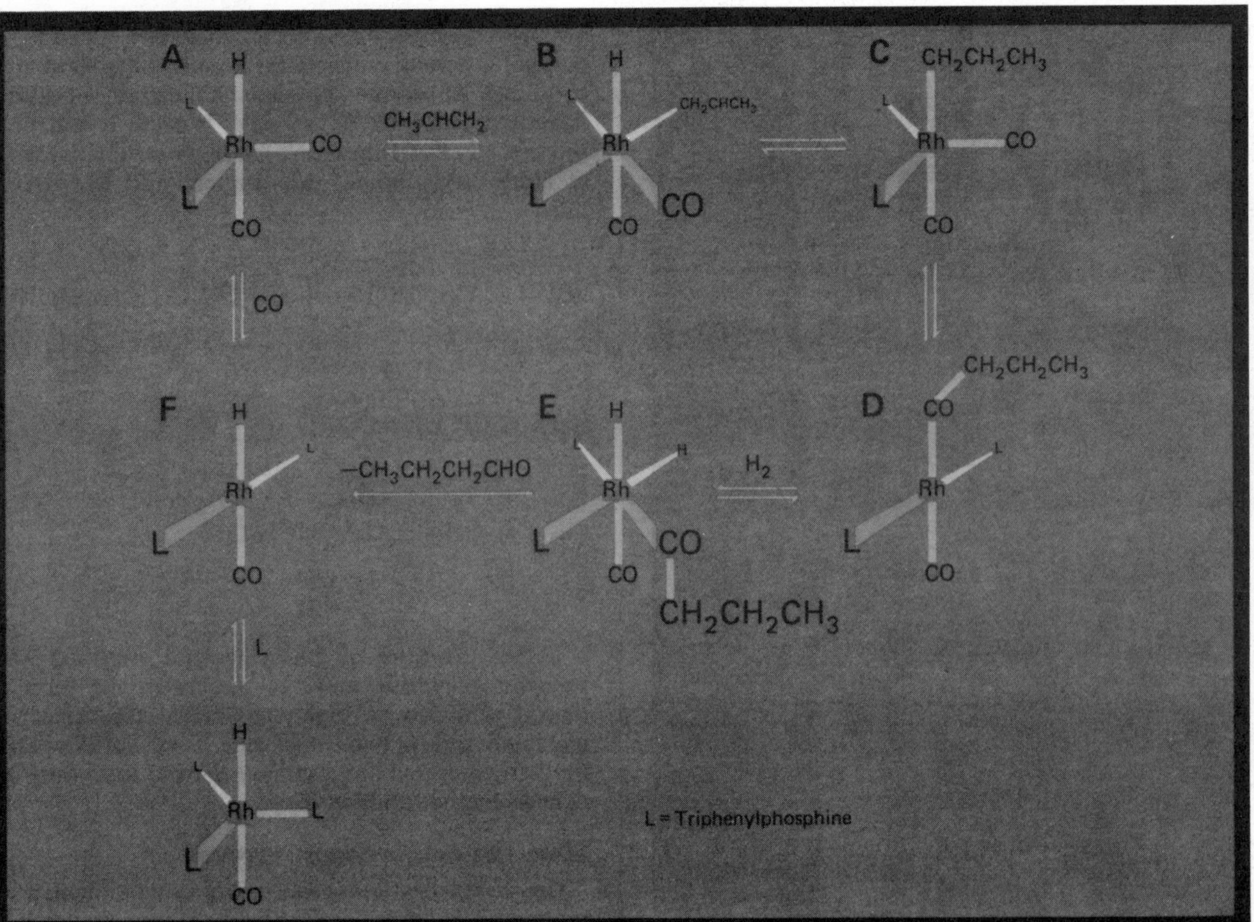

L = Triphenylphosphine

Rhodium-based catalyst in new oxo process may function as shown above and described in text below — Fig. 4

stream and in the bottoms from the final distillation column) at their rate of formation; (2) the rhodium catalyst remains in the reactor, with only minimal loss; (3) distillation of reaction solution for product recovery is avoided, so the catalyst need not be exposed to high temperatures (nor to oxygen that could leak in if the solution were vacuum-distilled).

The plant also includes equipment for periodic (and infrequent) makeup and removal of catalyst. Whenever catalyst must be added, an undisclosed convenient rhodium compound is dissolved along with triphenylphosphine into butyraldehyde, and this concentrated catalyst-precursor solution is pumped into the reactor. The system is designed in such a way that air cannot contact the catalyst solution and that operators are not exposed to malodorous aldehyde vapors.

Occasionally, a simple system concentrates a batch of the reaction solution, to recover catalyst at the end of its useful life. This is accumulated, then shipped to Johnson Matthey for rhodium recovery.

No squandering

A major advantage of the process is its efficient use of feedstock, which is especially valuable in light of current high valuations put upon propylene and synthesis gas.

Typical feedstock consumption is 750 kg of propylene (94 mole percent), and 740 normal m^3 of synthesis gas (99 mole percent), per 1,000 kg of *n*-butyraldehyde. The conventional oxo process, by contrast, is said to consume 930 kg of propylene and 1,200 m^3 of synthesis gas.

Catalyst usage is very efficient, too, because of the long catalyst life and the minimal loss during operation. Thus, catalyst cost is only a fraction of the total-product cost; and it is believed to be the same or slightly lower than that for the best of the conventional oxo processes.

Rhodium supply is not likely to be a problem, even though rhodium is a rare metal. Its use for oxo processes should have no significant impact on the overall rhodium market; and Johnson Matthey believes that reserves are adequate for 200 years. As noted above, the rhodium in the spent catalyst is recoverable.

How the catalyst works

Because the rhodium complex that provides the catalysis is formed under hydroformylation-reaction conditions, there is no need for complicated catalyst-synthesis and handling steps.

The exact reaction mechanism is not certain, but is believed to follow the sequence set out in Fig. 4. The starting point is the rhodium coordination complex shown at A, in which the rhodium atom carries five labile-bonded ligands: two that are triphenylphosphine, two that are carbon monoxide, and one hydrogen. In

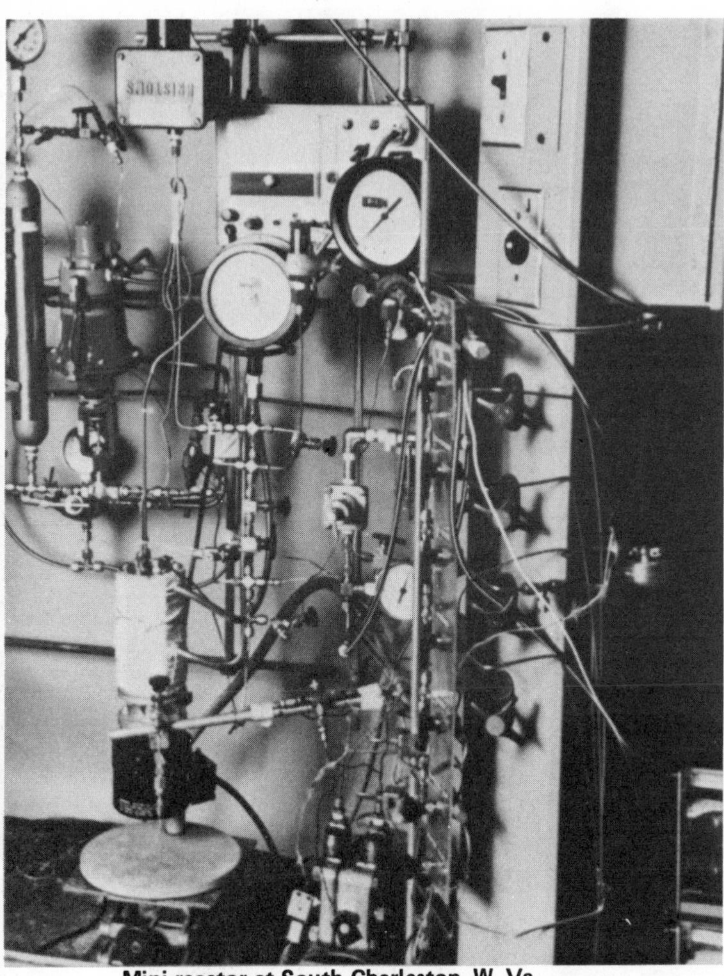

Mini-reactor at South Charleston, W. Va., generated data for reaction-rate model Fig. 5

the first step, this takes on propylene as an additional ligand, resulting in the structure shown at B.

This propylene complex rearranges to the alkyl complex shown at C, which undergoes carbon monoxide insertion to form the acyl complex illustrated at D. Oxidative addition of hydrogen gives the dihydroacyl complex shown at E. Finally, hydrogen transfers to the acyl group, and butyraldehyde is formed along with the species shown at F. Coordination with carbon monoxide returns the complex to the state shown at A (while coordination with triphenylphosphine also produces a "byproduct" complex having three triphenylphosphine ligands and only one carbon monoxide ligand).

The presence of excess triphenylphosphine, under the low-pressure conditions of the reaction, favors the high selectivity for normal aldehyde. This excess suppresses the dissociation of species A into one containing only a single phosphine ligand. By favoring the presence of species A in contact with the propylene, the steric effect of the two bulky triphenylphosphine ligands favors a high ratio of primary alkyl—if fewer such ligands were present, more propylene would form secondary alkyl groups, leading to more isobutyraldehyde.

Commercial significance

The high ratio of *n*-butyraldehyde to isobutyraldehyde is an asset because the former commands a better market. Each compound can be hydrogenated to the corresponding alcohol; and, of these two alcohols, *n*-butanol generally outperforms isobutanol in most end-uses, such as solvent applications. Further, *n*-butyraldehyde differs from its iso sister because it can be a precursor to 2-ethylhexanol, via this reaction sequence of aldol condensation, dehydration, and hydrogenation:

$$2CH_3CH_2CH_2CHO \longrightarrow CH_3CH_2CH_2\overset{\displaystyle OH}{\overset{|}{C}}H\underset{\underset{\displaystyle CH_3}{\underset{\displaystyle CH_2}{|}}}{C}HCHO$$

$$\xrightarrow{-H_2O} CH_3CH_2CH_2CH{=}\underset{\underset{\displaystyle CH_3}{\underset{\displaystyle CH_2}{|}}}{C}CHO$$

$$\xrightarrow{H_2} CH_3CH_2CH_2CH_2\underset{\underset{\displaystyle CH_3}{\underset{\displaystyle CH_2}{|}}}{C}HCH_2OH$$

Large quantities of 2-ethylhexanol esterified with phthalic anhydride move to market in the form of dioctyl phthalate, a highly important plasticizer for making flexible polyvinyl chloride. Total world markets for 2-ethylhexanol and butanols derived from butyraldehyde exceed 6 billion lb/yr.

How the achievement evolved

Commercialization of the LPO process climaxed an intensive collaborative effort by the three companies in chemistry and engineering, stretching back to 1964.

Early exploratory work by Union Carbide showed that the rhodium-catalyzed hydroformylation of propylene could be carried out at low pressure, with high selectivity for the normal aldehyde and with high enough catalyst productivity to justify a look at possible commercial use of the rhodium. The company obtained a basic patent (U.S. 3,527,809, issued Sept. 8, 1970) covering this achievement.

Based on a series of statistically designed experiments, the company developed kinetic models relating propylene hydroformylation and hydrogenation rates to the main process variables, namely, temperature, catalyst used, and CO, hydrogen and propylene concentrations. Models were also developed for the rate of formation of heavy byproducts resulting from the aldehyde-condensation reactions.

In England, meanwhile, Davy Powergas and Johnson Matthey were linking forces for similar study—the former recognizing a chance to develop unique processes, and the latter particularly interested in rhodium-catalyzed hydroformylation reactions. Publication of patents made all three parties realize that they had a common interest, so in 1971 they entered into a joint development program to convert the laboratory rhodium-oxo chemistry into a commercial process.

Drawing upon Union Carbide's mathematical models for reaction and byproduct-formation rates, Davy Powergas was able to optimize relationships among equipment size and cost, reactant concentrations, feedstock consumption, and rhodium inventory, seeking the

Processing unit of Union Carbide uses new oxo technology to make propionaldehyde from ethylene at Texas City Fig. 6

lowest possible production cost. The models showed that light byproducts as well as the heavy aldehyde-condensation products could be taken out of the reactor solution by the gas-recycle stream. This was a far more attractive approach than using several distillation columns, as proposed in an earlier flowscheme.

The high productivity promised by the rhodium catalyst made it important that the reactor be designed to avoid mass-transfer limitations. This was a key consideration in selecting the reactor type and preparing its conceptual design.

Late in 1972, Union Carbide decided to commercialize the process, by building the abovementioned 300-million-lb/yr plant at the company's Ponce, Puerto Rico, petrochemical complex. To lay more groundwork for this project, the company also decided to first erect a 200-ton/yr pilot unit at the same site, to test the process on the feedstocks available there and to provide scaleup data. Davy Powergas engineers helped to design and engineer this pilot unit.

Meanwhile, process engineering for the commercial unit was also done by Davy Powergas. This effort began at the time that the pilot plant started operating, and it drew upon the results that became available from that operation. Those results were helpful for, among other things, the final design of the reactors. Detailed engineering of the full-scale plant was by Union Carbide.

Initial startup in 1976 was so easy that, excluding outside interruptions, the plant was online for all but one hour in its first month of operation. During its first year, it was available for onstream operation (again excluding external interruptions) more than 99% of the time. This contrasts with about 90% availability for a conventional oxo plant, reports Union Carbide. The operation continues to be marked by unusual ease, stability and smoothness.

Design productivities, selectivities and feedstock-usage efficiencies have all been achieved, and product quality has met expectations. The ratio of *n*- to *iso* versions of butyraldehyde is normally controlled at around 10:1, but up to 16:1 has been achieved. Costs attributable to catalyst have been less than expected, and catalyst life exceeds one year.

A feature of the plant is its high degree of reaction control. Reaction temperature can be regulated to within ±0.5°C, and pressure to within ±0.8 psi, both computed over 8-h periods. The unit needs little operator attention.

From total shutdown, the unit can be brought back fully onstream in only 8 h. From outages due to dips in available power or similar causes, production can be restored in less than 45 min. Within 30 min of putting feedstock into the reactor, 80% of design output can be attained.

The LPO process has been licensed to Berol Kemi A.B., of Sweden, which is building a butyraldehyde plant scheduled to start up at Stenungsund in 1979. Meanwhile, Union Carbide itself is employing a variant of LPO to make propionaldehyde from ethylene and synthesis gas at Texas City, Tex., in a plant onstream since early 1975. It can operate at 150 million lb/yr, a rate which is about 50% above design capacity. (For further information, see pp. 188–189)

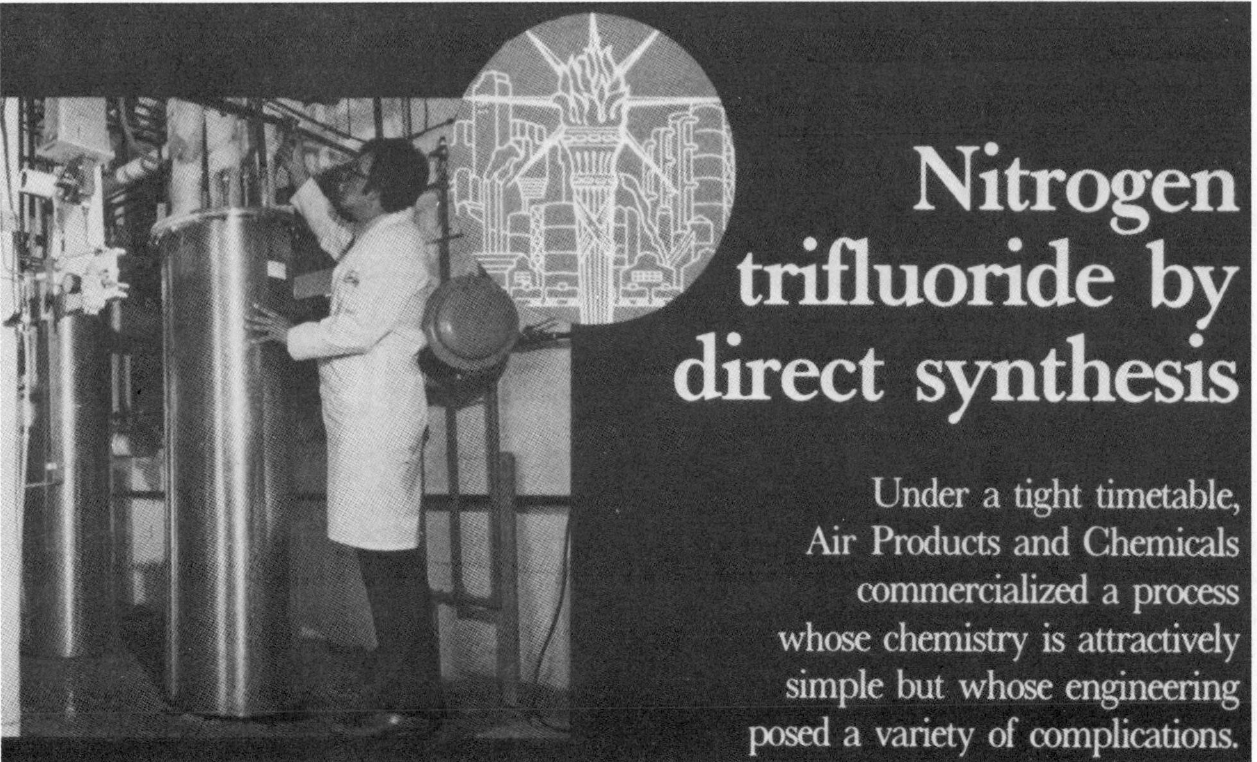

Nitrogen trifluoride by direct synthesis

Under a tight timetable, Air Products and Chemicals commercialized a process whose chemistry is attractively simple but whose engineering posed a variety of complications.

☐ In late 1975, the U.S. Air Force announced an expansion of its Chemical Laser Testing Program, which required large quantities of nitrogen trifluoride (NF_3) for use as a high-energy oxidizer for laser operation. At that time, however, production capacity did not exist. The state-of-the-art technology—electrolysis of ammonium acid fluoride (used commercially in the late 1950s and early 1960s)—had a host of drawbacks.

Previous attempts to fluorinate ammonia directly in the gas phase had resulted in the production of nitrogen. But a reaction based on the direct fluorination of ammonia in the presence of molten ammonium acid fluoride was tested by Air Products and Chemicals, Inc., and found to be feasible:

$$4NH_3 + 3F_2 \xrightarrow[260\,-\,300^\circ F]{NH_4F \cdot xHF} NF_3 + 3NH_4F$$

In commercializing this approach in a new plant at Hometown, Pa., Air Products succeeded in scaling up, by a factor of 500, a process involving very reactive and corrosive starting materials, and began producing a high-purity product having only trace amounts of impurities. Highlights of the NF_3 process are: (1) 20 to 25% energy savings over the electrolytic route; (2) high-purity product, greater than 99.5%; (3) marketable ammonium fluoride/ammonium acid fluoride by-product; (4) safety, because no hydrogen is evolved; and (5) 30 to 40% savings in capital costs, when onsite fluorine is available.

Early work in Air Products Specialty Gas Dept. laboratories confirmed that when NH_3 and F_2 were reacted in the gas phase, less than 5% of NF_3 was obtained. But the researchers theorized that the reaction might be feasible under other circumstances, if the intense heat associated with it could be dissipated. They decided to try using the ammonium acid fluoride in its molten state as a heat sink, and also as a transport medium for the NH_3. A laboratory reactor was built, and tests introducing F_2 and NH_3 directly into the salt showed a dramatic improvement in NF_3 yield. Modification of the operation parameters soon showed that NF_3 yields competitive with those achieved via electrolysis could be obtained.

A single mini-reactor was used for the pilot plant. The subsequent commercial design employed about 60 mini-reactors connected in parallel within multiple reactor vessels.

Reactor design

Pilot-plant work indicated the criticality of reactor design for optimum heat dissipation in the reaction zone. The engineers conceived a reactor based on the thermosiphon principle: F_2 gas is introduced via a packed tube submerged in the molten-salt bath. Sufficient heat-transfer area exists within the entire reactor to maintain overall temperature control of the bath, slightly above the salt's melting point. Mesh packing within the tube helps disperse the gas into very small bubbles, ensuring intimate contact of the gas with the molten salt, and thus spreading the heat of reaction over a large area.

Removing the thermal energy that accumulated in the molten salt posed some design problems. Because of the corrosive nature of the molten salt at operating temperature, unique and expensive materials would have been needed, and operating and maintenance problems encountered, if the salt were pumped through an external heat exchanger. Therefore, it was decided to use a jacketed vessel for a reactor.

A model, based on heat transfer in an electrolytic fluorine cell, was developed to select the optimum

Originally published December 5, 1977

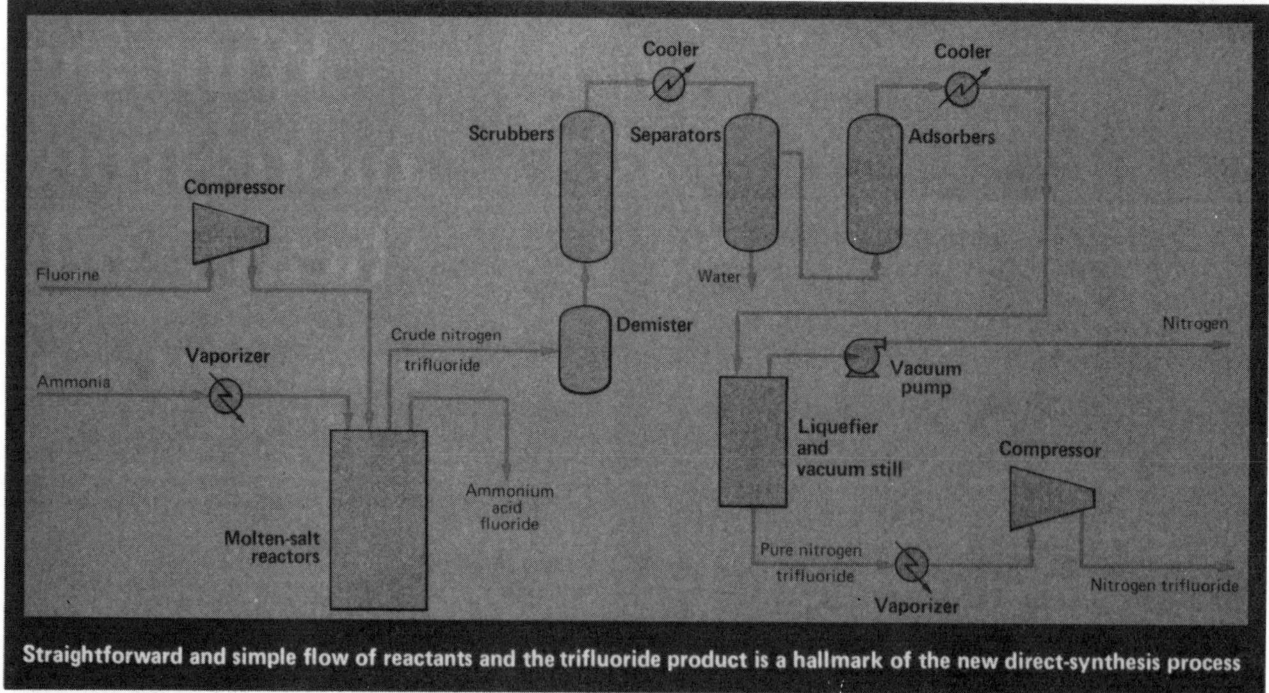

Straightforward and simple flow of reactants and the trifluoride product is a hallmark of the new direct-synthesis process

heat-transfer fluid. The jacketed NF_3 reactor resembled the cell (they contained a similar electrolyte). Both were mixed with about the same amount of gas. Although the molten salt was the controlling factor in the reactor, a 20% improvement in heat-transfer rate was attained by the employment of a material other than the electrolytic-cell coolant.

Process details, problems solved

Design of the final process was begun shortly after the process seemed feasible, and specification and design work proceeded on the equipment and systems for F_2 and NH_3 supply, heat transfer, purification, liquefaction, vaporization and compression.

The resulting process is shown above. Here is a summary of the performance of critical components in the system:

Reactors—Most of the details of the reactor are proprietary, but it can be said that the major difficulty in scaling up was the obtaining of an even distribution of F_2 to the parallel mini-reactors. The problems were accentuated by the hazardous nature of F_2 and the desirability of operating at low pressures. In the system devised, maldistribution is controlled to less than 10%.

Molecular-sieve adsorbers—A major early difficulty in the commercial plant was the capacity of the adsorbers, which remove traces of impurities (H_2O, CO_2, N_2O) from the process gas streams. When the plant began operating, the units worked at less than 10% capacity. This was because of a rise in temperature of the adsorbent, which eluted the adsorbed impurities prematurely. Actual temperatures in the plant exceeded those calculated from the heat of adsorption for these components.

Tests showed that pure NF_3 did not react with the adsorbent at the operating temperatures. But runs with other nitrogen fluorides indicated that even trace amounts would react, with the liberation of heat. These reactions were very temperature-dependent, and were triggered by hot spots in the adsorbers. Thus, pilot-plant studies were conducted to find a method of removing all other nitrogen fluorides except NF_3. A purification step was devised whereby these could be reduced to near zero.

This highly efficient scrubbing and purification system was required, not only to provide for complete removal of all active fluoride compounds but also to cope with large amounts of fluorine that could come from the reactors because of upsets. The system was designed as a combination spray and packed tower, with sufficient residence time to avoid production of semi-stable side-products.

Products recovery—A cryogenic liquefaction system recovers more than 99.7% of the NF_3. It employs continuous vacuum distillation of the trifluoride at liquid-nitrogen temperature, to minimize condensation of N_2 produced in side-reactions.

Process byproduct—The major byproduct of the reaction is ammonium acid fluoride, marketable as either ammonium bifluoride or as an aqueous solution of ammonium fluoride, for use in etching operations.

Safety—In designing the plant, operator exposure to the reactors and other equipment was held to a minimum by placing almost all controls external to the operation. Safety was also a key determinant in finding a process that did not produce H_2 as a byproduct, in order to prevent explosive reactions between it and nitrogen trifluoride.

Today, there is continuing effort to improve the process, solve corrosion problems and upgrade plant operation. The pilot unit continues to run in conjunction with the production plant to test new ideas and obtain technical data.

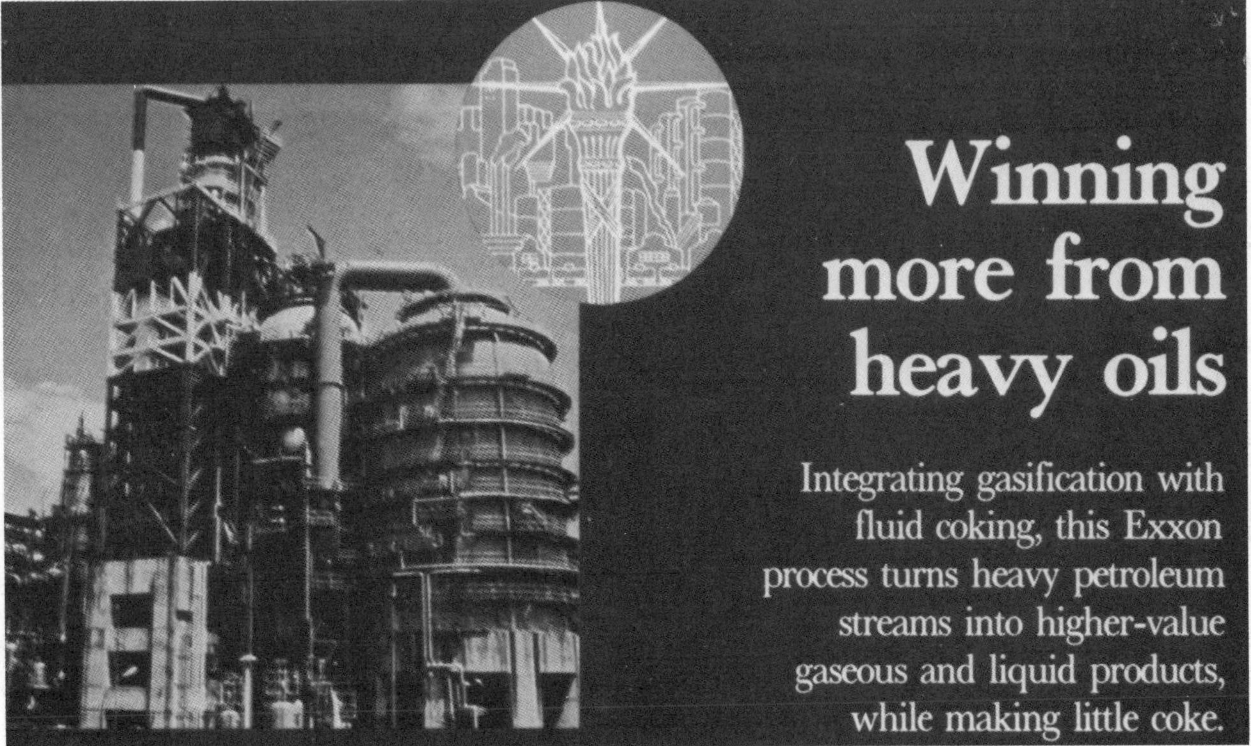

Winning more from heavy oils

Integrating gasification with fluid coking, this Exxon process turns heavy petroleum streams into higher-value gaseous and liquid products, while making little coke.

☐ The growing emphasis in the late 1960s on lessening sulfur emissions from combustion of hydrocarbons placed an increasing onus on heavy, high-sulfur, high-metals residua. Fluid coking could upgrade these bottoms into clean liquid fractions, such an naphtha and gas oil, but left a considerable quantity of sulfur- and metals-laden coke. Some refiners, reasoned Exxon Research and Engineering Co. (Florham Park, N.J.), would prefer the output of a readily desulfurized coke gas instead of the coke, which is sometimes hard to dispose of.

In 1969, Exxon R & E started work on such a task, and its efforts paid off in the Flexicoking process. The route went commercial in September of last year, when Japan's Toa Oil Co. started up a 21,300-bbl/stream-d unit at its Kawasaki refinery.

The now-proven process offers an environmentally acceptable, economical and energy-efficient method to convert vacuum residua and other high-gravity, dirty streams, even including Athabasca tar sands and heavy Venezuelan crudes, into lighter liquids and coke gas. However, Exxon R & E's achievement goes beyond this—the company also developed modifications to conventional furnace burners that will allow refineries to fire the low-heating-value gas in standard furnace designs as a replacement for natural gas or low-sulfur fuel oil.

Combining technologies

Flexicoking extends the capabilities of Exxon R & E's fluid-coking process. And, indeed, the new route draws heavily from the earlier technique, which was first commercialized in 1954. Flexicoking also takes advantage of well-established technology for gasification of coal and other carbonaceous solids.

However, several major areas of uncertainty posed serious challenges to coming up with a reliable process and mechanical design:

- Interactions between the coking and gasification sections could substantially affect critical process parameters in each section, and the properties of the recirculating coke particles.
- Corrosive slags could form from metals, such as vanadium and sodium, in the feedstream, and these could become concentrated in the solids residue from gasification.
- The coke's physical and chemical structure and the contaminants that are present differ from those of coal, so gasifier design know-how could not be straightforwardly transferred.
- Equipment components in the gasifier, heater and gasifier-overhead system would be exposed to different combinations of severe thermal, erosive and corrosive conditions from those encountered in previous, commercial plant designs.

Process development

In 1969, preliminary process and equipment evaluations began. These paper and laboratory studies led to one major revision of the original conceptual flowscheme. It had envisioned a combination coke-heater/gasifier unit. But the studies pointed to the desirability of separating the two functions into separate vessels, and of adding a controlled solids-recycle stream from the gasifier back to the heater. This latter feature substantially increases heat-balance-control flexibility, points out Exxon.

By 1972, work had progressed so far that a 750-bbl/d prototype unit got the green light. This facility was constructed at Exxon Co. U.S.A.'s Baytown, Tex., refinery, and it commenced a period of 16 months of trials in February 1974.

Better path to ethylbenzene
The new Mobil/Badger process, featuring a solid alkylation catalyst with high productivity and selectivity, avoids most of ethylbenzene makers' usual corrosion and pollution woes.

□ The prime accomplishment in the new Mobil/Badger ethylbenzene process is a novel catalyst that makes obsolete the well-established Friedel-Crafts alkylation for the manufacture of ethylbenzene.

But the availability of the catalyst also set off a chain of process developments that must be considered in total as an important and classical chemical-engineering achievement.

Briefly, Mobil and Badger engineers succeeded in:

- Developing a highly selective solid catalyst that avoids many of the corrosion and pollution problems of the previous Friedel-Crafts liquid-phase catalyst.
- Creating a complex, but efficient, catalyst-manufacturing method.
- Designing a novel vapor-phase, fixed-bed reactor with good product selectivity and excellent temperature control of the highly exothermic reaction. Good productivity and selectivity permit use of a relatively small reactor with low capital costs.
- Computer modeling the process, requiring close cooperation between research and process engineers, to develop optimum process performance.
- Designing an energy-recovery system that captures 95% of the process-heat input and the reaction heat, as useful low- and medium-pressure steam.
- Coming up with an optimum recycling method for polyalkylated reaction products that establishes steady-state reaction conditions, resulting in product yields over 99%.

Solid replaces liquid catalyst

Up until this process, the standard way to make ethylbenzene—an important intermediate in the manufacture of styrene—involved reacting ethylene with benzene, using a liquid-phase Friedel-Crafts catalyst (usually aluminum chloride). This procedure results in all sorts of problems: corrosion due to the nature of the liquid catalyst (requiring stainless-steel equipment), severe pollution problems in disposing of the spent catalyst, and energy losses in the exothermic reaction.

Mobil Oil scientists at Paulsboro, N.J.,* looking for a better way, developed a solid, acidic zeolite catalyst for this alkylation. Joining forces with The Badger Co., Inc., (Cambridge, Mass.), the team then worked out an optimum process scheme, based on pilot-plant tests and computerized process-simulation techniques.

Manufacturing the catalyst

While Badger concentrated on the process design, Mobil engineers worked on developing a commercially viable method to make the proprietary zeolite catalyst. The result was a complex, ten-step technique. Commercial test-batches were produced and used in a Badger-designed semiworks unit.

But all did not go smoothly at first. Tests on the early catalyst batches showed that the process would produce a relatively high yield of ethylbenzene byproducts, requiring additional distillation capacity. Mobil then came up with a modified catalyst that greatly reduced the amount of byproducts. This permitted production of high-purity ethylbenzene in a relatively low-cost plant.

Using dilute ethylene

A 40-million-lb/yr process-demonstration unit at Foster Grant Co.'s Baton Rouge, La., styrene monomer plant, using monomer-grade ethylene, confirmed all of the claims for the new process.

But subsequent laboratory work concentrated on dilute ethylene feedstock (as low as 15%) from such

*Catalyst Research & Development Group of the Mobil Research and Development Corp., working with Mobil Chemical Corp.'s Research & Development Div.

Originally published December 5, 1977

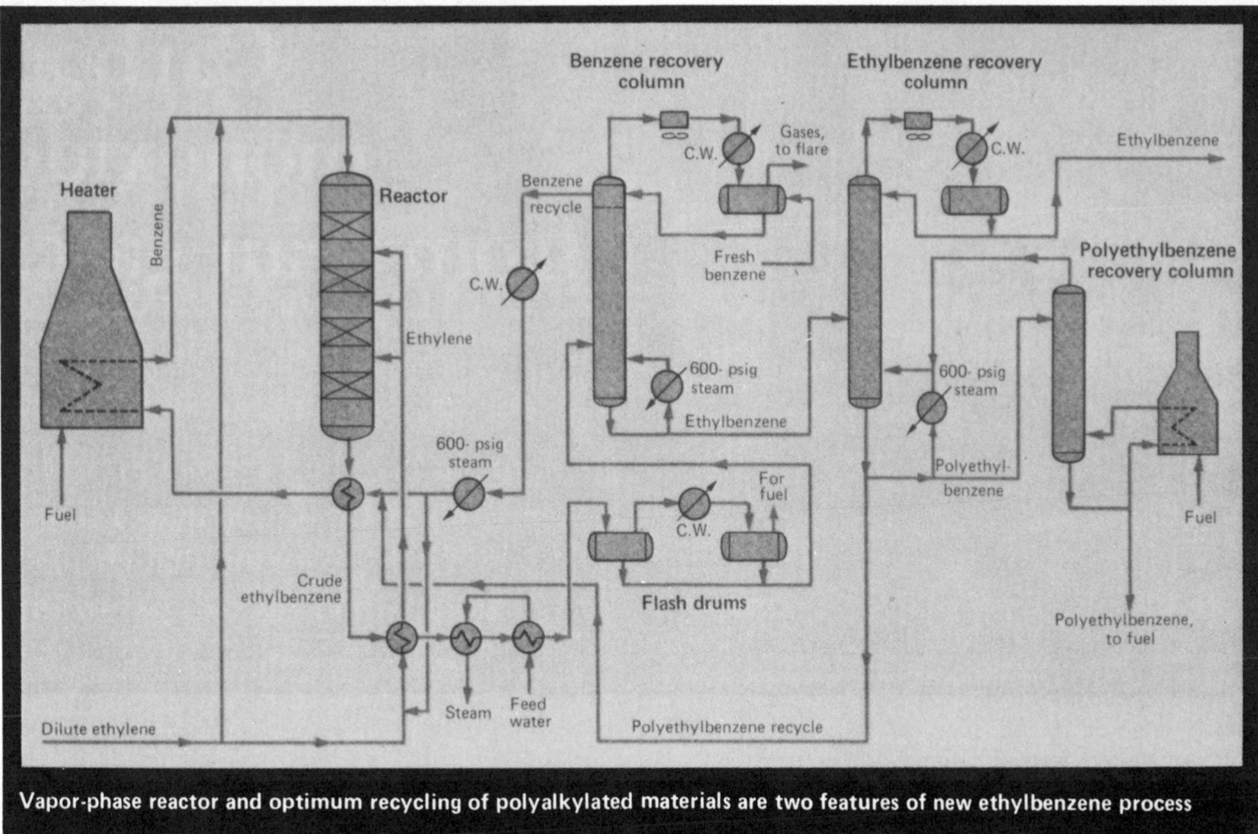

Vapor-phase reactor and optimum recycling of polyalkylated materials are two features of new ethylbenzene process

sources as raw coke-oven gas, and off-gas from catalytic cracking of gas oils.

This evaluation was carried out at Badger's Weymouth, Mass., laboratory. The results indicated that reactor operating conditions obtained with polymer-grade ethylene remained essentially the same for the dilute feedstock.

Confirmation of this was obtained upon converting Cosden Oil & Chemical's 40-million-lb/yr ethylbenzene unit at Big Spring, Tex., to the Mobil/Badger process (see flowsheet).

At Cosden, ethylene (from a fluidized catalytic gas-oil cracker) that has been treated to remove propylene and sulfur reacts with fresh plus recycled benzene in a low-chrome steel-alloy alkylation reactor containing the fixed-bed catalyst. Reaction takes place at about 800°F, and 200 to 400 psig.

Reactor effluent is cooled and condensed (generating medium-pressure steam), and the condensed hydrocarbons go to the distillation train. Benzene recovered here goes back to the reactor, as do recyclable nonselective aromatics. Meanwhile, high-purity ethylbenzene is recovered overhead.

Laboratory tests with pure polymer-grade ethylene indicated catalyst operating cycles of at least two weeks. Cycle lengths up to four weeks have been obtained in the Cosden operation.

Feedstock and operating conditions at Cosden's ethylbenzene plant

Feedstock (after sulfur, water and propylene removal):	
Hydrogen	11.6 vol. %
Nitrogen	13.5 vol. %
Carbon monoxide	3.9 vol. %
Methane	36.4 vol. %
Ethylene	17.9 vol. %
Ethane	16.7 vol. %
Propylene	10 ppm
Hydrogen sulfide	2 ppm
Operating conditions:	
Reactor configuration	4 beds in series, 1 reactor
Maximum temperature, °F	800
Outlet pressure, psig	200–300

Regenerating the catalyst

Mobil developed an adiabatic catalyst-regeneration model that successfully predicted regeneration performance (burning off the carbonaceous deposits) at Cosden. That company has only one reactor, requiring shutting down the plant during regeneration.* In the usual installation, one reactor will be onstream while another is being regenerated.

Catalyst regeneration, which takes place over 24 h, generally starts with low-oxygen gas. Oxygen is gradually increased as regeneration proceeds, to avoid developing high-catalyst-bed temperatures that could damage the catalyst.

*In the Cosden conversion, existing equipment was reused wherever possible, resulting in some sacrifices in process efficiency. An engineering study is now underway to revamp and expand the unit to take greater advantage of the Mobil/Badger technology.

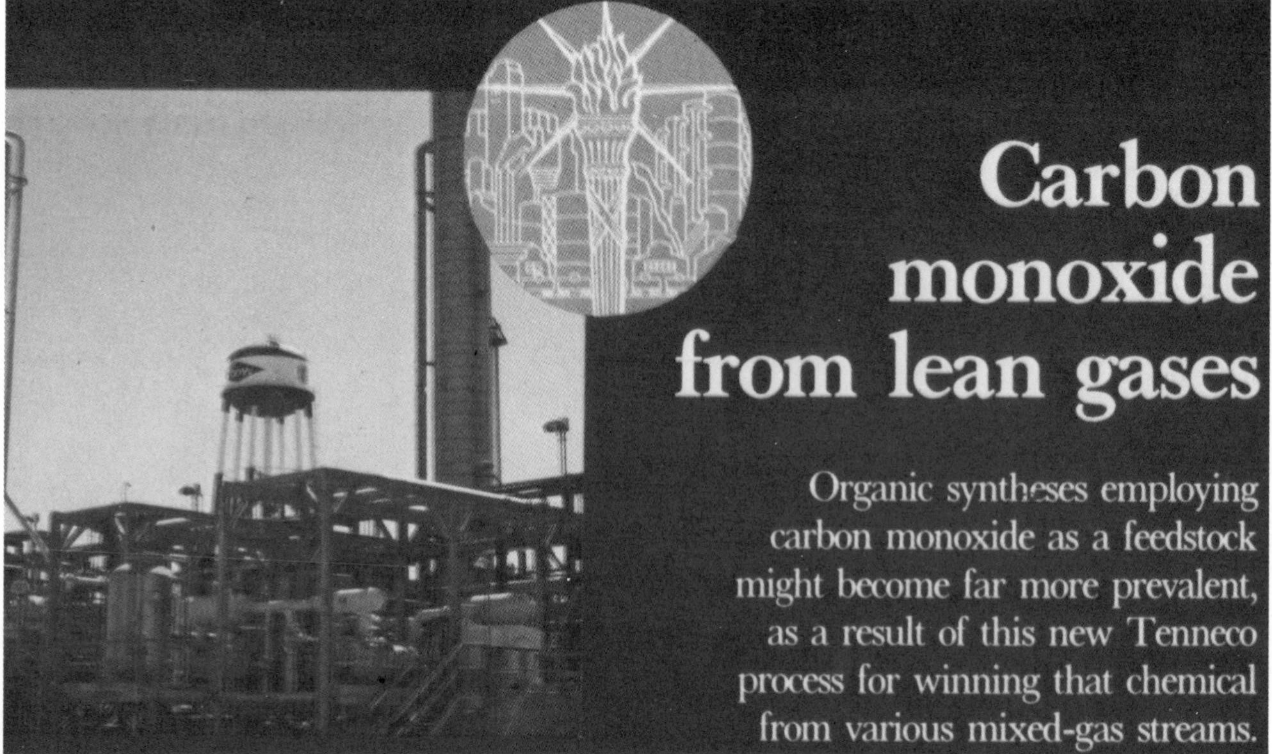

Carbon monoxide from lean gases

Organic syntheses employing carbon monoxide as a feedstock might become far more prevalent, as a result of this new Tenneco process for winning that chemical from various mixed-gas streams.

☐ Tenneco Chemicals, Inc.'s Cosorb technology significantly advances the chemical process industries' effort to conserve resources and eliminate pollution.

Described as the first process that can economically recover carbon monoxide from gas streams (including waste gases) at low pressure and ambient temperature, it opens up a low-cost source of CO for chemical synthesis. Not only is it effective on relatively rich gas mixtures coming from processes such as hydrocarbon reforming or partial oxidation; it can also cope with waste streams that contain nitrogen, such as basic-oxygen-furnace (BOF) gases, blast-furnace off-gases, and tail gas from carbon-black plants.

Rising hydrocarbon prices make CO more and more attractive as a less-costly source of carbon values for chemical synthesis. Cosorb becomes a vital link in such use because it economically recovers CO that might in many cases otherwise go to the atmosphere.

The importance of CO for chemical synthesis is already established in the manufacture of phosgene, toluene diisocyanate (TDI), and synthetic acids—including, rather recently, acetic acid. The monoxide is also employed in oxo syntheses (see article beginning on p. 110). Developments underway aim to extend its use to the production of terephthalic acid and *p*-cresol, and to use it as a co-monomer in thermoplastics.

Dow Chemical Co. was first to use the Cosorb process commercially, with a 75-million-lb/yr unit that started up in April 1976 at Freeport, Tex., recovering CO for manufacturing TDI and acrylates.

A second Tenneco licensee, Koor Chemicals, Ltd. (Israel), will buy a smaller Cosorb unit this fall, to purify CO generated by naphtha reforming. Another licensee, Prva Iska Baric (Yugoslavia), has recently initiated a TDI project that will include an upstream Cosorb unit producing approximately 20 million lb/yr of CO. And Kawatetsu Chemical (Japan) has been successfully piloting the recovery of CO from BOF gases via Cosorb.

Tenneco expects Cosorb to be used by the steel industry to improve blast-furnace efficiency via CO recycle, as well as in direct-reduction ironmaking. In coal gasification, Cosorb promises advantages in manufacturing relatively high-Btu fuel gas, because its ability to cope with nitrogen permits gasifying with air.

A variant of Cosorb has been demonstrated on pilot scale for recovering ethylene from dilute gas streams. As with CO, this technology provides recovery of a basic building block, ethylene, from gas streams now valued only as fuel. Commercialization of this so-called ESEP process awaits further pilot testing.

A deft absorption

The Cosorb process relies on selective absorption of CO by forming a chemical complex with cuprous aluminum tetrachloride ($CuAlCl_4$) in a toluene solvent. Most other components in the gas stream, such as nitrogen, hydrogen, carbon dioxide, oxygen and methane, are only slightly soluble and are not complexed. However, the feed gas must be free of water, hydrogen sulfide and sulfur dioxide because these compounds poison the solvent mixture.

Absorption by means of an inorganic salt in an organic solvent is unique, says Tenneco: practiced only in the laboratory before development of Cosorb. The absorbed gas is stripped out of the rich solvent at 99+% purity by a slight elevation of temperature.

Compared with other processes for recovering high-purity CO, Cosorb generally requires less investment and energy. The process based on copper-ammonia liquor, which has been operated commercially for some time, has high capital costs, uses a lot of energy because

Originally published December 5, 1977

of the poor absorption characteristics of the solvent and the high-pressure operation required, and suffers from both pollution and corrosion problems. Conventional cryogenic processing, on the other hand, cannot recover CO from gas streams containing nitrogen, due to the closeness of boiling points (nitrogen, −196°C; CO, −192°C).

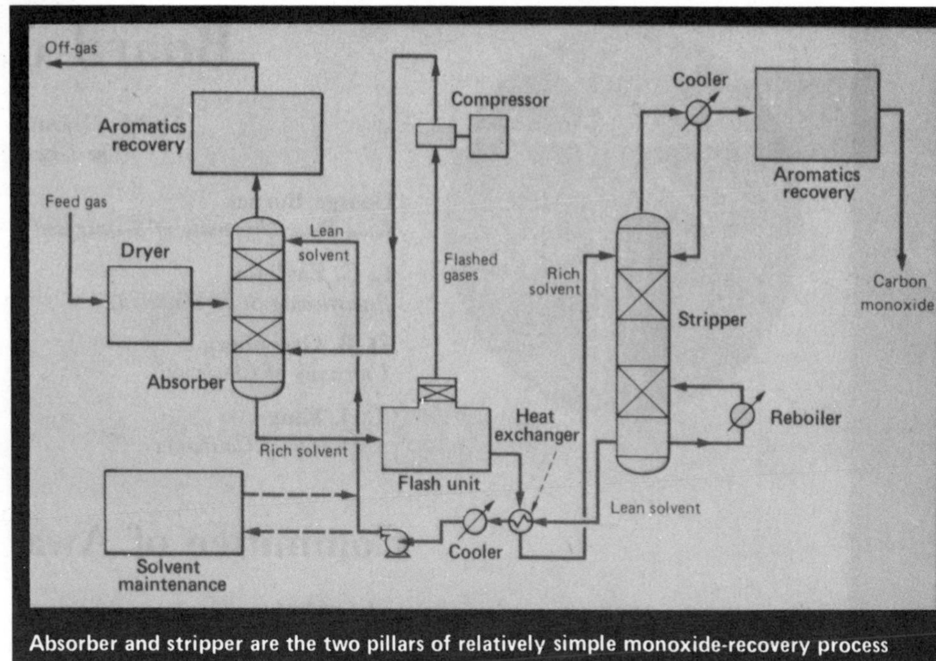

Absorber and stripper are the two pillars of relatively simple monoxide-recovery process

How it evolved

Tenneco's development of the Cosorb process grew out of a 1969 project that evaluated use of the $CuAlCl_4$ complex in toluene for recovering acetylene from gas streams. Earlier, Esso Research & Engineering had used this complex in studies for recovering ethylene and other compounds. That company (now Exxon Research & Engineering) owns the basic patents.

Although Tenneco found the acetylene recovery technically feasible, forecasted rises in natural-gas prices made investment in new acetylene facilities to use such technology economically unattractive. However, during the work on the acetylene-reactor product gas, which contains CO, the researchers also became more aware of the complex's advantages for absorbing the CO. Therefore, in 1970, Tenneco formed a research and engineering team to develop these into a new process to recover CO, under license from Exxon.

Initial laboratory work determined the optimum conditions for absorption and desorption, as well as the interaction of trace impurities in the gas with the solvent. Design information on physical properties, and on pertinent vapor-liquid equilibria, also had to be obtained since correlations were not available for this unusual system. Corrosion tests showed carbon steel to be satisfactory for most of the process.

A small-scale process demonstration in a glass, bench unit was followed by a 250,000-lb/yr unit at Pasadena, Tex., utilizing feed from a natural-gas steam reformer.

During operation of the Pasadena plant, corrosion was monitored carefully, energy consumption was minimized, and separation efficiencies and flooding data were determined for the packed absorption columns.

How it works

The process that evolved is a simple absorption-desorption operation. CO is absorbed selectively in a countercurrent tower at ambient temperature, then stripped at slightly elevated temperature by heat from toluene vapor out of the stripper reboiler.

The absorptivity of the solvent for CO is sufficiently high so that the process can operate at essentially 1 atm, minimizing the need for compression energy.

Before entering the process, feed gas must be thoroughly dried because water reacts with the solvent system, forming hydrogen chloride. The toluene solvent is essentially completely recovered by condensation and a followup adsorber loaded with activated carbon.

While degree of recovery of CO from the feedgas is affected by the balance established between the cost of that gas and the operating cost, levels as high as 99.5% can be achieved. Even though product purity is a function of feedgas composition, 99.5% is readily attainable in most cases. Any carryover of hydrogen chloride, formed by minute traces of water, is only in the parts-per-million range and can easily be reduced to 0.1 ppm.

When Dow decided to expand CO production at Freeport, it became interested in Cosorb because of dissatisfaction with corrosion problems and with ammonia pollution associated with operation of its copper-liquor process. Additional pilot work was done at Freeport by Tenneco and Dow engineers while the first commercial unit was being designed and built.

In the first year of operation after startup in April 1976, Dow experienced 98% onstream time. Operating experience at the Dow plant has verified the design criteria, based on engineering correlations developed from the laboratory and pilot-plant work. And the plant internals examined during a recent scheduled turnaround showed only minor pitting in a few areas subjected to the highest temperature level.

Tenneco Chemicals people feel that the Cosorb process generally provides the best operating economics and lowest capital investment for a plant to manufacture high-purity CO from CO-containing streams, such as synthesis gas from reforming of methane, LPG or naphtha. An economic comparison of Cosorb vs. cryogenic processing, based on a 100-million-lb/yr plant, indicates a final CO product cost of 7.3¢/lb for Cosorb and 9.4¢/lb for cryogenic. (The Cosorb investment in this study includes pressure-swing adsorption equipment to make high-purity hydrogen, while the cryogenic plant has an MEA unit to remove carbon dioxide.)

For further information, see pp. 73–75